# THE FUTURE FOR INSECTICIDES

**The Future for Insectides: Needs and Prospects**

*A Conference Held April 21-27, 1974 at the Rockefeller Foundation Study and Conference Center at Bellagio, Italy*

*Participants Standing:* Casida, Brooks, Nakajima, Elliott, Roelofs, Furtick, Smith, Narahashi, Barnes, Bowers, Wright, Oppenoorth, Spencer, Richardson. *Seated:* Wilkinson, Koehler, Boyce, Fukuto, Metcalf, McKelvey, Klein, Locke.

# Participants

DR. J. M. BARNES
MRC Toxicology Unit
Medical Research Council Laboratories
Carshalton, Surrey, England

DR. WILLIAM S. BOWERS
Department of Entomology
New York State Agricultural Experiment Station
Geneva, New York

DR. ALFRED M. BOYCE
College of Agriculture
University of California
Riverside, California

DR. GERRY T. BROOKS
Agricultural Research Council
Unit of Invertebrate Chemistry and Physiology
The Chemical Laboratory
The University of Sussex
Brighton, Sussex, England

DR. JOHN E. CASIDA
Department of Entomological Sciences
University of California
Berkeley, California

DR. MICHAEL ELLIOTT
Insecticides and Fungicides Department
Rothamsted Experimental Station
Harpenden, Herts, England

DR. T. ROY FUKUTO
Department of Entomology
University of California
Riverside, California

DR. WILLIAM F. FURTICK
FAO Regional Office for Europe
Geneva, Switzerland

DR. W. KLEIN
Institut für Okologische Chemie
München, Germany

DR. CARLTON S. KOEHLER
Department of Entomological Sciences
University of California
Berkeley, California

DR. MICHAEL LOCKE
Department of Zoology
The University of Western Ontario
London, Ontario, Canada

DR. JOHN J. MCKELVEY, JR.
The Rockefeller Foundation
New York, New York

DR. ROBERT L. METCALF
Department of Entomology
University of Illinois
Urbana, Illinois

PROF. M. NAKAJIMA
Department of Agricultural Chemistry
Kyoto University
Kyoto, Japan

PROF. TOSHIO NARAHASHI
Department of Physiology and Pharmacology
Duke University Medical Center
Durham, North Carolina

DR. F. G. OPPENOORTH
Laboratory for Research on Insecticides
Wageningen, The Netherlands

DR. RALPH W. RICHARDSON
The Rockefeller Foundation
New York, New York

DR. WENDELL ROELOFS
Department of Entomology
New York State Agricultural Experiment Station
Geneva, New York

DR. RAY F. SMITH
Department of Entomological Sciences
University of California
Berkeley, California

DR. E. Y. SPENCER
Research Branch, Agriculture Canada
London, Ontario, Canada

DR. CHRISTOPHER F. WILKINSON
Cornell University
Ithaca, New York

DR. J. W. WRIGHT
World Health Organization
Geneva, Switzerland

# INTRODUCTION TO THE SERIES

Advances in Environmental Science and Technology is a series of multiauthored books devoted to the study of the quality of the environment and to the technology of its conservation. Environmental sciences relate, therefore, to the chemical, physical, and biological changes in the environment through contamination or modification; to the physical nature and biological behavior of air, water, soil, food, and waste as they are affected by man's agricultural, industrial, and social activities; and to the application of science and technology to the control and improvement of environmental quality.

The deterioration of environmental quality, which began when man first assembled into villages and utilized fire, has existed as a serious problem since the industrial revolution. In the second half of the twentieth century, under the ever-increasing impacts of exponentially growing population and of industrializing society, environmental contamination of air, water, soil, and food has become a threat to the continued existence of many plant and animal communities of the ecosystem and may ultimately threaten the very survival of the human race.

It seems clear that if we are to preserve for future generations some semblance of the existing biological order and if we hope to improve on the deteriorating standards of urban public health, environmental sciences and technology must quickly come to play a dominant role in designing our social and industrial structure for tomorrow. Scientifically rigorous criteria of environmental quality must be developed and, based in part on these, realistic standards must be established, so that our technological progress can be tailored to meet such standards. Civilization will continue to require increasing amounts of fuel, transportation, industrial chemicals, fertilizers, pesticides, and countless other products, as well as to produce waste products of all descriptions. What is urgently needed is a total systems approach to modern civilization through which the pooled talents of scientists and engineers, in cooperation with social scientists and the medical profession, can be focused on the development of order and equilibrium among the presently disparate segments of the human environment. Most of the skills and tools that are needed already exist. Surely a technology that has created manifold environmental problems is also capable of solving them. It is our hope that the series in Environmental Science and Technology will not only serve to make this challenge more explicit to the established professional but will also help to stimulate the student toward the career opportunities in this vital area.

Finally, the chapters in this series of Advances are written by experts in their respective disciplines, who also are involved with the broad scope of environmental science. As editors, we asked the authors to give their "points of view" on key questions; we were not concerned simply with literature surveys. They have responded in a gratifying manner with thoughtful and challenging statements on critical environmental problems.

To facilitate communications with our contributors and readers, we are including below our addresses and telephone numbers.

JAMES N. PITTS, JR., Editor
Statewide Air Pollution Research Center
University of California
Riverside, CA 92502
Telephone: (714) 787-4584

ROBERT L. METCALF, Editor
Environmental Studies Institute
Departments of Biology and Entomology
University of Illinois
Urbana-Champaign, IL 61801
Telephone: (217) 333-3649

ALAN C. LLOYD, Associate Editor
Statewide Air Pollution Research Center
University of California
Riverside, CA 92502
Telephone: (714) 787-3852

# Preface

This book is rooted in a persistent concern which The Rockefeller Foundation and the profession of entomology share with the public: the need to improve the methods and the materials used for insect control in order to combat hunger, improve health, and provide an adequate quality of life for people. To address this issue specifically, the foundation first convened a meeting 6 years ago of entomologists versed in the entire gamut of methods employed for insect control. Four major programs grew out of that meeting, each multidisciplinary and multiuniversity in character. They deal with plant resistance to insect attack, juvenile hormones as sources of selective insecticides, pheromones for their potential impact on the population dynamics of insects, and biodegradable pesticides. These appeared to be the areas of research with strong prospects for yielding pioneering results to bring about effective and safe insect control.

Since pesticides constitute a group of chemicals that society cannot as yet do without, one obvious strategy for insect control is to accelerate the search for selective, biodegradable chemicals that will kill the target insect yet have little adverse effect on other, benign organisms in man's environment. A vital aspect of this strategy has been to enlarge the corps of young biochemists, toxicologists, entomologists, and related specialists capable of broadening the base of knowledge which is essential for designing selective biodegradable pesticides.

The chapters in this book point to the problems and show the recent progress that has been made in the development of such pesticides. But as the insecticides of today and those being fashioned for tomorrow lose their efficacy, or for other good reasons drop out of public favor, the never-ending search for new pesticides must continue.

In April 1974, The Rockefeller Foundation held a meeting at its Study and Conference Center in Bellagio for an updated assessment of the past, present, and future of the science and the art of the chemical approach to insect control. Leaders of the laboratories at which research is being conducted under the foundation's auspices met there with their colleagues from other laboratories. The essays embodied in this volume stem directly from that meeting. They vary in style, content, and point of view expressed, but are bound to one another by common points of view, namely: (1) the urgency to continue to search for chemicals that in makeup may approach the ideal insecticide, and (2) the realization that the complexity and the magnitude of the task call for human capabilities and financial resources beyond those that any one man, laboratory, institution, or country may possess.

We hope that this book will advance a clear and rational understanding of what insecticides may be expected to do, what the real and imagined hazards of their use may be. The ideas the authors have set forth should stimulate creative students and staff to step up their research on biodegradable pesticides and encourage governments, international agencies, industry, and academic institutions to see their way clear cooperatively to amplify their support for the basic research that is so essential to the discovery of better, safer insecticides. We hope further that the information in this volume will prompt the public and private sectors of nations to seek ways and means to put to judicious use the wealth of material and information at hand for exploiting, in the best sense of the word, existing chemicals as insecticides while it paves the way for the acceptance, registration, and expeditious production of new chemicals for insect control.

Robert L. Metcalf
John J. McKelvey, Jr.

# Acknowledgments

Many persons whose names do not appear in this book worked diligently and effectively to bring it into being. To them we are deeply grateful.

Dr. Carlton S. Koehler, Head of the Department of Entomology, Oregon State University and consultant to The Rockefeller Foundation, bore a major load in the production of this work. We are happy to express our appreciation to him for his signal contributions in the preparations leading to the conference held in Bellagio in 1974, in his participation in that conference, and in the work that followed the conference.

The editors are deeply indebted to Mrs. Dolores V. Tanno for her painstaking effort in final typing this volume of the Series.

# Contents

Page

I. INTERNATIONAL ASPECTS IN FOOD AND HEALTH

1. Insecticides in Food Production. William R. Furtick. . . . . . . . . . . . . . 1
2. Insecticides in Human Health. J. W. Wright . . 17

II. LIMITATIONS OF PRESENT DAY INSECTICIDES

3. Development of Resistance to Insecticides. F. J. Oppenoorth. . . . . . . . . . . . . . . . 41
4. Environmental Pollution by Insecticides. W. Klein. . . . . . . . . . . . . . . . . . . . 65
5. Selective Toxicity of Insecticides. G. T. Brooks. . . . . . . . . . . . . . . . . 97
6. Toxic Hazards to Man. J. M. Barnes . . . . . . 145

III. PROSPECTS FOR IMPROVEMENTS OF PRESENT DAY INSECTICIDES

7. Future Use of Natural and Synthetic Pyrethroids. M. Elliott. . . . . . . . . . . . . . . . . . . 163
8. Insecticide Synergism. C. F. Wilkinson . . . . 195
9. Organochlorine Insecticides, Survey, and Prospects. R. L. Metcalf . . . . . . . . . . 223
10. Insecticidal Activity and Biodegradability of Lindane Analogues. M. Nakajima . . . . . . . 287
11. Organophosphorus Insecticides. E. Y. Spencer . 295
12. Carbamate Insecticides. T. R. Fukuto . . . . . 313

IV. NOVEL CHEMICALS AND TARGETS

13. Prospects for New Types of Insecticides. J. E. Casida. . . . . . . . . . . . . . . . . . 349
14. New Targets of Insecticides in the Nervous System. Toshio Narahashi . . . . . . . . . . 371
15. Specialized Features of the Integument. M. Locke. . . . . . . . . . . . . . . . . . . . 397
16. Hormone Mimics. W. S. Bowers . . . . . . . . . 421
17. Pheromones. W. L. Roelofs. . . . . . . . . . . 445

V. PERSPECTIVES Page

18. Historical Aspects of Insecticide Development. A. M. Boyce. . . . . . . . . . . . . . . . . . . . 469
19. Insecticides and Integrated Pest Management. Ray F. Smith . . . . . . . . . . . . . . . . . . . 489

VI. STRATEGY FOR ACTION

20. Summary and Recommendations. R. L. Metcalf and John J. McKelvey, Jr.. . . . . . . . . . . . 509

Index . . . . . . . . . . . . . . . . . . . . . . . . . 513

# INTERNATIONAL ASPECTS IN FOOD AND HEALTH

# Insecticides in Food Production

WILLIAM R. FURTICK
The Food and Agriculture Organization of the United Nations
Rome, Italy

## I. INTRODUCTION AND PAST SITUATION

The trends and extent of growth for the production and use of pesticides since World War II are documented for the industrial countries, particularly in North America, Western Europe, and Japan. The figures are less readily available for the centrally planned societies of East Europe. The overall amount of pesticides sold to the developing countries is reasonably reliable, but the figures for individual countries, and even more the uses made of the pesticides imported, are not readily and reliably available.

Nearly all pesticide production is limited at least in part to the industrialized countries of Europe, North America, and Japan. Although much of the parent pesticidal compounds is formulated into final commercial form in the developing countries—in less frequent cases some steps in the synthesis of the parent compound being carried out in a developing country—there is very little production that includes all steps of synthesis and final formulation.

Although the rate of growth in pesticide use in developing countries has tended to be higher in recent years than in the developed countries, the small base from which this growth started still results in the total tonnage of use by all developing countries to be small in comparison with the countries with highly developed modern agriculture. In 1970 the United States consumed 45% of all pesticide production, Western Europe 23%, Eastern Europe 13%, Japan 8%, and the developing countries 7%, with the rest in Australia and other developed countries.

When only the insecticide sector is examined, it is apparent that the trends in insecticide use compared to other pesticides is considerably different between the developing and developed countries. The relative growth rate between insecticides and other pesticides has been much less in the developed world compared with the developing world. In the developed world, insecticides have declined in relative position among pesticides from being the dominant class of pesticide before 1960 to representing about one-third of total pesticide usage currently. This is in spite of large increases in total tonnage of insecticide use. The decline is primarily due to the rapid growth in the use of herbicides, which now represent the major portion of pesticides used and are still increasing in use at a more rapid rate than the others.

There is an indication that this trend may accelerate because of increased farmer substitution of herbicides for tillage due to much higher fuel and equipment costs and equipment shortages. Weed control is also of importance in preventing nutrient loss, and the increased prices and scarcity of fertilizer have a stimulating effect on weed control. The changing relationships between pesticides in the United States is illustrated in Table 1.

In the developing world, insecticides are the dominant class of pesticide used and are increasing at a rate that would appear to maintain their dominant position for some time to come. However, herbicide use is increasing rapidly where labor supply is depleted in many areas as a result of rapid urbanization.

International trends that have brought about greatly increased commodity prices in 1973 based on decreased stocks of basic agricultural commodities tended to further strengthen demand for pesticides. This was due to both the increase in crop acreage in the United States and the encouragement improved profitability gave to maximize production in the developed world. The result was to cause certain shortages of pesticides starting in 1974.

These same factors created pressure for increased food production in the developing countries, and the same trends were also apparent for large increases in pesticide, fertilizer, and machinery needs. This is clearly shown in Table 2.

The use pattern by crops for insecticides is fairly well known for the United States and some of the other developed countries, but is not well documented for the developing countries. These are illustrated in Table 3.

In the United States considerably more than half of all insecticide usage in agriculture is on cotton. Corn represents about 10% of all use. Of the fruit and vegetable crops, apples are one of the major crops for insecticide use. The breakdown in use by classes of compounds and projected levels in 1975 are found in Table 4.

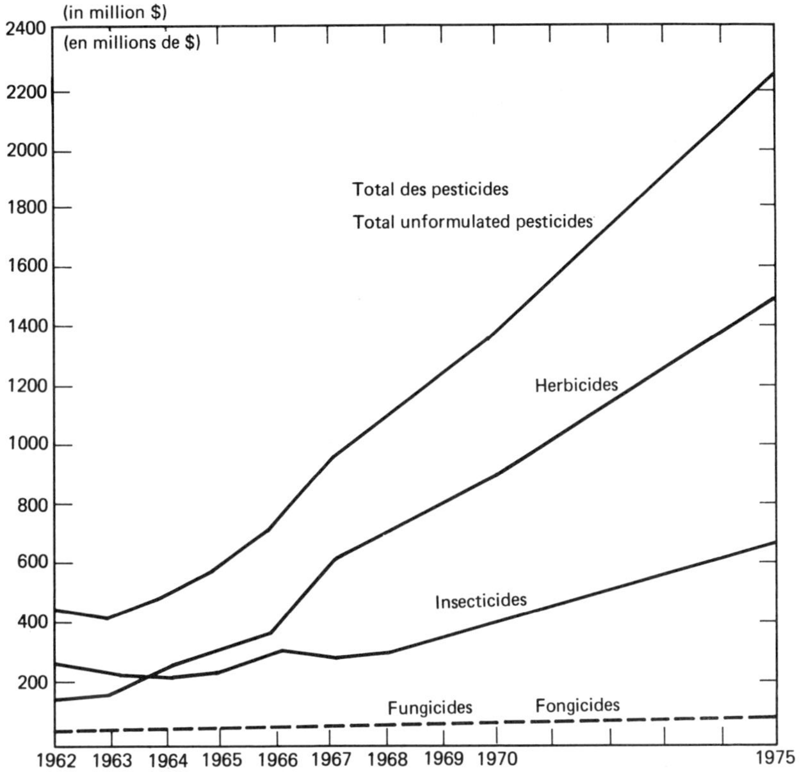
(in million $)
(en millions de $)
2400
2200
2000
1800
1600
1400
1200
1000
800
600
400
200
Total des pesticides
Total unformulated pesticides
Herbicides
Insecticides
Fungicides
Fongicides
1962
1963
1964
1965
1966
1967
1968
1969
1970
1975

TABLE 2. Insecticide Usage in 1972 and Projected Requirements for 1980 in Selected Developing Countries[a]

| Country | OCs | OPs | Carbs | Other | Total | Projected Total |
|---|---|---|---|---|---|---|
| Philippines | 482 | 203 | 267 | 12 | 924 | 24,499[b] |
| Indonesia | 2,263 | 266 | 71 | 109 | 2,709 | 21,382 |
| Thailand | 2,932 | 637 | 56 | 421 | 4,046 | 15,956 |
| India | 18,000 | 3,175 | 3,000 | 160 | 24,335 | 54,500 |

[a]In metric tons of active ingredients. Based on 1973 FAQ/UNIDO Survey.

[b]Based on extremely successful Masagsae Programs.

TABLE 3. Some Examples of Percentages of Crop Acres Treated, of Pesticide Amounts Used on Crops, and of Acres Planted to this Crop (USDA, 1968; USDA, 1970a)

| Crops | Insecticides | | Herbicides | | Fungicides | | % of Total Crop Acres |
|---|---|---|---|---|---|---|---|
| | % Acres | % Amount | % Acres | % Amount | % Acres | % Amount | |
| Nonfood | 1 | 50 | 0.5 | NA[a] | <0.5 | NA | 1.26 |
| Cotton | 54 | 47 | 52 | 6 | 2 | 1 | 1.15 |
| Tobacco | 81 | 3 | 2 | NA | 7 | NA | 0.11 |
| Food | 4 | NA | 11.5 | NA | <0.5 | NA | 98.74 |
| Field Crops | NA | NA | NA | NA | NA | 19 | NA |
| Corn | 33 | 17 | 57 | 41 | 2 | NA | 7.43 |
| Peanuts | 70 | NA | 63 | 3 | 35 | 4 | 0.16 |
| Rice | 10 | NA | 52 | 2 | 0 | NA | 0.22 |
| Wheat | 2 | NA | 28 | 7 | 0.5 | NA | 6.11 |
| Soybeans | 4 | 2 | 37 | 9 | 0.5 | NA | 4.19 |
| Pasture Hay & Range | 0.5 | 3 | 1 | 9 | 0 | NA | 68.40 |
| Vegetables | NA | 8 | NA | 5 | NA | 25 | NA |
| Potatoes | 89 | NA | 59 | NA | 24 | 12 | 0.16 |
| Fruit | NA | 13 | NA | NA | NA | NA | NA |
| Apples | 92 | 6 | 16 | NA | 72 | 28 | 0.07 |
| Citrus | 97 | 2 | 29 | NA | 73 | 13 | 0.08 |
| All Crops | 5 | 54 | 12 | 36 | 0.5 | 10 | NA |

[a]Not available.

TABLE 4. Breakdown of U.S. Market (User Level) - Insecticides

| Type of Compound | Value at User Level, $Million 1971 | 1975 (projected) |
|---|---|---|
| Arsenicals | 22.5 | 13.5 |
| Botanicals | 12.5 | 18.0 |
| Carbamates | 82.0 | 155.0 |
| Organo chlorines | 101.0 | 56.0 |
| Organo phosphorus | 281.0 | 342.8 |
| Total | 499.0 | 585.3[a] |

[a]Based on 1971 prices. (Source: Farm Chemicals.)

In developing countries, it is hard to evaluate easily the end use of pesticides imported, since they are used extensively in public health programs in addition to agricultural uses, and reliable figures on the relative use by sector and by specific crop are not available. It is clear that in the past nearly all use has been on cash crops, with cotton representing at least half of all use. Plantation crops such as sugar cane and tree crops have been major consumers, with vegetables, rice, and maize raised as cash crops of lesser but increasing importance. Use of insecticides to protect stored products is also of substantial importance.

## II. CURRENT INTERNATIONAL TRENDS AFFECTING AGRICULTURAL DEVELOPMENT

The disappearance of basic food stock reserves in 1972 led to a rapid increase in price for not only basic food commodities but also for industrial crops such as cotton and rubber. Some of the beverage and plantation crops such as cacao and sugar were also affected, though coffee and bananas had smaller price increases. The more favorable prices for farmers stimulated both production and demand for inputs such as fertilizer and pesticides. This factor led to shortages and rapidly escalating prices for these products, which produced greater emphasis by developing countries for increased local manufacture to the extent possible.

Inflation on a general basis resulted from the high food prices and shortages caused by rapidly increasing demand for many basic raw materials, such as petroleum, metals, and paper pulp. This created balance-of-payment problems for both industrial countries and developing countries exporting nonmajor raw materials. The balance-of-payment problems increased the desire for nations exporting agricultural products to increase their production as a means of easing payment problems, and the agricultural-importing countries to expand local production to reduce import requirements. Since both of these trends substantially increased pesticide demand, it had a major impact on the pesticide industry. This is particularly true because pesticides produce a high rate of return on money invested by the farmer. The rate of return is greatest in developing countries where low base productivity provides for greater cost-benefit potential.

The basic worldwide shortage of many raw materials that caused much of inflationary pressure made it difficult to expand production readily to meet the demand. The shortage and resultant rapid price rise in petroleum products affected the pesticide industry. Most pesticide manufacture involves petrochemical products, either as basic feed stocks in manufacture or as solvents and emulsifiers in formulation. In addition, the shortages of plastics, metals, and paper products used in packaging also affect production. In some cases shortages of transport for either raw materials or finished products slowed production. All these factors depressed supplies of both pesticides and fertilizers, which was a further factor that affected growth in agricultural production.

The depletion of the world food reserves with the result that food supplies were dependent on current production, which has the normal hazards of weather and other changing factors, created growing worry among world leaders both within and outside the agricultural field. The potential for disaster gave stimulation to policy shifts toward greater effort in agricultural production. This took place first in the industrial nations and led to reversal in United States agricultural policy. These changes in policy at the developing country level gave stimulus to activity in the modernization of agriculture, which is based on heavily increased use of pesticide fertilizer and machinery inputs.

One of the retardants to full modernization of agriculture in developing countries is capital shortages. The capital required to develop new sources of energy and raw materials and to modernize agriculture, all at the same time, strain the capability of world levels of capital formation. This creates very difficult choices among priority areas for expenditure of available capital resources.

## III. CURRENT TRENDS IN INSECT CONTROL AND PESTICIDE USAGE AND THEIR IMPLICATIONS

The growing concern for greater protection of the environment, especially in developed countries, has had major impact on insect control programs. This aspect is most heavily manifested in the efforts that were made to restrict or eliminate the use of pesticides that persist in the environment over long periods; it has had a major impact on the use of chlorinated hydrocarbons such as DDT, BHC, aldrin, and dieldrin.

The pressure from the environmental sector has been aimed not only at eliminating the use of pesticides that persist in the environment, but in minimizing the use of all pesticides. Some of the adverse consequences of widescale use of insecticides such as rapid development of insect resistance and elimination of natural control agents in the form of parasites and predators, have indicated that there is valid justification for restudy of insecticide-use patterns. It has also already led to greatly increased regulatory restrictions on the registration and use of all pesticides.

This trend has led to considerable interest in expanding the integrated control approach for insect management. In this approach an effort is made to keep insecticide use to the minimum and to rely to the extent possible on biological control and improved cultural practices and other methods to prevent the buildup of insect populations to levels that cause economic damage.

The interest in integrated control has caused the desire by entomologists for more specific, narrow-spectrum pesticides that could have the greatest possible impact on target species while at the same time causing minimum effects on other species in the environment. This interest has led not only to greater interest in developing highly specific insecticides, but also in isolating natural compounds that can be used to manipulate the insect population. These controls include such things as hormone or hormone mimics that can disturb the normal development pattern and pheromones that can be used as attractants or whose use can confuse normal behavior. This same trend has caused increased interest in artificial rearing of parasites, predators, insect disease, particularly insect viruses, and genetic manipulation of populations through the sterile male technique, or introduction of genetic lethals to the population.

Although the trends that affect the development of new compounds, the continued use of older compounds, and changing trends in the philosophy of insect control are readily apparent, they are fraught with many problems when it comes to maintaining adequate insect control capability in the future.

## IV. PROBLEMS IN THE DEVELOPMENT OF NEW INSECTICIDES, PARTICULARLY SPECIALTY COMPOUNDS AND NOVEL CHEMICALS USEFUL IN INSECT CONTROL

One of the biggest factors that creates difficulty in the development of any new insecticide is the rapidly increasing research cost required to develop a new product to the market stage. This cost has reached the level of more than $10 million to get a new product on the market and requires an average of 10 years to fulfill all the regulatory and performance requirements. Much of this cost is due to the greatly increased regulatory requirements that have resulted from heightened desire to safeguard man and the environment. There are requirements that are often clearly for the protection of the public sector and are unrelated to the safety and performance requirements needed to ensure an adequate agricultural market at reasonably assured levels of safety.

The fact that the development time from discovery to marketing has more than doubled in little more than a decade to an average time of 10 years has created a very short period of patent protection after marketing that does not give a very great incentive for such a large research investment. A thorough review of patent requirements appears justified.

Another problem in developing new compounds besides the research cost is the basic cost of manufacture. In the first place, the new discoveries consist nearly always of more complex molecules than those discovered earlier, and thus are much more expensive to manufacture as they usually require more steps in their synthesis. This cost is further increased because of much greater costs for new plant facilities and the greater complexity of plant requirements resulting from more complex synthesis.

The risk involved in developing and marketing a new compound that has much greater initial investment and thus requires a much higher price is further increased by the growing number of highly effective insecticides that are already marketed at a relatively low cost. Many of these now have patents that have expired, which means they are available for manufacture and sale on a wide basis at any price that is possible without recovery of research and developments costs (see Table 5).

TABLE 5. Cost per Pound of Commonly Used Insecticides[a]

| Compound (Technical Product-Unformulated) | Price $/lb. |
|---|---|
| Aldrin | 0.992 |
| Dieldrin | 1.665 |
| DDT | 0.179 |
| Heptachlor | 0.973 |
| Malathion | 0.900 |
| Parathion | 0.780 |
| Parathion (methyl) | 0.780 |
| Lindane | 1.406 |
| Chlordane | 0.602 |
| Endrin | 2.469 |
| Dichlorvos | 3.750 |

[a]Based on 1971 prices. From paper presented by R.A.E. Galley at UNIDO Pesticides Workshop, Vienna, 28 May to 1 June, 1973.

The factors already mentioned apply to all types of compounds, including the relatively broad-spectrum compounds which have been the primary interest of industry so far. When the narrow-spectrum specialized insecticides or novel chemistry such as hormones and pheromones are considered, then a further dimension of difficulty has been added. Unfortunately, the registration requirements, time schedule, and other development costs are essentially the same for narrow-spectrum compounds with a much more limited market and the broad-spectrum compounds with a wide market potential. These specialty compounds would normally require either market acceptance at a greatly increased price or a much longer market period to recover initial investment in order to be feasible commercially. This limitation is made more difficult than even the relatively high risk of broad spectrum compounds because of a lack of alternative markets in case of obsolescence due to new competing materials, buildup of resistance, or price competition changes and other factors.

As already mentioned, there is a world capital shortage, which means that each segment of a company is under even greater competition with every other sector for an insufficient capital supply. All the factors mentioned are large negative forces toward favorable capital considerations. They are somewhat counterbalanced by very satisfactory comparative returns from pesticides in the past in most companies. Unless this level of return on investment continues, the attitude of company management could shift very rapidly.

One of the major problems faced, particularly on an international basis, in justifying the risk involved in developing the specialized compounds desired by the entomologists is the lack of adequate data for reliable market analysis. The present and future magnitude of the various key insect problems, the types of parasites and predators needing protection, the price that could be justified for specialty compounds, the likelihood of major changes in production or agricultural practices that might affect use, and so forth, are frequently lacking.

There is also the growing problem that any new product may face increasing restrictions or expenses to gain access to developing-country markets. These restrictions are often placed for the following reasons:

1. To protect the products of local manufacture.

2. To force local manufacture or formulation.

3. To save foreign exchange.

4. To stimulate development of local institutions and research by:

    a. requiring local toxicological data;

    b. local residue data;

    c. local performance data beyond normal requirements.

Research and marketing activity are concentrated in the hands of the industrial sector for increased chemical research by governments, universities, and private institutions. At least in the Western industrial countries the ultimate marketing of compounds is limited to the industrial sector. The enormous research costs and time period required for development and the importance of adequate international patent protection make non-industry development nearly impossible. This hurdle is made more difficult by the conflict in desire by public sector research to make all discoveries widely available and the need of industry for complete control to protect a high investment and risk.

The risk to industry in devoting major resources to developing the type of compounds urged by the proponents of integrated control is greatly increased by the lack of both adequate research in the public sector and the trained staff required to shift successfully to integrated control on a major basis.

In spite of all these factors that are negative to major industrial participation in developing the requisites for sound pest management programs based on integrated control, there are other trends that are causing several of the major industrial concerns to consider major new initiatives. The restrictions on application leading toward prescription use of pesticides and the fact that most of the pest management programs being developed cannot be successful except on an area-wide basis under highly skilled supervision open the possibility of vertically integrated industrial entry into pest management. This could encompass the manufacture of chemicals, rearing of parasites and predators, and manufacture of virus and other pathogens and their coordinated use at the field level. If the industrial sector makes a decision to move in this direction, it would have major global implications for the whole field of insect control.

## V. POTENTIAL MEASURES TO OVERCOME THE PROBLEMS

1. Public Assumption of some Research Costs. It has already been indicated that a major portion of the development costs for new products has been levied through registration requirements designed entirely to give maximum protection to the public. The high degree of public benefit not only from the protection these requirements provide, but also from the assured supply of wholesome food at a reasonable price dependent on adequate insect control would justify public assumption of part of the cost. There are a number of ways this can be done. There are already examples in public partnership in weaponry development and other sectors of high public interest including atomic energy use.

2. Special Incentives for Research and Development on Desired Lines in the General Interest. A number of countries already provide incentives, from tax writeoffs to direct subsidies, to encourage specific types of research.

3. Changes in Patent Conventions. The long development period required in the public interest should lead to a general specified patent period of perhaps 10 years from date of registration rather than the present system based on filing date.

4. Internationally Adopted Registration Standards. A major added expense in the international development of new pesticides is the repetition of research required by differing standards and procedures for toxicology, residue analysis, and biological performance. There is no reason why internationally acceptable standards could not be developed to reduce both the cost and time required for development. It is recognized that it may be difficult to develop adequate agreement in all areas, but such international standards could be implemented in segments and would not require agreement in all areas before having an impact.

5. Internationally Developed Basic Information on Use Potentials, Requirements, and Related Statistics. The problem of lack of adequate information on which to develop market projections, research priorities, and capital allocations have already been discussed. Although developing adequate information would be a large task, it would appear a logical function of the international organizations and would be useful as developed without requiring complete information before having value.

6. Greatly Expanded International Research Network. One of the major costs in developing a compound for the various potential uses in the widely dispersed international markets is the cost of the biological and performance information under different ecosystems. Greater effort to develop a coordinated research network for this purpose which would provide not only research capability but improved speed of information exchange would speed up the time and reduce development costs immeasurably. The size of such an undertaking and the various problems involved are not to be minimized, but if the network were built step by step from the most important areas and problems first, it could have early impact long before full development.

7. Internationally Coordinated Intensification of the Necessary Training at all Required Levels to Implement Improved Pest Management Programs. One of the major limitations in implementing the improvements in insect control now possible with present technology, and in developing required new technology, is the lack of adequately trained staff at all levels, particularly in the developing countries. In the past, training

programs at the international level have concentrated primarily on training the research level staff. Training at this level needs to be intensified and broadened to include middle level staff and lower level technicians who are equally important in effectively implementing programs. A thorough study is needed of the most efficient methods of developing the needed training required.

## REFERENCES

(1) Association American Pesticide Control Officials. 1957. Pesticide Official Publication and Condensed Data on Pesticide Chemicals, Supplement, College Park, Maryland.

(2) Borkovec, A. B. 1970. Insect Chemosterilants, Wiley, New York.

(3) Council on Environmental Quality. 1972. Integrated Pest Management, Superintendent of Documents, Washington, D.C.

(4) Department of Agriculture. 1966. Suggested Guide for Use of Insecticides to Control Insects Affecting Crops, Livestock, Households, Stored Products, Forests, and Forest Products, Washington, D. C.

(5) Djerassi, C. 1973. Science, 181, 115.

(6) Fitzsimmons, K. P. 1972. Role of Industry in Advancing New Pest Control Strategies in Pest Control Strategies of the Future, National Academy of Sciences, Washington, D. C., pp. 352-361.

(7) Fukuto, T. R. 1957. Advances in Pest Control Research, vol. 1, Interscience, New York.

(8) Could, R. F., Ed. 1963. New Approaches to Pest Control and Eradication, Adv. Chem. Ser., No. 41, American Chemical Society, Washington, D. C. (1966). "Natural Pest Control Agents," ibid, No. 53.

(9) Jacobsen, M. 1965. Insect Sex Attractants, Wiley, New York.

(10) Kohll, C. F. 1973. The Formulation of Pesticides in Developing Countries, United Nations Industrial Development Organization.

Malaria thus remains a major public health problem in the tropics and subtropics. Although its eradication is within sight in many countries, in others representing a significant proportion of the world's population, long-term programs of control or the protection of communities exposed to the greatest risk are the best that can be expected.

In the absence of a vaccine against the disease and the difficulties associated with the implementation of mass chemotherapy operations, reliance will remain on the use of insecticides well into the foreseeable future. DDT is still the insecticide of choice in the majority of antimalaria programs, applied generally at 2 $g/m^2$ in the form of a water-dispersible powder. It has been estimated that the minimum quantity needed for the next 10 years is 40,000 tons of technical DDT per annum. Resistance to the chlorinated hydrocarbon insecticides has emerged in approximately 1% of the areas treated. The substitutes most commonly used are the organophosphorus materials malathion and fenitrothion and the carbamate propoxur (Wright *et al*., 1969). Total consumption of these compounds is currently in the vicinity of 2,000 tons of technical material per annum, but this will certainly increase as resistance to DDT spreads.

Consequently, a program of research will be essential during the next 20 years directed toward making the best use of compounds presently available, in the development of new and safe chemicals to control resistant species, and the design of an imaginative plan integrating all available measures into one overall operating procedure.

## III. FILARIASIS

It has been estimated that at least 250 million persons suffer from lymphatic filariasis; about 200 million of these are infected with *Wuchereria bancrofti*. This group of infections has a wide distribution affecting large tracts of Asia, Africa, and Latin America. Although there are indications that there may be a decrease in transmission of *W. bancrofti* in certain parts of the world such as Mauritius, Cape Verde, Réunion, Togo, Egypt, Japan, and Sri Lanka, in most situations there is a steady progression in both prevalence and distribution, with Asia and Africa being mainly affected. It has been estimated that the total population at risk in the world has doubled in 20 years: for example, in India the population exposed to the infection has risen from about 70 million in 1960 to 122 million in 1969. This increase has occurred in both urban and rural areas, but it has been particularly marked in the former, where there has been a great increase in actual and potential mosquito habitats.

Man is the only vertebrate known to harbor W. bancrofti, making him both reservoir and host. Transmission is through a mosquito, the most important species being Culex fatigans. This insect flourishes in unsanitary conditions, with massive breeding occurring in polluted drainage ditches and similar collections of water. There is no vaccine available to protect populations from acquiring the disease. Human filariasis can be controlled by drug therapy; diethyl carbamazine will destroy the majority of microfilarial and some of the adult worms. However, few mass-therapy campaigns have been successful, largely because of the adverse side effects of the drug. Neither biological nor genetical control is operationally practical at present. The only recourse open to public health authorities is, therefore, through the control of the vector (WHO Tech. Ser. No. 542).

The insecticidal approach has up to now been directed largely at the larval forms of the mosquito. In this, the organochlorines have not been completely successful, largely because of the inherent insensitivity to this group of chemicals of the average Culex fatigans population. In Rangoon it was found that before insecticidal application was begun a large proportion of the mosquito population was partially resistant to DDT and dieldrin, and that complete resistance could be anticipated if it were put under heavy pressure from these compounds (Rosen, 1967). Similar situations have been reported from Africa and Latin America (Brown and Pal, 1971). However, certain organophosphorus insecticides have brought about a satisfactory level of control in some situations. Of these the most successful have been fenthion, chloropyrifos, and chlorfenvinphos, not only because of their effectiveness and persistence in polluted water but also because of their ability to withstand the detoxifying effect of Bacillus subtilis and other organisms occurring naturally in contaminated water (Yasuno et al., 1965). In Rangoon City, Burma, fenthion has been used successfully on a routine basis, with weekly applications at 1 ppm since 1967, without resistance having emerged (Graham et al., 1972). Larviciding oils have been shown to be less effective and more expensive in comparison with these compounds (Self and Tun, 1970). Adulticiding programs have also been found to be expensive and have seldom produced the satisfactory results obtained with larvicides. However, for the control of Anopheles species that transmit filariasis in Africa, residual house spraying should be effective in areas where the mosquito populations have not yet developed resistance to the insecticide in use. The insect growth regulating compounds are being examined for their possible role in controlling the vectors of this disease, but it is too early to assess their operational potential.

Many governments, concerned with the increase in the incidence of filariasis, are planning or undertaking control programs, in which larviciding with one of the insecticides mentioned is being undertaken. These programs will inevitably increase in size and scope. It is anticipated that the resources to bring about major changes in the sanitary state of affected areas will be beyond the means of the average administration in the foreseeable future. There will, therefore, be an increasing demand for insecticides, and it may be assumed that this will continue for decades to come. It can also be anticipated that where insecticides are applied for long periods of time resistance will ultimately develop. The need, therefore, will not only be for increased quantities of existing compounds but also for effective substitutes. It is almost impossible at this stage to assess the quantities of insecticides and formulations that will be required during the next 10 to 20 years as the procurement of insecticides is normally a local rather than a national responsibility.

## IV. DENGUE HAEMORRHAGIC FEVER

Dengue fever has been an important health problem in many countries for a number of years. Large epidemics have occurred in widely separated geographic areas wherever the vector has occurred in high densities. Typical among these is that reported from Greece in 1927–1928. From about 1950 onward the situation was aggravated by severe outbreaks of haemorrhagic form of the disease in areas where dengue fever had been endemic. The basic cause is considered to be exposure of the human population to dengue viruses of different serotypes. These outbreaks occurred first in the Philippines, but have since occurred further to the west as far as the coast of India and Sri Lanka (Halstead, 1966). Several outbreaks have also been reported from the Americas. It is notable that dengue fever itself is now almost completely absent from the Mediterranean basin since the disappearance of *Aedes aegypti* from the area (Curtis, 1967).

As in filariasis, man is the reservoir and host. The main vector is *Ae. aegypti*. There is a marked relationship between urbanization and the incidence of dengue and dengue haemorrhagic fever. This species is highly domesticated and is seldom found far from man's habitation. It is this characterization that made its eradication feasible in certain countries in the Americas. With the crowding of man into the periphery of cities and towns the provision of water becomes a necessity. This is usually provided through strategically located standpipes. This in turn leads to extensive water storage in homes, which frequently gives rise to mosquito breeding. The situation is worsened by the ineffective disposal of discarded tires, bottles, cans, and other containers that can be transformed into breeding places. In certain cities, for example in Venezuela, up to as many as 2,000 potential breeding containers have been found around houses with garbage removal services (Gonzalez-Valdivieso, 1971).

Considerable success has been achieved in reducing or even eliminating mosquito populations by the application of larvicides. Until the widespread development of resistance in the Americas, the hydrocarbon compounds including DDT, dieldrin, and HCH gave excellent results. Because of resistance these have since been replaced by malathion, fenthion, and tetrachlorvinphos used as peripheral sprays (Schliessman and Calheiros, 1974). Where the greater part of the breeding occurs in containers holding water for domestic use, temephos (Abate) has been successful. This compound has the advantage of safety to man at the dosage used, lack of odor or taste, effectiveness, and persistence. It is the compound of choice in the majority of Asian and African towns and cities, used at a dosage of 1 ppm in the form of sand granules (Bang and Pant, 1972). The demand for insecticides of this nature as well as those which may be used for perifocal spraying will continue for an almost unlimited period of time. However, should resistance to temephos occur its replacement will represent a serious problem.

Adulticiding mainly through the application of compounds by the ULV technique has been demonstrated to be effective not only in reducing populations of *Ae. aegypti* during outbreaks of the disease but also in holding populations to a low level as a preventive measure. Experiments performed in Thailand applying malathion from the air at a dosage of 6 US fluid ounces per acre (438 ml/ha) 4 days apart over an urban area of 18 $km^2$ brought about a reduction in the adult *Ae. aegypti* population of between 95% and 99% after each spraying and an overall reduction of 80% to 99% for a period of 10 days (Lofgren *et al.*, 1970).

Experiments performed in Bangkok using ground application equipment for ULV were also successful. Applying fenitrothion through a nonthermal generator at a dosage of 438 ml/ha, in two treatments 2 days apart, brought about a 99% reduction of the mosquito population, which was maintained for 2 weeks. Large-scale applications using the same technique at a dosage of 511 to 1095 ml/ha gave sustained control for 4 to 5 months (Pant et al., 1973).

It would appear that the need for insecticides in Latin America for the eradication of Ae. aegypti through larviciding or peripheral spraying will remain high for at least the next 10 years. On the other hand, the control of the adult and larval forms will be a major public health problem for a much longer period in Asia, Africa, and parts of the Caribbean. Insecticides of unusual characteristics will be needed in both situations. In some cases compounds will be introduced into drinking water and in others insecticides will be applied in ULV quantities in densely populated urban areas. The greatest of care will have to be exercised to ensure that there are no adverse effects to those persons exposed to them.

No panacea exists with regard to this species through biological or genetical control. One cannot see techniques being developed for many years, and even when this is done these techniques will have to be integrated with insecticide usage.

## V. ONCHOCERCIASIS

Onchocerciasis is a parasitic disease of great social and economic importance, particularly in the savannah areas of Africa. It is not only a major cause of blindness and incapacity; it also prevents the human settlement and agricultural development of many fertile river valleys which at present lie uninhabited and unproductive.

The infection by the filarium Onchocerca volvulus is transmitted from man to man through the bite of the female blackfly Simulium spp. which breeds in the rapids of fast flowing streams and rivers. As no drug is currently available for the mass treatment of the disease and direct control of the adult fly is not practical, control is concentrated upon the destruction of the larvae in the circumscribed sites in which they mature. The physiological requirements of the larvae are such as to make this possible. Attached to grass, rocks, and other submerged supports, they depend on the turbulent water passing over them to provide oxygen and food. Consequently, particles suspended in the flowing water are consumed indiscriminately. It has been demonstrated in a number of programs in East and West Africa that the ingestion by the larvae of a suitable insecticide with this material results in their destruction.

One of the most important examples in international cooperation in recent times is the agreement reached on the strategy for controlling this disease in the Volta River Basin and the subsequent economic and agricultural development of the area. Seven countries, Dahomey, Ghana, Ivory Coast, Mali, Niger, Togo, and Upper Volta, are cooperating with the United Nations Development Program, the Food and Agriculture Organization, the International Bank for Reconstruction and Development, and the World Health Organization, supported financially by a group of donor countries, in carrying out a program of control extending over a period of 20 years. The area has a superficial surface of 700,000 square km, and contains almost 10 million inhabitants, of which more than 1 million are infected with the parasite; of these, 70,000 are considered to be "economically blind." Full-scale operations were scheduled to begin in late 1974, with WHO as the executing agency.

Temephos in the form of an emulsion concentrate applied at the rate of 1 ppm has been shown to be the most suitable toxicant. This choice has been based not only on its biological effectiveness but also on its biodegradability and its safety to man and other nontarget organisms. Application of the insecticide will be made to breeding places from the air using both fixed and rotary winged aircraft. Baseline data on the resistance of the blackfly to temephos have been completed, and trials are under way on the development of a substitute to this compound should resistance to it emerge. The difficulty in establishing colonies of this insect is a handicap in these studies, particularly when attempting to determine cross-resistance patterns.

Should this program be successful it will no doubt lead to the establishment of similar programs in other parts of Africa. The need for new insecticides and specialized formulations will therefore be required for a number of decades to come. However, it will be essential that this search for substitutes be done on the basis of the problems peculiar to the area and the ecology of the vector (WHO/OCP/73.1).

## VI. AFRICAN TRYPANOSOMIASIS

Human sleeping sickness and animal trypanosomiasis constitute major obstacles to the economic development of large areas of tropical Africa, with man and his animals being excluded from land suitable for stock raising and agricultural development by the presence of the tsetse fly *Glossina* spp. The situation has been aggravated in West Africa by persistent drought which has caused the death of large numbers of cattle and forced people to move from the dry savanna zone to the more humid tsetse fly belt in the south.

In the past a number of procedures have been used to control this disease, including chemotherapy, game destruction, and vegetation clearance. None of these have proved to be wholly satisfactory, and at present the only practical and economic procedure open to governments in Africa is the control or extermination of the fly through insecticide applications.

In the relatively open savanna the fly may be controlled either by the selective application of residual compounds such as DDT or dieldrin, or by three to five nonresidual, aerial applications made 2 to 3 weeks apart. In these, materials such as endosulfan, fenthion, or pyrethrins have been found to be effective.

In the dry savanna the riverine species of fly breed and rest under the shade of trees and bushes growing close to water courses. These can be controlled by a single residual application of DDT or dieldrin to the resting places, usually from the ground. Riverine species have also been controlled by residual applications of dieldrin from the air but to date success has been achieved only with very high dosages (0.5 to 2 kg/ha) (Pal and Wharton, 1974).

In humid environments the situation is more difficult. The distribution of *Glossina* is usually more widespread, at least during the rainy season, vegetation cover is more extensive, and residual applications are not effective for long periods of time for climatic and other reasons. In such situations dieldrin is usually more effective than DDT.

Serious environmental problems have arisen because of the treatment of large areas with blanket or even selective applications at high dosages of persistent and/or toxic insecticides such as dieldrin. Even large-scale applications at low dosages of endosulfan, an insecticide with a relatively low persistence, will be suspect until thorough studies on the possible hazards to man, animal, and other nontarget organisms have been completed.

An intensified search for compounds that will be effective in different environmental conditions is urgent. In this the examination of readily available and economically priced materials should be given preference. Parallel studies on persistence, improved formulation, more efficient application equipment, and the assessment of acute and long-term hazards to warm-blooded animals will be essential.

## VII. CHAGAS' DISEASE

Human cases of Chagas' disease have been recorded from almost every country in South America. It is a public health problem of the greatest importance in Brazil, Argentina, Chile, and Venezuela; in the latter country it has been estimated that some 2.8 million people are suffering from the disease. It is estimated that altogether about 7 million or more people are infected with the causative agent *Trypanosoma cruzi* and that of this number several million people suffer from heart or other damage as a result of the infection. Chagas' disease has been described as the most important cause of myocarditis in the world.

The vectors of the disease are blood-sucking bugs in the subfamily Triatominae, most species of which are found in the new world. There are some 95 potential vector species in the genera *Panstrongylus*, *Rhodnius*, and *Triatoma*. Transmission of the pathogen is usually by contact with infected feces of the bug. The rate of infection with *T. cruzi* in bugs in nature can reach as high as 75% in some areas of South America and in many rural locations 100% of the dwellings are infested with Triatominae. The bugs breed easily in areas of poor housing and are found in large numbers in cracks in the mud walls and in thatch walls and roofs.

At the present time, there is neither cure nor immunological protection for the disease. Thus an effective treatment of Chagas' disease represents a major therapeutic challenge. For the time being, prevention and control of the disease must depend on control of the vector bugs. The effectiveness of this control is not altogether satisfactory at present. There are several important control programs in the Americas that have as their objective the elimination of house infestation by means of insecticides and by housing improvement; little or no action has been taken against any extra-domiciliary vectors. Improvements in the socioeconomic status of the affected human populations, particularly as regards the quality of their housing and sanitary habits, would both reduce the degree of infestation and the probability of reinfestation. As this is a lengthy process, immediate control is being undertaken through the use of insecticides. The most effective control measure currently available consists of spraying the interior surfaces of dwellings with HCH (benzene hexachloride) or dieldrin to destroy the vector. DDT has not proven effective against Triatominae. In addition, the use of this insecticide in malaria eradication programs in rural areas not only had little effect on the bugs, but eliminated triatomid predators and parasitoids, thereby worsening the situation.

Very little is known about the biology and ecology of most peridomestic Triatominae, and relatively few extensive ecological studies have been carried out on purely domestic species. As a result of this, reliance has had to be made on nonselective insecticides providing the longest possible residual effect. This course raises operational costs and increases mammalian toxicity hazards.

As indicated, it is generally accepted at present in South America that the most effective insecticides against Triatominae are dieldrin and HCH; the use of dieldrin, however, has in the past caused many cases of pesticide intoxication among spraymen, some of which occurred in isolated areas where immediate treatment was not available, as well as many cases of poisoning and death in domestic animals. As a result, use of this compound has been withdrawn from most country programs. A recent development that will certainly give rise to serious consequences in the future is the appearance of insecticide resistance to HCH and dieldrin, that is, the cyclodiene insecticide group in certain populations of *Rhodnius prolixus* in Venezuela. Effective control of the bug with carbamate insecticides has been achieved in a number of limited trials; however, the data are as yet insufficient to justify large-scale operational use of these compounds (Valdivieso, 1968). Consequently, it is now imperative to accelerate the screening and field testing of possible alternative compounds to provide present and future operational programs with insecticides of proven effectiveness as soon as possible.

Systematic screening of compounds passing through the WHO program for the evaluation and testing of new insecticides for their effectiveness against Triatominae has now assumed a greater degree of urgency. After initial laboratory evaluation tests, it is intended that, in the future, candidate compounds against Triatominae will be tested in small- and large-scale field trials both by national groups collaborating with the WHO and by the newly established WHO Chagas' Disease Vector Research Unit in Acarigua, Venezuela.

## VIII. PLAGUE

The nature of bubonic plague as it exists today with natural foci spread throughout the world provides ample opportunity for the recrudescence of this disease. Outbreaks that have occurred in recent years in Indonesia and Vietnam demonstrate that plague can break out of these foci and affect large number of persons. Provided they can be instituted at an early stage, flea-control measures are still considered to be the most effective means of controlling outbreaks of flea-borne plague. Until recently DDT was used almost universally for rat flea control. However, resistance has necessitated the substitution of organophosphorus and carbamate compounds in many parts of the world. Here also insecticides will be required indefinitely. The biological control of fleas is not considered to be feasible.

Growing urbanization and the spread of single-crop culture have had unfortunate consequences, not least of which is the increase in both urban and rural rat populations in the United Kingdom, Denmark, and other countries.

The development of specific, effective rodenticides is urgently required, and this need will continue for an indefinite period of time.

## IX. TYPHUS FEVER

The control of body lice still depends almost entirely upon DDT. Although resistance to this compound has appeared in some areas, in almost all situations malathion has been a satisfactory substitute. It is unfortunate that the first emergence of resistance to malathion has been reported from Burundi. Effective alternatives among the carbamates with low toxicity to man have been found, and no concern is currently felt regarding the future control of this insect. There is no doubt that insecticides will be required indefinitely for lice control (Wright, 1971).

## X. ALTERNATIVE METHODS OF CONTROL

### A. Genetic Control of Insect Vectors of Disease

During recent years, the development and evaluation of genetic control techniques for the control of insects of medical importance have received increased attention in several parts of the world. For example, in Africa various genetic mechanisms are being considered and evaluated for application against *Aedes aegypti*; in Africa and in the United Kingdom the potential value of sterile hybrids within the *Anopheles gambiae* complex is being studied; development of the sterile male technique for use against tsetse flies is being considered; in 1972 a field release of sterile males of *An. albimanus* in El Salvador, Central America in an isolated area of 5 $mi^2$ successfully demonstrated population reduction; in Pakistan a large number of genetic stocks have been developed from *Culex tritaeniorhynchus*; and lastly, the sterile male technique is under study in relation to *Culex tarsalis* in the western United States (Pal and Whitten, 1974).

The research unit being conducted by WHO and the Indian Council for Medical Research on the genetic control of *Culex fatigans*, *Aedes aegypti*, and *An. stephensi* is carrying out studies in Delhi on the operational feasibility of genetic control on an extended scale. A variety of genetic mechanisms, that is, radio and chemosterilized males and males with cytoplasmic incompatibility factors and/or with translocations, are being investigated. Efforts are also being made to isolate and develop potentially useful genetic factors such as conditional lethals and genes conferring refractoriness for disease transmission. Pilot scale release experiments with chemosterilized *C. fatigans* and an integrated strain of this species with cytoplasmic incompatibility and sex-linked translocation complex have been carried out, and these have led to about 70% to 80% sterility of the eggs laid in the wild population, but a higher percentage of sterility has not been achieved for an extended period, and there has been no evidence of population reduction in view of the massive infiltration from adjoining areas. It appears that there is little to choose between the two methods and it is also evident that under the conditions of massive immigration and strong density dependent regulation of the population as in *C. fatigans*, it would be better to consider replacement of the population by strains carrying a refractory gene for filaria transmission.

In experiments with *A. aegypti*, it has been demonstrated in India that released mosquitos with a semi-dominant marker or translocation could inject this character into the wild population. The studies on this species have reached a stage where a large-scale genetic control experiment will be performed in 1975.

Although these experiments indicate that optimism for the eventually successful application of genetic control methods for vector species is well founded, a more detailed background of ecological information and a sophisticated level of development are required. In addition, it is doubtful whether the average developing country in which these techniques are most needed will be able to produce the highly skilled staff required to apply the technique. In these circumstances it would be unwise for health authorities to place any dependence on these procedures for 10 to 15 years.

## B. Biological Control of Disease Vectors

During the past decade WHO and a number of collaborating laboratories have made attempts to determine the possibility of using predators, parasites, microorganisms, and fungi for the control of vectors of public health importance. The present status of the work follows.

The nuclear polyhedrosis viruses and the granulosis viruses are currently receiving considerable attention, and their development has been stimulated by the pressures placed on the persistent chemical pesticides. In relation to vectors of public health importance, no virus has satisfactorily fulfilled the requirements of WHO for simulated trials and field testing. The potential safety hazards and the difficulty in characterizing these viruses and their host range make it highly unlikely that these will be used as biological control agents of any arthropod of medical importance within the near future. Their use in agriculture and forestry, of course, is already extensive but conclusive evidence is still needed concerning their safety to man and nontarget organisms (WHO/TRS.531).

*Bacillus thuringiensis* has not yet been demonstrated to be an effective practical agent for the control of insects of medical importance. Several new isolates of *B. sphaericus* are promising against certain mosquito species, and the requirements of primary testing are essentially complete. It is anticipated that limited field trials will shortly be conducted by WHO in one of its field research units. As promising as this new strain appears to be, several years will be necessary to obtain adequate information before its use as a functional insecticide can be established.

Repeated field trials have been conducted by a number of investigators, some with WHO support, in the use of *Coelomomyces* and *Lagenidium* and other fungi against mosquito larvae. Although these have sometimes been encouraging, the fungi have on the whole not performed as well as was originally hoped, so that research activity with them has been somewhat reduced in recent years. Recent work on infectivity and the method of sporulation has shown promise, but no field trials can be contemplated for the present.

Progress of studies with microsporidia, especially with *Nosema* for mosquito control, has been encouraging but only in the laboratory at present. Because of this possible cross-infectivity to mammals, considerable work is necessary on their safety and the practical application of these agents is not imminent.

Mermithid nematodes are presently the subject of several

extensive research programs as control agents for blackflies and possibly for tsetse flies. Although these organisms have performed well in the laboratory, mass culture creates a problem; this work may therefore be considered as being in the preliminary stages only.

Field trials have been conducted to evaluate the role of larvae of *Toxorhynchites* for *Aedes* mosquito control, but these have had only limited success. The major problems concern the sites of larval habitats plus the inherent problem of predator/prey interaction, which is usually only successful when the prey population is at its greatest. Telanomid wasps that parasitize the eggs of various species of triatomids are also being studied. Mass rearing, as would be required for control activities, has not been demonstrated, nor have there been any convincing tests under field conditions. It would appear that the general utility of introducing predators of insects of medical importance has yet to be demonstrated, although naturally occurring predators certainly play a considerable role in determining the bionomics of vector populations. However, there does not appear to be any way to enhance this natural predation to a great degree.

Various species of exotic fishes, particularly *Gambusia* and *Poecilia*, have been distributed around the world for approximately half a century, and under certain conditions have proved successful in the control of mosquito populations. On the other hand, an equilibrium with mosquito population is frequently reached, and their larvicidal effect is blunted. In addition, these aggressive and highly fecund species may often bring about undesirable changes in the fauna of the ecosystem. Their continued use is to be encouraged, but only under well-controlled conditions, especially in manmade habitats (wells, reservoirs, ditches, etc.) where their effect on other elements of the local fauna will be minimal. Studies are being conducted to evaluate the utility of various indigenous species of fishes in several places in the world. Although the use of these species may be more desirable for general and environmental reasons, their utility will only be shown should they be reared in very large numbers.

As far as vector species of human disease are concerned, the use of biological control agents as a routine operation cannot be considered feasible in the near future.

## XI. ALTERNATIVE COMPOUNDS

In an attempt to meet the challenge of insecticide resistance and environmental pollution, WHO established in 1960 a program for the systematic evaluation of new compounds that might be used in public health. Details of this scheme have been described by Wright (1971). However, it may be valuable to consider its structure in broad outline.

Newly developed compounds are submitted to WHO by the chemical industry (up to 45 companies have participated during the past 10 years) and a number of universities and independent institutes. Each compound is subjected to a series of seven evaluation stages, each succeeding stage demanding more exacting criteria of effectiveness and safety. Three of these stages are performed in the laboratory and four in the field. By meeting the criteria for each succeeding stage a compound advances to the next higher level of testing, until it finally qualifies for large-scale field testing in some cases, such as malaria, to determine whether the application of the compound is actually capable of interrupting transmission. Thus it moves from early assessment under controlled laboratory conditions to ultimate evaluation under normal stress of the field environment. The actual evaluation is performed by a complex of collaborating laboratories and WHO field research units, each playing its part in terms of agreed protocols.

The scheme has been of value to WHO from many viewpoints. The challenge of resistance has been met by developing compounds to meet emerging situations, to recognize and to influence trends in pesticide development, to obtain a sound knowledge of the toxicology and safety of different groups of compounds, and to approach environmental pollution with a sense of realism. No vector control program has yet failed because of resistance only; it has always been possible to produce an alternative material that was effective and safe.

Up to now almost 1800 compounds have been entered into the scheme, of which nearly 1500 have been actively examined. Between 1961 and 1965 the greatest number was forthcoming. Since 1965, coinciding with the first appearance of financial and other difficulties by companies, the numbers have declined considerably. At the same time contributions from nonindustrial sources have increased.

If one looks at the status of the scheme as it existed from October 1971 to date, one will see the degree to which the situation has deteriorated. There were 131 compounds received during this period of 2 and 1/2 years. Of these only 39 were recommended for examination at stages II and III or higher. Six were later withdrawn by the manufacturer; six were considered to be too toxic for field use; four were shown to lack sufficient biological effectiveness; fourteen were from nonindustrial sources, and to date few manufacturers have indicated their willingness to produce the compounds; nine are under study or are awaiting additional information from the producer. Only one appears to have promise for malaria control. Two are growth regulating agents and two are synthetic pyrethroids.

The reasons for the reluctance of industrial organizations to play a greater role in pesticide development are well known. However, events during the past 6 to 12 months are aggravating the situation. WHO's situation is further affected by the fact that in very few cases is there sufficient production potential in any compound to warrant its production for public health purposes only. WHO is compelled in most cases to await a decision by the manufacturer to proceed with production for agricultural purposes. Temephos is an important exception. The problem is no easier in the development of special formulations required in public health programs.

The widescale use of pesticides in agriculture has the effect of bringing great pressure on vectors of public health importance, the exposure of which is incidental. In practice it is frequently the agricultural use of a compound that brings resistance in vectors rather than public health application.

Finally, it is clear that many industrial organizations are approaching with great care the acquisition of patent rights in compounds produced by nonindustrial institutes. In no case but the pyrethroids and a limited number of analogues of DDT has WHO been able to obtain sufficient samples of undertake large-scale testing. It is WHO's policy not to proceed with stages V, VI, and VII testing of compounds not in general manufacture.

Thus the greater part of WHO's current program is related to compounds already in production—and for which WHO is attempting to find their most effective and integrated use by ecological, biological, chemical, and engineering investigations.

## REFERENCES

1. Bang, Y. H. and Pant, C. P. 1972. Bull. W.H.O. 46, 416

2. Brown, A. W. A. and Pal, R. 1971. "Insecticide Resistance in Arthropods." WHO Monograph Series 38.

3. Curtis, T. J. 1967. J. Med. Ent. 4 (1), 48-50.

4. Gonzalez-Valdivieso, F. E. 1971. Actual Situation of the Ae. aegypti Campaign in Venezuela. WHO unpublished document WHO/VBC/71.315.

5. Gonzalez-Valdivieso, F. E. Ensayo de Campo de la Accion del Insecticida OMS-33 Sobre *R. prolixus*. (Unpublished report).

6. Graham, J. E., Adbulcadar, M. H. M., Mathis, H. L., Self, L. S., and Sebastian, A. 1972. *Mosquito News* *32* (3), 399.

7. Halstead, S. B. 1966. *Bull. W.H.O.* *35*, 3.

8. Lofgren, C. S., Ford, H. R., Tonn, R. J., and Jatanasen, S. 1970. *Bull. W.H.O.* *42*, 15.

9. Pal, R. and Whitten, J. J. 1974. *The Use of Genetics in Insect Control*, Amsterdam, Elsevier/North Holland.

10. Pal, R. and Wharton, R. H. 1974. *Control of Anthropods*. Plenum, New York.

11. Pant, C. P., Nelson, M. J., and Mathis, H. L. 1973. *Bull. W.H.O.* *48*, 455.

12. Schliessman, D. J. and Calheiros, L. B. 1974. *Mosquito News* *34*, 1.

13. Rosen, P. 1967. *Bull. W.H.O.* *37*, 301.

14. Self, L. and Tun, M. M. 1970. *Bull. W.H.O.* *43*, 841.

15. Wright, J. W. 1971. *Bull. W.H.O.* *44*, 11-22.

16. Wright, J. W., Fritz, R. F., and Haworth, J. 1972. *Ann. Rev. Entom.* *17*, 75.

17. Wright, J. W., Fritz, R. F., Hocking, Kay S., Babione, R., Gratz, N. G., Pal, R., Stiles, A. R., and Vandekar, M. 1969. Bull. *W.H.O.* *40*, 67.

18. WHO Expert Committee on Filariasis. 1974. *W.H.O. Tech. Rep. Ser.* No. 542.

19. WHO. The Work of WHO. The Annual Report of the Director-General. Off. Rec. 213.

20. WHO unpublished document "Onchocerciasis Control in the Volta River Basin Area." OCP/73.1.

21. WHO. 1973. "The Use of Virus for the Control of Insect Pests and Disease Vectors." *W.H.O. Tech. Rep. Ser.* No. 531.

# DISCUSSION

Drs. Furtick and Wright have focused attention on present production and use of insecticides against a background of increasing consumption pressures, which are forcing inflation throughout the world. They have outlined clearly present trends and foreseeable problems in achieving required control of pests adversely affecting man's food supply and health. The prospects of controlling the major and minor plagues that so seriously affect man's health and well-being may be no brighter now than they were before the vast arsenal of modern compounds and greater scientific knowledge which we possess today became available.

Today a whole new set of factors influence pest control; they are the contemporary problems of society. They arise from changing life styles, new values related to environmental protection, uncontrolled human population growth, and a disturbed, faltering world economy. World needs for food, fiber, animal products, and health protection are increasing at unprecedented rates. Yet national restrictions are limiting the use of pesticides to increase these food and fiber needs and protect human health. Also, a shortage of investment capital and the costs and difficulties in meeting registration standards and criteria for use of new, improved pesticides are exacerbating the problems we already face.

The panel was thus confronted with several dilemmas as they opened discussion of these two papers. The speakers were asked for the state of knowledge concerning restrictions on the use of persistent pesticides in the developing nations. On a closely related issue, they were also asked how closely the developing nations monitor the use of insecticides and whether they are enacting legislation which will impose registration procedures similar to those in effect, for example, in the United States.

Serious efforts are being made in both FAO and WHO via contacts with national and international agencies to keep abreast of legislation in member nations as these relate to importation, application, residue levels permitted, and early testing of experimental materials. This wave of protective and/or restrictive legislation is relatively recent. For much of the past two decades, many developing nations have been testing grounds for experimental field application of the efficacy of new materials being produced in the developed nations; however, this pattern is changing rapidly. In fact, it was stated that nearly every country in the world now has some kind of pesticide law on the books or about to be passed by its legislative bodies. These comments tie in then to the need for trained personnel to enforce and implement laws and regulations as they may be established in various nations of the world. The technically trained manpower and equipment requirements necessary to establish fully effective pesticide residues and toxicology laboratories are very substantial. The discussion emphasized the need not only for training of skilled manpower for these sophisticated responsibilities but importantly, the increasing assistance in training and management required from the FAO and WHO international agencies.

The discussion then centered on the role of and continuing interest by major international pesticide manufacturing companies. It would appear that most of these great chemical corporations are still favorably inclined, from a management standpoint, to continue research on the development of new compounds and to make major new investments in research and development. At the same time, they are very alert to and are constantly reassessing the attractiveness of further investment for research and development in view of growing international concern for the environmental impact of pesticide use, both as reflected in damaging effects to the natural environment and to unforeseen hazards to human health. It was suggested that serious consideration of international public policy related to the production and distribution and regulation of pesticides for crop production may well be required in the immediate future.

The use of insecticides for the protection of public health reflects in general a very different set of concerns by national governments. The World Health Organization and its various committees on toxicology, formulation, and application are major forces in the development of recommendations to member nations regarding the importation and appropriate use of even the most controversial insecticides, many of which continue to be considered absolutely critical to the control of human diseases and the improvement of general public health. Many nations have come to depend heavily upon WHO in the human health field and FAO in the agricultural field to assist them in setting standards which have a degree of international neutrality, thus removing conflicts concerning their importation and use from the arena of political debate.

With regard to the potential measures suggested to overcome the problems listed in Dr. Furtick's paper, the panelists were then asked to consider the institutional capabilities of WHO and FAO in assuming greater responsibility in the development of internationally acceptable registration standards, in the development of basic information on use potentials and requirements, and in greatly expanding the international research network; also, the intensification of international strategies for the use, certification, and regulation of pest-control materials. In addition, discussion centered around the desirability of sharply increased research on new generations of pesticides that are biodegradable, and which are in some instances natural products or mimics of natural products that have no effect on nontarget organisms.

It was also pointed out that perhaps thousands of compounds have been developed over the past 20 years by the pesticide and chemical industries for control of specific organisms. Because they have lacked promise on the target problem, these materials have been discarded. It is believed that within this vast reservoir of materials many valuable products could still be salvaged for use in specific circumstances. Panelists discussed the problems associated with obtaining access to these compounds and mounting special testing programs to reevaluate them. Reference was made to the British National Research Development Corporation and the manner in which it functions as one model on a national basis for approaching this problem.

The discussion clearly pointed to the need for increasing and modifying the scope of activities of international agencies and/or developing a new institutional structure (international in character) that might deal with needs which are now global in nature, highly charged politically, and in many instances run counter to the industrial interests already active in the field.

# LIMITATIONS OF PRESENT DAY INSECTICIDES

# Development of Resistance to Insecticides

F. J. OPPENOORTH
Laboratory for Research on Insecticides
Wageningen, The Netherlands

## I. INTRODUCTION

A large and increasing number of insect species has developed resistance to insecticides as a result of the selective action of these compounds used for their control. An extensive account of this development and its biological background for insects of public health importance is given by Brown and Pal (1971); the situation for insects of agricultural importance is dealt with in a FAO monograph (Anonymous, 1970). The reader is referred to these publications and to the list of 80 reviews on resistance presented in the first one for more complete information.

The most disturbing aspect of resistance is the threat it poses for the future, since it is to be expected that the number of cases of resistance will continue to increase. This means that as long as insecticides constitute the main instrument for the control of insects, there is a continuous need for new insecticides that, in view of the extremely high costs of development, may not become available soon enough.

In this chapter we present a short description of the genetic and biochemical basis of resistance, since an understanding of the nature of the problem can be helpful in attempts to prevent or overcome resistance, as well as giving indications of properties that are desirable in new insecticides in order to reduce the danger of resistance development.

## II. GENETIC AND BIOCHEMICAL BACKGROUND

Resistance has invariably been found to be due to the selection of hereditary factors, and not to an adaptation of individual insects to the insecticide. Genes are selected that, alone or in combination, alter the physiology or biochemistry of an insect in such a way that higher amounts of the insecticide are required for killing them, even up to the point where an insect tolerates any given dose. These changes can also lead to resistance to other compounds than the one used for selection (cross-resistance). It is an important task for the student of resistance mechanisms to elucidate the basis of this cross-resistance and to attempt to predict what chemicals will be affected.

## A. Single Versus Complex Causes of Resistance

Resistance can, theoretically, be due to many changes and their interaction, such as reduced pick-up or penetration of the insecticide, increased detoxication, storage and excretion, reduced transportation to the site of action, and reduced action by alteration of this site. Two of these causes appear to be responsible for most cases of resistance: increased detoxication and altered site of action. In a number of cases these alterations have been shown to be brought about by changes in enzymes caused by mutants of single genes. A few examples follow.

### 1. Single Gene Alteration Causing Changes at the Site of Action.

Acetylcholinesterase (AChE) is the main site of action of two large groups of insecticides that have been widely used, the organophosphates and the carbamates. Moreover, it is one of the few sites of action of insecticides that are known, and it is therefore not surprising that examples of alterations of the site as a cause of resistance are to be found here.

There are now several species of acari (the spider mite, *Tetranychus urticae*, Smissaert, 1964), the cattle tick *Boophilus microplus* (Wharton and Roulston, 1970), as well as of insects (the housefly, *Musca domestica*, Tripathi and O'Brien, 1973), and the leafhopper, *Nephotettix cincticeps* (Hama and Iwata, 1971; Iwata and Hama, 1972) that have developed resistance because of an altered cholinesterase, which show a greatly reduced rate of reaction with the inhibitors. In hybrids of normal and mutant strains the two enzymes are present side by side. Besides a much slower reaction rate of the cholinesterase with the inhibitor, the mutant enzymes of some strains also show a reduced activity in hydrolyzing the substrate acetylcholine. This reduction in normal activity can be conceived to be an impairment that is necessary to achieve the adaptation to the presence of the insecticide. Apparently, some rare changes avoiding this impairment are possible, since Zahavi and Tahori (1970) found normal activity in some resistant strains of spider mites.

The single-gene resistance of this type can cause resistance factors as large as 2000 (Ballantyne and Harrison, 1967).

2. Single Gene Alteration Leading to More Efficient Detoxication Enzymes. Well-known examples of high degrees of resistance due to a single gene that produces an increased level of degradation are DDT resistance due to DDT-dehydrochlorinase (DDT-ase) (Lipke and Kearns, 1960) and malathion resistance due to a carboxylesterase (Welling and Blaakmeer, 1971), both found in the housefly. There is evidence that in both cases an enzyme with altered properties is produced (higher turnover for the insecticide), and not just a greater quantity of an enzyme (Oppenoorth, 1965). Both factors lead to practically complete immunity of the houseflies.

In other cases high degrees of resistance can only build up by the accumulation of more than one resistance gene. Some of these act independently, others show a clear interaction. The clearest examples of interaction are found where a factor for decreased penetration as well as for increased detoxication are present. If the insecticide enters slowly, even a moderate detoxication rate can prevent the buildup of a lethal concentration in the body, so that the two mechanisms together can lead to high degrees of resistance (Sawicki, 1970).

There is evidence that resistance due to several genes can develop when prolonged selection is exerted, and where genes causing high resistance are apparently absent (Georghiou and Hawley, 1971).

### B. Allelic Genes Causing Related Resistance Mechanisms

Generally the altered enzymes causing resistance of a certain type (e.g., DDT-ases, carboxylesterases, cholinesterases) differ between species and even between strains of the same species. The DDT-ase found in resistant houseflies can also degrade deutero-DDT, and not _o_-chloro-DDT, but the reverse is true for a DDT-ase found in _Aedes aegypti_ (Pillai _et al_., 1963). DDT-ases found in several strains of housefly differ in the relative rate of degradation of various DDT-analogs (Oppenoorth, 1965). Cholinesterases from strains of spider mites, _Tetranychus urticae_, vary with respect to their "insensitivity index" (the reciprocal of the rate of inhibition (Figure 1) and to their activity for the substrate-analog acetylthiocholine (Zahavi and Tahori, 1970). The cholinesterases in different strains of resistant cattle ticks show different insensitivity indexes that may partly explain the different cross-resistance patterns (Table 1). In the housefly one carboxylesterase has been found

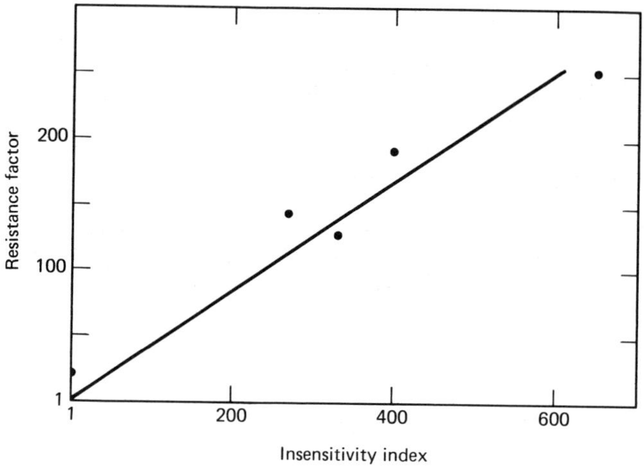


FIGURE 1. Relation between resistance to malathion and "insensitivity index" of acetylcholinesterase for malaoxon (the reciprocal of I 50) in seven strains of spider mites (from Zahavi and Tahori, 1970).

to be changed into detoxication enzymes attacking paraoxon or malaoxon in different strains (Oppenoorth and van Asperen, 1960).

There is evidence that the different forms of the enzymes are produced by alleles of the structural gene. Thus malathion and parathion resistance in the housefly were found to be allelic (Nguy and Busvine, 1960). In the housefly, DDT-ases of different activity were due to alleles (Oppenoorth, 1965). Several aberrant mutants of the cholinesterase in resistant spider mites were shown to be allelic (Schulten, 1968), and there is evidence that this also applies to those of the cattle tick (Stone, 1972).

The implication of this large variation in substrate specificity and sensitivity to inhibition for the resistance problem is that various strains can have different degrees of resistance, patterns of cross-resistance, and/or different sensitivity to synergists.

TABLE 1. Resistance Factors and Properties of AChE in Four Strains of Cattle Ticks

| Strain of *B. microplus* | AChE Activity as % of that of Yeerongpilly Strain | Insensitivity Index[a] of AChE | | Resistance Factor | | | |
|---|---|---|---|---|---|---|---|
| | | Coroxon | Diazoxon | Dioxathion | Coumaphos | Diazinon | Dimethoate |
| Yeerongpilly | 100 | 1 | 1 | 1 | 1 | 1 | 1 |
| Ridgelands | 14 | 2.9 | 21 | 7 | 2 | 10 | 400 |
| Biarra | 30 | 380 | 110 | 13 | 25 | 28 | 110 |
| Mackay | 27 | 1.1(8.3)[b] | 1.3 | 25 | 9 | 8 | 65 |

[a]Insensitivity index = reciprocal of bimolecular rate constant of inhibition, divided by that for the Yeerongpilly strain.

[b]Data obtained later, indicating a change in the strain.

Data from Wharton and Roulston (1970) and Schuntner and Smallman (1972).

## C. Different Causes for Resistance to a Single Insecticide

Resistance to an insecticide can be due to different causes in various strains of the same species. A well-known example is DDT-resistance in the housefly in which at least three genes can cause resistance due to different alterations (Grigolo and Oppenoorth, 1966). Two of these are detoxication mechanisms, DDT-ase and oxidation; the third is an insensitivity of the nerves due to gene kdr which produces either an altered site of action or a local barrier (Tsukamoto et al., 1965). It will be clear that the three mechanisms cause completely different cross-resistance patterns and sensitivity to synergists (Table 2). The three mechanisms not only provide a high degree of resistance to DDT applied in the usual way, but they also protect locally at the site of action, the sense organs. This is illustrated in Figure 2 which shows the degree of multiplicity of action potentials caused by DDT applied to labellar hairs of four strains of houseflies. Multiplicity in resistant strains is only obtained at very high dosages. In this figure the chromosome analysis of strain Fc indicates that the insensitivity is due to chromosomes 5 and 3, which are known to cause increased oxidative degradation and reduced penetration, respectively.

Another example is found in resistance to malaoxon in the housefly. Malaoxon can be degraded by a carboxylesterase as well as by a microsomal mixed function oxidase (Welling et al., 1974). Both mechanisms are present in strain G, only one in strain $E_1$, and this results in a different behavior with synergists (Figure 3).

## D. Increased Mixed Function Oxidase Activity as a Resistance Mechanism

This mechanism is a very important resistance mechanism since it affects insecticides from practically all chemical groups. We know less about the nature of the changes that cause this type of resistance, and it is becoming clear that it is more complex than other causes of resistance. Mixed function oxidases can degrade many lipophilic compounds also in nonresistant insects, where they are thought to play a role in the detoxication of many plant constituents (Krieger et al., 1971). A relationship was demonstrated between the feeding habits of Lepidoptera and the activity of their oxidase (oxidase activity in the species is decreasing in the order polyphagous<oligophagous<monophagous). The oxidation capacity was measured as the amount of aldrin that was converted to dieldrin by preparations

TABLE 2. Genetic and Toxicological Differentiation of Different Causes of DDT Resistance in Different Strains of Housefly

| Strain | DDT (10 μg) | DDT (1μg) +F-DMA (1μg) | DDT (5μg) +sesamex (10μg) | Chromosome Carrying Resistant Gene | Resistance Mechanism |
|---|---|---|---|---|---|
| L | 0 | 95 | 5 | 2 | DDT-ase |
| Fc | 0 | 0 | 98 | 5 | oxidation |
| kdr | 0 | 20 | 0 | 3 | ? not detoxication |

The figures represent percentage mortality in male houseflies upon topical application of DDT and synergists on the dorsal thorax. The LD 50 of DDT for susceptible flies is c. 0.15 μg.

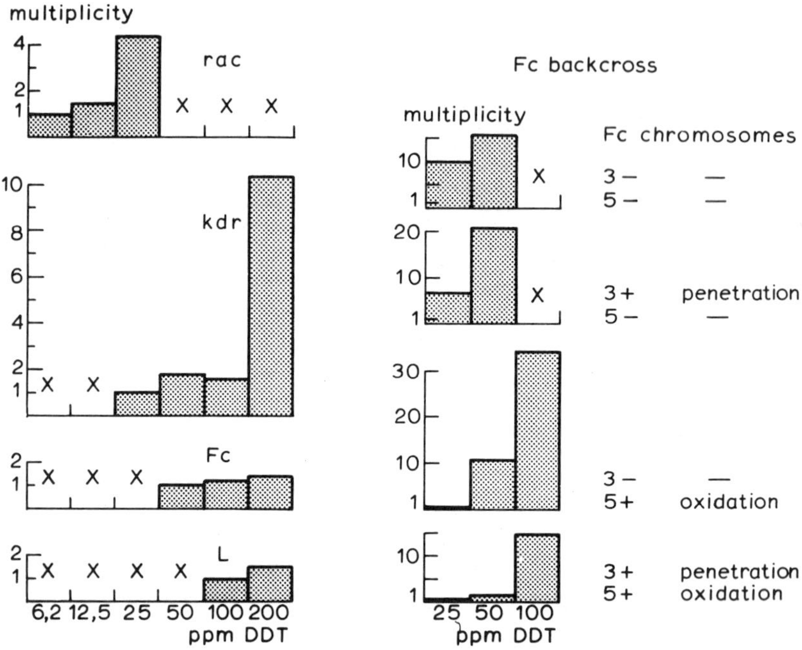


FIGURE 2. Protection of action of DDT on labellar sense organs in housefly strains with different resistance mechanisms. A labellar sense hair was submerged for one minute in DDT suspensions of different strengths. In the susceptible strain rac multiplicity of action potentials clearly occurs at 25 ppm DDT, in the resistant strains (kdr, Fc, and L) much higher concentrations are required. In strain FC two factors contribute to this resistance as shown in the chromosome analysis: chromosome 5 carrying a gene for oxidative degradation, and chromosome 3, causing reduced penetration which enhances the effect of the degradation (Van der Ven and Oppenoorth, unpublished results).

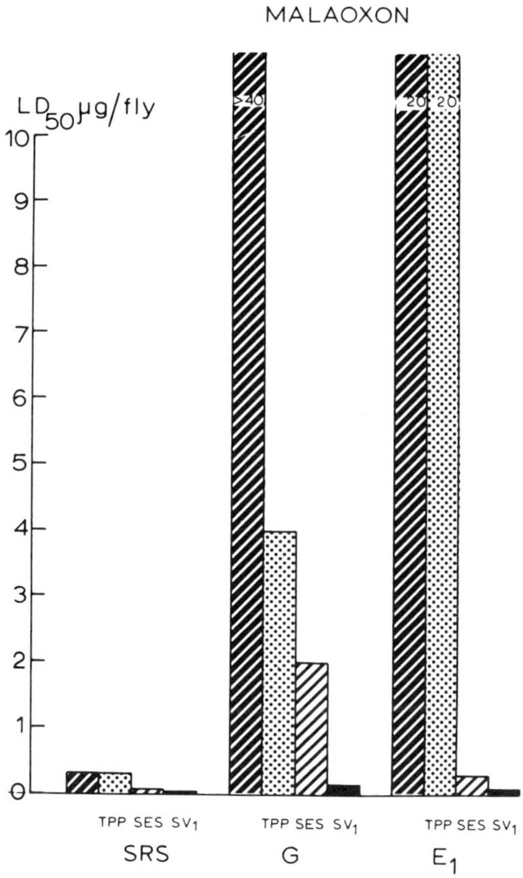


FIGURE 3. Different causes for malaoxon-resistance in various strains of the housefly. LD 50 of malaoxon upon topical application, alone or with synergists (10 µg TPP, triphenylphosphate, 2 µg sesamex, or 2 µg $SV_1$, O,O-diethyl, O-phenyl phosphorothionate). TPP blocks carboxylester attack, sesamex oxidative degradation, and $SV_1$ both (Data from Welling *et al.*, 1974).

of the guts of the caterpillars, and it was assumed that this activity, because of the low substrate specificity of the oxidases, would reflect the capacity to degrade a wide variety of plant constituents. This is in line with some observations in other species of Lepidoptera (Heliothis zea and H. virescens), where it was found that field strains had a higher capacity for oxidation of aldrin, probably as a result of selection with various insecticides (Williamson and Schechter, 1970). Also, in Trichoplusia ni, a parathion and a DDT-resistant strain showed a high capacity for oxidation of carbaryl (Kuhr, 1971). The genetic background of these strain differences is not known. It was shown that the gut of the strains contained an amount of cytochrome P-450/mg of protein, which was proportional to the oxidation capacity, indicating that in this case resistance may be due to a simple increase in the amount of enzyme or enzyme system.

The low degree of substrate specificity causes a wide spectrum of cross-resistance, affecting compounds from different chemical groups. Thus in the housefly a gene on chromosome 5 for resistance to DDT is also responsible for diazinon resistance. The situation is complex, however, since a gene on chromosome 2 causes diazinon resistance due to oxidation without cross-resistance to DDT (Oppenoorth, 1971). It is clear, therefore, that changes in substrate specificity of oxidases can be brought about by mutant genes, but it is not known what sites are changed. That the situation is a complex one further appears from the fact that there are several genes affecting the activity and substrate specificity of oxidases. Recently, for instance, a gene for increased oxidation of malaoxon (Welling et al., 1974) has been localized on chromosome 2 which is distinct from the one for diazinon resistance mentioned above (unpublished results). Georghiou found that the high capacity for epoxidation of aldrin in resistant strains, which is mainly due to genes on the second chromosome, is partly due to the other chromosomes. This again indicates that the genetic background determining the oxidation capacity is a complex one. In mammalian livers more than one oxidase have been shown to be present (Khandwala and Kasper, 1973; Comai and Gaylor, 1973). In insects where activity is generally measured in microsomes from whole insects or abdomens, the presence of more than one system is apparent from the effects of inhibitors on oxidations of various substrates and the different effects in various strains, as shown in Figures 4 and 5. A full understanding of the number of entities and their properties will only be obtained if it is possible to separate them and study them in vitro.

Resistance through increased mixed function oxidation is in particular threatening because of its versatility, resulting in

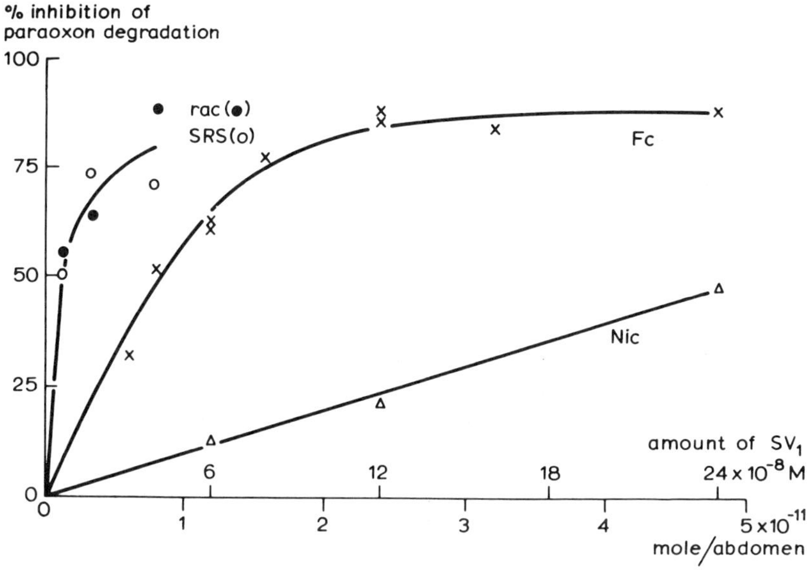


FIGURE 4. Inhibition of oxidative degradation of paraoxon by microsomes of different strains of housefly by $SV_1$, O,O-diethyl, O-phenyl phosphorothionate (Oppenoorth, unpublished).

adaptations to insecticides of widely different groups. Cross-resistance has been shown to new groups such as juvenile hormone and hormone-mimetics (Dyte, 1972; Cerf and Georghiou, 1972; Plapp and Vinson, 1973; Terriere and Yu, 1973) and although this cross-resistance has not been shown to be dependent on the oxidases, they are among the most likely candidates.

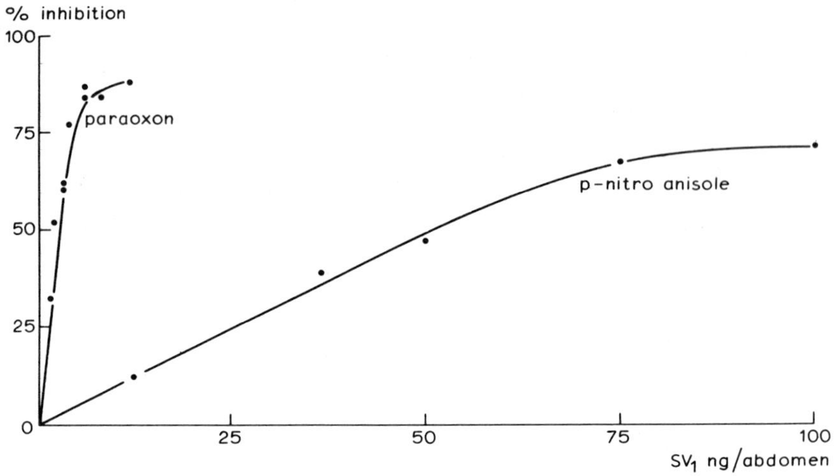


FIGURE 5. Inhibition of oxidative degradation of paraoxon and p-nitro anisole by microsomes of housefly strain Fc (Oppenoorth, unpublished).

## III. IS DEVELOPMENT OF RESISTANCE UNIVERSAL AND UNAVOIDABLE?

It is clear that there are certain cases of application of insecticides that need not lead to resistance. If treatments with insecticides are restricted to only a small part of an insect population, or are used only in cases of emergency, the chances that resistance genes will replace their wild-type alleles will be small, since the wild-type alleles are the ones that are having a selective advantage under normal circumstances. Thus it may be hoped that the selective treatment of anophelines by only applying insecticides to the inside of huts for malaria control will not cause resistance. A prerequisite for the success of this approach is that the rest of the population will not be under unnecessary pressure from the same insecticide because of its use for other purposes. For instance, if the breeding sites of mosquitos are contaminated by insecticides used in agriculture, development of resistance can be the result (Georghiou et al., 1973). The use of more selective and less persistent insecticides is of great importance in this respect, since this will decrease undue selection pressure on species to which the insecticides are not directed.

The question whether intensive and prolonged treatment with insecticides will always lead to resistance is more difficult to answer. Many examples of the development of resistance seem to answer this question in the affirmative, but the problem is that there are no lists of insects that did not develop resistance. Nevertheless, such cases exist. Control of spider mites in glass houses caused rapid resistance to a long list of insecticides, but Pentac® has been used on ornamental crops for at least 7 years now without signs of failure (in the United States McEnroe and Lakocy, 1969; in Holland Helle, personal communication).

From the fact that large quantities of insecticides to which resistance already developed in some species are continuously being used, it is obvious that they are still effective against other species. What we need, therefore, is balanced and quantitative information on questions such as: What is the "average" chance that a new insecticide will loose its value to a certain species in 1, 5, or 20 years? Such "averages" have little meaning for specific cases but still could be of some value in predicting cost-benefit relationships in the development of new compounds.

Theoretically, the development of resistance will depend on the intensity of selection, the number of individuals and generations involved, and more important for our problem, the presence of certain gene alleles that, alone or in combination with others, cause resistance. There is evidence that such genes can be extremely rare. Development of resistance to cholinesterase inhibitors in cattle ticks in Australia showed that it originated in certain sites and spread from there over the continent. A cholinesterase mutant in the housefly was only found in 1973, many years after introduction of the cholinesterase inhibitors as insecticides. As mentioned above, only a few species so far developed insensitive cholinesterases, and this again indicates that they are rare.

The chances that mutants will arise with biochemical adaptations to a new insecticide will, of course, be determined by the chemical structure of the compound. This applies to the vulnerability to detoxication as well as to the ease with which changes of the site of action that disable the insecticide will be possible. There is evidence that factors that favor survival in insects with acquired resistance resemble those existing in species that have always been tolerant. One example is the development of a carboxylesterase in several malathion-resistant insects, whereas this enzyme already occurred in the naturally tolerant species *Dermestes maculatus* (Dyte *et al.*, 1966). Another example is the development of DDT-ases in several species of DDT-resistant insects, and the occurrence of a similar enzyme in naturally tolerant Mexican bean beetles, *Epilachna varivestis* (Chattoraj and Kearns, 1958). This means that compounds acting on a large variety of species will be less vulnerable to development of resistance than more selective compounds. This is a very unpleasant situation in view of the need for selective insecticides that interfere as little as possible with natural enemies and other useful species of insects. It should be emphasized that selective insecticides (selectively killing some species of insects and not others) have thus two contradictory influences on the development of resistance: a favorable one, since undue selection of species to which the insecticide is not directed is avoided, and an unfavorable one, since the susceptible species can develop resistance by mimicking the properties of naturally tolerant species.

The answer to the question whether resistance development is universal and unavoidable, therefore, must necessarily be vague: development of resistance to any new compound is likely to take place upon prolonged and severe selection pressure in a number of insect species. How long this will take in different species and in how many of them cannot be predicted.

## IV. COUNTERMEASURES

Apart from the generally recognized fact that restriction in selection will delay or perhaps avoid development of resistance, there is no simple solution to the resistance problem. Since resistance as we have seen can depend on widely different genetic and biochemical backgrounds, the solution required is likely to be different for every case.

Solutions have been sought along the following lines.

### A. New Derivatives of Compounds That are no Longer Affected by the Resistance Mechanism

Development of compounds to which the resistance mechanism is not effective is a common approach to the problem. A well-known example forms the derivatives of DDT, synthesized to avoid the action of DDT-ase. Deutero-DDT and *o*-chloro-DDT, which cannot be attacked by the DDT-ases of *Aedes aegypti* and the housefly, respectively, have already been mentioned. Other DDT-analogs from which no HCl can be removed by DDT-ase are Prolan® and cyclopropane derivatives, and there is, of course, no cross-resistance to these compounds in insects where DDT-ase is the cause of resistance (Metcalf, 1955; Holan, 1971). The major stumbling block is the multiplicity of resistance as shown in Table 2, since other mechanisms for resistance can affect the analogs and confer cross-resistance to them. It becomes increasingly difficult to find compounds that are not affected by the accumulated resistance mechanisms present in "hardy" species that developed resistance to a whole series of insecticides. The rather obscure properties that determine the possibility of oxidative attack leave little but trial and error as a means to find suitable alternatives.

The altered cholinesterase in spider mites decreases the rate of inhibition by different inhibitors by a factor of $10 - 10^4$ (Smissaert *et al.*, 1970). No clear insight is available on the factors determining the magnitude of this change, although it was noticed that the presence of one bulky group in the acyl moiety of the inhibitor particularly impeded the reaction with the R enzyme.

## B. "Negatively Correlated" Insecticides

If "negatively correlated" insecticides, only toxic for the resistant insects and not for the susceptible ones, could be developed, it should be possible to shift the insect population back and forth by alternate use of the old and new insecticide (see Brown, 1958, for a review). It has never been possible to apply this in practice. Apart from the problem to synthesize suitable negatively correlated compounds, it is clear that the possibility of development of alternative resistance mechanisms to one of the compounds is detrimental to this solution. Nevertheless, the concept is a sound one, and too little rational efforts have thus far been made to test its applicability. That such efforts might well be successful is indicated by the fact that one of the compounds synthesized in an attempt to obtain suitable selective carbamates was found to be twice as toxic to a resistant strain of housefly than to the regular strain (Lee *et al.*, 1974).

## C. Synergists that Overcome Resistance by Blocking Detoxication

Detoxication mechanisms are among the most frequent and effective resistance mechanisms, and blocking the enzymes involved has been shown to be an effective means to restore susceptibility (see Wilkinson, 1968, for a review). Here again alternative resistance mechanisms can rapidly cause resistance to the mixture. For example, the housefly developed resistance to DDT synergized with WARF-antiresistant (Brown and Rogers, 1950). This synergist blocks DDT-ase, and the resistance to the mixture was probably of the kdr or oxidative type.

As discussed by Wilkinson (Chapter 8) synergists thus far have not been applied to any great extent, because of some inherent problems in the use of mixtures and because of the high cost of most of them. There is perhaps some greater promise in the use of so-called auto-synergizing compounds (Metcalf *et al*., 1971). These are compounds with built-in groups that block degradation enzymes, such as methylenedioxyphenyl or 2- or 3-propynyloxyphenyl groups. Such groups sometimes are present without having been introduced intentionally. It has been shown, for instance, that the PS group in malathion, which is known to block certain types of mixed function oxidases (Oppenoorth, 1971), blocks the enzyme for oxidative degradation of malaoxon (Welling *et al*., 1974). The result is that some strains of houseflies that are resistant to malaoxon are still susceptible to malathion.

Of course, mechanisms could be developed for the detoxication of the synergist or detoxication mechanisms insensitive to the inhibitor. This would also seem to apply to insecticides that are self-synergizing.

## V. CONCLUSIONS

It is apparent that there is no easy solution to resistance as a general phenomenon. There are ways, however, to control insects with certain types of resistance by alternative insecticides or by adding synergists. Just as in the case of the development of more selective insecticides where the rational approach based on the knowledge of the biochemical properties of different organisms is beginning to gain more impetus, a rational attempt to construct effective insecticides for dealing with specific resistant insects may well prove useful. This may only provide temporary solutions, however, and a continuous search for new compounds will therefore be required. Many species can adapt themselves rapidly but it should be clear that their resources of suitable mutants can still prove inadequate, as is already the case in many less successful species at present.

## REFERENCES

1. Anonymous. 1970. *Pest Resistance to Pesticides in Agriculture, Importance, Recognition, Countermeasures*. Food

and Agricultural Organization.

2. Ballantyne, G. H. and Harrison, R. A. 1967. Entom. Exp. Appl. 10, 231.

3. Brown, A. W. A. 1958. "Insecticide Resistance in Arthropods." W.H.O. Monograph Ser. 38.

4. Brown, A. W. A. and Pal, R. 1971. "Insecticide Resistance in Arthropods." W.H.O. Monograph Ser. 38.

5. Brown, H. D. and Rogers, E. F. 1950. J. Am. Chem. Soc. 72, 1864.

6. Chattoraj, A. N. and Kearns, C. W. 1958. Bull. Ent. Soc. Amer. 4, 95.

7. Cerf, D. C. and Georghiou, G. P. 1972. Nature (London) 239, 401.

8. Comai, K. and Gaylor, J. L. 1973. J. Biol. Chem. 248, 4947.

9. Dyte, C. E., Ellis, V. J., and Lloyd, C. J. 1966. J. Stored Prod. Res. 1, 223.

10. Dyte, C. E. 1972. Nature (London) 238, 48.

11. Georghiou, G. P., Breeland, S. G., and Ariaratnam, V. 1973. Environmental Ent. 2, 369.

12. Grigolo, A. and Oppenoorth, F. J. 1966. Genetica 37, 159.

13. Hama, H. and Iwata, T. 1971. Appl. Ent. Zool. 6, 183.

14. Holan, G. 1971. Bull. W.H.O. 44, 355.

15. Iwata, T. and Hama, H. 1972. J. Econ. Ent. 65, 643.

16. Khandwala, A. S. and Kasper, C. B. 1973. Biochem. Biophys. Res. Comm. 54, 1241.

17. Kuhr, R. J. 1971. J. Econ. Ent. 64, 1373.

18. Lee, A., Sanborn, J. R., and Metcalf, R. L. 1974. Pest. Biochem. Physiol. 4, 77.

19. Lipke, H. and Kearns, C. W. 1960. Adv. Pest Control Res. 3, 253.

20. McEnroe, W. D. and Lakocy, A. 1969. J. Econ. Ent. 62, 283.

21. Metcalf, R. L. 1955. Organic Insecticides, Interscience, New York.

22. Metcalf, R. L., Kapoor, I. P., and Hirwe, A. S. 1971. Bull. W.H.O. 44, 363.

23. Ngui, V. D. and Busvine, J. R. 1960. Bull. W.H.O. 22, 531.

24. Oppenoorth, F. J. 1965. Proc. 12th Intern. Congress Entomol. London, 240.

25. Oppenoorth, F. J. 1971. Bull. W.H.O. 44, 195.

26. Oppenoorth, F. J. and Asperen, K. van. 1960. Science 132, 298.

27. Pillai, M. K. K., Hennessy, D. J., and Brown. A. W. A. 1963. Mosquito News 23, 118.

28. Plapp, F. W. and Vinson, S. B. 1973. Pesticide Biochem. Physiol. 3, 131.

29. Sawicki, R. M. 1970. Pesticide Sci. 1, 84.

30. Schulten, G. G. M. 1968. Communs. R. Trop. Inst. Amsterdam 57, 1.

31. Schuntner, C. A. and Smallman, B. N. 1972. Pesticide Biochem. Physiol. 2, 78.

32. Smissaert, H. R. 1964. Science 143, 129.

33. Smissaert, H. T., Voerman, S., Oostenbrugge, L., and Renooy, N. 1970. J. Agric. Food Chem. 18, 66

34. Stone, B. F. 1972. Aust. Vet. J. 48, 345.

35. Terriere, L. C. and Yu, S. J. 1973. Pesticide Biochem. Physiol. 3, 96.

36. Tripathi, R. K. and O'Brien, R. D. 1973. Pesticide Biochem. Physiol. 3, 495.

37. Tsukamoto, M., Narahashi, T., and Yamasaki, T. 1965. Botyu-Kagaki 30, 129.

38. Welling, W. and Blaakmeer, P. T. 1971. Proceedings of the 2nd International IUPAC Congress of Pesticide Chemistry (A. S. Tahori, Ed.), Vol. II, 61, Gordon and Breach, New York.

39. Welling, W., Vries, A. W. de., and Boerman, S. 1974. Pesticide Biochem. Physiol. 4, 31.

40. Wharton, R. H. and Roulston, W. J. 1970. Ann. Rev. Ent. 15, 3081.

41. Wilkinson, C. F. 1968. World Rev. Pest Control 7, 155.

42. Williamson, R. L. and Schechter, M. S. 1970. Biochem. Pharmac. 19, 1719.

43. Zahavi, M. and Tahori, A. S. 1970. Biochem. Pharmac. 19, 219.

## DISCUSSION

Dr. Oppenoorth mentioned in his paper that examples of resistance due to an altered cholinesterase have been found in several species of mites, the cattle tick, the house fly, and a species of leafhopper. To further illustrate this phenomenon, the results of unpublished work of Georghiou (University of California, Riverside) with the Anopheline mosquito, *Anopheles albimanus*, were briefly discussed.

The original colony of *Anopheles albimanus* came from Guatemala where the population had been exposed to a variety of carbamate and organophosphorus insecticides. Under laboratory conditions the mosquitoes were subjected to selection pressure, first with a carbamate to which a very high degree of resistance developed. Then the same strain was subjected to selection pressure with an organophosphorus compound, resulting in an even higher degree of resistance. Examination of certain biochemical aspects clearly indicated that the basis for resistance is insensitivity of the ChE. When the susceptible strains are compared with the resistant ones, the ChE of the resistant mosquitoes ranged from 500- to 25,000-fold less sensitive to inhibition depending on the compound. A variety of organophosphorus and carbamate compounds have been evaluated with this strain of mosquitoes, and the results show that it is a general phenomenon. Apparently, this is the first recorded case of this phenomenon in mosquitoes. If it develops with other species, the control with carbamate and OP insecticides could become a very serious problem.

The great need to find a practical solution to the resistance problem was strongly emphasized. Aside from searching for new compounds, research should be done aimed at resuscitating some existing insecticides.

Oppenoorth is examining specific individual cases of resistance in great detail with the thought that learning the precise situation would provide clues to proceed toward a solution.

Synergists offer sufficient promise to warrant extensive investigations. In situations in which it is known that resistance is due to a certain metabolic mechanism, there are known inhibitors that, in theory, could be used to block those metabolic pathways. Theoretically, we should be able to use synergists to give that degree of flexibility to an insecticide formulation so that if an insect becomes resistant because of a certain mechanism, we should be able to formulate it with the right synergist. Conceivably, an insecticide might be formulated with several synergists, one for each type of resistance mechanism.

The actual determination of resistance mechanisms could lead to a practical solution. Anopheline mosquito resistance was mentioned as an example to consider. Since Anopheline resistance does not occur everywhere (it is localized in certain areas) the opinion was expressed that detailed examinations of the metabolism of the resistant mosquitoes should be made.

The question was asked, how is it possible that altered cholinesterase can hydrolyze acetylcholine and be acylated without there being some compound, either a carbamate or phosphate, which will inhibit it? The response was that it would be necessary to look at other types of compounds of related structure, but with systematic modification. However, this kind of study has not been done yet. As far as substrate reaction with acetylcholine, it is not much different than the normal cholinesterase. The $K_m$'s and the $k_2$'s and $k_3$'s are very similar, and yet the rates of inhibition are vastly different. Perhaps attention should be given to phosphinates and phosphonates, as well as reexamining phosphoramidates and certain other compounds that were discarded earlier. Mention was made of Smissaert's work, in which he used many inhibitors and poisons against spider mites and found that some are equally effective against resistant and susceptible ones, whereas others show fantastic differences.

There was discussion of the possible toxicity of synergists on the test insects when used alone. The three synergists shown in Oppenoorth's Fig. 3 were generally not toxic with the exception of SV-1, which is toxic to susceptible flies at high doses. Although SV-1 is an inhibitor of cholinesterase, it is not a good one. The conclusion is that generally the toxicity of synergists is negligible compared to the effective compound. However, in Bowers' work on synergizing juvenile hormone mimics he found that certain conventional synergists like piperonyl butoxide, sesamex, and others have definite juvenile hormone action when used alone. Synergists can be expected to have an effect because they tie up certain enzymes or change pathways in certain ways that make toxicants more poisonous, but their effects alone are not readily or usually observed.

In regard to DDT resistance in the housefly, the opinion was expressed that nerve insensitivity probably is not due to DDT detoxication inside the nervous system because nerve insensitivity is carried by a gene on the third chromosome, and detoxication is due to a gene on the second and fifth chromosomes. There is a possibility of finding a substitute for _p,p_'-DDT in some DDT analogs, since in terms of molecular mechanism some membrane receptors are probably slightly twisted or modified in such a way as to become insensitive to _p,p_'-DDT.

It was pointed out that a number of resistance factors involve changes within the nervous system. However, with many toxicants that work outside the nervous system, no resistance has appeared despite considerable use.

One of the most important problems affecting research on resistance is the difficulty in obtaining experimental compounds for research and for ultimate field development. Traditionally and necessarily, the chemical industries have been relied on to produce and market pesticides. In addition, until recent years research by industry has been necessarily relied on to develop new compounds. In general, only the very large chemical companies have been concerned with production and marketing of pesticides. Understandably, these companies are mainly interested in large-volume items that provide satisfactory profit, and primarily in compounds on which they have a sound patent position.

For instance, if a satisfactory synergist were developed, probably it would be difficult to find a company to produce it because of the relatively small quantity needed. In the United States, the costs of complying with EPA requirements for registration, that is, toxicology and pharmacology experiments, residues, and others, are enormous. In this particular regard, however, it was pointed out that perhaps registration problems might not be too great with certain compounds because a number of drugs in present-day use by humans are highly effective inhibitors of mixed-function oxidation. Some of the new compounds are imidazoles that are expected to be cleared soon for use in human therapy.

WHO is seeking ways to interest smaller local chemical companies in developing and/or producing insecticides for specific projects in localized areas. Recent experience indicated that certain emergency insecticide needs were supplied promptly by smaller companies, whereas the large companies could not make delivery for 18 months.

After considerable discussion of the subject of necessary reliance on the large chemical companies for research and developmental compounds, the statement was made that "industrial profits are perhaps not compatible with human survival."

A concluding suggestion was made to explore the possibility that the international organizations WHO, FAO, and AID could commission and supply specific compounds needed to support their particular programs.

# Environmental Pollution by Insecticides

W. KLEIN
Institute for Ecological Chemistry
Society for Radiation and Environmental Investigation
Munich, Germany

## I. INTRODUCTION

Pesticides, especially certain organochlorine insecticides, have been recognized as the first generation of environmental chemicals, together with inorganic air-pollutants and certain heavy metals. This paper attempts to put the potential for pollution by pesticides and by the sum total of human activities with chemicals into perspective. Changes in the chemical composition of the local, regional, and global environment have been occurring continuously and naturally ever since the earth began. Man has contributed to these changes from earliest history and this contribution has been increasing at a rate that is more than proportional to the increase in population. Chemical pollution has changed during history from local via regional to global proportions, which have become obvious only during the past decades.

"Environmental quality" will be used as a term of reference. Although changes of environmental quality are highly visible in certain areas, e.g., India, the basis for the following discussion is that changes of the chemical quality of the environment are undesirable. Any changes of the material composition of the environment resulting from the activities of man are changes of environmental quality, regardless of whether or not these changes represent a hazard.

## II. MAGNITUDE OF CHANGES OF ENVIRONMENTAL QUALITY

The parameters shown in Figure 1 represent a scale for environmental contamination by all chemicals, an indication whether there are local or long-term global contaminations, and the consequences that the use of the corresponding chemicals will have for man and his living environment. A useful weighing of the 6 parameters against one another could result in the

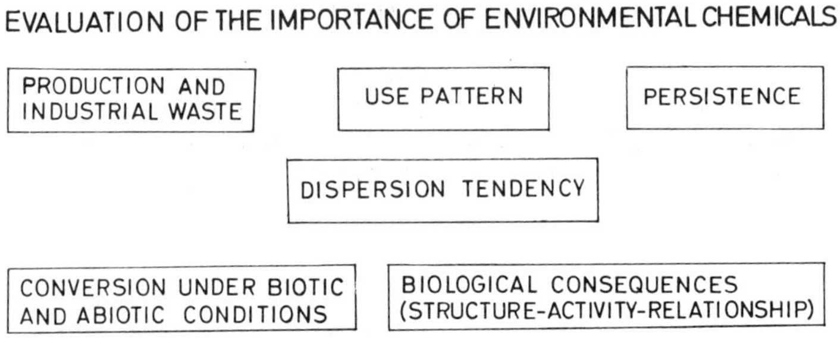


FIGURE 1. Parameters to be considered in environmental contamination by chemicals.

establishment of a priority list for environmental hazards due to chemicals. I would like to discuss these parameters by examples available - predominantly of pesticides - but applicable to any other chemical or drug.

As regards the world production of chemicals (Table 1), there are only estimates which I would like to mention here because of their significance for the situation with worldwide pollution. In 1950, the world production of synthetic organic chemicals amounted to a total of 7 million tons, in 1970 production was 63 million tons; for 1985, the author cited here expects an increase up to 250 million tons. This represents an increase by a factor of 2.5 for a period of ten years. Published production figures from the USA and the FRG show that the factor estimated here is realistic for organic chemicals.

The data given in Table 2 for the use of chemicals in the FRG show an average rate of increase by a factor of 2 for the past ten years. I would like to emphasize the high factor for chlorine, which is used in great part in the organic chemicals industry. All pesticides combined had a factor of 1.8, due mainly to the high increase in use of herbicides (Deutscher Bundestag, 1971). It is important to know from the amounts of chlorine used by chemical industry that proportion used in the synthesis of chemicals in order to predict from production and use-pattern which chemicals are or may become pollutants. Consequently registration requirements should be fixed according to expected production and use-pattern. A simple calculation shows the potential of present day human activities in changing the global environment.

TABLE 1. World Production of Industrial Chemicals

| INDUSTRIAL CHEMICALS<br>World Production | in $10^6$ t | Direct release in Env. | in $10^6$ t |
|---|---|---|---|
| 1950 | 7 | | |
| 1970 | 63 | | |
| | | solvents | 10 |
| | | detergents | 1.5 |
| | | pesticides | 1 |
| | | gaseous base chemicals | 1 |
| | | lubricating and industrial oil | 2-5 |
| | | miscellaneous | 7 |
| 1985 | 250 | | |
| NATURAL SOURCES | | | in $10^6$ t |
| | | methane | 1600 |
| | | terpene type hydrocarbons | 170 |

If we assume that the organic chemicals produced at present, 100 million tons per year (Table 3), were not decomposed or changed but were released into the environment quantitatively, there would be, with an even distribution over the total land surface of the earth, a load of 700 mg/m$^2$ or, in case of penetration into a soil layer of 10 cm depth, a concentration of 2.5 ppm.

Although this calculation contains some perhaps inadmissible simplifications it shows that, due to the industrial activity of man, the material global environment can be changed within a short period. With an example of a naturally occurring substance like mercury, the natural zero level should be added to the figures calculated.

If we consider under corresponding conditions - no conversion, evaporation or leaching - the possible environmental contamination from pesticides and nitrogen fertilizers, as compared to total organic chemicals, the result would be the contamination (in kg/ha) shown in Table 4, where the last line,

TABLE 2. Growth Factors for the Use of Chemicals in the Federal Republic of Germany

| Product | Consumption in 1969 ($10^3$ t) | Factor | Compared to |
|---|---|---|---|
| Chlorine | 1 749 | 2.5 | 1960 |
| Sodium hydroxide | 1 511 | 2.0 | 1960 |
| Hydrochloric acid | 600 | 2.4 | 1960 |
| Sulphur dioxide | 38 | 1.5 | 1963 |
| Carbon disulphide | 84 | 1.35 | 1960 |
| Sulphuric acid | 3 619 | 1.3 | 1960 |
| Nitric acid | 718 | 1.5 | 1962 |
| Lead oxides | 37.6 | 1.5 | 1960 |
| Lead carbonates | 0.9 | 0.4 | 1960 |
| Liquid ammonia | 2 106 | 1.8 | 1960 |
| Ethylene glycol | 171 | 2.0 | 1966 |
| Formaldehyde | 388 | 2.8 | 1960 |
| Tri- and tetrachloroethylene | 200 | | |
| Derivatives of phthalic acid (without plasticizers) | 165 | 2.0 | 1960 |
| Plasticizers on the basis of phthalic acid | 166 | 2.7 | 1960 |
| Synthetic rubber | 315 | 2.8 | 1960 |
| Plasticizers, total | 190 | 1.5 | 1966 |
| Detergents | 176 | 1.4 | 1966 |
| Fertilizers: | | | |
| Nitrogen | 1 011 | 1.5 | 1959/60 |
| Phosphorus pentoxide | 802 | 1.1 | 1959/60 |
| Insecticides, active materials | 1.0 | 1.8 (insecticides, herbicides, fungicides together) | 1960 |
| Herbicides, active materials | 8.8 | | |
| Fungicides, active materials | 4.7 | | |
| Mercury | 0.76 | | |
| Motor benzene | 14 084 | 2.8 | 1960 |
| Mineral oil products, total | 102 083 | 3.7 | 1960 |

TABLE 3. Maximum Global Concentrations of Organic Chemicals

MAXIMUM GLOBAL CONCENTRATIONS OF ORG. CHEMICALS

------------

| | |
|---|---|
| land surface of the earth | $149 \times 10^6$ $km^2$ |
| volume of oceans | $1.3 \times 10^9$ $km^3$ |
| weight of the atmosphere | $5.1 \times 10^{15}$ to |

------------

Org. chemicals production 1973, total — app. $100 \times 10^6$ to

Basis of calculation: no breakdown, dispersion of total amount in one medium only

------------

| | |
|---|---|
| Dispersion on total land surface | 700 mg / $m^2$ |
| or in 10 cm soil layer | 2.5 ppm |
| Dispersion in oceans | $0.8 \times 10^{-4}$ ppm |
| or in 1 m layer | 0.3 ppm |
| Dispersion in atmosphere | 0.02 ppm |

necessarily not containing data on the area of use, is of especial global importance. If all the pesticides used were distributed on the total land surface of the earth, there would result a contamination of 7 mg/$m^2$ equal to 1% of total organic chemicals; the nitrogen fertilizers would be present in concentrations of about 100 mg/$m^2$; all organic chemicals distributed on the surface would represent annually about 700 mg/$m^2$. For fertilizers similar quantities would be obtained with phosphorus, potash, etc. Apart from the fact that an even distribution does not occur under natural conditions, most of the substances produced are metabolized and/or changed to natural products, as e.g., fertilizers, so that the results of this mathematical example represent a maximum global contamination.

TABLE 4. Distribution of Organic Chemicals on the Land Surface of the Earth

| Land Surface of the Earth | $13,392 \times 10^9$ ha |
|---|---|
| agriculture | $1,424 \times 10^9$ ha |
| pastures | $3,001 \times 10^9$ ha |
| forests | $4,091 \times 10^9$ ha |

Soil Input per Year (no conversion, no evaporation or leaching)

| | world production Mio. tons/year | kg/ha in area of use locally | globally | kg/ha when distributed on total land surface |
|---|---|---|---|---|
| Pesticides | 1 | 2 - 4 | 0,12 (a/p/f) | 0,07 |
| Nitrogen fertilizers | 15 | 70 | 3,38 (a/p)<br>1,76 (a/p/f) | 1,12 |
| org. Chemicals total | 100 | | | 7,47 |

## III. MAJOR ENVIRONMENTAL CHEMICALS

Which chemicals in use today represent a hazard to environmental quality? The answer will be based on the knowledge we have for environmental chemicals in general, and for the model environmental chemicals "pesticides", respectively. First of all, we should consider for organic compounds, the amounts of those environmental chemicals which, due to their structure, are persistent, namely the chlorinated hydrocarbons. Today only that part of the pesticides which are the organochlorine compounds are included in discussions on environmental quality. Furthermore, even considering this, pesticides are relatively "small compounds" in production and use compared to figures given in Table 5, e.g., for $CH_3I$. The term "small compounds" is used in this paper in the industrial sense of low production and use.

All chemicals with undue persistence as a common characteristic should be evaluated along the same lines. Undue

TABLE 5.

| | |
|---|---|
| chlorinated paraffins | 25,793 |
| chloroform | 86,662 |
| carbon tetrachloride | 323,974 |
| vinyl chloride (monomer) | 1,100,302 |
| trichloroethylene | 222,444 |
| tetrachloroethylene | 241,973 |
| dichlorodifluoromethane | 82,726 |
| other halogenated hydrocarbons | 3,232,505 |

US - production of non-pesticidal halogenated chemicals, 1967 (tons)

persistence has been defined by IUPAC as follows:

Undue Persistence: A substance is unduly persistent whenever a measurable quantity thereof continues to exist in some discernable chemical form. Obviously, the most desirable chemical would be one that has a stability just sufficient to perform its function with no persistent residue. Since this is difficult to achieve, the rhythm of use in terms of function time must always be considered. Therefore, the "function time rhythm" of a chemical must take into consideration not only its stability, but also its persistence in terms of any residual terminal chemical definition.

Table 6 contains data on the US production of other chlorinated hydrocarbons which are produced in large quantities. "Other chlorinated compounds" is a frequently used term, which does not mean anything but is intended to suggest "nonpesticidal" and that furthermore nonpesticidal organochlorines would be harmless per se. These substances, however, especially the chlorinated paraffins, can enter the environment and come into contact with man following a similar route to that of pesticides and PCBs.

Apart from this situation regarding chlorinated environmental chemicals in general, a number of halogen-containing pesticides, e.g., the three active compounds shown in Figure 2, are not classified as halogenated hydrocarbon pesticides. Additional examples include 2,4-D, 2,4,5-T, pentachlorophenol,

TABLE 6.

Table VI Production of Organohalogen Compounds.

| Substance | Country | Year | 1000 tons | Remarks |
|---|---|---|---|---|
| Cl-paraffins | USA | 1971 | 25.8 | |
| Chloroform | USA | 1971 | 88.7 | |
| Carbon tetrachloride | | p.a. | 1000 | Industry |
| | | | 900 | Nature |
| Methyl chloride | BRD | 1969 | 121 | |
| Trichlorofluoromethane | USA | | 120 | World = 3X |
| | World | 1971 | 1000 | Integral |
| Dichlorofluoromethane | USA | | 180 | World = 3X |
| | World | 1971 | 1000 | Integral |
| Trichloroethylene | BRD | 1969 | 119 | |
| | USA | 1971 | 222 | |
| Tetrachloroethylene | BRD | 1969 | 82 | |
| | USA | 1971 | 242 | |
| Vinyl and vinylidene chloride | BRD | 1969 | 730 | |
| Vinyl chloride | USA | 1971 | 1100 | |
| Pentachlorophenol | USA | 1967 | 21 | |
| 2,4,5-Trichlorophenol | USA | 1967 | 12 | |
| 2,4-Dichlorophenoxy acetic acid | USA | 1967 | 41 | |
| 2,4,5-Trichlorophenoxy acetic acid | USA | 1967 | 15 | |
| _p_-Chloronitrobenzene | USA | 1967 | 50 | |
| _o_-Chloronitrobenzene | USA | 1967 | 17 | |
| Chlorobenzene | BRD | 1969 | 123 | |
| | USA | 1967 | 240 | |
| Methyl bromide | USA | 1967 | 10 | |
| 1,2-Dibromo-3-chloropropane | USA | 1967 | 2 | |
| Chloral | USA | 1967 | 26 | |
| Methyl iodide | World | p.a. | 40,000 | Nature |
| Other | USA | 1971 | 3232 | |

| | | | | |
|---|---|---|---|---|
| 2 | **o-chlorophenyl methylcarbamate** | 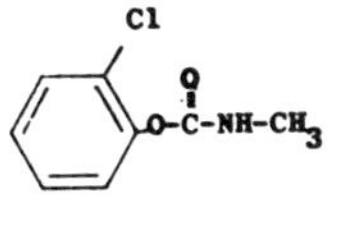 | Kumiai *Ins.* | AO M150 |
| 3 | 2-chlorophenyl-*N*-methylcarbamate | $C_8H_8ClNO_2$ | | |
| 5 | Hopcide® | | | |
| 6 | CPMC | | | |

| | | | | |
|---|---|---|---|---|
| 2, 3 | **4,4'-dichloro-*N*-methylben=zenesulfonanilide** | 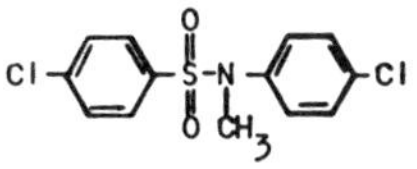 | Monte-catini *Syn.* | AO M2500<br>AD >1200<br>T NR |
| 6 | S-150 | $C_{13}H_{11}Cl_2NO_2S$ | | |

| | | | | |
|---|---|---|---|---|
| 1 | *bromophos* | 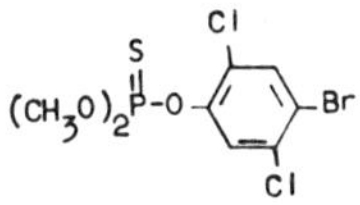 | Cela *Ins.* | AO 3750–7700<br>M2829–6000<br>Rb720<br>AD Rb2188<br>CO D120(m)<br>T NR |
| 2 | **O-(4-bromo-2,5-dichloro=phenyl) O,O-dimethyl phosphorothioate** | $C_8H_8BrCl_2O_3PS$ | | |
| 3 | *O,O*-dimethyl-*O*-2,5-dichloro-4-bromophenyl thionophosphate | | | |
| 4 | *O,O*-dimethyl-*O*-2,5-dichloro-4-bromophenyl phosphorothioate | | | |
| 5 | Nexion® | | | |
| 5 | Brofene® | | | |
| 6 | S-1942 | | | |
| 6 | ENT 27162 | | | |
| 6 | OMS 658 | | | |

FIGURE 2. Halogenated pesticides not classified as organochlorines.

dichlorvos (DDVP) and ronnel, etc. About 40 percent of all pesticide compounds contain halogens. The classification of pesticides as organophosphates, carbamates, etc., is based on the assumption that, after the first conversion, e.g., the saponification of the C-O-P- bonds in organophosphates, the so-called breakdown products no longer have any influence on environmental quality. According to our present very limited knowledge of intermediate degradation products in the environment, this assumption is not valid. It would be more reasonable to classify environmental chemicals according to chemical structures, independently of their use as pesticides, solvents, cosmetics, detergents, and for any other purpose.

With regard to DDT, as an example of the (so-called) pesticidal chlorinated hydrocarbons (organochlorines), we have considerable information regarding the parameters which can be used as basic for evaluation (FAO/WHO Monographs) (Figure 3).

| common trade name | DDT |
|---|---|
| formula: | 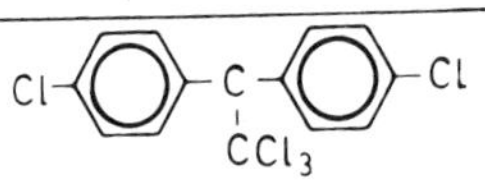 |
| purity: | 75-80%<br>main by-product o,p'-DDT |
| production level: | not exactly known |
| use pattern: | use against more than 100 different insects on large variety of crops, in public health, as moth-proofing agent<br>quantities unknown |
| metabolites: | mainly DDE, DDD, DDA, DCB |
| occurrence residues: | up to 7ppm including analogues |
| outside area of use: | concentration up to more than 10ppm in wildlife, fish including analogues<br>atmosphere: up to 200 pp$10^{12}$ |
| $LD_{50}$: | DDT rat - oral 250 mg/kg bodyweight<br>DDD " " 3400 " "<br>DDE " " 1000 " "<br>DDT monkey " >200 " " |
| chronic toxicity: | great number of data on DDT and analogues in animals and men - e.g. neoplastic disorders in mice (multigeneration)<br>no abnormalities in men after 11-19 years occupational exposure to around 18 mg/man/day |
| wildlife effects: | accumulation in food chains, reduction egg-shell thickness, certain bird species |

FIGURE 3. Summary of current status of knowledge about DDT.

Nevertheless, here too, some questions remain open, such as the question about quantitative use pattern. If we compare our knowledge about toxaphene, a chlorinated hydrocarbon, we notice that even the structures of most of its constituents are unknown (Figure 4). We have recently isolated from toxaphene

| common trade name | TOXAPHENE |
|---|---|
| formula: | average emp. $C_{10}H_{10}Cl_8$<br>appr. 25 camphene related 8Cl containing compounds<br>structures unknown |
| purity: | unknown |
| production level | unknown, however widely used in quantity and number |
| use pattern | agronomic, vegetable, fruit crops - ectoparasites on cattle, sheep, swine<br>quantities unknown |
| metabolites: | unknown |
| occurrence residues: | 0 - < 0.5 ppm |
| outside area of use: | unknown |
| $LD_{50}$: | guinea pig oral 365 mg/kg bodyweight<br>rat oral 60-120 " "<br>dog i.v. 5-10 " " |
| chronic toxicity: | rats - 25 ppm long term - no effects<br>monkeys - two years - 0.64 - 0.78 ppm in diet - no effects<br>volunteers - ten days during 30 min./day, 500 mg/m³ air - no effects |
| wildlife effects: | unknown |

FIGURE 4. Summary of current status of knowledge about toxaphene.

5 crystalline compounds, the sum of which is about 8% of toxaphene. They contain 7, 8, 9, and 10 chlorines respectively. A major compound could not be detected. The five compounds isolated so far all have an insecticidal activity comparable to the technical mixture.

We have better knowledge of parathion, insofar as the applied pesticide is concerned. The present data are rather limited as regards the influence of degradation products on environmental quality (Figure 5). Quantitative data on conversion products are limited and the final fate of the known major metabolites in the environment remains unknown as well.

| common trade name | PARATHION |
|---|---|
| formula: | 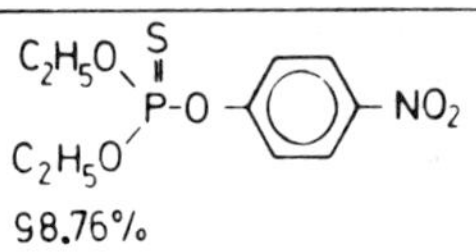 |
| purity: | 98.76% |
| production level: | 1966: 15 000 tons |
| use pattern: | wide spectrum insecticide, use on food crops, fruits vegetables, tea<br>quantities unknown |
| metabolites: | paraoxon, amino-parathion, amino-paraoxon, p-amino-phenol, p-nitro-phenol |
| occurrence residues: | 0,02-7ppm |
| outside area of use: | unknown |
| $LD_{50}$: | male rat - oral - 5.0-30.0 mg/kg bodyweight<br>guinea pig - " 9.3-32.0 " " |
| chronic toxicity: | no effect level on cholinesterase activity<br>rat: 0.05 mg/kg bodyweight<br>man: 0.05 " " /day |
| wildlife effects: | unknown |

FIGURE 5. Summary of current status of knowledge about parathion.

Finally, I would like to mention, as a last example of this comparison, the fungicide zineb which is a metal-organic compound for which there is only a little more information available than for toxaphene (Figure 6). The crucial point for having such widely varying knowledge for pesticides is based on the differing feasibility and detection limits in residue analysis. Nevertheless it is surprising that it is exactly those substances about which we have the best knowledge that

| common trade name | ZINEB |
|---|---|
| formula: | 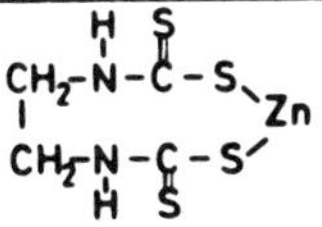 |
| purity: | unknown |
| production level: | unknown |
| use pattern: | fungicide on crop and fruit plants quantities unknown |
| metabolites: | ethylene-thiuram disulfide (ETD)<br>ethylene-thiuram monosulfide (ETM)<br>ethylene-thiurea (ETU) |
| occurrence residues: | 0-11 ppm, e.g. lettuce 3.6 ppm |
| outside area of use: | unknown |
| $LD_{50}$: | rat - oral - >5200 mg/kg bodyweight |
| chronic toxicity: | rat-2 years - 500 ppm goitrogenic effect<br>dog-1 year - 2000 ppm no effects<br>rat reproduction study-100 mg/kg/day orally resulted in sterility, resorption of fetuses and anomalous tails |
| wildlife effects: | unknown |

FIGURE 6. Summary of current status of knowledge about zineb.

are frequently discussed in public and that are submitted to strong legal restrictions of use, whereas there is hardly any interest in the environmental impact of less known substances. In this connection, besides DDT there should be mentioned the cyclodiene insecticides, and to some extent, PCBs as relatively well known models of organic environmental chemicals. For aldrin and dieldrin, for example, a large number of degradation products in various organisms is known, as well as their structures and concentrations in the total residues in the environment (Figure 7). Structures alone, however, are not adequate for predictions of environmental quality or for realistic evaluation of toxicological consequences. The amounts of the respective conversion products formed under practical conditions constitute an essential basis for evaluations.

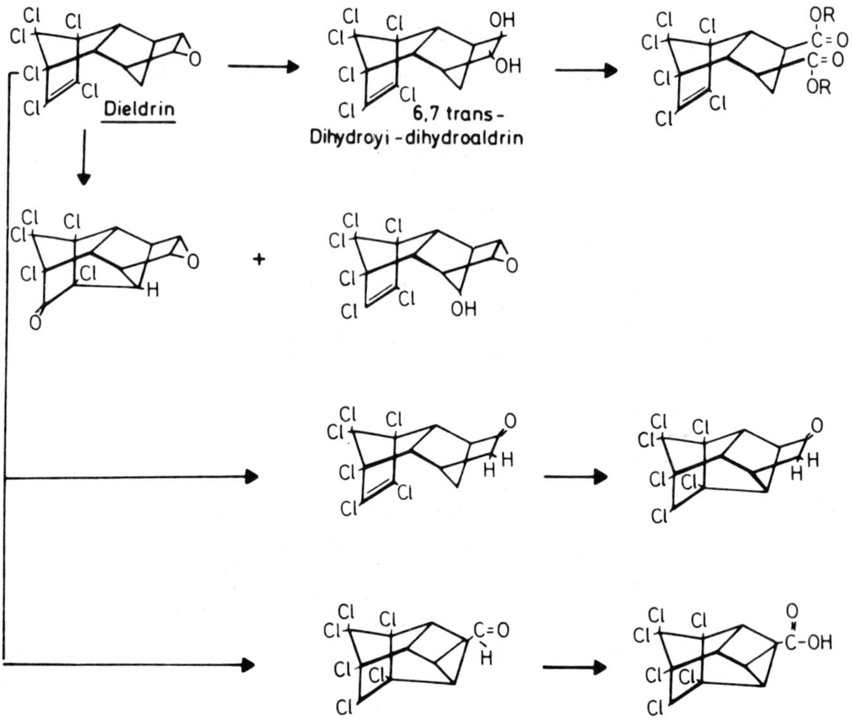


FIGURE 7. Metabolites of aldrin and dieldrin.

Figure 8 shows 4 dieldrin metabolites for which we have obtained semi-quantitative data in this regard. For the organisms given here the pentachloroketone does not have any significance - it is a major metabolite in urine of male rats, but urinary excretion is only 5% of the total in that animal. For any thorough estimates of the impact of a chemical on environmental quality we need at a minimum such data including, however, the effects of microorganisms and sunlight in degradations, and furthermore, we need to know the ultimate breakdown products.

Regarding the widely discussed accumulation of pesticides in mammals, it has been demonstrated in a number of examples with aldrin, dieldrin, and DDT, as well as with pesticides from

## Metabolism of Dieldrin in Mammals

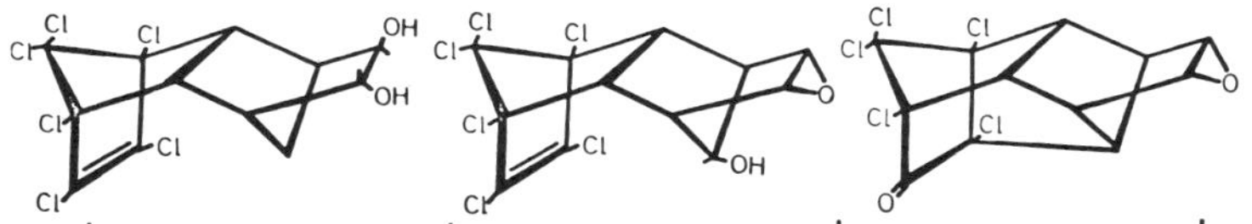


| animal | | applied dose | 9-hydroxy-dieldrin | 6,7-trans-dihydro-aldrin-diol | pentachloroketone | dieldrin | total excreted % of applied | excretion dieldrin + conversion products (ratio feces:urine) |
|---|---|---|---|---|---|---|---|---|
| mice | ♂ | 3 mg/kg | 8,4 | 14,0 | urine, trace | 5,7 | 28,0 | 95 : 5 |
| | ♀ | | 10,7 | 14,3 | | 2,7 | 32,6 | |
| rats | ♂ | 0,5 mg/kg | 8,8 | 2,3 | urine, major | 0,8 | 11,8 | 96 : 4 |
| | ♀ | | 4,6 | 2,4 | --- | 2,8 | 9,7 | |
| rabbits | ♂ | 0,5 mg/kg | --- | 1,4 | --- | 0,3 | 1,7 | 1 : 99 |
| | ♀ | | 0,2 | 2,0 | | 0,5 | 2,7 | |
| monkeys | ♂ | 0,5 mg/kg | 10,5 | 2,0 | --- | 9,5 | 22,0 | 75 : 25 |
| man | | --- | feces, major | urine, present | --- | | | --- |

FIGURE 8. Metabolism of dieldrin in mammals.

other chemical classes that there is a steady state of storage for each compound in mammals with the storage level depending on the daily dose. The time until this level is reached is specific for each chemical. Figure 9 shows an experiment with endrin in rats, where the steady state of storage is already reached after 6 days' application. In the organs of experi-

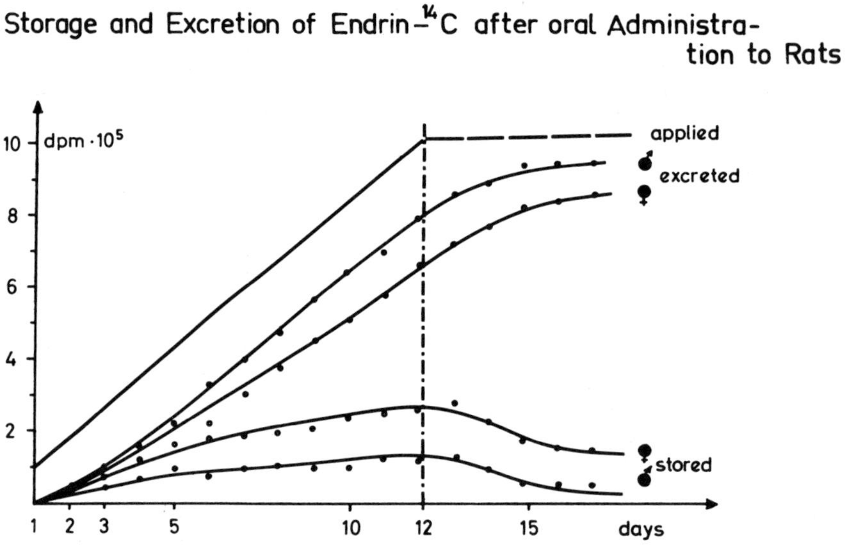


FIGURE 9. Storage and excretion of $^{14}C$ endrin after oral administration to rats.

mental animals, cyclodienes are mainly present as such, but other less "persistent pesticides" are present partly as metabolites. Since we have only a few balance studies of the fate of chemicals in different organisms, we can not judge the importance of this matter critically.

Another topic which must be considered with metabolites, is dispersion or translocation. Translocation of cyclodiene insecticides to the aerial parts of plants has been thoroughly investigated with the demonstration that both uptake by the roots and foliar excretion via the vapor phase occur and that there are differences in uptake depending on soil type, plant species, and even in varieties within one species (Lichtenstein *et al.*, 1970; Harris and Sans, 1967; Nash and Beall, 1970; Hulpke and Schuphan, 1970). The influence of the insecticide concentration on root absorption is proportional to the soil residues (Hulpke, 1970; Onsager *et al.*, 1970). Since applied insecticides in soil are exposed to sunlight, hydrolysis, and metabolic activity of soil organisms, growing plants may also absorb the resulting conversion products. It may be expected that these conversion products are absorbed at different rates than the unchanged insecticide depending on their solubility and/or the permeability of the plant membranes. Thus, the number of potential crop contaminants may increase with increasing time interval between (soil) treatment and harvest. Considering additionally the metabolism of parent compounds and conversion products in the plant itself, the number of expected residual chemicals is again increased. It would be desirable to establish qualitatively and quantitatively, the nature and amounts of all these chemical residues both for reasons of crop quality and for estimation of effects on global environmental quality.

## A. Open Air Experiments

To elucidate the long-term fate of environmental chemicals and their total residues in rotating crops, open air experiments were started in 1969 with aldrin-$^{14}C$. The site is shown in Figure 10.

*Aldrin*: To simulate practical conditions as far as possible, aldrin-$^{14}C$ was formulated and applied at a dose of approximately 3 kg/ha to soils which are used in the respective areas for growing the respective crops. Parallel experiments were carried out in England and Germany and partly in the U.S. and Spain. Crops used in the first year were potato, sugar beet, maize, and wheat. After harvest, the different crop parts and soil layers were analyzed separately. Figure 11 shows examples of residues and metabolic ratios in treated soil. In all crop samples, the residues of the conversion products, dieldrin, photodieldrin, and the dicarboxylic acid (Figure 12) have been positively identified and quantitized. In the soil samples, photoaldrin was found additionally. After one season following treatment, aldrin did not represent the main residue. The most significant residues were dieldrin and the already mentioned dicarboxylic acid. This acid is accompanied in most samples by two acidic byproducts which could not be identified due to the very low amounts present. Although the dihydro-

FIGURE 10. Experimental area for environmental studies with $^{14}C$ aldrin.

chlordene-1,3-dicarboxylic acid is a major hydrophilic conversion product found, it does not represent the terminal metabolite of aldrin in growing crops. After foliar application of the $^{14}C$-labeled acid to maize and oral administration to rats, we have recently demonstrated that besides conjugates, the compound is subject to further breakdown. As expected, photodieldrin was found to be only a minor residue in soils and plants, usually not more than 2-3% of the total residues. Photoaldrin which was found only in soil, amounts only to less than 1% of the total residues. There is one exception only: in the wheat soil samples from England photoaldrin represented 5% of the total residues. This result is astonishing since the other soil samples from the same site contained the same low relative amount of photoaldrin. We cannot explain this result, but it indicates that quantitative data resulting from absolutely equal experiments using the same soil can vary at least by the factor 5.

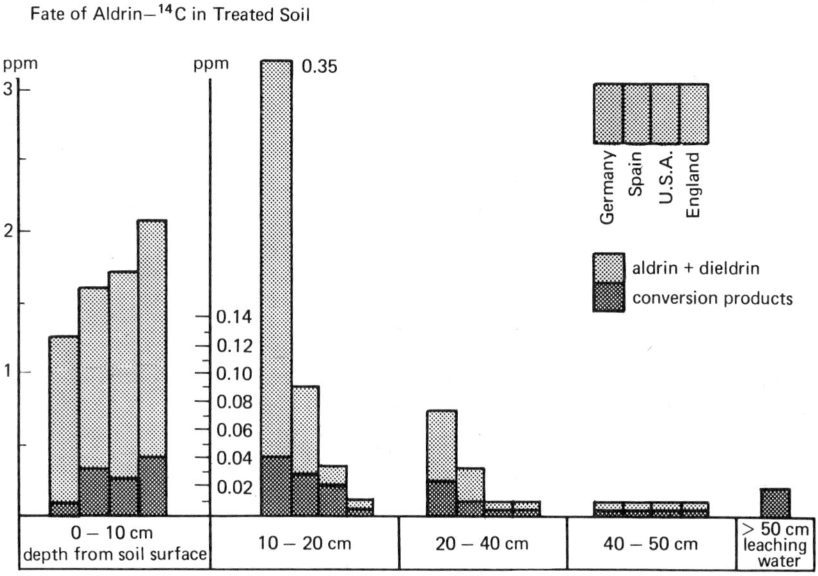


FIGURE 11. Fate of $^{14}C$ aldrin in treated soil.

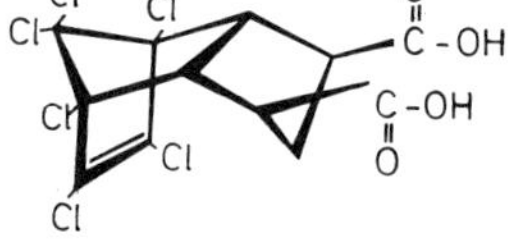


FIGURE 12. Dihydrochlordene-1,3-dicarboxylic acid.

Table 7 shows quantitative data on residues after dieldrin application. In 1970 it was applied as onion seed-dressing, in 1971 at 3 kg/ha to the same soil, in 1972 and 1973 there was no retreatment. Soil concentrations of total residues decreased by a factor of 3-4 over 12 months, and leaching amounts

TABLE 7. Balance of the Fate of Dieldrin in Plants and Soil

| | leaves (ppm) | | | peel (ppm) | | | peeled crop (ppm) | | | soil (0-10cm) (ppm | | | leaching water |
|---|---|---|---|---|---|---|---|---|---|---|---|---|---|
| | HEOD | PIPD | total | HEOD | PIPD | total | HEOD | PIPD | total | HEOD | PIPD | total | |
| 1970 onion seed dressing | 0.003 | <0.001 | 0.003 | 1.25 | 0.023 | 1.272 | 0.039 | 0.001 | 0.043 | 0.521 | 0.004 | 0.528 | n.d. |
| 1971 cabbage turnip soil treatment | 0.015 | 0.002 | 0.024 | 0.125 | 0.006 | 0.134 | 0.006 | 0.001 | 0.010 | 2.581 | 0.029 | 2.628 | 0.5ppb |
| 1972 carrots | <0.033 | | 0.035 | unpeeled crop 0.125 | 0.005 | 0.135 | | | | 0.634 | 0.035 | 0.690 | 0.3ppb |
| 1973 potatoes | 0.121 | 0.074 | 0.681 | 0.046 | 0.003 | 0.053 | 0.007 | 0.0004 | 0.009 | 0.204 | 0.011 | 0.232 | 0.2ppb |

Balance of the fate of dieldrin in plants and soil

decreased too. Since the absolute amount of residues in the cultivated plants is less than 5% of total residues in soil, and leaching is negligible in this respect, evaporation from soil must be the major route of disappearance of residues. These data are a beginning to give quantitative results. For long-term evaluation they have to be expressed in absolute quantities, e.g., mg per applied mgs. This is not only necessary for soil, plants, and food, but also for the atmosphere, where probably most organic chemicals of low molecular weight go, and which seems to be a sink for them. As far as the unintentional residues in the atmosphere are concerned, some data are known for the concentrations of the parent compounds.

Pesticides are introduced into the air in several ways: by drifting, from aeroplane application, by evaporation from plants and soil as well as via transpiration by plants. (For instance, four weeks after foliar application of 1.2 mg of $^{14}C$-labeled endrin per plant of white cabbage, residues in plants and soil amounted only to 6% and 94% had disappeared into the atmosphere.) In the atmosphere, pesticides may be converted photochemically. The ratio between the particle form and the gaseous form is not known. It can only be estimated that for DDT, owing to stronger adsorption onto particles of air pollutants, the particle form ratio is higher than that of dieldrin or lindane, and that in polluted areas this ratio is also higher than in unpolluted ones. On the other hand, it has also been estimated that the major part of pesticides in the atmosphere is present in gaseous form (Atkins and Eggleton, 1970). Nevertheless, it seems important to measure photochemical reactions of pesticides when adsorbed onto particles, since this may change the reaction velocity. Probably the type of reaction, too, is changed owing to chemical interaction with pollutants adsorbed on the particles. Up to now, we know of few investigations that have been carried out in this respect.

Concerning the mechanism of dispersion in the atmosphere, we have some data from extensive measurements of the distribution of radioactive isotopes. For this dispersion, besides numerous analyses, there are also mathematical models. We know, for instance, that for the troposphere of the Northern Hemisphere there are debris cycles with 12 to 25 day-periods and that there is a good mixture in the first circuit. On the other hand, transfer across the equator to the Southern Hemisphere is slow. There are some indications that the relatively constant levels of chlorinated hydrocarbons found in rainwater at different stations are representative for the Northern Hemisphere.

Figure 13 (Appleby, 1970) gives a survey of latitudes where DDT-residues are present in the human and wildlife populations and other parts of the environment. It reveals that residues prevail in zones where DDT has been extensively used in the past and in zones with dense population. In the Southern Hemisphere there may be a similar distribution pattern, but owing to the lack of data (there are sufficient data only for Australia and New Zealand) the graph is not complete for this half of the earth. What chemical changes do organochlorine compounds undergo after they have disappeared into the atmosphere? Owing to the difficulties involved in simulating the atmosphere, experiments in the laboratories allow for only

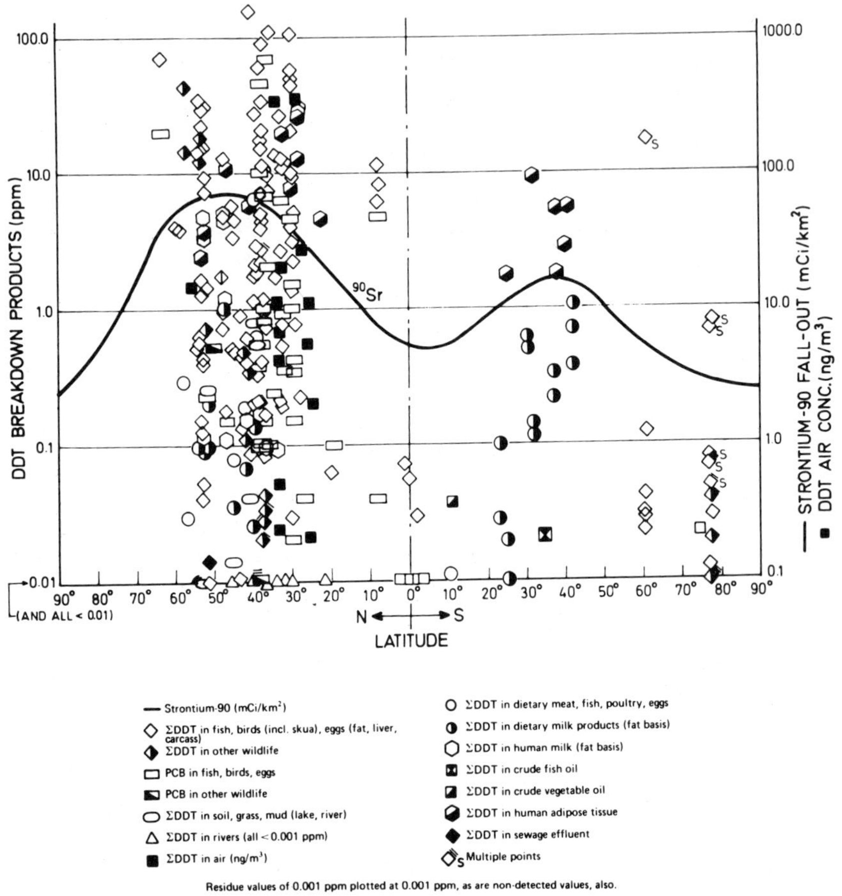


FIGURE 13. Relation of global latitude to presence of residues of DDT.

indirect conclusions as to the behavior of a chemical in the environment. However, preparative photochemical experiments lead to an understanding of the general reactivities of the environmental chemicals in question and to the synthesis and identification of photoproducts, which can occur in low concentrations. Photochemical reactions in solution reveal the conversion of these compounds when dissolved in plant waxes. The difficulties in predicting the conversions under atmospheric conditions can best be shown by the following example: it is generally known that wavelengths below 280 nm have no appreciable intensity at the earth's surface owing to absorption in the atmosphere. Therefore the irradiation experiments with UV lamps, which emit mainly 254 nm, permit no valid conclusions on the reactions in sunlight at the earth's surface. On the other hand, aldrin is easily photolyzed in sunlight, although it absorbs at considerably shorter wavelengths than 280 nm. So far it is not known whether this photolysis is due to the small portion of the sunlight of wavelengths below 280 nm which penetrates through the atmosphere, or to the photosensitization by singlet-oxygen or air pollutants. A catalysing effect of particulate matter in the atmosphere, which under practical conditions is possible, has also to be considered when the pesticides are adsorbed on to particulates.

Because pesticides can also reach higher layers of the atmosphere on the surface of various particles, irradiation experiments were carried out with wavelengths as low as 230 nm. Figure 14 shows the behavior of aldrin adsorbed on the surface of silica gel and irradiated with short wavelength UV-light. With short irradiation times, dieldrin (II) is the main product of the photochemical transformation. Under these conditions, very little photoaldrin is formed; with solid phase irradiation, however, photoaldrin (VII) is the main product. Dieldrin adsorbed on the surface of silica gel isomerizes during irradiation to photodieldrin (III), a relatively stable photoproduct. It could be shown, however, that it is converted to additional products at a finite rate. The chlorohydrin (IV) results from the addition to photodieldrin (III) of HCl, which has been eliminated from a second photodieldrin molecule. The elimination of HCl and the demonstration of numerous low molecular weight fragments are evidence for a photochemical degradation of photodieldrin. Both ketones V and VI are rearrangement products which are formed in the presence of acid catalysts. Besides the information on the abiotic conversion of the applied pesticides, substances are known which can result from some metabolites under abiotic conditions, as in Figure 15 for the main metabolites in higher plants. For this class of substances, at least, contributions to the estimation of the long-term alteration of the material environment are present today.

A further intensely discussed criterion used to classify organochlorine pesticides and PCBs as especially hazardous to the environment is the well known accumulation of residues in food chains, which is known to occur with DDT. Let me point out

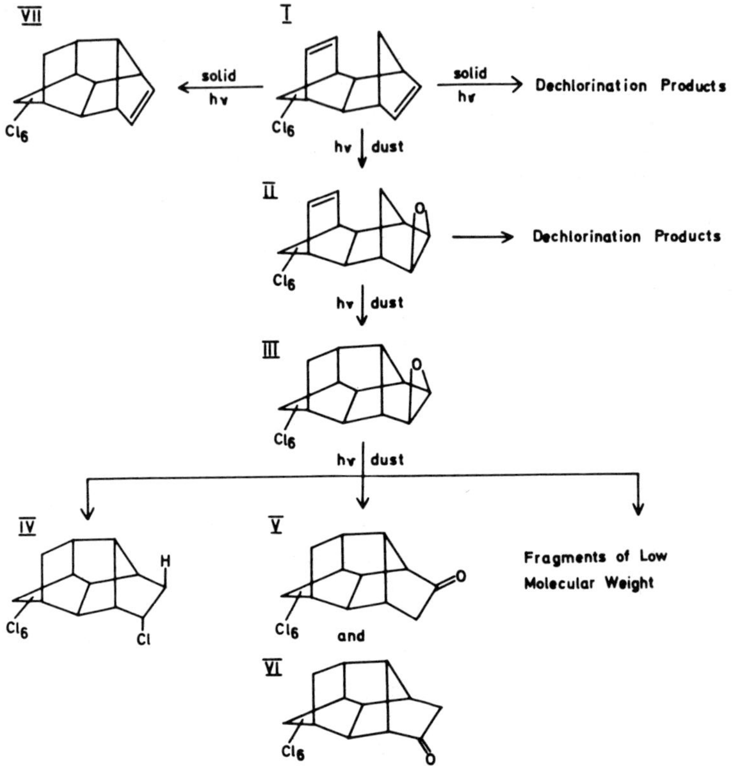


FIGURE 14. Photolysis of aldrin under short wavelength UV light.

that an enrichment in food chains of naturally occurring substances could take place as well.

Table 8 shows an example of the concentration of some elements in sea water compared to a marine organism, namely *Calanus finmarchicus*. As compared to its environment, the copepod contains some elements in lower concentrations, most elements, however, in a higher concentration than the sea water for instance phosphorus is 20 thousand-fold that of the environmental concentration.

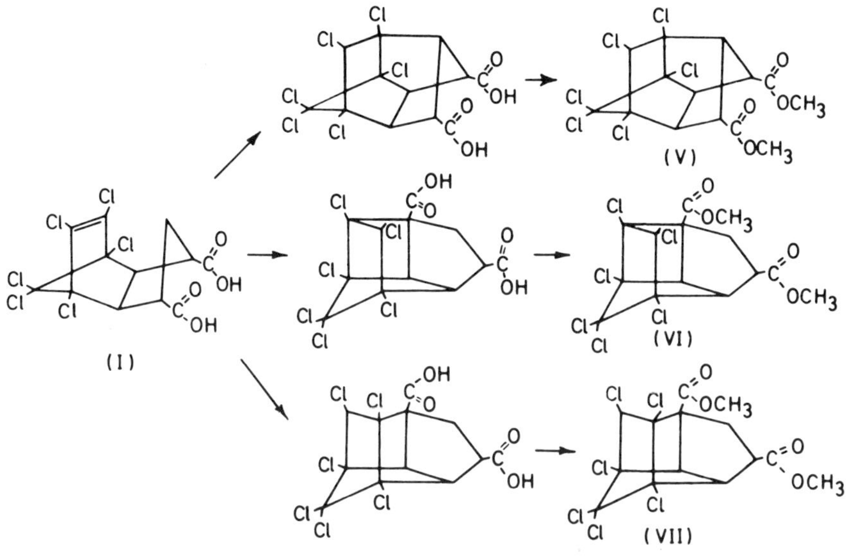


FIGURE 15. Photolysis products from dihydrochlordene dicarboxylic acid.

## IV. CONCLUSIONS

Continuous changes in chemical environmental quality by organic, natural, and man-made chemicals are facts to be reckoned with. Pesticides, despite their high acute toxicity as compared to many other chemicals, do not constitute the major impact in the concept of chemical environmental quality. The concept of constant environmental contamination despite increasing production appears at first to be an unrealistic approach. A necessary condition for realization of this ideal would occur if, on an international scale, it would be possible to bring all producers and users of industrial products to the point that they would publish their production levels, the use patterns of industrial products, and the quantities of the corresponding products used in various sectors. Then it would be possible for the first time to predict the alteration of the material environmental quality by means of the results of scientific work, in which radioiostopes play a central role, and we would be no

TABLE 8. Elements in Ocean Water as Compared to *Calanus finmarchicus* (% eight, total 100)

| | ocean water | Calanus finmarchicus | concentration factor |
|---|---|---|---|
| oxygen | 85.966 | 79.99 | 0.93 |
| hydrogen | 10.726 | 10.21 | 0.95 |
| chlorine | 1.935 | 1.05 | 0.54 |
| sodium | 1.075 | 0.54 | 0.50 |
| magnesium | 0.130 | 0.03 | 0.23 |
| sulfur | 0.090 | 0.14 | 1.6 |
| calcium | 0.042 | 0.04 | 1.0 |
| potassium | 0.039 | 0.29 | 7.4 |
| carbon | 0.003 | 6.10 | 2 000 |
| nitrogen | 0.001 | 1.52 | 1 500 |
| phosphorus | < 0.0001 | 0.13 | 20 000 |
| iron | < 0.0001 | 0.007 | 1 500 |

Elements in ocean water as compared to Calanus finmarchicus (% weight, total 100)

longer in the unsatisfactory situation of detecting the presence of unintended and/or undesired products only after the appearance of the negative effects which have been luckily, up to now, only local.

According to our present knowledge, the low global levels of these substances have no, or at least insignificant, effects on animals and plants. Effects of compounds present in ppb concentrations are known so far only in a few examples: the photosynthetic activity of phytoplankton is reduced by concentrations of about 100 ppb. This has been shown only for a few plankton species and, on the other hand, this is not the DDT-concentration in the world ocean. The potential synergistic action of low concentrations of foreign compounds nevertheless remains a problem to be considered. When saying that the chemical part of today's environmental quality has changed for the worse under special local conditions but shows no signs of deterioration globally, we must admit on the other hand that upon increasing use of synthetic products, the trend must be in the direction of a further and large-scale deterioration. Therefore, we could expect the known local effects to occur in larger areas in the future. This prediction of the trend would be changed

(a) if the consumed industrial products would break down quickly under environmental conditions, resulting in compounds which could be metabolized by organisms. That means then that relative to total production, the manufacture of persistent compounds should decrease. Considering the production-pattern of the chemical industry, the condition for this, namely the development of favorable substitutes according to an ecological point of view, would be an effort which can hardly be underestimated.

(b) in the case that man would recollect used products and handle them locally. They then should be recycled or should be returned into natural compound cycles by means of complete and controlled combustion. As regards waste, there is a principle difference between radioactive and other foreign compounds. Radioactive wastes with long half-lives should be disposed of exclusively under controlled conditions.

Today one can hardly imagine a real decrease in environmental contamination occurring within rather short periods because of increasing production; although in both sectors mentioned here, progress can be expected. However, it might be sufficient if, in spite of increasing world population and production due to the forced development of new technologies, environmental contamination by nonreusable waste could be kept at today's level.

## REFERENCES

1. Appleby, W. G. 1970. IUCN 11th Techn. Mtg. (New Delhi, 1969) 1, Morgues, Switzerland

2. Atkins, D. H. F. and Eggleton, A. E. J. 1970. In Symposium on the Use of Nuclear Techniques in the Measurement and Control of Environmental Pollution, Salzburg, Austria, 26-30 October, 1970; IAEA/SM-142.

3. Deutscher Bundestag. 1971. Materialien zum Umweltprogramm der Bundesregierung, zu Drucksache VI/2710, Verlag Heger, Bonn, Germany.

4. Harris, C. R. and Sans, W. W. 1967. J. Agr. Food Chem. 15, 861.

5. Hulpke, H. and Schuphan, W. 1970. Qual. Plant. Mater. Veg. XIX, 347.

6. Hulpke, H. 1970. Qual. Plant. Mater. Veg. XIX, 333.

7. Iliff, N. A. 1973. Environ. Qual. Safety 2, 64.

8. Lichtenstein, E. P., Schulz, K. R., Fuhremann, T. W., and Liang, T. T. 1970. J. Agr. Food Chem. 18, 100.

9. Nash, G. R. and Beall, M. L. 1970. Science 168, 1109.

10. Onsager, J. A., Rusk, H. W., and Butler, L. J. 1970. J. Econ. Entomol. 63, 1143.

## DISCUSSION

Dr. Klein was asked to explain the rationale for favoring the use of a small number of compounds with high production versus a large number of compounds with small production. From the pollution point of view alone it is much simpler and less expensive to monitor compounds in the environment when their number is relatively small. When there are a large number of compounds to consider, it is very difficult to detect and quantitatively measure all of them, and thus evaluation of each compound in respect to environmental quality is not possible. However, from the viewpoint of crop protection, a large number of compounds may be required so that the total amount of each occurring in the environment would be relatively small and probably would not have a big impact. Under these circumstances, perhaps less strict registration requirements for each compound are warranted. On the basis of present trends in crop protection and public health, a relatively large number of speciality compounds may develop.

Inquiry was made about any research to accelerate the decomposition of DDT after spraying. The response was that the breakdown of DDT and dieldrin on plants can be accelerated with certain photosynthesizers in the laboratory and on small-scale experiments outside. It is not known if large-scale tests out of doors have been made. Reference was made to a report by the Aerojet Corporation on research to degrade DDT in the soil, in which decomposition was accelerated by the addition of unspecified chemicals to the soil. The degradation products were not mentioned.

The question was asked if this group is satisfied with the public frame of mind as far as pollution is concerned on a variety of insecticides that are really quite viable not only in agricultural production but also for public health. From the discussion that followed it was clear that the group was not satisfied. It was mentioned that people tend to analogize persistence per se with hazard, and these are two quite different things. Apparently, most people generally believe that just because a compound is in the environment it constitutes a real risk. The question was posed, which was worse in terms of damaging the environment, to have small amounts of materials like DDT persisting in the environment or to apply an organophosphate like parathion once a week and wipe out most nontarget species? The former situation might be preferable if it were not for the fact that materials like DDT are biologically magnified. Any compound that is biomagnified is a potential hazard to something. However, certain compounds can be persistent without being biomagnified. Lipid solubility is another matter to consider in thinking about persistent compounds. These and many other potential deleterious factors, like carcinogenesis, are difficult to prove or disprove; hence they constitute a source of concern to people. The public is not aware of the fact that in the United States pesticides are only about 1% of the whole chemical production and about 10% of the persistent chemical production.

Studies on comparative amounts of DDT and PCBs in the world environment show that the amounts are about equal, and that PCBs came there as such and not as byproducts of decomposition of DDT. The distribution pattern is similar to the fallout of strontium 90 thus making it possible to compare strontium 90, DDT, and PCBs.

The effect of PCBs on the mink industry was mentioned. Mink reproduction is zero when fed on a diet containing 5 ppm of PCBs. All the fish in Lake Michigan now have above that level of PCBs. The fish formerly comprised about 40% of the commercial mink diet. Thus, in effect, the trace pollution of parts per trillion of PCBs has virtually wiped out the mink industry.

The recent destruction of several million chickens in the United States because of illegal dieldrin residues raised the question about the reality of tolerance standards in various foods. Some countries might use insecticide residue tolerance as a nontariff trade barrier. Perhaps consideration should be given to the establishment of a standard worldwide tolerance for insecticides in different foods. The dieldrin tolerance in chickens in the United States may be unrealistic. It was pointed out, however, that some of the chickens that were destroyed contained 3.5 ppm of dieldrin, which amounts to a health hazard.

Some of the problems with insecticides are due to over generalizations. Certain compounds can be used safely for one or more purposes but are not safe for other purposes. Research is needed to clarify just where certain compounds can and cannot be used with safety, and this information should be widely disseminated. A good example of over-generalization is the establishment of standards for the number of days after treatment of orchards and fields with certain insecticides before workers are permitted entry to do cultural work or harvest the crop.

Inquiry was made about the feasibility of considering an average optimum time of persistence for a given insecticide. Several examples in medical entomology were mentioned to show that it is not feasible to generalize on persistence time with any given insecticide--it depends on the particular problem.

If temephos (Abate®) is used in a river system for onchocerciasis control, a short persistence period is needed, long persistence must be avoided. However, when Abate® is used in drinking water to control *Aedes aegypti* a persistence of 3 months is needed. When fenitrothion is used on mud walls in Africa for control of malaria, the persistence period is 4 months--a longer period would be better, but when this compound is used in ULV equipment for control of *A. aegypti* a persistence period of from 6 to 12 hours only is wanted.

It was strongly emphasized that persistence problems are highly complex and there is urgent need for intensive study of the individual problems which will have the additional effect of putting an end to generalizations. The same comment applies to resistance and pollution.

# Selective Toxicity of Insecticides

G. T. BROOKS
Agricultural Research Council
Unit of Invertebrate Chemistry and Physiology
University of Sussex
Brighton, U. K.

## I. INTRODUCTION

Hitherto, the most economic insecticides have combined broad-spectrum insect toxicity with extended persistence in the environment. This combination is now regarded as undesirable, and broad-spectrum activity must be combined with limited persistence. However, more extended persistence may be acceptable if the toxic effect is selective for a limited number of pest species.

Selective toxicity in its broadest sense can be subdivided into ecological selectivity (Ripper et al., 1951) and physiological selectivity (Albert, 1965; Winteringham, 1969). In the former, selective toxic action is not a fundamental characteristic of the toxicant employed but arises because only the target pest organisms come into contact with it. Such selectivity is achieved by taking advantage of the particular behavioral characteristics of the target pest, as in the use of insecticidal baits against some species or systemic insecticides against others. The exploitation of behavioral differences in appropriate cases can be used to limit the spread of toxicants in the environment, and from all points of view, it should be the first principle of selectivity to expose only the target species wherever possible.

Physiological selectivity, on the other hand, becomes evident when marked differences in acute toxicity follow an equivalent contact between different organisms and the same toxicant. It is related to differences in the morphology and biochemistry of the organisms compared and, therefore, bears directly on the mode of action and design of insecticides. The present discussion concentrates on this aspect of selectivity.

## II. BIODYNAMICS AND SELECTIVITY

The most obviously desirable forms of insecticide selectivity are those between vertebrates (especially man) and insects, between beneficial parasites and predators and the insect pest, and between pollinators and pests. A thorough understanding of the biology and biochemistry of the species concerned is essential for the rational achievement of selectivity and the reasons for the selectivity between vertebrates and insects are probably best understood at present, this particular interspecific relationship has provided the principles by which the other relationships are examined.

To exert its action, an insecticide must cross membranes and become distributed between biophases that are fundamentally similar for all living tissue. Furthermore, broadly similar principles of metabolic transformation and excretion appear to operate in the biosphere. As in the case of drug action (Ariens, 1973), the toxic action of an insecticide as seen *in vivo* is the consequence of a pharmacokinetic phase followed by a pharmacodynamic phase. In the former phase, processes of partitioning into inert storage sites, metabolic toxication and/or detoxication, and excretion all compete for the toxicant which enters the organism through external (e.g., insect cuticle or vertebrate skin) or internal barriers (e.g., the alimentary tract). The amount of free toxicant remaining from this competition arrives eventually at the target organ and is the toxicant that is biologically available to interact with the target (site of action) during the pharmacodynamic phase. The overall situation is summarized in Figure 1, which shows that selective action may be a consequence of interspecific differences in either or both of these phases.

### A. Observed Toxicity and Intrinsic Activity

The overall intoxication process for insects (Figure 1) can be factored into various measurable parameters. Firstly, it is intrinsic toxicity of a toxicant at the site of action. For the anticholinesterase insecticides, the concentrations of toxicants required for 50% inhibition of the isolated enzyme ($I_{50}$) give a good indication of the relative toxicities they would have *in vivo* if they could interact directly with the enzyme without the intervention of the pharmacokinetic phase. The effect of this phase is often such that the relative intrinsic toxicities indicated in this way bear little relation to the relative toxicities ($LD_{50}$s) observed *in vivo*.

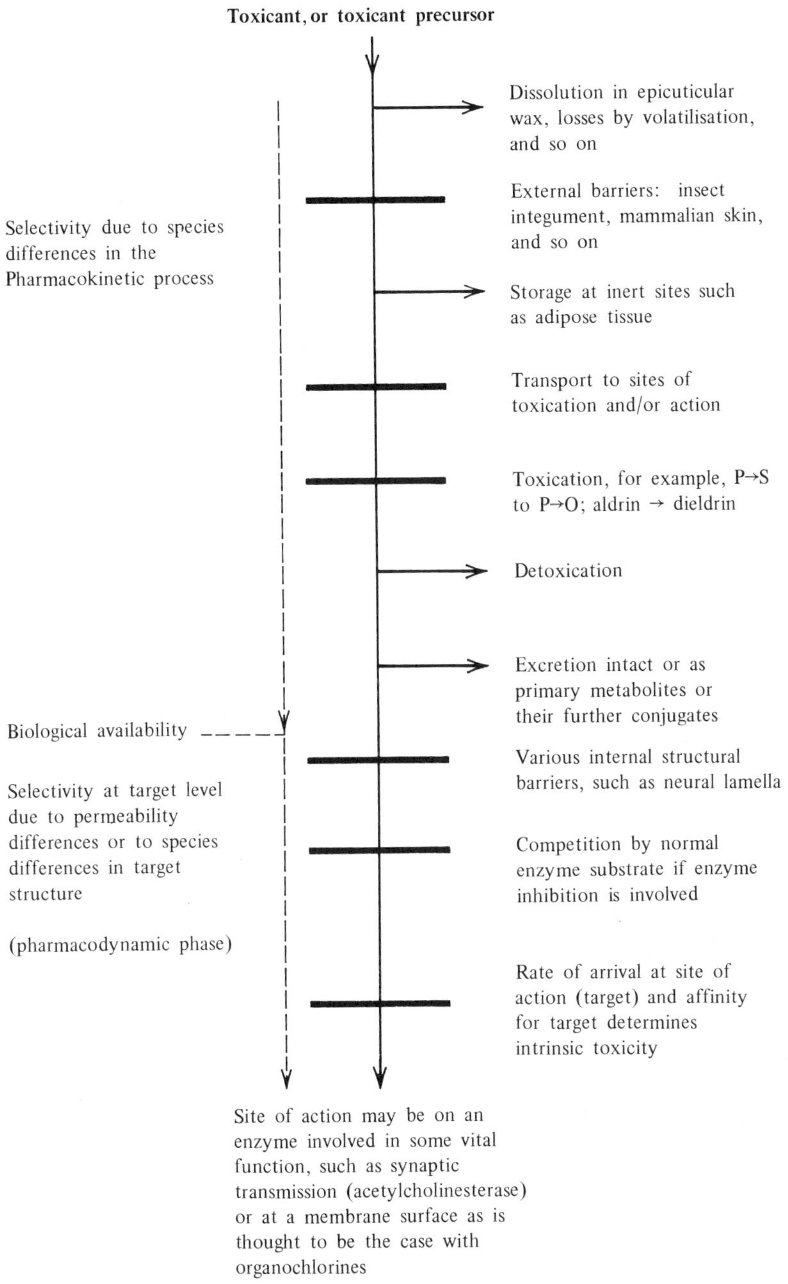
Toxicant, or toxicant precursor
Dissolution in epicuticular wax, losses by volatilisation, and so on
External barriers: insect integument, mammalian skin, and so on
Selectivity due to species differences in the Pharmacokinetic process
Storage at inert sites such as adipose tissue
Transport to sites of toxication and/or action
Toxication, for example, P→S to P→O; aldrin → dieldrin
Detoxication
Excretion intact or as primary metabolites or their further conjugates
Biological availability
Various internal structural barriers, such as neural lamella
Selectivity at target level due to permeability differences or to species differences in target structure
(pharmacodynamic phase)
Competition by normal enzyme substrate if enzyme inhibition is involved
Rate of arrival at site of action (target) and affinity for target determines intrinsic toxicity
Site of action may be on an enzyme involved in some vital function, such as synaptic transmission (acetylcholinesterase) or at a membrane surface as is thought to be the case with organochlorines

FIGURE 1. Dynamics of toxicant behavior in an insect or other animal (after Winteringham, 1969).

Thus, if a series of insecticides have similar $I_{50}$s for cholinesterase, but widely different toxicities *in vivo*, it is likely that the disparities result from differing partitioning properties between biophases and/or metabolic conversions. With the N-methylcarbamates of Table 1 for example, the disparities mentioned appear to be due mainly to metabolic detoxication, since the $LD_{50}$s (synergized $LD_{50}$; $SLD_{50}$) observed when the compounds are applied with a metabolic inhibitor (in this case the synergist piperonyl butoxide) are generally lower and more uniform, in accordance with the fairly uniform values of the $I_{50}$s. If the inhibitor used suppresses detoxication more or less completely, then departures from a straightforward correlation between $SLD_{50}$s and $I_{50}$s imply the additional influence of permeability and partitioning effects.

The total intoxication process of Figure 1 has now been factored into a parameter ($I_{50}$) indicating intrinsic toxicity and one ($SLD_{50}$) indicating the likely relative toxicity in the absence of metabolic detoxication. The influence of differing cuticle permeabilities may be assessed by observing the progress of poisoning following a topical application, and also by comparing the toxicities measured by different routes of application, such as injection and infusion (Sun *et al*., 1967; Sun and Johnson, 1969; Sun, 1971; Sun and Johnson, 1971, 1972). Thus if a topically applied insecticide produced toxic effects that increase over a long period (e.g., chlorinated insecticides), it is probable that slow penetration of the cuticle is combining with high toxicant stability to give a cumulative effect; there is ample opportunity for detoxication of the incoming poison for a slow penetrating but labile insecticide, and a toxic internal level may never be achieved.

In topical application in acetone, spray application in kerosene, and injection in acetone, dimethylsulfoxide gave similar $LD_{50}$s for several highly toxic organophosphorus insecticides, including parathion and organochlorine insecticides applied to houseflies (Sun, 1968; Sun and Johnson, 1972). For carbamates, including carbaryl, toxicities were higher and more uniform when measured by oil spray or injection, and the results suggest that these applications are equivalent, with kerosene acting to increase cuticle permeability; the synergistic action of dioctylphthalate with certain carbamates is attributed to a similar effect (Weiden, 1968). The implication is that an agent that greatly increases the speed of penetration of a labile toxicant should enable it to exert its effect before detoxication can intervene. Such measurements give some insight regarding the origins of selecticity and demonstrate the importance of biophysical factors in toxicity measurements.

TABLE 1. Detoxication and Synergism of Various Aryl N-Methyl Carbamates

| N-Methylcarbamate | $I_{50}M$ fly ChE | Topical $LD_{50}$ *Musca domestica* (μg/g) Alone | 1:5 p.b.[a] | SR |
|---|---|---|---|---|
| 2,3-Dimethoxyphenyl | $1.4 \times 10^{-5}$ | > 500 | 16.5 | > 30 |
| 2,4-Dimethoxyphenyl | $2.8 \times 10^{-5}$ | 155 | 13.0 | 11.9 |
| 2,5-Dimethoxyphenyl | $1.3 \times 10^{-5}$ | 13.0 | 4.9 | 2.7 |
| 3,4-Dimethoxyphenyl | $1.9 \times 10^{-5}$ | 400 | 12.0 | 33 |
| 3,5-Dimethoxyphenyl | $8.0 \times 10^{-6}$ | 11.0 | 4.4 | 2.5 |
| 3,4-Methylenedioxyphenyl | $1.3 \times 10^{-4}$ | 17.5 | 20.0 | 1.0 |
| 2-Propoxyphenyl | $8.7 \times 10^{-6}$ | 100 | 13.5 | 7.4 |
| 2-Propargyloxyphenyl | $2.9 \times 10^{-6}$ | 6.5 | 4.6 | 1.4 |
| 3-Propoxyphenyl | $1.6 \times 10^{-5}$ | 95 | 15.5 | 6.1 |
| 3-Propargyloxyphenyl | $4.0 \times 10^{-6}$ | 7.5 | 6.0 | 1.2 |
| 1-Naphthyl | $9.0 \times 10^{-7}$ | 900 | 12 | 75 |
| 5,6,7,8-Tetrahydro-1-naphthyl | $1.4 \times 10^{-6}$ | > 500 | 19 | > 26 |
| 5,8-Dihydro-1-naphthyl | $2.1 \times 10^{-6}$ | 167 | 33 | 5.1 |
| 4-Benzothienyl | $2.5 \times 10^{-7}$ | 18.5 | 8.0 | 2.3 |

[a]Piperonyl butoxide synergist. (From data of: Fukuto *et al.*, 1962; Metcalf, 1968.)

### B. Penetration Through the Integument and Gut

Insecticides mainly come in contact with the cuticle or skin of insects or mammals, or are absorbed from the alimentary tract following ingestion with the food. Lipophilic insecticides spread rapidly over the insect body following topical application or tarsal contact (Lewis, 1962; Quraishi and Poonawalla, 1969), and intake via the tracheolar system has been suggested as a major route of intake (Gerold, 1969). Passage through the cuticle and distribution via the haemolymph have previously been regarded as the main route of intoxication, and this subject has become controversial (Burt, 1970; Moriarty and French, 1971; Olson, 1973). In any case, the morphology of insects results in a relatively exposed nervous system, in contrast to the situation for mammals.

The difference between the flexible mammalian skin and the chitinized insect cuticle leads to an expectation of permeability differences between them toward insecticides, and comparisons between topical application and injection (measuring the effect of the integumental barrier) suggest that the cuticle is generally a less effective barrier than the skin (Winteringham, 1969). O'Brien, however, concluded (1967) that differential integument permeability itself is not a major cause of selective toxicity toward insects compared with mammals. This view is supported for cockroach cuticle and human skin by a theoretical treatment (Penniston et al., 1969) of the dependence of penetration rate (k) on oil-water partition coefficient (P). The relationship between the two parameters (in log-log plots) is parabolic rather than linear, so that as P increases, k is maximal for a particular value of P, which was found to be very similar for phosphate esters penetrating human skin and for various insecticides and other solutes penetrating American cockroach cuticle. Clearly, if the P values for optimum penetration through different integuments are similar, there is little reason to expect selective toxicity on the basis of penetration differences alone.

Nevertheless, small penetration differences may join with other factors in conferring overall selectivity. Certain insecticide-resistant strains of housefly, for example, have a genetically controlled factor that delays insecticide penetration; it normally confers little resistance to kill by itself but can greatly enhance the expression of other resistance mechanisms that may be present, especially if they involve metabolic detoxication (Sawicki and Farnham, 1968). Differing natural tolerances between different insect species, or between different stages of the same insect (e.g., in lepidopterous larvae), may well be partly due to permeability differences in the cuticle and/or internal membranes.

An illustration of the way in which successive small penetration barriers or their combination with other mechanisms, such as detoxication, can greatly alter sensitivity to an insecticide is provided by the discussions of Hewlett and Winteringham (cited in Winteringham, 1969). The effect depends on the fact that the amount of insecticide x entering an insect after a given time does not increase indefinitely as the applied dose z is increased, but approaches an upper limit a that can be defined by the equation $x = a(1-e^{-z/a})$. The shapes of the curves derived from this equation are such that mechanisms that independently effect modest reductions in the amount of insecticide penetrating or reaching the site of action can, when combined, reduce the attainable internal level a below that required for toxic effect; in other words, the insect becomes immune.

Interspecific differences in permeability of the gut may obviously have a role in the selective action of stomach poisons or other insecticides ingested with the food. For example, DDT penetration through the gut walls of the leaf roller *Argyrotaenia velutinana* and the grasshoppers *Melanoplus femur-rubrum* and *M. differentialis* is insignificant, and most of the ingested toxicant is excreted as DDT plue DDE by these tolerant insects. In contrast, large amounts of ingested DDT are absorbed through the guts of houseflies (Sternburg and Kearns, 1952). Carbaryl penetrates largely unchanged through the gut wall of mammals (Shah and Guthrie, 1971; Casper *et al*., 1973); in terms of penetration through the isolated gut, absorption is similar for mouse and tobacco hornworm (*Manduca sexta*) and slower for a cockroach (*Blaberus discoidalis*). It was also shown (Shah and Guthrie, 1971) that DDT and dieldrin penetrate the isolated guts only slowly, and are strongly retained in the gut tissue of all three species. In general, lower polarity favors retention by the gut tissue. With straight chain dialkoxy-analogues of dimethoate (except the dimethoxy-analogue), for example, gut penetration for the above three species increases with polarity (Shah *et al*., 1972), as found for insecticides penetrating the cuticle (Olson and O'Brien, 1963). The penetration pattern is inevitably complicated by metabolism in some cases (e.g., malathion), and although relatively small interspecific differences in penetration were found by Shah *et al*. (1972), they may contribute significantly to overall selectivity.

### C. Circulation and Distribution

The insect circulatory system is an open one, and the haemolymph directly bathes structures such as fat body, gut, and malpighian tubules that are involved in insecticide metabolism (Wilkinson and Brattsten, 1972) as well as the nervous system. In contrast, the mammalian nervous system is less exposed; entry from the gut is into the portal vein and results in transfer to the liver with its metabolizing enzymes, although entry into the systemic circulation via the lymphatics of the gut may avoid this route in some cases. Also, penetration through the skin or the mucosa of the mouth gives direct access to the circulation without passage through the liver. For both mammals and insects, the distribution of toxicants is greatly influenced by binding to soluble proteins and uptake by the cellular components of the blood (Moss and Hathway, 1964; Olson, 1973), which makes it difficult to relate penetration experiments with isolated integuments or gut preparations to the true situation *in vivo*.

Differences in the sizes of inert storage depots between different species of insects or mammals, between the sexes, between different ages of adults, or between the different stages of insect development may contribute to selective toxicity by altering biological availability. Thus for certain lepidopterous larvae (e.g., *Heliothis* spp.), the toxicities of lipophilic chlorinated insecticides decrease much more than those of more polar organophosphorus insecticides, as larval size (and fat depots) increases (Sun *et al*., 1967).

### D. Selectivity at the Site of Action

1. Metabolism. The possibility that metabolic toxication or detoxication may occur in the target tissue itself cannot be ignored, and the bio-available toxicant may be further modified at this level before it reaches the actual site of action. Clearly, interspecific differences in biotransformation at this point could result in selective toxicity.

2. Penetration. Penetration of toxicants into the mammalian and insect nervous systems appears to follow broadly similar principles. The vertebrate blood brain barrier retards the entry of all but highly lipophilic molecules, and there is a similar barrier in the insect central nervous system (CNS) (Treherne and Pichon, 1972) that appears to slow the penetration of ions 10 to 15 times (O'Brien, 1967). Unfortunately, inverse selectivity between insects and mammals is found with cholines-

terase inhibitors that are highly ionized at physiological pH, for these are frequently highly toxic to mammals and poorly toxic to insects. This is due to the presence in mammals of peripheral cholinesterase, which is critical for survival but is relatively unprotected against ionized anticholinesterases (O'Brien, 1967). Accordingly, the toxicities of ionizable inhibitors can be increased for insects relative to mammals by arranging for mostly the nonionized forms to be present at physiological pH.

This situation highlights the need to exploit other nerve targets that are peculiar to insects. Their noncholinergic neuromuscular junction is frequently discussed in this context, but no new developments in selective insect control have arisen from this difference so far.

3. Structure of the Site of Action. Interspecific differences in the topography of a particular site of action are a fundamental cause of selectivity. In anticholinesterases, the ratio of mammalian to insect cholinesterase inhibitory activity (e.g., $I_{50}$ for mammalian red cells, $I_{50}$ for housefly head cholinesterase) is frequently used as a measure of the relative affinities of a toxicant for the different enzymes. Examples of organophosphorus insecticides that show favorable mammalian toxicity and high $I_{50}$ ratios are dichlorvos (Table 2) ($I_{50}$ ratio 501), the phosphorothioate analogue ($I_{50}$ ratio 794) of the amidic phosphate ruelene [$(CH_3NH)(CH_3O)P(O)OPh$-2Cl, 4-$C(CH_3)_3$], Gardona® ($I_{50}$ ratio 300), and its corresponding diethyl ester ($I_{50}$ ratio 200). In the case of Gardona and its relative, the toxicities to rat (acute oral, 4000 to 5000 and 1100 mg/kg, respectively) are unexpectedly favorable, a finding attributed to partitioning characteristics that favor mammals, as well as to increased metabolic detoxication in the case of the dimethyl phosphate (Whetstone _et al._, 1966). The low toxicity of paraoxon to frogs, as compared with mice or American cockroaches, is due to the insensitivity of frog brain cholinesterase to this toxicant (O'Brien, 1967).

A now-classical example of selectivity between insect species is the high toxicity to houseflies and low toxicity to honeybees (_Apis mellifera_) shown by diisopropyl _p_-nitrophenyl phosphorothioate (isopropyl parathion). This compound and its P(O) analog are >212 and 17.8-fold, respectively, more toxic to houseflies, and housefly cholinesterase is about 36-fold more sensitive to the P(O) analogue than the bee enzyme is. The lower toxicity to honeybees appears related to a lower sensitivity of their cholinesterase to inhibition by the toxic phosphate, coupled with a low rate of its production from the

TABLE 2. Structures of Some Compounds Mentioned in the Text in Relation to Selective Toxicity

---

Parathion Group (Schrader, 1961; Metcalf and Metcalf, 1973)

| | |
|---|---|
| Parathion | $(C_2H_5O)_2P(S)O$ 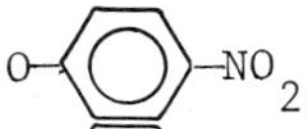 |
| Paraoxon | $(C_2H_5O)_2P(O)O$ 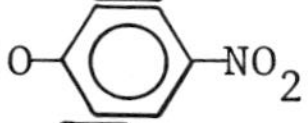 |
| Parathion-methyl | $(CH_3O)_2P(S)O$ 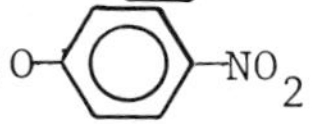 |

Derivatives of parathion-methyl having low mammalian toxicity

$(CH_3O)_2P(S)O$ 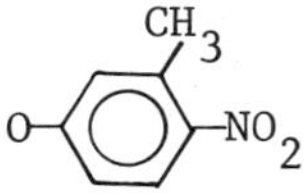
(Sumithion)

$(CH_3O)_2P(S)O$ 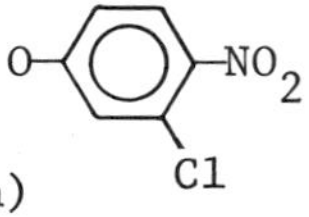
(Chlorthion)

$(CH_3O)_2P(S)O$ 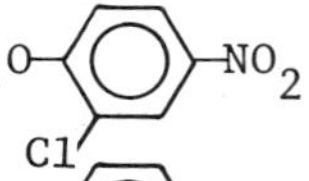

$(CH_3O)_2P(S)O$ 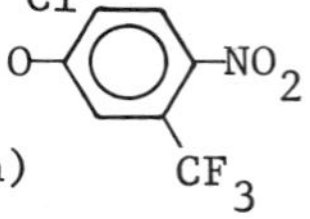
(Fluorthion)

(continued)

TABLE 2 (cont.)

| | | |
|---|---|---|
| Malathion Group (O'Brien, 1967) | | |
| Malathion | $(CH_3O)_2P(S)SCH(COOC_2H_5)CH_2COOC_2H_5$ (α at SCH, β at $CH_2$) | |
| Acethion | $(C_2H_5O)_2P(S)SCH_2COOC_2H_5$ | |
| Dimethoate | $(CH_3O)_2P(S)SCH_2CONHCH_3$ | |
| Vinylphosphate Group | | |
| Dichlorvos | $(CH_3O)_2P(O)OCH{=}CCl_2$ | (O'Brien, 1967) |
| Chlorfenvinphos | $(C_2H_5O)_2P(O)OC(=CHCl)-$ 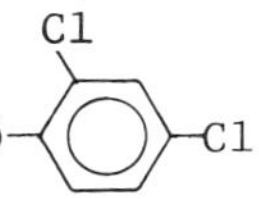 | (Whetstone et al., 1966) |
| Ethyl Tetrachlorvinphos | $(C_2H_5O)_2P(O)OC(=CHCl)-$ 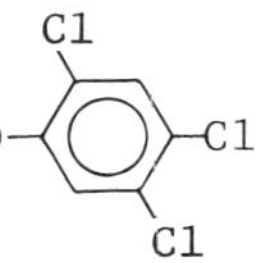 | |
| Tetrachlorvinphos (Gardona) | $(CH_3O)_2P(O)OC(=CHCl)-$ 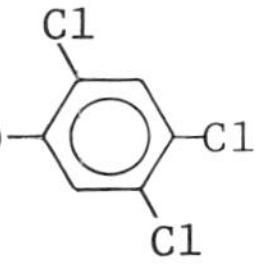 | |

(continued)

TABLE 2 (cont.)

DDT Group (See also Table 6; Holan, 1971a,b; Metcalf *et al.*, 1971; Hirwe *et al.*, 1972)

| | |
|---|---|
| Methoxychlor | $(4\text{-}CH_3OPh)_2CHCCl_3$ |
| DDD | $(4\text{-}Cl.Ph)_2CHCHCl_2$ |
| Prolan | $(4\text{-}CL.Ph)_2CHCH(NO_2)CH_3$ |
| Dianisylneopentane (DANP) | $(4\text{-}CH_3OPh)_2CHC(CH_3)_3$ |

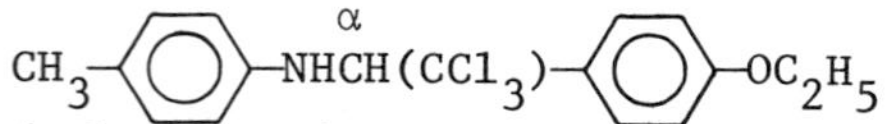


(substituted -trichloromethylbenzylanilines)

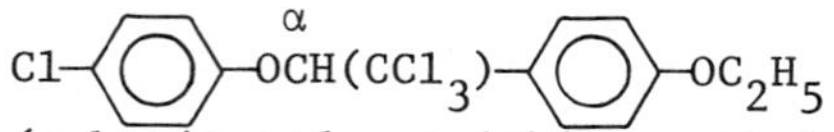


(substituted α-trichloromethylbenzyl phenyl ethers)

Cyclodiene Group

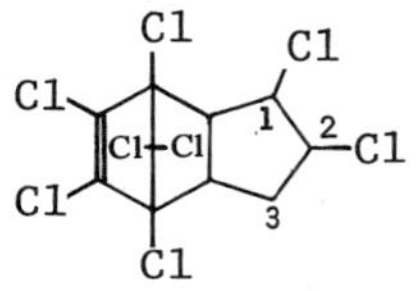


*endo*-*trans*-1,2-dichlorodihydrochlordene
*endo*-*cis*-1,2-dichlorodihydrochlordene
(Constituents of chlordane)

(continued)

TABLE 2 (cont.)

| | |
|---|---|
| 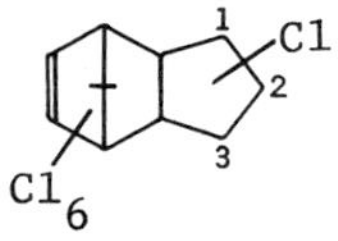 | 1-*exo*-chloro- = α-dihydroheptachlor<br>2-*exo*-chloro- = β-dihydroheptachlor<br>(Buchel *et al.*, 1964; Brooks and Harrison, 1967) |
| 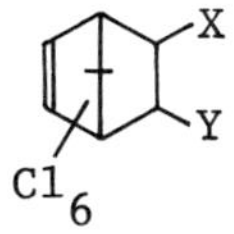 | X=$CH_2$Br, Y=H, Bromodan<br>X=Y=$CH_2$Cl, Alodan<br>(*endo*-substituents) |
| 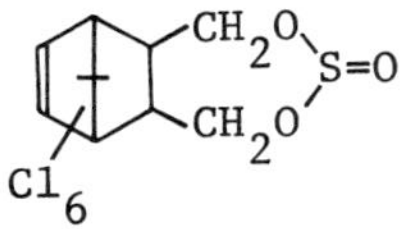 | α-endosulfan<br>β-endosulfan |
| 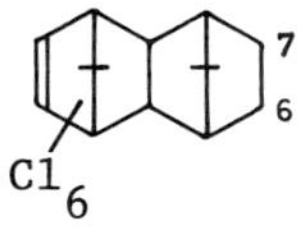 | 6-7-dihydroaldrin (Brooks and Harrison, 1969)<br>6-7-dihydroisodrin |
| 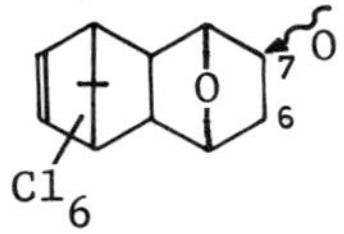 | 9-oxa-dihydroaldrin<br>(Brooks and Harrison, 1964; Brooks, 1972; El Zorgani *et al.*, 1970; Walker and El Zorgani, 1973; Walker, *et al.*, 1973) |

TABLE 2 (cont.)

---

<u>Cyclodiene Group</u> (cont.)

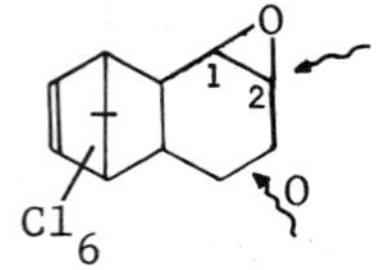


hydration (selective)

HCE

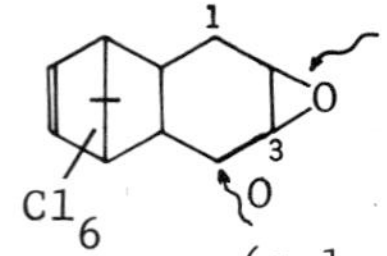


(selective)

hydration (selective)

HEOM

<u>Pyrethroid Group</u>

Pyrethrin I

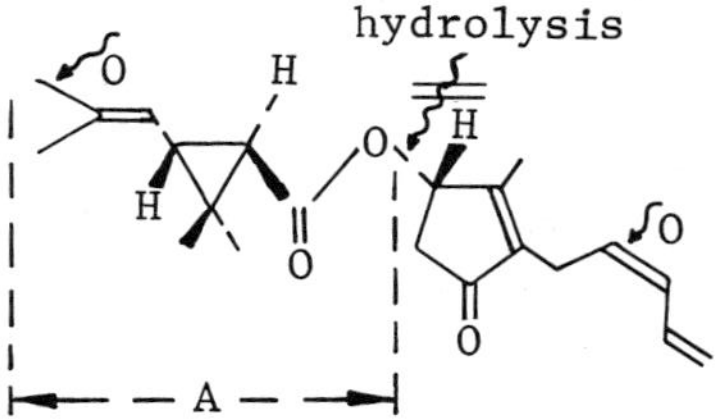


(Yamamoto <u>et</u> <u>al</u>., 1969; Miyamoto <u>et</u> <u>al</u>., 1971; Abernathy and Casida, 1973)

(continued)

TABLE 2 (cont.)

Pyrethroid Group (cont.)

Bioresmethrin

(+)-trans-isomer

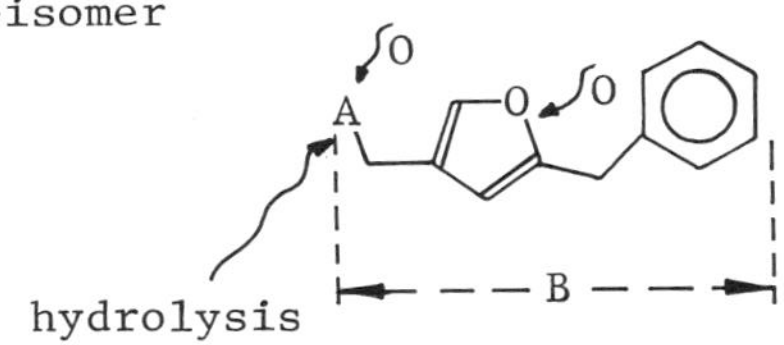

(+)-cis-isomer

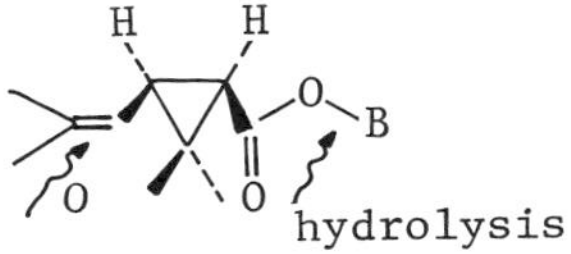

Cyclopentylidene analog

(+)-trans-isomer

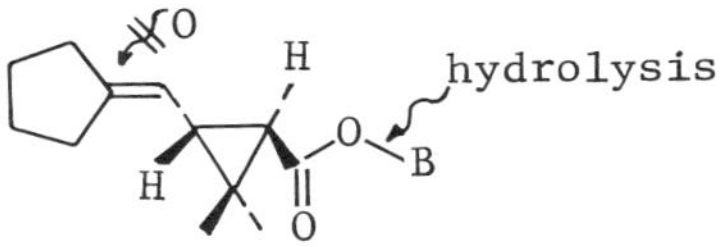

Nereistoxin Group (Sakai, 1970)

Nereistoxin  $(CH_3)_2NCH\begin{smallmatrix}CH_2S\\|\\CH_2S\end{smallmatrix}$

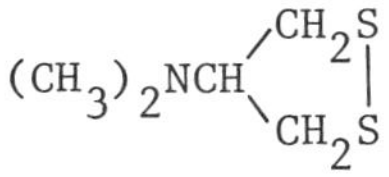

Dihydronereistoxin  $(CH_3)_2NCH(CH_2SH)_2$

(continued)

TABLE 2 (cont.)

| | |
|---|---|
| Nereistoxin Group (cont.) | |
| Cartap | $(CH_3)_2NCH(CH_2SCONH_2)_2$ |
| Insect Growth Regulators | |
| Cecropia juvenile hormone<br>JH1, $R=C_2H_5$<br>JH2, $R=CH_3$ | 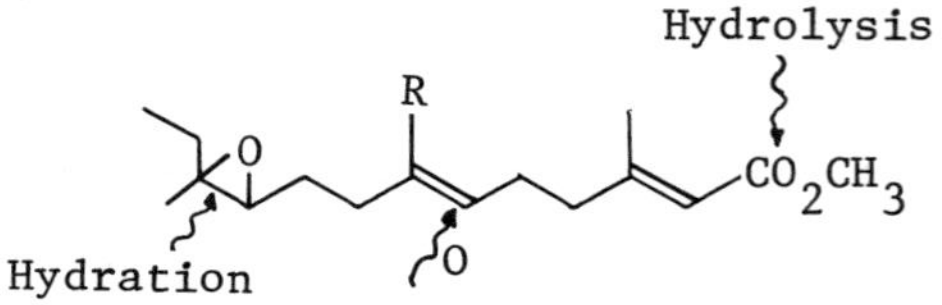 |
| Bowers' juvenile hormone mimic (Bowers, 1969) | 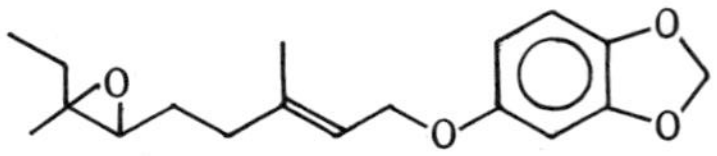 |
| Urea derivatives affecting cuticle deposition (Mulder and Gijswijt, 1973) | 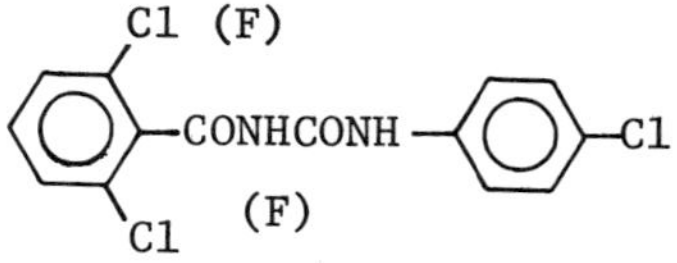 |

phosphorothioate; these phenomena may be due to steric effects associated with the isopropyl groups (Camp *et al.*, 1969).

It is usual to refer to the ratio of mammalian $LD_{50}$ to insect $LD_{50}$ for a particular insecticide, as the mammalian selectivity ratio (MSR), and there is a marked increase in the MSR when parathion-methyl (MSR, mouse-housefly = 71) is converted into Sumithion (MSR = 630). The 2-chloro (MSR >880), 3-chloro (MSR >140) and 3-$CF_3$ (MSR >650) derivatives of parathion-methyl (Table 2) also show this effect (Metcalf and Metcalf, 1973). Selectivity for Sumioxon has been attributed to relatively poor inhibitory activity toward mammalian cholinesterase (Hollingworth *et al.*, 1967) and reduced ability to enter the mammalian brain (Miyamoto, 1969). In support of the view that differential enzyme inhibition is an important factor in selectivity, fly head and bovine erythrocyte cholinesterases showed considerable differences in their susceptibility to inhibition by the phosphate analogs of compounds in this group, the greatest difference being found with the 3-$CF_3$ derivative, which itself has an MSR (mouse/housefly) of 92 (Metcalf and Metcalf, 1973).

The discovery of inhibitor-insensitive cholinesterases in resistant strains of the cattle tick *Boophilus microplus* provides an interesting example of target modification resulting in selectivity at the intraspecific level. Actually, the normal and resistant strains each contain both sensitive and insensitive forms of cholinesterase, but enzyme activity is lower in the resistant insects. In a coroxon (3-chloro-4-methyl-7-coumarinyl diethyl phosphate) resistant strain investigated by Nolan *et al.* (1972) it appears that the relatively insensitive components of the susceptible strain's enzymes are further modified to give an overall 500-fold difference in enzyme sensitivity to this insecticide. If the difference involves a change in the chemical or physical structure of the enzyme(s), then it might be allowed for by a suitable modification in toxicant structure.

The dimensions of the active site of cholinesterase in various mammalian and arthropod species have been explored with a view to designing selective inhibitors, and O'Brien (1967) deduced that the distance between anionic and esteratic sites is smaller for human red cell than for fly head cholinesterase. However, the design of molecules, specifically for target selectivity, must also allow for their need to survive the metabolic processes of the pharmacokinetic phase.

## III. METABOLISM AND SELECTIVITY

### A. Principles of Metabolism

Several recent reviews discuss the principles and pathways of insecticide biotransformation (Hathway et al., 1970, 1972; Dauterman, 1971; Brooks, 1972; Fukuto, 1972). The biotransformation (metabolism) of toxicants in vivo may produce equitoxic or more toxic products (bioactivation or toxication), less toxic products (detoxication), or mixtures arising from both types. The balance of these changes determines the amount of toxicant(s) within the organism, and, subject to the other components of the pharmacokinetic phase, determines the bioavailability of toxicant(s) at the target. Accordingly, biotransformation is a most crucial factor in the total intoxication process and differential metabolism is frequently responsible for both intraspecific (resistance) and interspecific selectivity.

The principles of biotransformation are rather similar in the animal species thus far examined, differences often being quantitative rather than qualitative. The enzymic reactions involved include oxidations (aromatic hydroxylation, aliphatic hydroxylation, N-dealkylation, O-dealkylation, epoxidation, desulfuration, thioether oxidation, oxidative dechlorination, etc.), reductions (e.g., $NO_2 \rightarrow NH_2$, reductive dechlorination), hydrolyses (e.g., ester hydrolysis, amide hydrolysis, epoxide ring hydration), and transfer reactions (including O-demethylation by glutathione and various conjugation reactions). The mixed-function oxidases (MFO) involved in many of the oxidations are located in the microsomal fraction (endoplasmic reticulum) derived from homogenates of tissues such as gut, fat body, and malpighian tubules of insects (Wilkinson and Brattsten, 1972), and from homogenates of vertebrate liver. These enzymes also occur in other vertebrate tissues, but liver is the most important site for this kind of toxicant metabolism, which is therefore rather more localized than in insects. Hydrolytic enzymes are located in both particulate and soluble fractions of cells, and this applies also to some of the conjugating enzymes. Glucuronyl transferases, for example, are located in the microsomal fraction of mammalian liver and kidney and catalyze the transfer of glucuronic acid from uridine diphosphate glucuronic acid (UDPGA) to reactive hydrogen atoms (e.g., in -OH, $-NH_2$, -COOH, -SH, etc.) produced in the primary oxidations, reductions, or hydrolyses.

The significance of a particular metabolic product may be difficult to assess when the metabolism is complex, but if antagonism is observed when an inhibitor of microsomal oxidations is coapplied with a toxicant *in vivo*, then an oxidative bioactivation is probably involved. Conversely, synergism may indicate that there is oxidative detoxication (Table 1). In such cases, toxicity may be altered by inhibiting or stimulating the enzymes involved, as appropriate, or by modifying the structures of those functional groups in the toxicant that are vulnerable to enzymic attack. Assuming that attack on a particular functional group results in detoxication, an alternative to inhibiting the enzyme involved is to look for an insect species to which the molecule is intrinsically toxic, and which has a low titer of the detoxifying enzyme. Thus Brattsten and Metcalf (1970) computed the toxicity of carbaryl, alone and with the synergist piperonyl butoxide, to 54 species of insects from 8 orders and 37 families. The variation in the ratio of the two $LD_{50}$s (synergistic ratio) gives some idea of the range of detoxicative microsomal oxidase activity towards molecules of this type. Also, the rate of aldrin epoxidation has been used to measure relative microsomal oxidase activity in the midguts of the 35 species of lepidopterous larvae (Krieger *et al*., 1971). From such information, one can pick out those species likely to be most susceptible to a toxicant which is attacked by microsomal oxidases.

## B. Influence of Molecular Structure

1. Organophosphorus Insecticides. a. *Parathion and relatives*. The conversion of this compound into paraoxon is an example of the oxidative desulfuration that is of almost universal occurrence *in vivo* and converts intrinsically nontoxic phosphorothioate esters into the corresponding highly toxic phosphates. The bioactivation requires time and allows opportunity (O'Brien, 1967) for enzymic degradation of both the P(S) and P(O) types to nontoxic products. The MSR is usually (though not invariably) higher for the P(S) than the P(O) analogue. Fortunately, the P(O) analogues themselves are often more toxic to insects than to mammals, and when the P(S) compound is applied, the effect seen *in vivo* is a result of the balance between activation and detoxication. In insects, this balance frequently lies in favor of more P(O) analogue in the tissues, as is very evident in the comparison between cockroaches and mice given by O'Brien (1967) for parathion.

For triesters of phosphorothioic or phosphoric acid, the possible detoxication mechanisms appear to be: (1) hydrolytic O-dealkylation or O-dearylation; (2) oxidative O-dealkylation or O-dearylation; and (3) O-dealkylation or O-dearylation by transfer to glutathione (GSH) or otherwise mediated by GSH. The P(S) phosphorus atom is much less electrophilic than the P(O) one, and hydrolytic deesterification appears insignificant for P(S) compounds. Oxidative dealkylation does not appear significant with P(S) esters, but the aryl moiety of parathion and a number of relatives is removed by a characteristic microsomal, mixed-function oxidase reaction requiring NADPH and oxygen, the products being a dialkyl phosphorothioic acid and the corresponding substituted phenol. This appears to be the most important detoxication reaction for parathion, although GSH-mediated O-dealkylation has also been reported (Nakatsugawa _et al._, 1968, 1969a, b, c). For paraoxon and other phosphates, hydrolysis of the acid anhydride link becomes significant and may mask the possible oxidative removal of this moiety; oxidative cleavage is said to occur for diazoxon in houseflies, but not in rats (Dauterman, 1971). Oxidative (Oppenoorth _et al._, 1971) and hydrolytic (Nolan and O'Brien, 1970) O-dealkylation of paraoxon has been indicated for houseflies and GSH-stimulated O-dealkylation for rats (Kojima and O'Brien, 1968).

A reduction in mammalian toxicity is achieved by converting parathion into the corresponding dimethyl ester (parathion-methyl; Table 2) and has been attributed to the GSH-mediated O-dealkylation, that removes one alkyl group and is much more active for dimethyl than for diethyl esters (Hollingworth, 1969). For the aromatic ring-substituted derivatives of parathion-methyl, such as Sumithion, lower enzyme inhibition by the P(O) analogues (Section II.D.3), coupled with the requirement for the P(S) → P(O) conversion, may enhance the value of O-demethylation. This mechanism operates against Sumithion and its P(O) analogue and in mice is very significant at nontoxic doses much exceeding those at which parathion-methyl is toxic. It is reinforced by the oxidative (for Sumithion) and hydrolytic dearylation (for Sumioxon) previously noted for parathion and paraoxon, respectively.

b. _Malathion, Acethion, and Dimethoate_. The selectivity of malathion (Table 2) is quite remarkable; it is toxic to many insects but generally has much lower toxicity to mammals and birds. Again, high toxicity to normal insects is associated with stability _in vivo_ leading to pronounced accumulation of the toxic P(O) analogue. In normal insects, relatively slow degradation involves primarily hydrolytic (phosphatase) cleavage of the methoxyl and thiol-ester groups. The difference between tolerant and susceptible species is that the former can also hydrolyze and oxidatively de-esterify the carboxyester groups. Mammals produce mainly a malathion monoacid that has been identified as the α-monoacid in the case of rats (Dauterman, 1971).

Human liver homogenates effect some O-demethylation and produce a large proportion of carboxyesterase products (about 90% of the added malathion); with rat liver there is relatively more O-demethylation (Matsumura and Ward, 1966). Malathion (Table 2) is a much better substrate for liver enzymes than parathion, which is degraded very slowly by both human and rat liver. The significance of the carboxyesterase detoxication is revealed by synergism of malathion toxicity to both mammals and some tolerant insects by the compound EPN® (ethyl p-nitrophenyl phenylphosphonothioate) that inhibits carboxyesterase activity *in vivo*.

The hide beetles *Dermestes maculatus* and *D. lardarius* differ ten-fold in natural tolerance to malathion, and the difference is overcome by triphenylphosphate (TPP), another carboxyesterase inhibitor (Dyte, 1969). Resistance to malathion in the flour beetle *Tribolium castaneum* is also frequently associated with enhanced carboxyesterase activity, although one strain has a mechanism that specifically O-demethylates malaoxon. Triphenyl phosphate can be used to detect the presence of the carboxy-esterase-resistance mechanism in *T. castaneum*.

The rather well-defined selectivity associated with the carboxyester group led to the preparation of some other organo-phosphates containing this moiety, the best-known one being acethion (O'Brien, 1967). This compound (Table 2) tends to show even more favorable mammalian toxicity than malathion, but has a narrow spectrum of insect toxicity. A variant on this theme is the introduction of an amide group, as in dimethoate, in the hope that vertebrates might also be differentially high in amidase activity. In this case, however, the selectivity between insects and vertebrates is not clear-cut, there being both sensitive and insensitive species in both classes. For insects, the differential selectivity of dimethoate appears to be polyfactorial. The high susceptibility of houseflies is rather obviously related to a combination of rapid penetration, sensitive cholinesterase, and a metabolic balance leading to a high internal level of the P(O) analogue, but the situation is less obvious with other insects. Unlike malathion, which is degraded in various vertebrate tissues, dimethoate is bio-transformed mainly in the liver, and although the balance between amidase and phosphatase attack varies between species, there is a marked positive correlation between total hydrolysis of dimethoate by liver homogenates and the tolerance of the species concerned. Thus biotransformation in the liver is a rather significant factor for vertebrate selectivity, which doubtlessly depends on the rate of hydrolytic detoxication exceeding that of oxidative activation to the toxic P(O) analog.

It is interesting to note that when dimethoate or its P(O) analog are converted into the corresponding free amides, toxicities increase more to houseflies than to mice, so that the MSR is improved in each case. The effect seems to be associated with a relative improvement in bioavailability for the insect target, since the P(O) amides are poorer inhibitors for both housefly head and vertebrate cholinesterase (Lucier and Menzer, 1970). Somewhat similar effects are seen when Bidrin® (3-hydroxy-N,N-dimethyl-*cis*-crotonamide dimethyl phosphate) is converted into the free amide (Menzer and Casida, 1965).

c. *Tetrachlorvinphos (Gardona®) and Relatives*. Sun (1972) found the MSR (mouse/housefly) to vary from 19 to 1060 (mean 376) for 14 analogues of tetrachlorvinphos and from 0.87 to 56 (mean 19.4) for 15 chlorinated insecticides, showing that these organophosphates are generally safer than the chlorinated insecticides tested. There is a progressive reduction in mammalian toxicity through the series chlorfenvinphos, "ethyl" tetrachlorvinphos, and tetrachlorvinphos (Table 2) without much change in insect toxicity. It appears that chlorination confers distribution properties that favor mammals, besides reducing the affinity for mammalian cholinesterase (Section II.D.3); in the dimethyl ester, these changes combine with an increased opportunity for GSH dependent dealkylation to optimize selectivity. Mammalian liver MFO de-ethylate chlorfenvinphos (Donninger *et al*., 1967) and such oxidation seems to involve alkyl esters of phosphates rather than phosphorothioates.

2. Carbamate Insecticides. The data of Table 1 show that metabolic detoxication has a very significant influence on the toxicity of these compounds. Metabolic differences may arise from structural changes that block or reveal preferred sites of oxidative attack, or alter the affinity for the detoxifying enzyme. Since some of the hydroxylation products of carbamate retain cholinesterase-inhibitory properties, the observed toxicity may also be influenced on this account. Comparison of the position isomers in the upper group of Table 1 reveals a substrate preference for the vicinal dimethoxy compounds that Metcalf (1968) has related to the preference for vicinal diphenols shown by oxidases such as tyrosinase. The substrate specificity found in natural tolerance is also evident in houseflies, resistant through selection with a particular carbamate.

Methylenedioxyphenyl- and propargyloxyphenyl-derivatives inhibit MFO, and carbamates containing such groups may inhibit their own metabolism (autosynergism); this may explain the low synergistic ratio for 3,4-methylenedioxyphenyl *N*-methylcarbamate compared with the 3,4-dimethoxy- analogue and for the propargyloxy-compounds compared with their propoxy-relatives (Metcalf, 1968).

Carbamates, in general, display considerable species selectivity. For example, only a few phenyl *N*-methylcarbamates are good miticides, whereas a variety of oxime carbamates have this activity. On the other hand, good activity against Mexican bean beetle larvae (*Epilachna varivestis*) is more often associated with phenyl *N*-methylcarbamates than with carbamoyloximes. Thiomethyl oxime carbamates such as methomyl (Lannate®) [$(CH_3S)(CH_3)C{=}NOCONHCH_3$] have outstanding activity to lepidopterous larvae such as *Heliothis*, and to a lesser extent, *Pieris* and *Prodenia* (Southern armyworm), whereas the corresponding alkoxy-series are particularly active against *Musca domestica*, and are less toxic to mammals than the thio-series (Weiden, 1968; Felton, 1968). However, the toxicities of members of both series to *Musca* become uniform in the presence of an MFO inhibitor, and selectivity can also be traced to differential detoxication in other cases.

Carbaryl (1-naphthyl *N*-methyl carbamate; Tables 1 and 3) provides an outstanding example of species selectivity due to detoxication. It is noteworthy that the honeybee appears to be deficient in MFO-attacking carbamates and is susceptible to carbaryl and numerous others (Table 3). Differences between insects in their ability to detoxify carbaryl are highlighted in the surveys by Brattsten and Metcalf (1970, 1973), and it is evident that interspecific selectivity will vary with sex and with age as the titer of MFO changes. The sulfur atom of aldicarb (Temik®) [$(CH_3S)C(CH_3)_2CH{=}NOCONHCH_3$] is biologically oxidized to give the corresponding sulfoxide and sulfone; these are more potent cholinesterase inhibitors than the parent, which may explain the high toxicity of aldicarb to both insects and mammals.

Fortunately, a number of the MFO inhibitors that are insecticide synergists are generally more stable in insects than in mammals. This differential metabolism in mammals means that the spectacular increase in toxicity seen, for example, when carbaryl is applied to houseflies with 2,3-methylenedioxynaphthalene or 1-naphthyl-2-propynyl ether is not seen with mice. Thus two chemicals that independently have low toxicity to housefly and mouse become highly toxic only to the housefly when combined, an almost ideal situation (Sacher *et al.*, 1969).

An interesting example of selectivity between insects and mammals arises when certain substituted phenyl *N*-methylcarbamates are converted into *N*-(acyl)methylcarbamates. For 2-*s*-butyl- (Table 4), 3-*s*-butyl-, and 3-isopropylphenyl *N*-methylcarbamates,

TABLE 3. Species Specificity of Some Aryl *N*-Methylcarbamates

| | $LD_{50}$ (μg/g; topical) | | | $LC_{50}$ (ppm) | |
|---|---|---|---|---|---|
| *N*-Methylcarbamate | Housefly | Honeybee | German Cockroach | Mosquito (*Culex*) | Oral $LD_{50}$ Rat (mg/kg) |
| 1-Naphthyl-(carbaryl) | 900 | 2.3 | >130 | 1.0 | 540 |
| 2-Isopropylphenyl- | 95 | 2.8 | >130 | 0.56 | 500 |
| 3-Isopropylphenyl- | 90 | 1.0 | 14.7 | 0.03 | 16 |
| 2-Isopropoxyphenyl-(Baygon®) | 25.5 | 0.8 | 11.3 | 0.35 | 250 |
| 4-Methylthio-3,5-xylenyl-(Mesurol®) | 24 | 1.1 | >130 | 0.23 | 100 |
| 4-Dimethylamino-3,5-xylenyl-(Zectran®) | 60 | 0.6 | >130 | 0.49 | 60 |

Data of Georghiou and Metcalf (1961); Metcalf and Fukuto (1965).

TABLE 4. Contact/Stomach Poison Activities of 2-*s*-Butylphenyl N-Acyl-N-Methylcarbamates

| Acyl Group | Oral $LD_{50}$ Mice (mg/kg) | Mean Lethal Concentration (ppm) | | |
|---|---|---|---|---|
| | | *Megoura viciae* | *Plutella maculipennis* | *Phaedon cochleariae* |
| None (parent) | 60–250 | 30 | 1500 | 900 |
| $COCH_3$ | >1000 | 50 | 3250 | >1000 |
| $COCH_2Cl$ | >1000 | 58 | 300 | 250 |
| $COCHCl_2$ | >1000 | 25 | 600 | >1000 |

Data of Fraser *et al*., 1968.

acylation (Fraser et al., 1968) produces remarkable reductions in mammalian toxicity. The spectrum of insecticidal activity is not increased, and toxicity may be reduced with simple alkanoyl groups. However, variants have been found that retain both the activity of the parent carbamate and the favorable mammalian toxicity. The acylcarbamates are poor inhibitors of insect cholinesterase *in vitro* but are inhibitory *in vivo* and the evidence suggests that deacylation to the parent carbamate occurs readily in insects (depending on the alkanoyl substituent) but not in mammals. Slow deacylation in mammals should give opportunities for other metabolic changes leading to detoxication. The propionyl and butyryl derivatives of 2-isopropoxyphenyl *N*-methylcarbamate show lower systemic activity than the parent, without much change in contact activity; the parent systemic activity is regained with methoxyacetyl, chloroacetyl, and dichloroacetyl groups, along with favorable reductions in mammalian toxicity. This pattern of systemic activities points to effects of the various acyl-substituents on distribution properties.

Veterinary insecticides such as carbaryl and Butacarb® (3,5-di-*t*-butyl *N*-methylcarbamate) have low mammalian toxicity without structural modification, but acylation may prove valuable in cases where potentially useful insecticidal properties are offset by high mammalian toxicity. Other types of proton substitution can also lead to favorable mammal-insect selectivity (Fahmy et al., 1970; Black et al., 1973). See Fukuto, Chapter XI, for further discussion.

3. Chlorinated Insecticides. a. DDT Group. In terms of acute toxicity, this group of compounds is remarkably safe to vertebrates, except fish. Nevertheless, the persistent nature of some members undoubtedly results in the once-prized broad spectrum insecticidal activity that can affect friend and foe alike. The group contains some remarkable and well-known examples of selectivity in favor of mammals. However, such selectivity is almost inevitably accompanied by a narrower spectrum of practical insecticidal activity. Methoxychlor, Prolan®, and DDD (Table 2) are classical examples and have found valuable applications against particular pests. DDD is frequently better than DDT as a mosquito larvicide and against insects in unexposed habitats. Methoxychlor is now increasingly used because of its biodegradability and low storage in animal tissues. Dianisylneopentane (Table 2), first described by Brown and Rogers (1950), is of theoretical interest since it was the first nonchlorinated yet insecticidal analogue of DDT.

Renewed interest in compounds of this type is reflected in the new groups of symmetrical and unsymmetrical analogues of DDT described by Holan (1971a, b) and by Metcalf et al. (1971). See Metcalf, Chapter 9. In addition, some α-trichloromethylbenzylanilines and α-trichloromethylbenzyl phenyl ethers show interesting insecticidal activity (Hirwe et al., 1972); the most active members of these two groups are shown in Table 2. In all these cases, synergistic effects with MFO inhibitors indicate that microsomal enzymes are responsible for detoxication, which involves the oxidation of aromatic alkoxyl or alkyl groups. The alkane, nitro-alkane, and oxetane moieties of some of these compounds provide additional points for enzymic attack, whereas the α-trichloromethylbenzylanilines suffer dehydrochlorination, followed by cleavage of the molecule at the NH-C bond to give the corresponding derivatives of aniline and acetophenone. The data of Table 5 show that combinations with an MFO inhibitor (sesamex) show increased toxicity to mice in some cases, but the effect is so much greater with the housefly that the ratio of synergised $LD_{50}$s (MSR) is actually increased in favor of the mouse. This situation is similar to that with the carbaryl/synergist combinations mentioned earlier.

b. Cyclodiene Insecticides. The insect toxicities in this group are generally higher than those of the DDT group, as are the mammalian toxicities. This combination produces MSRs not dissimilar to those of the DDT group, except for rather toxic compounds such as isobenzan and endrin. Endrin, however, is oxidatively transformed much more rapidly than its stereoisomer dieldrin, and is much less persistent in vertebrate tissues (Baldwin, 1971). The conversions of aldrin and heptachlor to their stable and toxic epoxides are good examples of microsomal oxidations producing toxic compounds. All the available evidence suggests that the precursors also are toxic (Brooks, 1974), but they are better insecticides from an environmental standpoint because their double bonds allow opportunities for chemical and for biological detoxications, apart from the widely occurring epoxidations.

Of the commercial cyclodienes, technical chlordane combines lowest mammalian toxicity (acute oral $LD_{50}$ for rats 457 to 590 mg/kg) with valuable insecticidal properties, although two other compounds, Alodan® and Bromodan® (Table 2) have remarkably low toxicities (>15,000 and 12,900 mg/kg, respectively) and have been used in the past to control ectoparasites. The halomethyl groups in these compounds are potentially vulnerable to hydrolytic or oxidative dechlorinations that should lead eventually to the corresponding carboxylic acids. In fact, unidentified acidic metabolites appeared in the urine of mammals given large doses of Alodan®, and Rahn (1963) attributed its low oral

TABLE 5. Selective Toxicity of Biodegradable DDT Analogues

| Compound | Housefly $LD_{50}$ (μg/fly) | | | Mouse $LD_{50}$ (mg/kg) | $\frac{LD_{50}\ Mouse}{LD_{50}\ Fly}$ (ppm) |
|---|---|---|---|---|---|
| | Alone | S[a] | SR[a] | | |
| DDT | 0.24 | 0.25 | 1 | 570 | 47 |
| 1,1-bis(_p_-ethoxyphenyl) 2,2-dimethylpropane | 1.92 | 0.32 | 6.0 | 5200 | 54 |
| 1,1-bis(_p_-ethoxyphenyl) 2-nitropropane) | 0.48 | 0.065 | 7.4 | 1150 | 48 |
| 1-(_p_-ethylthiophenyl)-1-(_p_-ethoxyphenyl)-2-nitropropane | 0.16 | 0.015 | 10.7 | 1040(360)[b] | 130(480)[c] |
| 1-(3,4-methylenedioxy-phenyl)-1-(_p_-ethoxy-phenyl)-2-nitropropane | 0.14 | 0.019 | 7.4 | >2000(>2000) | >286(>2000) |
| 1,1-bis-(_p_-ethoxyphenyl)-2-nitrobutane | 0.55 | 0.061 | 9.0 | 1160(980)[b] | 42(326)[c] |
| 2,2-bis(_p_-chlorophenyl)-3,3-dimethyloxetane | 1.27 | 0.66 | 1.9 | 3500 | 55 |

(continued)

TABLE 5 (cont.)

| Compound | Housefly $LD_{50}$ (μg/fly) Alone | $S^a$ | $SR^a$ | Mouse $LD_{50}$ (mg/kg) | $\frac{LD_{50}\ Mouse}{LD_{50}\ Fly}$ (ppm) |
|---|---|---|---|---|---|
| 2,2-bis(p-ethoxyphenyl)-3,3-dimethyloxetane | 0.52 | 0.010 | 52.0 | 1200(380)[b] | 46(760)[c] |

[a] S = 5 μg sesamex/fly applied immediately after the insecticide; SR = $LD_{50}$ without synergist/$LD_{50}$ with synergist.

[b] Figures in parentheses are $LD_{50}$ with synergist (sesamex) at 1:1 or 2:1 ratio with the insecticide.

[c] Figures in parentheses are ratio of synergised $LD_{50}$s (mouse $LD_{50}$s by i.p.).

(Data of Holan, 1971a, b).

toxicity to a combination of detoxication and poor absorption from the gut.

Technical endosulfan (Table 2) is a mixture of α- and β-isomers, of which the β-isomer is about three-fold less toxic to rats (orally) and somewhat less toxic to insects than the α-isomer (Maier-Bode, 1968). This toxicity change is in the same direction as that between endrin and dieldrin, which sterically resemble the α- and β-isomers, respectively. Endosulfan sulfate, formed from both isomers _in vivo_, is as toxic to rats as α-endosulfan, and as toxic to houseflies as β-endosulfan. The mammalian toxicity of technical endosulfan appears similar to that of aldrin; it is less toxic to birds but more toxic to fish. In particular, it combines a useful spectrum of insecticidal activity with relatively low persistence due (in addition to sulphate formation in some situations) to its conversion to several derivatives related to the innocuous endosulfan-diol.

By suitable molecular modifications, cyclodiene insecticides can be made increasingly vulnerable to detoxication, subject usually to loss of the broad-spectrum insecticidal activity. For example, reduction of the reactive double bonds in isodrin, aldrin, and heptachlor makes the resulting dihydro-compounds vulnerable to oxidative detoxication (hydroxylation), rather than toxicative epoxidation. Of the three known monochlorodihydrochlordenes (dihydroheptachlor isomers), the β-isomer (β-DHC; Table 2) showed good toxicity to houseflies, mosquito larvae, _Pieris brassicae_ larvae, and bedbugs, and remarkably low toxicity to mice, rats, dogs, and hens, with the oral $LD_{50}$ range for the vertebrate group between 2000 and 9000 mg/kg (Buchel _et al_., 1964). Its toxicity and that of the α-isomer (derived from heptachlor) is synergized against houseflies by an MFO inhibitor such as sesamex (Brooks and Harrison, 1967). The dihydroheptachlor isomers have not found practical application.

The toxicities of 6,7-dihydroisodrin and 6,7-dihydroaldrin to houseflies are increased ten-fold, and that of the 9-oxa-analog of dihydroaldrin (Table 2) about twenty-fold by sesamex, which inhibits detoxicative hydroxylation at the 6 position formerly occupied by a double bond in the precursors of these molecules (Brooks and Harrison, 1969). Interesting variations in biodegradability are obtained by altering the position of the epoxide ring in the 9-oxa-analog. Thus the epoxide ring in the isomeric 1,2-epoxide HCE (Table 2) is stable in insects that, however, can hydroxylate the cyclohexane ring to an extent dependent on the species. In the housefly, inhibition of this detoxication by sesamex results in a considerable increase in toxicity (45-fold) toward that of dieldrin. Tsetse flies

(*Glossina* spp.) and stable flies (*Stomoxys calcitrans*) appear to be deficient in microsomal oxidases, and for these, the unsynergized toxicities of HCE are only three- to four-fold and nine-fold lower, respectively, than dieldrin. In contrast to HCE, the symmetrical 2,3-epoxide HEOM suffers rapid epoxide ring hydration to the corresponding transdiol in several species (e.g., housefly, blowfly, flour beetle) and is nontoxic to them. That this epoxide has intrinsic toxicity to some insects is shown, however, by its toxicity to *Glossina* spp. and to *S. calcitrans* (12- to 14-fold and 60-fold less than dieldrin, respectively) which are also deficient in epoxide hydrase. A two-fold synergism of HEOM with sesamex against *G. austeni* (*S. calcitrans* not tested) is indicative of the moderate inhibitory effect of sesamex on epoxide hydrase (Brooks, 1972, 1973a). Clearly, the toxicity of HEOM depends largely on the absence of the rather widespread epoxide hydrases, and its toxicity to tsetse flies represents rather an extreme case of selectivity.

Vertebrates, generally speaking, are able both to hydroxylate the cyclohexane rings of HCE and HEOM, and to hydrate the epoxide rings (Brooks *et al*., 1970; Brooks, 1972; El Zorgani *et al*., 1970; Walker and El Zorgani, 1973), so that more detoxication routes are available to them than to the insects thus far examined. However, the capacity for enzymic hydration varies considerably among species; the order is fish $<$ birds $<$ mammals; however, the rat and the rook (*Corvus frugilegus*) are exceptional in having lower and higher hydrase activity, respectively, than expected. HCE has been most studied in vivo as being the more interesting insecticide; products of cyclohexane ring hydroxylation (oxidation) predominate, but the trans-diol is excreted by rat, rabbit, and quail. The relatively rapid *in vitro* metabolism of HCE is paralleled by rapid excretion *in vivo*, the approximate half-times of elimination following intraperitoneal injection (10 to 30 mg/kg; $^{14}$C-HCE at 0.14 μCi/μmole) being $\ngtr$5, 2, 4, and 7 days, respectively, for rat, rabbit, quail, and pigeon. For the birds especially, these values compare favorably with dieldrin, which is also oxidized and hydrated *in vivo* (in mammals). It is interesting to note that the shag (*Phalocrocoras aristotelis*) has proved to be an efficient accumulator of dieldrin, and its liver microsomes have poor ability to hydroxylate or hydrate even these degradable structural analogs, although they epoxidize aldrin at a moderate rate (Walker *et al*., 1973). *In vitro*, the pH optimum is higher (8.0 to 9.0) for epoxide hydrase than for hydroxylation (pH 7.0 to 7.5), and the balance between the two activities can be shifted by pH change as well as by omitting the cofactors (NADPH and oxygen) required for oxidation. This has to be remembered when attempts are made to predict the likely situation *in vivo* from experiments in vitro.

Interesting interspecific differences in toxicity are found when the hexachloronorbornene nucleus of cyclodiene insecticides is progressively dechlorinated (Soloway, 1965). Together with the selectivity found in simple derivatives of hexachloronorbornene such as Alodan and Bromodan, this observation suggests that useful properties might reside in cyclodiene isosteres having methyl or chloromethyl groups instead of chlorine atoms (Brooks, 1973b). Toxaphene is a mixture of polychloronorbornanes, polychloronorbornenes, and related compounds that may be expected to contain interesting biodegradable and selective components of this type, and it is currently being examined on this basis (Casida, 1974).

c. *Lindane*. The γ-isomer of hexachlorocyclohexane combines excellent insecticidal properties with intermediate mammalian toxicity (acute oral $LD_{50}$ 90 to 200 mg for several species), low chronic toxicity, and moderate environmental persistence (see Ulmann, 1973, for review). Because of the strict conformational requirement for toxicity, it is difficult to devise structural modifications that retain insecticidal activity. Nevertheless, Nakajima and his colleagues (Kurihara, 1973) have prepared mixed halogen analogs having the γ-conformation that are more toxic than lindane, and a γ-methoxypentachloro analog that is nearly equitoxic with lindane to mosquitoes. The presence of groups such as methoxyl in lindane isosteres is likely to provide opportunities for biodegradation, additional to those already present in lindane, although the spectrum of insecticidal activity may be narrowed, as in other cases (see Nakajima, Discussion, Chapter IX).

4. Derivatives of Botanicals and Natural Toxins. a. *Pyrethroids*. The excellent insect/mammal selectivity of natural pyrethrins (acute oral $LD_{50}$, rat, ranging from 260 to >1400 mg/kg) is well known. However, this selectivity disappears when the compounds are administered intravenously ($LD_{50}$s 1 to 5 mg/kg). Likewise, the toxicities of the synthetic pyrethroids bioresmethrin (5-benzyl-3-furylmethyl(+)-*trans*-chrysanthemate; Table 2) and its (+)-*cis*-isomer are, respectively, 8000 mg/kg and 168 mg/kg orally, 340 mg/kg and 6 to 7 mg/kg intravenously (Verschoyle and Barnes, 1972). It appears that neither insects nor mammals can readily hydrolyze the esters of secondary alcohols found in natural pyrethrins and allethrin, so that detoxication is mainly oxidative (Yamamoto *et al*., 1969; Casida *et al*., 1971). In contrast, the esters of primary

alcohols used in the new synthetic pyrethroids are both hydrolyzed and oxidized by mammals and probably also by insects. Thus a combination of oxidative and hydrolytic detoxication accounts for the low toxicity of bioresmethrin to mammals (Miyamoto *et al.*, 1971; Abernathy and Casida, 1972). Furthermore, the higher toxicity to mice of the (+)-*cis*-isomers of bioresmethrin and other pyrethroids containing primary ester functions is accompanied by lower hydrolytic rates (in liver microsomes) than those for the (+)-*trans*-isomers.

The methyl groups in the isobutenyl group of chrysanthemic acid are major sites of oxidative attack that can be protected with a consequent increase in both insect and mammalian toxicity, by, for example, conversion into a cyclopentylidenemethyl moiety (Table 2). Inhibitors of mixed function oxidases (sesamex, piperonyl butoxide) do not alter the toxicity of bioresmethrin or its cyclopentylidenemethyl analogue to mice, and the esterase inhibitor DEF (S,S,S-tributyl phosphorotrithioate) synergizes only the latter, although it inhibits the hydrolysis of both *in vitro*. This suggests that if either the oxidative or the hydrolytic detoxication mechanism for bioresmethrin is inhibited, the remaining route is still adequate for protection (Abernathy and Casida, 1972). In the cyclopentylidenemethyl analogue, the oxidative route is suppressed by the structural change; ester hydrolysis becomes vital for detoxication and its inhibition results in marked synergism. This situation is reminiscent of that noted with the degradable dieldrin analogs; it illustrates the value of synergists in these studies and the potential for the manipulation of selectivity ratios by suitable molecular modifications (see Elliot, Chapter VII, for further discussion.

b. *Rotenone*. Microsomal mixed-function oxidases of fish liver, mammalian liver, and insect tissues convert rotenone into rotenolones I and II, 8'-hydroxyrotenone, and 6',7'-dihydroxyrotenone, together with the products of combinations of these routes and water soluble products. However, water soluble products (probably conjugates) appear more extensively in vertebrate than in insect tissues, and their formation is apparently stimulated by factors present in the soluble fraction of vertebrate liver but not of insect cells. In consequence, an insect such as the American cockroach tends to accumulate the primary hydroxylation products (some of which retain significant intrinsic toxicity) as well as unchanged rotenone (Fukami *et al.*, 1969). Deficiencies in conjugating mechanisms may well contribute to interspecific selectivity, for even if the primary products of MFO activity are inactive, their removal from the microsomes by conjugation might in some cases be rate limiting for continued oxidation of the parent toxicant. Any suppression of conjugation would then lead to accumulation of unchanged toxicant in the tissues. This possibility has been examined by Dorough *et al.* (1972) as an alternative to inhibiting the microsomal oxidases.

c. *Nereistoxin*. This toxin (4-dimethylamino-1,2-dithiolane; Table 2), isolated from the marine annelid *Lumbrineris heteropoda*, causes ganglionic blocking and consequent paralysis by interacting with the acetylcholine receptors. There are species differences in response; lepidopterous larvae are rather sensitive, cockroaches recover from the paralysis, and houseflies are unaffected. Dihydronereistoxin, the corresponding propane-1,3-dithiol (Table 2) is also active and has been regarded as the active form of nereistoxin. However, nereistoxin in formed from various propane-1,3-dithiol esters *in vivo* and according to Sakai (1970), there is a direct correlation between the amount of nereistoxin formed *in vivo* and ganglionic-blocking action in isolated nerve cords. Since the formation of dihydronereistoxin and nereistoxin requires ester hydrolysis, or hydrolysis followed by dehydrogenation, respectively, "opportunity factors" for selective detoxication can be incorporated by making esters of appropriate stability. The 1,3-bis-carbamoyl ester (cartap) has favorable mammalian toxicity (rat oral $LD_{50}$ 500 to 750 mg/kg) and is in practical use as an insecticide.

5. Juvenile Hormone and Juvenile Hormone Mimics. Since natural juvenile hormones (JH) act at the gene level, they represent a new dimension in insect control, and the possibilities for interspecific differences in terms of selective behavioral changes as well as morphogenetic effects appear almost unlimited. During the life of an insect, JH may act on a variety of target tissues in yet undefined ways, thus the question of selective action due to target differences is complex and lies in the future. For their use in insect control, JH or its mimics currently have to be applied in the same way as classical insecticides. Their biological availability at the target(s) is therefore governed by the pharmacokinetic principles of Figure 1; however, they are subject to any additional constraint an insect may apply to a compound that may be its own natural hormone. Accordingly, current research emphasizes identification, metabolism, and biosynthesis, with the last area in particular providing possible new targets for selective action.

The compounds thus far fully examined, including JH mimics (JHM), have low acute toxicity to mammals, have not shown activity toward mites, ticks, or spiders (Bagley and Bauernfeind, 1972), and often show remarkable specificity between different insects. Although numerous compounds of synthetic origin and some of natural origin show JH activity, the identification of such hormones in insects is limited to verification of the presence or absence of the two known *cecropia* hormones [methyl *trans,trans,cis*-10,11-epoxy-7-ethyl-3,11-dimethyltrideca-2,6-dienoate (JH1) and its 7-methyl analog (JH2)], or their $C_{16}$ relative *trans,trans*-methyl epoxyfarnesoate (JH3) in extracts. The JH2 and JH3 are present in the tobacco hornworm, *Manduca sexta*, (Judy *et al.*, 1973), and JH3 is present in the May bug, *Melolontha melolontha*, but none of these compounds are in larvae of the lepidopteran, *Vanessa io*, the larvae or adults of the meal beetle (*Tenebrio molitor*), or the larvae of houseflies (Trautmann *et al.*, 1974). However, another investigation indicated the presence of JH1 in houseflies (Bieber *et al.*, 1971).

Terpenoid JH and JHM are known to be metabolized by one or more of ester hydrolysis, epoxide hydration, double bond epoxidation or double bond hydration, depending on structure, and there are interspecific variations in the extent of these conversions (Slade and Zibitt, 1972; White, 1972; Ajami and Riddiford, 1973) that suggest that some part of the selectivity phenomena may be related to differential metabolism. If this is the case, shifts in the activity spectra should occur in the presence of appropriate inhibitors of the metabolizing enzymes, just as in the case of classical insecticides. Bowers' discovery (1968) that certain MFO inhibitors such as sesamex have JH activity, and the subsequent preparation of terpenoid JHM containing the methylenedioxyphenyl group (Bowers, 1969), emphasized this point. Inhibition of the epoxide hydrases and carboxyesterases involved in JH and JHM metabolism has been shown recently (Brooks, 1973a; Slade and Wilkinson, 1973), as well as synergistic effects on JH activity (Solomon *et al.*, 1973). The situation is complicated because in some cases there may be a requirement for activation of JHM to "proximate" juvenoids. Thus there are indications that the diols formed by epoxide hydrase action may be inactive in some insects but active in others. In some cases the high activity of JHM containing terminal alkoxy-groups, compared with the corresponding epoxy-analogs (Sarmiento *et al.*, 1973), may relate to greater stability of the alkoxy-group in vivo. Another possibility is that diol formation from the epoxide is deactivating, whereas tertiary alcohol formation by alkoxyl cleavage is activating; these questions await further elucidation.

## IV. SUMMARY AND CONCLUDING COMMENTS

The development of selective insecticides falls into two main areas: (1) the search for new classes of insecticides that may affect targets with characteristics peculiar to insects, e.g., insect neuromuscular junction, cuticle synthesis, or hormone biosynthesis; and (2) the accumulation of information about metabolic and excretory processes so that the fate of existing insecticide groups can be manipulated by appropriate structural modification or enzyme inhibition.

In the first category are the formamidine acaricides related to chlorphenamidine (Galecron®). This compound is a monoamine oxidase inhibitor (Beeman and Matsumura, 1973) having low mammalian toxicity. It may interfere with the regulation of biogenic amines, but the precise mode of action is uncertain. Also of interest is a new group of 1-(2,6-dichlorobenzoyl)-3-phenylureas with low mammalian toxicity and broad-spectrum insecticidal activity related to their interference with cuticle deposition (Mulder and Gijswijt, 1973). Such compounds, having new modes of action, should be toxic to insects that are resistant to existing insecticides through alterations in target sensitivity (e.g., cattle tick resistance to organophosphates and insect resistance to cyclodienes).

Research in the second category would be aided by methods for the rapid assessment of the activity levels of the major metabolizing enzymes in a wide spectrum of target and nontarget species in vivo. An example is the assessment of mixed function oxidase levels by measuring synergistic ratios with appropriate combinations of insecticides and synergists. *In vitro* measurements can give valuable information, but their limitations must be recognized, especially in regard to the additional influence of conjugation *in vivo*. The insect excretory mechanisms themselves may afford a target for selective attack, whereas compounds that interfere with the conjugating enzymes might be synergistic for some of the existing insecticides.

## REFERENCES

1. Abernathy, C. O. and Casida, J. E. 1973. *Science* *179*, 1235.

2. Ajami, A. M. and Riddiford, L. M. 1973. *J. Insect Physiol.* *19*, 635.

3. Albert, A. 1965. *Selective Toxicity*, Methuen, London.

4. Ariens, E. J. 1973. In Drug Design, (E. J. Ariens, Ed.), vol. IV, Academic, New York and London, p. xi.

5. Bagley, R. W. and Bauernfeind, J. C. 1972. In Field Experiences with Juvenile Hormone Mimics, J. J. Menn and M. Beroza, Eds., Academic, New York and London, pp. 113-151.

6. Baldwin, M. K. 1971. The Metabolism of the Chlorinated Insecticides Aldrin, Dieldrin, Endrin, and Isodrin. Ph.D. Thesis, University of Surrey.

7. Beeman, R. W. and Matsumura, F. 1973. Nature (London) 242, 273.

8. Bieber, M. A., Sweeley, C. C. and Faulkner, D. J. 1971. Cited in Trautmann et al., 1974.

9. Black, A. L., Chiu, Y., Fahmy, M. A. H., and Fukuto, T. R. 1973. J. Agr. Food Chem. 21, 747.

10. Bowers, W. S. 1968. Science 161, 895.

11. Bowers, W. S. 1969. Science 164, 323.

12. Brattsten, L. B. and Metcalf, R. L. 1970. J. Econ. Entomol. 63, 101.

13. Brattsten, L. B. and Metcalf, R. L. 1973. Pestic. Biochem. Physiol. 3, 189.

14. Brooks, G. T. 1966. World Rev. Pest Control 5, 62.

15. Brooks, G. T. 1972. Pathways of Enxymatic Degradation of Pesticides. In Environmental Quality and Safety, F. Coulston and F. Korte, Eds., Thieme Verlag Stuttgart, Academic, New York, vol. 1, pp. 106-163.

16. Brooks, G. T. 1973a. Nature (London) 245, 382.

17. Brooks, G. T. 1973b. The Design of Insecticidal Chlorohydrocarbon Derivatives. In Drug Design, E. J. Ariens, Ed., Academic, New York and London, vol. IV, pp. 379-444.

18. Brooks, G. T. 1974. Chlorinated Insecticides, vol. 2, Chemical Rubber Company, Cleveland.

19. Brooks, G. T. and Harrison, A. 1964. J. Insect Physiol. 10, 633.

20. Brooks, G. T. and Harrison, A. 1967. Life Sci. (Oxford) 6, 1439.

21. Brooks, G. T. and Harrison. A. 1969. Biochem. Pharmacol. 18, 557.

22. Brooks, G. T., Harrison, A., and Lewis, S. E. 1970. Biochem. Pharmacol. 19, 255.

23. Brown, H. D. and Rogers, E. F. 1950. J. Amer. Chem. Soc. 72, 1864.

24. Büchel, K. H., Ginsberg, A. E., Fischer, R., and Korte, F. 1964. Tetrahedron Lett. 33, 2267.

25. Burt, P. E. 1970. Pestic. Sci. 1, 88.

26. Camp, H. B., Fukuto, T. R., and Metcalf, R. L. 1969. J. Agr. Food Chem. 17, 249.

27. Casida, J. E., Kimmel, E. C., Elliott, M., and Janes, N. F. 1971. Nature (London) 230, 326.

28. Casida, J. E., Holmstead, R. L., Khalifa, S., Knox, J. R., and Ohsawa, T. 1974. Personal communication.

29. Casper, H. C., Pekas, J. C., and Dinusson, W. E. 1973. Pestic. Biochem. Physiol. 2, 391.

30. Dauterman, W. C. 1971. Bull. W.H.O. 44, 133.

31. Donninger, C., Hutson, D. H., and Pickering, B. A. 1967. Biochem. J. 102, 26.

32. Dorough, H. W., Mehendale, H. M., and Lin, T. 1972. J. Econ. Entomol. 65, 960.

33. Dyte, C. E. 1969. Proc. 5th Br. Insectic. Fungic. Conf. 2, 393.

34. El Zorgani, G. A., Walker, C. H., and Hassall, K. A. 1970. Life Sci. (Oxford) 9, 415.

35. Fahmy, M. A. H., Fukuto, T. R., Myers, R. O., and March, R. B. 1970. J. Agr. Food Chem. 18, 793.

36. Felton, J. C. 1968. Insecticidal Activity of Some Oxime Carbamates. In Symposium on Pesticidal Carbamates, J. Sci. Food Agr. Suppl., pp. 32-38.

37. Fraser, J., Harrison, I. R., and Wakerley, S. B. 1968. Synthesis and Insecticidal Activity of N-Acyl-N-Methyl-carbamates. In Symposium on Pesticidal Carbamates, J. Sci. Food Agr. Suppl, pp. 8-12.

38. Fukami, J., Shishido, T., Fukunaga, K., and Casida, J. E. 1969. J. Agr. Food Chem. 17, 1217.

39. Fukuto, T. R. 1972. Drug Metabol. Rev. 1, 117.

40. Fukuto, T. R., Metcalf, R. L., Winton, M., and Roberts, P. 1962. J. Econ. Entomol. 55, 341.

41. Georghiou, G. P. and Metcalf, R. L. 1961. J. Econ. Entomol. 54, 231.

42. Hathway, D. E., Brown, S. S., Chasseaud, L. F., and Hutson, D. H. (Reporters). 1970. Foreign Compound Metabolism in Mammals, vol. 1, A Specialist Periodical Report, The Chemical Society, London.

43. Hathway, D. E., Brown, S. S., Chasseaud, L. F., Hutson, D. H., Moore, D. H., Sword, I. P., and Welling, P. G. (Reporters). 1972. Foreign Compound Metabolism in Mammals, vol. 2, A Specialist Periodical Report, The Chemical Society, London.

44. Hirwe, A. S., Metcalf, R. L., and Kapoor, I. P. 1972. J. Agr. Food Chem. 20, 818.

45. Holan, G. 1971a. Nature (London) 232, 644.

46. Holan, G. 1971b. Bull. W.H.O. 44, 355.

47. Hollingworth, R. M. 1969. J. Agr. Food Chem. 17, 987.

48. Hollingworth, R. M., Fukuto, T. R., and Metcalf, R. L. 1967. J. Agr. Food Chem. 15, 242.

49. Judy, K. J., Schooley, D. A., Dunham, L., Hall, M. S., Bergot, B. J., and Siddall, J. B. 1973. Proc. Nat. Acad. Sci. 70, 1509.

50. Kojima, K. and O'Brien, R. D. 1968. J. Agr. Food Chem. 16, 574.

51. Krieger, R. I., Feeny, P. P., and Wilkinson, C. F. 1971. Science 172, 579.

52. Kurihara, N. 1973. Personal communication.

53. Lewis, C. T. 1962. Nature 193, 904.

54. Lucier, G. W. and Menzer, R. E. 1970. J. Agr. Food Chem. 18, 698.

55. Maier-Bode, H. 1968. Residue Rev. 22, 1.

56. Matsumura, F. and Ward, C. T. 1966. Arch. Environ. Health 13, 257.

57. Menzer, R. E. and Casida, J. E. 1965. J. Agr. Food Chem. 13, 102.

58. Metcalf, R. A. and Metcalf, R. L. 1973. Pestic. Biochem. Physiol. 3, 149.

59. Metcalf, R. L. 1968. The Role of Oxidative Reactions in the Mode of Action of Insecticides. In Enzymatic Oxidations of Toxicants, E. Hodgson, Ed., Raleigh, North Carolina State University, pp. 151-174.

60. Metcalf, R. L. and Fukuto, T. R. 1965. J. Agr. Food Chem. 13, 220.

61. Metcalf, R. L., Kapoor, I. P., and Hirwe, A. W. 1971. Bull. W.H.O. 44, 363.

62. Miyamoto, J. 1969. Residue Rev. 25, 251.

63. Miyamoto, J., Nishida, T., and Ueda, K. 1971. Pestic. Biochem. Physiol. 1, 293.

64. Moriarty, F. and French, M. C. 1971. Pestic. Biochem. Physiol. 1, 286.

65. Moss, J. A. and Hathway, D. E. 1964. Biochem. J. 91, 384.

66. Mulder, R. M. and Gijswijt, M. J. 1973. Pestic. Sci. 4, 737.

67. Nakatsugawa, T., Tolman, N. M., and Dahm, P. A. 1968. Biochem. Pharmacol. 17, 1517.

68. Nakatsugawa, T., Tolman, N. M., and Dahm, P. A. 1969a. Biochem. Pharmacol. 18, 685.

69. Nakatsugawa, T., Tolman, N. M., and Dahm, P. A. 1969b. Biochem. Pharmacol. 18, 1103.

70. Nakatsugawa, T., Tolman, N. M., and Dahm, P. A. 1969c. J. Econ. Entomol. 62, 408.

71. Nolan, J. and O'Brien, R. D. 1970. J. Agr. Food Chem. 18, 802.

72. Nolan, J., Schnitzerling, H. J., and Schuntner, C. A. 1972. Pestic. Biochem. Physiol. 2, 85.

73. O'Brien, R. D. 1967. Insecticides, Action and Metabolism, Academic, New York.

74. Olson, W. P. 1973. Pestic. Biochem. Physiol. 3, 384.

75. Olson, W. P. and O'Brien, R. D. 1963. J. Insectic. Physiol. 9, 777.

76. Oppenoorth, F. J., Voerman, S., Welling, W., Houx, N. W. H., and Oudenweyer, J. W. V. D. 1971. Nat. New Biol. 233, 187.

77. Penniston, J. T., Beckett, L., and Bentley, D. L. 1969. Mol. Pharmacol. 5, 333.

78. Quraishi, M. S. and Poonawalla, Z. T. 1969. J. Econ. Entomol. 62, 988.

79. Rahn, H. W. 1963. Arch. Int. Pharmacodyn. Ther. 144, 126.

80. Ripper, W. E., Greenslade, R. M., and Hartley, G. S. 1951. J. Econ. Entomol. 44, 448.

81. Sacher, R. M., Metcalf, R. L., and Fukuto, T. R. 1969. J. Agr. Food Chem. 17, 551.

82. Sakai, M. 1970. Nereistoxin and its Derivatives. In Biochemical Toxicology of Insecticides, R. D. O'Brien, and I. Yamamoto, Eds., Academic, New York and London, pp. 13-19.

83. Sarmiento, R., McGovern, T. P., Beroza, M., Mills, G. D., Jr., and Redfern, R. E. 1973. Science 179, 1342.

84. Sawicki, R. M. and Farnham, A. W. 1968. Bull. Entomol. Res. 59, 409.

85. Shah, A. H. and Guthrie, F. E. 1971. Pestic. Biochem. Physiol. 1, 1.

86. Shah, P. V., Dauterman, W. C., and Guthrie, F. E. 1972. Pestic. Biochem. Physiol. 2, 324.

87. Slade, M. and Wilkinson, C. F. 1973. Science 181, 672.

88. Slade, M. and Zibitt, C. H. 1972. Metabolism of Cecropia Juvenile Hormone in Insects and Mammals. In. Insect Juvenile Hormones, J. J. Menn and M. Beroza, Eds., Academic, New York and London, pp. 155-176.

89. Solomon, K. R., Bowlus, S. B., Metcalf, R. L., and Katzenellenbogen, J. A. 1973. Life Sci. (Oxford) 13, 733.

90. Soloway, S. B. 1965. Adv. Pest Control Res. 6, 85.

91. Sternburg, J. and Kearns, C. W. 1952. J. Econ. Entomol. 45, 497.

92. Sun, Y. P. 1968. J. Econ. Entomol. 61, 649.

93. Sun, Y. P. 1971. J. Econ. Entomol. 64, 624.

94. Sun, Y. P. 1972. J. Econ. Entomol. 65, 632.

95. Sun, Y. P. and Johnson, E. R. 1969. J. Econ. Entomol. 62, 1130.

96. Sun, Y. P. and Johnson, E. R. 1971. J. Econ. Entomol. 64, 75.

97. Sun, Y. P. and Johnson, E. R. 1972. J. Econ. Entomol. 65, 349.

98. Sun, Y. P., Schaefer, C. H., and Johnson, E. R. 1967. J. Econ. Entomol. 60, 1033.

99. Trautmann, K. H., Schuler, A., Suchy, M., and Wipf, H. K. 1974. Z. Naturforsch. (in press).

100. Treherne, J. E. and Pichon, Y. 1972. The Insect Blood-Brain Barrier. In Advances in Insect Physiology, J. E. Treherne, M. J. Berridge, and V. B. Wigglesworth, Eds., Academic, London and New York, pp. 257-312.

101. Ulmann, E., Ed. 1973. Lindane, Monograph of an Insecticide, revised 2d Edn. Freiburg im Breisgau, Verlag K. Schillinger.

102. Verschoyle, R. D. and Barnes, J. M. 1972. Pestic. Biochem. Physiol. 2, 308.

103. Walker, C. H. and El Zorgani, G. A. 1973. Life Sci. (Oxford) 13, 585.

104. Walker, C. H., El Zorgani, G. A., Craven, A. C. C., Kenny, J. D. R., and Kurukgy, M. 1973. Studies of Comparative Metabolism Using Dieldrin Analogues. In Proceedings of the FAO/IAEA Symposium on Nuclear Techniques in Comparative Studies of Food and Environmental Contamination, Otaniemi, Finland.

105. Weiden, M. H. J. 1968. Insecticidal Carbamoyloximes. In Symposium on Pesticidal Carbamates, J. Sci. Food Agr. Suppl., pp. 19-31.

106. Whetstone, R. R., Phillips, D. D., Sun, Y. P., and Ward, L. F. 1966. J. Agr. Food Chem. 14, 352.

107. White, A. F. 1972. Life Sci. (Oxford) 11(II), 201.

108. Wilkinson, C. F. and Brattsten, L. B. 1972. Drug Metabolism Rev. 1, 153.

109. Winteringham, F. P. W. 1969. Ann. Rev. Entomol. 14, 409.

110. Yamamoto, I., Kimmel, E. C., and Casida, J. E. 1969. J. Agr. Food Chem. 17, 1227.

## DISCUSSION

There was expression of great concern about the tactics of the pesticide industry in causing the public to be exposed to very toxic, nonselective insecticides, when in most cases highly selective compounds could be made and sold. Methyl parathion versus fenitrothion is a typical example. Efforts were made twice by a participant to convince a major chemical company to produce fenitrothion in the United States but both times, when the matter reached the board of directors, it was turned down because they could sell all of the methyl parathion that could be produced. It is unnecessary to use these extremely toxic compounds in agriculture. Through visionary management, WHO has kept highly toxic compounds from operational use. It should be possible through FAO and other agencies to get a counterpart of this kind of philosophy involved in agricultural pest control. The strong urging for the development of this philosophy by the group of participants in this conference is regarded as one of the most valuable things that can be done.

Inquiry was made about the effectiveness in control of mosquitoes with DDT in solution in oil, compared to a wettable powder in suspension in water. In WHO's malaria-eradication program, most of the DDT is used as a 75% water dispersable powder--the DDT having a particle size between 10 and 20 μ. Therefore, the exposure of the mosquitoes is to DDT in the solid form. Because the insect cuticle has a waxy covering, it might be expected that the DDT-oil preparation would be more effective. However, since the DDT in the water suspension is a lipid water liquid crystal and the structure of the cuticle allows both water and lipid penetration, toxic action is relatively rapid. Apparently, no quantitative data are available to provide a precise answer to the original question.

The DDT-oil question prompted comment on the common belief that the solvents used with insecticides are biochemically inert. Actually, they are not inert. Now it is known that various oils and certain other solvents interact nicely with the microsomal oxidases. Some of these solvents are metabolized, and they probably go through an inhibitor phase so they could be acting as a built-in synergist to some extent; another effect is enzyme induction. Work with naphthas and various other kerosene-type compounds shows that they are highly effective enzyme-inducing agents with certain insects. Conceivably, then, an insecticide intended to kill an insect is dissolved in a material that allows the insect to fight the insecticide. Such interactions warrant investigation since this might explain some of the varying results in effectiveness obtained with different formulations. Industrial research on insecticides has not been oriented toward selectivity. A survey of the insecticide screens of most major companies showed that essentially all industrial screens are designed to pick up broad-spectrum, highly potent compounds. Reference was made to herbicide screening that was formerly conducted on the same basis as insecticides. A university researcher on herbicides obtained from most major companies all compounds that had shown above a very low level of activity and he worked with them on a large number of plant species. This work resulted in a large number of selective compounds that had been bypassed as the result of the limited industrial screens. On the basis of this work, a small company completely revamped its screening program, designing it to get maximum breadth, with the result that a large number of specialized compounds have been developed. The belief is that similar results are likely to be obtained if insecticide screens are greatly broadened. It is hoped that the favorable experience with herbicides will convince concerned companies to give attention to development of selective insecticides.

Screening for synergists also should be done on a broad-spectrum basis. Here again, in the herbicide area a good example was cited. Antagonism between carbaryl and particular materials being used for selective herbicides on rice became evident, so the chemical company's researchers decided that they were dealing with a carbamate-type herbicide that had limited persistence because the plants metabolized it too quickly. Therefore, they started on the reverse and systematically screened thousands of compounds and synergists and have found three already, one of which appears very likely to eliminate the metabolic pathway to break down.

There was concern expressed about the undesirability of continuing work on developing compounds that are ChE inhibitors. This was prompted by the fact that cholinesterase would be at the top of a list of similarities between vertebrates and insects. Perhaps differences between vertebrates and insects should be the theoretical basis for developing new insecticides. Surface-to-volume ratio is one difference that can be stated another way, that vertebrates have $10^{13}$ cells and insects have $10^{6}$. The circulation times are very similar—it takes 5 to 10 seconds for anything to get around the hemocoel and about the same time to get around the vertebrate system. However, the main difference is in the mass of cells that provide a buffer in vertebrates that makes them very susceptible to chronic effects and very resistant to acute effects. Insects are sensitive to acute poisons because of their small number of cells; therefore, it is not appropriate to compare the $LD_{50}$s for insects and vertebrates. Fundamentally, it will be necessary to work out the real differences using biochemical evolution as a base—insects and vertebrates developed about 600 million years ago. When these differences are determined, certain of them can be magnified in the development of insecticides that are not dangerous to vertebrates.

An interesting example in another field was mentioned, namely, the use of arsenical compounds to control trypanosomiasis. Arsenic in rudimentary form has been known to be toxic to humans for over a century, but during a relatively recent period, sophisticated arsenical compounds were developed that are virtually innocuous to man in his bloodstream but lethal to trypanosomes.

In further discussion of $LD_{50}$s in relation to number of cells in insects and vertebrates, it was pointed out that still another factor must be considered with human beings, namely, that the human body is a very highly sophisticated machine. These complicated mechanisms are sensitive to disruption of just one or more very small groups of cells over a very long period of time. There was strong sentiment by several participants against the use of $LD_{50}$s for insecticides on vertebrates on the basis of a lack of safety. Nevertheless, some guidelines on the effect of insecticides on vertebrates are necessary. The present system of extrapolation provides a crude comparison between insects and vertebrates; $LD_{50}$s or other types of toxicity measurements on human beings are not possible.

# Toxic Hazards to Man

J. M. BARNES
MRC Toxicology Unit, Medical Research Council Laboratories
Woodmansterne Road, Carshalton, Surrey, U. K.

## I. INTRODUCTION

In considering needs and prospects for insecticides in the future, the question of safety can only be based on lessons from the past. The introduction of insecticides with modes of action that differ from those with which we are now familiar, or the development of new methods of application, will each present a problem requiring special consideration if safe use is to be guaranteed. Under all foreseeable circumstances the people who will be most heavily exposed to an insecticide will be those who make or apply the concentrated formulations. Residues in food or water arising from normal operational application will result in only a very small exposure. If more attention had been paid in the recent past to following up those most heavily exposed to the insecticides introduced on a wide scale since 1947, we would now be in a better position to assess the likelihood of exposure leading to any serious long-term effects.

For the purpose of this paper, toxic hazards may be considered under three main heads:

(1) Immediate effects - related to the principal toxic action of the compound.

(2) Late effects - teratogenesis, carcinogenesis, mutagenesis.

(3) Other effects - identifiable or hypothetical, and not the direct result of the principal toxic action.

For the purpose of illustrating these points, those insecticides in current use that fall into the four main groups (pyrethroids, organophosphorus anticholinesterases, carbamate anticholinesterases, DDT, dieldrin, and other organochlorines) will be considered. Where compounds with a novel mode of action are proposed for use, the hazards they might present should in the first instance be considered in the same way.

## II. IMMEDIATE TOXIC EFFECTS

These are related to the main mode of action of the insecticide, and if a compound is going to be dangerous to use, this probability will be revealed in conventional laboratory toxicity tests. Tests on rats have proved to be a reliable index of the comparative hazards from members of a group having a similar mode of action. Where a compound is toxic to insects but not to rats, it is wise to ensure that other mammals are also insensitive before assuming that man will also be insensitive. A great deal is known about the manner and degree of exposure during the prevailing methods of insecticide application (Wolfe, Durham, and Armstrong, 1967). The main route of exposure is the skin, hence the importance of establishing in test animals the ratio of acute oral and acute dermal toxicity (Gaines, 1969). For some operations exposure can be kept low by protection or the use of special equipment. For conventional indoor residual spraying exposure is inevitably heavy (Durham and Wolfe, 1961). The most toxic and hazardous materials can be applied safely if appropriate measures are taken, but it is important to bear in mind the practical limitation of any proposed safe method of use, particularly where extensive personal protection is needed. Measurements of exposure can be made, provided that facilities are available for properly conducted introductory trials such as the World Health Organization (WHO) has carried out on insecticides proposed for use in public health (Wright, 1971). There is a need for similar facilities to be established, so that agricultural insecticides may be safely introduced into developing countries. Only in the case of organophosphorus compounds has a suitable method for measuring exposure by a specific biochemical reaction been devised that makes it possible to withdraw an individual from further exposure before serious poisoning develops. Despite their similar mode of action, this has not proved possible in the case of carbamate insecticides (Vandekar et al., 1968). However, knowledge of the biochemical characteristics of the action of carbamates on enzymes means that the first toxic symptoms, which are incapacitating, have provided an adequate early-warning system to prevent dangerous overexposure.

Experience over many years has shown that DDT is safe when used under a wide range of conditions; dieldrin, although safe for outdoor application, may cause severe but not lethal toxic effects in men heavily exposed by indoor spraying over prolonged periods. Endrin has produced lethal effects occasionally, even in outdoor application. This pattern can be related to the comparative toxicity of these insecticides to mammals, coupled with their behavior on repeated ingestion.

The pyrethrins have toxic effects on rats similar to those of DDT (Verschoyle and Barnes, 1971), but have not yet produced any toxic effects in man. Exposure to pyrethrins has probably never been either as severe or as prolonged as it has to the organochlorine insecticides. However, the safety of pyrethrins seems to be determined by a detoxification mechanism operating in mammals but not in insects.

For any new insecticide the hazards can only be assessed on the basis of some knowledge of the compound's mode of action and metabolism in mammals. Any developments that increase selectivity either in the mode of action on the target species or in restricting exposure to the target species will obviously reduce hazards to man.

In summary, acute immediate toxic hazards to man from new insecticides are probably not difficult to predict. The exact threat they may present can best be assessed by a properly phased introduction, as currently used for vector control agents. These should be emulated for important crop protective agents proposed for use in developing countries, in order to increase yields from commercially important crops.

## III. LATE EFFECTS

### A. Teratogenesis, Carcinogenesis, Mutagenesis

Active teratogens and carcinogens can readily be detected by tests on mammals. Appropriate procedures have been proposed by several regulatory bodies concerned with the safety of drugs and food additives. Among the most recent and detailed have been those of the Canadian Government (1973).

The main problems center on the interpretations of the findings in these tests.

**1. Teratogenesis.** If any particular compound is thought to present a teratogenic hazard, it may be circumvented by avoiding exposure of women of child-bearing age. However, so many compounds can in high doses lead to teratogenic effects in rodents that it is necessary to limit consideration to compounds that are teratogenic in doses that produce no obvious adverse effects on the mother. If a woman was poisoned by an insecticide and subsequently found to be pregnant, therapeutic abortion might be envisaged.

It should be borne in mind that teratogens are likely to exert their effects only during the very early weeks of a human pregnancy. Very little is known about the mode of action of teratogens, though the well-known alkylating agents used in cancer therapy are teratogenic to rodents.

2. Carcinogenesis. Cancer in people who have been occupationally exposed to certain toxic substances is a well-recognized hazard. Within the past 3 years, two widely used industrial chemicals, bis-chloromethyl ether and vinyl chloride, have been added to the list. Probably the best-known chemical carcinogen presenting a hazard to man is 2-nathphylamine. Men with very limited exposure to this compound have developed bladder cancer. A few such cases have even been attributed to the handling of the rat poison alpha-naphthyl thiourea, in which 2-naphthylamine was present at 2% as an impurity.

No compound that produced cancer readily in laboratory animals should ever be considered for possible use as an insecticide. The problem of the hazards presented by compounds like DDT and dieldrin which produce liver tumors in mice but not in rats, remains unresolved at the present time. The fact that phenobarbitone, to which man has been exposed in large numbers, produces similar tumors in mice (Thorpe and Walker, 1973) suggests that the response in mice may not be a reflection of a true carcinogenic response.

There is no generally accepted biochemical lesion that is known to initiate the development of cancer. It is widely assumed that a chemical carcinogen must initiate the lesion by reacting with the DNA of cells in the target organ. However, the inevitability of a cancer arising as a result of damage to DNA is complicated by the existence of DNA repair mechanisms.

Although certain chemical carcinogens will, at a high dose, produce tumors in 100% of animals well before the end of their life span, other compounds produce few tumors even in animals maintained for their life span on maximum-tolerated doses. There is a fairly steep dose-response curve for active carcinogens, but few experiments have been done on large numbers of animals exposed to low doses of carcinogens.

However, it seems probable that the ultimate development of cancer that kills the host depends on more than the chance collisions between molecules of a carcinogen and DNA. At the present time, there is no agreed formula for deriving a no-effect level of exposure to an accepted chemical carcinogen.

Species vary in their response to the same carcinogen; sometimes this has been shown to be related to the metabolic treatment of the carcinogen. However, it should be assumed that a compound that is carcinogenic for rats will also present a carcinogenic hazard for man.

The best recognized reaction between DNA and chemical carcinogens is alkylation of guanine, hence the anxiety about methylating and other alkylating agents. The present biochemical situation may be summed up by saying that reactions with DNA *in vitro* are irrelevant. If changed DNA can be isolated from tissues of treated animals, it might arouse suspicion, but the failure to detect methylated DNA in a tissue could not be used as evidence that no carcinogenic response was to be expected. The best that can be expected is that conventional long-term feeding tests will identify powerful carcinogens. Any claim that an insecticide might present a hazard by reacting with DNA must be considered on its merits and investigated by any practical means (see below).

3. Mutagenesis. Whereas cancer is essentially a disease of the whole animal, mutagenic effects are produced in single cells that may be self-contained; in other words, cells or organisms in culture or part of a more complex organism. Many compounds can produce mutagenic effects on bacteria including, for example, formaldehyde, which is an intermediate metabolite in mammals.

It is not difficult to demonstrate a carcinogenic effect in laboratory animals, but is is very difficult to demonstrate a mutagenic effect. The only mutagenic effects that are important are the nonlethal ones. There is some evidence that mutagens can act as carcinogens, presumably because the basis of their effects is an initial reaction with DNA.

Although the threat of a carcinogenic effect is a serious one for the exposed person, he or she is the only one who suffers from the toxic effect. A mutagen might act as a carcinogen, but if it only affected the germ cell, it would be the next generation that would be affected.

At present various government bodies are trying to formulate recommendations for tests to demonstrate mutagenic activity and provide evidence on which a realistic appraisal of hazards to man could be derived. A compound that can react with, and particularly that can alkylate, DNA in the intact organism will come under suspicion as being something likely to present a mutagenic hazard.

## IV. OTHER EFFECTS

### A. Organophosphorus Insecticides

1. Phosphorylation of Protein Other than Acetylcholinesterase (ChE). This is known to occur with many O-P insecticides. The only reaction of recognized importance is the phosphorylation

by some O-Ps of the protein in the mammalian nervous system, as a result of which extensive damage to neurones can occur (Johnson, 1970). This delayed neurotoxic effect is only produced by some compounds and can be recognized by appropriate animal studies. Except under special circumstances, compounds with this property should not be developed commercially. Recent work has suggested that nerve cell damage only follows when the phosphorylated protein undergoes the further reaction known as "aging" so that a simple test for phosphorylation of this particular esterase is not in itself enough evidence that the O-P compound will be neurotoxic.

Chymotrypsin and other proteins can be phosphorylated, but the extent and significance of this in the whole animal is unknown. A curious effect noted first with demeton and then with demeton methyl may be related to such an action (Barnes and Denz, 1954; Vandekar, 1958). Rats at an early stage of chronic poisoning show a poor food utilization in that growth rate is not commensurate with food intake.

The O-P compounds show teratogenic effects injected into fertile birds' eggs and also inhibit tissue growth. These effects may be due to phosphorylation of unidentified proteins.

Some O-Ps have pronounced anaesthetic effects. The basis for this is unknown (Vandekar, 1957).

2. Alkylation. Methyl phosphates, including some anticholinesterases, can act as methylating agents *in vitro* and will react with DNA (Bedford and Robinson, 1972). *In vivo* trimethyl phosphate, which has a low acute toxicity, will damage DNA in the intact mouse tests (Elling *et al*., 1968). There is no evidence that dichlorvos, in the largest doses that can be tolerated either in the diet or by inhalation, will produce carcinogenic effects. There was no evidence of damage to testicular function in mice fed dichlorvos (Dean and Thorpe, 1972). Thus despite evidence that dichlorvos could methylate DNA *in vitro* and exert some mutagenic effects on bacteria at concentrations several orders of magnitude greater than that tolerable to the animal, whole animal studies failed to reveal any evidence of significant effects from any DNA damage that may have been produced *in vivo*. Since any effective O-P insecticide will have an affinity for ChE several orders of magnitude greater than that for DNA, it seems probable that these thorough experiments with dichlorvos can be used as a basis for assuming that any methyl phosphate anticholinesterase will not present a hazard as an alkylating agent. This generalization will doubtless be challenged.

## B. Carbamate Anticholinesterases

There is no evidence that these compounds will carbamylate proteins other than cholinesterase in the whole animal. Some noninhibitory carbamates that can be tolerated by mammals in large doses will carbamylate the protein in the nervous system that is related to the neurotoxic lesion produced by certain O-P compounds (Johnson, 1970). The esterase activity of this protein can be reduced to very low levels and maintained there for days, but no nervous damage follows. Thus carbamylation of this particular protein does not produce the same extent of tissue injury as does phosphorylation and subsequent aging of the same protein.

On balance, carbamate anticholinesterases could appear to carry fewer risks of unexpected side effects because their capacity to carbamylate seems much more limited than that of O-P compounds to phosphorylate. Furthermore, the reaction of carbamylation of ChE and at least one other protein seems to entail a lesser risk of permanent damage to the protein.

However, one must always be prepared to meet surprise challenges. It has recently been shown that if carbaryl is reacted with nitrous acid, then N-nitroso-methyl-naphthyl carbamate is produced, and this is a potent mutagen for bacteria and probably a carcinogen (Elespurn and Lijinsky, 1973).

Since nitrite is present in many foods and carbaryl residues could presumably be present in other foods derived from crops treated with the insecticide, a reaction could occur in the consumer's stomach. However remote such a possibility might be, this kind of observation does suggest the desirability of looking into other possibilities of contact between nitrate or nitrite and N-methyl carbamates, such as might lead to the production of an N-nitroso compound on a larger scale. It is sometimes better to anticipate trouble by an appropriate investigation.

## C. DDT

Rats on diets containing levels of DDT that do not produce obvious signs of poisoning may show changes in their gait pattern that is dose-related (Khairy, 1959). The physiological basis of this has not been investigated, but presumably is a reflection of aberrations in the proprioceptive responses to the stimuli from walking. No changes of this nature have been detected in spraymen examined in the WHO long-term studies in Brazil and India.

A claim that the hearts of animals chronically exposed to DDT showed an abnormal reaction to pitressin (Lokaneva, 1968) has been reexamined in two species and not upheld (Jeyaratnam, 1974). The control observations in the study were inadequate, nor did the hearts of these animals, in which high concentrations of DDT were present, show any increased sensitivity to catecholamines. As DDT is a chlorinated hydrocarbon, this was considered to be a possibility. Many years ago it was shown that even in the highest attainable levels in rat tissue fat, DDT had no effect on the metabolism of the fatty tissue *in vitro* (Parker, 1960).

DDT does, however, have a marked effect on the mixed function oxidase enzyme system in rodent livers. The stimulation of these enzymes probably reflects the attempt to metabolize DDT and convert it to the acid that can be excreted. There is evidence that these enzyme systems are also stimulated in men heavily exposed to DDT (Poland *et al*, 1970). Although, by the indices selected, the formulators studied had higher levels of activity than matched controls, the values found did not fall outside the range encountered within healthy members of the general population.

The possibility that DDT or compounds with this effect could produce evidence of Vitamin D disturbance as a result of the stimulated microsome activity should not be lost to sight. Such effects have been described in epileptic patients with a long history of drug taking (Dent *et al*, 1970).

It is the disturbance in liver microsome activity in birds that is probably responsible for the adverse effects on reproduction. The metabolism of their endogenous hormones is thrown out of balance. It has been suggested that DDT acts as an inhibitor of carbonic anhydrase in the bird's oviduct to inhibit eggshell formation. The significance of such a reaction in terms of other possible side effects in birds and mammals is not known.

## D. Dieldrin and Aldrin

A comprehensive study of factory workers with several years exposure, including some that led to acute convulsions, revealed no evidence of other adverse effects including studies of liver function (Jaeger, 1970), although specific tests of liver microsome activity as in the DDT workers was not undertaken.

Evidence of some impairment in muscle function in chronically exposed rats probably reflects some disturbance in function of the CNS (Khairy, 1960) but this was not detected in the factory workers.

### E. Pyrethroids

The toxic effects of these compounds seem to be limited to their action on nerve membranes. Unfortunately, there is a limited amount of the findings in conventional long-term animal toxicity tests that have been published. What there is on compounds like allethrin provides no evidence of other toxic activity. As new compounds of this family are produced, particularly those with halogen groups and those of greater stability, complications from long-term exposure could arise. However, if the products of degradation either by hydrolysis or the action of the mixed-function oxidases are water soluble and unreactive, it is likely that they will be rapidly excreted by mammals.

## V. CONCLUSIONS

The immediate toxic effects of new active insecticides should present no insuperable barrier to their safe use, at least under some circumstances. As long as there is always a temptation to allow substances of low immediate toxicity, for example DDT, to be used without any precautions, it is probably wiser under all circumstances to insist that all compounds are used in such a way as to reduce human exposure to the minimum by insisting on good practices. It is only after 40 years of widescale use that a compound like vinyl chloride has been found to carry a risk of an unpleasant form of cancer in some people exposed to it.

The more thoroughly the biological activity of molecules is studied, the better will become our predictive powers, particularly for the problem of delayed toxic effects. At the present time, the behavior of so few compounds have been comprehensively studied, and the biochemical basis of these long-term effects are so ill-understood that one must continue to rely on empirical animal tests as the only feasible early warning system for likely toxic hazards. However, there should be at least some working hypothesis about the mode of action of any new compound proposed for use as an insecticide, together with some information on its breakdown products and their reactivity, so that some predictions on the basis of existing knowledge can be proposed. In the absence of any such information for a compound that is toxic though capable of being applied, a widescale introduction should be postponed until more information becomes available.

The stimulus to some further research may come from expressions of doubt about safety from both well- and ill-informed quarters. The first line of defense against all such criticism is the ability to be able to claim safe practices in application. Research to prove a negative, as in the case of dichlorvos, can be long and expensive, but for a valuable material may prove a sound investment. The final evidence of safety, that will always be extremely difficult to obtain in the case of pesticides, is that those who by their occupation are the most heavily exposed to a compound, have a health record within the normal range.

## REFERENCES

1. Barnes, J. M. and Denz, F. A. 1954. Br. J. Ind. Med. 11, 11-19.

2. Bedford, C. T. and Robinson, J. 1972. Xenobiotica 2, 307-337.

3. Canadian Government. 1973. The Testing of Chemicals for Carcinogenicity, Mutagenicity, and Teratogenicity. Health Branch, Ministry of National Health and Welfare, Ottawa, Canada.

4. Dean, B. J. and Thorpe, E. 1972. Arch. Toxicol. 30, 51-59.

5. Dent, C. E., Richens, A., Rowe, D. J. F., and Stamp, T. C. B. 1970. Br. Med. J. 4, 69-72.

6. Durham, W. F. and Wolfe, H. R. 1961. Bull. W.H.O. 26, 75-91.

7. Elespurn, R. K. and Lijinsky, W. 1973. Food Cosmet. Toxicol. 11, 807-817.

8. Elling, U. H., Cumming, R. B. and Malling, H. V. 1968. Mutat. Res. 5, 417.

9. Gaines, T. R. 1969. Toxic. Appl. Pharmac. 14, 515-534.

10. Jager, K. W. 1970. Aldrin, Dieldrin, Endrin, and Telodrin. An Epidemiological and Toxicological Study of Long-Term Occupational Exposure. Elsevier, Amsterdam.

11. Jeyaratnam, J. 1974. To be published.

12. Johnson, M. K. 1970. Biochem. J. 120, 523-531.

13. Khairy, M. 1959. Q. J. Exp. Psychol. 11, 84-91.

14. Khairy, M. 1960. Br. J. Ind. Med. 17, 146-148.

15. Lokaneva, A. M. 1968. Abstract of the Fourth Scientific Conference of Soviet Union on Hygiene and Toxicology of Pesticides, Kiev 11-14 June, 1968, Kiev, pp. 731-738.

16. Parker, V. H. 1960. Biochem. J. 77, 74.

17. Poland, A., Smith, D., Kuntzmann, R., Jacobson, M., and Conney, A. H. 1970. Clin. Pharmac. Ther. 11, 724-732.

18. Thorpe, E. and Walker, A. I. T. 1973. Food Cosmet. Toxicol. 11, 433-442.

19. Vandekar, M. 1957. Nature 179, 154.

20. Vandekar, M. 1958. Br. J. Ind. Med. 15, 158-167.

21. Vandekar, M., Hedayat, S., Plestina, R. and Ahmady, G. 1968. Bull. W.H.O. 38, 609-623.

22. Wolfe, H. R., Durham, W. F., and Armstrong, J. F. 1967. Arch. Envir. Health 14, 622-633.

23. Wright, J. W. 1971. Bull. W.H.O. 44, 11-22.

## DISCUSSION

Dr. Barnes has just provided us with an excellent factual summary of some of the acute and chronic hazards (either known or hypothetical) associated with the major groups of insecticides in use today. I can add little to the scientific facts he has given us, but I would like to take this opportunity to emphasize a few general, perhaps more philosophical, points which Dr. Barnes touched on. These relate mainly to some of the problems encountered in assessing the toxic hazards to man, not only of insecticides, but of any type of chemical we utilize.

It is an unfortunate fact that the average layman is under the erroneous impression that insecticides pose a very real and serious threat to his or her safety. This attitude has arisen largely through the extensive publicity given to irresponsible statements or written articles by numerous misinformed individuals during the last decade, but has unfortunately been amply confirmed, at least in the layman's eyes, by tragedies such as that associated with the drug thalidomide. There is currently a general antiinsecticide sentiment in the public sector. Since public opinion can be translated into votes, votes to politicians, and politicians to restrictive pesticide legislation, it is of considerable importance that we make every effort to educate the layman as to the real extent of the hazard associated with pesticide usage. If we fail to do this we run the risk of having future pesticide policy ultimately determined by misinformed public opinion. Because this conference is concerned with insecticides for the future, the problem of how to win public opinion seems most appropriate for discussion. The major problem is that the average layman is only too willing to accept the benefits of pesticide usage but is unwilling to accept that the use of any chemical is always associated with a certain amount of risk. How can we really assess this risk and give the public the straight answers it demands?

There is usually little or no problem associated with the assessment of an acute toxic hazard. This is a relatively simple task for the toxicologist, and although there is always the problem of extrapolating animal data to man, it can usually be accomplished with a fair degree of accuracy. The risk of acute toxic exposure is restricted to a relatively small number of individuals working directly with the concentrated insecticides, or, of course, to those involved in accidents or suicides. Deaths and serious illnesses resulting from acute accidental exposures, though not large, could undoubtedly be markedly reduced by replacing some of the more toxic insecticides now in use with other currently existing materials. This raises the question: Should we continue to register insecticidal compounds with high acute mammalian toxicity if other safer materials are available?

It is an interesting reflection of human psychology, however, that most people are not overly concerned about compounds known to be highly toxic. They are much more concerned about materials that just might cause some effect over a long period of low-level exposure, particularly carcinogenic, mutagenic, or teratogenic effects. It is in this area of evaluating possible long-term effects of very small doses of materials where most difficulty exists. The assessment of this type of hazard has become a central problem for the modern toxicologist, and yet it has been called nontoxicology. The problem arises quite simply, of course, from the fact that it is never possible to prove conclusively that some effect will not occur. The toxicologist is, therefore, in an uncomfortable and unique position in the life sciences, because while his colleagues are busily examining real phenomena and analyzing changes in them, he is faced with attempting to determine the absence of changes and to demonstrate something negative.

The practical problems encountered in this area are quite different from those involved in assessing acute effects, since it is not a matter of trying to demonstrate an effect on an individual but usually of attempting to assess the frequency of occurrence of an event in a population, for example, tumor formation or chromosome aberrations. In order to truly demonstrate significant effects at high levels of confidence, experiments would have to be conducted with impossibly large numbers on animals. Clearly, compromises have to be made, and consequently the level of confidence actually attained in these tests is usually quite low. As a result, the final interpretation of the data becomes more subjective than scientific, the experts sometimes disagree, and the layman is understandably confused and disturbed. Dr. Weinberg has termed the evaluation of this type of hazard "trans-science" and Dr. Coulston has described it as "part science and part intuitive art." Clearly this is a problem, and it is unfortunate that the success or failure of many insecticides will ultimately depend on this kind of pseudoscientific situation. How can we adequately interpret such data and what are some of the guidelines that should be used in obtaining the data? There is still a great deal of disagreement with respect to the appropriate tests to use to satisfactorily demonstrate carcinogenic, teratogenic, and mutagenic effects; the period of time or number of generations over which the tests should be conducted; and the species or strains of animals that should be employed.

How can we reduce to a minimum the risk of chronic toxicity to humans and how can we educate the public to accept the fact that the use of any chemical is fundamentally inseparable from some degree of risk?

# PROSPECTS FOR IMPROVEMENTS OF PRESENT DAY INSECTICIDES

# Future Use of Natural and Synthetic Pyrethroids

MICHAEL ELLIOTT
Department of Insecticides and Fungicides
Rothamsted Experimental Station
Harpenden, Hertfordshire, England

## I. INTRODUCTION

Pyrethrum is the only naturally occurring insecticide used extensively today (for reviews see Gnadinger, 1936, 1945; Elliott, 1951; Metcalf, 1955; West, 1959, Crombie and Elliott, 1961; Matsui and Yamamoto, 1971; Casida, 1973; Elliott and Janes, 1973). In many ways, the mixed active constituents of pyrethrum, the natural pyrethrins, are ideal insecticides (Potter, 1973) because they act rapidly against a wide range of insect species, have no action on mammals under normal conditions (Casida, 1973a), and, being quickly rearranged or decomposed in air and light to inactive products (for references and data see Chen and Casida, 1969; Bullivant and Pattenden, 1971, 1972, 1973a, b), do not leave harmful residues. Furthermore, in practice, few insects have developed serious levels of resistance to the pyrethrins (but see Lloyd and Parkin, 1963; Collins, 1973; Farnham and Sawicki, 1974; Keiding, 1974).

## II. THE NATURAL PYRETHRINS

The pyrethrins are useful where many persistent contact insecticides are inappropriate, for example in houses, restaurants, and aircraft, on food and for preharvest treatment of expensive food crops, and in public health (Stevenson, 1963). As films in heavy white oil they persist adequately in dark warehouses to control insect infestations of grain (Potter, 1935, 1938). Their rapid action against insects ("knockdown") is particularly valuable in some applications (Page et al, 1949; Sawicki, 1962a, b), and with relatively inexpensive synergists (Casida, 1970; Yamamoto, 1973) such as piperonyl butoxide (Brown, 1970, 1971), sulphoxide (Synerholm et al, 1947), or tropital (Hopkinson and Maciver, 1965; Maciver, 1966), which lower the amount of active constituent required. The natural product has a firmly established and annually increasing range of applications.

However, regular supplies of the natural pyrethrins depend on favorable climatic conditions for growing Chrysanthemum cinerariaefolium and on manual labor for harvesting, and relatively small quantities are produced (Glynne Jones, 1973) compared with organophosphorus compounds, organochlorine compounds, and carbamates. About 60% of the present need for pyrethroids is satisfied by the natural compounds, usually formulated with a synergist. High cost, instability, and the presence of numerous noninsecticidal but potentially physiologically active constituents in pyrethrum extract (Heat, 1973), as well as possible complications from formulations with synergists, must limit the prospects for expanded use of the natural product. Pyrethrum has outstanding activity against many agricultural and forest pests, including lepidopterous larvae (Miskus and Lyon, 1973), and advances have been made in applying ultra-low-volume sprays (Brett, 1974; Maas and Zindel, 1974) and in stabilizing formulations (Miskus and Andrews, 1972). Nevertheless, the problems of the inherent instability of the natural pyrethrins and of synthetic pyrethroids available commercially so far have not been adequately overcome. Recent ingenious methods of stabilization, such as cyclodextrin inclusion compounds (Yamamoto and Katsuda, 1974) would inevitably add to the overall cost of the insecticidal formulation. It seems improbable, therefore, that the total world consumption of natural pyrethrins will rise greatly above twice the present level in the foreseeable future. Present production is about 23,000 metric tons of dried flowers, equivalent to about 350 tons of active ingredients (Glynne Jones, 1973).

## III. SYNTHETIC PYRETHROIDS

Allethrin was the first synthetic pyrethroid produced commercially (Schechter, Green, and LaForge, 1949; Sanders and Taff, 1954; Roark, 1952, 1955) and the quantity used annually still exceeds that of any other synthetic pyrethroid at least tenfold. It is an effective substitute for natural pyrethrins in some applications, especially aerosols, and is extensively used in mosquito coils, where heat stability and volatility are an advantage. However, it is sometimes less well synergized than the natural compounds, so formulations are not necessarily less expensive. Tetramethrin (neopyramin) (Fig. 12) was introduced in 1964 (Kato et al., 1964) but like allethrin (Elliott et al, 1950) which also has excellent rapid action against flying insects, it is inferior as a killing agent to the natural product against many insect species. (Fales *et al.*, 1972; Winney, 1973).

The potential for developing insecticides based on the structure of the pyrethrins was demonstrated with resmethrin and bioresmethrin not only more active against many insect species than the natural compounds, but also of even lower oral and intravenous toxicity to mammals (Elliott, 1971; Verschoyle and Barnes, 1973; Abernathy, *et al.*, 1973). In this respect, bioresmethrin, more than 30,000 times more potent to the housefly than to the rat, remains an outstandingly potent and safe insecticide (Elliott, 1971). Tetramethrin, resmethrin (NRDC 104), and bioresmethrin (NRDC 107), with compounds developed more recently such as cismethrin (NRDC 119, Lhoste *et al.*, 1971), K-Othrin (RU 11,679, Velluz *et al.*, 1969), bioallethrin (Lhoste *et al.*, 1967, 1968), S-bioallethrin (Rauch *et al.*, 1972), proparthrin (kikuthrin, Nakanishi *et al.*, 1970), prothrin (Katsuda *et al.*, 1967), and kadethrin, RU 15,525 (a recently developed knockdown agent) (Martel and Buendia, 1974; Lhoste and Rauch, 1974) could, given approval by appropriate authorities, satisfy the present market for pyrethroids. The total cost, with large-scale economic synthesis and less reliance on synergists, would probably be lower than at present.

## IV. FUTURE DEVELOPMENTS OF PYRETHROIDS

New insecticides are needed to replace compounds currently considered unsuitable and to combat resistant strains of insects. High cost and insufficient stability in air and light are important factors limiting the scope of pyrethroids at present. It is appropriate now to assess the prospects for developing new pyrethroids without these disadvantages and perhaps with additional favorable characteristics.

### A. Relative Potencies of Pyrethroids and Other Insecticides

Great potency against insects offsets the high cost of an expensive insecticide and the smaller amounts needed diminish the risk of environmental contamination. To evaluate the pyrethroids in these respects, Fig. 1 shows the insecticidal activities as median lethal doses in milligrams per kilogram of insect bodyweight for representative examples of four groups of insecticides against five species of insect. The values, which are for unsynergized compounds, were all obtained by topical application and are therefore comparable both within and between

FIGURE 1. $LD_{50}$ Values of Insecticides. The data for *Anopheles stephensi*, *Glossina austeni*, and *Stomoxys calcitrans* are by courtesy of F. Barlow and A. B. Hadaway.

Key

| Organophosphates | Pyrethroids | Carbamates | Organochlorines |
|---|---|---|---|
| 1. Dichlorvos | 1. NRDC 161 | 1. Propoxur | 1. DDT |
| 2. Dimethoate | 2. NRDC 143 | 2. Carbaryl | 2. Methoxychlor |
| 3. Malathion | 3. Phenothrin | 3. Promecarb | 3. Lindane |
| 4. Fenthion | 4. RU 11,679 | 4. Mobam | 4. Dieldrin |
| 5. Fenitrothion | 5. Cismethrin | 5. Landrin | 5. Aldrin |
| 6. Tetrachlorvinphos | 6. Bioresmethrin | 6. Carbofuran | 6. α-Endosulfan |
| 7. Phoxim | 7. Pyrethrin I | | |
| 8. Bromophos | | | |
| 9. Iodofenphos | | | |
| 10. Chlorpyrifos | | | |
| 11. Phenthoate | | | |
| 12. Parathion | | | |
| 13. Diazinon | | | |

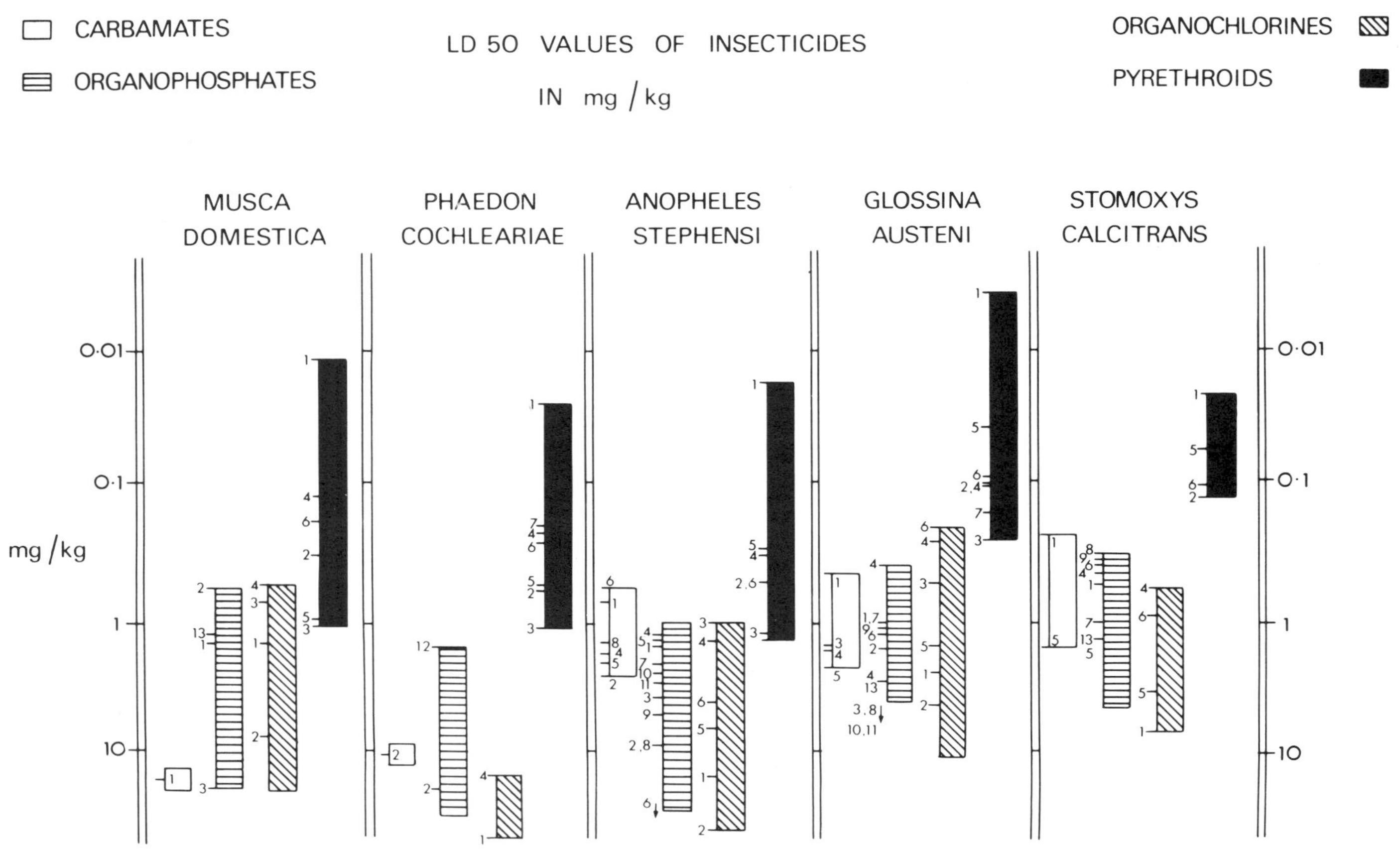
CARBAMATES
ORGANOPHOSPHATES
LD 50 VALUES OF INSECTICIDES
IN mg / kg
ORGANOCHLORINES
PYRETHROIDS
MUSCA DOMESTICA
PHAEDON COCHLEARIAE
ANOPHELES STEPHENSI
GLOSSINA AUSTENI
STOMOXYS CALCITRANS
0·01
0·1
1
10
mg / kg

species. Synergists would increase the activity of pyrethroids (Farnham, 1973; Elliott, 1971) and of some of the other insecticides by about one further order of magnitude under appropriate conditions. The figure shows the pyrethroids to be more active than other groups of insecticides; tests against other insect species give similar results. Therefore, efficient control can often be achieved with quantities of pyrethroids smaller than those needed with other classes of insecticides.

## B. Structural Features of Insecticidal Esters

To assess the prospects for developing less expensive, more stable pyrethroids that still retain the favorable properties of low mammalian toxicity and great potency to insects, it is appropriate to start by considering the structure of pyrethrin I (Fig. 2), the most important of the six, recognized insecticidal constituents of pyrethrum extract. Pyrethrin I, one of the most

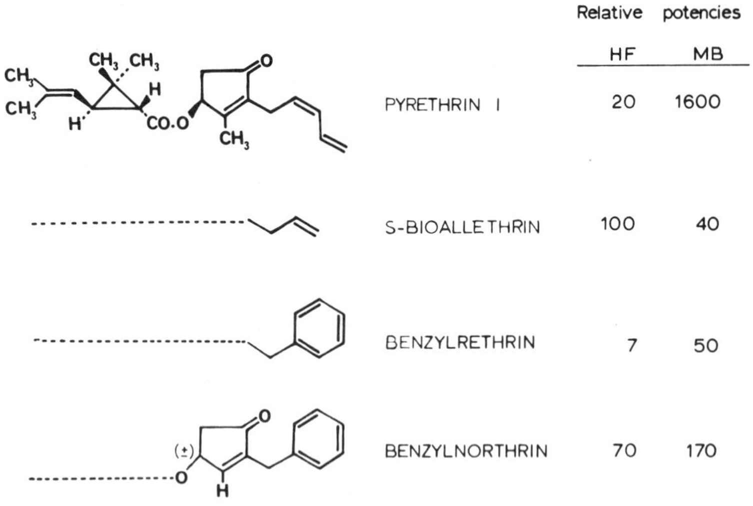


FIGURE 2. Structural Requirements for Alcohol (1).

powerful insecticides known against many insect species, may be considered the prototype of the group, with all the structural features essential for high activity. These will be analyzed by considering the insecticidal activities of related compounds, compared with the synthetic pyrethroid bioresmethrin (Elliott *et al.*, 1967) as standard and assigned the potency of 1000 for two species: the housefly, *Musca domestica*, L., and the mustard beetle, *Phaedon cochleariae*, Fab. The merits of these two species have been discussed (Elliott, 1969).

Many synthetic compounds related to the pyrethrins have now been reported in the scientific literature and especially in patents since the first examples were described by Staudinger and Ruzicka (1924). To limit this chapter to a reasonable length, only modifications producing a useful change in the properties of the compounds will be considered, and regrettably, much elegant and distinguished work must therefore be omitted.

Figure 2 shows that pyrethrin I is very effective against mustard beetles but relatively weak against houseflies, which are apparently able to diminish the activity by detoxification processes (Casida, 1970). In the presence of a synergist such as sesamex under special conditions (Farnham, 1973) pyrethrin I becomes nearly 400 times more potent to houseflies, whereas synergists have little effect on mustard beetles.

The activity, with or without synergist, depends on the intact ester and derivatives of either acidic or alcoholic component not related in structure to the parent compound are inactive (Elliott, 1954, 1969, 1971). All isosteric replacements of the ester link reported so far produce a loss of activity; the most active nonester is only one-twentieth as active as the related ester (Berteau and Casida, 1969). Methyl groups in a definite steric relation to the carboxyl group of the acid component are present in the most active esters so far described, and greatest potency is attained when they are supported by a cyclopropane ring, as in chrysanthemic acid (Pyrethrin I). Esters from tetramethylcyclopropane carboxylic acid are more effective insecticides with some alcoholic components such as 5-benzyl-3-furylmethylalcohol (below) (Matsui and Kitahara, 1967; Berteau and Casida, 1969; Barlow *et al.*, 1971) but are relatively inactive with others (3-phenoxybenzylalcohol). Greatest activity is attained in esters from dimethylcyclopropanecarboxylic acids with unsaturated groups at C-3 on the cyclopropane ring cis or trans to the carboxyl function, for example, *trans*-isobutenyl as in pyrethrin I. Further examples are discussed below (Section IV.D). Recently, esters of substituted (S)-α-isopropylphenylacetic acids which have structural features in common with (+)-*trans* (natural)

[1R,3R]-chrysanthemate esters were shown to have considerable insecticidal activity (Ohno et al., 1974) but the examples described so far are less potent than esters derived from the most effective cyclopropane constituents (Elliott et al., 1973a, b, 1974).

## C. Influence of Alcohol Structure

Unsaturation in the side chain of the alcohol (e.g., the cis-butadienyl group in pyrethrin I) seems essential for the highest insecticidal activity against a wide range of insect species (Elliott, 1969), but the data in Fig. 2 for S-bioallethrin against houseflies show that the conjugated system of the natural ester is not essential. Allethrin, more stable to heat than pyrethrin I, is thus an effective constituent of mosquito coils but is less active against mustard beetles and other insect species (Elliott et al., 1950). The related compound benzylrethrin (Chen and Barthel, 1953) demonstrates that compounds with aromatic rather than aliphatic unsaturation in the side chain can also be effective insecticides. Benzylrethrin was relatively difficult to synthesize and insufficiently active as an insecticide to justify commercial production. Benzyl northrin (Elliott et al., 1971c) showed that the ring system could be modified and simplified with improved insecticidal activity. This stimulated syntheses of many variations (Fig. 3) in which the bridging structure to the acid, via the alcoholic group, was varied while the terminal aromatic ring, a center of unsaturation, was retained. One alcohol discovered as a result of these investigations was 5-benzyl-3-furylmethyl alcohol (Elliott et al., 1967, 1971a,b), still the most readily available and effective constituent of pyrethroid esters for kill (Elliott et al., 1973a) and knockdown (e.g., Kadethrin, RU 15,525; Lhoste and Rauch, 1974). Comparing this alcohol with (+)-pyrethrolone shows the relative spatial dispositions of the unsaturated side chains (phenyl and cis-butadienyl) and the alcoholic groups (primary and secondary) to be very similar in the two compounds. The great insecticidal potency of bioresmethrin showed for the first time that activity against insects much greater than that of the natural pyrethrins could be attained, and that it was possible to replace an asymmetric alcohol, (+)-pyrethrolone, with a symmetrical, more easily synthesized alcohol (5-benzyl-3-furylmethyl) of related structure. When justified by biological activity, however, manufacture of pyrethroid esters derived from single optical isomers of acid and alcohol is feasible, as demonstrated by the

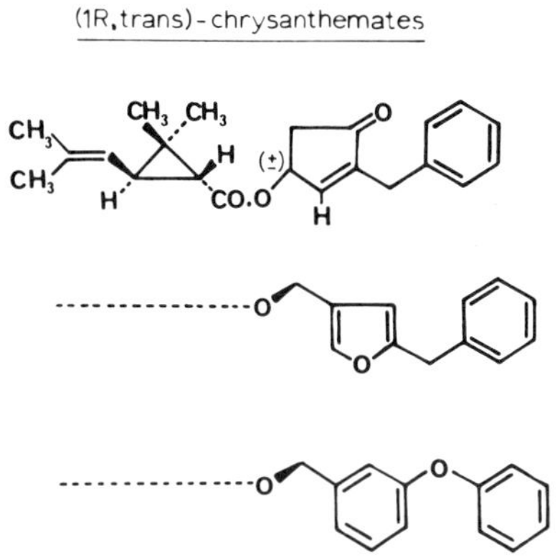


| | Relative potencies HF | MB |
|---|---|---|
| BENZYLNORTHRIN | 70 | 170 |
| BIORESMETHRIN | 1000 | 1000 |
| PHENOTHRIN | (310) | 700 |

FIGURE 3. Structural Requirements for Alcohol (2).

elegant commercial synthesis of S-biollethrin by Martel and coworkers (Rauch et al., 1972).

A more recent advance has been the discovery of active esters from 3-phenoxybenzyl alcohol (Fujimoto et al., 1971, 1973; Elliott, 1971) which, although usually less potent than those from 5-benzyl-3-furylmethyl alcohol, are more easily synthesized and, as discussed below, more stable in air and light. In three effective alcohols, therefore, [(+)-pyrethrolone, 5-benzyl-3-furylmethyl alcohol, and 3-phenoxybenzyl alcohol] an unsaturated center in the side chain is maintained in an appropriate stereochemical situation by one of three ring systems, acting as bridging structures. The potency of esters of these alcohols provides a standard by which other compounds may be assessed.

All reported attempts to increase insecticidal activity by introducing a substituent (Fig. 4) at various positions on either synthetic alcohol have failed, with one exception. Introducing an α-cyano group onto 3-phenoxybenzyl alcohol gives a cyanhydrin that produces esters of exceptional potency with appropriate acids, as discussed below (Section IV.D). No other compounds with cyclic bridging structures are of sufficient practical importance to be considered here; furthermore, when the bridge is replaced by an acyclic structure, less effective compounds are generally obtained. Three such variations are shown in Fig. 5. The compounds, especially butethrin (Sota,

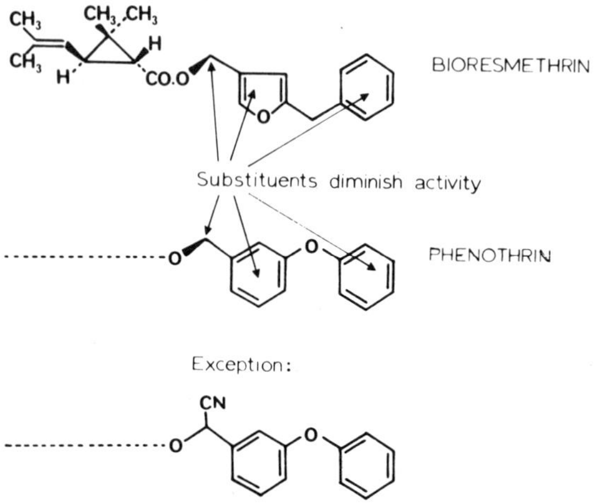


FIGURE 4. Effect of Substitution.

*et al*., 1971) (the chloro ester) are relatively effective against houseflies but where tests have been done have proved weak against mustard beetles.

No guiding principle seems yet to have been discerned to predict the most effective combination from a range of promising pyrethroid acids and alcohols; direct testing is still essential.

## D. The Acid Side Chain

With all other structural features unchanged, the activity of bioresmethrin was increased by rearranging the isobutenyl side chain of chrysanthemic acid to *cis*-butenyl (Fig. 6) (Elliott *et al*., 1973). Further, with appropriate alcohols, the *cis*-butadienyl substituted acid gave exceptionally potent esters, which were, however, relatively difficult to synthesize (Elliott *et al*., 1975) and potentially unstable. The ethanochrysanthemate (RU 11679) of which a practicable synthesis has been described, also shows exceptional activity.

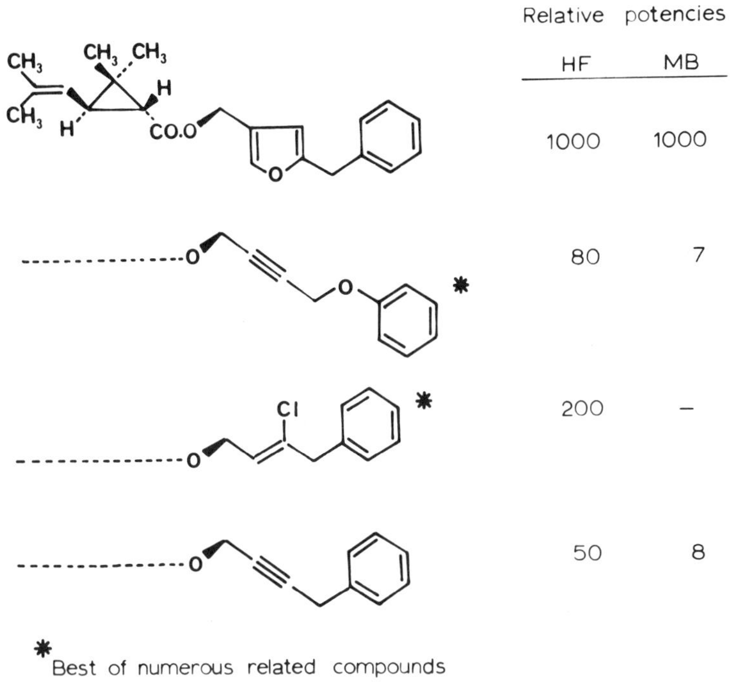


FIGURE 5. Esters of Acyclic Alcohols.

In another approach, the monochloro-isostere of an ester with a propenyl side chain was prepared and found to have greater activity than the parent ester. This led to an extended examination of halopyrethroids isosteric with chrysanthemates (Elliott *et al*., 1973a, b, 1974a) - the ester of (±)-allethrolone with the (±)-*trans*-dihalovinyl acid had been synthesized by Farkas *et al*. in 1959. A (±)-*trans*-dibromovinyl ester was reported independently (Brown *et al*., 1973). Esters with the *trans*-dichlorovinyl side chain were generally more effective than those with difluoro (not shown) or dibromo substituents and in this series the *cis*-dihalovinyl cyclo-propane esters were usually more active than the *trans* (Fig. 7). Cismethrin (NRDC 119) with a *cis*-isobutenyl side chain is similarly more effective than bioresmethrin against many agricultural and public health pests and in other applications (Barlow and Hadaway, 1974).

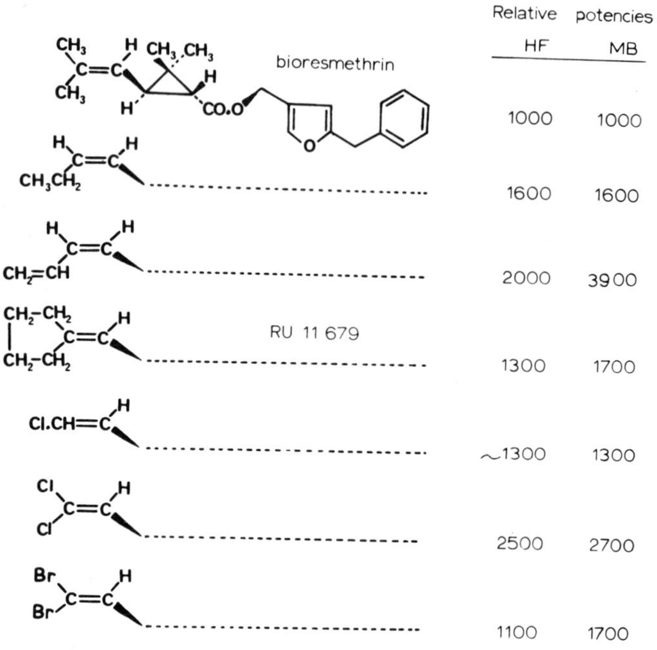


FIGURE 6. Effective Sidechains at C-3 of Acid.

The (+)-cis-dibromovinyl acid (corresponding in stereochemistry to (+)-cis-chrysanthemic acid) gives highly crystalline esters; this is a considerable advantage because it permits separation and purification of one optical isomer (temporarily coded as NRDC 161) of the ester with 3-phenoxybenzaldehyde cyanhydrin (Fig. 8) (Elliott *et al.*, 1974). The insecticidal potency of this particular isomer demonstrates the outstanding activity possible within the pyrethroid class. To at least five species of insect (Fig. 1) it is considerably more potent than bioresmethrin, itself much more active than DDT (Fig. 8). Furthermore, by pretreatment of houseflies with sesamex, the $LD_{50}$ for NRDC 161 is diminished by a factor of 15 to the remarkably small value of about 0.002 mg/kg (Elliott *et al.*, 1974). This compound appears, therefore, to be the most active pyrethroid yet discovered.

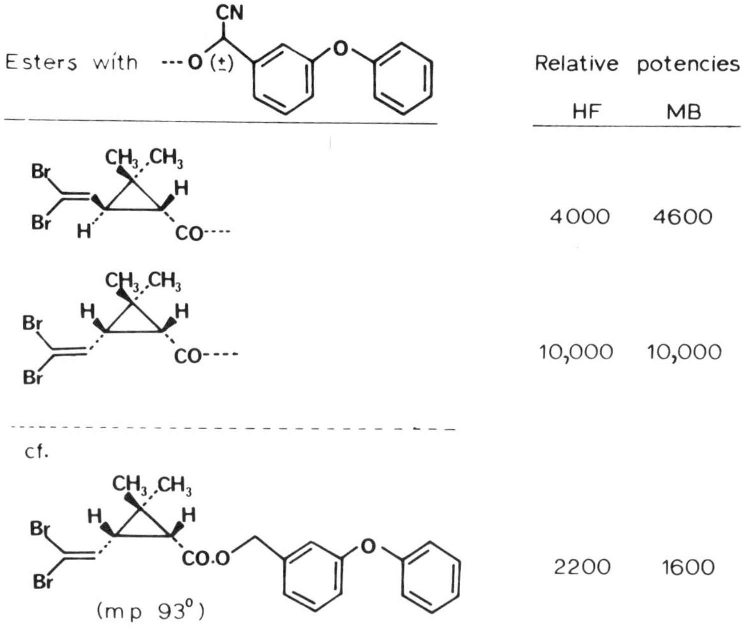


FIGURE 7. Effective Combinations (1).

## V. PROPERTIES OF PYRETHROIDS

### A. Photostability

Esters of dihalovinylcyclopropane carboxylic acids thus demonstrate the great insecticidal potency of this group of compounds. Another advantage of these acids is that with appropriate alcohols, the esters are considerably more stable in air and light than previous pyrethroids.

Pyrethroids are traditionally photolabile compounds. For many applications this is an advantage; without persistent residues, for example, there is less danger to predators and other beneficial insects and no risk of long-term environmental contamination. The natural pyrethrins and synthetic compounds such as bioresmethrin have at least two centers of instability to light. These are the side chain in the acid (Chen and Casida, 1969) and either the pentadienyl group in pyrethrolone

Active isomer, NRDC 161:

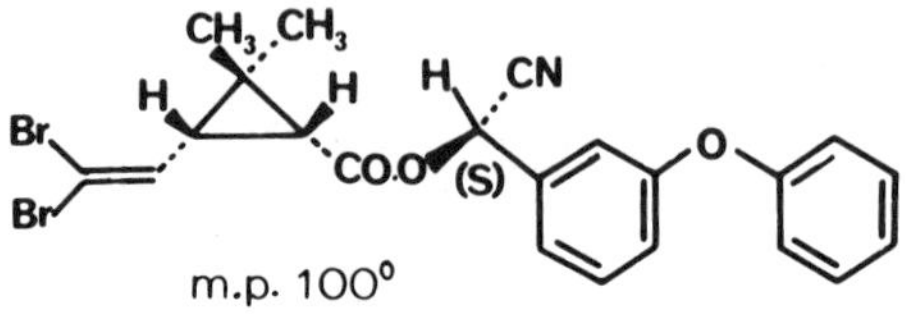


m.p. 100°

Insecticidal activity:

| | LD50 (ng/insect) | | |
|---|---|---|---|
| | NRDC 161 | bioresmethrin | DDT |
| Musca domestica (HF) | 0.3 | 5 | 40 |
| Phaedon cochleariae(MB) | 0.3 | 4 | 500 |
| +Anopheles stephensi | 0.04 | 1 | 50 |
| +Stomoxys calcitrans | 0.9 | 2 | – |
| +Glossina austeni | 0.08 | 3 | 90 |

+Data from F. Barlow and A.B. Hadaway

FIGURE 8. Effective Combinations (2).

or the furan ring in bioresmethrin (Elliott et al., 1973c; Ueda et al., 1974a). 3-Phenoxybenzyl chrysanthemate (phenothrin; Fujimoto et al., 1973) lacks the alcoholic center of instability but is a less effective insecticide than resmethrin. Much of this loss of insecticidal activity in the chrysanthemate is restored in the dichlorovinyl ester, which is relatively more effective, especially against mustard beetles, than would have been predicted from comparable results with alternative acidic and alcoholic components. In this 3-phenoxybenzyl ester of the dichlorovinyl acid (the (±)-cis, trans ester is coded NRDC 143 and provisionally named permethrin) the groups essential for high insecticidal activity have been retained in a molecular framework in which the photolabile centers have been replaced by more stable ones. Figure 9 shows representative data for the relative stabilities and insecticidal activities of the four combinations of acid and alcohol. The esters containing the furan ring are little more stable than the natural pyrethrins or allethrin.

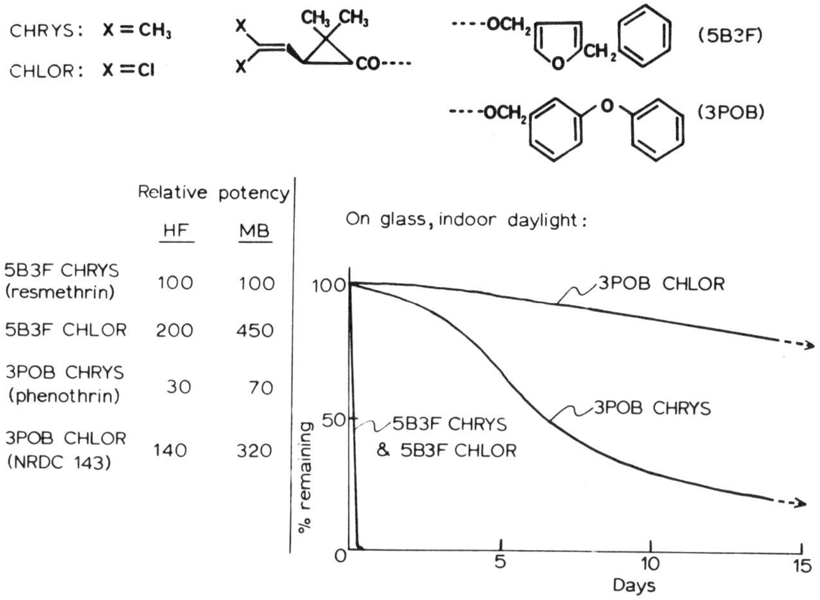


| | Relative potency HF | Relative potency MB |
|---|---|---|
| 5B3F CHRYS (resmethrin) | 100 | 100 |
| 5B3F CHLOR | 200 | 450 |
| 3POB CHRYS (phenothrin) | 30 | 70 |
| 3POB CHLOR (NRDC 143) | 140 | 320 |

FIGURE 9. Insecticidal Activity and Photostability.

Recent results from field tests have confirmed the photostability of permethrin indicated in the laboratory and performance against a wide range of pests important in horticulture and agriculture, as well as in the accepted applications of pyrethroids, has been promising. It is important, therefore, to consider the implications of widespread use of more stable pyrethroids in situations where previously other classes of insecticides including the chlorinated hydrocarbons (aldrin, dieldrin, DDT, lindane, toxaphene, etc.) have been used, but where these are now less acceptable for environmental reasons, such as high mammalian toxicity and undue persistence, or because insects are resistant.

## B. Polarity and Physical Characteristics

In this context, it is relevant to consider the polarities of representative groups of insecticides. Figure 10 shows the

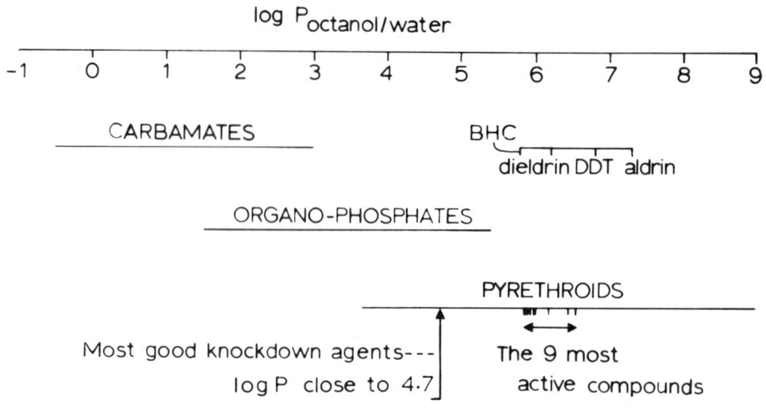


FIGURE 10. The Polarity Spectrum of Insecticides.

polarities of various groups of insecticides estimated by their octanol-water partition coefficients (Briggs *et al.*, 1974b). The pyrethroids are much more lipophilic than the majority of organophosphorus compounds and carbamates. Of the 300 pyrethroids studied, the nine most potent insecticides fall within a very narrow range of polarities close to that of the organochlorine compounds. For both organochlorine compounds and pyrethroids, lipophilicity is probably an important factor determining their ability to penetrate to and act intracellularly in the nervous systems of insects (Burt and Goodchild, 1974). With the less stable pyrethroids used to the present time, this lipophilicity is not a disadvantage from the environmental standpoint, because the compounds are known to be rapidly metabolized in mammalian systems (Elliott *et al.*, 1972; Casida *et al.*, 1974; Ueda *et al.*, 1974b) and to be excreted as polar metabolites. The products of photodecomposition are also relatively harmless. Unlike the organochlorine compounds, therefore, neither the pyrethroids nor their metabolites or photodecomposition products are stored in mammalian body fat and thus do not pass into food chains, nor contaminate the environment in other ways. Although some of the latest pyrethroids are more stable in air and light, they, too, have low mammalian toxicities (Barnes and Verschoyle, 1974; Elliott *et al.*, 1973b, c). Preliminary results indicate that they are also very rapidly metabolized and excreted from mammalian systems (Elliott *et al.*, 1974b) and that their photodecomposition products, when formed, are innocuous.

Knowledge of the relative polarities of pyrethroids also helps to interpret the influence of structure on another valuable property of pyrethroids, their rapid knockdown action. A direct indication that the nature of the group at C-3 on the cyclopropane ring influences the knockdown action of pyrethroids is the better activity in this respect of pyrethrin II compared with pyrethrin I (Sawicki and Thain, 1962) (see Fig. 11). Thus much of the knockdown efficiency of pyrethrum

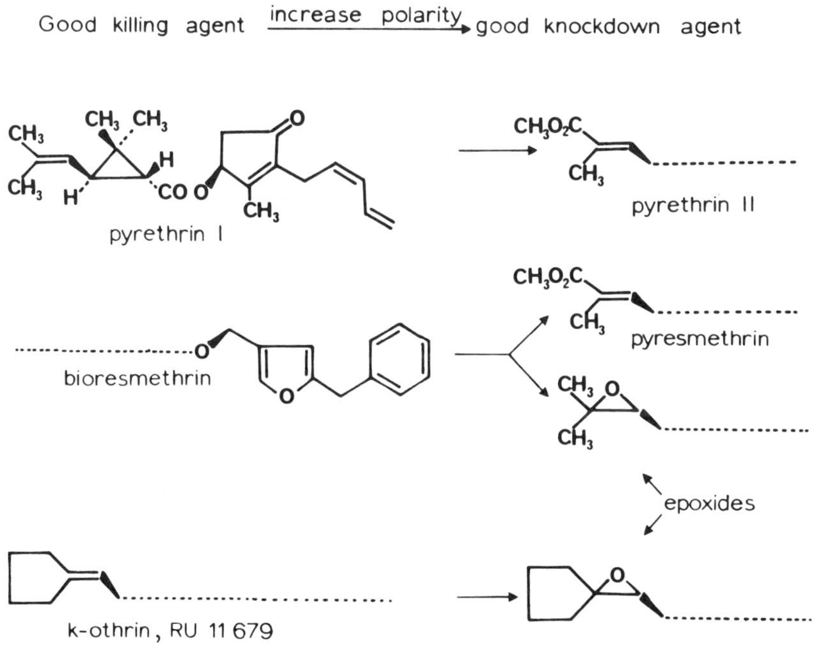


FIGURE 11. Polarity and Knockdown Efficiency.

extract comes from the content of pyrethrin II (Sawicki, 1962a, b); similarly the ester of 5-benzyl-3-furylmethyl alcohol with pyrethric acid (pyresmethrin) has more rapid knockdown action, although poorer kill, than bioresmethrin. Knockdown compounds are generally more polar than those with good killing action (Briggs et al., 1974a). The difference in speed of action appears related to the greater speed of penetration to the central nervous system associated with rapid knockdown action (Burt and Goodchild, 1974). The epoxides of bioresmethrin and of K-Othrin, RU 11679 provide two further examples of how insecticidal properties are modified by changes in molecular structure which influence polarity. Both these epoxides give better knockdown but poorer kill than the unsaturated compounds from which they are derived. Such modifications illustrate the possibility of designing pyrethroids with physical properties appropriate for specific applications. Bioallethrin (Fig. 12)

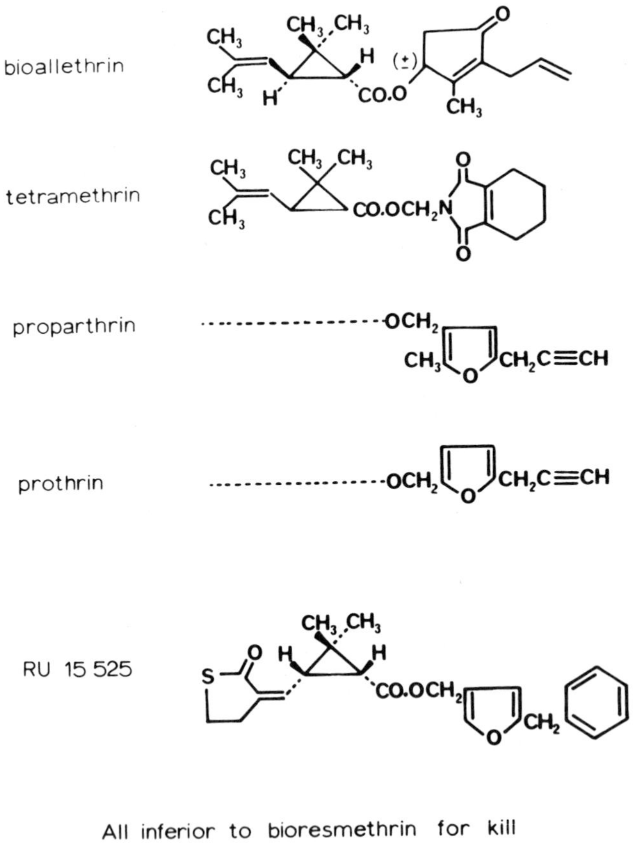


FIGURE 12. Practical Knockdown Agents.

and especially S-bioallethrin (Fig. 2) (Rauch et al., 1972) are even better knockdown agents than pyrethrin I, consistent with their slightly greater polarity. Tetramethrin (Fig. 12) (Kato et al., 1964) is a powerful knockdown agent with a polarity in the region deduced (Briggs et al., 1974a) to be best for knockdown, but it has relatively poor killing action against many species of insects. Its epoxide (not shown) has neither knockdown nor kill, perhaps because it is now too polar even for knockdown (Briggs et al., 1974a). RU 15,525, kadethrin, is even more effective for knockdown than neopynamin or bioallethrin (Lhoste and Rauch, 1974) and was designed specifically for rapid action from knowledge of structure activity relationships (Martel and Buendia, 1974). Similarly, proparthrin (Nakanishi et al., 1970) and prothrin (Katsuda et al., 1969), pyrethroid esters, with relatively low molecular weight and being more volatile, are especially suitable for intermittent or continuous thermal dispersion.

## C. Mammalian Toxicity

The very low mammalian toxicity of most pyrethroids is a very important factor influencing their future development for more general use. As with insecticidal activity, recent investigations (Elliott, 1971; Verschoyle and Barnes, 1972; Abernathy et al., 1973; Miyamoto, 1974a, b; Casida et al., 1974; Barnes and Verschoyle, 1974; Ueda et al., 1974) define more precisely the structural features necessary for low toxicity to mammals combined with high insecticidal activity. Some of the information is summarized in Fig. 13. The most important difference between insect and mammalian detoxifying systems is the ability of the latter to cleave the esters to acidic and alcoholic products (Abernathy and Casida, 1973) which are rapidly eliminated before or after further modifications or conjugation (Ueda et al., 1974; Miyamoto et al., 1974a). The available evidence suggests that unmetabolized esters, the insecticidal agents, are not stored significantly in any part of the mammalian system. Compounds with an ester link susceptible to cleavage as well as sites at which oxidative metabolism can occur show low mammalian toxicity, for example, bioresmethrin and pyresmethrin (male rats: oral $LD_{50}$ 8000 mg/kg or higher; intravenous, 340 mg/kg) (Elliott, 1971; Verschoyle and Barnes, 1972; Abernathy et al., 1973). Modifications to either acidic or alcoholic components that cause steric hindrance to ester cleavage increase mammalian toxicity. Thus saturated and unsaturated groups (as in the tetramethylcyclo-

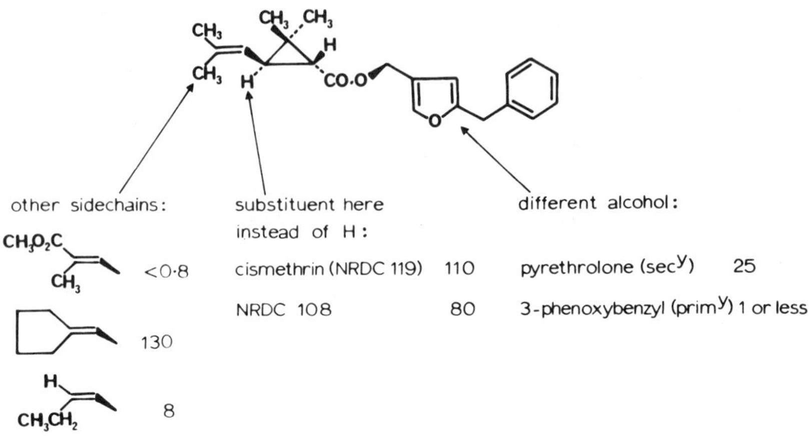


FIGURE 13. Changes in Mammalian Toxicity (approximate factors given are for oral toxicity to rats, relative to bioresmethrin).

propane carboxylate NRDC 108 and cismethrin, NRDC 119, respectively) cis to the cyclopropane carboxylate function give higher mammalian toxicity. Similarly esters of secondary ((+)-pyrethrolone, allethrolone, and α-cyano-3-phenoxybenzyl alcohol) rather than primary (5-benzyl-3-furylmethyl and 3-phenoxybenzyl alcohols) alcohols are more toxic to mammals, particularly by the intravenous route (Verschoyle and Barnes, 1972; Barnes and Verschoyle, 1974).

3-Phenoxybenzyl alcohol is a particularly favorable alcoholic component, esters from it being readily cleaved and metabolized by oxidative attack (Miyamoto et al., 1974a, b; Abernathy et al., 1973; Casida et al., 1974). Dihalovinyl esters of this alcohol are considerably less toxic to mammals than are those of 5-benzyl-3-furylmethyl alcohol (Elliott et al., 1973b, c; Barnes and Verschoyle, 1974). Absence of the trans methyl group in the isobutenyl side chain (e.g., in the cis-butenyl compound, Fig. 6) or substitutions of halogens for the isobutenyl methyl groups in chrysanthemic acid are not necessarily associated with higher mammalian toxicity, although this was easily recognized as a site of metabolic attack in both insects and in mammals.

## VI. CONCLUSIONS

The foregoing discussion has indicated that insecticidal activity is maintained through numerous structural variations, and that there is wide scope for manipulating physical properties and chemical structure in pyrethroids. Design of compounds with selective action between insect species or with short, medium, or long persistence on various surfaces or in special situations should be possible, and appropriate compounds, if manufactured on a large scale, should be less expensive than current examples. Much information is now available about the factors that affect both mammalian toxicity and insecticidal potency. However, if the more stable compounds developed recently are found suitable for widespread application and in situations from which other pyrethroids are excluded by instability, it is essential to anticipate and plan for possible development of resistant insect species; this situation has recently been envisaged and possible approaches for designing appropriate strategies for insecticide management formulated by Sawicki (1974). With such foresight, the prospects for efficient and economic insect control with pyrethroids, which are active at lower concentrations than most other insecticides, are good.

## ACKNOWLEDGMENTS

I thank all of my colleagues at Rothamsted Experimental Station, especially Norman Janes, David Pulman, Andrew Farnham, and Paul Needham, for continued productive collaboration and Ian Graham-Bryce and Norman Janes for much help and constructive criticism in preparing this manuscript.

## REFERENCES

1. Abernathy, C. O. and Casida, J. E. 1973. Science 179, 1235.

2. Abernathy, C. O., Ueda, K., Engel, J. L., Gaughan, L. C., and Casida, J. E. 1973. Pestic. Biochem. Physiol. 3, 300.

3. Barlow, F., Elliott, M., Farnham, A. W., Hadaway, A. B., Janes, N. F., Needham, P. H., and Wickham, J. C. 1971. Pestic. Sci. 2, 115.

4. Barlow, F. and Hadaway, A. B. 1974. Environmental Quality and Safety. Special Issue, Proceedings of the Third International Congress of Pesticide Chemistry, Helsinki (in press).

5. Barnes, J. M. and Verschoyle, R. D. 1974. Environmental Quality and Safety. Special Issue, Proceedings of the Third International Congress of Pesticide Chemistry, Helsinki (in press).

6. Berteau, P. E. and Casida, J. E. 1969. J. Agr. Food Chem. 17, 931.

7. Brett, R. L. 1974a. Pyrethrum Post 12(3), 103.

8. Briggs, G. G., Elliott, M., Farnham, A. W., and Janes, N. F. 1974a. Pestic. Sci. (in press).

9. Briggs, G. G., Elliott, M., Farnham, A. W., Janes, N. F., Needham, P. H., Pulman, D. A., and Young, S. R. 1974b. Environmental Quality and Safety. Special Issue, Proceedings of the Third International Congress of Pesticide Chemistry, Helsinki (in press).

10. Brown, D. G., Bodenstein, O. F., and Norton, S. J. 1973. J. Agr. Food Chem. 21, 767.

11. Brown, N. C. 1970. Report A 28/52, The Wellcome Research Foundation, Berkhamsted, Herts., United Kingdom.

12. Brown, N. C. 1971. Pyrethrum Post 11(2), 66.

13. Bullivant, M. J. and Pattenden, G. 1971. Pyrethrum Post 11(2), 72.

14. Bullivant, M. J. and Pattenden, G. 1972. J. Chem. Soc. D, 864.

15. Bullivant, M. J. and Pattenden, G. 1973a. Tetrahedron Lett., 3679.

16. Bullivant, M. J. and Pattenden, G. 1973b. Pyrethrum Post 12(2), 64.

17. Burt, P. E. and Goodchild, R. E. 1974. Pestic. Sci. 5 (in press).

18. Casida, J. E. 1970. J. Agr. Food Chem. 18, 753.

19. Casida, J. E. 1973. Pyrethrum, the Natural Insecticide, Academic Press, New York.

20. Casida, J. E. 1973a. In Pyrethrum, the Natural Insecticide, (J. E. Casida, Ed.), Chap. 5, p. 101, Academic, New York.

21. Casida, J. E., Ueda, K., Gaughan, L. C., Jao, L. T., and Soderlund, D. M. 1974. Environmental Quality and Safety. Special Issue, Proceedings of the Third International Congress of Pesticide Chemistry, Helsinki (in press).

22. Chen, Y-L. and Barthel, W. F. 1953. J. Amer. Chem. Soc. 75, 4287.

23. Chen, Y-L. and Casida, J. E. 1969. J. Agr. Food Chem. 17, 208.

24. Collins, W. J. 1973. J. Econ. Entomol. 66, 44.

25. Crombie, L. and Elliott, M. 1961. Fortschritte der Chemie organischer Naturstoffe 19, 120.

26. Elliott, M., Needham, P. H., and Potter, C. 1950. Ann. Appl. Biol. 37, 490.

27. Elliott, M. 1951. Pyrethrum Post 2(3), 18.

28. Elliott, M. 1954. J. Sci. Food Agric. 11, 505.

29. Elliott, M., Farnham, A. W., Janes, N. F., Needham, P. H., and Pearson, B. C. 1967. Nature (London) 213, 493.

30. Elliott, M. 1969. Chem. Ind. (London), 776.

31. Elliott, M. 1971. Bull. W.H.O. 44, 315.

32. Elliott, M., Janes, N. F., and Pearson, B. C. 1971a. J. Chem. Soc. C, 3551.

33. Elliott, M., Janes, N. F., and Pearson, B. C. 1971b. Pestic. Sci. 2, 243.

34. Elliott, M., Janes, N. F., and Payne, M. C. 1971c. J. Chem. Soc. (London), 2458.

35. Elliott, M., Janes, N. F., Kimmel, E. C., and Casida, J. E. 1972. J. Agr. Food Chem. 20, 300.

36. Elliott, M. and Janes, N. F. 1973. In Pyrethrum, the Natural Insecticide, (J. E. Casida, Ed.), Chap. 4, p. 55, Academic, New York.

37. Elliott, M., Farnham, A. W., Janes, N. F., Needham, P. H., and Pulman, D. A. 1973a. Nature 244, 456.

38. Elliott, M., Farnham, A. W., Janes, N. F., Needham, P. H., Pulman, D. A., and Stevenson, J. H. 1973b. Nature 246, 169.

39. Elliott, M., Farnham, A. W., Janes, N. F., Needham, P. H., Pulman, D. A., and Stevenson, J. H. 1973c. Proceedings of the 7th British Insecticide and Fungicide Conference, 721.

40. Elliott, M., Farnham, A. W., Janes, N. F., Needham, P. H., and Pulman, D. A. 1974a. Nature 248, 710.

41. Elliott, M., Ueda, K., Gaughan, L. C., and Casida, J. E. 1974b. Unpublished results.

42. Elliott, M., Janes, N. F., and Pulman, D. A. 1975. J. Chem. Soc. Perkin I 21, 2470.

43. Fales, J. H., Bodenstein, O. F., Waters, R. M., Fields, E. S., and Hall, R. P. 1972. Soap Cosmetics Chem. Spec. 46(5), 60.

44. Farkas, J., Kourim, P., and Sorm, F. 1959. Coll. Czech. Chem. Comm. 24, 2230.

45. Farnham, A. W. 1973. Pestic. Sci. 4, 513.

46. Farnham, A. W. and Sawicki, R. M. 1974. Environmental Quality and Safety. Special Issue, Proceedings of the Third International Congress of Pesticide Chemistry, Helsinki (in press).

47. Fujimoto, K., Okuno, Y., Itaya, N., Kamoshita, K., Mizutani, T., Kitamura, S., Nakai, S., and Kameda, N. 1971. Japan Patent, 21473.

48. Fujimoto, K., Itaya, N., Okuno, Y., Kadota, T., and Yamaguchi, T. 1973. Agric. Biol. Chem. 37, 2681.

49. Gnadinger, C. B. 1936. Pyrethrum Flowers, McLaughlin Gormley King Co., Minneapolis.

50. Gnadinger, C. B. 1945. Pyrethrum Flowers, Supplement to 2d Edition (1936-1945). McLaughlin Gormley King Co., Minneapolis.

51. Glynne Jones, G. D. 1973. In Pyrethrum, the Natural Insecticide, (J. E. Casida, Ed.), Chap. 2, p. 17, Academic, New York.

52. Head, S. W. 1973. In Pyrethrum, the Natural Insecticide, (J. E. Casida, Ed.), Chap. 3, p. 25, Academic, New York.

53. Hopkins, L. O., and Maciver, D. R. 1965. Pyrethrum Post 8(2), 3.

54. Kato, T., Ueda, K., and Fujimoto, K. 1964. Agric. Biol. Chem. 28, 914.

55. Katsuda, Y., Chikamoto, T., Ogami, H., Hirobe, H., and Kunishige, T. 1969. Agric. Biol. Chem. 33, 1361.

56. Keiding, J. 1974. Environmental Quality and Safety. Special Issue, Proceedings of the Third International Congress of Pesticide Chemistry, Helsinki (in press).

57. Lhoste, J., Lambert, J., and Rauch, F. 1967. C. R. Acad. Agr. France, 686.

58. Lhoste, J., Rauch, F., and Lambert, J. 1968. Phytiatrie - Phytopharmacie 2, 143.

59. Lhoste, J., Martel, J., and Rauch, F. 1971. Meded. Fac. Landbouwwetensh Rijksuniv. Gent. 36, 978. (Chem. Abstr. 1972, 77, 30321 n).

60. Lhoste, J., and Rauch, F. 1974. Environmental Quality and Safety. Special Issue, Proceedings of the Third International Congress of Pesticide Chemistry, Helsinki (in press).

61. Lloyd, C. J. and Parkin, E. A. 1963. J. Sci. Food Agric. 14(9), 655.

62. Maas, W. and Zindel, E. E. 1974. Pyrethrum Post 12(3), 110.

63. Maciver, D. R. 1966. Pyrethrum Post 8(3), 3.

64. Martel, J., and Buendia, J. 1974. Environmental Quality and Safety. Special Issue, Proceedings of the Third International Congress of Pesticide Chemistry, Helsinki (in press).

65. Matsui, M. and Kitahara, T. 1967. Agric. Biol. Chem. 31, 1143.

66. Matsui, M. and Yamamoto, I. 1971. In Naturally Occurring Insecticides, (M. Jacobson and D. G. Crosby, Eds.), Chap. 1, p. 3, Dekker, New York.

67. Metcalf, R. L. 1955. In Organic Insecticides, Their Chemistry and Mode of Action, Chap. 3, p. 37, Interscience, New York.

68. Miskus, R. P. and Andrews, T. L. 1972. J. Agr. Food Chem. 20, 313.

69. Miskus, R. D. and Lyon, R. L. 1973. In Pyrethrum, the Natural Insecticide, (J. E. Casida, Ed.), Chap. 15, p. 281, Academic, New York.

70. Mijamoto, J., Suzuki, T., and Nakae, C. 1974a. Pestic. Biochem. Physiol. 4, 438.

71. Mijamoto, J., Suzuki, T., and Kadota, T. 1974b. Environmental Quality and Safety. Special Issue, Proceedings of the Third International Congress of Pesticide Chemistry, Helsinki (in press).

72. Nakanishi, M., Mukai, T., Inamasu, S., Yamanaka, T., Matsuo, H., Taira, S., and Tsuruda, M. 1970. Botyu-Kagaku 35(3), 87.

73. Ohno, N., Fujimoto, K., Okuno, Y., Mizutani, T., Hirano, M., Itaya, N., Honda, T., and Yoshioka, H. 1974. Agric. Biol. Chem. 38, 881.

74. Page, A. B. P., Stringer, A., and Blackith, R. D. 1949. Ann. Appl. Biol. 36, 225.

75. Potter, C. 1973. Development Forum, (United Nations Center for Economic and Social Information, Geneva, Switzerland, pub.) 1, (1).

76. Potter, C. 1935. Ann. Appl. Biol. 22, 769.

77. Potter, C. 1938. Ann. Appl. Biol. 25, 836.

78. Rauch, F., Lhoste, J., and Birg, M. L. 1972. Mededelingen Fakulteit Landbouwwetenschappen Gent 37(2), 755.

79. Roark, R. C. 1952. U. S. Dept. Agr. Bur. Entomol. Plant Quar. Agric. Res. Admin., E 846 (Sept.)

80. Roark, R. C. and Nelson, R. H. 1955. U. S. Dept. Agric., Agric. Res. Serv. ARS-33-12.

81. Sanders, H. J. and Taff, A. W. 1954. Ind. Eng. Chem. 46, 414.

82. Sawicki, R. M. 1962a. J. Sci. Food Agric. 13, 283.

83. Sawicki, R. M. 1962b. J. Sci. Food Agric. 13, 591.

83. Sawicki, R. M. and Thain, E. M. 1962. J. Sci. Food Agric. 13, 292.

84. Sawicki, R. M. 1974. Environmental Quality and Safety. Special Issue, Proceedings of the Third International Congress of Pesticide Chemistry, Helsinki (in press).

85. Schechter, M. S., Green, N., and LaForge, F. B. 1949. J. Amer. Chem. Soc. 71, 3165.

86. Sota, K., Amano, T., Aida, M., Noda, K., Hayashi, A., and Tanaka, I. 1971. Agric. Biol. Chem. 35, 968.

87. Staudinger, H. and Ruzicka, L. 1924. Helv. Chim. Acta 7, 390.

88. Stevenson, J. H. 1963. Roy. Soc. Health 83(4), 207.

89. Synerholm, M. E., Hartzell, A., and Cullman, V. 1947. Contrib. Boyce Thompson Inst. 15, 35.

90. Ueda, K. 1974. J. Agr. Food Chem. 22, 212.

91. Ueda, K., Gaughan, L. C. and Casida, J. E. 1975. J. Agr. Food Chem. 23, 106.

92. Velluz, L., Martel, J., Monime, G. 1969. C. T. Hedb. Seanc. Acad. Sci. Paris 268, 2199.

93. Verschoyle, R. D. and Barnes, J. M. 1972. Pestic. Biochem. Physiol. 2, 308.

94. West, T. F. 1959. J. Roy. Soc. Arts 108, 423.

95. Winney, R. 1973. Pyrethrum Post 12(1), 2.

96. Yamamoto, I. 1973. In Pyrethrum, the Natural Insecticide, (J. E. Casida, Ed.), Chap. 10, p. 195, Academic, New York.

97. Yamamoto, I. and Katsuda, Y. 1974. Environmental Quality and Safety. Special Issue, Proceedings of the Third International Congress of Pesticide Chemistry, Helsinki (in press).

## DISCUSSION

One hundred and fifty years of investigations of pyrethrum and pyrethrins laid the background for the last 25 years of work on synthetic pyrethroids. Within the past 25 years there have been many exciting discoveries, including one or more compounds with each of the following desirable features: improved knockdown activity; high potency in the absence of synergists; insect-mammalian selectivity of a magnitude not previously encountered; greater potency than achieved with any other type of insect toxicant; and simplified and more economical structures, often a single compound rather than an isomer mixture. Many of these developments have occurred without sacrificing the safety for humans, wildlife, or the environment. By 1980 we may see a doubling in the number of commercial synthetic pyrethroids and expanded uses in horticulture and agriculture, potentially at gram-per-acre levels. These are exciting times for the pyrethroid field. This state of progress and the excitement are largely due to the long and creative efforts of Dr. Michael Elliott and his colleagues at Rothamsted who have maintained for many years that pyrethroids are the best insecticides. Dr. Elliott's presentation of their research program certainly supports this conviction.

The investigations on pyrethroids illustrate how very long and intensive the study of any group of compounds has to be before we can define the most appropriate derivatives for ultimate development. A similar situation exists with some of the chlorinated hydrocarbons where we go back 100 years and still have much to learn. Accordingly, it is essential to have a continuity of funding, because it often takes a good many years before there is any practical impact from studies as fundamental as these.

It was suggested that the use of a single synthetic pyrethroid in lieu of the complex natural mixture of the pyrethrins might result in more rapid onset of insecticide resistance. However, on the basis of reactions at the molecular level, there seems to be no reason to expect that a mixture of closely related compounds should induce less resistance than a single component. The metabolic fate of pyrethrins I and II are now well understood, and there is no indication for interactions of any large magnitude between isomeric components in their metabolism. The manner in which the compounds are metabolized depends largely on the unique properties of the various alcohol and acid moieties.

It was pointed out that although it would be easier and cheaper to use synthetic pyrethroids containing mixtures of the isomeric acid components, the most potent compounds should be the ones to be introduced into the environment so that the total amount used will be as small as possible. Fundamental understanding of structure-activity relationships often makes it possible to replace asymmetric compounds or components such as pyrethrolone with symmetrical ones (5-benzyl-3-furylmethanol) of comparable structure. These two components apparently fulfill a very similar structural function in the molecule for insecticidal action. Thus where optical and geometrical isomerism are important for insecticidal activity, the symmetrical alcoholic component may be advantageous.

Discussion about the potential cost of the new pyrethroids suggested that these would fall into the general range of about \$50 to \$55 per pound, or about that of the natural pyrethrins. Some of the synthetic pyrethroids already on the market are less expensive than the natural pyrethrins. Others under development might be even cheaper under large-scale production. The newer compounds are also more effective and may not require the special formulations important with the pyrethrins. An important feature of the economics of the use of the newer synthetic pyrethroids is that very low dosages, perhaps just a few grams per acre may be effective against insect pests, particularly in public health programs.

In response to a question about the most promising of the new synthetic pyrethroids, this was stated as likely to be NRDC 143 or permethrin which is photostable and relatively cheap. It is a mixture of isomers with a m.p. of 28 to 38°C and can potentially be formulated in any of the ways in which polar, water-imiscible compounds are normally formulated. Ultra-low volume (ULV) formulations are likely to be of particular interest. The most potent compound discussed in the chapter is NRDC 161, whose commercial possibilities remain to be evaluated.

The final discussion centered around the production of the naturally occurring pyrethrins from *Chrysanthemum* flowers. Can the agricultural policies of countries where these are produced be changed to lead to much greater production? It appears that in Kenya, which produces about 70% of the crop, a very large increase cannot be effected while maintaining a reliable product. The standards are set by the Pyrethrum Marketing Board of Kenya both for production of the crop and the marketing of extracts. Kenya is able to sell all the pyrethrum it produces. There is an intensive program to increase both production and yields of the active ingredient. The normal average is 1.2% and yields of up to 3% have been achieved in some cases, proving a significant increase in production.

The annual production of pyrethrum is about 200 metric tons of active constituents, valued at about \$25 million. This is a very low figure relative to the organochlorine and organophosphorus insecticides. If natural pyrethrum production were doubled, this might significantly change the balance in uses of natural pyrethrins versus synthetic pyrethroids in present day use. However, over the next 20 to 50 years and with the possibility of expanding uses to wider spheres such as agriculture, there will not be an adequate supply to fulfill all future needs. Labor costs in Kenya are increasing, and there has been little success to date in developing mechanical methods of harvesting. There is still dependence on good weather for consistent crop production. Production in other countries seems uncertain. Ecuador has some interest, and Nepal has just set up a pyrethrum board to try and get into the business.

# Insecticide Synergism

C. F. WILKINSON
Department of Entomology
Cornell University
Ithaca, New York

## I. INTRODUCTION

The biological activity of a chemical can often be considerably modified by the prior or simultaneous exposure of the test organism to another chemical agent. When this occurs it is commonly referred to in toxicology as an interaction. If the net result of the interaction is to decrease biological activity, the effect is said to be antagonistic, whereas if activity is enhanced, it is described as being synergistic. Insecticide synergists, therefore, are compounds that enhance the toxicity of insecticides with which they are combined, and in most cases the effect is observed at synergist concentrations which per se elicit no toxic response. Synergism can theoretically occur through interaction of the synergist with any of the several processes that determine the ability of the insecticide to penetrate the organism and to be subsequently transported to the target site at a concentration sufficient to cause a toxic effect. Most known cases of insecticide synergism, however, result primarily from the ability of the synergist to interfere with the metabolic detoxication of the insecticide.

All insecticides are metabolized to some extent by the insect pests at which they are directed as well as by man and other nontarget species which are inadvertently exposed. It is now well established that the degree and duration of action of many biologically active compounds are closely associated with their rate of metabolism, and consequently variations in the susceptibility of different organisms to insecticides (selective toxicity) can often be attributed directly to differences in detoxication capacity. The important practical problem of insect resistance to insecticides is also known to result in many cases from selection by the insecticide of those members of the pest population with a gene complement causing enhanced detoxication capability. In view of the important role of metabolism in these and other areas, it is not surprising that synergists often have dramatic effects on the spectrum and intensity of insecticidal action.

Much has been written in recent years concerning the chemistry, mode of action and use of insecticide synergists (Brooks, 1968; Casida, 1970; Hodgson and Philpot, 1974; Metcalf, 1967; Wilkinson, 1968a, b, 1971a, b), and another comprehensive review of the subject is not warranted at this time. Instead, this chapter attempts to summarize the current state of knowledge in the field and endeavors to assess the practical potential for the future development and use of synergists in insect control programs.

## II. INSECTICIDE METABOLISM

The apparent complexity of the metabolic pathways of insecticides in living organisms is more a reflection of the diversity of the chemicals themselves than of the systems catalyzing their biotransformation. In reality most insecticides, like other lipophilic "foreign" compounds, are metabolized by relatively few general enzyme types catalyzing a few major types of reactions. These enzyme systems appear to have evolved to enable living organisms to transform potentially harmful lipophilic "foreign" compounds into more polar, hydrophilic materials that can be more readily removed from their bodies. The fact that such metabolic transformations usually lead to a decrease in the toxicity of a given chemical is really secondary to their overall function of increasing hydrophilicity and probably results solely from the often rigid structural specificity involved in the interaction of the parent chemical with its biological receptor. The metabolism of a typical insecticide occurs in two major steps often referred to as primary and secondary detoxication (Fig. 1). Primary detoxication constitutes the initial metabolic attack on the molecule and usually involves a biotransformation reaction in which a polar, reactive group is either added to or uncovered in the chemical. The products of primary metabolism are sometimes directly excreted, but more often than not they undergo one of several secondary reactions that result in their conjugation with a variety of endogenous materials such as glucose, glucuronic acid, sulfate, phosphate, or amino acids. The reactions of primary metabolism are usually those initially responsible for decreasing biological activity, and the enzymes catalyzing these reactions are therefore rate limiting with respect to toxicity. All known cases of synergism occur through synergist interactions with the enzymes of primary detoxication.

The major types of reactions by which primary detoxication is effected are indicated in Fig. 1 and include oxidation,

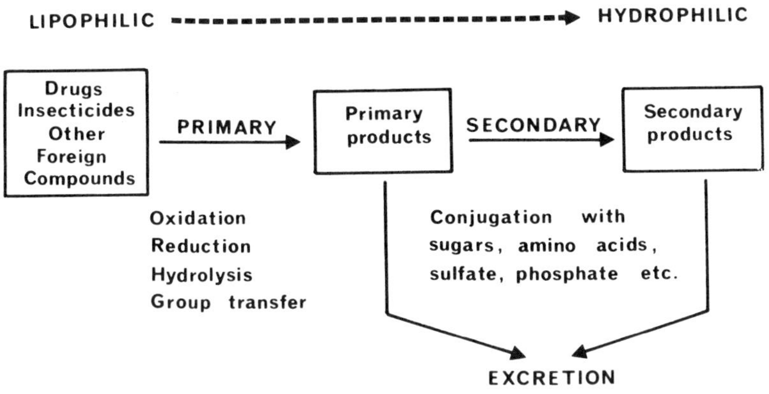


FIGURE 1. Metabolism of lipophilic foreign compounds.

reduction, hydrolysis, dehydrochlorination, and group transfer. Oxidation must be considered of maximum overall importance and is effected mainly by the mixed-function oxidase (mfo) system which appears to play a ubiquitous role in detoxication throughout the plant and animal kingdom (Parke, 1968). The mfo system is located mainly (though not exclusively) in the liver of mammals, fish, and birds, and occurs in several insect organs and tissues (gut, fat body, and Malpighian tubules) depending on the species concerned (Wilkinson and Brattsten, 1972). It occurs in the microsomal particulate fraction of tissue homogenates which is derived from the membranous endoplasmic reticulum of the intact cell. The mfo system is well suited to its role in the detoxication of lipophilic "foreign" compounds because it exhibits a broad substrate specificity and can catalyze several types of biotransformation involving numerous functional groups. Among these reactions are hydroxylation of aromatic, aliphatic, and alicyclic compounds; dealkylation of ethers and substituted amines; epoxidation of double bonds; oxidation of thioethers to sulfoxides and sulfones; and desulfuration of phosphorothionates to phosphates. With the exception of the latter, which is responsible for the insecticidal activation of phosphorothionates to phosphates, these reactions constitute important detoxication pathways, and at least one of them is crucial in the metabolism of almost all groups of organic insecticides. The reactions all rely on a common electron transport pathway (Fig. 2) that transfers electrons

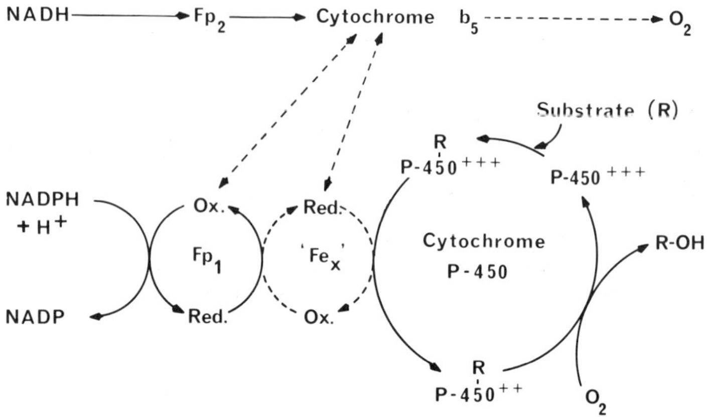


FIGURE 2. Microsomal electron transport chain.

from NADPH through a flavoprotein (NADPH cytochrome c reductase) to a unique cytochrome, cytochrome P-450, which is the terminal oxidase of the chain. Insecticides, drugs, and other foreign compounds form a complex with the oxidized form of the cytochrome, and the complex, after reduction by electrons passing down the chain from NADPH, reacts with and activates molecular oxygen. During the enzymatic reaction one atom of oxygen is incorporated into the insecticide substrate and the other is reduced to water.

The other major type of reaction by which primary detoxication occurs is ester hydrolysis. This is of most general importance with the organophosphate insecticides, but even with these compounds esterase action is often accompanied by oxidative metabolism. There are, of course, many types of esterases in living organism, but those of most importance in organophosphate metabolism are the A-esterases which cleave the phosphate ester (P-O-X) bond and the B-esterases, often called aliesterases or carboxyesterases (Dauterman, 1971). The latter are responsible for cleaving one of the carboxyester groups of malathion, a reaction which results in the rapid detoxication of this insecticide in mammals and in many strains of resistant, though not susceptible, insects.

Although most of the natural pyrethroids and the earlier synthetics are metabolized largely by microsomal oxidation, ester hydrolysis has been shown to be an important detoxication pathway with the newer synthetic materials such as resmethrin and tetramethrin (Abernathy and Casida, 1973).

## III. SYNERGISTS AND THEIR ACTIVITY

The chemical structure of a synergist for any given insecticide depends on the nature of the enzyme(s) responsible for its detoxication. Attention is focused on materials known to interfere with the mfo system, because this is of maximum overall importance in insecticide metabolism and these compounds are the only ones that have attained commercial significance.

The most important group of oxidase inhibitors is the derivatives of methylenedioxybenzene (1,3-benzodioxole) (Fig. 3). The activity of these compounds was first recognized as a

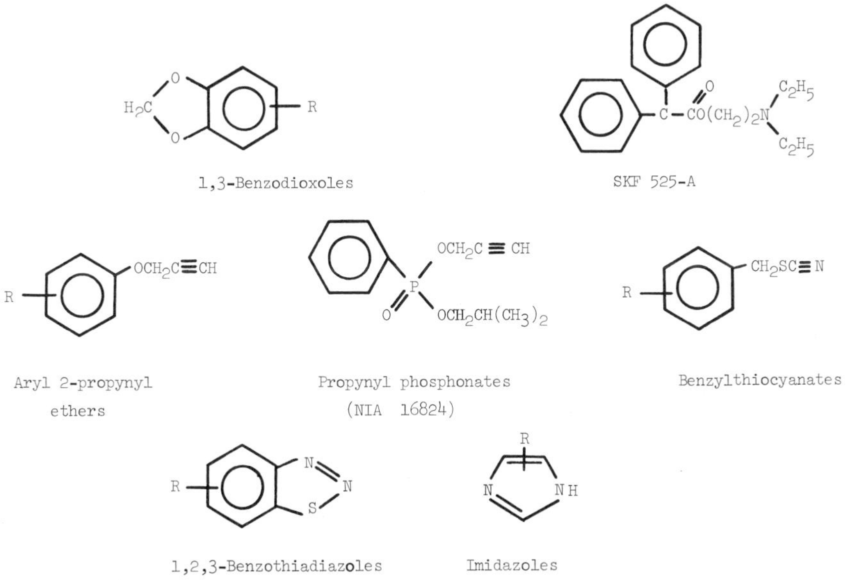


FIGURE 3. Insecticide Synergists.

result of the discovery that the ability of sesame oil to synergize the insecticidal action of the pyrethrins was due to the fact that it contained two 1,3-benzodioxoles, sesamin, and sesamolin. Subsequent studies with numerous natural and synthetic 1,3-benzodioxoles have established the general ability of these compounds to enhance the toxicity of a variety of insecticidal chemicals. Structure-activity studies have clearly established the importance of the intact 1,3-aryldioxole ring, and even small structural modifications in this usually cause a marked decrease or complete loss of synergistic activity. The nature of the substituents on the aryl ring is less critical for activity and probably serve mainly to promote the necessary lipophilicity which allows the molecule to penetrate the insect cuticle (Wilkinson, 1971b). Several of the more polar compounds which are not active by topical application show synergistic activity when incorporated into baits (Weiden and Moorefield, 1965). Of the many hundreds of compounds of this type tested for synergistic activity only four, piperonyl butoxide, sulfoxide, propyl isome, and Tropital® (Fig. 4), currently regis-

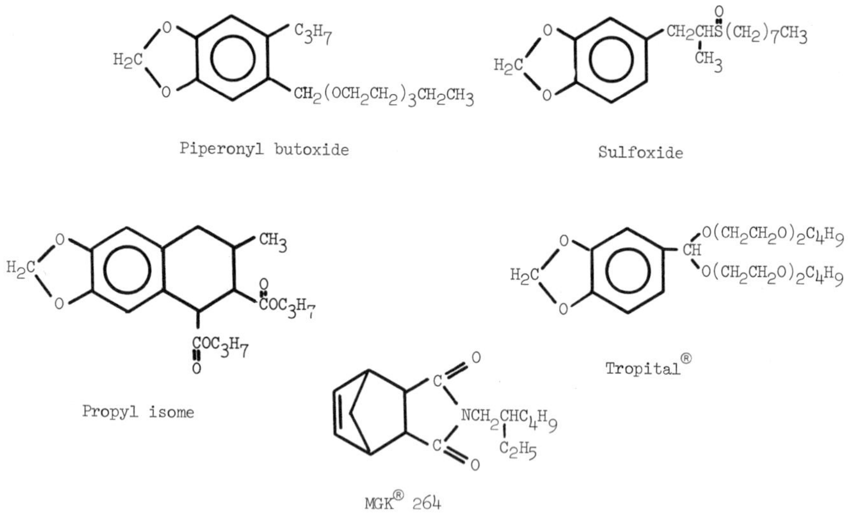


FIGURE 4. Insecticide synergists registered for commercial use.

tered for commercial use in the United States, are used in aerosol formulations of the pyrethrin insecticides.

For some time it appeared that synergistic activity was uniquely associated with the 1,3-benzodioxoles, but in recent years several other groups of compounds have been shown to possess similar properties. These include the alkylamines such as SKF 525-A [2-diethylamino-ethyl 2,2-diphenylpentanoate] (Fig. 3), long known as a drug potentiator in mammals; several materials containing acetylenic bonds, such as the aryl 2-propynyl ethers (Sacher et al., 1968) (Fig. 3); oxime ethers (Hennessy, 1970); the organothiocyanates (Bakry et al., 1964) (Fig. 3); and propynyl phosphonate esters like NIA 16824 [O-(2-methylpropyl)-O-(2-propynyl) phenylphosphonate] (Casida, 1970) (Fig. 3). Attempts to register compounds of the latter type for commercial use were initiated but were subsequently discontinued because of their irritant properties. Another material, MGK 264 (N-octyl bicycloheptene dicarboximide) (Fig. 4) is currently registered for commercial use as a synergist.

More recently there has been considerable interest in the synergistic activity of compounds with nitrogen containing rings, particularly the substituted 1,2,3-benzothiadiazoles (Felton et al., 1970; Gil, 1973) and imidazoles (Wilkinson et al., 1974a, b) (Fig. 3). Regression analysis has established that the synergistic activity of a large number of substituted 1,2,3-benzothiadiazoles with carbaryl is correlated with the hydrophobic parameter ($\pi$) and the homolytic free radical contant ($\sigma^{\bullet}$) (Gil, 1973). Hydrophobic character also appears to be of dominant importance in determining the synergistic activity of the imidazoles (Wilkinson et al., 1974a). Although a large number of 1- and 4 (5)-substituted imidazoles are among the most potent in vitro inhibitors of microsomal oxidation yet discovered, only the 1-substituted compounds are active as in vivo synergists for carbaryl against houseflies (Wilkinson et al., 1974a, b).

Because these compounds inhibit microsomal oxidation, they are capable of synergizing to some extent the action of all insecticides metabolized by this system. These include the pyrethrins, the chlorinated hydrocarbons, the organophosphates, and the carbamates. Synergistic activity varies with both the synergist and the insecticide, the most dramatic effects being observed with the phenyl N-methylcarbamates where synergistic ratios (ratio of $LD_{50}$ of insecticide alone to $LD_{50}$ of synergized insecticide) of several hundred are not uncommon (Table 1). Since many of the commercial organophosphates are phosphorothionates (P=S) and require microsomal oxidation for conversion (activation) to the highly potent phosphate (P=O) antichlines-

TABLE 1. Synergism of Carbaryl Toward Susceptible Houseflies by Compounds Representing Different Synergist Groups[a]

| Synergist | Topical $LD_{50}$ of carbaryl in 1:5 ratio with synergist (μg/g) | Synergistic ratio |
|---|---|---|
| None | 900 | --- |
| Piperonyl butoxide | 12.5 | 72 |
| 1,2-Methylenedioxynaphthalene | 4.0 | 225 |
| *p*-Nitrobenzylthiocyanate | 77.5 | 12 |
| 2-Propynyl-4-chloro-2-nitrophenyl ether | 4.2 | 214 |
| 2-(Diethylamino)ethyl 2,2-diphenyl pentanoate (SKF 525A) | 58.5 | 15 |
| 5,6-Dichloro-1,2,3-benzothiadiazole | 14.5 | 62 |
| 1-(2,3-Dimethylphenyl) imidazole[b] | 2.5 | 360 |

[a]Data taken from Gil (1973); Wilkinson (1971); Wilkinson *et al.*, 1974b).

[b]Flies pretreated with 50 μg synergist 1 hour prior to carbaryl.

terases, the toxicity of these compounds is often antagonized when combined with a "synergist." This is not always the case, however, since the microsomal enzymes both activate and detoxify phosphorothionates and the net result of combining them with oxidase inhibitors is often difficult to predict (Wilkinson, 1971b).

With most insecticides the degree of synergism observed is directly related to the rate at which the insecticide is metabolized, and this depends, among other factors, on the structure of the insecticide and the metabolic capacity of the insect concerned. The optimum dose ratio of the synergist to the insecticide often varies considerably with different materials and depends on factors related to the potency of the synergist itself and its ability to penetrate to the site of detoxication. Certain synergists such as naphtho [2,3-d]-1,3-dioxole (Metcalf *et al.*, 1966) and some of the aryl 2-propynyl ethers (Barnes and Fellig, 1969) are extremely effective at low ratios with carbaryl.

A theoretically attractive way of utilizing the synergistic activity of certain groups (synergophores) is to incorporate them directly into the structure of the insecticide molecule. Attempts to develop this "self-synergism" concept have to date had only marginal success, however, possibly because the synergist group often appears to have a deleterious effect on the interaction of the insecticide with its target receptor. With the carbamates, some degree of self-synergism is suggested by the fact that inclusion of the 1,3-dioxole and 2-propynyl ether moieties in the phenyl ring yields potent nonsynergizable insecticides (Metcalf *et al.*, 1966), but little or no activity is observed with carbamates containing the thiocyano or 1,2,3-thiadiazole groups (Wilkinson, 1971b) (Table 2 ). It is probable that the pyrethroid, Barthrin, owes some of its insecticidal activity to the presence in its structure of the 1,3-benzodioxole grouping.

Examples of synergism resulting from esterase inhibition are limited mainly to those with malathion. The B-esterase responsible for cleavage of one of the carboxyester groups of malathion is highly susceptible to inhibition by compounds such as EPN (*O*-ethyl-*O*-*p*-nitrophenyl phenylphosphonothioate), TOCP (tri-*o*-cresyl phosphate), DEF (*S*,*S*,*S*-tributyl phosphorotrithioate), and numerous other trisubstituted aliphatic and aromatic phosphates and noninsecticidal carbamates. Although many of these compounds have been shown to enhance substantially the toxicity of malathion to resistant strains of insects in the laboratory, they have not achieved any real practical significance.

TABLE 2. Toxicity of Methylcarbamates Containing Synergistic Groups, and the Degree to Which They can be Synergized[a]

| Methylcarbamate | Topical $LD_{50}$ (μg/g, female fly) Alone | With P.B. | Synergistic ratio |
|---|---|---|---|
| 3,4-Methylenedioxyphenyl | 17.5 | 20.0 | 1.0 |
| 2-(2-Propynyloxy)phenyl | 6.5 | 4.6 | 1.4 |
| 3-(2-Propynyloxy)phenyl | 7.5 | 6.0 | 1.2 |
| 4-Thiocyanatophenyl | 1000[b] | 1000[b] | 1.0 |
| 6-Methylcarbamoyl-1,2,3-benzothiadiazole | c | --- | --- |

[a]Data from Wilkinson (1971b).

[b]$LD_{50}$ values (ppm); data of Weiden and Moorefield (1965), measured by incorporation of piperonyl butoxide in bait.

[c]Little or no activity (Kirby, personal communication, 1971).

It is, however, probable that synergism by esterase inhibition will be of importance with some of the newer synthetic pyrethroids such as resmethrin and tetramethrin which are detoxified more by ester hydrolysis rather than oxidation. Materials currently being evaluated for this purpose include 1-naphthyl N-propylcarbamate, DEF, and some of the propynyl phosphonates (Jao and Casida, 1974; Yamamoto, 1973).

## IV. MODE OF ACTION OF OXIDASE INHIBITORS

In attempts to obtain information that might lead to the rational design of new and more effective insecticide synergists, attention in recent years has been focused on studies to elucidate the mechanism by which a variety of materials exert their inhibitory activity on the microsomal enzyme couple. Most of these studies have involved the 1,3-benzodioxoles. Although several theories have been proposed (Casida, 1970; Wilkinson, 1971a) we still only have a hazy understanding of the precise mode of action of these compounds.

When the microsomal enzyme system is provided simultaneously with a combination of two oxidizable substrates, one will competitively inhibit the metabolism of the other by a process termed alternative substrate inhibition. This is a commonly observed phenomenon with various combinations of drugs and other foreign compounds. If the two drugs bind to the same site, the degree of inhibition observed in any particular case will vary in a manner reflecting the relative enzyme affinities and turnover numbers of the two compounds involved. If binding occurs at two different, but closely associated sites, inhibition will probably involve steric or allosteric interactions, but unfortunately we still have little knowledge concerning the nature of the substrate binding sites at cytochrome P-450. Because it has been established that the 1,3-benzodioxoles are themselves microsomal enzyme substrates, and are oxidatively metabolized to the corresponding catechols by this system, it has been suggested that their inhibitory activity occurs mainly by the alternative substrate mechanism (Casida et al., 1966). Although this mechanism may in part account for the activity of the 1,3-benzodioxoles, and indeed may explain their varying efficacy with different insecticides, the often rigid structural requirements for microsomal enzyme inhibition and synergism strongly suggest that other, more specific, interactions are involved. Because of the established importance of the intact 1,3-dioxole ring, the required integrity of the methylenic hydrogens of this ring, and the decreased activity of the corresponding dideuterated methylene derivatives, the several other mode of action theories that have been proposed have emphasized inhibitory interactions involving this part of the molecule (Fig. 5).

Hennessy (1965) proposed that the electrophilic benzodioxolium ion formed by hydride ion transfer from the methylene group of the 1,3-dioxole ring might interact by ligand addition or displacement with the hemochrome of cytochrome P-450. Hansch (1968), however, as a result of regression analysis of *in vivo* structure-activity relationships in a series of 1,3-benzodioxoles, concluded that the mechanism of inhibition involved the formation of a homolytic free radical by hydrogen abstraction from the methylene group of the ring. The most recent theory is that of Ullrich (1973) who has proposed that the interaction of the 1,3-benzodioxoles with cytochrome P-450 might result in the formation of a reactive carbanion following the liberation of methylenic proton. We, therefore, have theories covering all possible types of hydrogen transfer. It is likely that a complete understanding of the inhibitory mechanism will have to await identification of the active oxygen species generated at cytochrome P-450, although it is now generally accepted that microsomal oxidation proceeds by a free radical mechanism. Recent studies by Marshall and Wilkinson (1973) have established that the 1,3-benzodioxoles will effectively inhibit the epoxidation of aldrin in certain nonenzymatic model systems such as Fenton's reagent (EDTA, $Fe^{+2}$ and $H_2O_2$) which are known to generate mostly hydroxyl radicals (OH·). It appears that the 1,3-benzodioxoles compete with aldrin for the available OH· radicals, and it is of interest and possible relevance to their inhibition of microsomal oxidation that they are also converted to the corresponding catechols by the Fenton's reagent (Marshall and Wilkinson, 1973). Since the initial reaction of the 1,3-benzodioxoles with a OH· radical would probably result in the formation of a homolytic radical through methylenic hydrogen abstraction, these results lend some support to Hansch's proposal. It is also of interest that further reaction of the homolytic radical with another OH· radical would lead to formation of the 2-hydroxy-1,3-benzodioxole suggested as an intermediate in the microsomal metabolism of the 1,3-benzodioxoles to the corresponding catechols (Casida *et al.*, 1966).

Spectrophotometric studies on the binding of 1,3-benzodioxoles to microsomal cytochrome P-450 have established that two distinct types of binding occur. The addition of 1,3-benzodioxoles to oxidized microsomal suspensions leads to the formation of a Type I optical difference spectrum (Fig. 6) with a trough at 420 nm and a peak at 385 nm (Hodgson and

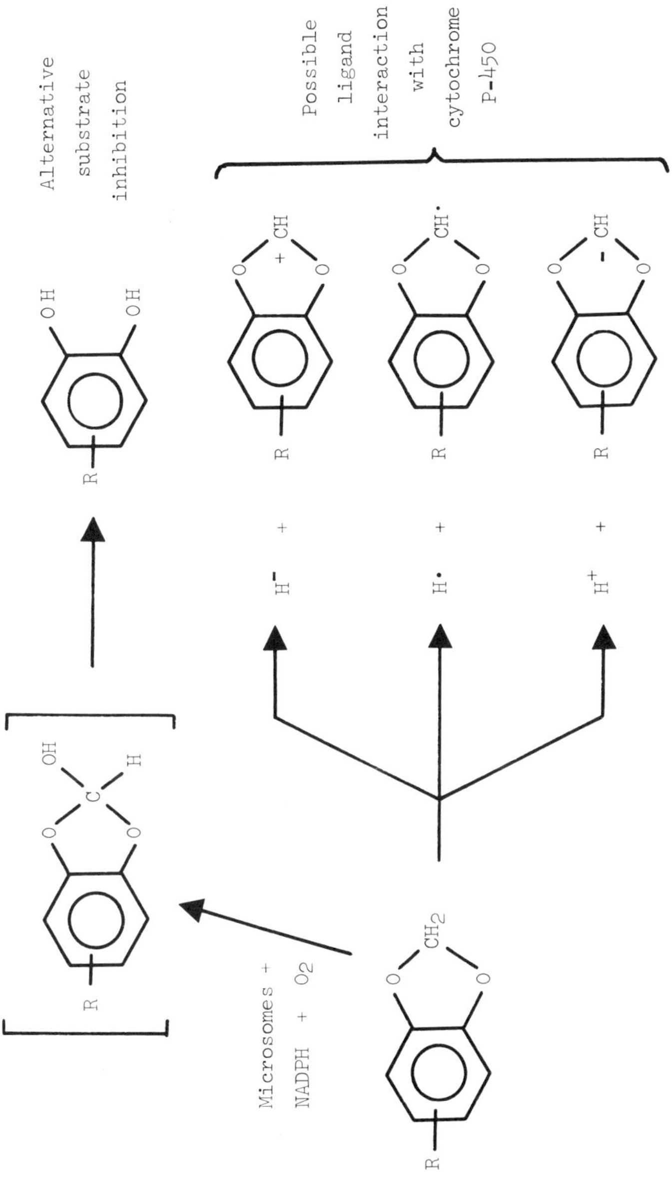


FIGURE 5. Theories for mode of action of 1,3-benzodioxole synergists.

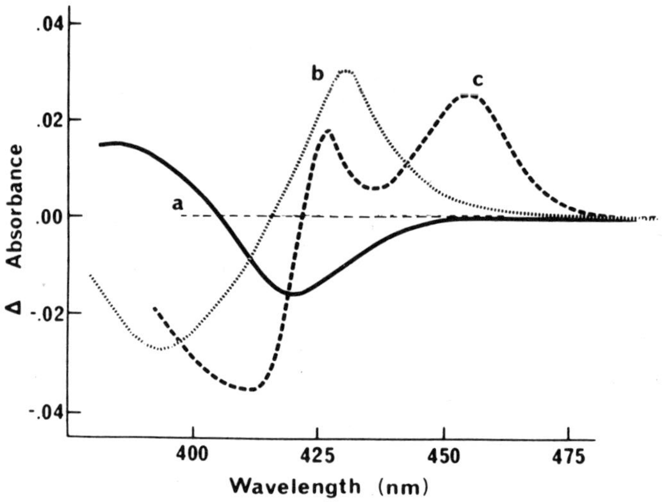


FIGURE 6. Microsomal optical difference spectra. a = Type 1; b = Type II; c = Type III.

Philpot, 1974). Such difference spectra are commonly observed with many foreign compounds and are thought to represent formation of the enzyme substrate complex. Difference spectra observed in NADPH - reduced microsomes, however, are quite different and consist of double Soret peaks at 455 and 427 nm which exist in a pH-dependent equilibrium and are termed Type III (Hodgson and Philpot, 1974) (Fig. 6). Although not yet established, it is possible that the Type III difference spectra result from ligand binding by either the benzodioxolium, the homolytic radical, or the carbanion resulting from removal of hydrogen atom from the methylene group of the 1,3-benzodioxole ring.

The microsomal binding characteristics of the 1,2,3-benzothiadiazoles and imidazoles are different from the 1,3-benzodioxoles and clearly indicate that the inhibitory activity of these nitrogen-containing compounds occurs by a different mechanism.

Although the substituted imidazoles are the most recently discovered group of microsomal enzyme inhibitors and synergists, we already have a better understanding of their probable mode of action than we do with other groups of compounds. The imidazoles exhibit a distinct Type II optical difference spectrum (Fig. 6) with a peak at 430 to 431 nm and a trough at 390 to 393 nm in both oxidized and NADPH-reduced microsomal suspensions. This is typical of a variety of nitrogen-containing compounds such as aniline or pyridine, and is usually interpreted as indicating direct ligand interaction between the nonbonded electrons on the nitrogen atom and the fifth or sixth ligand of the heme moiety of cytochrome P-450. This is considered to be the same ligand responsible for the binding of oxygen and CO to the cytochrome (Fig. 7), and this is supported

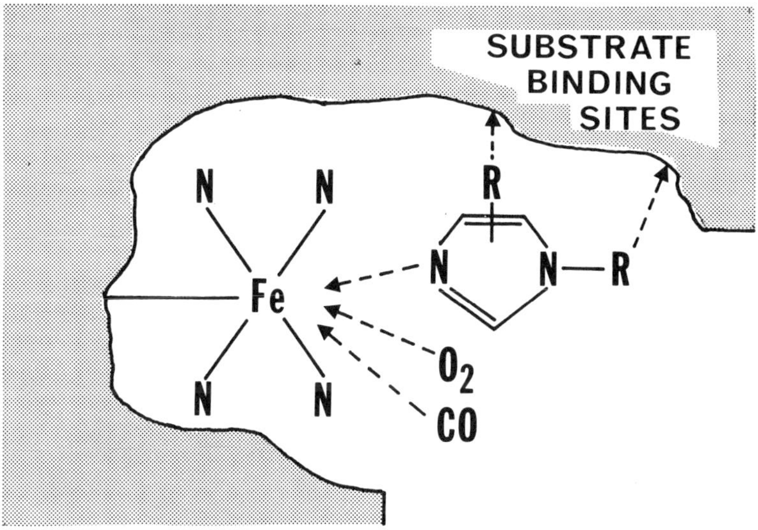


FIGURE 7. Postulated mode of action of substituted imidazoles.

by the fact that in NADPH-reduced microsomes the Type II imidazole spectrum is readily displaced by CO to produce the typical CO optical difference spectrum with peaks at 450 and 420 nm. Since as shown in Table 3 the molar $I_{50}$ values for microsomal enzyme inhibition by the imidazoles closely parallel their spectral dissociation constant ($K_s$) (a measure of their binding affinity) it appears that the ability of the compounds to bind to cytochrome P-450 is the major factor responsible for their inhibitory action. Structure-activity studies suggest

TABLE 3. Molar $I_{50}$ Values and Spectral Dissociation Constants ($K_s$) for Some Substituted Imidazoles

General Structure: 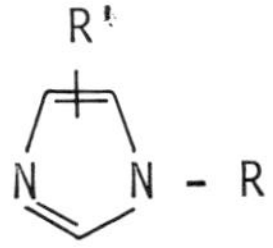


| Substituents R | Substituents R' | $I_{50}$ (M), Epoxidase activity (rat liver microsomes) | $K_s$ (M) rat liver microsomes |
|---|---|---|---|
| Phenyl | H | $1 \cdot 5 \times 10^{-6}$ | $6 \cdot 8 \pm 0 \cdot 8 \times 10^{-6}$ |
| 2-Isopropyl phenyl | H | $1 \cdot 2 \times 10^{-6}$ | $5 \cdot 3 \pm 0 \cdot 8 \times 10^{-6}$ |
| H | Phenyl | $4 \cdot 6 \times 10^{-6}$ | $4 \cdot 6 \pm 0 \cdot 5 \times 10^{-6}$ |
| H | Naphthyl | $4 \cdot 0 \times 10^{-7}$ | $4 \cdot 2 \pm 0 \cdot 4 \times 10^{-7}$ |

that both binding and inhibition are dependent on steric parameters (related to the accessibility of the nonbonded nitrogen electrons) and to the overall lipophilic character of the molecule.

The 1,2,3-benzothiadiazoles also interact to produce a Type II difference spectrum with oxidized microsomes, but under reduced conditions a peak displacement is observed (from 422 to 444 nm, Gil, 1973). Whether this is indicative of the interaction of an intermediate metabolite of 1,2,3-benzothiadiazole or whether it simply reflects a change in conformation of the reduced ligand complex is not known at this time. Under reduced conditions similar optical difference spectra are observed with a series of substituted phenyhydrazines that also inhibit microsomal enzyme activity.

In summary, it appears that we are quite close to understanding the mode of action of many insecticide synergists, and it is likely that with further *in vitro* studies and structure-activity investigations with a variety of inhibitors of microsomal oxidation in preparations from both mammals and insects, we shall be able to develop new and potentially more effective insecticide synergists.

## V. MAMMALIAN TOXICOLOGY OF SYNERGISTS

The acute and subacute toxicities of the insecticide synergists currently registered for use under specified conditions are shown in Table 4 and have been discussed in previous reviews (Casida, 1970; Hodgson and Philpot, 1974; Mrak, 1969). In general, their innate toxicity is low, and they present little or no acute hazard to man. Because the compounds are relatively short-lived, they are currently exempt from tolerance requirements when used preharvest on growing crops according to good agricultural practice. Tolerances ranging from 8 to 20 ppm, however, have been established for piperonyl butoxide used in postharvest application to a variety of vegetables, grains, nuts, and fruits.

At high concentrations some of the 1,3-benzodioxoles have been shown to possess tumorigenic properties in rats and mice (Casida, 1970; Hodgson and Philpot, 1974), but the results are often somewhat equivocal. There is a considerable need for further study in this area with greater emphasis on the route of administration by which humans are most frequently exposed (i.e., inhalation of aerosol formulations).

In view of the basic similarities between the microsomal detoxication system in insects and mammals, synergists do, of course, present a potential hazard in synergizing the acute

TABLE 4. Acute and Subacute Mammalian Toxicity of Synergists Registered for Commercial Use[a]

| Synergist | Acute $LD_{50}$ (mg/kg) | | | No effect level in rat diet for 2-24 months (ppm) |
|---|---|---|---|---|
| | Rat (oral) | Mouse (i.p) | Rabbit (dermal) | |
| Piperonyl butoxide | 7,500-12,800 | >640 | 1,880 | >5,000 |
| Sulfoxide | 2,000 | >640 | >9,000 | >2,000 |
| Propyl isome | 15,000 | >640 | 375 | >5,000 |
| Tropital | >4,000 | >640 | >10,000 | >300 |
| MGK 264 | 2,800 | --- | 470 | >5,000 |

[a]Data from Casida (1970).

toxicity of drugs, insecticides, and other environmental chemicals to man and other animals. That this can occur is shown by the prolongation of barbiturate sleeping time that has been observed in mammals exposed to high levels of several commercially used 1,3-benzodioxole synergists (Casida, 1970; Hodgson and Philpot, 1974; Mrak, 1969). For similar reasons synergists may also stabilize and potentiate the activity of carcinogenic materials such as benzo[α]pyrene in mammals and may consequently be considered potentially cocarcinogenic. The latter has been observed in mice exposed to high levels of piperonyl butoxide (Casida, 1970; Hodgson and Philpot, 1974; Mrak, 1969), but recent findings that the carcinogenic activity of some compounds occurs only after microsomal activation dictates that the overall effect of microsomal enzyme inhibitors on carcinogen potency be subject to a much more thorough evaluation.

It should be emphasized that the possibility of such hazards occurring as a result of the present usage patterns of commercial insecticide synergists is negligible. The 1,3-benzodioxoles are in general rapidly degraded by mammals (probably the major reason for their low activity), and the total level of human exposure to natural and synthetic compounds of this type is in any case very low.

The potential for hazardous synergistic interactions in man is, however, a justifiable cause for concern in the development of new types of insecticide synergists and will prove particularly important with those exhibiting a greater degree of metabolic and/or environmental stability than existing compounds. The 1,2,3-benzothiadioxoles, for example, produce a particularly long-lasting prolongation of barbiturate sleeping time in mice and many of the substituted imidazoles are effective at very low levels of exposure; indeed, a slight effect on hexobarbital sleeping time in mice has been observed with concentrations of some of the substituted imidazoles as low as 1 mg/kg (Wilkinson et al., 1974b). Despite the inhibitory potency of some of the substituted imidazoles, it is of interest that at least one such material, Burimamide®, N-methyl-N'[4-(4(5)-imidazolyl)butyl]thiourea, is currently being considered for use as an antagonist of the histamine $H_2$--receptors which control the secretion of acid by the stomach in humans. The compound is a potent potentiator of hexobarbital sleeping time and an inhibitor of microsomal oxidation.

It should also be pointed out that the interaction of many of these synergists with the hepatic microsomal enzyme system is biphasic in nature and that the initial inhibitory effects are often followed by an "induction" phase which causes microsomal enzyme activity to increase above the normal, control level.

The phenomenon of microsomal enzyme induction is encountered following the exposure of mammals to many drugs and insecticides (particularly the chlorinated hydrocarbons), and although it is sometimes viewed with concern, its toxicological implications for man have not been fully established except in certain acute cases involving human multidrug therapy.

In summary, it is extremely important that in developing new groups or synergists for commercial use, considerable attention be given not only to establishing their innate mammalian toxicity, but also to evaluating their activity on mammalian liver microsomes and to determining possible acute and chronic interactions with a variety of other potentially hazardous environmental chemicals. Residue tolerances of stable synergists will require careful evaluation.

## VI. COMMERCIAL USE AND DEVELOPMENT OF SYNERGISTS

The ability of insecticide synergists to enhance the toxicity of insecticides with which they are combined and therefore to reduce substantially the amount of toxicant required for effective insect control, has several rather obvious practical advantages and consequences.

One of these, of course, related directly to cost. To date, the use of insecticide synergists has been considered economically justified only with the expensive pyrethroid insecticides, and even here the choice of available materials is very limited. Only five compounds, MGK 264, piperonyl butoxide, sulfoxide, propyl isome, and Tropital®, are currently registered for commercial use. Tropital®, the last commercial compound to be developed, was registered in 1966; thus for almost 8 years now, no effective new materials have been registered. Furthermore, four of the five compounds available are 1,3-benzodioxoles which are all obtained from the natural product safrole. This material which has been available in the amount of only about 4 million pounds per year is currently in short supply, and during the next few years this will probably substantially decrease the production and increase the cost of piperonyl butoxide, the most important commercial synergist. There is, therefore, a need for new types of pyrethroid synergists that are not so dependent on the supply of safrole. Some of the newer synthetic pyrethroids are extremely potent insecticides, but since they are also very expensive materials, it is probably that the availability of suitable synergists will play an important role in their successful commercial development. As discussed earlier, it is likely that in many cases these will be esterase inhibitors rather than the existing microsomal enzyme inhibitors. Most of the other insecticide chemicals we have employed during the past three decades have been relatively cheap materials and, except in cases in which insect resistance has developed, they have proved effective when used alone. There has, therefore, been little need or incentive for industry to attempt to synergize these materials, and in any case the cost of the available synergists has been prohibitive. For one reason or another many of the insecticide chemicals we have used are now considered unacceptable, and new materials will have to be found to replace those presently being phased out. The more rigorous requirements for environmental compatibility and greater selectivity will undoubtedly increase the cost of developing and registering these new and more sophisticated chemicals, and it is probable that the economic advantages to be gained from the use of synergized formulations might become a more significant factor in the future.

There are, of course, considerations other than those simply related to economics. Clearly, the use of synergized insecticide formulations can be beneficial in reducing environmental contamination by insecticides and could enable the use of materials which when used alone are too metabolically unstable, biodegradable or nonpersistent. Synergists can also extend the effectiveness of certain insecticides and allow the control of species or strains of insects that are either naturally tolerant or that have developed some degree of resistance to the material alone. Indeed, the use of synergized formulations to control resistant insects has considerable potential, since when resistance is due to enhanced metabolism, an appropriate material can theoretically be used selectively to counteract the particular metabolic pathway involved. Different synergists could be employed for different resistance mechanisms and could consequently provide a considerable degree of insecticide flexibility.

If, however, synergists are to be employed on a broad agricultural or public health front, it will be necessary to develop new types of relatively cheap, effective, safe, and environmentally sound materials. It is this author's belief that this can be done, but there are numerous problems that must be resolved. In view of the several new types of potential synergist groups now being discovered, it is probable that new compounds will be found that will be relatively cheap and easy to produce and which will prove to be effective under laboratory testing conditions. There are, however, several technical difficulties in developing effective formulations of insecticide-synergist combinations for field use and it is probably for this reason that in the past many synergized formulations that have looked promising in the laboratory have often proved disappointing when evaluated under field conditions. For a new compound to be commercially viable it must be active at low ratios relative to the insecticide, and it is probable that a material with similar physical and chemical properties to the insecticide would be advantageous for the preparation of a compatible formulation. Clearly, a highly selective synergist would be advantageous from the viewpoint of safety, but the overall similarities between the detoxication systems in insects and mammals dictate this to be a rather remote possibility in most cases. Some degree of selectivity can be achieved by using synergists that are more rapidly degraded in mammals than in insects, as is the situation with the 1,3-benzodioxoles. The question of environmental persistence of the synergist is of considerable practical importance, and indeed some degree of persistence is essential if the synergist is to be employed in residual spray formulations. Most existing synergists are environmentally unstable, and their period of effectiveness in the field is consequently too brief to allow them to be effective. More stable synergists can be developed, but if they are too stable they will pose potentially serious problems of environmental or human hazard. The presence of residues of stable, highly potent synergists in the environment or in marketable food produce would be a justifiable cause for concern, and it is highly unlikely whether registration could ever be obtained on such materials. It is, however, interesting to speculate that the insecticidal persistence of a material could in theory be regulated by combining it with specific synergists with varying rates of biodegradation.

With the exception of its application with the pyrethroid insecticides, we are still a long way away from developing the synergist concept into a general practical reality. It remains as an attractive possibility, however, and if some of the problems discussed can be satisfactorily overcome insecticide synergism could indeed be a valuable asset in our future pest control programs.

## REFERENCES

1. Abernathy, C. O. and Casida, J. E. 1973. Science 179, 1235.

2. Bakry, N., Metcalf, R. L., and Fukuto, T. R. 1964. J. Econ. Entomol. 57, 478.

3. Barnes, J. R. and Fellig, J. 1969. J. Econ. Entomol. 62, 86.

4. Brooks, G. T. 1968. Meded. Rijksfac. Landbouwetensch. Gent. 33, 629.

5. Casida, J. E. 1970. J. Agric. Food Chem. 18, 753.

6. Casida, J. E., Engel, J. L., Esaac, E. G., Kamienski, F. X., and Kuwatsuka, S. 1966. Science 153, 1130.

7. Dauterman, W. 1971. Bull. W.H.O. 44, 133.

8. Felton, J. C., Jenner, D. W., and Kirby, P. 1970. J. Agric. Food Chem. 18, 671.

9. Gil, L. D. 1973. Ph.D. dissertation, Cornell University.

10. Hansch, C. 1968. J. Med. Chem. 11, 920.

11. Hennessy, D. J. 1965. J. Agric. Food Chem. 13, 218.

12. Hennessy, D. J. 1970. In Biochemical Toxicology of Insecticides, (R. D. O'Brien and I. Yamamoto, Eds.), Academic, New York.

13. Hodgson, E. and Philpot, R. M. 1974. Drug Met. Rev. (in press).

14. Jao, L. T. and Casida, J. E. 1974. Pest. Biochem. Physiol. 4, 465.

15. Marshall, R. S. and Wilkinson, C. F. 1973. Pest. Biochem. Physiol. 2, 425.

16. Metcalf, R. L. 1967. Ann. Rev. Entomol. 12, 229.

17. Metcalf, R. L., Fukuto, T. R., Wilkinson, C. F., Fahmy, M. A., El-Aziz, S. A., and Metcalf, E. R. 1966. J. Agric. Food Chem. 14, 555.

18. Mrak, E. M. (Chairman). 1969. "Report of the Secretary's Commission on Pesticides and their Relationship to Environmental Health." U.S. Dept. Health, Education, and Welfare, Washington, D. C.

19. Parke, D. V. 1968. The Biochemistry of Foreign Compounds, Pergamon, Oxford.

20. Sacher, R. M., Metcalf, R. L., and Fukuto, T. R. 1968. J. Agric. Food Chem. 16, 779.

21. Schenkman, J. B., Remmer, H., and Estabrook, R. W. 1967. Mol. Pharmacol. 3, 113.

22. Ullrich, V. and Schnabel, K. H. 1973. Drug Met. Disp. 1, 176.

23. Weiden, M. H. and Moorefield, H. H. 1965. J. Agric. Food Chem. 13, 200.

24. Wilkinson, C. F. 1968a. World Rev. Pest Control 7, 155.

25. Wilkinson, C. F. 1968b. In Enzymatic Oxidation of Toxicants, (E. Hodgson, Ed.), North Carolina State University, Raleigh.

26. Wilkinson, C. F. 1971a. In Pesticide Chemistry, (A. S. Tahori, Ed.), Proceedings Second International IUPAC Congress Pesticide Chemistry, vol. II, Gordon and Breach, New York.

27. Wilkinson, C. F. 1971b. Bull. W.H.O. 44, 171.

28. Wilkinson, C. F. and Brattsten, L. B. 1972. Drug Met. Rev. 1, 153.

29. Wilkinson, C. F., Hetnarski, K., Cantwell, G. P., and Di Carlo, F. J. 1974a. Biochem. Pharmacol. (in press).

30. Wilkinson, C. F., Hetnarski, K., and Hicks, L. J. 1974b. Pest. Biochem. Physiol. 4, 299.

31. Yamamoto, I. 1973. In Pyrethrum the Natural Insecticide, (J. E. Casida, Ed.), Academic, New York.

# DISCUSSION

The discussion centered on the feasibility of using synergists in insect control. Thus far, synergists have been of great value as research tools for investigations on the pathways of insecticide metabolism and mode of action, but, apart from their classical use with the pyrethrins, they have not found economic application in practical situations. Synergists are potentially useful in one or more of the following applications:

1. To increase the efficiency of compounds that have a broad spectrum of insecticidal activity but only moderate toxicity due to detoxication.

2. To extend the activity spectrum of selective toxicants when selectivity depends on detoxication that can be inhibited by a suitable synergist.

3. To combat resistance when this is due to enhanced detoxication of an insecticide by the resistant strain.

4. To reduce the amount of an expensive insecticide that is actually required for control. Any reduction in insecticide application rates effected thereby would be helpful from an environmental standpoint; the synergists in use thus far not having been themselves shown to constitute any environmental hazard.

The major problem in the practical exploitation of synergists are as follows:

1. The cost of the synergist has hitherto been greater than that of the insecticide to be synergized. Synergism of the natural pyrethrins is the classical exception, four methylenedioxyphenyl compounds (piperonyl butoxide, sulfoxide, propylisome, and Tropital®) and MGK 264 being currently registered for this use. It seems possible that appropriate synergists might be similarly used to potentiate insect juvenile hormone analogs, which are likely to remain expensive in practical application.

2. Commercial synergists are often combined with the insecticide in 10:1 ratio; only rarely is the ratio 1:1 used, or feasible. Even with the larger ratio, the effect is generally to increase the insecticidal activity of poor or moderate insecticides to an acceptable rather than a superior level; the toxicities of the most effective insecticides are rarely improved. Nevertheless, the increase in activity of moderately toxic compounds does mean that more insecticidal combinations are potentially available.

3. Because of the differing physical and chemical properties of insecticide and synergist, there is great difficulty in developing formulations that will keep the two together under practical conditions. For example, mixtures of carbaryl and various synergists have been evaluated as residual sprays for the control of houseflies and mosquitoes. The deposits give good control for several weeks, then fail, due apparently to selective disappearance of the synergist.

4. The action of synergists appears to be largely species specific, the best examples being found with diptera, especially houseflies.

On the question of cost, there is an evident need to move away from the methylenedioxyphenyl (1,3-benzodioxole) synergists, which still depend on increasingly expensive safrole for their production. Several new types of synergistic mixed-function oxidase inhibitors have been examined during the last few years (see Chapter VIII), and it seems likely that relatively cheap but effective compounds may eventually result from such investigations. Low-cost inhibitors of esterases and epoxide hydrases are also of interest, because the detoxication of some of the new pyrethroids depends significantly on esterase activity, while both esterase and epoxide hydrase activity are involved in the deactivation of molecules related to <u>cecropia</u> juvenile hormone.

Formulation is evidently a major problem, and satisfactory solutions might permit, among other benefits, the use of lower synergist insecticide ratios. One way to ensure that the physical properties of insecticide and synergist are as similar as possible is to look for synergistic structural analogs, as has been done in the case of DDT. For example, nontoxic or poorly toxic analogs in a series containing the insecticide to be synergized may be of interest in this context. Even with similar physical properties, differing chemical properties may lead to selective deterioration of the synergist in the mixture. This selective removal of the synergist can, however, have some advantages; the relatively favorable mammalian selectivity of insecticide/NDP synergist combinations evidently depends on the more rapid metabolism of synergist in the mammal compared with the insect and its consequent removal from the mixture.

There is a great potential for the use of synergized insecticides in vector control, and synergists that would be effective with larvicides in water would be especially valuable. Microencapsulation of the insecticide/synergist combination might be helpful, because there is evidence from experiments with microencapsulated hormone analogs that capsules of appropriate size are actually ingested by mosquito larvae. With this technique it may be possible to ensure release of synergized toxicants inside the gut.

The incorporation of synergistic (synergophoric) groups such as methylenedioxyphenyl or 2-propynyloxyaryl moieties into insecticide molecules of appropriate structure avoids the formulation problem but may have a deleterious effect on insecticidal activity through a change in polarity leading to altered pharmacokinetics and/or an altered interaction with the target. However, the concept appears to work in the case of 2-propargyloxyphenyl- and 3,4-methylenedioxyphenyl-N-methylcarbamates. The latter compound shows quite good insecticidal activity toward houseflies, despite relatively low anticholinesterase activity, and it may be that low affinity for the enzyme is offset by the increased bioavailability of the stabilized molecule at the target. Inexpensive synthetic methods are required for the incorporation of synergophores if this technique is to be of any value, and the 2-propynyl group, for example, has the advantage over the 3,4-methylenedioxy-group in that it can usually be readily attached provided the molecule contains a single phenolic hydroxyl group.

The question of species specificity of synergized combinations merits further exploration. Some idea of the possibility for the carbaryl/piperonyl butoxide (1:5) mixture is given in the survey of 54 species of insects from eight orders carried out by Brattsten and Metcalf. Considerations should also be given to synergists acting in ways other than by inhibiting metabolism, for example, by increasing cuticular penetration.

The insecticide synergists currently registered for use present little or no acute hazard to man. Their short life in the environment is a disadvantage from the standpoint of satisfactory formulation with insecticides but advantageous from the toxicological standpoint. Compounds such as the imidazoles and benzothiadiazoles may prove to be rather more stable environmentally, and their ability to inhibit the metabolism of drugs and other environmental chemicals in man as well as insects must be kept in mind. Much depends on determining the pattern of residues remaining after their application with insecticides under practical conditions, but it may be expected that the residue levels ingested by mammals as a result of such applications would be negligible compared with the levels of similar compounds already ingested by humans in some therapeutic situations.

In summary, insecticide synergism is a useful concept that is difficult to bring into field practice. Because two compounds are so much more difficult to formulate than one, it seems unlikely, except for special cases, that synergized combinations will come into universal use so long as new and highly efficient insecticides continue to be synthesized. However, new insecticides are becoming increasingly difficult to synthesize and develop. Research on synergists is valuable from a theoretical point of view and against the day when we really may have to make the most of insecticidal compounds already available.

# Organochlorine Insecticides, Survey, and Prospects

ROBERT L. METCALF
Department of Entomology
University of Illinois
Urbana-Champaign, Illinois

## I. INTRODUCTION

Organochlorine insecticides, including DDT, lindane, the cyclodienes, and toxaphene have been important factors in applied ecology since World War II. The unprecedented environmental persistence of these compounds, together with their broad spectrum insecticidal efficiency and their relatively low cost, effected a total revolution in insect control that has continued for nearly three decades. This revolution has seen the number of synthetic pesticide chemicals increase from about 30 to more than 900, and the annual volume of insecticide production in the United States increase from about 80 million pounds in 1947 to about 1,200 million pounds in 1971 (Pesticide Review).

Organochlorine is a generic term applied to this familiar group of insecticides which structurally resemble one another principally because they are polychlorinated hydrocarbons (Fig. 1). The important insecticidal compounds have become familiar household words over the past 30 years either because of their success and popular appeal in insect pest control or because of their persistence and ubiquitous presence as micropollutants in the total environment. A very full account of these organochlorine insecticides has been published recently (Brooks, 1973).

DDT or 2,2-bis-(p-chlorophenyl)-1,1,1-trichloroethane was first synthesized by Othmar Zeidler (1874) and its insecticidal properties discovered by Paul Müller (1939). Because of its extraordinary persistence, high insecticidal efficacy, and low cost this compound opened up vast new possibilities in insect control techniques and became the prototype of the other organochlorine insecticides. At the height of its production in the United States in 1963 approximately 176 million pounds (81.3 x $10^6$ kg) were produced (Pesticide Review), and world cumulative production is estimated at about 4.4 x $10^9$ pounds (M.I.T., 1970).

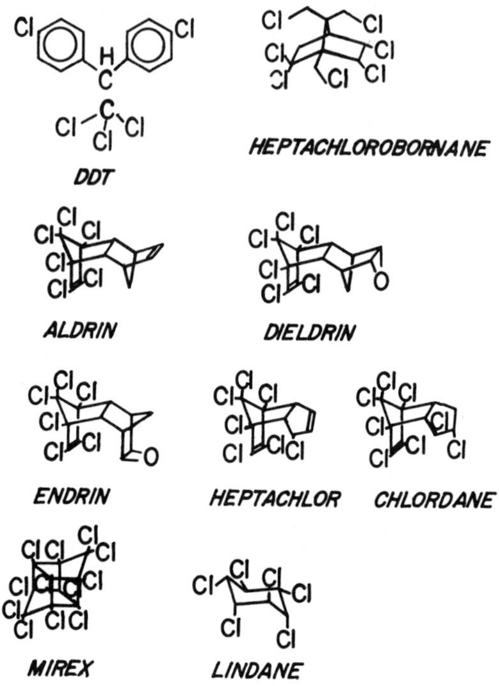


FIGURE 1. Chemical structures of important organochlorine insecticides.

Technical DDT has contained variously from 48 to 80% p,p-isomer m.p. 108 to 109°C; 11 to 29% o,p'isomer m.p. 74°C; and 0.1 to 1% o,o'-isomer m.p. 92.5 to 93.0°C together with up to 37% DDD or 2,2-bis-(p-chlorophenyl)-1,1-dichloroethane, m.p. 112° (Metcalf, 1955).

DDD has been marketed as an insecticide in its own right with a toxicity about 0.08 that of DDT (Table 1). Another analog methoxychlor or 2,2-bis-(p-methoxyphenyl)-1,1,1-trichloroethane, m.p. 89° is used as a very safe nonaccumulative insecticide (Table 1). United States production is estimated as 10 million pounds (Lawless et al., 1972). Perthane® or 2,2-bis-(p-ethylphenyl)-1,1-dichloroethane m.p. 56 to 57° is of very low mammalian toxicity (mouse oral $LD_{50}$ 6600 mg/kg) and is used to a limited extent as a specialty insecticide.

TABLE 1. Nontarget Toxicity of Organochlorine Insecticides (Pimentel, 1971; Hayes, 1963)

| | Oral $LD_{50}$ mg/kg | | | | |
|---|---|---|---|---|---|
| | Rat ♂ | Rat ♀ | Mallard[a] | Pheasant[b] | Rainbow Trout[c] |
| DDT | 113 | 118 | > 2240 | 1296 | 0.012 |
| DDD | 3400 | | | | 0.009 |
| Methoxychlor | 6000 | | > 2000 | | 0.074 |
| Aldrin | 39 | 60 | 150(520) | 16.8 | 0.0061 |
| Dieldrin | 46 | 46 | 40–75 (385) | 79 | 0.0031 |
| Endrin | 17.8 | 7.5 | 5.6 | 1.8 | 0.0028 |
| Heptachlor | 100 | 162 | 200–590 ( 2000) | 180 | 0.013 |
| Chlordane | 335 | 430 | 1200 | 340–550 | 0.022 |
| Endosulfan | 43 | 18 | 33(310) | 620–850 | 0.0032 |
| Toxaphene | 90 | 80 | 70.7 | 40 | 0.004 |
| Lindane | 88 | 91 | > 2000 | 500–600 (75) | 0.018 |
| Mirex | 312 | | > 2400 | | |

Continued

TABLE 1 (cont.)

| | $LC_{50}$ ppm | | | |
|---|---|---|---|---|
| | Bluegill[d] | Fairy Shrimp[e] | Stonefly[f] | Waterflea[g] |
| DDT | 0.005 | 0.0047 | 0.041 | 0.0036 |
| DDD | 0.056 | 0.0056 | 3.0 | 0.0032 |
| Methoxychlor | 0.083 | 0.0047 | 0.030 | 0.0008 |
| Aldrin | 0.010 | 45 | 0.008 | 0.028 |
| Dieldrin | 0.015 | 1.4 | 0.006 | 0.240 |
| Endrin | 0.0008 | 0.0064 | 0.004 | 0.020 |
| Heptachlor | 0.026 | 0.150 | 0.008 | 0.042 |
| Chlordane | 0.095 | 0.160 | 0.170 | 0.029 |
| Endosulfan | | 0.0092 | 0.024 | 0.24 |
| Toxaphene | 0.0066 | 0.180 | 0.018 | 0.015 |
| Lindane | 0.10 | 0.120 | 0.012 | 0.460 |

[a] *Anas platyrhynchos*.

[b] *Phasianus colchicus torquatus*.

[c] *Salmo gairdnerii*.

[d] *Lepomis macrochirus*.

[e] *Gammarus lacustris*.

[f] *Pteronarcys californica*.

[g] *Daphnia pulex*.

Benzene hexachloride (BHC) or hexachlorocyclohexane was first synthesized by Michael Faraday in 1825. The crude chlorination product is a mixture of six stereoisomers, and the singular insecticidal properties of the γ-isomer (lindane) were discovered independently and almost simultaneously by Dupire in France about 1941 (Dupire and Raucourt, 1943) and by Thomas of ICI in England in 1942 (Slade, 1945). Crude BHC contains 10 to 18% of the active γ-isomer m.p. 112 to 113°C (configuration aaaeee); in addition 53 to 80% α-isomer m.p. 159 to 160° (a mixture of enantiomers aaeeee and aeeeea); 3 to 14% β-isomer m.p. 309 to 310° (eeeeee); 2 to 16% δ-isomer m.p. 138-139° (aeeeee); 3 to 5% ε-isomer m.p. 219 to 220° (aeeaee); and a trace of ζ-isomer m.p. 88 to 89° (aaeaee), which are insecticidally inactive (Metcalf, 1955; Melnikov, 1971). Production of crude BHC in the United States reached a maximum of 117 million pounds ($5.3 \times 10^7$ kg) in 1951. Cumulative world production is difficult to estimate, but it probably exceeds 2.5 billion pounds ($1.1 \times 10^9$ kg). Because of severe off-flavors resulting from the use of crude BHC and the objectionable environmental properties of the inactive isomers, especially the extremely stable β-isomer, 99 to 100% pure γ-isomer or lindane is now produced commercially in substantial quantities (e.g., about 1 million pounds annually in the United States). A monograph of lindane as an insecticide has recently appeared (Blaquiere *et al.*, 1972).

Toxaphene, a polychloroterpene, was first described in 1948 as chlorinated camphene containing 67 to 69% chlorine and approximating $C_{10}H_{10}Cl_8$. Toxaphene has recently been shown by Casida *et al.* (1974) to contain as many as 175 individual components, most of which are apparently polychlorbornanes. The chief active ingredients are compounds such as 2,2,5-*endo*,6-*exo*-8,9,10-heptachlorobornane (Fig. 1). Mixtures similar to toxaphene have been produced in the United States as Strobane® (66% chlorine) and in the Soviet Union as chloropinene (66 to 68% chlorine) (Melnikov, 1971). Official production figures are not available, but it is estimated that toxaphene is currently being produced in the United States at about 50 million pounds annually (Lawless *et al.*, 1972). Cumulative world production has been estimated at about 1 billion pounds ($4.5 \times 10^8$ kg) (Casida *et al.*, 1974).

Cyclodiene insecticides are Diels-Alder condensation products of hexachlorocyclopentadiene, and the widely used insecticides are the complex multiple ring structures chlordane, heptachlor, aldrin, dieldrin, endrin, endosulfan, alodan, and mirex (Fig. 1). Chlordane was first synthesized by Julius Hyman in 1944 (U.S. Patents 2,509,160 and 2,606,910) and its insecticidal properties described by Kearns *et al*. (1946). Chlordane is principally the *cis* (m.p. 106.5 to 108°) and *trans* m.p. 104.5 to 106°) isomers of 2,3,4,5,6,7,8,8-octachloro-2,3,3a-4,7,7a-hexahydro-4,7-*endo*-methanoindene and was also independently discovered in Germany by Riemschneider (1947) as M-410. Heptachlor or 1,4,5,6,7,8,8-heptachloro-3a,4,7,7a-tetrahydro-4,7-methanoindene m.p. 95 to 96°, is a minor constituent of technical chlordane but also has been developed as a commercial insecticide (U.S. Patent 2,576,666). Aldrin or 1,2,3,4,10,10-hexachloro-1,4,4a,5,8,8a-hexahydro-1,4-*endo*,*exo*-5,8-dimethano-naphthalene m.p. 104° (U.S. Patent 2,635,977), and dieldrin the 6,7-epoxide of aldrin m.p. 176 to 177° (U. S. Patent 2,676,547) are two additional hexachlorocyclopentadiene adducts developed as insecticides by Hyman. Endrin, m.p. 245° dec. is the *endo*, *endo*-isomer of dieldrin (Fig. 1) (U.S. Patent 2,676,132).

Endosulfan or 6,7,8,9,10,10-hexachloro-1,5,5a,6,9,9a-hexahydro-6,9-methano-2,4,3-benzodioxathiepin-3-oxide, two isomers α-m.p. 108 to 109° and β-m.p. 206 to 208°, was first described by Finkenbrink (1956) (U.S. Patent 2,799,685). Mirex or dodecachloro-octahydro-1,3,4-metheno-2H-cyclobuta-(cd)-pentalene, m.p. 485° is described in U.S. Patent 2,671,043. Its 4-keto analog is chlordecone or Kepone® 1,2,3,5,6,7,8,9,10,10-decachloro-[5.2.1.$0^{2,6}$ .$0^{3,9}$ .$0^{5,8}$]-decano-4-one, m.p. 349° dec. used in baits for cockroaches and ants. Isobenzan or 1,3,4,5,6,7,8,8 -octachloro-3a,4,7,7a-tetrahydro-4,7-methanophthalan, m.p. 122°, has been used commercially but was withdrawn because of its high mammalian toxicity (rat oral $LD_{50}$ 4.8 to 5.5 mg/kg). Alodan or 1,2,3,4,7,7-hexachloro-bis-5,6-chloromethylbicyclo-(2.2.1)-2-heptene, m.p. 105° has been used to control animal ectoparasites because of its low mammalian toxicity (rat oral $LD_{50}$ 15,000 mg/kg). Bis-(pentachlorocyclopentadiene-2,4-yl) m.p. 122° is a cyclodiene acaricide (Metcalf, 1971).

Production figures for the cyclodienes are incomplete. However, in the United States in 1971, 81 million pounds (3.7 x $10^7$ kg) of cyclodienes, including toxaphene, were produced (Pesticide Review). Lawless *et al*. (1972) have estimated the current United States production as chlordane, 25 million pounds; aldrin, 10 million pounds; heptachlor, 6 million pounds; dieldrin, <1 million pounds; endosulfan, <1 million pounds; and endrin, <1 million pounds. World cumulative production must have exceeded 600 million pounds (2.7 x $10^8$ kg).

Several widely used organochlorine acaricides (miticides) should be included in this survey. Chlorofenthol or DMC is 1,1-bis-(*p*-chlorophenyl) ethanol, m.p. 69.5 to 70° (U.S. Patent 2,430,586). Dicofol or Kelthane® or 1,1-bis-(*p*-chlorophenyl)-2,2,2-trichloroethanol, m.p. 78.5 to 79.5° is the most widely used acaricide (U. S. Patent 2,812,280). Chlorobenzilate a closely related compound, is ethyl 4,4'-dichlorobenzilate, b.p. 141° at 0.06 mm (U. S. Patent 2,745,780), and chloropropylate is isopropyl 4,4'-dichlorobenzilate, m.p. 73 to 75° (U. S. Patent 2,745,780). Ovex is p-chlorophenyl p-chlorobenzene sulfonate, m.p. 86.5° (U. S. Patent 2,528,310), and its analogs are genite or 2,4-dichlorophenyl benzene sulfonate m.p. 45 to 47° (U. S. Patent 2,618,583), and fenson or p-chlorophenyl benzene sulfonate, m.p. 61 to 62°. Tetradifon is 2,4,5,4'-tetrachlorodiphenyl sulfone, m.p. 148° (U. S. Patent 2,812,281). Tetrasul is the corresponding 2,4,5,4'-tetrachlorodiphenyl sulfide, m.p. 88° (U. S. Patent 3,054,719). Sulphenone® is p-chlorophenyl phenyl sulfone, m.p. 98° (U. S. Patent 2,593, 001), and chlorobenside is p-chlorobenzyl p-chlorophenyl sulfide, m.p. 74° (British Patent 713,984). The properties of these compounds are described by Metcalf (1971).

## II. USES OF ORGANOCHLORINE INSECTICIDES

### A. Organochlorine Insecticides and Public Health

The discovery of the insecticidal properties of DDT by Paul Müller on September 25, 1939, introduced a new era in the control of the insect vectors of human diseases. The story has been told many times (Müller, 1955; Metcalf, 1972); we will only summarize it here. DDT's insecticidal properties were discovered just at the beginning of World War II, and it was quickly found to be nearly the ideal insecticide for the control of louse-borne epidemic typhus which had played such a significant role in World War I. It soon became evident that DDT was nearly as toxic to the human body louse *Pediculus humanus* as the *Chrysanthemum* derived pyrethrins, would persist for months on inert surfaces, and was so safe to man and animals that they could be dusted with it (Fig. 2) or live in houses where all the rooms were sprayed, without any ill effects. Moreover, DDT could be produced at a cost of less than 1/100 that of the pyrethrins, previously the most effective weapon against insect vectors.

1. DDT and Typhus. A 5% DDT louse powder was evaluated for typhus control in 1942, and The Rockefeller Foundation Typhus Team in Algiers in 1943 developed techniques for the mass delousing of humans (Fig. 2). The use of DDT was given the ultimate practical test in Naples in July 1943, when more than 3,265,786 individual applications were made to the city's inhabitants over a 7-month period, and an epidemic of typhus was

FIGURE 2. Use of DDT dusting powder for typhus control in Afghanistan (from World Health Organization).

terminated with only 1,403 cases (Soper *et al.*, 1947). DDT went on to play a notable role in controlling typhus among inmates liberated from German concentration camps and in preventing major epidemics throughout Europe. Bayne-Jones (1950) concluded "for the first time in the history of typhus in wartime, epidemics of the disease were brought under control before they had run their previously customary course." DDT became the principal public health weapon against typhus used by WHO until widespread development of resistant human lice reduced its effectiveness (Brown and Pal, 1971).

2. DDT in Malaria Control. It was discovered during World War II that DDT applied to the interior of dwellings at 1 to 2 g/m$^2$ would remain toxic to adult *Anopheles* mosquitoes for 6 months to a year (Metcalf *et al.*, 1945). The adult female *Anopheles* in search of a blood meal habitually invades human habitation about every 2 days. During feeding she approximately doubles her body weight, and her chances of absorbing a toxic dose of DDT are very great if all interior surfaces of houses are routinely sprayed. Thus the transmission of malaria can be interrupted at the weak link between formation of gametocytes in human blood and the development of infective sporozoites in the mosquito salivary glands.

DDT residual house spraying became one of the simplest of all public health measures and was applied to the eradication of malaria for the first time by the Italian entomologist Missiroli (1947) to houses in the Latina Province of Italy. Spraying of all dwellings was begun in the Pontine marshes area on June 5, 1945, using 5% DDT in kerosene, applied at about 2 g/m$^2$ of wall surface. After 2 years of spraying, the malaria vectors *Anopheles labranchiae* and *A. sacharovi* had nearly disappeared and the *Plasmodium* parasite index had declined from 10.32% in March 1946 to 0.49% in 1947 and to 0.24% in 1948. No more cases of malaria were reported after midsummer 1949.

The International Health Division of The Rockefeller Foundation began an attack on malaria in Sardinia under the direction of John Logan (1953). DDT treatment of all dwellings and other adult resting sites and larval breeding areas was begun on October 1, 1945 and the total cases of malaria were reduced from 75,447 before treatment to 39,303 the next year, and after 5 years to 9 cases. There was no evidence of the transmission of malaria after 1950. The disappearance of malaria from Sardinia has brought about dramatic economic changes, and the island has changed from a sheep- and goat-herding economy into a grain-growing agriculture. It has now become a prosperous winter playground of the Mediterranean area (Levi, 1968).

These spectacular successes demonstrated that malaria could be completely controlled by thorough residual house spraying at a cost of 11 to 44 cents per person per year. Programs were established in the malarious countries of the Americas and the Pan American Sanitary Bureau in 1950 recommended that a coordinated plan be developed to achieve continental eradication of malaria. The development of DDT resistance in several species of anophelines expedited planning, and in 1955 the World Health Assembly proposed that WHO should undertake the worldwide eradication of malaria (Soper *et al.*, 1961). An immense program was begun that has now eradicated malaria in 36 countries with a population of 710 million, is controlling malaria in another 27 countries, and is underway in yet another 53 countries. By 1971, 1329 million people of 1802 million living in originally malarious areas of 146 countries were protected (WHO, 1971). According to Bruce-Chwatt (1970) the malaria eradication program has decreased world malaria mortality from 2.5 million per annum to 1 million. The balance is fragile, however, and premature termination of DDT spraying in Ceylon in 1963, when cases of malaria had been reduced from several million to only 17, resulted in a colossal epidemic in 1968 with more than 2.5 million cases (Metcalf, 1970).

3. Other Organochlorine Insecticides in Public Health. The spectacular successes of DDT spurred research with other organochlorine insecticides. BHC was employed in malaria control in Central Africa against *Anopheles gambiae* using residual house spraying with dosages as low as 0.25 to 0.5 $g/m^2$. Its unpleasant odor and higher vapor pressure made it less residually effective than DDT, and it was soon replaced by the more residual dieldrin, which became widely used for malaria eradication at 0.2 $g/m^2$ of wall surface in Africa, Indonesia, India, and the Middle East. Unfortunately, both lindane (BHC) and dieldrin exhibit cross resistance, and in general the use of dieldrin for adult mosquito control has rapidly selected for resistant strains (Section III). Dieldrin, unlike DDT, has produced a high incidence of poisoning in spray workmen, with 47 to 100% of those with clinical poisoning having epileptiform convulsions up to 120 days after the last exposure, together with symptoms of psychic disturbances and loss of memory (Patel and Rao, 1958). Dieldrin as a residual spray was also highly effective in controlling Chagas' disease or American trypanomiasis vectored by Triatominae bugs. Prolonged exposure of workmen in this program resulted in 10 to 20% developing clinical symptoms of poisoning from 4 to 8 months after exposure. More than half had convulsions, backaches,

blurred vision, dizziness, slight involuntary muscular movements, sweating, difficulty in sleeping, nausea, and general malaise. In 51 clinical cases from 285 spraymen in Venezuela there were severe epileptiform convulsions with as many as 30 seizures in one case. These occurred as long as 84 to 105 days after the last exposure to dieldrin and tended to be chronic and diagnosable by abnormal encephalograms, which were also found in one-third of the spray workers not clinically ill. This condition was associated with alterations of reflexes, disorientation, personality changes, tachycardia, and arrhythmia. Thus dieldrin has been almost entirely withdrawn for vector control programs.

Chlordane was used initially with great success in housefly control programs in Italy in 1948 with a substantial decrease in gastroenteric disease, but subsequent chlordane resistance reversed this favorable trend (Brown, 1958). Initially, dieldrin was the most spectacular of the organochlorine insecticides in fly control, with marked reductions in enteric diseases in Georgia, but resistance rapidly supervened and public health gains were lost (Brown, 1958).

When DDT resistance in human body lice became an important phenomena in typhus control programs, the *Pediculus humanus* vectors were found to be highly susceptible to 0.5% lindane dusting powder, and this material has been used extensively for typhus control, although louse resistance is now common (Brown and Pal, 1971). Lindane has remained as the most effective control measure for the mange mite or itch mite *Sarcoptes scabiei*.

## B. Organochlorine Insecticides in Agriculture

The widespread use of the organochlorine insecticides has improved agricultural production of many crops, such as potatoes, apples, and sweet corn. As an example, after the first use of DDT in 1946 the yields of potatoes in New York and Wisconsin were increased from 56 to 68% over those previously obtained with the best production from lead arsenate and bordeaux mixture (Metcalf, 1965). DDT became the standard remedy for the control of such major pests as the codling moth, *Laspeyresia pomonella*, attacking deciduous fruits, and the pink bollworm, *Pectinophora gossypiella*, attacking cotton. At the height of its popularity, DDT was registered for use on 334 agricultural commodities, and was without rival as the predominant insecticide in worldwide use. However, the development of resistance by many important crop pests and increasing concern about environmental pollution have led to substantially decreased use and outright banning of DDT in many areas. Table 2 shows the approximate number of registered agricultural uses for the organochlorine insecticides in the United States at the peak of their use (1969), the range of residue tolerances, and major agricultural uses.

Methoxychlor is slowly developing its own areas of use but has had most application for livestock pests and home gardens.

Toxaphene was a close rival in popularity to DDT, giving good control of the cotton bollweevil, *Anthonomus grandis*, where DDT is mediocre. Toxaphene has been widely used in dips to control the cattle tick, *Boophilus annulatus*, lice, and other cattle pests. With the banning of the use of DDT, toxaphene has become the predominant organochlorine pesticide in the United States and is registered on approximately 168 agricultural commodities (USDA, 1969).

Benzene hexachloride was used initially as a soil insecticide and for the control of cotton pests. Its use in soil has been abandoned because of severe off-flavors produced in root crops. BHC has also been used extensively in the Middle East and Africa for the control of migratory locusts by aerial spraying of swarms with 10% oil solutions. Lindane, the highly purified *gamma*-isomer, is widely used worldwide as a seed treatment at 0.33 to 5.33 oz/100 lb of seed. It is also used for general insect control on fruits and vegetables.

Cyclodiene insecticides have had a great impact on worldwide agriculture. Chlordane became the first successful soil poison against ants and termites and is still preeminent. Aldrin, dieldrin, and heptachlor have been widely used for the control of grasshoppers and migratory locusts and became the most successful soil and seed treatments against wireworms, rootworms, white grubs, and the like. Dieldrin was for a time unique in the protection of sheep against the black blowfly *Phormia regina* until excessive residues in mutton and resistance ruled it out. Endosulfan has been widely used for the control of insect pests of fruits and vegetables. These uses are summarized in Table 2.

Mirex is widely used as a stomach poison for the fire ant *Solenopsis saevissima richteri* in oil or fat baits at 0.3%.

## C. Organochlorine Insecticides in Household and Home Garden

Nearly 20% of the United States pesticide use is in home operations (Lawless *et al.*, 1972), and 42 individual insecticide chemicals are sold including aldrin, BHC, chlordane, DDT,

TABLE 2. Uses of Organochlorine Insecticides in Agriculture (USDA, 1969)

| | Registered Uses | Residue Tolerances (ppm) | Major Uses |
|---|---|---|---|
| DDT | 240 | 1-7 | Cotton--pink bollworm, bollworms, forest defoliators, deciduous fruits, vegetables |
| DDD | 60 | 7 | Deciduous fruits, vegetables, tobacco |
| Methoxychlor | 87 | 2-100<br>0 (milk) | Fruits, vegetables, forage, cattle |
| Aldrin | 72 | 0.02-0.1 | Soil and seed treatment |
| Dieldrin | 71 | 0.02-0.1 | Soil and seed treatment, moth-proofing |
| Endrin | 25 | 0 | Postharvest, mouse control |
| Heptachlor | 28 | 0-0.1 | Soil and seed treatment |
| Chlordane | 81 | 0.3 | Fruits, vegetables, soil insects, ants, termites |
| Endosulfan | 53 | 0.2-2 | Fruits, vegetables |
| Toxaphene | 168 | 2-7 | Cotton, bollweevil, fruits, vegetables, cattle dips and sprays |
| Lindane | 123 | 4-10 | Fruits, vegetables, cattle dips and sprays, timberbark beetles, seed treatment |

DDD, dieldrin, heptachlor, dicofol (Kelthane®), chlordecone (Kepone®), methoxychlor, ovex (Tedion®), and toxaphene. These organochlorine insecticides formed the largest use pattern for at least 20 years from 1945 to 1965. DDT was used in the majority of garden dust and spray preparations and had a wide vogue as a "moth proofing" agent, used at 0.5% in volatile petroleum solvent. Lindane was also widely used in garden sprays and in vaporizers and shelf papers for fly and cockroach control. Chlordane was the material of choice for ant and roach control and has been used routinely for termite control as a 1% solution or emulsion for soil treatment under and around foundations. Dieldrin at 0.05% in volatile petroleum solvent has been very widely used in mothproofing. Aldrin and heptachlor have had more limited use in termite "proofing" and some use along with toxaphene as garden insecticides. Chlordecone is widely used as a bait for cockraoches and ants. Dicofol and ovex are included as acaricides in garden preparations.

Today, with increasing insect resistance and restrictions, methoxychlor has become the most widely used of the organochlorines in home and garden formulations.

## III. INSECT RESISTANCE TO ORGANOCHLORINE INSECTICIDES

The selection of genetic resistance to the organochlorine insecticides in important insect vectors and agricultural pests has been a major factor in the decline in the use of these compounds. This factor is the subject of Chapter III in this book and has recently been ably reviewed for insect vectors and pests of man and animals by Brown and Pal (1971) and for agricultural pests by Brown (1971). Therefore, it is considered but briefly here.

Insect resistance to insecticides came of age along with the use of DDT after World War II. Housefly control with DDT was initially so spectacular that Time magazine (1947) described a DDT offensive as "No Flies in Iowa" and the Reader's Digest (Arundel, 1948) had a piece "Entire Towns Abolish Flies."

However, these promises were never realized, because within 2 years after the widespread use of DDT for fly control, resistant races were reported from Sweden, Denmark, Italy, and the United States. Development of resistance in the rest of the world was not far behind. BHC, chlordane, and dieldrin were successively introduced as substitutes for fly control, but resistance developed even more rapidly, a single cycle of spraying with BHC or dieldrin often producing nearly complete immunity as in California and Georgia (Brown and Pal, 1971). By 1950 to 1951 housefly populations in California, Florida, Denmark, and Egypt were almost totally resistant to all the organochlorine insecticides. At latest documentation (Brown, 1971) a total of 98 species of economically important insects have developed immunity to DDT and 140 species to the cyclodienes. Major pests resistant to DDT include 14 *Anopheles* vectors of malaria; the human body louse, *Pediculus humanus*; the oriental rat flea, *Xenopsylla cheopis*; vector of plague, *Culex pipiens fatigans*; vector of filariasis, *Culex tarsalis*; vector of western equine encephalitis, *Aedes aegypti*; vector of yellow fever, the bedbugs *Cimex lectularius* and *C. hemipterus*; the pink bollworm, *Pectinophora gossypiella* and the bollworms, *Heliothis zea* and *H. virescens*, of cotton; the cabbage looper *Trichoplusia ni*; the imported cabbageworm, *Pieris rapae*; and the diamondback, *Plutella maculipennis*, the codling moth, *Laspeyresia pomonella*; the Colorado potato beetle, *Leptinotarsa decemlineata*; the beet armyworm, *Spodoptera exigua*, and the spruce budworm, *Choristoneura fumiferana*. This list includes many of the world's most destructive insects, and it is certain that DDT resistance has virtually eliminated the successful area chemical control of the housefly, jeopardized the global malaria eradication program (Bruce-Chwatt, 1971), seriously affected programs to control other vector-borne diseases, and hampered control programs for many important agricultural and forest pests.

Resistance to BHC and lindane and to the cyclodienes, which represent a single class of compounds as far as cross-resistance is concerned (Brown, 1971), has been nearly as devastating. It is widespread in the housefly, in *Anopheles gambiae* and 36 other species of *Anopheles* (BHC and dieldrin); the human body louse, *Pediculus humanus* (lindane); the bedbugs (BHC); and most species of pest mosquitoes (all organochlorines). BHC and cyclodiene resistance has developed in most important cotton pests, including the cotton boll weevil, *Anthonomus grandis* (toxaphene); cotton leafworms, *Alabama argillacea* and *Spodoptera littoralis* (toxaphene); cotton bollworm, *Heliothis virescens* (endrin); spiny bollworm, *Erias insulana* (endrin); cabbage looper, sugarcane borer, *Diatraea saccharalis* (endrin); rice stem borer, *Chilo suppressalis* (BHC); corn rootworms, *Diabrotica* spp. (aldrin, heptachlor); Colorado potato beetle (BHC); alfalfa weevil, *Hypera postica* (heptachlor); cotton stainer, *Dysdercus peruvianus* (BHC); cocoa capsid, *Distantiella theobroma* (BHC); root maggots, *Hylemya* spp. (dieldrin); grain weevils, *Sitophilus* spp. (BHC); and the german cockroach, *Blattella germanica* (chlordane).

## A. Cross Resistance

Insect pest control problems resulting from the development of resistance have been dealt with, over the past 25 years, by introducing a new insecticide to which the insects are susceptible. As an example, the floodwater mosquito *Aedes nigromaculis* in California has been controlled since 1945 with DDT, BHC, toxaphene, aldrin, EPN, parathion, methyl parathion, malathion, fenthion, chlorpyrifos, temephos (Abate®) (Brown and Pal, 1971), *m*-*sec*-butylphenyl *N*-methyl, *N*-thiophenyl carbamate, and most recently the growth hormone methoprene.

Resistance to the organochlorine insecticides is of two major types: (1) to DDT and to its analogs DDD and methoxychlor, and (2) to lindane and the cyclodienes. Either type can arise independently of the other, and each can be superimposed on the other to give multiresistance. These possibilities, together with the relatively permanent or stable nature of the resistant races in insect populations, (Brown, 1971) have substantially limited the remedial use of alternative organochlorine insecticides. Indeed, the problems of insecticide resistance are so serious that some form of pest management program that can break up the seemingly hopeless cycle of development of resistance to one insecticide after another by reliance on ecologically sound tactics (Chapter XVIII), seems absolutely essential to continued success in insect pest control.

## B. GENETICS OF INSECTICIDE RESISTANCE

These resistances are discussed in detail in Chapter III. In summary, DDT resistance in the housefly is controlled by at least two mechanisms: (1) a dormant gene on chromosome II which controls the production of an enzyme DDT'ase that dehydrochlorinates DDT to the noninsecticidal DDE, and (2) a gene on chromosome V which determines the microsomal oxidation of DDT to polar metabolites such as $\alpha$-hydroxy-DDT or dicofol (Fig. 3). Related mechanisms exist in other resistant species and behavioristic resistance and decreased target site sensitivity have also been demonstrated as resistance mechanisms.

Lindane resistance has been shown to result from increased levels of detoxication producing water-soluble metabolites (Bradbury and Standen, 1959), and cyclodiene resistance is presently very poorly understood (Brown and Pal, 1971).

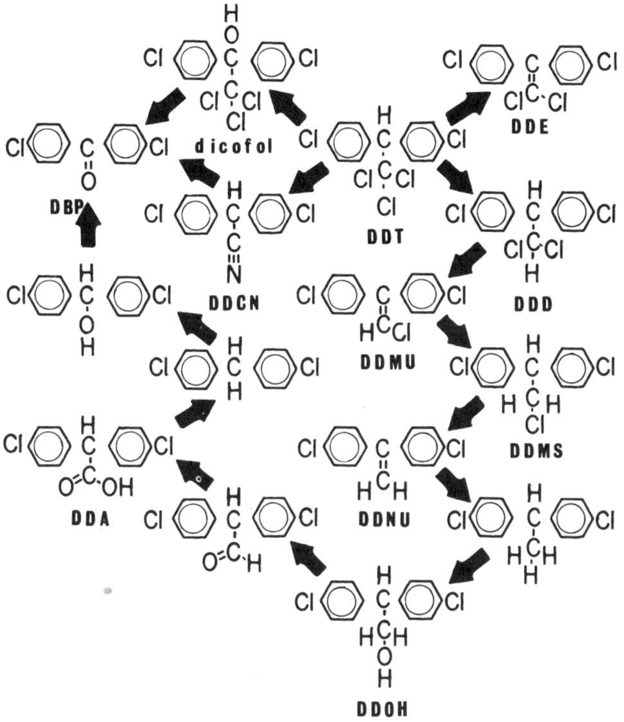


FIGURE 3. Degradative pathways of DDT in the environment (from Metcalf, J. Agricultural and Food Chemistry 1973, copyright American Chemical Society).

## C. DDT'ase

A major step in understanding insecticide resistance was the discovery of the enzyme DDT'ase in the DDT-resistant housefly (Sternburg *et al.*, 1954). This enzyme is a simple protein, molecular weight 36,000, isoelectric point 6.5, pH optimum 7.4, and temperature optimum 35 to 37°, which requires glutathione as a cofactor (Lipke and Kearns, 1959). DDT'ase catalyzes the dehydrochlorination of DDT and related compounds by an $E_2$-type elimination controlled by the availability of electrons at the α-carbon. This enzyme reaction is similar to the $OH^-$ catalyzed dehydrochlorination of DDT as shown by a deuterium isotope effect resulting from replacement of α-H with α-D to form α-deutero-DDT (II). In $OH^-$ catalyzed dehydrochlorination, α-D-DDT ($K \times 10^5 = 8.9 \times 10^2$ l/min/mol) reacted only 0.14 as rapidly as DDT ($k \times 10^5 = 6.1 \times 10^3$) for a deuterium isotope effect of 6.8 (Metcalf and Fukuto, 1968). *In vivo* in the housefly, the deuterium isotope effect as determined by the relative $LD_{50}$ values for DDT and α-D-DDT was 1.25 to 1.5 (Barker, 1960; Moorefield *et al.*, 1962) and was 1.7 for 2,2-bis-(*p*-bromophenyl)-1,1,1-trichloroethane and 3.0 for DDD (Metcalf and Fukuto, 1968).

## D. Resistance-Proof DDT Derivatives

*Deutero-DDT* has been suggested as a practical insecticide for *Aedes aegypti* and other mosquito larvae resistant to DDT (Pillai *et al.*, 1964), although it was ineffective against DDT-resistant houseflies (Table 3).

However, α-deutero-methoxychlor (IV) is substantially more toxic than methoxychlor to both DDT-resistant flies and to mosquitoes (Table 3). This shows that the decreased reactivity of the C-D bond acting in conjunction with the higher electron density of the α-C produced by the inductive and mesomeric effects of the *p,p*-$CH_3O$ groups (Fig. 4) has reduced DDT'ase attack to a level permitting a substantial portion of the compound to reach the site of action (Metcalf and Fukuto, 1968).

*α-Fluoro Analogs*. The substitution of the α-C-F bond with an energy of 116 kcal per mole for the weaker α-C-H bond, energy 98 kcal per mole, inhibits the attack of DDT'ase and has cleverly been used as a resistance-proofing mechanism. Although α-F-DDT or 2,2-bis-(*p*-chlorophenyl)-2-fluoro-1,1,1-trichloroethane (fly $LD_{50}$ 125 μg/g) is of low toxicity because of adverse steric factors at the site of action, Hennessey and Villarica (1969) found that α-F-DDD (VII) or 2,2-bis-(*p*-chlorophenyl)-2-fluoro-1,1-dichloroethane was about 5 times more toxic than DDD to S flies and had a resistance ratio (RR) of 4 between $R_{SP}$ and $S_{NAIDM}$ flies as compared with a value of >25 for DDD (Table 3). As an example of the amazing effect of the α-F atom in decreasing DDT'ase activity, the $-CHBr_2$ analog of DDT is such an excellent substrate that it has an $LD_{50}$ of >500 to the $S_{NAIDM}$ housefly. However, its α-F analog 2,2-bis-(*p*-chlorophenyl)-2-fluoro-1,1-dichloroethane (VII) had an $LD_{50}$ of 7 μg/g to the $S_{NAIDM}$ housefly and a RR of 4.4 (Metcalf and Fukuto, 1968).

TABLE 3. Effects of Variations in DDT Structure on Susceptible and Resistant Insects (Metcalf and Fukuto, 1968)

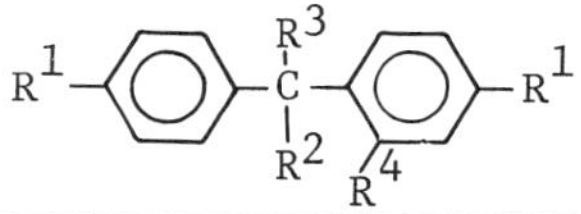


| | $R^1$ | $R^2$ | $R^3$ | $R^4$ | Topical $LD_{50}$ μg/g *Musca domestica* Susceptible | Resistant | $LC_{50}$ ppm *Culex pipiens fatigans* |
|---|---|---|---|---|---|---|---|
| I | Cl | $CCl_3$ | H | H | 2.0 | >500 | 0.07 |
| II | Cl | $CCl_3$ | D | H | 2.3 | >500 | 0.016 |
| III | $CH_3O$ | $CCl_3$ | H | H | 9.0 | >500 | 0.067 |
| IV | $CH_3O$ | $CCl_3$ | D | H | 8.0 | 34 | 0.033 |
| V | Cl | $HCCl_2$ | H | H | 19.5 | >500 | 0.038 |
| VI | Cl | $HCCl_2$ | D | H | 6.5 | >500 | 0.02 |
| VII | Cl | $HCCl_2$ | F | H | 4.1 | 16.5 | 0.024 |
| VIII | Cl | $CCl_3$ | H | F | 4.6 | >500 | 0.041 |
| IX | Cl | $CCl_3$ | H | Cl | 11.0 | 36.0 | 0.054 |
| X | Cl | $CCl_3$ | D | F | 4.30 | 33.0 | 0.020 |
| XI | Cl | $CCl_3$ | D | Cl | 9.5 | 22.0 | 0.023 |
| XII | Cl | $HC(CH_3)NO_2$ | H | H | 8.5 | 45.0 | 0.064 |
| XIII | $CH_3O$ | $C(CH_3)_3$ | H | H | 26.0 | 145 | 0.43 |

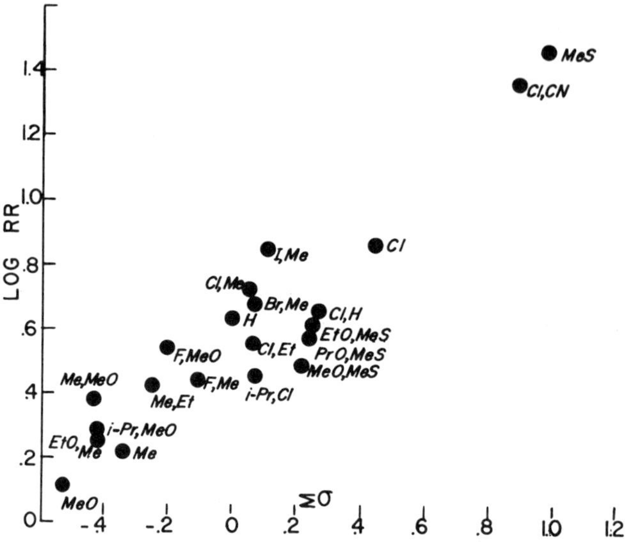


FIGURE 4. Relationship of summed sigma values for para substituents of DDT analogs to resistance ratio (RR or $LD_{50}$ R/S houseflies). MeS substituents are plotted as MeSO primary oxidation product. (From Metcalf, J. Agricultural and Food Chemistry 1973, copyright American Chemical Society.)

o-Chloro-DDT. As shown in Table 3, the incorporation of an o-chloro atom into the DDT molecule to form 2-(4-chlorophenyl)-2-(2',4'-dichlorophenyl)-1,1,1-trichloroethane (IX) produces a compound highly effective against R-DDT flies (Hennessey et al., (1961). The bulky ortho-substituent interferes sterically with dehydrochlorination as it undergoes $OH^-$ catalyzed dehydrochlorination ($k \times 10^5 = 1.07 \times 10^3$ 1/mol/min) only 0.17 as rapidly as p,p'-DDT($k \times 10^5 = 6.1 \times 10^3$) (Metcalf and Fukuto, 1968). The ortho-chloro-DDT had a RR value of 3.3 (Table 3) and simultaneous incorporation of α-D and o-Cl as in 2-(4-chlorophenyl)-2-(2',4'-dichlorophenyl)-2-deutero-1,1,1-trichloroethane (XI), improved the toxicity further with a RR value of 2.3 (Table 3). Decreasing the size of the ortho substituent as in o-fluoro-DDT (VIII) reduced the activity against the $R_{SP}$ housefly (Table 3). The impressive action of o-chloro-DDT against R flies must be due to a decrease in the affinity of the molecule for DDT'ase as well as to increased stability of the α-C-H bond, as dehydrochlorination in o-chloro-DDT proceeds at about 0.17 the rate of DDT. In α-deutero-DDT which was not toxic to the R flies, the rate was 0.14 that of DDT (Metcalf and Fukuto, 1968).

Altered Para-Substituents. The dehydrochlorination mechanism of DDT and its analogs is an $E_2$-type elimination, controlled by the availability of electrons at the α-carbon (Cristol, 1945). Thus the rate of $OH^-$ catalyzed dehydrochlorination of various p,p'-disubstituted DDT analogs depends on the polar nature of the substituents and the overall difference in rates between 2,2-bis-(p-nitrophenyl)-1,1,1-trichloroethane ($k \times 10^5 = 6.9 \times 10^6$ 1/mol/min) with electron withdrawing groups, and 2,2-bis-(p-methoxyphenyl)-1,1,1-trichloroethane ($k \times 10^5 = 29.5$) with electron-donating groups, is about $2 \times 10^5$ fold. This effect is also shown in the action of DDT'ase in the housefly as shown in Fig. 4 where the RR values ($LD_{50}$ R flies/$LD_{50}$ S flies) are plotted against the summed Hammett sigma valies for a variety of diaryltrichloroethanes. The sigma values represent a quantitative measure of electron-withdrawing (positive values) and electron-donating (negative values). This explains the inactivity of the p,p'-$NO_2$ analog of DDT, which is too unstable to reach the site of action and the superior activity of the p,p'-$CH_3O$ analog (methoxychlor) against R flies. Methoxychlor (III) has performed effectively against field strains of DDT-resistant houseflies which have moderate levels of DDT'ase activity (March and Metcalf, 1949), and even when the DDT'ase level is very high, α-deutero-methoxychlor (IV) had an RR value of 4.2 to $R_{SP}$ flies compared to >200 for α-deutero-DDT (II) (Table 3).

## IV. ORGANOCHLORINE INSECTICIDES AS ENVIRONMENTAL POLLUTANTS

It is ironic that the same properties that made the organochlorine insecticides so successful for insect control (their environmental stability and persistence, their water insolubility and lipid solubility, and their low cost) have resulted in their present position as the most discussed and ubiquitous of the environmental micropollutants. Some of these properties are illustrated in Table 4. World comprehension of environmental quality is totally different today from that of 30 years ago. In the 1940s it was generally assumed that dilution in the environment was a satisfactory answer to pesticide pollution. Today there is abundant evidence that the organochlorine compounds can persist in soil and water for periods of years to

TABLE 4. Some Environmental Properties of Organochlorine Insecticides

| | Solubility (ppm 20-25°) | | Vapor Pressure 20-25° |
|---|---|---|---|
| | $H_2O$[a] | Benzene | |
| DDT | 0.0012 | $7.8 \times 10^5$ | $1.9 \times 10^{-7}$ |
| DDD | 0.002 (ca) | | |
| DDE | 0.0013 | | |
| Aldrin | 0.027-0.20 | $8.3 \times 10^5$ | $2.3 \times 10^{-5}$<br>$4.0 \times 10^{-5}$ |
| Dieldrin | 0.05-0.25 | $5.6 \times 10^5$ | $1.8 \times 10^{-7}$ |
| Endrin | 0.06-0.23 | $1.38 \times 10^5$ | $2.0 \times 10^{-7}$ |
| Heptachlor | 0.056 | $1.06 \times 10^5$ | $3.0 \times 10^{-4}$ |
| Heptachlor epoxide | 0.35 | | |
| Chlordane | 0.009 | | $1.0 \times 10^{-5}$ |
| Endosulfan | | $4.5 \times 10^5$ | |
| Toxaphene | 0.4-3 | $>4.5 \times 10^5$ | $1.0 \times 10^{-6}$<br>$3.3 \times 10^{-5}$ |
| Lindane | 7.3 | $2.9 \times 10^5$ | $1.6 \times 10^{-4}$<br>$9.4 \times 10^{-4}$ |

(continued)

TABLE 4 (cont.)

| | Average U.S. levels (ppm) | | | | |
|---|---|---|---|---|---|
| | Cropland soil[b] | Water[c] ppb | Human fat[d] | Human milk[e] | Food[f], daily intake mg |
| DDT | 0.17 | 0.008–0.144 | 2.3 | 0.022 | 0.037 |
| DDD | | 0.004–0.080 | | 0.0049 | 0.016 |
| DDE | 0.06 | 0.002–0.011 | 8.0 | 0.041 | 0.024 |
| Aldrin | 0.02 | <0.001–0.006 | | | 0.002 |
| Dieldrin | 0.03 | 0.008–0.122 | 0.29 | 0.0073 | 0.006 |
| Endrin | <0.01 | 0.008–0.214 | | | |
| Heptachlor | <0.01 | 0–0.002 | <0.03 | | |
| Heptachlor epoxide | <0.01 | <0.001–0.008 | 0.24 | 0.0027 | 0.003 |
| Chlordane | 0.04 | | | | |
| Endosulfan | <0.01 | | | | |
| Toxaphene | 0.07 | | | | |
| Lindane | <0.01 | 0.003–0.022 (BHC) | 0.60 (BHC) | 0.0024 (BHC) | 0.004 |

[a]Metcalf *et al.*, 1973.

[b]Wiersma *et al.*, 1972.

[c]Breidenback *et al.*, 1967.

[d]Durham, 1969.

[e]Curley and Kimbrough, 1969.

[f]Duggan, 1969.

decades (Table 5), and that they can be biomagnified in the tissues of invertebrates, fish, birds, and mammals from ppt levels in the aquatic environment by as much as $10^5$ to $10^7$-fold to ppm levels in living animals. As an example, the levels of

TABLE 5. Soil Persistence of Organochlorine Insecticides

| Insecticide | Years for | |
|---|---|---|
| | 50% loss[a] | 95% loss[b] |
| DDT | 3-10 | 4-30 (10) |
| Aldrin | 1-4 | 1-6 (3) |
| Chlordane | 2-4 | 3-5 (4) |
| Dieldrin | 1-7 | 5-25 (8) |
| Endrin | 4-8 | --- |
| Heptachlor | 7-12 | 3-5 (3.5) |
| Lindane | 2 | 3-10 (6.5) |
| Toxaphene | 10 | |

[a]Menzie, 1972.

[b]Edwards, 1964.

DDT and dieldrin in lake trout, _Salvelinus namaycush_, of Lake Michigan (Table 6) show not only the accumulative properties from the average concentrations in the lake of DDT 6 ppt and dieldrin 2 ppt, but also the progressive increase in level of accumulation with age (EPA, 1972). It appears that these poikilothermic animals, especially in cold water, have little if any protection against continuous uptake and storage of these difficultly degradable xenobioties. Fish and other animals so contaminated have been widely judged by regulatory agencies to be "unfit for human consumption."

TABLE 6. Bioaccumulation of DDT(T) and Dieldrin in Lake Trout, *Salvelinus namaycush*, Lake Michigan (EPA, 1972)

| Trout Size (in.) | Body Residue (ppm) | |
|---|---|---|
| | DDT(T) | Dieldrin |
| 2.0 - 5.9 | 0.89 | 0.03 |
| 6.0 - 9.9 | 2.24 | 0.12 |
| 10.0 - 15.9 | 6.00 | 0.14 |
| 16.0 - 21.9 | 8.00 | 0.21 |
| 22.0 - 26.9 | 14.62 | 0.23 |
| 27.0 - 32.9 | 19.23 | 0.26 |

## A. Food Web Transport

The movement of the highly stable lipid partitioning organochlorine insecticides and their degradation products, for example, DDE from DDT, dieldrin from aldrin, and heptachlor epoxide from heptachlor, through food webs, represents a most disturbing feature of environmental pollution by these compounds. The phenomenon is well known, and only a few examples need be given. Hunt and Bishoff (1960) first described the ecological magnification of DDD resulting from the treatment of Clear Lake, California, with this chemical at 0.014 to 0.02 ppm to control the Clear Lake gnat, *Chaoborus astictopus*. After 13 months, DDD residues of 10 ppm were found in plankton, 903 ppm in fat of phytophagous fish, 2690 ppm in fat of carnivorous fish, and 2133 ppm in fat of a fish-eating bird, the Western grebe, for a $10^5$ ecological magnification. Harrison *et al.* (1970) describe the ecological magnification of DDT in Lake Michigan, from 0.000002 ppm in water, to 0.410 in amphipods, to 3-6 ppm in fish, and to 99 ppm in herring gulls. Woodwell *et al.* (1967) describe an estuary of Long Island, New York, where a DDT concentration of 0.00005 ppm became successively magnified to 0.04 ppm in plankton, 0.16 to 0.42 ppm in invertebrates, 0.17 to 2.07 ppm in fish, and 3.15 to 75.5 ppm in predatory birds. The DDT levels increased about tenfold at each trophic level and were magnified about 1.2 x $10^5$. Hunt (1966) records the results of an application of toxaphene to Big Bear Lake, California, at 0.2 ppm to control rough fish (goldfish). Toxaphene accumulated to 73 ppm in plankton, 200 ppm in goldfish, and 1700 ppm in the fat of a fish-eating pelican. In a terrestrial ecosystem of Missouri, where cornfields were treated with aldrin at an accumulative total of 25 lb/A over 15 years, Korschgen (1970) found 0.06 ppm aldrin-dieldrin in the soil, 1.49 ppm in earthworms, 9.67 ppm in the predaceous beetle *Poecilus*, 4.60 ppm in the toad *Bufo americanus*, and 10.3 to 14.4 ppm in the garter snake *Thamnophis sirtalis*, an overall magnification of about 200 fold.

An interesting comparison of the ecological magnification of several organochlorine insecticides in Lake Poinsett, South Dakota, is shown in Table 7 (Hannon *et al*., 1970). Hamelink *et al*. (1971) suggest that the bioaccumulation of lipid-partitioning organochlorine compounds results from a series of partitionings and is roughly proportional to their solubility in fat/solubility in water (Table 4). They estimated partition coefficients of DDT of about $10^6$, dieldrin 5 x $10^4$, heptachlor epoxide 3 x $10^4$, and lindane $10^3$ which are remarkably close to the actual levels of concentration observed in nature. Verification of this idea has been found from laboratory model ecosystem studies where there was an excellent correlation ($r = >0.95$) between levels of ecological magnification in the fish *Gambusia affinis* and water solubility, and lipid/water partition coefficient (Metcalf *et al*., 1974).

## B. Environmental Chemistry

DDT is commonly thought of as a nondegradable or "hard" insecticide. This is not strictly true, as seen from the degradative pathways shown in Fig. 3. It is remarkable that although DDT has had very extensive usage over the past 35 years, knowledge of its environmental chemistry has accumulated slowly, and major discoveries are still being made. The microsomal oxidation of DDT at the α-carbon to dicofol or 1,1-bis-(*p*-chlorophenyl)-2,2,2-trichloroethanol was first described in *Drosophila melanogaster* by Tsukamoto (1959). Reductive dechlorination to form DDD or 2,2-bis-(*p*-chlorophenyl)-1,1-dichloroethane was first elucidated in yeast (Kallman and Andrews, 1963) and later in the rat (Datta *et al*., 1964; Peterson and Robinson, 1964). Only in 1972 was a major new anaerobic degradation product DDCN or bis-(*p*-chlorophenyl) acetonitrile identified in sewage sludge (Albone *et al*., 1972; Jensen, *et al*., 1972).

TABLE 7. Organochlorine Insecticides in Lake Poinsett Ecosystem (Hannon _et al_., 1970) (concentration in ppm)

| | Water | Crayfish | Plankton | Insects | Carp | |
|---|---|---|---|---|---|---|
| DDT-T | 0.00008 | 0.002 | 0.005 | 0.919 | 0.151[a] | 1.417[b] |
| Aldrin-Dieldrin | 0.00002 | 0.001 | 0.0007 | 0.164 | 0.030 | 0.268 |
| Heptachlor & epoxide | 0.00006 | 0.001 | 0.0011 | 0.312 | 0.026 | 0.174 |
| Lindane | 0.00003 | 0.001 | 0.0002 | 0.001 | 0.004 | 0.042 |
| Toxaphene | 0.001 | 0.02 | 0.005 | 0.02 | 0.176 | 1.152 |

[a]Tissue.

[b]Fat.

The major environmental problem with DDT results from its dehydrochlorination to form DDE or 2,2-bis-(*p*-chlorophenyl)-1,1-dichloroethylene which has essentially the same lipophilic tendencies as DDT but is much more stable in living systems. Thus DDE becomes the major degradation product stored in animal tissues and increases in percentage with the trophic level of the organism (Woodwell *et al*., 1967).

DDT entering the bodies of animals is stored in lipids and slowly eliminated by successive dechlorinations and dehydrochlorinations, as shown in Fig. 3, to form DDA or bis-(*p*-chlorophenyl) acetic acid which is substantially more water partitioning and can be excreted in urine. The half-life for this conversion and excretion in humans is about 0.5 years (Durham, 1969).

*Aldrin-Dieldrin*. Knowledge of the environmental fate of aldrin and dieldrin has only become relatively well understood by 1973, or 25 years after the introduction of these compounds as insecticides in 1948. The dominant reaction is the epoxidation of the double bond of aldrin to form the 6,7-epoxide, dieldrin, was first demonstrated in the rat (Bann *et al*., 1956) and American cockroach, *Periplaneta americana* (Giannotti *et al*., 1956). Subsequently, it was shown that this is a universal reaction (Fig. 5) catalyzed by microsomal oxidases microbiologically in the soil (Gannon and Bigger, 1958), photochemically on the surfaces of plants (Gannon and Decker, 1958), and in the vertebrate liver. Because of the ease with which this reaction occurs, the highly stable dieldrin (v.p. $1.8 \times 10^{-7}$ mm Hg at 20°) appears everywhere in the environment as the major product following the use of aldrin (v.p. $2.3 \times 10^{-5}$ mm Hg at 20°) and has become a ubiquitous micropollutant of soil, water, food, and human tissues (Table 4). Further photochemical reaction of dieldrin produces photodieldrin (Fig. 5), or 10-oxa-3,6-*exo*-4,5,13,13-hexachloro-$(6.3.1.1^{3,6}.1^{9,11}.0^{2,7}.0^{5,12})$-tridecane, a cagelike compound (Crosby, 1972). Photodieldrin is also formed biologically from dieldrin by microorganisms and in vertebrates (Matsumura *et al*., 1970). Both dieldrin and photodieldrin are substantially more toxic to animals than aldrin and much more environmentally persistent, so that these environmental chemical transformations are highly deleterious to environmental quality.

The fate of $^{14}C$ aldrin in soil and crops has been studied by Klein *et al*. (1973), using 2.9 to 3.2 kg/ha applied to soil in which potatoes were grown. At harvest, more than 60% of the total dosage was recovered from soil and the layer 0 to 10 cm deep contained aldrin 0.58 to 0.59 ppm, dieldrin 0.40 to 0.62 ppm, and more hydrophilic metabolites, principally dihydro-

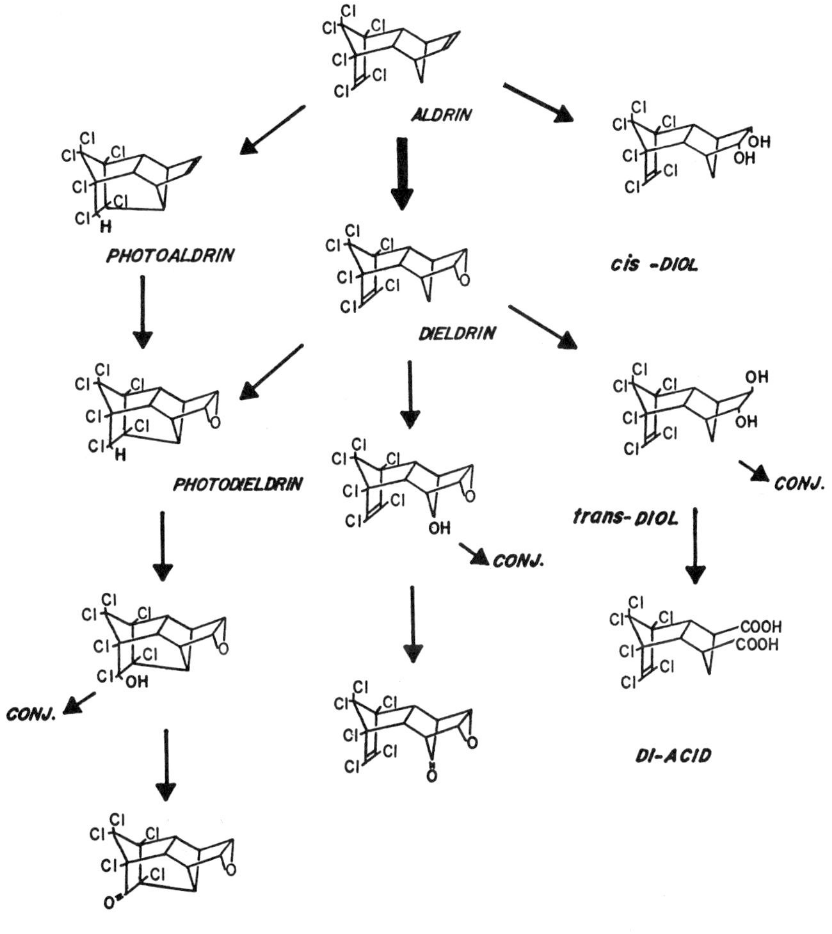


FIGURE 5. Degradative pathways of aldrin in the environment.

chlordene dicarboxylic acid (Fig. 5) (1,2,3,4,8,8-hexachloro-1,4,4a,6,7,7a-hexahydro-1,4-*endo*-methyleneindene-5,7-dicarboxylic acid) at 0.11 to 0.74 ppm. Photodieldrin (Fig. 5) comprised 1% of the total recovered radioactivity. Potato tubers grown in the treated soil contained 0.20 to 0.58 ppm aldrin and 1.68 to 2.26 ppm dieldrin together with 0.38 to 0.76 more hydrophilic metabolites.

The degradative pathways for aldrin-dieldrin in animals have been partially elucidated only after years of study. Biological conversion to more hydrophilic compounds is relatively slow in most organisms, and substantial amounts of intact dieldrin are excreted. For example, the rat excretes about 10 times as much $^{14}C$ from dieldrin in feces as in urine. In the female this is largely intact dieldrin. The principal fecal metabolite in the male is keto-photodieldrin (Fig. 5) or Klein's metabolite (which has been reported as more toxic than dieldrin). Aldrin *trans*-diol (Fig. 5) was the most abundant polar metabolite in the feces, together with a hydroxy compound, probably 5-hydroxydieldrin. Urinary excretion in male rats was largely 9-ketodieldrin together with traces of aldrin *trans*-diol (Fig. 5) and lesser amounts of dieldrin (Matthews *et al.*, 1971). Dieldrin metabolism in sheep was similar (Feil *et al.*, 1970). The hydroxylated metabolites are generally present as conjugates with glucuronic acid and the like. Plant tissues can hydroxylate aldrin directly to aldrin *cis*-diol and produce a monohydroxy and monoketo derivative as well (McKinney and Mehendale, 1973).

Heptachlor behaves rather like aldrin and is readily epoxidized to heptachlor epoxide or 1,4,5,6,7,8,8-heptachloro-3a,4,7,7a-tetrahydro-2,3-epoxy-4,7-methanoindene, m.p. 159 to 160°. This microsomal reaction occurs in animal tissues (Davidow and Radomski, 1953), in and on plant surfaces where it may form from atmospheric oxygen (Gannon and Decker, 1958), and in the soil (Gannon and Bigger, 1958). Heptachlor epoxide, which is considerably more acutely toxic (rat oral $LD_{50}$ ♂ 47, ♀ 61 mg/kg) than heptachlor (rat oral $LD_{50}$ ♂ 100, ♀ 162 mg/kg) is found nearly everywhere as a persistent pollutant in the living and nonliving environment (Table 4). Heptachlor differs from aldrin, however, in ease of hydrolysis to form 1-hydroxychlordene or 1-hydroxy-4,5,6,7,8,8-hexachloro-3a,4,7,7a-tetrahydro-4,7-methanoindene m.p. 194 to 196°. This compound (rat oral $LD_{50}$ >2400 mg/kg) is a true detoxication product and is readily converted to 1-hydroxy-2,3-epoxychlordene or 1-hydroxy-4,5,6,7,8,8-hexachloro-2,3-epoxy-1,2,3a,4,7,7a-hexahydro-4,7-methanoindene, m.p. 115 to 130° subl., which is an excretory metabolite in rat and rabbit (Menzie, 1969) and is formed in soils by microbial action (Miles *et al.*, 1969).

*Chlordane*. Both the *cis*- and *trans*-chlordanes also form a single epoxide, oxychlordane or 1-*exo*, 2-*endo*-4,5,6,7,8,8-octachloro-2,3-*exo*-epoxy-2,3,3a,4,7,7a-hexahydro-4,7-methanoindene, m.p. 99 to 101°, *in vivo* in mammals (Schwemmer *et al.*, 1970). Oxychlordene accumulates in animal tissues as 50 to 90% of the residue after ingestion or absorption of chlordane. It has been found in cow's milk, and appears to be the major terminal residue of chlordane in animal tissue (Street and Blan, 1972; Barnett and Dorough, 1974). Oxychlordane is more toxic than the chlordane isomers *cis*-(rat oral $LD_{50}$ 392 mg/kg) and *trans*-(rat oral $LD_{50}$ 327 mg/kg). There is also evidence of the formation of hydrophilic degradation products such as chlordene-dihydrodiol and 1-hydroxy-2-chlorodihydrochlordene (Korte, 1967).

*Endosulfan*. Both α and β isomers are oxidized in biological systems to form endosulfan sulfate, or 6,7,8,9,10,10-hexachloro-1,5,5a,6,9,9a-hexahydro-6,9-methano-2,4,3-benzodioxathiepin-3,3-dioxide, m.p. 180°. Endosulfan sulfate is the terminal oxidative product from endosulfan and is formed in plant surfaces and in the tissues of animals. It is excreted in milk after the ingestion of endosulfan (Maier-Bode, 1968; Gorbach *et al.*, 1968). Degradation products of endosulfan excreted in mammalian urine include endosulfan alcohol or 1,4,5,6,7,7-hexachloro-2,3-bis-(hydroxymethyl)-bicyclo-2.2.1-heptene-5 and 1-α-hydroxy endosulfan ether or 1-hydroxy-4,5,6,8,8-hexachloro-1,3,3a,4,7,7a-hexahydro-4,7-methanoisobenzofuran (Gorbach *et al.*, 1968).

*Endrin*. This cyclodiene exists in the *endo-endo* configuration which is inherently less stable than the *endo-exo* configuration of its stereoisomer dieldrin. In the presence of light, endrin isomerizes to form Δ-keto endrin or 1,8-*exo*-9,10,11,11-hexachloropentacyclo-$(6.2.1.1.^{3,6},0^{2,7},0^{4,10})$-dodecan-5-one, a caged structure analogous to keto photodieldrin (Fig. 5). The Δ-ketoendrin is also formed from endrin by microorganisms and plants. In the rat, endrin is considerably less persistent than dieldrin and is degraded largely to 9-ketoendrin (on methylene bridge), which is found in tissues and urine, and to 9-hydroxyendrin, which is excreted largely in feces (Baldwin *et al.*, 1970). Lesser amounts of another mono-hydroxyendrin, presumed to be 5-hydroxyendrin, are also found in the rat. Presumably these hydroxy-endrins also form conjugates.

Mirex. This perchlorocompound is almost impervious to degradation, and when fed to the rat is stored and slowly excreted virtually unchanged (Gibson *et al.*, 1972).

Lindane. In animals, lindane is degraded initially by dehydrochlorination to form several stereoisomers of pentachlorocyclohexene. This can be hydroxylated by microsomal enzymes forming the excretory metabolite 2,3,4,5,6-pentachloro-2-cyclohexen-1-ol, but the pentachlorocyclohexenes are also subsequently degraded further by dehydrochlorination to form a series of polychlorobenzenes, of which the major products are 1,2,4-, 1,3,5-, 1,2,3,4-, and 1,2,3,5-. The chlorobenzenes are then hydroxylated to form polychlorphenols, of which the major products are 2,3,5-, 2,4,5-, 2,4,6-, 2,3,4,5-, and 2,3,4,6-, which are excreted as urinary conjugates (Freal and Chadwick, 1973; Kurihara and Nakajima, 1974). Conjugation of γ-pentachlorocyclohexene with glutathione also occurs in the rat and housefly, and after dehydrochlorination, 2,4-dichlorophenyl mercapturic acid has been found as a metabolite in rat urine (Grover and Sims, 1965). This compound may hydrolyze to 2,4-dichlorobenzenethiol. Degradation of lindane in plants apparently follows similar pathways, that is, through pentachlorocyclohexene (Casida and Lykken, 1969).

Toxaphene. Nothing definitive has yet been published on the degradation pathways of the mixture of compounds found in toxaphene.

## C. Microsomal Induction

An inherent problem with the widespread environmental contamination of the persistent organochlorine insecticides is the effect of their bioconcentration in organisms on the induction of liver microsomal enzyme activity (Street, 1969). The organochlorine insecticides, along with other lipid soluble compounds, enhance the development of extra protein in the endoplasmic reticulum of the liver, together with increased cytochrome P-450 and drug-metabolizing enzymes. The more refractory the organochlorine insecticide is to degradation, the more effective its inductive properties. Induction has been demonstrated in mammals following the feeding or injection of ppm quantities of DDT, DDD, DDE, chlordane, heptachlor, heptachlor epoxide, aldrin, dieldrin, endrin, α-, β-, γ-benzene hexachloride, toxaphene, mirex, and chlordecone (Street, 1969; Conney and Burns, 1972). The more biodegradable compounds such as methoxychlor and Perthane® have borderline effects.

The biochemical consequences of exposure to organochlorine micropollutants have been reviewed by Conney and Burns (1972). Human beings exposed to DDT and lindane in pesticide factories have been found to metabolize drugs such as antipyrine and steroids about twice as fast as a control population, because of increased levels of microsomal oxidases. It is not known whether DDT storage in the general population has any marked effect on the metabolism of drugs or endogenous steroids, and indeed controls without DDT would be impossible to obtain. However, levels of the organochlorine insecticides stored in the body tissues of carnivorous mammals and birds at the upper ends of food webs are high enough (Section IV.A.) to have appreciable effects. DDT fed to animals markedly accelerated the metabolism of dieldrin and reduces its storage in adipose tissues. Organochlorine insecticides are potent stimulators of estrogen metabolism, and their presence in the diet decreases estrogen effects on the rat uterus. They also alter pathways of steroid metabolism in man.

Birds seem particularly vulnerable to these organochlorine inducers, and DDT fed at 10 ppm enhances hepatic metabolism of 17-β-estradiol and other steroids (Peakall, 1970) which control the deposition of medullary bone, the chief source of calcium during eggshell formation. Thus the decline in population of the peregrine falcon and other raptorial birds appears to be associated with thin eggshells, and this syndrome has a highly significant correlation with the ppm levels of DDE in the egg (Cade *et al.*, 1971).

There is no doubt that the stimulation of microsomal metabolism by trace levels of organochlorine insecticides can affect drug metabolism in man and higher animals and can also play a role in the development of insecticide resistance in invertebrates. This microsomal induction can affect adversely the endogenous metabolism of steroids, which can have effects on the fertility of animals. These complicated biochemical effects of micropollution from persistent organochlorine insecticides remain as one of the most compelling reasons for replacing these materials with biodegradable substitutes.

## D. Carcinogenicity

The area of greatest concern about the persistent organochlorine insecticides in the environment is that of induction of malignancy. The Mrak Commission (1969) after intensive study has judged aldrin, dieldrin, DDT, mirex, Strobane®, heptachlor, and chlorbenzilate as positive for tumor induction in one or more species, and has recommended that exposure of human beings to these substances be minimized. As we have seen, residues of these chemicals are so persistent in water and soil (Tables 4 and 5) and so easily carried through food chains that human tissue cells everywhere are exposed. Relatively recent studies with DDT have shown that the BALB/C strain of mice develops malignant tumors when fed with DDT at levels comparable to those found in foods and in human tissues. Tarjan and Kemany (1969) showed a generalized increase in frequency of tumors in several generations of mice after feeding DDT at 3 ppm, and Faur and Kemen (1968) found increased numbers of malignancies when DDT was fed to mice at 0.3 to 0.6 mg/kg. WHO has repeated these studies and found that DDT fed to mice at 0.3 mg/kg body weight per day for a lifetime produced a significant increase in liver tumors in males. It was concluded "the Committee did not regard this new evidence in mice as providing them with an adequate basis for recommending the withdrawal of DDT where its continued use for disease control and for protecting food and crops could be life saving" (WHO, 1973).

Dieldrin has also been shown in recent studies to be highly carcinogenic to mice. Walker et al. (1972) found that dieldrin fed at levels as low as 2.5 ppm in the diet shortened the lifetime of CAR/1 mice and was an inducer of liver hepatomas in male mice at the lowest dosage fed, 0.1 ppm. The incidence of liver hepatomas in male mice rose from 7% in controls to 28% at 0.1 ppm dieldrin and to 53% at 10 ppm. In females the figures were control 4%, 0.1 ppm dieldrin 30%, and 10 ppm dieldrin 62%. Experiments in rats and dogs were less definitive (Walker et al., 1969), and again there was ambivalence among the experts, an EPA Advisory Committee concluding "The committee did not feel that the balance of data indicated a carcinogenicity hazard" (O'Brien, 1973). However, after exhaustive restudy and a public hearing, EPA concluded that the carcinogenic properties of aldrin and dieldrin constituted an "imminent hazard to human health" (Carter, 1974).

Heptachlor and its major metabolite heptachlor epoxide appear to be even more carcinogenic in mice than aldrin-dieldrin (Carter, 1974), and their ban, along with chlordane, has been proposed by EPA, August 1976.

If these data indicate a potential human health hazard, there is little to be done about it. Human body residues of DDE and dieldrin will persist in appreciable amounts for a generation or more. For example, dieldrin in the human body has an estimated half-life of 369 days (Mrak, 1969). DDT in human adipose tissue is lost with a half-life estimated at about 1 year, but the half-life of DDE is about 8 years (Morgan and Roan, 1974). Therefore, it must be concluded that the only way that these persistent organochlorine insecticides, shown to be carcinogenic for laboratory animals, will not be a threat to human health is to prevent these substances from entering the environment.

## V. DEVELOPMENT OF SUBSTITUTES

The problem of environmental pollution resulting from the tremendous production and use of the organochlorine insecticides during the past 25 years and steadily increasing concern about chemical carcinogens (WHO has estimated that 85% of human cancers result from environmental exposure to chemicals) have brought about severe legal restrictions on the use of these "hard" insecticides. The Mrak Commission (1969) recommends that we should "eliminate within 2 years all uses of DDT and DDD in the United States except those essential for the preservation of human health or welfare" and "restrict the usage of certain persistent pesticides in the United States to specific essential uses which create no known hazard to human health or to the quality of the environment." These include aldrin, dieldrin, heptachlor, chlordane, benzene hexachloride, and lindane. It was recommended that the use of toxaphene receive close surveillance.

Acting on these recommendations, the Environmental Protection Agency (EPA) cancelled the remaining uses of DDT effective January 1, 1973, and is presently reviewing the uses of all these organochlorine insecticides. Aldrin and dieldrin registrations were cancelled October 1, 1974.

In the Soviet Union the use of aldrin, dieldrin, and endrin is not permitted, and residues of DDT and heptachlor are not permitted in milk, butter, grain, meat, and other food products (Melnikov, 1971). The use of these organochlorine compounds is restricted or eliminated in England, Japan, Switzerland, West Germany, Norway, and Sweden. These increasing restrictions will improve the quality of the environment but will have had the unpleasant effect of greatly increasing the use of highly toxic organophosphorus insecticides such as methyl and ethyl parathion (Spencer, Chapter X). Thus the production of DDT in the United States has declined from 164.2 million pounds in 1960 to 59.3 million pounds in 1970, whereas the production of methyl and ethyl parathion has increased from 18.2 million pounds in 1960 to 56.6 million pounds in 1970 (Pesticide Review). This changing use pattern has had serious effects on human health. As an example, it is reported that the annual use of about 4.8 million pounds of the parathions in El Salvador caused

2028 cases of poisoning and 30 deaths in 1972. WHO has estimated that there are some 500,000 cases of pesticide poisoning worldwide with more than 5000 deaths (C&E News, 1974).

As related elsewhere in this book, there will be urgent demands over the next 25 years in public health programs (Wright, Chapter II) and in food production for persistent, safe, and relatively inexpensive insecticides. These demands cannot be fully accommodated by any other group of insecticides presently available or foreseen. Therefore, in the remainder of this chapter we will examine modifications that can be made in existing organochlorine structures to develop compounds which still produce the typical biochemical lesions in insects yet which may be useful as persistent, biodegradable insecticides of low toxicity to man and higher animals. Whether these substitutes may still properly be termed organochlorine insecticides is questionable.

## A. DDT-Type Compounds

DDT is a particularly promising target for structural optimization because of the ease of synthesis of a wide variety of analogs with varying aryl and alkyl moieties in the generalized formula:

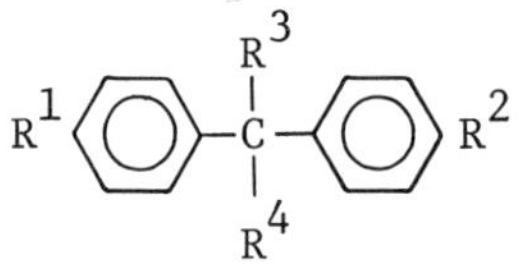

A considerable variety of these compounds had been shown to be insecticidally active by 1955 (Müller, 1955), and a much wider range of types have been critically evaluated by WHO, Programme for Evaluation and Testing of New Insecticides (Metcalf and Fukuto, 1968). Two cardinal principles are involved in the development of environmentally suitable DDT-type compounds: (1) critical fit at the site of action, and (2) incorporation of suitable degradophores (groups serving as substrates for mixed function oxidase enzymes) to replace the environmentally stable aryl chlorines of DDT, DDD, and DDE. These principles have been reviewed by Metcalf _et al_ (1972) and Metcalf (1973).

1. Site of DDT Action. Despite 30 years of study, there is little certain knowledge about the critical biochemical lesion produced in DDT poisoning. However, a great deal has been

learned from structure-activity relationships that can be interpreted into a rational theory about the design of effective structural analogs. DDT is clearly a nerve poison in both invertebrates and vertebrates, producing symptoms of intoxication marked by hyperexcitability, tremors of increasing severity, and finally convulsions, prostration, and death. The tremors are the result of initiation of repetitive after-discharge in the nerve, a multiplication of nerve impulses so that a single sensory impulse produces a prolonged volley of afferent impulses. These actually stimulate the insect victim to a death that may result from metabolic exhaustion or endogenously produced neurotoxins (Sternburg, 1963). The site of DDT action is in the nerve axon and is associated with increasing negative after-potential and prolongation of the action-potential, perhaps due to specific inhibition of the $K^+$ efflux (Narahashi and Haas, 1968).

Theories of the mode of action of DDT have been reviewed recently (Metcalf, 1973). The most promising are those of Mullins (1956) and Holan (1969), which suggest a specific cavity or receptor site within the axonal structure that is highly complementary in structure to the DDT-type compound, which fits into the receptor because of its lipophilicity and the effect of Van der Waals' forces, producing distortion or blocking and subsequent ion leaks.

Mullins (1956) proposed a generalized receptor site in the regular interstices of the cylindrical lipoprotein strands of the membrane lattice of the nerve axon, which were estimated to be about 40 Å in diameter. When these are packed into hexagonal array with a 2 Å separation, hypothetical pores result into which DDT-type molecules should fit snugly in end-on position, distorting the membrane and presumably producing ion leaks to cause nerve excitation.

Holan (1969) conceived of the DDT-type molecule acting as a molecular wedge, entering into the lipoprotein of the nerve membrane so that the dichlorodiphenylmethane moiety is locked into the overlying protein layer by formation of a molecular complex, while the trichloromethyl group fits into a pore in the membrane, the "sodium gate" keeping it open to $Na^+$ ions and delaying the falling phase of ion potential. Holan concluded from study of the size and shape of a number of toxic DDT analogs that the molecular wedge should have an effective Van der Waals' diameter of 60 ± 0.5 Å or close to that of the hydrated sodium ion, and that the maximum allowable distance between _p,p_'-substituents of the aryl rings should be about 14.0 Å.

Even more recently, ATP'ase has been suggested as the primary target site for DDT action. Matsumura and Patil (1969) found DDT ($I_{50}$ 3 x $10^{-7}$ M) inhibited the $Na^+$, $K^+$, and $Mg^{++}$ ATP'ases of rat brain homogenate and suggested that the involvement of ATP'ase in $Na^+$ and $K^+$ transport through nerve membranes might explain the mode of action. Cutkomp et al. (1971) showed that the oligomycin-sensitive $Mg^{++}$ ATP'ase of mitochondria was most sensitive to DDT both in cockroach nerve cord ($I_{50}$ 6.4 x $10^{-7}$ M) and coxal muscle ($I_{50}$ 2.7 x $10^{-7}$ M). Toxic DDT analogs of varying structures were also active as ATP'ase inhibitors (Desaiah et al., 1974).

2. Critical Fit at DDT Receptor. The present views of the mode of action of DDT, as described above, imply a precise physical action between the insecticide and macromolecular element of the nerve axon, the "DDT receptor" (Gunther et al., 1954; Mullins, 1956; and Holan, 1969). Analogs isosteric with DDT and with appropriate lipophilicity and biochemical stability produce the characteristic symptoms of axonal instability. Therefore, the $LD_{50}$ to various insect species, as determined under uniform conditions, can be used as a precise quantitative measure of the fit of individual DDT analogs at the receptor site (Gunther et al., 1954; Metcalf and Fukuto, 1968). Where groups substituted for the Cl atoms of DDT are degradophores attacked by the microsomal oxidase enzymes, prior inhibition of these enzymes by administration of piperonyl butoxide (Metcalf et al., 1971) or sesamex (Holan, 1971) gives a synergized $LD_{50}$ which is a better estimation of intrinsic activity than the $LD_{50}$ of the compound alone (Tables 8 and 9).

In general, typical DDT-like activity is found in all isosteric combinations of Cl (Van der Waals' radius 1.75 Å) and $CH_3$ (radius 1.97 Å). Thus the completely methylated DDT isostere 1,1-bis-(p-tolyl)-neopentane (LXXIV) ($LD_{50}$ Musca 1250, synergized $LD_{50}$ 35, $LC_{50}$ Culex 1.0), approaches the toxicity of DDT (synergized $LD_{50}$ 5.5) and produces indistinguishable, toxic symptoms (Metcalf et al., 1972). Replacement of an aryl Cl with $CH_3$ (XL) decreases its toxicity to Musca to about 0.25 but does not affect it to Phormia or Culex, while replacement of both aryl Cl atoms with $CH_3$ groups (XVIII) decreases the toxicity to Musca to about 0.14 and also decreases it to Phormia and Culex (Table 9). The decreases in toxicity are to be expected from the biodegradability of the $CH_3$ group (Section V.B.) and are minimized by synergism with piperonyl butoxide. Replacement of $-CCl_3$ by $C(CH_3)_3$ also decreases its toxicity to Musca by about 0.25 and to Culex by about 0.16 (Table 3), but the synergized $LD_{50}$ to Musca is only about 2 times that of DDT. Replacement of a single Cl in the $CCl_3$ group, ($CCl_3$ by $C(CH_3)Cl_2$) decreases the $LD_{50}$ to Musca to about 0.3 but increases the toxicity to Culex 3 times.

TABLE 8. Toxicity of Symmetrical DDT Analogs with Altered Aromatic Substituents (Metcalf *et al*., 1971, 1972)

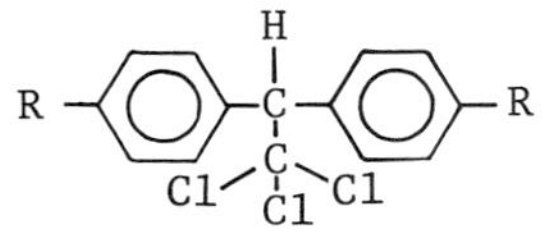

| | R | Topical $LD_{50}$ μg/g | | | | $LC_{50}$ ppm |
|---|---|---|---|---|---|---|
| | | *Musca domestica* | | *Phormia regina* | | *Culex pipiens fatigans* |
| | | Alone | p.b.[a] | Alone | p.b.[a] | |
| XIV | H | 2900 | 575 | 1250 | 1250 | 1.1 |
| XV | F | 30.5 | 16.5 | | | 0.074 |
| I | Cl | 14.0 | 5.5 | 11.5 | 8.25 | 0.07 |
| XVI | Br | 27.0 | 10.5 | | | 0.074 |
| XVII | I | 35.0 | 14.0 | | | 1.4 |
| XVIII | $CH_3$ | 100 | 17.5 | 61.2 | 21.5 | 0.081 |
| XIX | $C_2H_5$ | 70 | 29.0 | | | 0.18 |
| XX | $C_3H_7$ | 55 | 45 | 250 | 250 | 0.145 |
| XXI | $(CH_3)_2CH$ | >500 | | | | >10 |
| XXII | $(CH_3)_3C$ | >500 | | | | >10 |
| III | $CH_3O$ | 45 | 3.5 | 10.0 | 4.6 | 0.067 |
| XXIII | $C_2H_5O$ | 7.0 | 1.75 | 6.9 | 7.4 | 0.04 |
| XXIV | $C_3H_7O$ | 125 | 18.5 | 250 | 61.2 | 0.6 |
| XXV | $C_4H_9O$ | >500 | >500 | >250 | >250 | 1.0 |
| XXVI | $(CH_3)_2CHO$ | >500 | 175 | >250 | >250 | 1.0 |
| XXVII | $HC{\equiv}CCH_2O$ | >500 | 28.0 | >250 | >250 | 0.07 |
| XXVIII | $CH_3S$ | 225 | 17.0 | 36.4 | 14.5 | 0.21 |
| XXIX | $CH_3SO$ | >500 | | | | >10 |

(continued)

TABLE 8 (cont.)

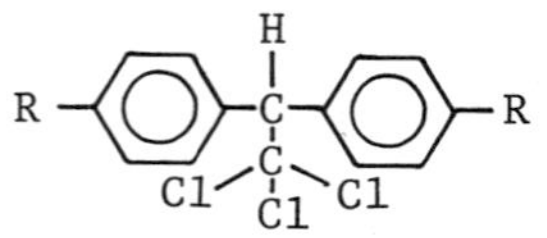

| | R | Topical $LD_{50}$ µg/g | | | | $LC_{50}$ ppm |
|---|---|---|---|---|---|---|
| | | Musca domestica | | Phormia regina | | Culex pipiens fatigans |
| | | Alone | p.b.[a] | Alone | p.b.[a] | |
| XXX | $CH_3SO_2$ | >500 | | | | >100 |
| XXXI | $NO_2$ | >500 | 37 | | | >100 |
| XXXII | $NH_2$ | >2500 | 1500 | | | >100 |
| XXXIII | CN | >2500 | 1500 | | | 9 |
| XXXIV | CHO | >2500 | 1500 | | | >100 |
| XXXV | COOH | >2500 | 1500 | | | >100 |
| XXXVI | OH | >2500 | 1500 | | | >100 |

[a]Piperonyl butoxide synergist applied at 50 µg/insect 1 hr before application of compound.

TABLE 9. Toxicity of Asymmetrical DDT Analogs with Alkyl and Alkoxy Substituents (Metcalf *et al*., 1971, 1972)

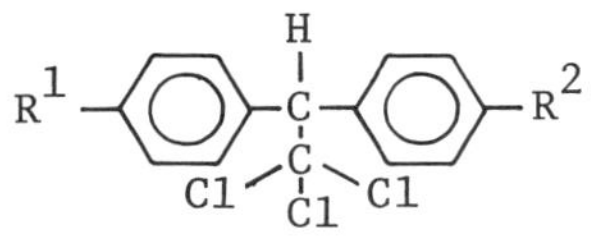

| | $R^1$ | $R^2$ | Topical $LD_{50}$ µg/g *Musca domestica* Alone | *Musca domestica* p.b.[a] | *Phormia regina* Alone | *Phormia regina* p.b.[a] | $LC_{50}$ ppm *Culex pipiens fatigans* |
|---|---|---|---|---|---|---|---|
| XXXVII | Cl | H | 160 | 4.0 | 125 | 122.5 | 0.12 |
| XXXVIII | $CH_3$ | H | >500 | 33 | >250 | >250 | 0.61 |
| XXXIX | $CH_3$ | F | 250 | 25 | 47.5 | 25 | 0.059 |
| XL | $CH_3$ | Cl | 62.5 | 6.5 | 11 | 9.2 | 0.032 |
| XLI | $CH_3$ | Br | 31.5 | 1.75 | 9.2 | 4.7 | 0.04 |
| XLII | $CH_3$ | I | 120 | 22.5 | 40 | 29.0 | 0.042 |
| XLIII | $CH_3$ | $C_2H_5$ | 11.0 | 3.0 | | | 0.082 |
| XLIV | $CH_3$ | $C_3H_7$ | 52.5 | 37.0 | 112.5 | 92.5 | 0.075 |
| XLV | $CH_3$ | $CH_3O$ | 23.5 | 4.9 | 12.2 | 9.0 | 0.085 |
| XLVI | $CH_3$ | $C_2H_5O$ | 9.0 | 1.7 | 5.2 | 2.1 | 0.13 |
| XLVII | $CH_3$ | $C_3H_7O$ | 20.0 | 1.85 | 11.3 | 9.7 | 0.042 |
| XLVIII | $CH_3$ | $C_4H_9O$ | 33.0 | 1.95 | 70 | 28.7 | 0.063 |
| XLIX | $C_2H_5$ | Cl | 245 | 65 | 16 | 16 | 0.047 |
| L | $C_2H_5$ | Br | 47 | 22 | 11.5 | 9.2 | 0.044 |
| LI | $C_2H_5$ | $C_2H_5O$ | 18.5 | 3.5 | 9.7 | 5.2 | 0.036 |
| LII | $(CH_3)_2CH$ | Cl | 215 | 95 | | | 0.15 |
| LIII | $(CH_3)_2CH$ | $CH_3O$ | 155 | 67.5 | | | 0.32 |
| LIV | $CH_3$ | COOH | >5000 | | | | 4.0 |
| LV | $CH_3O$ | F | 47.0 | 11.5 | 15 | 9.25 | 0.47 |

(continued)

TABLE 9 (cont.)

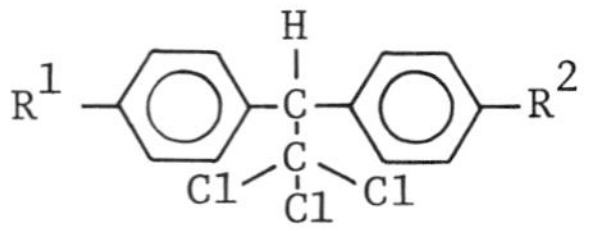


| | $R^1$ | $R^2$ | Topical $LD_{50}$ μg/g *Musca domestica* Alone | *Musca domestica* p.b.[a] | *Phormia regina* Alone | *Phormia regina* p.b.[a] | $LC_{50}$ ppm *Culex pipiens fatigans* |
|---|---|---|---|---|---|---|---|
| LVI | $CH_3O$ | Cl | 41.5 | 7.0 | | | 0.058 |
| LVII | $CH_3O$ | $C_2H_5O$ | 16.0 | 3.7 | 10 | 10 | 0.039 |
| LVIII | $CH_3O$ | $C_3H_7O$ | 24.0 | 8.5 | 30 | 27.5 | 0.18 |
| LIX | $CH_3O$ | $C_4H_9O$ | 21.0 | 20.5 | 107 | 75 | 0.18 |
| LX | $CH_3O$ | $C_5H_{11}O$ | 160 | 26 | 185 | 130 | >1 |
| LXI | $CH_3O$ | $C_6H_{13}O$ | 500 | 72.5 | >250 | >250 | >1 |
| LXII | $CH_3O$ | $C_8H_{17}O$ | >500 | >500 | >250 | >250 | >1 |
| LXIII | $CH_3O$ | $(CH_3)_2CHO$ | 160 | 67.5 | | | 0.32 |
| LXIV | $CH_3O$ | $HC{\equiv}CCH_2O$ | 34.0 | 3.3 | 9.5 | 8.5 | 0.050 |
| LXV | $CH_3O$ | $-OCH_2O-$ | 22.0 | 4.5 | 65 | 24.5 | 0.036 |
| LXVI | $C_2H_5O$ | $C_3H_7O$ | 9.5 | 3.3 | 7.5 | 5.25 | 0.036 |
| LXVII | $C_2H_5O$ | $C_4H_9O$ | 21.5 | 3.75 | 9.7 | 3.7 | 0.074 |
| LXVIII | $C_2H_5O$ | $-OCH_2O-$ | 13.5 | 3.5 | 11.5 | 10.5 | 0.055 |
| LXIX | $CH_3S$ | $CH_3O$ | 32 | 4 | 13 | 6.7 | 0.11 |
| LXX | $CH_3S$ | $C_2H_5O$ | 32 | 2.8 | 15.5 | 13 | 0.70 |
| LXXI | $CH_3S$ | $C_3H_7O$ | 35.5 | 2.75 | 14.5 | 7.25 | 0.11 |
| LXXII | $CH_3O$ | $C_2H_5$ | 31.0 | 3.5 | 9.7 | 6.7 | 0.049 |
| LXXIII | $C_2H_5O$ | OH | >5000 | | | | 3.7 |

[a]Piperonyl butoxide synergist applied at 50 μg/insect 1 hr before application of compound.

Despite this seeming flexibility in structure, there is an optimum size for both the DDT-type molecule and for its constituent parts. Fahmy _et al_. (1973) have synthesized a series of silicon analogs of DDT including 1,1-bis-(_p_-chlorophenyl)-_tert_-butylsilane (LXXV), 1,1-bis-(_p_-chlorophenyl)-dichloromethyl silane (LXXVI), and bis-(_p_-chlorophenyl)-methyltrimethylsilane (LXXVII) which were nearly inactive as insecticides to _Musca_ and _Culex_. This suggests that the nearly 50% larger size of silicon (Si-C bond 1.86 Å versus C-C 1.48 Å) causes an increase in size of the molecule in all tetrahedral dimensions so that interaction with the DDT receptor is prevented; for example, the Van der Waals' radius of $-Si(CH_3)_3$ is 3.3 Å versus 2.79 Å for $-C(CH_3)_3$ (Charton, 1969).

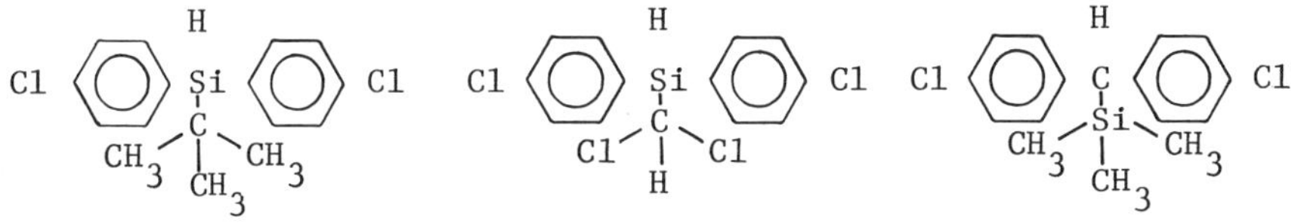


LXXV LXXVI LXXVII

There is, then, a relatively precise range of physical size for the active DDT-type molecule, so that the summation of Van der Waals' volumes for each of the substituents must be considered. For example, 2,2-bis-(_p_-chlorophenyl)-2,1,1,1-tetrachloroethane is completely inactive, but 2,2-bis-(_p_-chlorophenyl)-2-fluoro-1,1,1-trichloroethane is moderately toxic ($LD_{50}$ _Musca_ 125, $LC_{50}$ _Culex_ 0.092), where the atomic radius of Cl is 1.75 Å versus 1.47 Å for F. Further reduction of crowding around the α-carbon using DDD or 2,2-bis-(_p_-chlorophenyl)-1,1-dichloroethane (V) as a model ($LD_{50}$ _Musca_ 19.5, $LC_{50}$ _Culex_ 0.038) permits insertion of α-F (VII) with increased toxicity ($LD_{50}$ _Musca_ 4.1, $LC_{50}$ _Culex_ 0.024) and even insertion of α-Cl ($LD_{50}$ _Musca_ 18.5, $LC_{50}$ _Culex_ 0.24) without loss of activity (Metcalf and Fukuto, 1968).

3. Essential Configuration. The biological action of DDT depends on an optimum configuration of the phenyl rings and trichloromethyl group about the central α-carbon. For maximum toxic action the rings must assume a "butterfly" or trihedralized configuration (Rogers _et al_., 1953). This essential feature is demonstrated by the low insecticidal activity of 2,7-dichloro-9-trichloromethyl-9,10-dihydroanthracene (LXXVIII) ($LD_{50}$ _Musca_ >500, $LC_{50}$ _Culex_ 0.35) in which the aryl rings are

planar (Vingiello and Newallis, 1960; Metcalf, 1973). The diphenylmethane moiety is not essential for DDT-like activity, and an atom of N,O, or S can be interposed, as in α-trichloromethyl-p-chlorobenzyl-N-(p-chloroaniline (LXXIX) ($LD_{50}$ *Musca* 115, $LC_{50}$ *Culex* 0.026) and α-trichloromethyl-p-chlorobenzyl p-chlorophenyl ether (LXXX) ($LD_{50}$ *Musca* 90, $LC_{50}$ *Culex* 0.035) (Hirwe *et al*., 1972). α-Trichloromethyl-p-ethoxybenzyl-N-p-ethoxyaniline ($LD_{50}$ *Musca* 15.5, $LC_{50}$ *Culex* 0.19) and α-trichloromethyl-p-ethoxybenzyl p-ethoxyphenyl ether ($LD_{50}$ *Musca* 27, $LC_{50}$ *Culex* 0.11) were generally more effective insecticidally than the p,p'-dichloro-analogs. These compounds produce characteristic DDT-like symptoms of poisoning, and observations of Fischer-Hirschfelder-Taylor molecular models show that the staggered alignment of aryl-X-C-aryl permits the aryl rings and p,p'-substituents to assume configurations almost identical to that of DDT. Clearly they fit the DDT-receptor site.

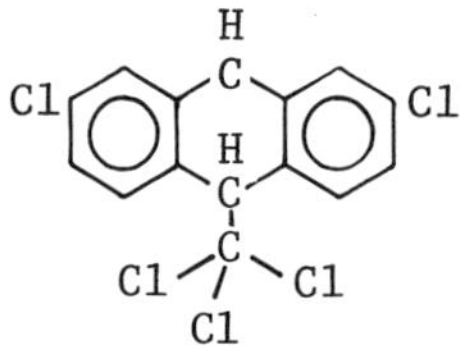


LXXVIII

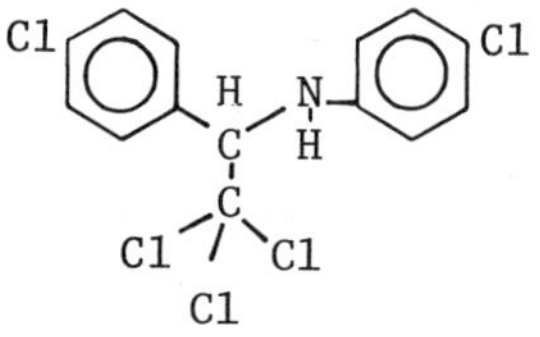


LXXIX

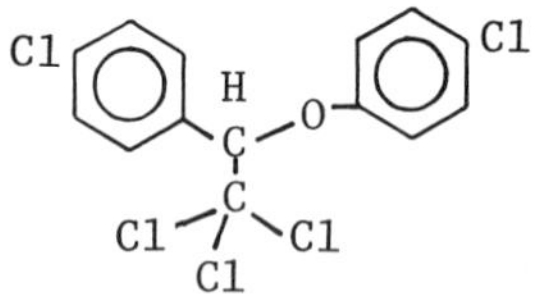


LXXX

4. Altered Aromatic Substituents. As discussed in Section III-D, the nature of the substituents of the aromatic rings of the DDT-type molecule has a pronounced effect on the stability of the compound to dehydrochlorination. As shown in Table 8, strong electron-withdrawing groups such as $NO_2$ and CN produce essentially inactive compounds. Inactivity also results from incorporation of highly polar substituents such as COOH and OH. Acceptable p,p' substituents that produce active analogs are F, Br, $CH_3$, $CH_3O$, and $C_2H_5O$ (Table 8). Of these compounds 2,2-bis(p-fluorophenyl)-1,1,1-trichloroethane (XV) has been produced commercially in Germany during World War II as "Gix" (Metcalf, 1948). It has a much shorter residual life than DDT, is considerably more expensive, and is not biodegradable. Methoxychlor or 2,2-bis-(p-methoxyphenyl)-1,1,1-trichloroethane (III) has been the most successful of the DDT substitutes commercially, as it is of very low toxicity to mammals and is biodegradable. The analog 2,2-bis(p-tolyl)-1,1,1-trichloroethane (XVIII) is the safest and most biodegradable of all the trichloroethanes but is of somewhat decreased insecticidal activity unless synergized (Table 8). Perthane® or 2,2-bis-(p-ethylphenyl)-1,1-dichloroethane, an analog of DDD, has also been produced commercially as a low toxicity, safe, household insecticide.

5. Asymmetrical Analogs. Earlier investigators of DDT structure versus activity concluded that symmetry of structure is an essential feature for high insecticidal activity (O'Brien, 1967). The pronounced activity of a variety of asymmetrical analogs, as shown in Table 9, proves that this is not so. A number of these compounds are nearly as effective as DDT to a variety of insects. Furthermore, where the pairs of p,p'-substituents are carefully chosen to provide degradophores readily susceptible to attack by microsomal oxidase enzymes, these compounds become highly biodegradable and of very low toxicity to higher animals (Metcalf *et al.*, 1971, 1972). Among the most promising of these asymmetrical analogs as DDT-substitutes are 2-(p-ethoxyphenyl)-2-(p-tolyl)-1,1,1-trichloroethane (XLVI) and 2-(p-methoxyphenyl)-2-(p-tolyl)-1,1,1-trichloroethane (XLV). A remarkable feature of such compounds is the degree of asymmetry that can be incorporated without losing insecticidal action. For example, 2-(p-methoxyphenyl)-2-(p-butoxyphenyl)-1,1,1-trichloroethane (LIX) is a highly active insecticide (Table 9). Incorporation of the $CH_3S$ group as in 2-(p-methylthio)-2-(p-methoxyphenyl)-1,1,1-trichloroethane (LXIX) $LD_{50}$ *Musca* 32, $LC_{50}$ *Culex* 0.11 produced an active compound with short-lived residual action.

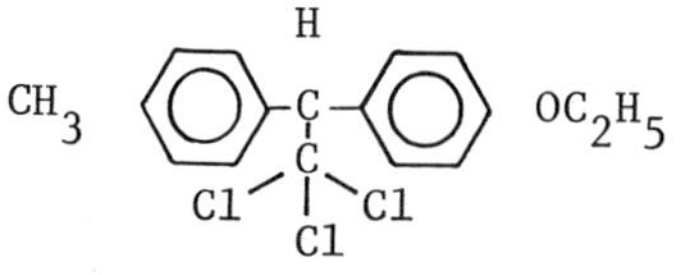

XLVI

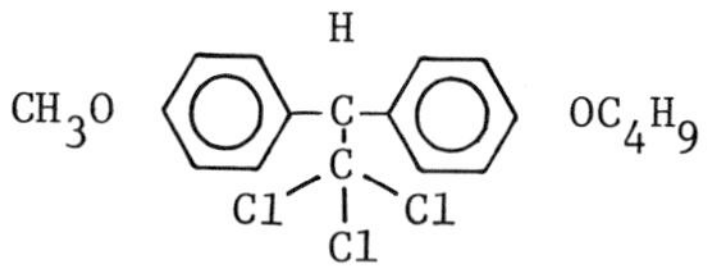

LIX

Holan (1971) has applied the asymmetry principal to the diphenyl-2-nitropropane series and has found that 1-(*p*-ethoxyphenyl)-1-(*p*-ethylthiophenyl)-2-nitropropane ($LD_{50}$ *Musca* 8.0) and 1-(*p*-ethoxyphenyl)-1-(3,4-methylenedioxyphenyl)-2-nitropropane ($LD_{50}$ *Musca* 7.0) were more active than the highly effective 1,1-bis-(*p*-ethoxyphenyl)-2-nitropropane ($LD_{50}$ *Musca* 24.0). Recent study in this laboratory has shown the Prolan® type diphenyl-2-nitropropane moiety has more flexibility for asymmetry than the DDT-type diphenyl-trichloroethane (Section V.A.2.). Thus 1-(*p*-methoxyphenyl)-1-(*p*-hexyloxyphenyl)-2-nitropropane ($LD_{50}$ *Musca* 37, $LC_{50}$ *Culex* 0.19) was a highly toxic molecule and even 1-(*p*-methoxyphenyl)-1-(*p*-octyloxyphenyl)-2-nitropropane ($LD_{50}$ *Musca* 75, $LC_{50}$ *Culex* 1) had substantial insecticidal activity.

6. Altered Aliphatic Substituents. Within the isosteric and configurational requirements previously discussed, aliphatic structures resembling $-CCl_3$ may be incorporated into compounds having strong DDT-like activity. Some of these alterations have already been discussed under "resistance proofing" in Section III.D. and illustrated in Table 3. The insecticide DDD, 2,2-bis-(*p*-chlorophenyl)-1,1-dichloroethane (V) has had considerable commercial use. An analog Perthane® or 2,2-bis-(*p*-ethylphenyl)-1,1-dichloroethane, also incorporating the $-CHCl_2$ moiety has been used as a safe household insecticide. An interesting and effective modification incorporates the dichlorocyclopropane moiety, such as 1,1-bis-(*p*-ethoxyphenyl)-2,2-dichlorocyclopropane (LXXXII) ($LD_{50}$ *Musca* 2.75, *Culex* 0.048) (Holan, 1969).

Among the most active groups of compounds are the nitropropanes with $-CH(CH_3)NO_2$. of which Prolan® or 1,1-bis-(*p*-chlorophenyl)-2-nitropropane (XII) has been most intensively investigated (Haas *et al.*, 1951) (Table 3). The geometry of the nitropropane group must closely complement that of the DDT receptor. The bis-(*p*-chlorophenyl)-methylene moiety is environmentally persistent (Section V.B.), and a more satisfactory analog 1,1-bis-(p-ethoxyphenyl)-2-nitropropane (LXXXI) ($LD_{50}$ *Musca* 5.0, *Culex* 0.045), which is biodegradable, has been developed by Holan (1971).

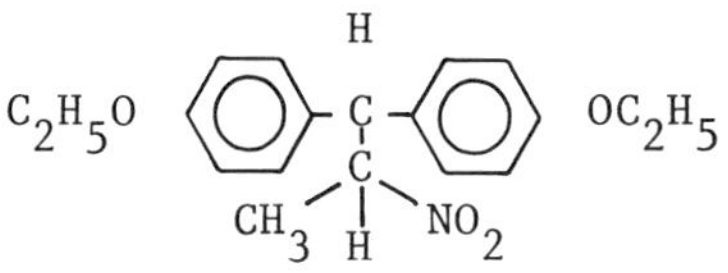

LXXXI

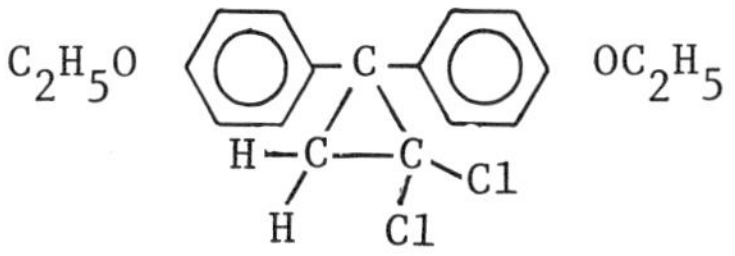

LXXXII

Holan (1971) has also introduced a readily biodegradable moiety into the aliphatic portion, for example, 1,1-bis-(p-chlorophenyl)-3,3-dimethyl oxetane. Again the p,p'-$C_2H_5O$ analog or 1,1-bis(p-ethoxyphenyl)-3,3-dimethyloxetane (LXXXIII) ($LD_{50}$ *Musca* 26, $LC_{50}$ *Culex* 0.048) is superior.

The isosteric substitution of $C(CH_3)_3$ produces interesting compounds, the most effective of which are the p,p'-$CH_3O$ analog or 1,1-bis-(dianisyl)-neopentane (XIII) and the p,p'-$C_2H_5O$ analog or 1,1-bis-(p-ethoxyphenyl)-neopentane (LXXXIV) ($LD_{50}$ *Musca* 37.5, $LC_{50}$ *Culex* 0.27) (Holan, 1971). These two analogs, like ethoxy Prolan®, are notable for not containing any chlorine atoms. DDT analogs with $CBr_3$ and $CF_3$ substituted for $CCl_3$ are nearly inactive.

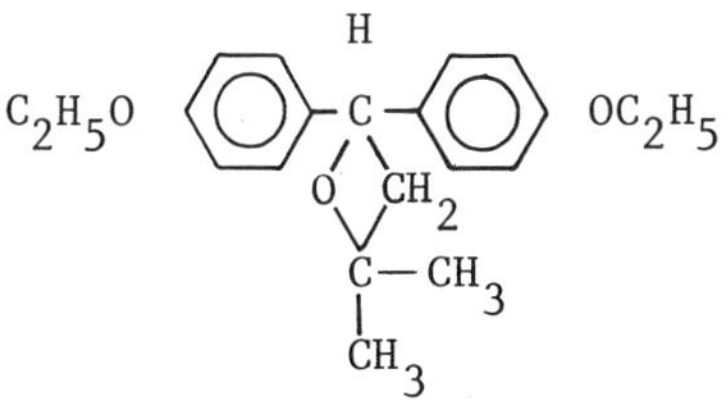

LXXXIII

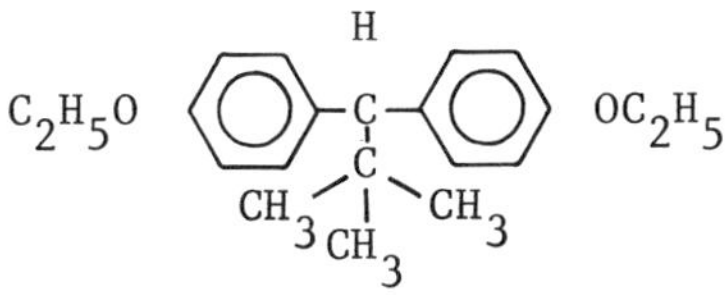

LXXXIV

## B. Degradation of DDT-Substitutes

There is clearly no point in using analogs as DDT substitutes for insect control unless they are substantially less hazardous to environmental quality. Thus it has been important to investigate the environmental fate of a variety of DDT analogs with altered aromatic and aliphatic substituents, including insecticides as old as methoxychlor and Prolan® which have been available for 28 and 23 years, yet about whose degradation little or nothing was known. The laboratory-model ecosystem provides a useful way of comparing quantitatively the environmental fate of these compounds using radiolabeled preparations (Metcalf et al., 1971). Using this technique, investigations have been made of the environmental fate of most of the changes in aromatic or aliphatic moieties of the DDT-type molecule, which are insecticidally promising (Kapoor et al., 1970, 1972, 1973; Coats et al., 1974; Hirwe et al., 1972, 1974). The degradative pathways of analogs with altered aromatic substituents are shown in Fig. 6, and of those with altered aliphatic substituents in Fig. 7. A summary of the model ecosystem results from these studies is given in Table 10.

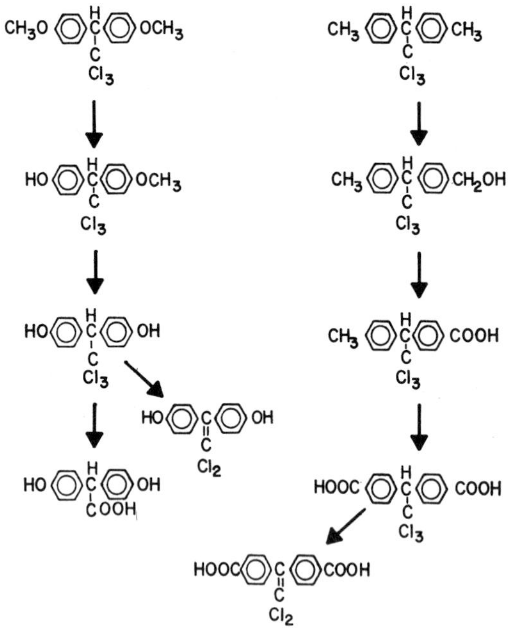

FIGURE 6. Degradative pathways of DDT analogs with degradophores on the aromatic rings (after Metcalf, 1973).

TABLE 10. Laboratory Model Ecosystem Studies of Biodegradability of DDT Analogs[a]

| | Water Solubility (ppm) | *Gambusia affinis* (fish) | |
|---|---|---|---|
| | | Ecological Magnification[b] | Biodegradability Index[c] |
| $ClC_6H_4HC(CCl_3)C_6H_4Cl$ (DDT) | 0.0012 | 84,500 | 0.015 |
| $CH_3OC_6H_4HC(CCl_3)C_6H_4OCH_3$ (methoxychlor) | 0.62 | 1,545 | 0.94 |
| $C_2H_5OC_6H_4HC(CCl_3)C_6H_4OC_2H_5$ | 0.16 | 1,536 | 2.69 |
| $CH_3C_6H_4HC(CCl_3)C_6H_4CH_3$ | 2.21 | 140 | 7.14 |
| $CH_3SC_6H_4HC(CCl_3)C_6H_4SCH_3$ | 0.57 | 5.5 | 47 |
| $CH_3C_6H_4HC(CCl_3)C_6H_4Cl$ | 0.10 | 1,400 | 3.43 |
| $CH_3C_6H_4HC(CCl_3)C_6H_4OC_2H_5$ | 0.028 | 400 | 1.20 |
| $CH_3OC_6H_4HC(CCl_3)C_6H_4SCH_3$ | 0.19 | 310 | 2.75 |
| $CH_3OC_6H_4HC(CMe_3)C_6H_4OCH_3$ | 0.69 | 1,636 | 1.04 |
| $ClC_6H_4HCCH(Me)NO_2C_6H_4Cl$ | 0.003 | 112 | 3.27 |

[a]Data from Kapoor *et al.* (1973), Coats *et al.* (1974), Hirwe *et al.* (1974).

[b]Concentration of parent compound in fish/concentration in water.

[c]Ratio of polar/nonpolar metabolites.

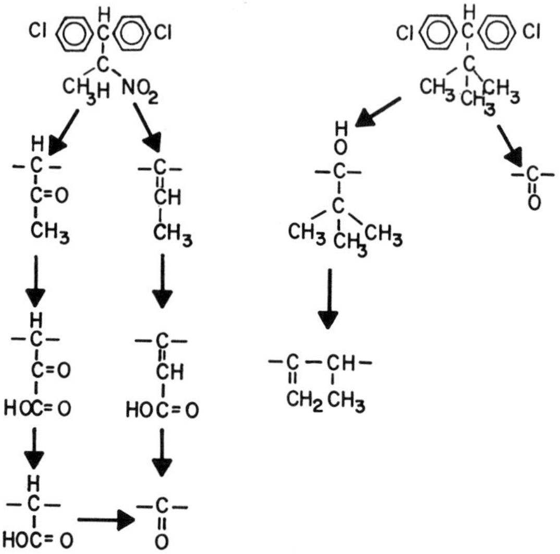


FIGURE 7. Degradative pathways of DDT analogs with altered aliphatic moieties (after Coats *et al.*, 1974 and Hirwe *et al.*, 1975).

1. Aryl Degradophores. The presence of aryl substituents on the DDT-type molecule, which can serve as substrates for microsomal oxidation, provides degradophores that can be converted *in vivo* into more water-partitioning moieties, and in elimination of these xenobiotics from living organisms. Methoxychlor serves as an excellent example. It is well known that methoxychlor does not accumulate in animal lipids. When fed to the rat at 100 ppm it was stored at only 1 ppm (Kunze *et al.*, 1950) in contrast to DDT, which when fed at 1 ppm was stored at 13 to 18 ppm (Laug *et al.*, 1950). The degradation pathways for methoxychlor proceed largely by O-dealkylation of $CH_3O$-aryl to form mono- and bis-OH derivatives (phenols) which increases the water solubility from 0.62 ppm to 76 ppm (Kapoor *et al.*, 1970). Methoxychlor is therefore conjugated and eliminated from the body rather than stored in lipids. Methoxychlor forms an environmental hazard only in organisms like the snail *Physa* (Table 10) and coldwater fish such as salmon, where O-dealky-

lation takes place only very slowly and allows the compound to accumulate. Other alkoxy-derivatives of DDT such as ethoxychlor or 2,2-bis-(p-ethoxyphenyl)-1,1,1-trichloroethane are also detoxified by O-dealkylation (Table 10) and form a substantially reduced environmental hazard compared to the stable DDT and its degradation products DDE and DDD (Fig. 3) which are stored in tissues rather than eliminated from the animal body (Table 10).

Methyl and other alkyl substituents of the aryl rings of DDT can also serve as effective degradophores and are rapidly oxidized in higher animals to alcohols and to mono- and bis-COOH derivatives (carboxylic acids). In methylchlor or 2,2-bis-(p-methylphenyl)-1,1,1-trichloroethane, this microsomal oxidation increases the water solubility from 2.21 ppm to 50 ppm, thus favoring elimination rather than storage (Kapoor et al., 1972). The presence of a single $CH_3$-aryl group on the DDT-type molecule, acting as a degradophore in 2-(p-methylphenyl)-2-(p-chlorophenyl)-1,1,1-trichloroethane substantially aids in decreasing tissue storage and food chain transfer (Table 10) (Kapoor et al., 1973). An especially interesting effect on biodegradability was obtained by incorporating two different degradophores, for example, $CH_3$- and $C_2H_5O$-, as in 2-(p-methylphenyl)-2-(p-ethoxyphenyl)-1,1,1-trichloroethane. This is a highly active compound (Table 9) with improved degradability (Table 10), proceeding through two different microsomal reactions, O-dealkylation and side chain oxidation.

The $CH_3S$-aryl group can also serve as a degradophore by microsomal oxidation to $CH_3SO$- and $CH_3SO_2$ (Table 10). These oxidations increase the polarity and water solubility of the molecule, for example from 0.57 ppm for methiochlor or 2,2-bis-(p-methylthiophenyl)-1,1,1-trichloroethane to 29 ppm for 2,2-bis-(p-methylsulfinylphenyl)-1,1,1-trichloroethane (Kapoor et al., 1970) and aid in elimination from the animal body.

2. Alkyl Degradophores. Suitable degradophores on the aliphatic portion of the DDT molecule are more difficult to incorporate into appropriate stereochemical complementarity to the DDT receptor. The methyl isosteres of the $-CCl_3$ group are surprisingly stable to microsomal oxidation, and 1,1-bis-(dianisyl) neopentane, the methoxychlor isostere, has almost the same properties of biodegradability as methoxychlor (Table 10) (Coats et al., 1974). Microsomal degradation centers on O-demethylation of the aryl rings to form mono- and bis-OH derivatives which aid in the elimination of the compound from the animal body. The tert-butyl group was recalcitrant to microsomal attack, although small amounts of the benzophenone

derivatives were found. Dianisylneopentane was also susceptible to microsomal oxidation at the α-C to form 1,1-bis-(p-methoxyphenyl)-1-hydroxyneopentane, which rearranged to form 2,3-dianisyl-3-methylbut-1-ene.

The nitropropane analogs such as Prolan® are also resistant to microsomal oxidation, and degradation takes place largely through elimination of the nitro group to form 1,1-bis-(p-chlorophenyl)-1-propene and by oxidation to 1,1-bis-(p-chlorophenyl)-2-propanone (Fig. 7, Hirwe et al., 1974). There is also some reduction of the $NO_2$- group to form 1,1-bis-(p-chlorophenyl)-2-aminopropane. The improved biodegradability of Prolan® over DDT (Table 10) is largely the result of inability of the former to form a stable degradation product such as DDE, because both Prolan® and DDT form the same degradation products DDA and DDBP (Figs. 3 and 7).

The degradation of other DDT analogs may involve substantially greater structural fragmentation. The α-trichloromethyl p-ethoxybenzyl N-(p-ethoxyaniline) is degraded both by O-dealkylation to mono- and bis-OH derivatives similar to ethoxychlor, and by an elimination of HCl, rearrangement, and cleavage of the two rings to form p-ethoxyaniline and p-ethoxybenzoic acid which are further degradable (Hirwe et al., 1972). The corresponding α-trichloromethyl p-ethoxybenzyl p-ethoxyphenyl ether is more stable and only degrades by O-dealkylation.

The 1,1-bis-(phenyl)-3,3-dimethyloxetanes are stated to undergo degradation by electrocyclic-fission forming formaldehyde and 1,1-bis-(phenyl)-2,2-dimethylethylene (Holan, 1971).

The aryl-chlorine bond is highly resistant to enzymatic cleavage (Section V.B.1.). Even with suitable aliphatic degradophores, DDT analogs formed from chlorobenzene are most likely to persist in the environment as bis-(p-chlorophenyl) methylene moieties which like DDE may be bioaccumulative. Therefore, any serious efforts to produce new DDT-type analogs should concentrate on the introduction of aryl degradophores such as alkyl, alkoxyl, and alkylthio which will aid in the total degradation of the compound. Some of the compounds designed in this way, such as 1,1-bis-(p-ethoxyphenyl)-neopentane and 1,1-bis-(p-ethoxyphenyl)-2-nitropropane, no longer contain chlorine and can scarcely be categorized as organochlorine insecticides, although they act biologically in a manner indistinguishable from DDT.

### C. Biodegradable Analogs of Other Organochlorine Insecticides

Little systematic study has been given to the design of biodegradable derivatives of lindane or the cyclodienes. Nakajima (Discussion) reports on lindane derivatives incorporating degradophores such as $CH_3$-, $CH_3O$-, and $CH_3S$- in place of one or more chlorine atoms, which preserved high insecticidal activity and yet were degradable by microsomal enzymes, indicating the potentially fruitful nature of this approach, in other types of organochlorine compounds.

There is reason to believe that the incorporation of degradophores, such as $CH_3$-, into suitable cyclodiene molecules should produce active and biodegradable insecticides, as suggested by Brooks (Chapter V). Endosulfan (Fig. 1) incorporates the sulfite degradophore (O-SO-O) which is both oxidizable and hydrolyzed to diols and is apparently considerably more biodegradable than the other cyclodienes (Section IV.B.) although more positive proof is desirable. The epoxide forming aldrin and heptachlor are particularly undesirable because of the very high stability imparted as in dieldrin and heptachlor epoxide, and because of the implications of epoxides as carcinogens (Section IV.D.). Therefore, the introduction of cyclodiene insecticides that do not contain readily oxidizable double bonds would seem to be an environmental improvement. Büchel _et al_. (1964) have described β-dihydroheptachlor or 2-_exo_-4,5,6,7,8,8-heptachloro-4,7-methano-4,4a,7,7a-tetrahydroindane, which is highly insecticidal ($LD_{50}$ _Musca_ 3.75, $LC_{50}$ _Culex_ 0.079), yet has an oral $LD_{50}$ to the rat and mouse of >5000 mg/kg. β-dihydroheptachlor and other cyclodienes lacking reactive double bond are susceptible to oxidative detoxication and biodegradation by hydroxylation, as discussed by Brooks (Chapter V). It seems remarkable that no systematic effort has been made to develop them as practical insecticides.

Now that the structures of the active ingredients of toxaphene are being elucidated (Casida _et al_., 1974), it seems possible that biodegradable compounds may be found or devised by structural modification.

## VI. SUMMARY AND CONCLUSIONS

The organochlorine insecticides originating in the post-World War II period have been the mainstay of insect pest control both in agriculture and public health for more than 25 years. The success of these cheap persistent chemicals in insect control has made it possible for the first time to control vector-borne diseases such as typhus and malaria and to obtain previously unheralded increases in crop production and timber protection. At the same time the rapid increase in insect resistance to these compounds and their extraordinary environmental persistence has forced the realization that they cannot be used indefinitely and in exponentially increasing quantities without grave consequences to agriculture and public health, and to the quality of the environment. Therefore, most of the present-day organochlorine insecticides are being withdrawn from use.

Although substitute compounds have been developed from the organophosphorus, carbamate, and pyrethroid compounds, there will remain for the next 25 to 50 years a demand for cheap, persistent, yet biodegradable insecticides to solve increasingly serious problems in agriculture and public health. There is, therefore, a great opportunity to develop satisfactory new insecticides with target-site action similar to the present-day organochlorine insecticides. These new compounds must incorporate the principles of biodegradability, for as Abelson (1970) has stated "Companies producing fat-soluble nonbiodegradable organic chemicals should give careful attention to the question of what they may responsibly set loose in the environment." Because the incorporation of degradophores into the organochlorine structures inevitably decreases the in vivo stability of the compounds, and also affects their environmental persistence, it would appear that some sacrifice in long-term persistence and in intrinsic insecticidal activity may be required in order to protect environmental quality. This sort of choice, for example, the substitution of methoxychlor for DDT or endosulfan for dieldrin, will greatly improve the benefit/risk involved in the use of insecticides.

## REFERENCES

1. Abelson, P. 1970. Science 170, 495

2. Albone, E. S., Eglinton, G., Evans, N. C., and Rhead, M. M. 1972. Nature 240, 420.

3. Anonymous. April 14, 1947. Time Magazine 49, 75.

4. Arundel, G. E. June, 1948. Reader's Digest 52, 22.

5. Baldwin, M. K., Robinson, J., and Parke, O. U. 1970. J. Agr. Food Chem. 18, 1117.

6. Bann, J. M., De Cino, T. J., Earle, N. W., and Sun, Y. P. 1956. J. Agr. Food Chem. 4, 937.

7. Barker, R. J. 1960. J. Econ. Entomol. 53, 35.

8. Barnett, J. K. and Dorough, H. W. 1974. J. Agr. Food Chem. 22, 612.

9. Bayne Jones, S. 1950. Yale J. Biol. Med. 22, 483.

10. Blaquiere, C., Bodenstein, G., Demozay, D., Herbst, M., Marechal, G., and Sieper, H. 1973. Lindane--Monograph of an Insecticide, K. Schillinger, Freiburg, Germany.

11. Bradbury, F. R. and Standen, H. 1959. Nature 183, 983.

12. Breidenback, A. W., Gunnerson, C. G., Kawahara, F. K., Lichtenberg, J. S., and Green, K. S. 1967. Publ. Health Repts. 82, 139.

13. Brooks, G. T. 1973. Chlorinated Insecticides, CRC Press, Cleveland, Ohio.

14. Brown, A. W. A. 1958. "Insecticide Resistance in Arthropods," WHO, Geneva.

15. Brown, A. W. A. 1971. Pesticides in the Environment, (White-Stevens, R., Ed.), vol. I, pt. II, p. 457, Marcel-Decker, New York.

16. Brown, A. W. A. and Pal, R. 1971. "Insecticide Resistance in Arthropods," WHO, Geneva.

17. Bruce-Chwatt, L. J. 1970. Misc. Pub. Entomol. Soc. Amer. 7(1), 7.

18. Bruce-Chwatt, L. J. 1971. Bull. W.H.O. 44, 419.

19. Büchel, K. H., Ginsburg, A. E., Fischer, R., and Korte, F. 1964. Tetrahedron Lett. 33, 2267.

20. Cade, T. J., Lincer, J. L., White, C. M., Roseneau, D. G., and Swartz, L. G. 1971. Science 172, 955.

21. Carter, L. J. 1974. Science 186, 239.

22. Casida, J. E., Holmstead, R. L., Khalifa, S., Knox, J. R., Ohsawa, T., Palmer, K. J., and Wong, R. Y. 1974. Science 183, 520.

23. Casida, J. E. and Lykken, L. 1969. Ann. Rev. Plant Physiol. 20, 607.

24. Charton, M. 1969. Amer. Chem. Soc. 91, 615.

25. Chemical and Engineering News. April 22, 1974. p. 33.

26. Coats, J. R., Metcalf, R. L., and Kapoor, I. P. 1974. Pest. Biochem. Physiol. 4, 201.

27. Conney, A. H. and Burns, J. J. 1972. Science 178, 576.

28. Cristol, S. J. 1945. J. Amer. Chem. Soc. 67, 1494.

29. Crosby, D. R. 1972. Environmental Photooxidation of Pesticides, p. 260. In "Degradation of Synthetic Organic Molecules in the Biosphere," Nat. Acad. Sci., Washington, D. C.

30. Curley, A. and Kimbrough, R. 1969. Arch. Environ. Health 18, 156.

31. Cutkomp, L. K., Yap, H. H., Vea, E. V., and Koch, R. B. 1971. Life Sci. 10, 1201.

32. Datta, P. R., Laug, E. P., and Klein, A. K. 1964. Science 145, 1052.

33. Davidow, B. and Radowski, J. 1953. J. Pharmacol. Exp. Therap. 107, 259.

34. Desaiah, D., Cutkomp, L. K., and Koch, R. B. 1974. Pest. Biochem. Physiol. 4, 232.

35. Duggan, R. E. 1969. Ann. N. Y. Acad. Sci. 160, 173.

36. Dupire, A. and Raucourt, M. 1943. Compt. Rend. Acad. Agr. France 29, 470.

37. Durham, W. F. 1969. Ann. N. Y. Acad. Sci. 160, 183.

38. Edwards, C. A. 1964. Soils and Fertility 27, 451.

39. E. P. A. 1972. "The Pollution Potential in Pesticide Manufacture," Office of Water Programs, Pesticide Study Series 5, TS-00-72-04, Washington, D. C.

40. Fahmy, M. A., Fukuto, T. R., and Metcalf, R. L. 1973. J. Agr. Food Chem. 21, 585.

41. Faur, N. and Kemen, T. 1968. Hygiena Sanit. 2, 86.

42. Feil, V. J., Hedde, R. D., Zaylskie, R. G., and Zachrison, C. H. 1970. J. Agr. Food Chem. 18, 120.

43. Finkenbrink, W. 1956. Nachbl. Deutsch. Pflsch, Dienst. 8, 183.

44. Freal, J. J. and Chadwick, R. W. 1973. J. Agr. Food Chem. 21, 424.

45. Gannon, N. H. and Bigger, J. H. 1958. J. Econ. Entomol. 51, 1.

46. Gannon, N. H. and Decker, G. C. 1958. J. Econ. Entomol. 51, 8.

47. Giannotti, O., Metcalf, R. L., and March, R. B. 1956. Ann. Entomol. Soc. Amer. 49, 588.

48. Gibson, J. R., Ivie, G. W., and Dorough, H. W. 1972. J. Agr. Food Chem. 20, 1246.

49. Gorbach, S. G., Christ, O. E., Kellner, H. M., Kloss, G., and Borner, E. 1968. J. Agr. Food Chem. 16, 950.

50. Grover, P. L. and Sims, P. 1965. Biochem. J. 96, 521.

51. Gunther, F. A., Blinn, R. C., Carman, G. E., and Metcalf, R. L. 1954. Arch. Biochem. Biophys. 50, 504.

52. Haas, H., Naher, M., and Blickenstaff, R. 1951. Ind. Eng. Chem. 43, 2875.

53. Hamelink, J., Waybrandt, R. C., and Ball, R. C. 1971. Trans. Am. Fish. Soc. 110, 207.

54. Hannon, M. R., Greichus, Y. A., Applegate, R. L., and Fox, A. C. 1970. Trans. Amer. Fish. Soc. 99, 496.

55. Harrison, H. L., Loucks, O. L., Mitchell, J. W., Parkhurst, D. F., Tracy, C. R., Watts, D. G., and Yannacone, V. J. Jr. 1970. Science 170, 503.

56. Hayes, W. J. Jr. 1963. "Clinical Handbook on Economic Poisons," U. S. Dept. Health, Education, and Welfare, Public Health Service Communicable Disease Center, Atlanta.

57. Hennessey, D. J., Fratantoni, J., Hartigan, J., Moorefield, H. H., and Weiden, M. J. 1961. Nature 190, 341.

58. Hilton, J. G. 1971. Report of the DDT Advisory Committee to the Administrator, Environmental Protection Agency. Final report.

59. Hirwe, A. S., Metcalf, R. L., and Kapoor, I. P. 1972. J. Agr. Food Chem. 20, 818.

60. Holan, G. 1969. Nature 221, 1025.

61. Holan, G. 1971. Bull. W.H.O. 44, 355.

62. Hunt, E. G. 1966. Biological Magnification of Pesticides. In "Scientific Aspects of Pest Control," Nat. Acad. Sci. Publ. 1402, 251.

63. Hunt, E. G. and Bischoff, A. I. 1960. Calif. Fish Game 46(1), 91.

64. Ivie, G. W., Gibson, J. R., Bryant, H. E., Begin, J. J., Barnett, J. R., and Dorough, J. W. 1974. J. Agr. Food Chem. 22, 646.

65. Jensen, S., Göthe, R., and Kindstedt, M. O. 1972. Nature 240, 442.

66. Kallman, B. J. and Andrews, A. K. 1963. Science 141, 1050.

67. Kapoor, I. P., Metcalf, R. L., Nystrom, R. F., and Sangha, G. K. 1970. J. Agr. Food Chem. 18, 1145.

68. Kapoor, I. P., Metcalf, R. L., Hirwe, A. S., Lu, P. -Y., Coats, J. R., and Nystrom, R. F. 1972. J. Agr. Food Chem. 20, 1.

69. Kapoor, I. P., Metcalf, R. L., Hirwe, A. S., Coats, J. R., and Khalsa, M. S. 1973. J. Agr. Food Chem. 21, 310.

70. Kearns, C. W., Ingle, L., and Metcalf, R. L. 1946. J. Econ. Entomol. 38, 661.

71. Khalifa, S., Mon, T. R., Engel, J. L., and Casida, J. E. 1974. J. Agr. Food Chem. 22, 653.

72. Klein, W., Kohli, J., Weisberger, I., and Korte, F. 1973. J. Agr. Food Chem. 21, 152.

73. Korschgen, L. J. 1970. J. Wildlife Manag. 34, 186.

74. Korte, F. 1967. Botyu-Kagaku 32, 46.

75. Kunze, F. M., Laug, E. P., and Prickett, C. S. 1950. Proc. Soc. Exp. Biol. Med. 75, 415.

76. Kurihara, N. and Nakajima, M. 1974. Pest. Biochem. Physiol. 4, 220.

77. Laug, E. P., Nelson, P., Arthur, A., Fitzhugh, O. G., and Kunze, F. M. 1950. J. Pharmacol. Exp. Therap. 98, 268.

78. Lawless, E. W., von Rümker, Rosemarie, and Ferguson, T. L. 1972. E.P.A. Pesticide Study Series 5, "The Pollution Potential in Pesticide Manufacturing," Technical Studies Report TS-00-72-04.

79. Levi, G. 1968. World Health Mag. 13 (April).

80. Lipke, H. and Kearns, C. W. 1959. J. Biol. Chem. 234, 2123.

81. Logan, J. 1953. The Sardinian Project - An Experiment on the Eradication of an Indigenous Malaria Vector, Johns Hopkins Press, Baltimore.

82. Maier-Bode, H. 1968. Residue Rev. 22, 1.

83. March, R. B. and Metcalf, R. L. 1949. Bull. Calif. State Dept. Agr. 38, 93.

84. M. I. T. 1970. "Man's Impact on the Global Environment," Massachusetts Institute of Technology, Cambridge.

85. Matsumura, F. and Patil, K. C. 1969. Science 166, 121.

86. Matsumura, F., Patil, K. C., and Boush, G. M. 1970. Science 170, 1206.

87. Matthews, H. B. and Matsumura, F. 1969. J. Agr. Food Chem. 17, 845.

88. McKinney, J. D. and Mehendale, H. M. 1973. J. Agr. Food Chem. 21, 1079.

89. Melnikov, N. N. 1971. Chemistry of Pesticides, Springer-Verlag, Berlin.

90. Menzie, C. M. 1969. "Metabolism of Pesticides," U. S. Dept. of the Interior, Washington, D. C.

91. Menzie, C. M. 1972. Ann. Rev. Entomol. 17, 199.

92. Metcalf, R. L. 1948. J. Econ. Entomol. 41, 416.

93. Metcalf, R. L. 1955. Organic Insecticides: Their Chemistry and Mode of Action, Interscience, New York.

94. Metcalf, R. L. 1965. Methods of Estimating Effects, p. 17. In Research in Pesticides, (C. O. Chichester, ed.), Academic Press, New York.

95. Metcalf, R. L. 1970. Amer. Zool. 10, 583.

96. Metcalf, R. L. 1971. Chemistry and Biology of Pesticides, p. 1. In Pesticides in the Environment, ( R. White-Stevens, Ed.), vol. 1, pt. 1, Dekker, New York.

97. Metcalf, R. L. 1972. DDT Substitutes, p. 25. In CRC Critical Reviews in Environmental Control, CRC Press, Cleveland, Ohio.

98. Metcalf, R. L. 1973. J. Agr. Food Chem. 21, 511.

99. Metcalf, R. L. and Fukuto, T. R. 1968. Bull. W.H.O. 38, 633.

100. Metcalf, R. L., Hess, A. D., Smith, G. E., Jeffrey, G. M., and Ludwig, G. L. 1945. Pub. Health Rep. 60, 753.

101. Metcalf, R. L., Kapoor, I. P., and Hirwe, A. S. 1971. Bull. W.H.O. 44, 363.

102. Metcalf, R. L., Kapoor, I. P., and Hirwe, A. S. 1972. Development of Persistent Biodegradable Insecticides Related to DDT, p. 244. In "Degradation of Synthetic Organic Molecules in the Biosphere," Nat. Acad. Sci., Washington, D. C.

103. Metcalf, R. L., Kapoor, I. P., Lu, P. -Y., Schuth, C. K., and Sherman, P. 1973. Environ. Health Perspectives, June, 35.

104. Metcalf, R. L., Sanborn, J. R., Lu, P. -Y., and Nye, D. 1975. Arch. Environ. Contamination Toxicol. 3, 151

105. Miles, J. R. W., Tu, C. M., and Harris, C. R. 1969. J. Econ. Entomol. 62, 1334.

106. Missiroli, A. 1947. Riv. Parassitol. 8, 141.

107. Moorefield, H. H., Weiden, M. H., and Hennessey, D. J. 1962. Contrib. Boyce Thompson Inst. 21, 481.

108. Morgan, D. P. and Roan, C. C. 1974. Essays Toxicol. 5, 39.

109. Mrak, E. F., Chairman. 1969. "Secretary's Commission on Pesticides and Their Relationship to Environmental Health," U. S. Dept. Health, Education, and Welfare, Washington, D. C.

110. Müller, P. 1955. Das Insekitzid Dichlorodiphenyl-trichloroethane und Seine Bedeutung, vol. I, Birkhauser-Verlag, Basel.

111. Mullins, L. J. 1956. Amer. Inst. Biol. Sci. Publ. 1, 123.

112. Narahashi, T. and Haas, H. G. 1968. J. Gen. Physiol. 51, 177.

113. O'Brien, R. D. 1967. Insecticides, Action, and Metabolism, Academic Press, New York.

114. O'Brien, R. D. 1973. Science 182, 222.

115. Peakall, D. 1970. Science 168, 592.

116. Pesticide Review. 1972. U.S.D.A. Washington, D. C.

117. Patel, T. B. and Rao, U. N. 1958. Brit. Med. J. 1, 919.

118. Peterson, J. E. and Robinson, W. H. 1964. Toxicol. Appl. Pharmacol. 6, 321.

119. Pillai, M. K. K., Abedi, Z. H., and Brown, A. W. A. 1963. Mosquito News 23, 112.

120. Pimentel, D. 1971. "Ecological Effects of Pesticides in Nontarget Species," Executive Office of the President, Office of Science and Technology.

121. Rogers, E. F., Brown, H. D., Rasmussen, I. M., and Neal, R. E. 1953. J. Amer. Chem. Soc. 75, 2991

122. Schwemmer, B., Cochrane, W. P., and Polen, P. B. 1970. Science 169, 1087.

123. Slade, R. E. 1945. Chem. Ind. 314.

124. Soper, F. L., Andrews, J. A., Bode, K. F., Coatney, G. R., Earle, W. C., Keeney, S. M., Knipling, E. F., Logan, J. A., Metcalf, R. L., Quarterman, K. D., Russell, P. F., and Williams, L. L. 1961. Am. J. Trop. Med. Hyg. 10, 451.

125. Soper, F. L., David, W. A., Markham, F. S., and Riehl, L. A. 1947. Amer. J. Hyg. 45, 305.

126. Sternberg, J. 1963. Ann. Rev. Entomol. 8, 19.

127. Sternberg, J., Kearns, C. W., and Moorefield, H. H. 1954. J. Agr. Food Chem. 2, 1125.

128. Street, J. C. 1969. Ann. N.Y. Acad. Sci. 160, 274.

129. Street, J. C. and Blan, S. E. 1972. J. Agr. Food Chem. 20, 395.

130. Sullivan, W. 1972. New York Times, July 13.

131. Tarján, R. and Kemány, T. 1969. Food Cosmet. Toxicol. 7, 215.

132. Terracine, B. 1973. Indust. Med. 42, 19.

133. Tsukamoto, M. 1959. Botyu-Kagaku 24, 141.

134. U.S.D.A. 1969. "Summary of Registered Agricultural Pesticide Chemical Uses," Ed. III, vol. III.

135. Vingiello, F. A. and Newallis, P. E. 1960. J. Org. Chem. 25, 905.

136. Walker, A. I. T., Stevenson, D. E., Robinson, J., Thorpe, E., and Roberts, M. 1969. Toxicol. Appl. Pharmacol. 15, 345.

137. Walker, A. I. T., Thorpe, E., and Stevenson, D. E. 1972. Food Cosmet. Toxicol. 11, 415.

138. Wiersma, G. B., Tai, H., and Sand, P. F. 1972. Pest. Monit. J. 6, 194.

139. Weisgeiber, I., Kohli, J., Kaul, R., Klein, W., and Korte, F. 1974. J. Agr. Food Chem. 22, 609.

140. Woodwell, G. M., Wurster, C. F. Jr., and Isaacson, P. A. 1967. Science 156, 821.

141. W.H.O. Tech. Rept. Series No. 513. 1973. Safe Use of Pesticides.

142. W.H.O. 1971. Vector Control, WHO Chron. 25(5).

143. Zeidler, O. 1874. Ber. Deut. Chem. Ges. 7, 1180.

# Insecticidal Activity and Biodegradability of Lindane Analogues

MINORU NAKAJIMA
Department of Agricultural Chemistry
Kyoto University
Kyoto, Japan

## I. INTRODUCTION

Lindane, the γ-isomer of BHC, is a potent insecticide. It has a very simple and yet very specific structure; the three vicinal chlorine atoms are axial, whereas the other three are in equatorial configurations. We have been synthesizing a number of lindane analogs in which one or two of the six chlorine atoms are replaced by other substituents, that is, H, F, Br, I, Me, SMe, or alkoxy. The compounds are broadly classified according to the position of substituents, other than chlorine, as *meso*-type and *dl*-type analogs. They were synthesized by the routes classified as A-F in Table 1. Their structures and configurations were confirmed by elementary analyses for C, H and halogen, and pmr spectroscopy.

## II. INSECTICIDAL ACTIVITY

The $LD_{50}$ values against the mosquito (*Culex pipens pallens*), housefly (*Musca domestica*), and German cockroach (*Blattella germanica*) were determined by the topical application method. For compounds having an Me, alkoxy, or alkylthio group, that would be susceptible to microsomal oxidation, the synergized $LD_{50}$ values were also determined by the application of piperonyl butoxide (PB) prior to the insecticide application. As shown in Tables 2 and 3, the analogs exhibit various degrees of activity. In general, the *meso*-type analogs are more active than the *dl*-type analogs. It is remarkable that some of the *meso*-type analogs are almost as active as lindane, especially against mosquitoes.

The synergistic ratios (SR) for activity against the mosquito and German cockroach (except for the compound VI) are generally low and show no significant variation with structural modification. This indicates that the variation in insecticidal

TABLE 1. Synthetic Methods of Lindane Analogs

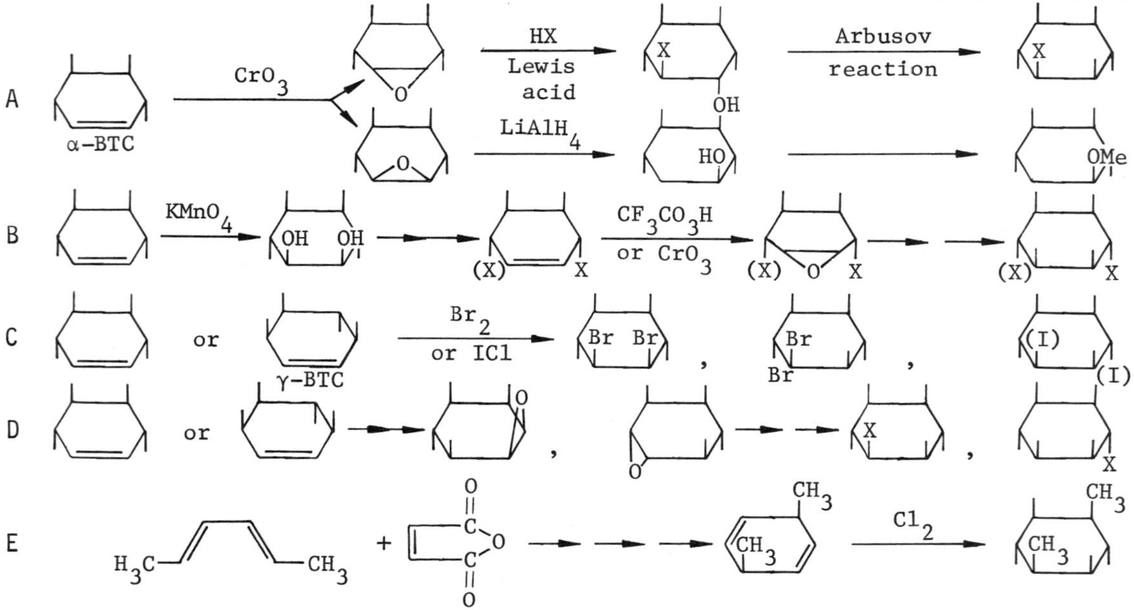


F Methylation of *muco*-inositol with $(CH_3)_2SO_4$

The short vertical lines represent Cl substituent.

TABLE 2. Lindane Analogs and Their Insecticidal Activity Against *C. pipiens pallens*

| Number | Compound | m.p.°C | Synthetic Method | $LD_{50}$ [a] |
|---|---|---|---|---|
| *Meso-type* | | | | |
| I | lindane | 112 | - | 1.5 |
| II | 3-Br | 106 | B | 0.59 |
| III | 3,6-di-Br | 106 | B | 0.93 |
| IV | 3-I | 129 | C | >600 |
| V | 3-OMe | 97 | B | 3.9 |
| VI | 3-SMe | 142 | D | 1.5 |
| VII | 3-OH | 130 | D | >590 |
| VIII | 3-OAc | 117 | D | >510 |
| *DL-Type* | | | | |
| IX | 1-H | 109 | A | 66 |
| X | 1-F | 125 | A | 6.6 |
| XI | 1,2-F,Br | 105 | A | 12 |
| XII | 1-Br | 121 | A | 29 |
| XIII | 1,2-di-Br | 123 | C | 170 |
| XIV | 1,6-di-Br | 114 | C | 1200 |
| XV | 1,2,3-tri-Br | 128 | B | >250 |
| XVI | 1-I | 106 | C | 1600 |
| XVII | 1,2-H,OMe | 83 | A | - |
| XVIII | 1-OMe | 72 | A | 15 |
| XIX | 1-OEt | 90 | A | 68 |

(continued)

TABLE 2 (cont.)

| Number | Compound | m.p.°C | Synthetic Method | $LD_{50}$[a] |
|---|---|---|---|---|
| DL-type (cont.) | | | | |
| XX | 1-Me | 74 | A | 5.1 |
| XXI | 1,4-di-Me | 136 | E | 100 |
| XXII | 1-CN | 159 | A | >570 |
| XXIII | 1-$CONH_2$ | 178 | A | I[b] |
| XXIV | 1-$CO_2H$ | 160 | A | I[b] |
| XXV | 1-$CH_2OH$ | 70 | A | I[b] |
| XXVI | hexa-OMe | syrup | F | >610 |
| XXVII | α-isomer | 158 | - | 1000 |

[a] Topical $LD_{50}$ μmole x $10^4$/mosquito 3 to 5-day-old female adult; reared in laboratory of Dainippon Jochugiku Co., Ltd. after 24-hour motality at 26 ± 1°C.

[b] I = Inactive.

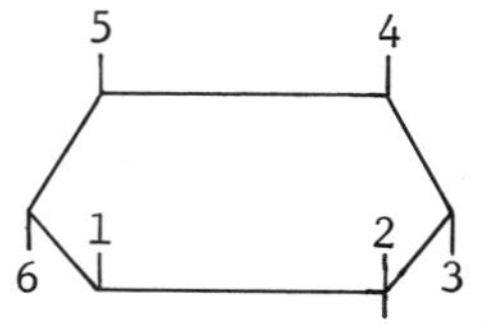

TABLE 3. Insecticidal Activity of Biodegradable Lindane Analogs

| | | Topical $LD_{50}$ (μmole x $10^4$/insect) | | | | | | | | |
|---|---|---|---|---|---|---|---|---|---|---|
| | | *C. pipiens pallens*[a] | | | *Musca domestica* ($S_{NAIDM}$)[b] | | | *Blattella germanica*[c] | | |
| No. | Compound | Alone | With P.B. | SR | Alone | With P.B. | SR | Alone | With P.B. | SR |
| Meso-type | | | | | | | | | | |
| I | lindane | 1.5 | 0.68 | 2.2 | 0.83 | 0.14 | 5.7 | 18 | 4.8 | 3.8 |
| V | 3-OMe | 3.9 | 2.4 | 1.7 | 37 | 0.77 | 48 | 35 | 12 | 2.9 |
| VI | 3-SMe | 1.5 | 0.91 | 1.6 | 30 | 0.26 | 120 | 790 | 6.6 | 120 |
| DL-type | | | | | | | | | | |
| IX | 1-H | 66 | 42 | 1.6 | 16 | 4.3 | 3.6 | 160 | 98 | 1.6 |
| XVII | 1,2-H,OMe | - | - | - | 200 | 50 | 4.0 | >1600 | ≒1600 | >1 |
| XVIII | 1-OMe | 15 | 10 | 1.4 | 120 | 3.8 | 31 | 200 | 37 | 5.4 |
| XIX | 1-OEt | 68 | 53 | 1.3 | 280 | 30 | 9.2 | 930 | 330 | 2.8 |
| XX | 1-Me | 5.1 | 3.2 | 1.6 | 17 | 0.48 | 35 | 160 | 27 | 5.8 |
| XXI | 1,4-di-Me | 100 | 67 | 1.5 | 130 | 4.8 | 28 | >1600 | 260 | 6.2 |
| XXVI | hexa-OMe | >610 | >300 | - | >1500 | >1500 | - | >1500 | >1500 | - |

[a] 3 to 5-day-old female adult; reared in laboratory of Dainippon Jochugiku Co., Ltd., after 24-hour mortality at 26±1°C. P.B.; piperonyl butoxide, 0.25 μg/mosquito was applied as 0.1% acetone solution 1 hr before the application of insecticides. SR; synergistic ratio = $LD_{50}$ alone/$LD_{50}$ with P.B.

[b] 4 to 5-day-old female adult; reared in laboratory of Sumitomo Chemical Co., Ltd., after 24-hour mortality at 25±1°C. P.B.; 25 μg/fly was applied as 5% acetone solution 1 hr before the application of insecticides.

(continued)

TABLE 3 (footnotes cont.)

[c] 2 to 3-week-old male and female (1:1); reared in laboratory of Sumitomo Chemical Co., Ltd., after 72-hr mortality. P.B.; 50 μg/cockroach was applied as 5% acetone solution 1 hr before the application of insecticides.

activity among the analogs is not critically controlled by the oxidative degradation in these insects. The exceptionally high SR value for compound VI with the German cockroach may be due to the use of metabolic pathways other than microsomal oxidation in the transformation of the SMe group. However, the SR values observed for houseflies show significant variation with a change in structure. The high SR values of compounds V, VI, XVIII, XX, and XXI suggest a potent metabolic activity for the housefly mixed-function oxidase (mfo) system, as well as a high biodegradability for these compounds. The rates of detoxication can be ranked as: *meso*-SMe(VI) > *meso*-OMe(V) > *dl*-Me(XX) ≒ *dl*-OMe(XVIII) ≒ *dl*-$Me_2$(XXI) > *dl*-OEt(XIX) > lindane(I) ≒ *dl*-H(IX) ≒ *dl*-H,OMe(XVII). This sequence is very similar to, but not quite the same as, that observed for the symmetrical DDT analogs by Metcalf (see Chapter IX). This is probably because a conformational specificity might not be critical in the metabolism of susceptible groups. It shows that the mfo enzymes possibly catalyze sulfide oxidation, *O*-dealkylation, and side-chain oxidation. Here, the *meso*-SMe compound (VI) again shows a very high SR value that may be attributed in part to causes similar to those observed for the value with the German cockroach.

## III. MICROSOMAL METABOLISM

To confirm the structure of possible metabolites, *in vitro* metabolism experiments were performed according to the method of Tsukamoto and Casida using microsomal mfo preparations from susceptible female housefly abdomens. The results are shown in Table 4. The metabolites were identified with authentic samples synthesized independently by electron capture gas chromatography. The figures in Table 4 are the percentages of applied compounds metabolized to the identified metabolite only. The metabolites from 3-OMe(V), 3-SMe(VI), and 1,4-$Me_2$(XXI) analogs could not be identified, although there was evidence for significant degradation in their gas chromatograms. Therefore, the figures should not be taken directly as representing the ease of degradation.

TABLE 4. Microsomal Metabolism of Lindane Analogs

| Substrates | | Transformation | % Metabolism/mg Protein |
|---|---|---|---|
| XVII | 1,2-H,OMe | OMe → OH | 6.0 |
| XVIII | 1-OMe | OMe → OH | 8.3 |
| XIX | 1-OEt | OEt → OH | 7.7 |
| XX | 1-Me | Me → $CH_2OH$ | 0.43 |
| | Aldrin | Dieldrin | 4.6 |

Microsomal source; $S_{NAIDM}$, 3 to 5-day-old female. A standard incubation mixture; 1.92 x $10^{-7}$ mole substrate, 1-ml microsomal suspension (20 abdomens/ml), 3-ml incubation medium (pH.7.48, 0.25M-sucrose-0.1M-phosphate buffer), 4 μmole NADPH. Total volume 4 ml in 50-ml Erlenmeyer flask. Reactions were carried out aerobically, with shaking at 31 to 32°C for 60 min. The protein concentration was determined by the method of Lowry _et al_. using BSA as standard.

# Organophosphorus Insecticides

E. Y. SPENCER
Research Institute, Agriculture Canada
London, Ontario, Canada

## I. INTRODUCTION

Four years ago it was reported that the estimated loss to British agriculture and animal husbandry would be of the order of $1200 million if pesticides were abolished (Bruce-Chwatt, 1971). In the light of today's prices, this figure would be larger. There are areas in developing countries where, in the absence of protective agents, some crops are a total loss. We may deplore the use of large quantities of toxic substances in the environment, but the need to meet food shortages by high-yielding monoculture crops requires their protection from destruction by insect pests, disease, and weeds. Thus the minimum use of pesticides in an integrated program of crop protection, where the chemical is applied at the optimum time and most efficient manner so as to take advantage of natural predators, will be with us for some time to come. Accepting this as a fact, in spite of "alternatives" that are promoted from time to time with only occasional success, the challenge must be to improve the effectiveness of present-day insecticides as well as to look for modified or alternative methods of control.

The importance of insecticides in relation to food and health has been discussed already in this symposium. Their deficiencies, such as toxic hazards, pollution of the environment, development of resistance, and their variable selectivity, have also been reviewed. Therefore, because of the continued need for pesticides, the aim must be to reduce their deficiencies. The state of the art and possibilities for improvement of the pyrethroids and organochlorine insecticides have already been discussed, together with the function and potential of synergists. My topic is concerned with the organophosphorus insecticides, to be followed by a separate chapter on the carbamate insecticides.

It is rather remarkable that the main developments of four groups of organochlorine insecticides, together with the organophosphorus and carbamate antichlolinesterase insecticides, appeared during the decade beginning in 1940 and therefore added a great impetus to chemical pest control. The relatively stable, long-lasting, broad-spectrum organochlorine insecticides appeared as the answer to the control of many undesirable insects. They were most effective, but these favorable qualities of stability and broad-spectrum activity as well as lipid solubility were disadvantageous with respect to deleterious effects on the environment and brought most of them into disfavor. Subsequently, public opinion forced a great restriction in their use, often before adequate replacements were available. This has resulted in a vacuum that had to be filled in part by organophosphorus and carbamate insecticides. We are, therefore, presented with a new challenge associated with a broader use of these insecticides and the resultant increase in efficacy requirements.

## II. DEVELOPMENT OF ORGANOPHOSPHORUS INSECTICIDES

Although organophosphorus chemistry dates back to the middle of the nineteenth century, the development of a search for organophosphorus compounds with insecticidal activity did not begin until 1936. In that year Schrader, the chemist, and Kükenthal, the biologist, started their systematic search of esters and esteramides of phosphoric and thiophosphoric acids and pyrophosphates. Two products that were developed commercially at that time were tetraethylpyrophosphate (TEPP) and octamethyl pyrophosphoramide (schradan), the latter exhibiting the unique property of systemic activity, that is, being readily translocated through the plant. Later, there followed parathion, which is still in use although of high mammalian toxicity. In fact, its use is now increasing because of restrictions on organochlorine insecticides.

In the meantime, the biochemical lesion of these and related organophosphorus toxicants was being investigated and shown to be the inactivation of cholinesterase with the resultant accumulation of acetylcholine producing the characteristic cholinergic symptoms. From these studies it was soon demonstrated that there were unexplained differences between *in vitro* and *in vivo* activity in some cases. When highly pure toxicants were used, *in vitro* activity for schradan and parathion was relatively low by contrast with TEPP. It therefore appeared that enhanced *in vivo* activity for some organophosphorus insecticides must be due to an activation by the organism. The active metabolite from parathion was soon shown to be the oxygen analog, paraoxon. In the case of schradan the characterization of the active metabolite was more difficult (Heath *et al.*, 1955; Spencer *et al.*, 1957; Spencer, 1957) and eventually it was proved to be the hydroxymethyl derivative. It is of interest to note that the activation and subsequent oxidative demethylation demonstrated then (almost 20 years ago) is the same as the form currently demonstrated as one of the key metabolic pathways of activation and degradation.

This same reaction demonstrated the influence of functional groups and spatial configuration on the activity of closely related compounds. Thus by merely oxidizing one methyl group of schradan to a hydroxymethyl, the anticholinesterase activity was increased over 100,000 times (Spencer et al., 1957).

The three organophosphorus insecticides mentioned are highly toxic to mammals as well as insects. It was therefore a major development when a short time later an organophosphorus insecticide appeared with low mammalian toxicity, namely malathion. It was not long before the basis for this selectivity was explained. Likewise followed a ready explanation for the potentiation in mammalian toxicity when malathion was used with certain other organophosphorus compounds (O'Brien, 1967).

These rapid developments led to a ready explanation of toxic action by enzyme inhibition and metabolic detoxification by hydrolysis. The reaction of the phosphorus ester is determined by the magnitude of the electrophilic character of the phosphorus atom, the strength of the P-X bond in $(RO)_2P(O)$-X, and the steric effects of the substituents, particularly about the X group. Since the strength of the P-X bond is affected by these substituents, the range in activity and number of compounds possible becomes enormous. Metcalf, reviewing the impact of the development of organophosphorus insecticides on basic and applied science, stated in his presidential address to the Entomological Society of America, "the principles relating chemical structure to toxic action are so well understood today that a skillful chemist can produce a compound with some degree of biological activity at nearly every attempt," (Metcalf, 1959). The success at this time, in elucidating the basis for the selective toxicity of malathion together with the specific inhibition of cholinesterase, led to assumptions about the "rational development" of new organophosphorus insecticides (O'Brien et al., 1958). Although the trial-and-error as well as the imitative method had been highly profitable in the development of many commercial compounds, it was now assumed that this "rational" approach would be more fruitful. In fact, the basic triester phosphoric ester hypothesis with slight variations and correlation with cholinesterase inhibition, together with the "rational" approach, actually restricted the outlook in searching for promising leads. On the other hand, by following this approach many "irregularities" were found that did not fit the expected pattern and therefore spurred on further research to find a plausible explanation (Spencer, 1961). This rigid general formula, as a requisite for toxicological activity, also led to the incorrect conclusion that the primary path of degradation was via a hydrolase, accounting for any dealkylation of the phosphoric esters. It is now known that organophosphorus insecticides are metabolized by several enzymes in addition to the hydrolases, namely mixed-function oxidases, transferases, and others (Dauterman, 1971).

This assortment of biological processes, that vary in their relative contribution between organisms as well as their specificity for the particular organophosphorus insecticide, are major factors in contributing to differences in toxicity. Further variables include the rate of penetration and path to the target site, as well as the reactivity of the specific cholinesterase. These factors are distinct from some outside forces that also have an influence, some of which will be mentioned later.

In view of the complexity of these biological systems, it is not surprising that progress in the development of the large number of organophosphorus insecticides has been due to a mixture of trial and error, coupled with "educated" leads and a touch of serendipidity. Let us look at some examples to illustrate this and note consistencies, irregularities, and amazing specificity in the search for selective toxicity.

An illustration of unpredictable toxicity of related compounds with different substituents was recently reported for the phosphorylated pyridazinyl compound O-methyl O-[2-methyl-3-oxo-4-chloro-5(2H)-pyridazinyl] N-isopropyl-amidothiophosphate (Konecny, 1973). Replacement of P=S by P=O, isomerization of P=S to $S-CH_3$ or replacement of isopropylamido by an ethyl ester all reduced insecticidal activity unexpectedly.

An example of striking differences in toxicity to both the mouse and housefly, with minor changes in chemical substituents of dimethyl phenyl phosphates and phosphorothioates is shown in Table 1 (Metcalf and Metcalf, 1973).

Of the many derivatives examined, the MSR ranged from 1.0 to >880. Factors that contributed to these differences include differential enzyme inhibition, rates of O-demethylation, side chain oxidation, P=S desulfuration, selective penetration, and nucleophilic displacement. None of these could be predicted with any precision, so that in summation, when these factors are linked favorably for the mammal and lethally for the insect, a high selectivity factor results.

The relatively high activity of carboxyesterases in mammals, by contrast with their low level in insects, had earlier

TABLE 1.

| Compound | $LD_{50}$ | | $MSR^{a}$ |
|---|---|---|---|
| | Mouse | Housefly | |
| 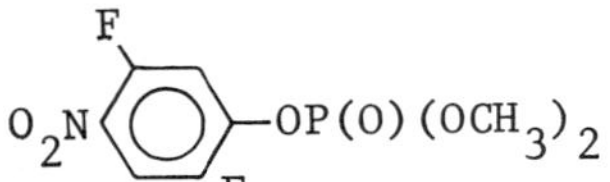 | 46 | 18 | 2.5 |
| 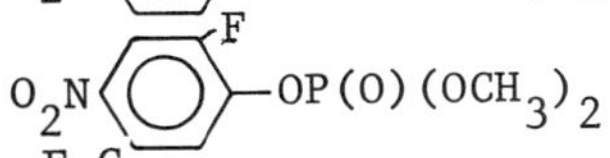 | 12 | 1.4 | 8.8 |
| 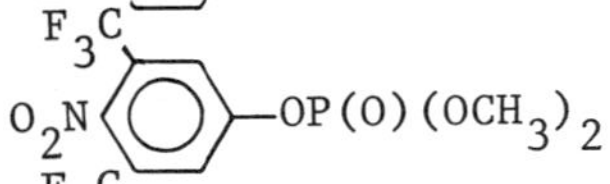 | 291 | 3.15 | 92 |
| 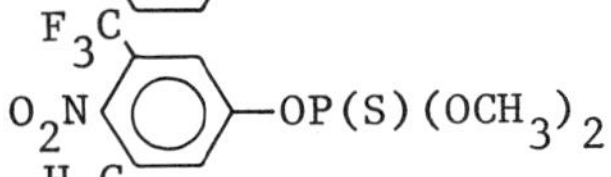 | >2000 | 3.1 | >650 |
| 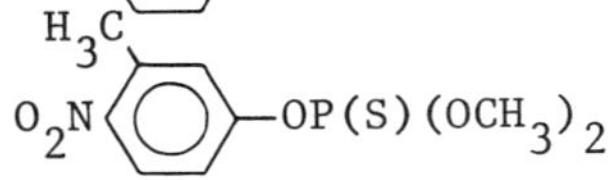 | 1500 | 2.4 | 630 |
| 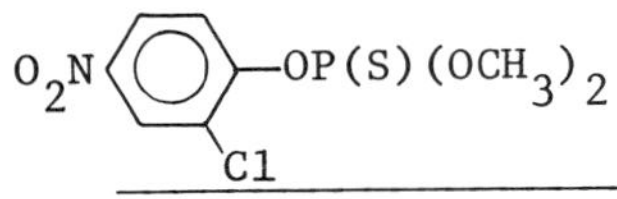 | >1500 | 1.7 | >880 |

[a] Mammalian selectivity ratio or $LD_{50}$ mouse/$LD_{50}$ fly.

accounted for the basis of the selective toxicity of such carboxyester-containing organophosphorus insecticides as malathion and the loss of this property when malathion was used together with carboxyesterase inhibitors (O'Brien *et al.*, 1958). In examining this hypothesis further, with carboxyester-containing vinyl phosphates and phosphorothioates, it was found that these esters were inert to carboxyesterases (Morello *et al.*, 1968). However, the hypothesis was supported to the extent that no potentiation in toxicity was found with agents that were effective with malathion. Although this illustrates the influence of steric configuration on enzyme specificity, the recent report of malaoxon monoacid in resistant houseflies arising from microsomal oxidation and not carboxyesterase activity, emphasizes the need for caution in proposing enzyme mechanisms from isolated metabolites (Welling *et al.*, 1974).

The nonconformity in the metabolism of some vinyl phosphate isomers is again shown in an examination of a more recently established metabolic pathway, namely the alkyl transferases with glutathione as the acceptor (Fukami *et al.*, 1966). This was shown to be an important detoxifying pathway, particularly with many dimethyl phosphorus ester insecticides, including the *cis* isomer of mevinphos. However, the *trans* isomer of this vinyl phosphate as well as both geometrical isomers of closely related vinyl phosphate Bomyl, were unaffected by this system but were hydrolyzed by a phosphatase to yield dimethyl phosphate (Morello *et al.*, 1968).

The mixed-function oxidases play a large role in the metabolism of organophosphorus insecticides and also show considerable specificity. They activate phosphorothionates and phosphorodithioates by desulfuration to yield the more toxic phosphates and phosphorothiolates, respectively. Oxidative *N*-dealkylation occurs to yield the intermediate *N*-hydroxymethyl or *N*-α-hydroxyethyl derivatives of *N*-methyl and *N*-ethyl compounds, respectively. In some cases, this is a large activation as cited earlier for the phosphoramide, schradan, whereas with many the intermediates formed differ only slightly in toxicity and subsequently split out formaldehyde or acetaldehyde, resulting in *N*-dealkylation. Oxidative dealkylation, on the other hand, results in inactivation of the toxicant with the production of the desalkyl derivative for organophosphates, but not for organophosphorothioates in the insect and mammalian systems thus far examined (Dauterman, 1971). By contrast, oxidative dearylation appears to be an important degradative step in the cleavage of the aryl-phosphate bond, particularly for phosphorothionates.

For those organophosphorus insecticides containing thioether components, it is assumed that mixed-function oxidases also play a key role in their activation by rapid conversion to the sulfoxide, and then more slowly to the sulfone. Mixed-function oxidases are probably also responsible for the oxidation of aliphatic side chains. In some cases, this may not be significant as far as any change in toxicity is concerned, one such example being the substituents of the aromatic component of diazinon. However, in others there is a tremendous increase in toxicity. Oxidation of one methyl substituent in tri-*ortho*-cresyl phosphate results in the formation of a cyclic phosphate with a 12,000,000 times increase in anticholinesterase activity.

Apart from the significance of mixed-function oxidases in various metabolic functions that affect the activity of the toxicant, the level of mixed-function oxidase activity in an organism can be a major factor in determining the ultimate toxicity of a chemical. For example, in the metabolism of the vinyl phosphate insecticide chlorfenvinphos by oxidative deethylation, the rate varied from 10% for the rat to almost 100% for the dog within a 2-hour period, with the rate for the rabbit and the mouse in between (Donninger, 1971). This was in direct correlation with the toxicity: $LD_{50}$ mg/kg: rat 10, mouse 100, rabbit 500, and dog >1200. Since this enzyme system is inducible by specific agents, the toxicity can be reduced by prior treatment of the animals before exposure to the toxicant.

Because the mixed-function oxidase levels show a large variation in activity in the life cycle of insects, there is a potential for increase in organophosphorus insecticide effectiveness by taking advantage of low levels in the life cycle to maximize toxicity for those compounds not requiring activation (Hollingworth, 1971). In other words, these low levels could represent a stage of enhanced susceptibility for some materials.

There are many examples of nonbiological modifications of insecticides by light and heat, often to more toxic compounds (Dauterman, 1971). A common early example was the inadvertent production of S-ethyl parathion by heat isomerization from parathion in attempting purification by fractional distillation. Recently, however, an *in vivo* isomerization has been reported with the isolation of S-methyl fenitrothion from pine seeds after several days incubation with fenitrothion (Hallet *et al.*, in press). The mechanism has not been determined, but the more active isomer is not an artifact and therefore broadens somewhat the performance of the compound originally applied.

In attempts to develop insecticides of greater specificity with low mammalian toxicity, the combined effects of carbamates and phosphoric esters have been investigated by derivitization of insecticidal carbamate esters with dimethyl or diethyl phosphoryl or phosphorothioyl groups. Since both insecticidal carbamates and organophosphorus compounds react with acetylcholinesterase, it was anticipated that the new joint compounds might have unusual activity. However, in one series examined none was found to be a superior inhibitor or insecticide (Hentnarski *et al*., 1972). Another hypothesis suggested that such compounds should have differential toxicity since they would have an O-C bond susceptible to cleavage in mammalian systems to nontoxic fragments, whereas the P-N bond would be cleaved by the insect to regenerate the original toxic carbamate (Fukuto, 1971). A differential toxicity was indeed found with an $LD_{50}$ of 32 for houseflies and 760 for the mouse. Activity against mosquitoes was sufficient so that further testing appeared warranted. In an earlier investigation, substituted dihydrobenzofuranyl groups were coupled with the organophosphorus moiety (Fahmy *et al*., 1970). They generally proved to be highly insecticidal and less toxic to the mouse than the corresponding carbamate. One of these (2,2-dimethyl-2,3-dihydrobenzofuranyl-7 *N*-methyl-*N*-dimethoxy-phosphino-thioyl) carbamates, a highly effective selective insecticide, when treated with *m*-chloroperbenzoic acid yielded on rearrangement a carbamoyl disulfide (Fahmy *et al*., 1974). However, in an attempt at preparation by an alternative route, the reaction of the carbamate (carbofuran) with sulfur monochloride in pyridine resulted in the *N*-substituted biscarbamoyl *mono*-sulfide. This compound retained its insecticidal properties yet was less toxic to the white mouse. Thus via a truly serendipitous route, another group of compounds with selective toxicity was discovered!

Because of the accepted general structural requirements for an organophosphorus anticholinesterase mentioned earlier, the emergence of methamidophos (Monitor®) (*O*-methyl *S*-methyl phosphoramidothioate) as an effective insecticide was a surprise because of its relatively simple structure. To further the astonishment, the high toxicity to flies is in contrast to the moderate anticholinesterase activity, and the reduced activity of the *N*-methyl derivatives was unexpected. The low mammalian toxicity of the *N*-acetyl derivative, acephate (Orthene®)-rats $LD_{50}$ 900 mg/kg, by contrast with that of 20 for Monitor®, is also difficult to explain. Finally, in connection with this unique compound and some derivatives, it has just been reported that in an attempt to establish physical and chemical parameters as a basis for the systemic movement of these materials, the following conclusions were made: "the foregoing results indicate that even within relatively restricted series of compounds, systemic movement in the cotton plant is not directly correlated with any single chemical or physical parameter," (Hussain et al., 1974). It therefore appears in the rather unpredicted behavior of this phosphoramidothioate, there are still many variables not accounted for.

Another unusual compound is phoxim (diethoxy-thiophosphoryloxyimino)-phenyl acetonitrile, in that its low mammalian toxicity is inconsistent with a high rate of mammalian cholinesterase inhibition, indicating that there must be factors other than target enzyme inhibition to explain its selectivity (Vinopal et al., 1971). However, to be an effective insecticide in the field, its volatility has to be reduced, which is done by incorporation into the soil during application (Harris, personal communication).

Although the main aim is to develop chemicals that are reasonably specific in their activity against undesirable organisms, the detrimental effect on nontarget organisms must not be ignored. It has been observed that chlorfenvinfos (2-chloro-1-(2,4-dichlorophenyl)-vinyl diethyl phosphate) reduced plant yield. Further study of this narrow-spectrum soil insecticide indicated that the plant cell membrane is the principal site of action. It inhibits water uptake and active transport, yet high concentrations inhibit membrane permeability and thus permit leakage of the cell contents. The effect is dependent on the state of plant growth and the presence of divalent cations, suggesting the possibility of interference of other chemicals in the soil with the phytotoxicity of the insecticide. The cis form is more active than the trans, whereas closely related vinyl phosphate, tetrachlorvinphos (2-chloro-1-(2,4,5-trichlorophenyl)-vinyl dimethyl phosphate), and phosphamidon (2-chloro-2-diethyl carbamoyl-1-methyl vinyl dimethyl phosphate) are without effect, emphasizing again the complexity of the structure-activity relationship (Lee and Wilkinson, 1973).

Other nontarget organisms of particular concern are beneficial soil animals such as earthworms, Collembola, and Carabidae. Biomass reductions for earthworms (Lumbricus terrestris L.) varied from severe for phorate and a closely related thioether, Counter®, to moderate (45%) for leptophos (Tomlin, personal communication). Collembola and Carabidae showed a similar variation in response to these toxicants.

Where the restrictions on the use of organochlorine insecticides for soil treatment has forced a consideration of their replacement with organophosphorus toxicants, new factors modifying the effectiveness of the toxicant have to be taken into account. For example, it has been shown that the availability of fensulfothion, a sulfoxide, when used on mineral soils, is greatly affected by cationic content. Especially $Fe^{+++}$ and $Al^{+++}$ increase the adsorption and therefore reduce the availability (Bowman, Annual Report, 1973, Research Institute, Agriculture Canada, London). The sulfoxides and sulfones of phorate and the closely related thioether, Counter®, behave in a similar manner, being more persistent than the parent material. Thus soil type influences the persistence, residue content, and effectiveness so that recommended application levels have to be varied accordingly. In searching for organochlorine replacements as soil insecticides, highly residual candidates, such as Cyolane® (AC47470-cyclic propylene (diethoxyphosphinyl) dithioimido-carbonate), have to be discarded in favor of moderately and slightly residual ones (Harris and Hitchon, 1970).

The improvement in effectiveness of organophosphorus insecticides to adequately replace the residual organochlorine insecticides has required a greater knowledge of the behavior of the insect, as the time of application is often more critical. The European corn borer in southwestern Ontario used to be controlled by DDT, and the timing of application was not critical. Subsequent attempts to use parathion as a replacement showed its effectiveness to be erratic (McLeod, personal communication). Investigations indicated that the variation in temperature within the region was such that time of emergence was influenced sufficiently to be critical for precise timing of application. In addition, the temperature difference at its extremes resulted in one- and two-generation areas. In an attempt to allow for this variable, day degree recorders were used at check points. A plot of numbers of moths caught in light traps versus day degrees for 1972 and 1973, gave excellent coincidence to the curves by contrast with plots on the calendar base. It is thus hoped by this method to be able to determine more precisely the timing of parathion application and so enhance its effectiveness and reduce the amount required for control.

Further improvement in performance of present-day organophosphorus insecticides can be made in some cases by modifying formulation and application. For example, control of the fruit fly with phorate in corn was greatest with top dressing, whereas furrow treatment was less effective (Walker and Turner, 1973). Another development that should reduce environmental contamination yet extend the field life of nonpersistent chemicals is the encapsulation technique. The toxicant is enclosed in minute capsules by materials that break down on exposure to sunlight, moisture, and air; thus hazard is reduced in handling and a nonpersistent toxicant is gradually released.

In summing up the prospects for improvement of present-day organophosphorus insecticides, it would seem that they are good. Increased knowledge of the biology of the insect can assist in more critical timing of application, with the resultant improved effectiveness and reduction in environmental contamination as well as the maximum use of natural predators. Improved methods of formulation will greatly assist in this. The combination of increased knowledge of the biochemistry of the organisms, their pathways of metabolizing toxicants, and structure-activity correlations will enable the development of even more effective organophosphorus insecticides that will include the use of educated guesses and chance discovery. In addition, with the main target enzyme acetylcholinesterase shown to be more complex than originally assumed, there is potential for the development of new types of inhibitors to attack a different part of the molecule of one of the more critical isoenzymes, and it might even be a different organophosphorus compound from the standard pattern (O'Brien, 1974).

The term "organophosphorus" insecticides has purposely been used throughout, rather than "phosphate," so that it would include phosphonates, of which there are several recently in commercial use, and other nonphosphate compounds. The monocyclic phosphates are of interest not only as anticholinesterases, but also as alkylating agents (Eto and Ohkawa, 1970). On the other hand, a bicyclic phosphorus ester has just been reported that has high toxicity without cholinesterase inhibition (Bellett and Casida, 1973).

## REFERENCES


1. Bellet, E. M. and Casida, J. E. 1973. Science 182, 1135.

2. Bowman, B. T. 1973. Annual Report, Research Institute, Agriculture Canada, London.

3. Bruce-Chwatt, L. J. 1971. W.H.O. Bull. 44, 419.

4. Dauterman, W. C. 1971. W.H.O. Bull. 44, 133.

5. Donninger, C. 1971. W.H.O. Bull. 44, 265.

6. Eto, M. and Ohkawa, H. 1970. Biochemical Toxicology of Insecticide, Academic, New York, p. 93-104.

7. Fahmy, M. A. H., Chiu, Y. C., and Fukuto, T. R. 1974. J. Agr. Food Chem. 22, 59.

8. Fahmy, M. A. H., Fukuto, T. R., Myers, R. D., and March, R. B. 1970. J. Agr. Food Chem. 18, 793.

9. Fukami, J. I. and Shishido, T. 1966. J. Econ. Entomol. 59, 1338.

10. Fukuto, T. R. 1971. W.H.O. Bull. 44, 411.

11. Hallett, D. J., Greenhalgh, R., Weinbeyer, P., and Prasad, R. Can. J. Forest. Res., in press.

12. Harris, C. R., personal communication.

13. Harris, C. R. and Hitchon, J. L. 1970. J. Econ. Entomol. 63, 2.

14. Heath, D. F., Lane, D. W. J., and Park, P. O. 1955. Philos. Trans. Roy.

15. Hentnarski, B. and O'Brien, R. D. 1972. J. Agr. Food Chem. 20, 543.

16. Hollingworth, R. M. 1971. W.H.O. Bull. 44, 155.

17. Hussain, M., Fukuto, T. R., and Reynolds, H. T. 1974. J. Agr. Food Chem., 22, 225.

18. Konecny, V. 1973. Pestic. Sci. 4, 775.

19. Lee, T. T. and Wilkinson, C. E. 1973. Pestic. Biochem. Physiol. 3, 341.

20. McLeod, D. G. R., personal communication.

21. Metcalf, R. L. 1959. Bull. Entomol. Soc. Am. 5, 3.

22. Metcalf, R. A. and Metcalf, R. L. 1973. Pestic. Biochem. Physiol. 3, 149.

23. Morello, A., Vardanis, A., and Spencer, E. Y. 1968. Can. J. Biochem. 46, 885.

24. O'Brien, R. D. 1967. Insecticides. Action and Metabolism, Academic, New York.

25. O'Brien, R. D. 1974. Amer. Chem. Soc., Los Angeles.

26. O'Brien, R. D., Thorn, G. D., and Fisher, R. W. 1958. J. Econ. Entomol. 51, 714.

27. Spencer, E. Y. 1957. Chem. Soc. Spec. Publ. 8, 171.

28. Spencer, E. Y. 1961. Can. J. Biochem. Physiol. 39, 1790.

29. Spencer, E. Y., O'Brien, R. D., and White, R. W. 1957. J. Agr. Food Chem. 5, 123.

30. Tomlin, A. D., personal communication.

31. Vinopal, J. H. and Fukuto, T. R. 1971. Pestic. Biochem. Physiol. 1, 44.

32. Walker, P. T. and Turner, C. R. 1973. Br. Insectic. Fungic. Conf., p. 183, Brighton.

33. Welling, W., deVries, A. W., and Voerman, S. 1974. Pestic. Biochem. Physiol. 4, 31.

## DISCUSSION

The discussion that followed Dr. Spencer's position paper on organophosphorus insecticides focused on problems associated with the use of present-day materials and on future needs with respect to agricultural use and protection of public health.

Concern of major dimension was expressed over the continued use of organophosphorus insecticides that are hazardous to man, domestic animals, and wildlife. There was general agreement that materials with high acute mammalian toxicity, for example parathion, should be replaced with compounds of lower toxicity, particularly in view of the large number of safe materials that have been discovered in recent years. However, far greater concern was expressed over the use of materials that produce insidious side effects in mammals, leading to eventual death after prolonged periods, even though the material was reasonably safe on an acute basis. A case in point is the phosphonothioate insecticide Phosvel®(leptophos), a compound that has been shown experimentally to produce delayed neurotoxic effects in chickens. This material has been accused of being responsible for the death of 1500 water buffalo that were inadvertently exposed to it in the upper Nile delta in 1971. A temporary permit allowing the application of this compound on tomato crops was recently issued in the United States in spite of evidence suggesting that it is a potentially dangerous compound. Apparently, EPA is well aware of the water buffalo incident in Egypt but is unable to rule against the use of leptophos because of the general absence of data affirming neurotoxic activity. Nevertheless, there was agreement among the conference participants that leptophos should be examined further before its use is allowed.

According to Dr. Barnes, other neurotoxic organophosphorus insecticides are being used in different parts of the world. However, they are not used in the United Kingdom because of general policy against any material for which delayed neurotoxic effects have been demonstrated. Some of these were introduced a few years ago as substitutes for dieldrin against soil insects. It was mentioned also that Ciba-Geigy voluntarily withdrew a promising new compound after it was discovered to be a demyelinating agent, even though this compound was in heavy demand because of its outstanding effectiveness in controlling sheep blow flies in Australia.

With respect to unexpected side effects produced by certain organophosphorus insecticides, it was suggested that studies be conducted to determine the specificity of these compounds in inhibiting different types of enzymes. The ability that organophosphorus esters have to phosphorylate biological tissue, for example, inhibition of chymotrypsin or binding to other protein besides cholinesterase, appears to vary from one compound to another, and studies of this nature are needed to anticipate and understand side effects produced by these compounds.

From the discussion that followed, it was clear that new compounds will be needed in the near future that are effective in controlling insects of public health significance. The types of compounds desired for this purpose were summarized by Mr. Wright, who described work conducted by WHO in their search for a substitute for DDT in the malaria eradication program. When this organization began searching for a DDT substitute, it was quickly discovered that the organophosphate fenthion was a very attractive material, and its attraction was that it could withstand the detoxyfying effects of mud walls in addition to being highly effective against adult mosquitoes. Of all the organophosphorus compounds examined to date, fenthion is the most effective in terms of long-term residual activity. This material is effective for as long as 4 months on mud walls but, unfortunately, cannot be used because it is one of the few compounds that has a cumulative toxic effect on the people living in the villages. The other extreme is malathion, a very safe material but one that when sprayed on mud walls lasts for about 2 weeks. Thus there is a difference of 4 months for fenthion and 2 weeks for malathion. Fortunately, however, further screening showed that fenitrothion possessed the virtues of both fenthion and malathion in that it was effective for 3 months on mud, about 4 months on thatch and other cellulose-based materials, and is so safe that it is not necessary to take more precaution than the normal hygienic ones. Fenitrothion is obviously the compound that will have the most effect on malaria in Africa. However, the major problem anticipated with this compound is that mosquitoes undoubtedly will develop resistance to it, not necessarily because of the high usage of it in mosquito control, but because fenitrothion obviously also will be used for agricultural purposes. All of this together will bring resistance faster. In the future suitable alternate compounds of the quality of fenitrothion, or some means, possibly through synergism, of finding a way to extend the life of a compound like fenitrothion from about 8 years, the expected period of usefulness, to about 12 years, should be developed.

There are approximately 20 companies collaborating with WHO at the present time in research concerned with the control of insect-disease vectors. Frequent visits are made to WHO by scientific staff members of these companies, where they are informed of the major problems that face WHO so that they can determine which of their compounds under development might fit into the WHO program. After the compounds are placed into the program, the companies are encouraged to go into the field to observe the trials, because methods of application and formulation sometimes are as important as the compound itself. Field trials usually are repeated three times, and attempts are made, with the company's cooperation, to improve the formulation each time. In this way the companies are able to obtain a clear picture of the problems that occur in the field. It is necessary that the field trials be conducted by WHO since no government will allow a company to apply experimental compounds in villages and huts where people live. WHO is prepared to test any compound without cost to a company and determine whether it might fit into the disease-vector control program.

The possibility of developing an insecticidal organophosphate that is a solid under field conditions was discussed. Evidently, solid organophosphates seem to stand up longer in the field than do liquids and, in some cases, are easier to formulate. However, very few organophosphate insecticides are solids, and this is attributable to the general physical properties of organophosphorus compounds that tend to be liquids.

The need for new materials for agricultural use also was discussed, with emphasis placed on needs related to insect-pest management. With the changeover in philosophy in recent years from a predominantly chemical to an integrated approach to pest control (pest management), obviously, new types of organophosphorus insecticides will be necessary that will fit into this kind of program. However, the need with respect to integrated pest management in agriculture are much more difficult to define than the need in public health protection. In developing new insecticides for pest-management purposes, full consideration should be given to the protection of natural enemies of insect pests. However, agricultural situations are extremely complex, and a wide array of natural enemies must be considered. Natural enemies that exist in one crop are different in other crops, and often differences exist within a single crop. For example, natural enemies in cotton in California may differ from those in Texas. Therefore it is not possible, except in a very general way, to indicate exact needs for the future. In general, compounds should be developed that provide differential toxicity between broad groups of natural enemies such as parasitic hymenoptera, certain diptera, certain hemiptera, coccinelids, and a few other coleoptera. It should be added that not all natural enemies of insect pests need to be protected, but those that are worth protecting are essential in a pest-management program.

Because of changes in insect-control practices, it was suggested that some of the older compounds, particularly those sitting on industrial shelves, be reexamined. In addition, old compounds that are not patented and, therefore, considered public property, should also be considered for development if they prove to be useful insecticides within the framework of current control methods. A good example of this is isopropyl parathion, a compound that a number of years ago was shown to be about 100-fold more toxic to a number of pests than it is to bees and hymenoptera parasites. It is also significantly safer to mammals than its analogs, parathion or methyl parathion. However, attempts to persuade the chemical industry to develop this compound have not been successful. This is attributed in part to industry's reluctance to invest large sums of money into a compound that they do not have proprietary control over.

It was pointed out that within the past 18 months, two well-founded corporations have been established that are prepared to consider old compounds for development. These corporations are evaluating compounds whose patents have expired and are prepared to manufacture and sell these materials, provided the material has been registered for usage. The high cost of developing information to satisfy registration requirements is a major stumbling block to the development of new compounds and is actually a deterrent to the development of old non-patentable materials. Compounds such as isopropyl parathion, because of their favorable properties of selectivity, are needed for use in insect-pest management, and it is unfortunate that industry is unwilling to develop materials for which patent protection is not available.

Compared to other classes of insecticides, the organophosphorus esters represent the single most important group of chemicals used in insect control. As a group, they are highly versatile materials, and organophosphorus esters are available for a variety of uses, for example, as contact insecticides, stomach poisons, systemic insecticides in both plants and animals, and fumigants. Despite the wide spectrum of materials available, the chemical industry has continued in its efforts to synthesize and evaluate organophosphates, and several new compounds are currently being examined for eventual development and sale. Interest in organophosphates still remains high in institutions outside of industry, where much of the fundamental work on the metabolism and mode of action of these compounds is conducted. Information originating from outside sources is of immense value to industry, and any fundamental breakthrough is usually quickly followed up. Unfortunately, and for obvious reasons, important discoveries made in industry often are slow in reaching print and there is a need for more informational flow.

# Carbamate Insecticides

T. R. FUKUTO
Department of Entomology
Division of Toxicology and Physiology
University of California, Riverside

## I. INTRODUCTION

Although enormous advances have been made during the past two decades in the development of carbamate esters as useful insecticides, relatively little effort has been channeled specifically for the design of carbamate insecticides with favorable properties of selectivity. For the most part, particularly in the early period of their development, the approaches taken for the design and synthesis of new carbamate insecticides have been empirical. Although the practice of analog synthesis and screening has resulted in the discovery of a number of useful insecticides, several of the commercially more prominent carbamate insecticides in current use possess the undesirable property of also being highly toxic to warm-blooded animals. Examples of these are carbofuran (2,2-dimethyl-2,3-dihydrobenzofuranyl-7 methylcarbamate) and aldicarb [2-methyl-2-(methylthio)propionaldehyde _O_-methylcarbamoyl oxime] with toxicity to mice ($LD_{50}$ or median lethal dose) of 2 and 0.3 to 0.5 mg/kg, respectively (Fahmy _et al._, 1970).

With increasing sophistication in approaches used for the discovery of new insecticidal chemicals, attributable mainly to increased information on the metabolism and mode of action of insecticides, greater emphasis has been placed on the design of materials less toxic to mammals, although attainment of high insecticidal activity remains the primary objective.

This chapter is concerned with recent attempts to discover selectively toxic carbamate insecticides, in other words, carbamates that are toxic to insects but relatively nontoxic to mammals, based on differences in rates and routes of metabolic transformations that different chemical moieties undergo in insects and mammals. Particular attention is given to our attempts to convert highly toxic methyl-carbamate insecticides into "safe" materials by derivatization studies. Discussion on the general aspects of the toxicology and mode of action of carbamate insecticides will be avoided and the reader is referred to recent review articles (Metcalf, 1971; Knaak, 1971; Fukuto, 1972) for background information.

## II. DERIVATIZED CARBAMATES

### A. *N*-Dimethoxyphosphinothioyl-*N*-Methylcarbamate Esters

The initial discovery that *N*-acylation of the methylcarbamyl moiety produced derivatives with reduced mammalian toxicity, while still retaining effectiveness against insects (Fraser *et al*., 1965; Fahmy *et al*., 1966), stimulated our interest several years ago in the examination of derivatized carbamates as potential insecticides. Derivatization of such common carbamate insecticides as carbaryl, Zectran®, Landrin, and propoxur with different aliphatic acyl moieties produced compounds that were much less toxic to mice than the parent carbamates. The acyl derivatives were quite effective insecticides, although, in general, they were less active than their respective carbamates. Of the acyl groups examined, derivatization with acetyl gave products with highest insecticidal activity. Because of the poorer anticholinesterase activity of the acyl derivatives relative to the parent carbamates, it was suggested that the carbamate, generated *in vivo* from the derivative, was responsible for intoxication. Support for this hypothesis was provided in a subsequent study (Miskus *et al*., 1969) in which the metabolic conversion of *N*-acetyl Zectran® in the spruce budworm into relatively large amounts of Zectran® was demonstrated. In contrast, *N*-acetyl Zectran® was detoxified in the mouse through hydrolysis of the carbamate ester moiety to nontoxic products.

The starting point for our approach was based on the selectivity of the safe organophosphorus insecticide malathion, for which it had been suggested that slow *in vivo* oxidation of malathion to the anticholinesterase malaoxon provided the opportunity for detoxifying enzymes in mammals, probably a carboxyesterase, to degrade malathion or malaoxon to nontoxic products (Krueger and O'Brien, 1959). In contrast, in malathion-resistant houseflies phosphotriesterase appeared to be one of the detoxifying enzymes responsible for the degradation of organophosphorus insecticides (Matsumura and Dauterman, 1964). With these peculiar differences in the metabolism of malathion between houseflies and mice in mind, it was expected that substitution of the hydrogen on the carbamyl nitrogen atom of toxic carbamate esters with a dialkoxyphosphinothioyl moiety would provide derivatives of reduced mammalian toxicity, owing to the opportunity in mammals for the occurrence of metabolic reactions leading to nontoxic products. The rationale for the approach is illustrated in the postulated metabolic pathways presented in Scheme 1.

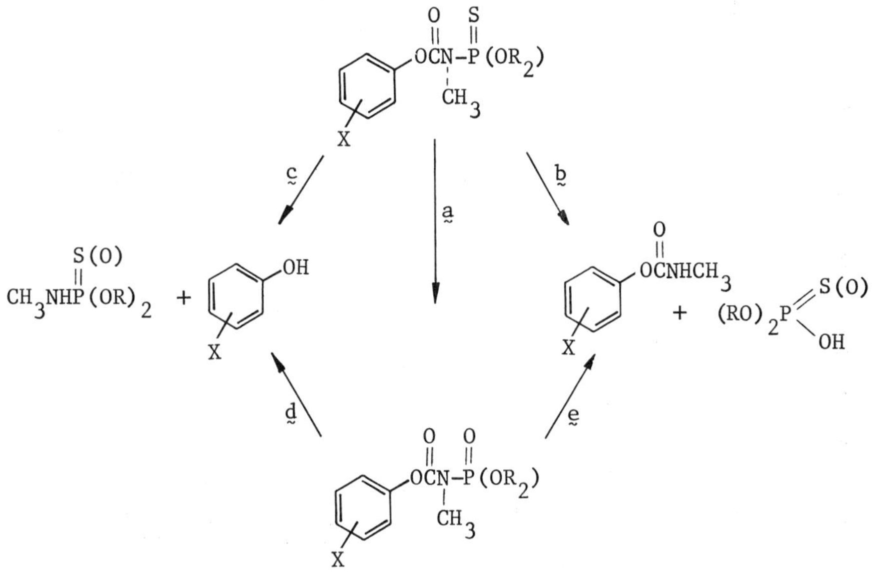

Scheme 1

It was anticipated that in houseflies, because of phosphotriesterase activity, pathways b or a-e would be the major routes for the metabolism of the N-dialkoxyphosphinothioyl derivative to produce in vivo lethal quantities of the original methylcarbamate insecticide. In the mouse, owing to the carboxyesterase activity, pathways c or a-d were expected to be of greater significance, resulting in the production of the nontoxic phenol and phosphoramidothioate or phosphoramidate. On the basis of this rationale, a number of N-dialkoxyphosphinothioyl derivatives were prepared and examined for toxicity to insects and mice. Evidence supporting this rationale is found in the toxicological data presented in Table 1. Included are $LD_{50}$ values (median lethal dose) for susceptible ($S_{NAIDM}$) and carba-

TABLE 1. Toxicological Properties of N-Dialkoxyphosphinothioyl)-N-Methylcarbamate

| Structure Number | R | $I_{50}$ (M) | $LD_{50}$ mg/kg | | |
|---|---|---|---|---|---|
| | | | Housefly | | Mouse |
| | | Fly AChE | $S_{NAIDM}$ | $R_{MIP}$ | (oral) |
| | m-Isopropylphenyl-OC(=O)N($CH_3$)R | | | | |
| 1 | H (MIP) | $3.7 \times 10^{-7}$ | 41 | 125 | 16 |
| 2 | $P(S)(OCH_3)_2$ | $6.6 \times 10^{-5}$ | 32.5 | 33.5 | 760 |
| 3 | $P(S)(OC_2H_5)_2$ | $2.9 \times 10^{-6}$ | 67.5 | 165 | 400-550 |
| | 2-Isopropoxyphenyl-OC(=O)N($CH_3$)R | | | | |
| 4 | H (propoxur) | $6.9 \times 10^{-7}$ | 22 | 45 | 24 |
| 5 | $P(S)(OCH_3)_2$ | $8.4 \times 10^{-6}$ | 32 | 37 | 1400 |
| 6 | $P(S)(OC_2H_5)_2$ | $1.4 \times 10^{-5}$ | 37 | 38 | 1700 |
| | $CH_3SC(CH_3)_2CH{=}NOC(=O)N(CH_3)R$ | | | | |
| 7 | H (aldicarb) | $8.4 \times 10^{-5}$ | 5.5 | 30 | 0.3-0.5 |
| 8 | $P(S)(OCH_3)_2$ | $2.9 \times 10^{-4}$ | 95 | 170 | 170 |

(continued)

TABLE 1 (cont.)

| Structure Number | R | $I_{50}$ (M) Fly AChE | $LD_{50}$ mg/kg Housefly $S_{NAIDM}$ | $R_{MIP}$ | Mouse (oral) |
|---|---|---|---|---|---|
| | 2,2-Dimethyl-2,3-dihydrobenzofuranyl-7-OC(=O)N($CH_3$)R | | | | |
| 9 | H (carbofuran) | $2.5 \times 10^{-7}$ | 6.7 | - | 2 |
| 10 | P(S)($OCH_3$)$_2$ | $9.3 \times 10^{-6}$ | 13 | - | 150-190 |

mate-resistant ($R_{MIP}$) strains of houseflies, the white mouse, and anticholinesterase activity ($I_{50}$ or molar concentration to inhibit 50% of the enzyme in 15 min) for housefly-head acetylcholinesterase. The results clearly show that the *N*-dialkoxyphosphinothioyl-*N*-methylcarbamate derivatives were substantially less toxic to the white mouse and at the same time, except for the aldicarb analog, were approximately of equal toxicity to houseflies as the parent carbamates. In addition, in some cases the derivatized carbamates were more toxic than the methylcarbamates to a strain of houseflies ($R_{MIP}$) that by selection pressure had developed resistance to *m*-isopropylphenyl methylcarbamate (MIP). Derivatization with the dimethoxyphosphinothioyl moiety reduced mouse toxicity of MIP 48-fold, propoxur 58-fold, aldicarb about 400-fold, and carbofuran about 80-fold.

The anticholinesterase activities of the derivatized carbamates against housefly-head acetylcholinesterase (HAChE) were consistently lower than the parent carbamates, indicating that the derivatives per se were not directly responsible for housefly toxicity. To support the rationale that was used for the design of these methyl-carbamate derivatives, the comparative metabolism of a representative compound, 2,2-dimethyl-2,3-dihydrobenzofuranyl-7 *N*-dimethoxyphosphinothioyl-*N*-methylcarbamate (**10**), was examined in the white mouse and houseflies, using materials labeled with $^{14}C$ in the ring or *N*-methyl moiety and with $^{32}P$ (Krieger *et al*., 1974). The major metabolites recovered from the internal parts of the housefly after topical treatment were carbofuran and 3-hydroxy-carbofuran, along with minor amounts of other secondary oxidative metabolites of carbofuran. In comparison, the white mouse and rat treated orally with $^{14}C$ ring-labeled **10** produced as major products recoverable from the urine the conjugates of the respective phenols derived from carbofuran, 3-keto-carbofuran and 3-hydroxy-carbofuran. These three phenolic conjugates accounted for approximately 75% of the administered dose. Overall, the data strongly supported the hypothesis that derivatized carbamates such as **10** are degraded in mammalian systems to nontoxic products, and the toxic methylcarbamate is generated *in vivo* in insects.

When the mouse was treated with $^{32}P$-labeled **10**, the primary urinary product, isolated in quantities approaching 70% of the administered dose, was dimethyl methylphosphoramidate. This, coupled with the almost equivalent amounts of phenolic derivatives that were recovered with $^{14}C$ ring-labeled **10**, indicated the pathway below (analogous to *a*–*d* in Scheme 1) as the principal route for the metabolism of **10** in the mouse.

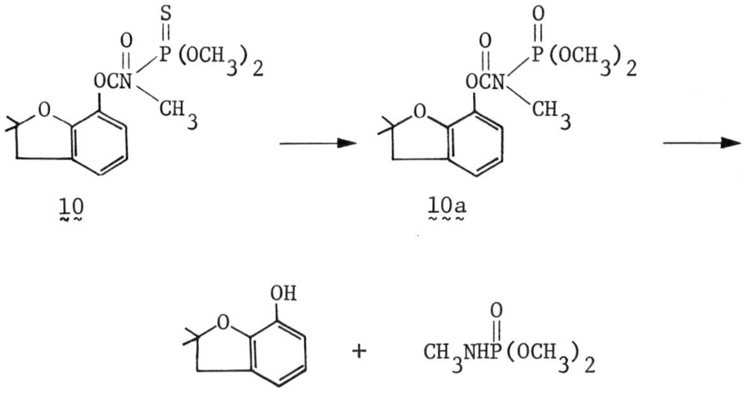


Scheme 2

## B. Biscarbamyl Sulfides

The proposed intermediacy of 2,2-dimethyl-2,3-dihydrobenzofuranyl-7 *N*-dimethoxyphosphinyl-*N*-methylcarbamate (10a) in the metabolism of 10 to the phenol and dimethyl methylphosphoramidate necessitated the synthesis of 10a to establish whether it coincided chromatographically with other metabolites isolated from the mouse after treatment with 10. Attempts to synthesize 10a by analogous condensation procedures used to prepare 10 (Fahmy *et al.*, 1970) resulted in failure, and the synthesis of 10a was accomplished by oxidation of 10 with *m*-chloroperbenzoic acid (Fahmy and Fukuto, 1972). When 10 was treated with *m*-chloroperbenzoic acid in methylene chloride, the products observed were 10a (40%), carbofuran (9) (13.7%), and an unexpected rearrangement product, the corresponding *N*-(dimethoxyphosphinylthio)carbamate (10b) (35%). The products are indicated in Scheme 3.

The rearrangement product 10b was unstable on silica gel thin-layer plates and decomposed after standing overnight to carbofuran and biscarbofuran disulfide (10c) according to Scheme 4.

In an attempt to prepare 10c by an independent method, 2 equivalents of carbofuran were treated with 1 equivalent of sulfur monochloride in the presence of pyridine. Unexpectedly, the product obtained was not 10c, and instead, the reaction

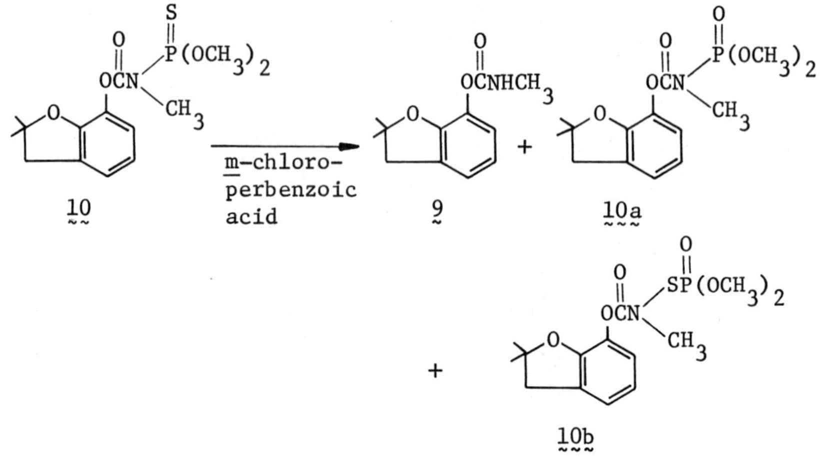


Scheme 3

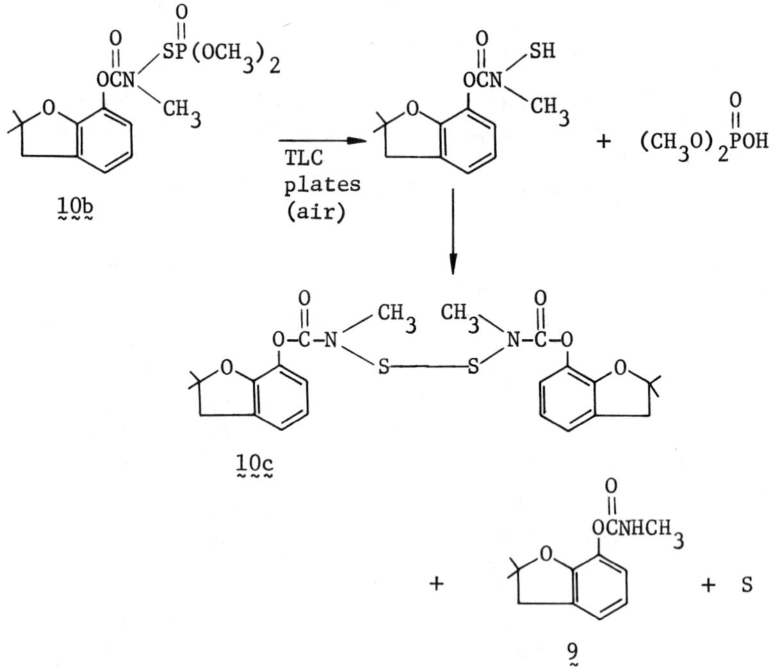


Scheme 4

biscarbofuran sulfide (10d) as indicated in Scheme 5. To our

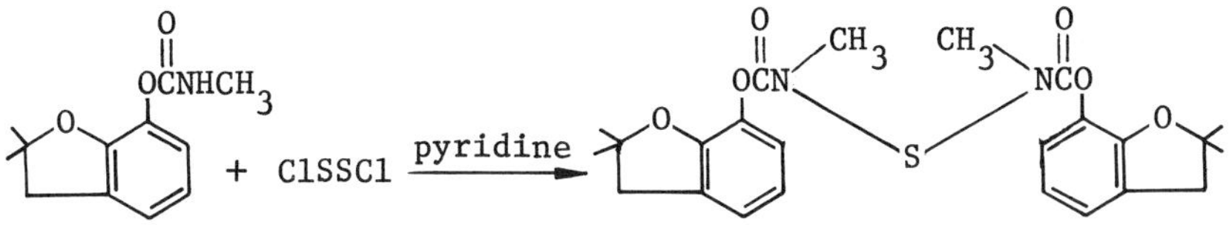


9

10d

Scheme 5

surprise, 10d proved to be quite insecticidal (comparable in activity to carbofuran) but possessed the added feature of being significantly less toxic to the white mouse than carbofuran.

Because of the favorable properties of selectivity exhibited by 10d, other biscarbamyl sulfides were synthesized and examined for toxicological properties (Table 2). From examination of the data in Table 2, it is apparent that the biscarbamyl sulfides retained much of the insecticidal activity exhibited by the corresponding methylcarbamate esters. Except for bisaldicarb sulfide (7d), the biscarbamyl sulfides were seven- to eight-fold more toxic to mosquito larvae (*Culex pipiens quinquefasciatus*) than the respective methylcarbamates. Further, as a systemic insecticide bisaldicarb sulfide (7d) proved to be substantially more effective in laboratory tests against the cotton aphid (*Aphis gossypii*), perforator (*Bucculatrix thurberiella*), and mite (*Tetranychus cinnabarinus*) than aldicarb. For example, at the approximate dosage of 1.2 lb of actual material per acre, 7d gave virtually complete control of cotton aphids and perforator for over 21 weeks, as compared to 15 weeks for aldicarb.

The mouse toxicity data in Table 2 show that compared to the original methylcarbamates, the biscarbamyl sulfides were from 5- to 50-fold less toxic to the white mouse after oral treatment. The largest difference in mouse toxicity was observed between carbofuran (10) and biscarbofuran sulfide (10d) (25- to 50-fold), and the smallest difference was between aldicarb (7) and bisaldicarb sulfide (7d) (five-fold).

On an overall basis, the biscarbamyl sulfides were less effective anticholinesterases than the present methylcarbamates

TABLE 2. Toxicological Properties of Insecticidal Methylcarbamate Esters and Their Corresponding N-Substituted Biscarbamyl Sulfides

| | | $LD_{50}$ (mg/kg) | | | | Anticholinesterase Activity $k_i$ ($M^{-1}$ $min^{-1}$) | |
|---|---|---|---|---|---|---|---|
| | | Housefly | | | | | |
| Structure Number | Compound | Alone | +P.B.[a] | Mosquito Larvae | Mouse (oral) | Bovine Erythrocyte | Fly-head |
| 9 | Carbofuran | 6.7 | 0.9 | 0.052 | 2 | $1.9 \times 10^6$ | $1.3 \times 10^7$ |
| 10d | Biscarbofuran sulfide | 19 | 1.2 | 0.007 | 50-100 | $2.5 \times 10^4$ | $9.2 \times 10^4$ |
| 1 | 3-Isopropylphenyl methylcarbamate (MIP) | 41 | 1.6 | 0.038 | 16 | $7.5 \times 10^5$ | $7.7 \times 10^5$ |
| 1d | Bis(MIP)sulfide | 85 | 4.6 | 0.0056 | 200 | $2.7 \times 10^4$ | $2.3 \times 10^4$ |
| 4 | Propoxur | 22 | 0.9 | 0.33 | 24 | $4.3 \times 10^4$ | $1.2 \times 10^6$ |
| 4d | Bispropoxur sulfide | 35 | 2.5 | 0.041 | 700 | $4.6 \times 10^3$ | $2.8 \times 10^4$ |
| 7 | Aldicarb | 5.5 | 1.0 | 0.16 | 0.3-0.5 | $3.5 \times 10^4$ | $2.0 \times 10^4$ |
| 7d | Bisaldicarb sulfide | 8.5 | 3.2 | 0.17 | 1.6-2.5 | $7.4 \times 10^3$ | $2.1 \times 10^3$ |

[a]Piperonyl butoxide (P.B.) was applied at a constant dose of 40 μg/fly in combination with varying doses of insecticide.

against both bovine erythrocyte AChE (BAChE) and housefly-head AChE (HAChE). In the case of BAChE, in other words an AChE from a mammalian source, the reduction in anticholinesterase activity of the biscarbamyl sulfides was more or less in agreement with reduction in mouse toxicity. In comparison, the differences in anticholinesterase activity between the methylcarbamate and biscarbamyl sulfide against HAChE were substantially greater than the differences in housefly toxicity. In general, the relationship between AChE inhibition and toxicity was similar to that observed for the N-dimethoxyphosphinothioyl-N-methylcarbamate derivatives, suggesting that factors other than cholinesterase inhibition are responsible for the selective properties of the biscarbamyl sulfides.

Recently completed studies (Chiu *et al.*, 1974) on the comparative metabolism of biscarbofuran sulfide (**10d**) in houseflies and the white mouse revealed that **10d** was rapidly degraded into nontoxic phenolic products in the mouse, whereas substantial amounts of carbofuran were generated in houseflies. The results indicate that the selective action of these carbamate derivatives is attributable primarily to differences in the metabolic reactions that take place in insects and mammals.

## C. Sulfenylated Carbamate Esters

The successful achievement of the desired order of selectivity by substitution of the methylcarbamyl hydrogen atom with a dimethoxyphosphinothioyl moiety prompted the examination of carbamate derivatives similarly substituted with other functional groups. Of the various derivatized carbamates that were examined, the N-arylsulfenyl and N-alkylsulfenyl derivatives proved to be particularly effective insecticides, in addition to being less toxic to mammals (Black *et al.*, 1973a). These compounds are conveniently prepared by the reaction between the methylcarbamate ester and sulfenyl halide in the presence of pyridine.

Toxicological properties of a wide variety of N-arylsulfenyl and N-alkylsulfenyl derivatives of *m*-isopropylphenyl methylcarbamate (MIP), propoxur, aldicarb, and carbofuran are presented in Table 3. According to the data, the sulfenylated derivatives are substantially less toxic to the white mouse than the original methylcarbamates. Substitution of the nitrogen proton with an arylsulfenyl moiety reduced mouse toxicity by 10- to 50-fold, and substitution with alkylsulfenyl reduced toxicity by 5- to 17-fold. Overall, substitution with the 4-*t*-butylphenylsulfenyl moiety provided derivatives with the largest improvement in mouse toxicity.

TABLE 3. Toxicological Properties of N-Arylsulfenyl and N-Alkylsulfenyl Derivatives of Insecticidal Methylcarbamates

| Structure Number | R | Housefly $LD_{50}$, μg/g Alone | Housefly $LD_{50}$, μg/g 1:5 P.B. | *Culex fatigans* $LC_{50}$, ppm | Mouse (oral) mg/kg |
|---|---|---|---|---|---|
| | 3-Isopropylphenyl-OC(=O)N($CH_3$)SR | | | | |
| 1 | MIP | 41.0 | 11.0 | 0.038 | 16 |
| 11 | Phenyl | 75.0 | 16.0 | 0.006 | 150-200 |
| 12 | 4-Tolyl | 92.5 | 19.5 | 0.005 | 50-100 |
| 13 | 4-*t*-Butylphenyl | 92.4 | 18.5 | 0.005 | 150 |
| 14 | 2-Me-4-*t*-butylphenyl | 59.0 | 18.0 | 0.009 | 100-150 |
| 15 | 4-Cl-phenyl | 145 | 17.5 | 0.006 | 125 |
| 16 | $CH_3$ | 240 | 19.5 | 0.028 | 50-100 |
| | 2-Isopropoxyphenyl-OC(=O)N($CH_3$)SR | | | | |
| 4 | Propoxur | 24.0 | 5.6 | 0.33 | 24 |
| 17 | Phenyl | 36.0 | 8.0 | 0.039 | 300 |
| 18 | 4-Tolyl | 36.0 | 3.9 | 0.028 | 350-400 |
| 19 | 3-Tolyl | 23.5 | 6.7 | 0.020 | 300-400 |

(continued)

TABLE 3 (cont.)

| Structure Number | R | Housefly $LD_{50}$, μg/g Alone | 1:5 P.B. | *Culex fatigans* $LC_{50}$, ppm | Mouse (oral) mg/kg |
|---|---|---|---|---|---|
| **20** | 2-Tolyl | 24.0 | 3.9 | 0.024 | 400 |
| **21** | 2,4-Xylyl | 27.5 | 7.8 | 0.014 | 400 |
| **22** | 2-Isopropylphenyl | 28.5 | 8.5 | 0.013 | 500 |
| **23** | 4-*t*-Butylphenyl | 9.0 | 4.5 | 0.013 | 750-1000 |
| **24** | 2-Me-4-*t*-butylphenyl | 24.5 | 6.5 | 0.014 | >850 |
| **25** | 4-Br-phenyl | 26.5 | 3.7 | 0.010 | 300-400 |
| **26** | 4-Br-2-Me-phenyl | 37.0 | 5.6 | 0.020 | 400 |
| **27** | $CH_3$ | 65.0 | 4.5 | 0.105 | 300-400 |
| | $CH_3SC(CH_3)_2CH{=}NOC({=}O)N(CH_3)SR$ | | | | |
| **7** | Aldicarb | 5.5 | 3.4 | 0.16 | 0.3-0.5 |
| **28** | 2-Tolyl | 12.5 | 8.8 | 0.084 | 3-5 |
| **29** | 4-*t*-Butylphenyl | 7.5 | 3.2 | 0.014 | 10 |

(continued)

TABLE 3 (cont.)

| Structure Number | R | Housefly $LD_{50}$, μg/g Alone | 1:5 P.B. | *Culex fatigans* $LC_{50}$, ppm | Mouse (oral) mg/kg |
|---|---|---|---|---|---|
| | 2,2-Dimethyl-2,3-dihydrobenzofuranyl-7-OC(=O)N($CH_3$)SR | | | | |
| 9 | Carbofuran | 6.7 | 2.5 | 0.052 | 2 |
| 30 | Phenyl | 9.3 | 5.0 | 0.0045 | 25-50 |
| 31 | 4-Tolyl | 9.0 | 3.3 | 0.0045 | 100-125 |
| 32 | 3-Tolyl | 6.5 | 3.0 | 0.004 | 25-50 |
| 33 | 2-Tolyl | 3.7 | 1.8 | 0.004 | 100-125 |
| 34 | 2,4-Xylyl | 9.0 | 4.0 | 0.003 | 50-100 |
| 35 | 4-*t*-Butylphenyl | 2.7 | 2.3 | 0.0025 | 75 |
| 36 | 2-Me-4-*t*-Butylphenyl | 7.5 | 3.3 | 0.002 | 75-125 |
| 37 | 2-Isopropylphenyl | 8.3 | 4.0 | 0.003 | 50-75 |
| 38 | 4-Cl-phenyl | 12.5 | 4.6 | 0.002 | 50 |
| 39 | 4-Br-phenyl | 9.0 | 4.6 | 0.004 | 50-75 |
| 40 | 4-Br-2-Me-phenyl | 11.3 | 3.3 | 0.0025 | 100-150 |
| 41 | Penta-Cl-phenyl | 26.5 | 6.0 | 0.0085 | 50 |
| 42 | 4-MeO-phenyl | 7.8 | 3.0 | 0.006 | 25-50 |

(continued)

TABLE 3 (cont.)

| Structure Number | R | Housefly $LD_{50}$, μg/g Alone | 1:5 P.B. | *Culex fatigans* $LC_{50}$, ppm | Mouse (oral) mg/kg |
|---|---|---|---|---|---|
| 43 | 3,4-$OCH_2O$-phenyl | 5.0 | 3.3 | 0.0065 | 10–25 |
| 44 | 4-CN-phenyl | 22.5 | 3.9 | 0.022 | 25–50 |
| 45 | 2-Naphthyl | 8.0 | 4.3 | 0.0032 | 40 |
| 46 | 2-Benzothiazolyl | 8.5 | 4.4 | 0.0095 | 50 |
| 47 | $CH_3$ | 4.9 | 1.65 | 0.026 | 20 |
| 48 | $C_2H_5$ | 12.8 | 6.0 | 0.024 | 10–15 |

Although the sulfenylated carbamates were variable in their toxicity to houseflies and mosquito larvae, they were, on the whole, quite effective insecticides and in many cases were more effective than the parent methylcarbamate. Without exception, the derivatives were all more toxic to mosquito larvae, and virtually all of the arylsulfenyl carbofuran and MIP derivatives were outstanding as larvicides. In every case, introduction of a hydrophobic substituent to the benzenesulfenyl moiety increased larvicidal activity, suggesting that lipophilicity of the carbamate derivative plays an important role in determining effectiveness against mosquito larvae. Of the various arylsulfenyl ring substituents, the 4-*t*-butyl moiety produced derivatives with highest insecticidal activity, for example, compounds **23**, **29**, and **35**. Although no obvious relationship between structure and insecticidal activity was apparent, in general, ring substituents that were electron withdrawing gave compounds with poorest activity.

Sulfenylated derivatives are not necessarily restricted to arylsulfenyl or alkylsulfenyl moieties. Dialkylaminosulfenyl chlorides also were used as derivatizing agents to produce dialkylaminosulfenyl derivatives to methylcarbamate esters, but in these cases a lower level of selectivity was achieved (see Table 4). Although the toxicological properties of the limited number of these derivatives are less favorable than those of the arylsulfenyl derivatives, the results point out the range of modifications that can be made without significantly affecting insecticidal activity.

Anticholinesterase data, in other words, bimolecular inhibition constants ($k_i$) against BAChE and HAChE, for a restricted number of methylcarbamates and their sulfenylated derivatives are presented in Table 5. Kinetic analysis of the inhibition process showed that the sulfenylated derivatives behaved typically as carbamylating agents, similar to the behavior of methylcarbamate esters.

With the exception of two propoxur derivatives (**17** and **20**), the sulfenylated carbamates were poorer anticholinesterases than the parent methylcarbamates. Except for **17**, reduction in the rate constants for the inhibition of BAChE with derivatization was in approximate agreement with mouse toxicity. However, a similar relationship was not observed between inhibition of HAChE and toxicity to houseflies, and in fact, the three 4-*t*-butylphenylsulfenyl derivatives (**12**, **23**, and **35**) showed the poorest anticholinesterase activity, even though these compounds were generally the most effective insecticides.

To establish the basis for the selective toxicity of these compounds, the comparative metabolism of a representative deri-

TABLE 4. Toxicological Data for Aminosulfenyl Derivatives of Aldicarb and Carbofuran

| Structure Number | R | Housefly $LD_{50}$, μg/g Alone | 1:5 P.B. | *Culex fatigans* $LC_{50}$, ppm | Mouse $LD_{50}$ (oral) mg/kg |
|---|---|---|---|---|---|
| | $CH_3SC(CH_3)_2CH{=}NOC(O)N(CH_3)SR$ | | | | |
| 7 | Aldicarb | 5.5 | 3.4 | 0.16 | 0.3-0.5 |
| 49 | $N(CH_3)_2$ | 12.0 | 5.5 | 0.13 | - |
| 50 | Morpholino | 7.6 | 5.5 | 0.17 | - |
| | 2,2-Dimethyl-2,3-dihydrobenzofuranyl-7-$OC(O)N(CH_3)SR$ | | | | |
| 9 | Carbofuran | 6.7 | 2.5 | 0.052 | 2 |
| 51 | $N(CH_3)_2$ | 15.0 | 4.1 | 0.027 | 10-20 |
| 52 | Morpholino | 10.5 | 4.2 | 0.048 | 10-20 |

TABLE 5. Anticholinesterase Activities of Some N-Arylsulfenyl Derivatives of Aryl Methylcarbamates

| Structure Number | R | $k_i$, BAChE ($M^{-1}$ $min^{-1}$) | $k_i$, HAChE ($M^{-1}$ $min^{-1}$) |
|---|---|---|---|
| | *m*-Isopropylphenyl-OC(=O)N($CH_3$)SR | | |
| **1** | MIP | $7.5 \times 10^5$ | $7.7 \times 10^5$ |
| **11** | Phenyl | $1.2 \times 10^5$ | $3.7 \times 10^4$ |
| **13** | 4-*t*-Butylphenyl | $1.8 \times 10^4$ | $6.5 \times 10^4$ |
| | 2-Isopropoxyphenyl-OC(=O)N($CH_3$)SR | | |
| **4** | Propoxur | $4.3 \times 10^4$ | $1.2 \times 10^6$ |
| **17** | Phenyl | $2.9 \times 10^4$ | $9.8 \times 10^6$ |
| **20** | 2-Tolyl | $7.0 \times 10^4$ | $1.5 \times 10^5$ |
| **23** | 4-*t*-Butylphenyl | $7.4 \times 10^3$ | $5.2 \times 10^4$ |
| | 2,2-Dimethyl-2,3-dihydrobenzofuranyl-7-OC(=O)N($CH_3$)SR | | |
| **9** | Carbofuran | $1.9 \times 10^5$ | $1.3 \times 10^7$ |
| **30** | Phenyl | $8.1 \times 10^5$ | $9.8 \times 10^5$ |
| **33** | 2-Tolyl | $1.0 \times 10^5$ | $1.7 \times 10^5$ |
| **35** | 4-*t*-Butylphenyl | $7.8 \times 10^4$ | $6.7 \times 10^5$ |

vative, 2,2-dimethyl-2,3-dihydrobenzofuranyl-7 *N*-methyl-*N*-(2-toluenesulfenyl)carbamate (**33**) or *N*-(2-toluenesulfenyl)-carbofuran was examined in the white mouse and housefly (Black *et al*., 1973b). Quantitative data for the recovery of metabolic products isolated from susceptible houseflies treated with carbonyl-$^{14}C$ and benzofuranyl aromatic ring-$^{14}C$ **33** at 2.85 μg/g dose are given in Table 6. The principal metabolites found internally in the housefly were carbofuran (**9**) and 3-hydroxy-carbofuran (**9a**). Other minor metabolites were 3-keto-carbofuran (**9b**) and 3-hydroxy-*N*-$CH_2$-OH-carbofuran (**9c**). The substantial amounts of carbofuran (**9**) recovered from the internal extracts of houseflies, corresponding to 0.68 μg/g and 0.32 μg/g for the carbonyl and ring labels, respectively, strongly implicate this product

TABLE 6. Metabolic Products Recovered from Susceptible Houseflies after Topical Application of 2.85 μg/g N-(2-Toluenesulfenyl) Carbofuran (33)

| Structure Number | Total Recovery[a] Analysis[b] | Ring Label % 90.4 | CO Label % 93.7 |
|---|---|---|---|
| | Cage Rinse | | |
| 33 | N-(2-Toluenesulfenyl)carbofuran | 1.6 | 2.3 |
| 9 | Carbofuran | 0.6 | 0.5 |
| | Conjugates | 2.5 | 1.5 |
| | Body Rinse | | |
| 33 | N-(2-Toluenesulfenyl)carbofuran | 26.1 | 22.8 |
| 9 | Carbofuran | 1.2 | 2.9 |
| 9a | 3-Hydroxy-carbofuran | 5.9 | 7.2 |
| | Internal Extracts | | |
| 9 | Carbofuran | 17.2 | 37.3 |
| 9a | 3-Hydroxy-carbofuran | 8.1 | 9.4 |
| 9c | 3-Hydroxy-N-$CH_2$OH-carbofuran | 1.6 | trace |
| | Conjugates | | |
| 9a | 3-Hydroxy-carbofuran | 28.2 | 3.5 |
| 9b | 3-Keto-carbofuran | 3.4 | 1.4 |
| 9c | 3-Hydroxy-N-$CH_2$OH-carbofuran | 1.2 | 1.3 |
| | Unextractables | 2.4 | 5.3 |
| | $CO_2$ | - | 4.6 |
| | TOTALS | 100.0 | 100.0 |

[a]Based on administered radioactivity.

[b]Based on recovered radioactivity.

as the agent responsible for the insecticidal activity of 33. Such amounts are comparable or even greater than the amounts recovered when houseflies are treated at a similar dosage with carbofuran itself (Dorough, 1968). In light of the high anti-cholinesterase and insecticidal activity of carbofuran, there is little doubt that the toxicity of 33 is attributable to the production of lethal amounts of carbofuran *in vivo*. The scheme for the metabolism of 33 in houseflies is presented in Figure 1.

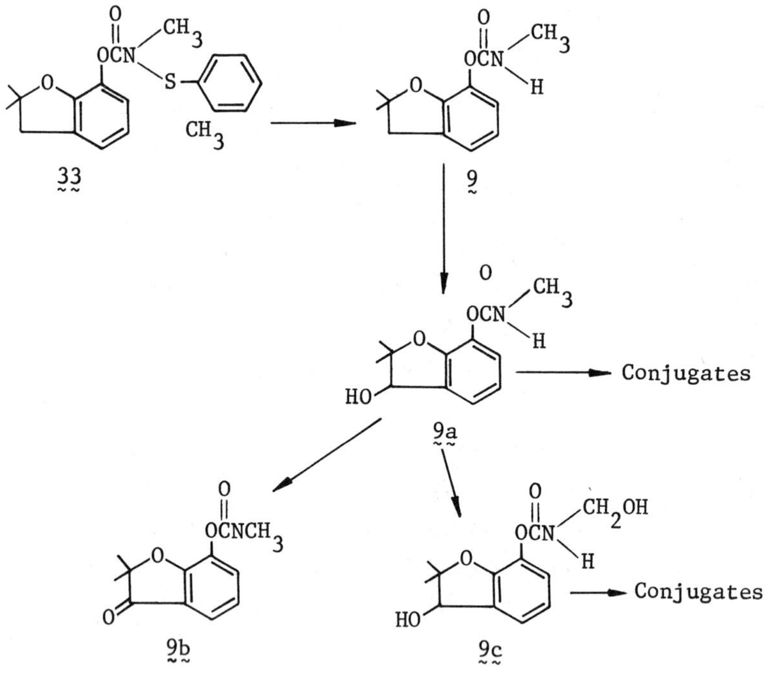


FIGURE 1. Metabolism of *N*-(2-Toluenesulfenyl) Carbofuran in the Housefly.

Table 7 summarizes data for the distribution, identity, and relative amounts of the various metabolites recovered after oral treatment of the mouse at a dose of 15 mg/kg. The greatest amount of radioactivity appeared in the urine as water-soluble conjugates within 24 hours after treatment. Compared to the housefly, metabolism of N-(2-toluenesulfenyl)-carbofuran (33) in the mouse was quite complex, and the bulk of the applied radioactivity was recovered as water-soluble conjugates of 3-hydroxy-carbofuran (9a), 3-hydroxy-N-$CH_2OH$-carbofuran (9c), or as conjugates of the various phenolic products (9d, 9e, and 9f). Noteworthy was the fact that little, if any, carbofuran was isolated. Of added significance was the detection of relatively large amounts of 3-hydroxy-N-$CH_2OH$-carbofuran (26.8 to 31.0%), either in the conjugated or unconjugated form. This compound previously was reported as a minor metabolite of carbofuran in rats and houseflies (Dorough, 1968). The other major metabolite, 3-hydroxy-carbofuran (9a), also was recovered predominantly in the form of a water-soluble conjugate. Although 9a is quite toxic to mice, in other words, mouse oral $LD_{50}$ of 7 mg/kg (Black et al., 1973b), evidently it is transformed into a conjugate as it is produced in vivo and eliminated before attaining a critical level in the mouse. Conjugation reactions doubtlessly play an important role in the relative safety of 33 to the mouse.

The various pathways that may be suggested for the metabolism of 33 in the white mouse are presented in Figure 2. The primary route of metabolism is a series of detoxication steps and probably proceeds via the key intermediate N-(2-toluenesulfenyl)-3-hydroxy-carbofuran (in brackets). Although this intermediate was not isolated, evidence for its formation rests with the recovery of significant amounts of N-(2-toluenesulfenyl)-3-keto-carbofuran (33b) and the relatively large amounts of 9a and 9c.

The results of the metabolism study indicate that the selective toxicity of 33 and related sulfenylated derivatives is a consequence of different metabolic pathways in insects and mammals. Apparently, the arylsulfenyl group on the methylcarbamyl moiety allows the mouse to carry out metabolic reactions leading to less toxic products that are rapidly conjugated, whereas in houseflies the toxic parent carbamate is produced.

Although a reasonable case, based on difference in intoxication and detoxication processes between insects and animals, is made for the selective action of the sulfenylated derivative, more recent results indicate that other factors also may be involved. To establish the intermediacy of the key N-(2-toluene-

TABLE 7. Metabolic Products Recovered from the White Mouse After Oral Administration of 15 mg/kg N-(2-Toluenesulfenyl) Carbofuran (33)

| Structure Number | Total Recovery | Ring Label % 98.7 | CO Label % 90.0 |
|---|---|---|---|
| | Organo-soluble[a] urinary metabolites | | |
| 33b | N-(2-Toluenesulfenyl) 3-keto-carbofuran | - | 0.2 |
| 9a | 3-Hydroxy-carbofuran | 1.0 | 3.3 |
| 9b | 3-Keto-carbofuran | 0.3 | 0.3 |
| 9c | 3-Hydroxy-N-$CH_2OH$-carbofuran | 5.9 | 10.6 |
| | Water-soluble conjugates of urinary metabolites | | |
| 9d | Carbofuran phenol | 12.9 | - |
| 9e | 3-Hydroxy-carbofuran phenol | 11.7 | - |
| 9f | 3-Keto-carbofuran phenol | 4.5 | - |
| 9a | 3-Hydroxy-carbofuran | 29.2 | 13.3 |
| 9c | 3-Hydroxy-N-$CH_2OH$-carbofuran | 20.9 | 20.4 |
| | Unextractables | 10.4 | 11.8 |
| | Fecal metabolites | | |
| 33 | N-(2-Toluenesulfenyl)carbofuran | 0.5 | 0.3 |
| 33a | N-(2-Toluenesulfenyl)carbofuran | - | 1.4 |
| 33b | N-(2-Toluenesulfenyl) 3-keto-carbofuran | - | 2.7 |
| 9 | Carbofuran | 1.0 | - |
| | Unextractables | 1.7 | 1.2 |
| | $CO_2$[a] | - | 34.5 |
| | TOTALS | 100.0 | 100.0 |

[a]Based on total recovered radioactivity.

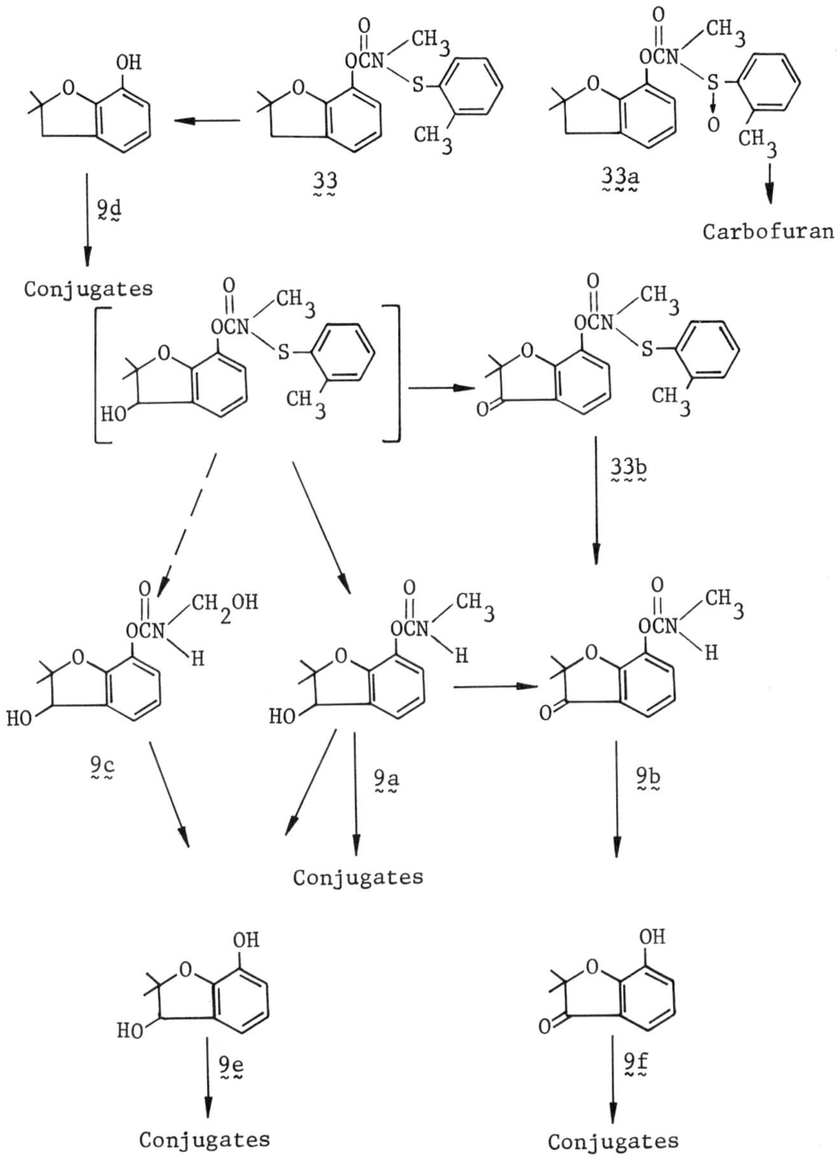


FIGURE 2. Metabolic Pathways of N-(2-Toluenesulfenyl) Carbofuran in the White Mouse.

sulfenyl)-3-hydroxycarbofuran in the metabolic pathway in mice, attempts were made to isolate this material by incubating N-(2-toluenesulfenyl)-carbofuran with different subcellar fractions of mouse liver, with and without NADPH cofactor. The fractions examined were liver homogenate, microsome, and supernatant. However, after repeated experimentation, evidence for the proposed intermediate could not be obtained, and further, carbofuran was the major product isolated, particularly in the absence of NADPH. The results presented in Table 8 show that each of the indicated subcellular fractions of mouse liver rapidly cleaved the N-S bond to generate carbofuran. In the presence of NADPH, the carbofuran that was produced was converted further to other oxidation products, for example, 3-hydroxycarbofuran (9a). Although *in vitro* results do not necessarily reflect actual processes that take place within an animal, nevertheless, the facility with which mouse liver tissue was able to catalyze the cleavage of the N-S bond with resultant liberation of carbofuran cannot be disregarded. Further study is required to solve this dilemma.

TABLE 8. Products Obtained after 10-min Incubation of N-(2-Toluenesulfenyl)-carbofuran (33) with Subcellular Fractions of Mouse Liver at 30°C

| Fraction | Products (% of Total Radioactivity) | | |
|---|---|---|---|
| | 33 | Carbofuran | Other |
| Buffer (control) | 100 | 0 | 0 |
| Homogenate | 11 | 84 | 5 |
| Homogenate + NADPH | 8 | 52 | 40 |
| Microsome | 26 | 72 | 2 |
| Microsome + NADPH | 22 | 48 | 30 |
| Boiled microsome[a] + NADPH | 0 | 100 | 0 |
| Supernatant | 8 | 91 | 1 |
| Supernatant + NADPH | 9 | 89 | 2 |

[a]The microsomal fraction was boiled for 10 minutes prior to incubation with 33.

## III. INCORPORATION OF SELECTOPHORES

### A. Alkoxyiminomethylphenyl Methylcarbamates

The rationale used in the design of methylcarbamate esters appropriately derivatized on the methylcarbamyl moiety is based on expected differences in the metabolic route that the derivatives undergo in insects and mammals. A similar approach, but based more on fundamental organic chemistry, was used in the recent discovery of selectively toxic methylcarbamate esters containing the alkoxyiminomethyl moiety attached to the aromatic nucleus as indicated in the structure below (Sanborn *et al.*, 1974).

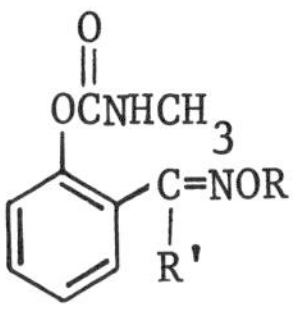


Toxicological data for a representative number of these compounds are given in Table 9. Corresponding isomers substituted in the *meta* or *para* positions were substantially less effective insecticides and therefore were omitted from the table.

The data in Table 9 show that all of these compounds are good housefly toxicants and, on the whole, are quite safe to the white mouse. An added feature of these compounds is their relatively high toxicity to insecticide resistant houseflies ($R_{SP}$) and one of the compounds (**53**) was actually more toxic to resistant than to susceptible houseflies.

The basic premise used in the examination of these carbamate esters for selective toxicity originates from the anticipated stability in the mouse of the alkoxyiminomethyl moiety to Beckmann rearrangement. An earlier study (Lee *et al.*, 1974) revealed that the *bis*-methylcarbamate of salicylaldoxime is highly toxic to the mouse. This compound rearranged to the corresponding nitrile under acidic conditions, and it was expected that this rearrangement also would take place in the acid stomach of the mouse. Since compounds that were susceptible to rapid rearrangement were toxic to the mouse, it was predicted that replacement of the methylcarbamyl oxime moiety with an alkoxyiminomethyl moiety would result in compounds with lower mouse toxicity, owing to the increased stability of this group to the Beckmann rearrangement. Support for this rationale is found in the favorable mouse toxicities of the compounds listed in Table 9.

TABLE 9. Toxicological Data for Alkoxyiminomethylphenyl Methylcarbamates

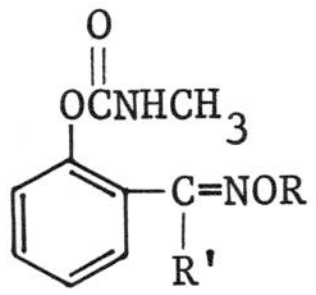

| Structure Number | R | R' | Housefly $LD_{50}$ (μg/g) $S_{NAIDM}$ Alone | $S_{NAIDM}$ 1:5 P.B. | $R_{SP}$ | Mouse $LD_{50}$ (oral) mg/kg |
|---|---|---|---|---|---|---|
| 49 | $CH_2C{\equiv}CH$ | H | 8.7 | 2.0 | 10.3 | 50 |
| 50 | $CH_2C{\equiv}CH$ | $CH_3$ | 11.5 | 1.5 | 14.4 | 200–300 |
| 51 | $C_2H_5$ | H | 13.4 | 1.7 | 12.9 | 150–200 |
| 52 | $\underline{n}$-$C_3H_7$ | H | 17.1 | 1.9 | 19.9 | 300–400 |
| 53 | i-$C_3H_7$ | H | 15.9 | 1.8 | 20.4 | 400–500 |

Anticholinesterase data for the same compounds (Table 10) show that they are much more effective in inhibiting BAChE than HAChE. Although a quantitative correlation between enzyme inhibition and toxicity to the housefly or mouse was not apparent, the results indicate that, on an overall basis, toxicity is related to anticholinesterase activity. Whether the actual reason for the selective properties of these compounds is attributable to enzyme sensitivity or to stability to the Beckmann rearrangement must await further study.

## IV. SUMMARY

Evidently, selectively toxic carbamate esters may be designed by exploiting differences between the fundamental biochemistry of insects and mammals. The successful achievement of converting toxic carbamates into safe derivatives by appropriate substitution on the methylcarbamyl moiety suggests that, in the future, all new methylcarbamate insecticides should be similarly derivatized. The examples provided in this review by no means represent all of the modifications that can be incorporated into toxic carbamates. For example, derivatization by groups such as the aminosulfenyl moiety has barely been explored. Derivatization with functional groups that can change the physical properties of the molecule also is worthy of consideration, for example, introduction of perhaps the ethylthioethylsulfenyl moiety to produce compounds that may be effective systemic insecticides. The success of this approach depends on our imagination and the appropriate use of available information concerning intoxication and detoxication processes in insects and mammals.

From the favorable toxicological properties observed for the alkoxyiminomethylphenyl methylcarbamates, it is clear that further studies are warranted on the incorporation of selectophoric groups in the carbamate molecules. This approach has been widely exploited in the development of selectively toxic organophosphorus but has been neglected with carbamate esters.

A recent incident in which some 2400 wild ducks died near San Jacinto, California after they had been exposed to carbofuran is believed to be one of the worst single incidents of wildlife deaths involving pesticides in the United States. In this particular incident, carbofuran was used to treat a field for the control of the Egyptian alfalfa weevil and pea aphids. This carbamate insecticide is highly effective against these insects and preferred by farmers over other materials, even though it is highly toxic to warm-blooded animals. It is

TABLE 10. Anticholinesterase Activity of the Alkoxyiminomethylphenyl Methylcarbamates for BAChE and HAChE

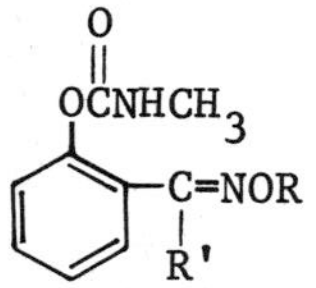


| Structure Number | R | R' | $I_{50}$ (M) HAChE | $I_{50}$ (M) BAChE | Ratio $I_{50}$ BAChE/HAChE |
|---|---|---|---|---|---|
| 49 | $CH_2C{\equiv}CH$ | H | $1.17 \times 10^{-6}$ | $1.31 \times 10^{-4}$ | 112 |
| 50 | $CH_2C{\equiv}CH$ | $CH_3$ | $9.70 \times 10^{-7}$ | $2.28 \times 10^{-4}$ | 297 |
| 51 | $C_2H_5$ | H | $8.40 \times 10^{-7}$ | $1.27 \times 10^{-4}$ | 151 |
| 52 | $\underline{n}$-$C_3H_7$ | H | $1.26 \times 10^{-6}$ | $7.70 \times 10^{-5}$ | 61 |
| 53 | $\underline{i}$-$C_3H_7$ | H | $5.00 \times 10^{-7}$ | $5.62 \times 10^{-4}$ | 1124 |

conceivable that these deaths could have been avoided had one of the safer carbofuran derivatives been used in place of carbofuran. This unfortunate incident serves to remind us of the continuing need for the development of new materials with favorable properties of selectivity.

## REFERENCES


1. Black, A. L., Chiu, Y.-C., Fahmy, M. A. H., and Fukuto, T. R. 1973a. J. Agr. Food Chem. 21, 747

2. Black, A. L., Chiu, Y.-C., Fukuto, T. R., and Miller, T. A. 1973b. Pestic. Biochem. Physiol. 3, 435.

3. Chiu, Y.-C., Musson, D. E., Black, A. L., and Fukuto, T. R. 1974. Unpublished work.

4. Dorough, H. W. 1968. J. Agr. Food Chem. 16, 319.

5. Fahmy, M. A. H. and Fukuto, T. R. 1972. Tetrahedron Lett. 41, 4245.

6. Fahmy, M. A. H., Fukuto, T. R., Myers, R. O., and March, R. B. 1970. J. Agr. Food Chem. 18, 793.

7. Fahmy, M. A. H., Metcalf, R. L., Fukuto, T. R., and Hennessy, D. J. 1966. J. Agr. Food Chem. 14, 79.

8. Fraser, J., Clinck, G. C., and Ray, R. C. 1965. J. Sci. Food Agr. 16, 615.

9. Fukuto, T. R. 1972. Drug Metab. Revs. 1, 117.

10. Knaak, J. B. 1971. Bull. W.H.O. 44, 121

11. Kreiger, R. I., Lee, P. W., Fahmy, M. A. H., and Fukuto, T. R. 1974. Unpublished work.

12. Krueger, H. R. and O'Brien, R. D. 1959. J. Econ. Entomol. 52, 1063.

13. Lee, A., Sanborn, J. R., and Metcalf, R. L. 1974. Pestic. Biochem. Physiol. 4, 77.

14. Matsumura, F. and Dauterman, W. C. 1964. Nature (London), 202, 1356.

15. Metcalf, R. L. 1971. Bull. W.H.O. 44, 43.

16. Miskus, R. P., Andrews, T. L., and Look, M. L. 1969. J. Agr. Food Chem. 17, 842.

17. Sanborn, J. R., Lee, A., and Metcalf, R. L. 1974. Pestic. Biochem. Physiol. 4, 67.

## DISCUSSION

The discussion centered around desirable requisites for mammalian selectivity or safety of the carbamate insecticides. The carbamates, as the most recent class of insecticides to be exploited on a large scale, include carbaryl produced in very large amounts (about 55 million pounds) in the United States in 1972, as a broad-spectrum agricultural insecticide; carbofuran (produced at about 8 million pounds), which is the most generally effective insecticide of the group and is widely used against soil pests; aldicarb (produced at about 2 million pounds), as a successful soil and systemic insecticide; and methomyl (produced at about 2 million pounds), as a foliar insecticide. Propoxur is a useful household insecticide and a promising replacement for DDT in the WHO malaria control programs. Other carbamates used in lesser amounts include (Bux®), a soil insecticide, landrin, 4-dimethylamino-3,5-xylenyl N-methylcarbamate (Zectran®), 4-dimethylamino-3-tolyl N-methylcarbamate (Matacil®), 4-methylthio-3,5-xylenyl N-methylcarbamate (Mesurol®), m-tolyl N-methylcarbamate (Tsumacide®), o-chlorophenyl N-methylcarbamate (CMPC), and 3,5-di-tert-butylphenyl N-methylcarbamate (butacarb).

As a group the carbamates have many desirable properties including cyrstalline structure and pleasant odor, and are well adapted to persistence on surfaces. However, they suffer from lack of high mammalian selectivity. In fact, adjudged by their mammalian selectivity ratios (MSR) or rat oral $LD_{50}$/housefly topical $LD_{50}$ (see Brooks, Chapter V), they are generally close to 1:1. MSR values for the commonly used compounds include carbaryl 0.6, carbofuran 1.0, aldicarb 0.175, methomyl 10, and propoxur 4.5. The redeeming feature with these compounds is their generally poor penetration through mammalian skin and the reversible nature of cholinesterase inhibition. Nevertheless, incidents of poisoning have occurred as in the WHO experimental use of m-isopropylphenyl N-methylcarbamate which caused illness in workers spraying houses in Nigeria, carbofuran which poisoned a substantial number of field workers detasseling corn in Indiana in 1974, methomyl which poisoned field workers in vegetable crops in Canada in 1972, and o-chlorophenyl N-methylcarbamate which caused poisoning in factory workers in Taiwan. Thus the margin of safety in the use of many of the carbamates is narrow, particularly with the highly toxic carbofuran and aldicarb which are among the most toxic compounds in use as pesticides. These compounds also pose substantial hazards to wildlife as carbaryl and carbofuran are extremely toxic to honeybees, and carbofuran recently killed 2400 wild ducks in an alfalfa field in California.

It was considered that the principal reason for lack of selectivity in the carbamate compounds is the absence of a delay factor such as that provided in certain organophosphorus compounds by P=S to P=O *in vivo* activation. Therefore, the advances detailed by Dr. Fukuto in the Carbamate Position Paper involving derivatized carbamates that do provide such a delay factor with consequent increase in safety (MSR values for derivatized carbofuran of 30 or more) were generally regarded as quite important.

In the WHO malaria control program the carbamates have become the most promising substitutes for DDT because they kill adult Anopheline mosquitoes with contact exposure of 30 minutes or less, whereas DDT may require 2 hours or more. In addition, propoxur, for example, provides a mosquito knockdown without direct contact on a treated surface as a result of airborne suspended particles that may be effective under eaves of treated houses for 3 to 4 weeks after application. More new carbamates have shown promising activity as DDT replacements in the WHO programs than any other class of compound, and considerable experience with spray applications in experimental houses has shown that if the rat acute oral $LD_{50}$ is above 100 mg per kg, these compounds do not have adverse cholinergic effects on humans in sprayed houses. For the reasons advanced above, the carbamates are particularly attractive in public health entomology because it appears unlikely that properly chosen compounds will ever be lethal to spraymen or inhabitants since transitory and incapacitating illness would result in discontinuation of exposure.

Discussion of the practical significance of derivatized carbamates such as the sulfenylated derivatives described in the Position Paper indicated that these would probably remove the hazards already experienced with carbofuran. In the insect the sulfenylated carbamate forms an *N*-methylcarbamate *in vivo* and inhibits cholinesterase, whereas in higher animals they are detoxified by other metabolic pathways. Thus derivatized carbamates could be important in protecting both field workers and wildlife.

Carbofuran is urgently needed for control of the alfalfa weevil, but its high toxicity to humans and wildlife is a problem. Reduced dosages down to 0.25 lb per acre have been very effective for alfalfa weevil control, but labeling and grower advertising has been directed at a dosage of 1 lb/acre. It was pointed out that the United States chemical industry had requested licensing for two of Dr. Fukuto's patents on derivitized carbofuran and that there was rejuvenated interest in safer carbamates because of the work reported. This was cited as a convincing example of the influence of the Rockefeller Foundation research program on "Selective and Nonpersistent Insecticides" on the world chemical industry.

The derivatized carbamates also offer a substantial advantage in protection of the carbamates in water, in other words, in their use as mosquito larvicides. The carbamates alone have very disappointing larvicidal activity yet the sulfenylated derivatives are among the most effective mosquito larvicides yet discovered in the WHO "Program for the Evaluation of Insecticides for Vector Control." Discussion suggested that lipoid/water partition values for the carbamates alone were too low (on the polar side) for adsorption by larvae and that the sulfenylated compounds were highly lipophilic and readily absorbed by the mosquito.

As a final subject for discussion, the question was posed as to which area of insecticide development under discussion, in other words, organochlorines, organophosphorus, carbamates, pyrethroids, insect growth regulators, or new botanical compounds, offered the most promise in terms of research payoff in the development of new insecticides suitable for use in agriculture and public health. Several views were expressed in the discussion. Perhaps the most important was the necessity for new types of screening programs for insecticidal activity that went beyond the conventional 24-hour acute toxicity test. The feeling was expressed that there were probably thousands of compounds that had been bypassed in such screening activities that in total life cycle screens or in other novel evaluations could become useful products.

The discussion closed with an expression of views on the role of the world chemical industry vis-a-vis integrated control or pest management programs. Perhaps it is "putting the cart before the horse" to develop compounds by empirical screening and then to try to fit them into the rapidly developing large-scale pest-management programs. It would be far more prudent to seek to develop the types of compounds needed for such pest-management programs by more elaborate and better organized testing programs. It was recognized that these programs would be expensive, but with the extent of worldwide emergencies in pest control in both agriculture and public health, there may be few if any alternatives. The chemical industry has been slow to move in this direction both because of uncertain markets and because of present preoccupation with EPA regulations and product safety.

It was concluded that the role of the Rockefeller program was congruent with that of the Bellagio Conference in seeking to develop new approaches and ideas for pest control materials that could be exploited by the world chemical industry. Hopefully the Bellagio Conference publication will be useful in aiding this objective.

# NOVEL CHEMICALS AND TARGETS

# Prospects for New Types of Insecticides

JOHN E. CASIDA
Division of Entomology and Parasitology
University of California
Berkeley, California

## I. INTRODUCTION

Three major groups of insecticide chemicals (chlorinated hydrocarbons, organophosphorus compounds, and methylcarbamates, referred to here as the "Big Three") provide effective control of most important pest insects. Other types of insecticides employed prior to the 1940s have either been displaced because of their relative ineffectiveness, or continue in use but on a limited basis.

A great variety of new insecticidal compounds has been discovered during the past 35 years. These discoveries have resulted from advances in four areas: (1) natural products research; (2) modification of known toxicants and random screening; (3) modification of screening procedures to seek new types of useful activity; (4) comparative biochemistry and physiology. However, with a few notable exceptions, new compounds resulting from these advances have not reduced the importance of the Big Three in pest control.

This review attempts to bring into perspective the progress made and prospects for future success in finding alternative insecticides to the Big Three. Little or no attempt is made to consider pyrethroids, synergists, hormones, pheromones, attractants, repellents, and chemosterilants, most of which are dealt with in other chapters of this book. Many very old types of compounds and some relatively new ones are discussed. They are lethal or repellent agents or in other ways deter feeding or damage by insects and mites. Emphasis is given to the great variety of biochemical lesions already known to be useful in insect control toxicants.

The references cited represent only a small fraction of the pertinent literature. Citations are placed after each compound name or designation since many of them are reviews relating to more than one aspect of studies on the chemical.

## II. NATURAL PRODUCTS

There is a continuing search for natural materials that affect insects in a manner that might be useful in pest control. Some organisms appropriate as source materials have been known to biologists for decades or centuries. Other organisms and sources have been discovered only recently as a result of screening programs. The desired response is standardized as a bioassay for isolation and characterization of the biologically active component. This process allows a sifting among the seemingly infinite variety of natural chemicals for new insecticides that can be used directly in insect control or as leads for analog synthesis. Candidate insect control agents have been discovered in plants, microorganisms, insects, and other invertebrates.

### A. Botanical Origin (Figure 1)

A tremendous variety of ecologically significant insect-plant interactions are known that involve plant components that are toxic to insects or deter insect attack. In other cases the insecticidal component provides no survival benefit to the plant. Thus both keen biological observations and screening programs are necessary to discover new botanical insecticides.

Relatively complex natural products can often be simplified for increased potency and economical synthesis. There are two excellent examples with the botanicals. Pyrethrin I (Casida, 1973) was the starting point for all pyrethroid insecticides and physostigmine (Metcalf, 1971) for the methylcarbamate insecticides.

Rotenone (Fukami and Nakajima, 1971; Crombie, 1972; Gosalvez and Merchan, 1973) continues to be an important, non-persistent insecticide for control of caterpillars and leaf-chewing beetles. Structure-activity studies on its naturally occurring analogs, derivatives, and a few related chromano-chromanones from synthesis have not led to simplification of the molecule with maintenance of potency. The biosynthesis route leading to rotenone is now well established. Rotenone is converted on metabolism and photodecomposition to a great variety of less toxic products, many of which have been identified. Although certain tests indicate a possible carcinogenic activity, others under more realistic use conditions show no such effect with rotenone.

Rotenoids are not the only insecticidal botanicals that lack a nitrogen atom. The clerodendrins (Kato *et al.*, 1972)

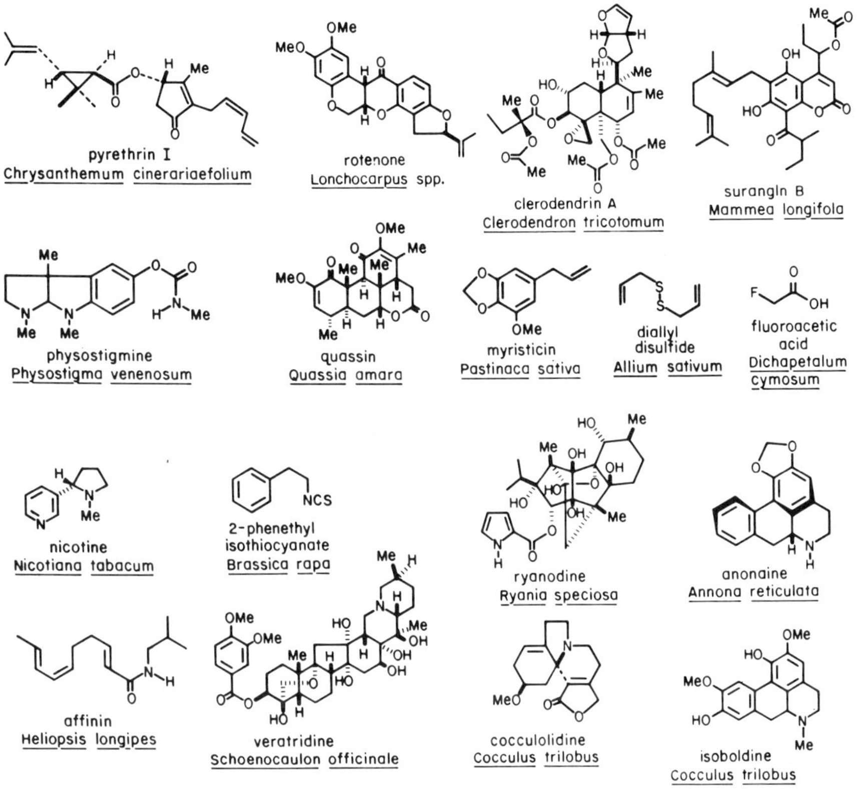


FIGURE 1. Insecticidal Compounds of Botanical Origin.

are antifeeding compounds for larvae of polyphagous insects. 5,7-Dihydroxycoumarins such as surangin B (Crosby, 1971; Crombie, 1972) require an acetoxy residue in the 1'-position of the coumarin side chain for high insecticidal activity, a conclusion based on studies with at least 27 natural compounds of this general type. The amaroid quassin (Crosby, 1971; Valenta, 1972) and the benzodioxole myristicin (Lichtenstein and Casida, 1963) also have a limited spectrum of insecticidal activity. A variety of benzodioxoles are known to be morphogenetic agents and developmental inhibitors in insects, but the commercial importance of this class of compounds is restricted to their use as pyrethroid synergists. Diallyl disulfide (Amonkar and Banerji, 1971), an insecticidal component of garlic, lacks the appropriate properties for general use as an insecticide. Fluoroacetic acid (Metcalf, 1966), a natural toxicant in plants, is probably the intermediate metabolite in conversion of various insecticidal fluoroethyl ethers, fluoroacetamides, and fluoroacetate esters to fluorocitric acid, a respiratory blocking agent. Several of these derivatives have been extensively tested as acaricides and served as the subject for detoxification and mode of action studies. Although it has been difficult to achieve the desired species specificity, studies are continuing in this area with organofluorine compounds.

There are a large number and great variety of insecticidal plant alkaloids, but not any one of them is used extensively at the present time for insect control. Nicotine (Schmeltz, 1971) ranks historically as the first and most important insecticidal alkaloid. Its action on the cholinergic nervous system, metabolism, biosynthesis, and photochemistry are quite well defined. Structure-activity studies have led to simpler nicotinoids with improved insecticidal activity, but even the best compounds to date are not competitive with other available insect control agents. 2-Phenethyl isothiocyanate (Lichtenstein *et al.*, 1962) is a simple botanical in which the isothiocyanate is undoubtedly the toxophoric group. Unsaturated isobutylamides such as affinin (Jacobson, 1971) and others are potent insect knockdown and killing agents, but they are also pungent and unstable compounds, features that appear to be associated with the same structural properties that confer insecticidal activity. However, it is possible that analog syntheses will partially overcome these limitations.

Most of the other insecticidal alkaloids from plants are not attractive candidates for analog synthesis because of their complexity, photolability, and sometimes the possibility of high mammalian toxicity. Examples of such compounds are the highly insecticidal sabadilla and hellebore alkaloids, represented by veratridine (Crosby, 1971), many of which are also highly toxic to mammals; the pentacyclic polyol ester of pyrrolecarboxylic acid, ryanodine (Crosby, 1971; Valenta, 1972), which again is highly toxic to mammals when injected; the insecticidal aporphine alkaloids such as anonaine (Crosby, 1971); insecticidal and antifeeding compounds such as cocculolidine (Wada and Munakata, 1968) and isoboldine (Wada and Munakata, 1968). Studies of these alkaloids should continue on the possibility that a structural mimic of only a portion of the molecule may retain the activity and therefore prove useful.

Many other botanical products are also insecticidal, but some of them have undesirable properties such as carcinogenic or cocarcinogenic activity.

## B. Microbial Origin (Figure 2)

Some strains of Bacillus thuringiensis produce the β-exotoxin (Angus, 1971; Bond et al., 1971; Norris, 1972; Bond and Somerville, 1973; Faust, 1973) that inhibits nucleotidases and

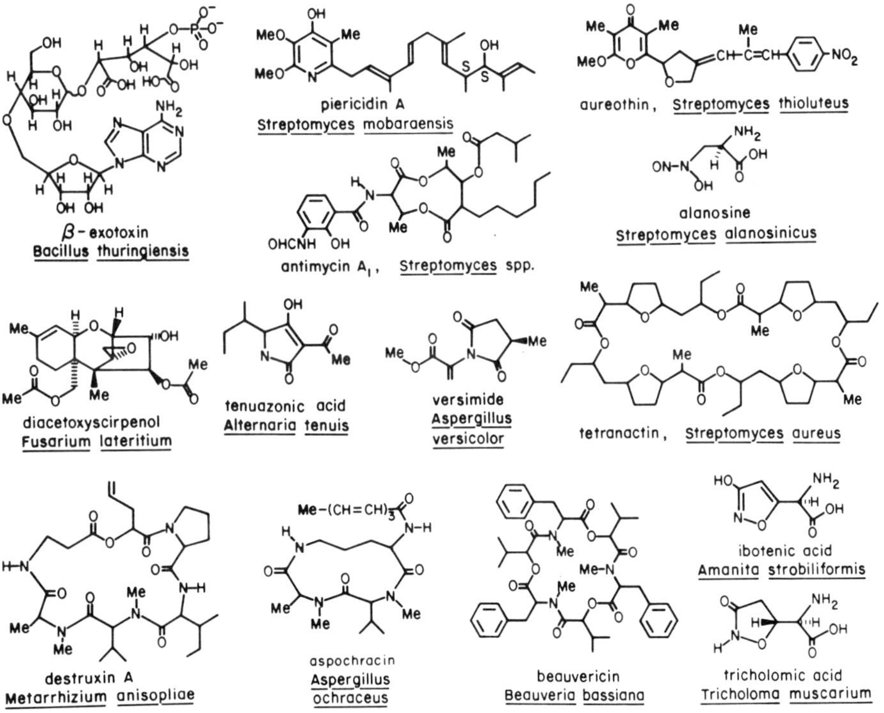


FIGURE 2. Insecticidal Compounds of Microbial Origin.

DNA-dependent RNA polymerase involved with ATP, leading to morphogenetic effects in many insects and mites. However, little if any β-exotoxin is present in any commercial strains of this Bacillus used in the United States. The commercial success of *B. thuringiensis* depends on the presence of a crystalline protein parasporal body (δ-endotoxin) that is formed by all strains and, following activation by gut protease, produces paralysis and death in lepidopterous larvae. The toxic moiety from the δ-endotoxin is believed to interfere with membrane transport systems of the midgut epithelium and with synaptic transmission of nerve impulses.

A variety of compounds from *Streptomyces* have insecticidal activity. Piericidin A (Tamura and Takahashi, 1971; Suzuki and Tamura, 1972), aureothin (Oishi *et al*., 1969) and antimycin $A_1$ (Kido and Spyhalski, 1950; van Tamelen *et al*., 1961) are highly toxic to mammals as well as insects. Alanosine (Kenaga, 1969) and related hydroxynitrosamino compounds inhibit insect reproduction. Prasinons A and B (Box *et al*., 1973) from *Streptomyces prasinus* are high molecular weight, nitrogen-containing larvicidal stomach poisons of moderate toxicity to mammals. Tetranactin (Sagawa *et al*., 1972) is more selective, being an effective acaricide with low mammalian toxicity. Other fungi produce relatively simple insecticidal compounds that either lack nitrogen (diacetoxyscirpenol) (Cole and Rolinson, 1972) or contain this element (tenuazonic acid, versimide) (Cole and Rolinson, 1972). Some nitrogen-containing insecticidal products from fungi contain 12- to 19-membered ring systems; these include destruxin A (Tamura and Takahashi, 1971; Suzuki and Tamura, 1972), which is one of six related compounds from the same fungus, aspochracin (Tamura and Takahashi, 1971; Suzuki and Tamura, 1972), and beauvericin (Hamill *et al*., 1969; Ovchinnikov *et al*., 1971). The insecticidal activity of the *Beauveria* organism appears to be due to the invasive properties of the fungus overwhelming the insects' defense mechanism rather than to the exotoxin. The largest amount of structure-activity information is available for the piericidins, destruxins, and alanosine analogs, but the scope of such studies is still very restricted. There are also other insecticidal fungal products including the very potent carcinogens, the aflatoxins.

Ibotenic (Takemoto *et al*., 1964) and tricholomic acids (Iwasaki *et al*., 1965) from mushrooms are muscicidal amino acids containing the isoxazole and isoxazolidinone skeletons, respectively. The isoxazole has served as the starting point for a useful fungicide and, with incorporation of a phosphoryl moiety, of insect control agents. The simplicity of these ring systems suggests that further attention be given to this area of development.

## C. Animal Origin (Figure 3)

Arthropods elaborate several venoms and defense secretions. Many of the arthropod venoms are highly insecticidal materials, but they are not of interest in the present context since they are only active on injection, a task for which the insect is admirably suited but the economic entomologist is not (Beard,

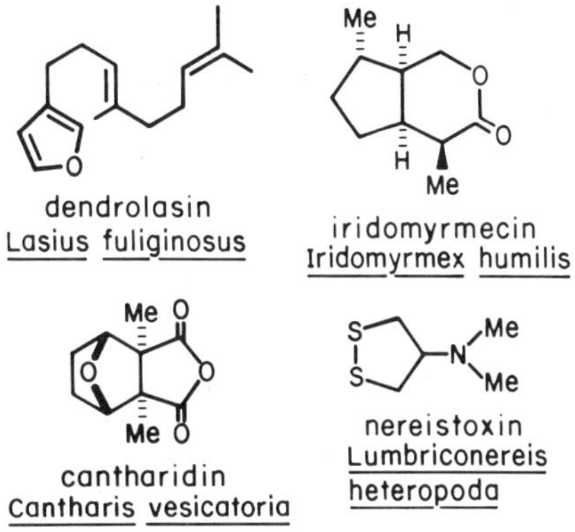


FIGURE 3. Insecticidal Compounds of Animal Origin.

1971). The defense secretions from ants with insecticidal activity include the terpenoid dendrolasin (Cavill and Clark, 1971) and the iridolactone iridomyrmecin (Cavill and Clark, 1971). Cantharidin (Cavill and Clark, 1971) from beetles of the family Meloidae also has high biological activity, including in insects. However, the insecticidal activity of these and other arthropod-derived toxicants is generally not very high, and no useful insect control agents have resulted as yet from analog synthesis.

Nereistoxin (Konishi, 1972; Sakai and Sato, 1972), from a marine annelid, has served as a starting point for a very useful insecticide, cartap, the bis-thiocarbamate derivative, that acts as a synaptic blocking agent. A high degree of structural specificity exists for this action so new compounds will probably only be alternative precursors that degrade in the insect or the environment to release dihydronereistoxin or nereistoxin.

Many other toxicants are known to occur in aquatic and terrestrial invertebrates other than the arthropods and in some vertebrates, but they have not been shown to have high insecticidal activity or, if so, they are of complex structure and are highly toxic to mammals.

## III. MODIFICATION OF KNOWN TOXICANTS AND RANDOM SCREENING (Figure 4)

Perusal of pharmacological and toxicological literature reveals a great variety of toxicants, some of which are relatively simple compounds. The list of materials known to be toxic to mammals, microorganisms, and other life forms is growing at a rapid rate. The similarity of biochemical pathways in different organisms indicates that a compound active on one or a few organisms can be modified to act, perhaps lethally, on other organisms. Other types of pesticides are often the base point for discovery of a new insecticide chemical. This is due in part to the industrial practice of screening on a great variety of pest organisms. It is not surprising to find that modifications on nematocides, anthelmintics, and acaricides provide useful insecticides. It is more surprising but still common to find powdery mildew-control activity associated with acaricidal activity. Often a small structural modification alters a herbicidal compound to an insecticide without phytotoxic properties. Thus compounds synthesized in anticipation of one type of pesticidal activity often prove to be inactive in this area but useful in control of other types of pests. Many of our current insecticides have resulted from a half century of studies on modification of suitable toxicants with the goal of obtaining selective insecticidal activity.

New compounds of unusual structure, and perhaps resulting from novel synthesis procedures, are sometimes found by random screening to have useful biological activity. This is the only basis, other than natural products research, for the discovery of totally new types of insect control agents. These tests are often made with mixtures of chemicals with the subsequent finding that an unanticipated impurity or degradation product accounts for the insecticidal activity.

Considerable progress has been made in the modification of known bioactive materials with acidic OH or acidic NH groups in the search for insecticidal and acaricidal compounds. These studies also involve derivatizing the acidic moiety as an ester or carbamate in order to achieve preferential release of the active toxicants in the pest organisms. The oldest examples

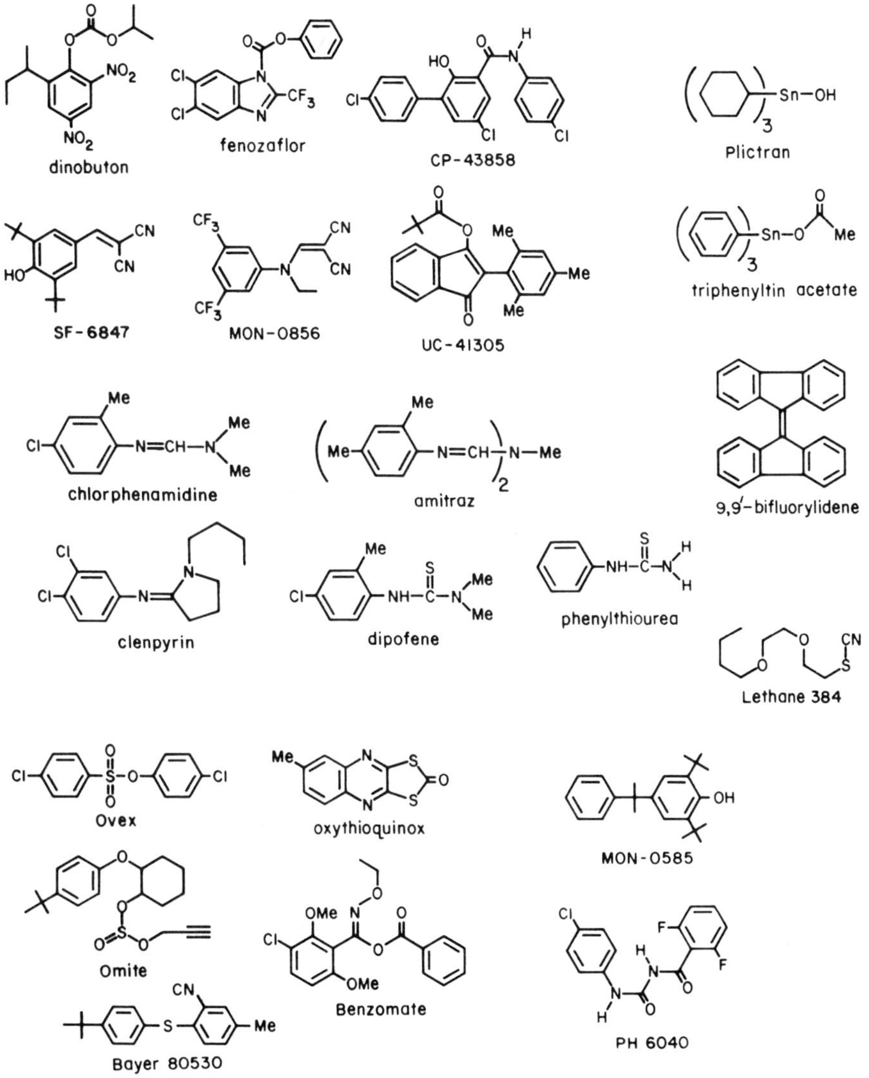


FIGURE 4. Insecticidal Compounds from Modification of Known Toxicants and Random Screening.

are the 2,4-dinitrophenols, optimized with the 6-*sec*-butyl chain and an isopropyl carbonate protective group in the development of dinobuton (Pianka and Smith, 1965) acaricide. Other examples are the trifluoromethylbenzimidazoles (fenozaflor) (Saggers and Clark, 1967), salicylanilides (CP-43858) (Cantu and Wolfenbarger, 1973), dicyanovinylphenols (SF-6847) (Horiuchi *et al*., 1971; Muraoka and Terada, 1972), dicyanovinylanilines (MON-0856) (Darlington *et al*., 1972; Cantu and Wolfenbarger, 1973), 3-acyloxy-2-arylindones (UC-41305) (Sousa *et al*., 1973), and dicyano- and α-acyl-α-cyanocarbonylphenylhydrazones (Büchel and Draber, 1972). It is likely that this area of investigation will have increasing impact on pest-control practices.

Triorganotins, known as biocides for many decades, are becoming one of the most important classes of acaricide chemicals. Although plictran (Luijten, 1972) is the most important commercial mite-control agent of this type, several other triorganotins are currently under development. The nature of the organo substituents is very important in conferring selectivity in this series, and the anion involved also has some significance, possibly relative to solubility characteristics. Triphenyltin acetate (Ascher and Ishaaya, 1973) is an antifeeding compound for some insects, reducing the activity of their digestive enzymes, and it also inhibits insect reproduction. It is anticipated that triorganotins will become increasingly important as insect- and mite-control agents, replacing other types of compounds in some cases.

Formamidines such as chlorphenamidine (Knowles *et al*., 1972; Knowles and Roulston, 1973) are potent acaricides and ovicides, and some of them also affect insects by inducing deleterious alterations in their behavior. Related compounds that are also highly active include a diamidide (amitraz) (Harrison *et al*., 1973; Knowles and Roulston, 1973), a cyclic amidine (clenpyrin) (Enders *et al*., 1973), and a thiourea (dipofene) (Knowles and Roulston, 1973). Chlorophenamidine and amitraz undergo oxidative and hydrolytic activation, respectively, each forming the corresponding *N*-methylformamidine. Dipofene is probably also activated by metabolic oxidation. Chlorphenamidine has already become a very important acaricide and insecticide, and the related compounds are of great current interest, particularly in tick control.

The finding that phenylthiourea (Ogita, 1958) is toxic to DDT-resistant fruit fly strains suggests that structural modifications in the thiourea series may provide useful insecticides. 9,9'-Bifluorylidene (Moorefield, 1965) is a very effective stomach poison for several pest insects, having many desired properties such as selective toxicity but lacking in one feature in that it is a highly colored dye. Knowledge of its mode of action may provide openings for synthesis of effective, noncolored replacements. The organothiocyanates such as Lethane 384 (Ohkawa *et al.*, 1972) are useful knockdown agents, poisoning as a result of *in vivo* liberation of cyanide, but they lack the potency, selectivity, and odor characteristics sought in ideal insect control agents. However, some of these deficiencies may be overcome with a revitalization of research interest in this area.

Acaricides discovered in screening programs [for example, ovex (Knowles *et al.*, 1972), omite (Knowles *et al.*, 1972), Bayer 80530 (Hammann *et al.*, 1969), oxythioquinox (Aziz and Knowles, 1973), and benzomate (Soeda *et al.*, 1972)] are varied in type, generally lack insecticidal activity, and are usually of unknown mode of action. Progress in improving on these compounds will be facilitated by increased knowledge of their mode of action.

Among a great variety of developmental inhibitors, two are of particular interest. MON-0585 (Sacher, 1971) blocks the developmental cycle of mosquito larvae at the pupal stage regardless of when the exposure takes place. On the other hand, a benzoylurea, PH 6040 (Mulder and Gijswijt, 1973), blocks chitin deposition at or near the terminal stage in chitin polymerization, and the insects invariably die in the apolytic stage of ecdysis.

## IV. MODIFICATION OF SCREENING PROCEDURES

The screening process is an essential feature in the search for new insecticides because it is the only procedure for optimizing the structure for a particular type of biological activity. The answers obtained from the screening process are directly related to the questions posed, in other words, the nature of the screen. Thus the screening process has progressed through several conceptual stages. In early work the compounds passing the screen acted quickly and at low dose by either stomach or contact routes of entry. Then the screens were expanded to test for ovicidal as well as larvicidal and adulticidal effects. Later, systemic properties and selective toxicity were emphasized. Then the degree of feeding was evaluated, leading to the discovery of antifeeding compounds (clerodendrins, isoboldine, organotins). More recently the insects have been held for several developmental stages or through a life cycle, a more expensive screen but one selecting out reproduction inhibitors (chemosterilants, alanosine, organotins), hormones (morphogenetic agents), and developmental inhibitors acting in various ways or at various stages (MON-0585 and PH 6040). These latter developments raise the question of whether practitioners of chemical pest control will tolerate the use of compounds that permit insect damage to continue for a few days or weeks before mortality is effected. Slow kill of immature insects may be acceptable when the adults rather than the larvae are damaging, but a rapid kill is often necessary to prevent intolerable levels of economic damage by the pest.

## V. SELECTIVE TOXICANTS BASED ON KNOWLEDGE OF COMPARATIVE BIOCHEMISTRY AND PHYSIOLOGY

Proposals for academic research continue to promise that the knowledge of comparative biochemistry and physiology will soon permit the synthesis and development of highly selective insecticides, specific to insects and safe for man. This idealized situation has yet to be accomplished on any major scale.

A novel mode of action (biochemical or physiological lesion) is generally discovered as the result of a new compound that elicits the unusual response. These compounds are natural products or come from industrial screening of synthetic compounds as discussed above. Once a new mode of action is found, the major industrial effort is placed in defining the limits of the response by the synthesis of analogs, since the fruits of the effort may find direct utility in pest control. The academic research concentrates on mode of action, sometimes redefining or further defining structure-function relations and thereby supplementing industrial efforts. The ultimate feedback of academic knowledge stimulates further industrial research, leading to an interdependent and yet reasonably efficient overall effort.

Model biochemical and physiological systems as well as the organisms themselves are useful for structure-activity determinations, the lack of correlation being attributed to detoxification, metabolic activation, or physical-chemical properties. Classical examples of this approach are investigations on: inhibitors of acetylcholinesterase (organophosphorus compounds and methylcarbamates); uncouplers of oxidative phosphorylation (surangin B, a variety of synthetic compounds with acidic OH or NH groups, and possibly the formamidines); inhibitors of oxidative phosphorylation (antimycin $A_1$, triorganotins); inhibitors of NADH oxidase enzymes (rotenone, piericidin A) that bind with both the NADH dehydrogenase and lipid components of this system; monoamine oxidase inhibitors (chlorphenamidine); ATPase inhibitors (chlorinated hydrocarbons and pyrethroids); neurophysiological systems sensitive to chlorinated hydrocarbon and pyrethroid insecticides; the unique nerve-muscle junction of insects with L-glutamate as the transmitter substance at excitatory synapses; the excretory system that is often under the control of neurohormones and diuretic hormones, one or both responding to insecticide shock. Compounds such as PH 6040 that block chitin deposition, or more generally those that interfere with the biochemistry of cuticle formation and maintenance, are particularly attractive since they would presumably affect only those organisms that rely on a chitinous exoskeleton. Compounds with certain types of hormonal activity would give the same end result but possibly involve more complex and less stable molecules. Other chemicals that alter molting by interference at any step between hormone action and ultimate tanning may be useful. It is clear that the use of _in vitro_ "mode of action" assays and screens will become of increasing importance in future research designed to discover new insecticide chemicals.

## VI. SUMMARY

The diverse compounds of nature and of the chemical laboratory provide many selective toxicants that, once recognized by screening techniques, can be optimized for pest control. The process of optimization is often facilitated by careful consideration of stereochemical features, chemical reactivity, and hydrophobic properties. It is usually limited by the practicalities and economics of synthesis, and it is likely that the availability of intermediates will become of ever-increasing importance in the future.

The shift to insecticides of entirely new types and modes of action has both advantages and disadvantages. They may be suitable replacements for the Big Three in some areas where these compounds, for one reason or another, no longer do the required job. New compounds may provide temporary solutions to resistance problems and provide techniques for pest control not previously envisaged. This may require a reeducation of farmers, pest control operators, and public health workers to effectively use the new compounds. These alternative insecticides may appear to be safer than others previously used, in part because the undesirable features are frequently uncovered only after years of research by many laboratories. When a new mode of action is encountered, it is often difficult to evaluate the short- or long-term potential of damage for man from the compound. In fact, this potential cannot be fully evaluated until the mode(s) of action is defined at a molecular level. However, many of these questions are resolved in routine toxicology, teratogenesis, mutagenesis, and carcinogenesis tests that must be made prior to registering the compound for use.

No single type of insect control agent can be expected to solve all problems involved in production of food and maintenance of high public health standards. We clearly need an arsenal of insect-control weaponry of multiple types or potential. This cannot be achieved only through further studies of analogs of known compounds, such as the Big Three; it requires discoveries or scientific breakthroughs of fundamental importance and eventual practical application. It appears likely that within the next 25 years there will be as large a changeover in the insect-control agents in use as there has been since the beginning of man's efforts to control pest insects. The insecticides of the future will be characterized by their diversity, both as to chemical type and mode of action. This is the only way we will maintain a lead over our insect competitors for the available food and fiber supply.

## REFERENCES

1. Amonkar, S. V. and Banerji, A. 1971. Science 174, 1343.

2. Angus, T. A. 1971. Chapter 10 in Naturally Occurring Insecticides, M. Jacobson and D. G. Crosby, Eds., Marcel Dekker, New York, pp. 463-497.

3. Ascher, K. R. S. and Ishaaya, I. 1973. Pestic. Biochem. Physiol. 3, 326.

4. Aziz, S. A. and Knowles, C. O. 1973. J. Econ. Entomol. 66, 1041.

5. Beard, R. L. 1971. Chapter 6 in Naturally Occurring Insecticides, M. Jacobson and D. G. Crosby, Eds., Marcel Dekker, New York, pp. 243-270.

6. Bond, R. P. M. and Somerville, H. J. 1973. Microbial Toxins, Paper No. 1, Vth International Colloquium on Insect Pathology and Microbial Control, VIth Annual Meeting of the Society of Invertebrate Pathology, Oxford.

7. Bond, R. P. M., Boyce, C. B. C., Rogoff, M. H., and Shieh, T. R. 1971. Chapter 12 in Microbial Control of Insects and Mites, H. D. Burges and N. Hussey, Eds., Academic, New York, pp. 275-303.

8. Box, S. J., Cole, M., and Yeoman, G. H. 1973. Appl. Microbiol. 26, 699.

9. Büchel, K. H. and Draber, W. 1972. Adv. Chem. Ser. 114, 141.

10. Cantu, E. and Wolfenbarger, D. A. 1973. J. Econ. Entomol. 66, 527.

11. Casida, J. E., Ed. 1973. Pyrethrum the Natural Insecticide, Academic, New York.

12. Cavill, G. W. K. and Clark, D. V. 1971. Chapter 7 in Naturally Occurring Insecticides, M. Jacobson and D. G. Crosby, Eds., Marcel Dekker, New York, pp. 271-305.

13. Cole, M. and Rolinson, G. N. 1972. Appl. Microbiol. 24, 660.

14. Crombie, L. 1972. In Insecticides, Vol. I, Proceedings of the Second International IUPAC Congress of Pesticide Chemistry, A. S. Tahori, Ed., Gordon and Breach Science, New York, pp. 101-118.

15. Crosby, D. G. 1971. Chapter 5 in Naturally Occurring Insecticides, M. Jacobson and D. G. Crosby, Eds., Marcel Dekker, New York, pp. 177-239.

16. Darlington, W. A., Ludvik, G. F., and Sacher, R. M. 1972. J. Econ. Entomol. 65, 48.

17. Enders, E., Stendel, W., and Wollweber, H. 1973. Pestic. Sci. 4, 823.

18. Faust, R. M. 1973. Bull. Entomol. Soc. Amer. 19, 153.

19. Fukami, H. and Nakajima, M. 1971. Chapter 2 in Naturally Occurring Insecticides, M. Jacobson and D. G. Crosby, Eds., Marcel Dekker, New York, pp. 71-97.

20. Gosalvez, M. and Merchan, J. 1973. Cancer Res. 33, 3047.

21. Hamill, R. L., Higgens, C. E., Boaz, H. E., and Gorman, M. 1969. Tetrahedron Lett, 4255.

22. Hammann, I., Hoffman, P., Unterstenhoefer, G., Kleimann, H., Marquarding, D., Offermann, K., and Ugi, I. 1969. South African patent 69 01,922.

23. Harrison, I. R., Kozlik, A., McCarthy, J. F., Palmer, B. H., Wakerley, S. B., Watkins, T. I., and Weighton, D. M. 1973. Pestic. Sci. 4, 901.

24. Horiuchi, F., Fujimoto, K., Ozaki, T., and Nishizawa, Y. 1971. Agr. Biol. Chem. 35, 2003.

25. Iwasaki, H., Kamiya, T., Oka, O., and Ueyanagi, J. 1965. Chem. Pharm. Bull. 13, 753.

26. Jacobson, M. 1971. Chapter 4 in Naturally Occurring Insecticides, M. Jacobson and D. G. Crosby, Eds., Marcel Dekker, New York, pp. 137-176.

27. Kato, N., Takahashi, M., Shibayama, M., and Munakata, K. 1972. Agr. Biol. Chem. 36, 2579.

28. Kenaga, E. E. 1969. J. Econ. Entomol. 62, 1006.

29. Kido, G. S. and Spyhalski, E. 1950. Science 112, 172.

30. Knowles, C. O. and Roulston, W. J. 1973. J. Econ. Entomol. 66, 1245.

31. Knowles, C. O., Ahmad, S., and Shrivastava, S. P. 1972. In Insecticides, Vol. I, Proceedings of the Second International IUPAC Congress of Pesticide Chemistry, A. S. Tahori, Ed., Gordon and Breach Science, New York, pp. 77-98.

32. Konishi, K. 1972. In Insecticides, Vol. I, Proceedings of the Second International IUPAC Congress of Pesticide Chemistry, A. S. Tahori, Ed., Gordon and Breach Science, New York, pp. 179-189.

33. Lichtenstein, E. P. and Casida, J. E. 1963. J. Agr. Food Chem. 11, 410.

34. Lichtenstein, E. P., Strong, F. M., and Morgan, D. G. 1962. J. Agr. Food Chem. 10, 30.

35. Luijten, J. G. A. 1972. Chapter 12 in Organotin Compounds, Vol. 3, A. K. Sawyer, Ed., Marcel Dekker, New York, pp. 931-974.

36. Metcalf, R. L. 1966. Chapter 7 in Pharmacology of Fluorides. Handbook of Experimental Pharmacology, Vol. 20, Part I, F. A. Smith, Ed., Springer-Verlag, Berlin, pp. 355-386.

37. Metcalf, R. L. 1971. Bull. W.H.O. 44, 43.

38. Moorefield, H. H. 1965. In Research in Pesticides, C. O. Chichester, Ed., Academic, New York, pp. 41-51.

39. Mulder, R. and Gijswijt, M. J. 1973. Pestic. Sci. 4, 737.

40. Muraoka, S. and Terada, H. 1972. Biochim. Biophys. Acta 275, 271.

41. Norris, J. R. 1972. In Insecticides, Vol. I, Proceedings of the Second International IUPAC Congress of Pesticide Chemistry, A. S. Tahori, Ed., Gordon and Breach Science, New York, pp. 119-139.

42. Ogita, Z. 1958. Nature 182, 1529.

43. Ohkawa, H., Ohkawa, R., Yamamoto, I., and Casida, J. E. 1972. Pestic. Biochem. Physiol. 2, 95.

44. Oishi, H., Hosokawa, T., Okutomi, T., Suzuki, K., and Ando, K. 1969. Agr. Biol. Chem. 33, 1790.

45. Ovchinnikov, Y. A., Ivanov, V. T., and Mikhaleva, I. I. 1971. Tetrahedron Lett., 159.

46. Pianaka, M. and Smith, C. B. F. 1965. Chem. Ind. (London), 1216.

47. Sacher, R. M. 1971. Proc. Brit. Insectic. Fungic. Conf. 6th, 611.

48. Sagawa, T., Hirano, S., Takahashi, H., Tanaka, N., Oishi, H., Ando, K., and Togashi, K. 1972. J. Econ. Entomol. 65, 372.

49. Saggers, D. T. and Clark, M. L. 1967. Nature 215, 275.

50. Sakai, M. and Sato, Y. 1972. In Insecticides, Vol. I, Proceedings of the Second International IUPAC Congress of Pesticide Chemistry, A. S. Tahori, Ed., Gordon and Breach Science, New York, pp. 455-467.

51. Schmeltz, I. 1971. Chapter 3 in Naturally Occurring Insecticides, M. Jacobson and D. G. Crosby, Eds., Marcel Dekker, New York, pp. 99-136.

52. Soeda, Y., Kato, S., Takiguchi, D., Sakimoto, R., and Ohkuma, K. 1972. J. Agr. Food Chem. 20, 936.

53. Sousa, A. A., Durden, Jr., J. A., and Stephen, J. F. 1973. J. Econ. Entomol. 66, 584.

54. Suzuki, A. and Tamura, S. 1972. In Insecticides, Vol. I, Proceedings of the Second International IUPAC Congress of Pesticide Chemistry, A. S. Tahori, Ed., Gordon and Breach Science, New York, pp. 163-177.

55. Takemoto, T., Nakajima, T., and Yokobe, T. 1964. J. Pharm. Soc. Japan 84, 1232.

56. Tamura, S. and Takahashi, N. 1971. Chapter 11 in Naturally Occurring Insecticides, M. Jacobson and D. G. Crosby, Eds., Marcel Dekker, New York, pp. 499-539.

57. Valenta, Z. 1972. In Insecticides, Vol. I, Proceedings of the Second International IUPAC Congress of Pesticide Chemistry, A. S. Tahori, Ed., Gordon and Breach Science, New York, pp. 191-205.

58. van Tamelen, E. E., Dickie, J. P., Loomans, M. E., Dewey, R. S., and Strong, F. M. 1961. J. Amer. Chem. Soc. 83, 1639.

59. Wada, K. and Munakata, K. 1968. J. Agr. Food Chem. 16, 471.

## DISCUSSION

Naturally occurring toxicants are obviously a fertile source of leads for the synthesis of new types of insecticides. Dr. Casida in the position paper has outlined the synthetic processes necessary for structural optimization for insecticidal activity, for improvement of selectivity ratio, suitability for pest management, and finally for economic registration before you have a marketable compound. The greatest problem with naturally occurring toxicants is their lack of persistence. However, they are especially useful for models and for mode of action studies leading to alternatives to organochlorine, organophosphorus, and carbamate insecticides.

Nereistoxin from a Japanese marine annelid is a unique case of the discovery of a useful insecticide derived from the toxin of a marine animal. Last year 600 to 1000 tons of cartap, a practical derivative produced by this optimization process, were produced and used for the control of the rice-stem borer and other agricultural pests. Unfortunately, cartap is also very toxic to silkworms that were sometimes poisoned on mulberry leaves by very small particles of cartap that drifted from agricultural areas. This illustrates the need for highly selective toxicants for specific insects.

The isoxazole compounds that have insecticidal and fungicidal activity are another interesting example. Tricholomic acid and ibotenic acid are flycidal toxins isolated from mushrooms, having isoxazolidine and isoxazoline structures, respectively. Both toxins are derivatives of amino acids with sweet tastes even stronger than monosodium glutamate. Iwai in 1965 developed a convenient synthesis for isoxazole compounds and discovered a new soil-fungicide 3-hydroxy-5-methyl isoxazole, named hymexazole. It exhibits remarkable activity against soil-borne diseases of many crops and is now widely used to control damping-off diseases of rice and sugarbeets in their seedling stages. Hymexazole has low toxicity to mammals, birds, and fish and when administered to rats is rapidly excreted in urine. Its soil persistence is relatively short since a large portion of the molecule is metabolized to acetoacetamide and finally to carbon dioxide by soil microorganisms. The fact that removal of the amino acid moiety of the isoxazole compounds destroys flycidal activity but not insect fungicidal activity suggests a possible interaction with the neuromuscular junction perhaps like glutamic acid.

These examples demonstrate the possibilities for the discovery of "safe pesticides" by optimization of the structures of endogenous substances such as amino acids, carbohydrates, fatty acids, nucleic acids, and their derivatives. Only about 1% of the marine toxins and venoms have been identified, so many potential sources remain.

It was pointed out that many of the five-membered ring compounds containing oxygen and nitrogen are potent inhibitors of microsomal oxidases, for example, oxazoles that are about as active as imidazoles. Such compounds are a good starting place to look for active biocides as the imidazoles, for example, are fungicides, protozoacides, insecticides, and synergists. The thiadiazoles include herbicides, fungicides, insecticides, and synergists. All of these activities must be interrelated and dependent on the nitrogen-containing rings. It is unusual for such compounds to have only one mode of action, and their microsomal-enzyme interactions will have important bearing upon their use.

Ferns were suggested as sources for new toxicants because of their reputed immunity to insect attack and the isolation of ecdysones from them. However, it was pointed out that a literature search showed that over 200 species of insects feed in ferns and survive.

The question of the possible role of toxic substances in plants and animals as defense secretions and whether or not these signified a chemical basis for biological control, was considered. In plants it appears that such substances including nicotine and the pyrethrins have little or no defense significance. They would never have been discovered if someone had not ground up or extracted the organism and offered it to an insect. In microbial toxins such as *Bacillus thuringiensis*, however, the picture is less clear. This product, B.T. insecticide, has shown increased toxicity with each new strain of *Bacillus*. The differences may not relate to amino acid composition but rather to changes in tertiary structure that would allow different release rates of toxic subfragments. In general, the marketing and acceptance of this product has increased although problems have been experienced with the very narrow spectrum activity and the toxicity to silkworms in Japan.

In regard to insect toxins, these must often be directed at the host, and it was concluded that every chemical intermediate used biologically to carry information or messages, that is, pheromones, can potentially be useful in insect control once its chemistry and the ecological aspects of its use are understood. Very many of the drugs used in medicine were optimized from natural products, and we must conclude that "Nature knows better than we," and is certainly more varied in chemical expression than any chemist can ever be.

Organometallic compounds as insecticides were discussed as potential new insecticides. There are reservations about the use of such substances as the organotin compounds because of possible environmental pollution. WHO, for example, carried out studies with organotins for snail control in schistosomiasis programs. However, because of possible toxicity to fish and other organisms, such studies have been terminated for the present. However, the mode of action of organotins as inhibitors of oxidative phosphorylation is well understood, and these compounds break down in the environment to inorganic tin that is without biological activity. Thus it was concluded that organotins were perhaps little different than organophosphorus compounds as environmental pollutants. However, organomercury and organolead compounds are much more hazardous because of the intrinsic toxicity of the metals themselves. It would seem that the ultimate scarcity of these particular metals would preclude their use in pesticides because of cost and their value to industry.

# New Targets of Insecticides in the Nervous System

TOSHIO NARAHASHI
Department of Physiology and Pharmacology
Duke University Medical Center
Durham, North Carolina

## I. INTRODUCTION

It has been well established that the nervous system is the target site of a wide variety of insecticides. However, the potency of any particular insecticide is not determined solely by its effect on the target site. Insecticide has to enter the body via cuticle or other routes, may be converted into a more or less active form, may be stored in nontarget sites such as adipose tissues, and may be excreted intact or after detoxication. Only after reaching the target site can insecticide exert its toxic action either in its original form or in an activated form. These processes of insecticidal action are illustrated in Figure 1.

To elucidate the mechanism of action of insecticide, the study at the target site is of critical importance. Three objectives are conceivable in such study at the target site:

The first and most straightforward objective is to clarify the detailed mechanisms of action of insecticide on the target site. For certain insecticides, the target site is an enzyme or an enzyme system. For instance, most organophosphorus and carbamate insecticides inhibit cholinesterases thereby disturbing the nervous function. Kinetic studies of cholinesterase inhibition by a variety of agents have led to a model for the receptor site on cholinesterase molecule (see Koelle, 1970; Holmstedt, 1959; Wilson, 1959). For some other insecticides, the target site is not an enzyme but neural membranes. These insecticides presumably bind to certain macromolecules in the nerve membrane thus causing effects as manifested by functional changes of the neuron. Electrophysiological analyses have proved highly successful in elucidating the mechanism whereby insecticides modify the function of nerve membranes.

The second objective in the study of target site is the structure-activity relationship. There are two kinds of structure-activity relationships. One is an apparent or prac-

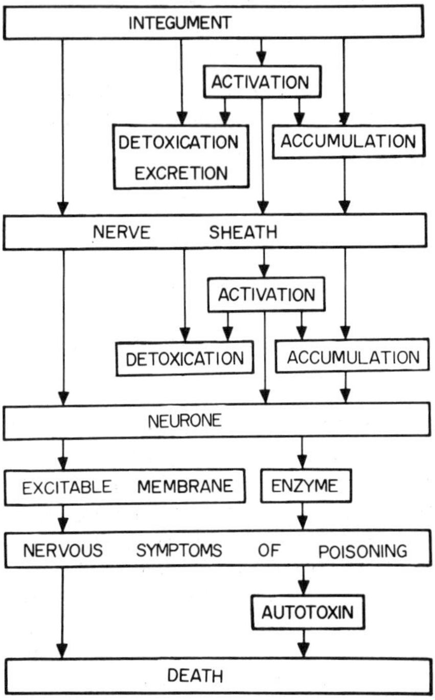


FIGURE 1. Toxic Process of Insecticide (Narahashi, 1964).

tical structure-activity relationship derived from the measurements of insecticidal potencies of various analogs and derivatives. The insecticidal potency is a final manifestation of a series of processes and reactions including cuticle penetration, detoxication, and action at the target site, so that it does not represent the interaction of insecticide with the target site. Despite this, the apparent structure-activity relationship based on the measurements of insecticidal potencies is a useful parameter to screen various compounds for their insecticidal activity.

The true structure-activity relationship can be obtained only by measurements of potencies at the target site. Experiments are simplest and most straightforward if the target site is isolated *in vitro* in a purified form. The best example is acetylcholinesterase which is inhibited by organophosphorus and carbamate insecticides (O'Brien, 1960, 1967). It is available in a purified and crystallized form and is easy to handle for in vitro experiments. Another example is seen in NADH oxidase which is inhibited by rotenone (Fukami, 1961; Fukami and Tomizawa, 1956, 1958a, b; Lindahl and Öberg, 1961; Öberg, 1961). ATPases are inhibited by DDT and other chlorinated hydrocarbon insecticides (Matsumura and Narahashi, 1971; Koch, 1969; Koch et al., 1969, 1971), but the significance of such inhibition in the toxic action remains to be studied.

For most other insecticides, the target site has not been isolated in vitro. For example, DDT, cyclodienes, and pyrethroids act on the nerve membranes through direct interactions with the macromolecules associated with ionic channels (Narahashi, 1971a, 1974b). Their effects are not exerted by inhibitions of enzymes. Therefore, the only way to study the direct action of such insecticides on the target site is to use isolated nerve preparations.

The third objective is to clarify the mechanism whereby various factors influence the insecticidal potency (Narahashi, 1971a). For instance, temperature has a profound effect on the insecticidal activity of certain compounds such as DDT and pyrethroids, and the temperature dependency cannot be explained on the basis of cuticle penetration or detoxication alone. The effects of these insecticides on the target site (nerve) are greatly influenced by temperature. Similar examples can be found in the mechanism of insecticide resistance. In some cases insecticide resistance can be accounted for in terms of detoxication, but in some other cases other factors are involved also. The sensitivity of the target nervous tissue is greatly diminished in resistant strains of insects.

## II. NERVE EXCITATION AND DRUG ACTION

### A. Mechanism of Nerve Excitation

Before discussing the mechanism of action of insecticides on nerve, it would be appropriate to review the current model for nerve excitation and the rationale for the study of drug action in general. The scheme of excitation is illustrated in Figure 2. There are concentration gradients for sodium and potassium across the nerve membrane, sodium concentration being higher outside than inside and potassium concentration being higher inside than outside. Since the nerve membrane is permeable to potassium but only sparingly so to other ions including sodium, chloride, and calcium, the resting membrane

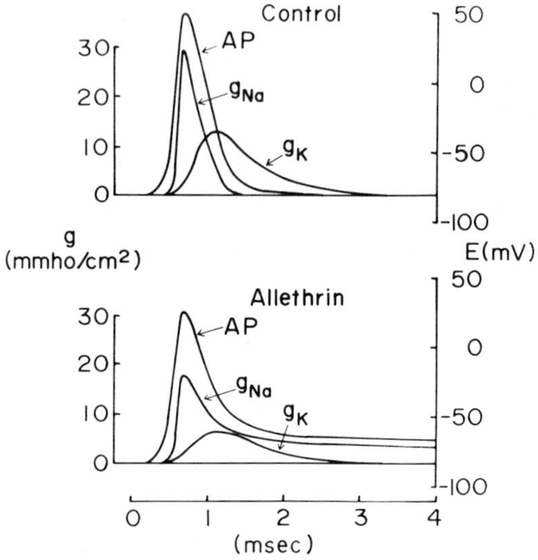


FIGURE 2. Schematic Drawings of the Time Course of Changes in Sodium Conductance (Permeability) ($G_{Na}$) and Potassium Conductance (Permeability) ($G_K$) During an Action Potential (AP) Before and After Application of Allethrin. (Narahashi, 1971b).

potential ($E_{RP}$), inside negative with respect to the outside, is close to the equilibrium potential for potassium ($E_K$) defined by the Nernst equation

$$E_{RP} \simeq E_K = \frac{RT}{F} \ln \frac{[K]_o}{[K]_i} \tag{1}$$

where R, T, and F refer to the gas constant, the absolute temperature, and the Faraday constant, respectively, and $[K]_o$ and $[K]_i$ refer to the external and internal potassium concentrations. respectively. When the membrane potential becomes less negative (depolarized) beyond a certain level (threshold) by a stimulating outward current across the membrane, the sodium permeability of the membrane undergoes a large and fast generative increase. Thus the membrane becomes permeable largely to sodium, and the membrane potential approaches the sodium equilibrium potential ($E_{Na}$) given by the equation

$$E_{AP} \simeq E_{Na} = \frac{RT}{F} \ln \frac{[Na]_o}{[Na]_i} \qquad (2)$$

where $E_{AP}$ is the membrane potential at the peak of the action potential, and $[Na]_o$ and $[Na]_i$ refer to the external and internal sodium concentrations, respectively. Since the external sodium concentration far exceeds the internal concentration, the membrane potential is reversed in polarity during the action potential. The increased sodium permeability starts decreasing soon, and the potassium permeability starts increasing beyond its resting level so that the membrane becomes almost exclusively permeable to potassium bringing back the membrane potential toward the resting level. Thus the falling phase of the action potential is produced. As a result of these permeability changes, sodium ions enter and potassium ions leave during the action potential.

It should be emphasized that these permeability changes during nervous activity are not directly supported by metabolic energy. Even after injection of metabolic inhibitors such as azide, dinitrophenol, or cyanide into the giant axon, a number of action potentials can still be produced as long as the proper concentration gradients of ions are maintained across the nerve membrane (Hodgkin and Keynes, 1955). However, metabolic energy plays an important role after excitation, pumping out the extra sodium that has entered and absorbing potassium. This mechanism called sodium-potassium pump, is stimulated by an increase in internal sodium concentration. It also works at rest, keeping the proper concentration gradients of sodium and potassium.

## B. Mechanism of Action of Insecticides on Nerve

Most neuroactive agents affect nervous function through interactions with the ionic permeability mechanisms rather than inhibiting metabolism. This is also true for insecticides with a few exceptions including rotenone and cyanide. The mechanisms of the changes in nerve excitability brought about by drugs may be classified as follows:

1. Change in Membrane Potential. Since the ability of the nerve membrane to undergo ionic permeability changes on stimulation is a function of the membrane potential, an increase or decrease in membrane potential (hyperpolarization or depolarization) should change excitability. For example, a large depolarization blocks nervous conduction because no sodium permeability increase can take place on stimulation as is observed with batrachotoxin (Narahashi *et al.*, 1971; Albuquerque *et al.*, 1973) and with grayanotoxins (Seyama and Narahashi, 1973; Narahashi and Seyama, 1974). No insecticide has been known to block conduction primarily by depolarization. Allethrin causes a slight depolarization of nerve membranes, but this is only partially responsible for conduction block (Narahashi, 1962).

2. Inhibition of Sodium Permeability Increase. Certain neuroactive agents are known to inhibit the sodium-permeability increase that occurs on stimulation. The most dramatic examples are tetrodotoxin (TTX) and saxitoxin (STX), both of which exert a highly selective inhibitory effect on this mechanism without any effect on other permeability parameters or on the resting membrane potential (Narahashi, 1972; Narahashi *et al.*, 1960, 1964, 1967a; Nakamura *et al.*, 1965; Hille, 1968). This effect is exerted at a very low concentration (with the apparent dissociation constant in the order of nanomolars) and only from outside of the nerve membrane (Narahashi *et al.*, 1967b). Local anesthetics block nervous conduction by the inhibition of the sodium permeability increase, but they inhibit the potassium permeability increase as well (Taylor, 1959; Shanes *et al.*, 1959; Narahashi *et al.*, 1969a; Blaustein and Goldman, 1966).

Certain insecticides block nervous conduction by this mechanism but with a less specific affinity for sodium. At relatively high concentrations (10 to 30μM), allethrin suppresses the sodium permeability increase, thereby blocking the action potential (Narahashi and Anderson, 1967; Wang *et al.*, 1972; Narahashi, 1971b). However, the potassium permeability increase (that produces the falling phase of the action potential) is also inhibited by allethrin (Figure 2). Aldrin-*trans*-diol also inhibits the sodium permeability increase with a smaller inhibitory effect on the potassium permeability increase (van den Bercken and Narahashi, 1974).

3. Inhibition of Sodium Permeability Decrease and Potassium Permeability Increase. Since the sodium permeability decrease (sodium inactivation) and the potassium permeability increase are responsible for the falling phase of the action potential,

an inhibition of either or both of them should prolong that phase. Typical examples are seen with DDT and allethrin. DDT inhibits both permeability mechanisms thereby increasing and prolonging the negative (depolarizing) after-potential that following the spike phase of the action potential (Figure 3)

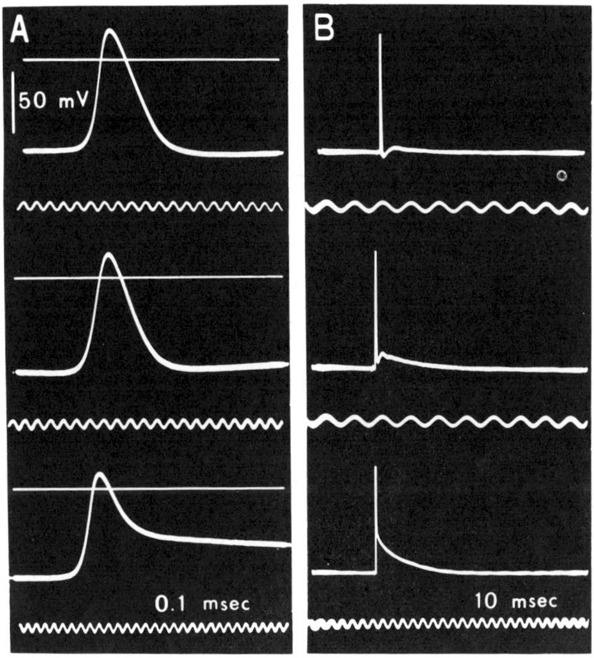


FIGURE 3. Effects of 1 x $10^{-4}$ M DDT on the action potential of the cockroach giant axon. A and B were recorded at different sweep speeds. Top records, control; middle record, 38 min after application of DDT; bottom records, 90 min after (Narahashi and Yamasaki, 1960).

(Narahashi and Haas, 1967, 1968; Hille, 1968). As a result of the large negative after-potential, repetitive after-discharges are often induced (Figure 4) (Narahashi and Yamasaki, 1960).

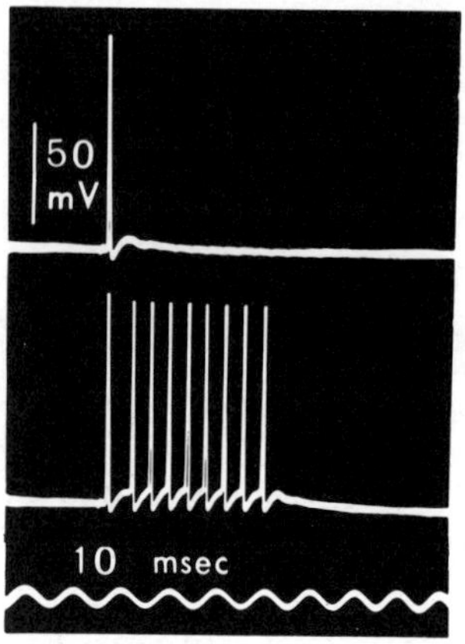


FIGURE 4. Repetitive discharges induced by a single stimulus in the cockroach giant axon exposed to 1 x $10^{-4}$ M DDT. Upper and lower records are taken 20 and 22 min after treatment with DDT (Narahashi and Yamasaki, 1960).

Allethrin also exerts similar effects as DDT at relatively low concentrations but inhibits the sodium permeability increase also at higher concentrations (Figure 2) (Narahashi and Anderson, 1967; Wang et al., 1972; Narahashi, 1971b).

There are noninsecticidal chemicals that also inhibit the sodium permeability decrease with a small inhibitory effect on the potassium permeability increase. Typical examples are Condylactis toxin (Narahashi et al., 1969b) and pronase (Armstrong et al., 1973). They both prolong the falling phase of the action potential. Tetraethylammonium inhibits the potassium permeability increase thereby prolonging the falling phase of the action potential (Tasaki and Hagiwara, 1957; Armstrong and Binstock, 1965; Hille, 1967).

## III. SYNAPTIC TRANSMISSION AND DRUG ACTION

### A. Mechanism of Synaptic Transmission

When an action potential arrives at the nerve terminal, a transmitter substance is released from the terminal. The postsynaptic membrane has a special affinity for the transmitter and is depolarized as a result of the transmitter-receptor interaction. The depolarization of the postsynaptic membrane stimulates the postsynaptic neuron or the effector organ, thus completing synaptic transmission. The released transmitter may be quickly hydrolyzed by an enzyme; for instance, acetylcholinesterase is present on or near the cholinergic postsynaptic membrane and hydrolyzes acetylcholine (ACh) into choline and acetic acid. In inhibitory synapses the released transmitter prevents the effect of the excitatory presynaptic fibers on the same postsynaptic element.

A variety of transmitter substances have been identified. In vertebrate animals ACh is the transmitter in skeletal neuromuscular junctions, synapses in the ganglia, cholinergic neuroeffector junctions, and certain synapses in the spinal cord and brain. Other excitatory or inhibitory transmitter substances identified or suspected in vertebrates include norepinephrine, dopamine, l-glutamate, $\gamma$-aminobutyric acid (GABA), glycine, 5-hydroxytryptamine (5-HT), and substance P (Krnjević, 1974). In insects no substance has so far been firmly identified as the transmitter substance, but there is some evidence to indicate that ACh, l-glutamate, GABA, and possibly norepinephrine, dopamine, and 5-HT are the excitatory or inhibitory transmitters in certain synapses in the ganglia (Gerschenfeld, 1973; Pitman, 1970; Boistel, 1968; Treherne, 1966). l-Glutamate is possibly the excitatory transmitter and GABA the inhibitory transmitter in insect neuromuscular junctions (Gerschenfeld, 1973).

## B. Effects of Drugs on Synaptic Transmission

Since synaptic transmission involves multiple steps, several sites of action are conceivable for any particular drug. Conduction at the presynaptic nerve terminal may be affected. Tetrodotoxin blocks the conduction thereby impairing synaptic and neuromuscular transmission (Furukawa *et al.*, 1959). DDT causes repetitive firing in the presynaptic fibers, and this action is thought to be responsible for the facilitation of synaptic transmission (Narahashi, 1974).

Certain drugs are known to inhibit or disrupt the synthesis of the transmitter substance in the presynaptic fiber. For example, hemicholinium 3 inhibits the uptake of choline from the presynaptic nerve terminal, causing a depletion of ACh store (MacIntosh, 1963; Narahashi, 1974a).

The mechanism by which the transmitter is released is vulnerable to various drugs and experimental conditions. Botulinum toxin is highly effective in blocking the transmitter release (Simpson, 1971; Narahashi, 1974a). Calcium ions are essential for the transmitter to be released, but magnesium ions inhibit the release (Katz, 1962, 1966; Hubbard, 1970). Guanidine and phenols facilitate the transmitter release (Otsuka and Endo, 1960). Some insecticides are suspected of affecting this mechanism, but none of the studies so far performed is documented with clear-cut experimental evidence. For instance, the action of dieldrin in producing synaptic after-discharges may be due to a stimulation of transmitter release (Shankland and Schroeder, 1973; Akkermans _et al._, 1974). Recent observations strongly support the notion that aldrin-_trans_-diol, a metabolite of dieldrin, is one of the active forms of dieldrin (Wang _et al._, 1971; van den Bercken and Narahashi, 1974; Akkermans _et al._, 1974). γ-BHC has a potent action in inducing synaptic after-discharges (Yamasaki and Ishii, 1954b), and it is quite possible that it facilitates transmitter release, though no attempt has been made to prove this notion experimentally.

The mechanism whereby the released transmitter is removed or inactivated is also important as the target of a variety of drugs. Acetylcholinesterase (AChE) is of particular importance in connection with insecticides since a number of organophosphate and carbamate insecticides inhibit AChE causing facilitation and blockage of cholinergic synaptic transmission. It is well established that certain insecticides inhibit AChE only after being activated. The kinetics of AChE inhibition by various inhibitors have been studied in detail, and effective reactivators of AChE have successfully been developed as a result of such studies (see Koelle, 1970; O'Brien, 1967). Pralidoxime is a typical example and is being used as an effective antidote of organophosphorus insecticide poisoning.

The postsynaptic membrane is affected by a variety of agents. Two types of effects are known: one is to reduce the sensitivity of the postsynaptic membrane to the transmitter substance, and the other is to depolarize the postsynaptic membrane. The former type of drugs are often called nondepolarizing blocking agents, and the latter depolarizing block agents. One typical example of the nondepolarizing blocking agent is _D_-tubocurarine which competes with ACh for the receptor site.

Several insecticides have recently been shown to exert their blocking action on the vertebrate skeletal neuromuscular junction by the mechanism similar to that of _D_-tubocurarine. Nereistoxin blocks the neuromuscular transmission of the frog primarily by a reduction of the sensitivity of the end-plate membrane to ACh (Deguchi _et al._, 1971). Figure 5 depicts the peak amplitude of end-plate depolarization induced by iontophoretic application of ACh as a function of the current intensity applied to ACh pipette. The ACh-induced depolarization is

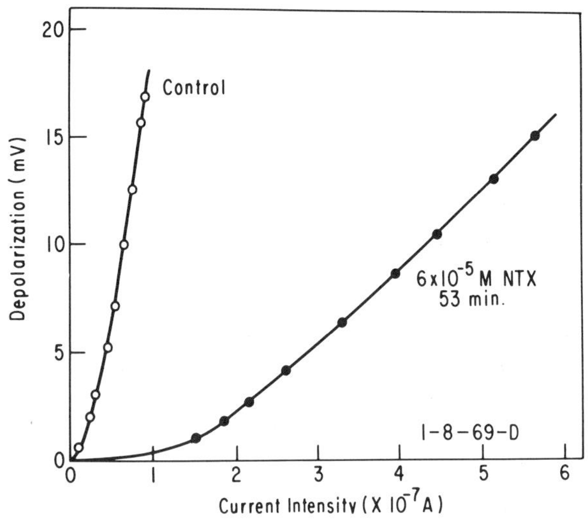


FIGURE 5. Effects of $6 \times 10^{-5}$ M nereistoxin (NTX) on the peak depolarization of the frog end-plate membrane produced by iontophoretic application of acetylcholine. The amplitude of depolarization is plotted against the intensity of current (45 sec duration) applied to the acetylcholine pipette (Deguchi et al., 1971).

greatly diminished by nereistoxin. Similar mechanism has been suggested for the block of synaptic transmission in the cockroach ganglion (Sakai, 1967, 1969). Chlordimephorm blocks the frog neuromuscular transmission in a similar manner (Wang et al., 1974). It has been found recently that nicotine, at concentrations insufficient for the classical depolarizing neuromuscular block ($5 \times 10^{-6}$ M), suppresses the sensitivity of the frog end-plate membrane to ACh (Wang and Narahashi, 1972).

The end-plate membrane undergoes increases in conductances to sodium and potassium as a result of ACh-receptor interactions. The sodium and potassium conductance changes can be measured by voltage clamp techniques. When the end-plate membrane potential is clamped at the potassium equilibrium potential (-100 mV), an end-plate current (EPC) that is carried exclusively by sodium current can be recorded by nerve stimulation. Likewise, an EPC carried by potassium current can be recorded at the sodium equilibrium potential (+50 mV). Figure 6 depicts families of EPCs

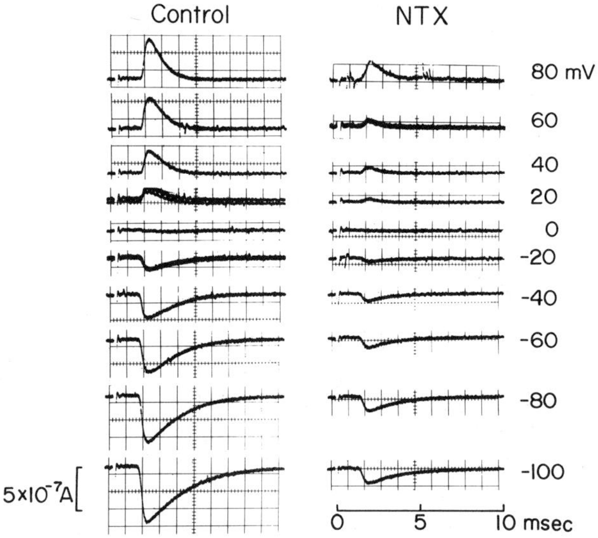


FIGURE 6. End-plate currents recorded at various holding membrane potentials before and 12 min after application of 4 x $10^{-5}$ M nereistoxin (NTX) to the frog sartorius muscle. Downward deflections indicate the inward end-plate currents (Deguchi et al., 1971).

recorded at various membrane potentials indicated on the right-hand side. The EPC reverses its polarity near 0mV, which is the equilibrium potential. Nicotine and nereistoxin have been found to inhibit both conductance increases equally (Deguchi et al., 1971; Wang and Narahashi, 1972). An example of nereistoxin experiment is shown in Figure 8. However, this is not the case for some other blocking agents. For example, procaine inhibits both conductances but in different manners; the sodium EPC is accelerated in its initial falling phase and is followed by a slow terminal phase, whereas the potassium EPC is slightly slowed (Deguchi and Narahashi, 1971).

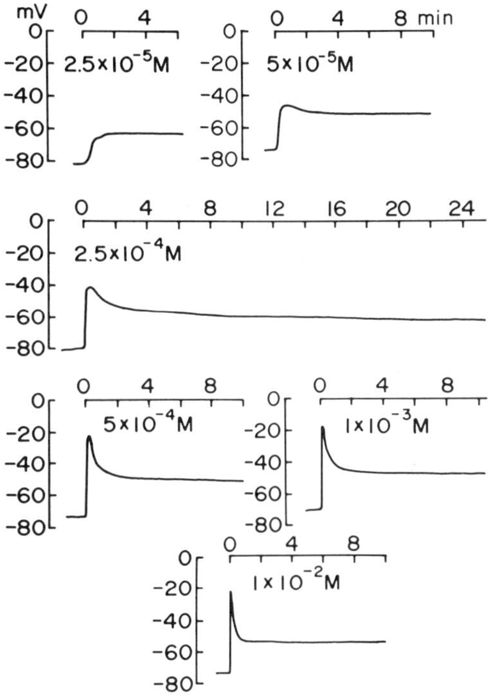


FIGURE 7. Effects of various concentrations of nicotine on the resting membrane potential of the end-plate of the frog sartorius muscle (Wang and Narahashi, 1972).

Certain agents depolarize the postsynaptic membrane thereby first stimulating and then blocking synaptic transmission. For instance, ACh, carbamylcholine, and decamethonium depolarize the cholinergic end-plate membrane. However, the depolarization is not maintained but is followed by a partial repolarization in the continuous presence of the drugs during which the end-plate membrane remains insensitive to the transmitter ACh. This phase is called desensitization block. Nicotine at relatively high concentrations (in the order of $5 \times 10^{-5}$ M) exerts a depolarization-repolarization block (Thesleff, 1955; Wang and Narahashi,

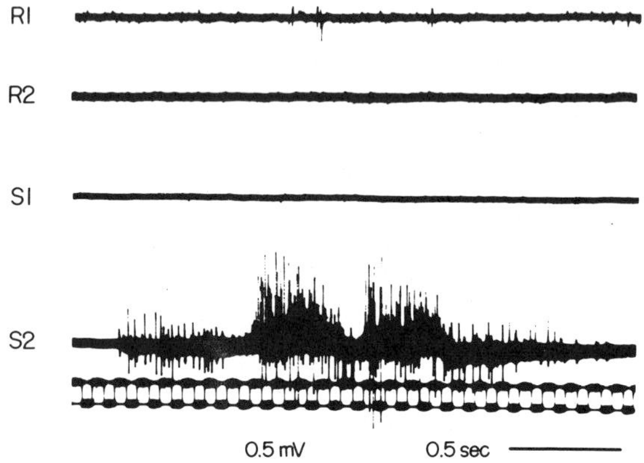


FIGURE 8. Discharges of impulses recorded from the femur muscle before (R1 and S1) and 9 min after (R2 and S2) application of 2.8 x $10^{-5}$ M DDT directly to the exposed thoracic ganglia of the housefly. R, DDT-resistant strain (bwb; ocra; ar; ac); S, susceptible strain (Lab) (Tsukamoto _et al._, 1965).

1972). Figure 7 illustrates the time course of depolarization and repolarization induced by nicotine. It should be noted that sodium and potassium conductances of the end-plate membrane are affected differently by nicotine under this condition. The potassium conductance is increased and decreased during application of nicotine, whereas the sodium conductance is increased and only partially decreased (Wang and Narahashi, 1972). The residual sodium conductance is responsible for the residual depolarization in the presence of nicotine.

## IV. SELECTIVE TOXICITY DUE TO FACTORS OTHER THAN TARGET SITE

Selective toxicity of insecticides could be due to factors other than the target site. Similarly some chemicals may not possess any insecticidal potency simply because they cannot reach the target site for various reasons. Penetrability of insecticide through the cuticle is the first factor to be considered. A number of toxic compounds are useless as insecticides owing to their lack of penetrability. DDT derivatives in which chlorines at the *para*-positions on the benzene rings are replaced by hydroxy or *amino* groups have no insecticidal activity although they are toxic when directly applied to the nerve (Wu *et al*., 1974).

Activation and detoxication are also important factors when selective toxicity is considered. Parathion is highly toxic to both mammals and insects because neither of them is superior to the other in the efficiency of enzymatic detoxication. On the contrary, malathion is much less toxic to mammals than insects, because of the presence in mammals of the efficient detoxifying enzyme carboxyesterase which is low in activity in insects (O'Brien, 1967). A variety of insecticide-resistant strains of insects have potent detoxifying enzymes that are absent or weak in normal susceptible strains (O'Brien, 1966, 1967; Brown, 1964).

## V. SELECTIVITY OR POTENCY DIFFERENCE DUE TO FACTORS AT TARGET SITE

Target site plays an important role in the selective toxicity. Examples known for insecticides may be divided into three categories: (1) due to intrinsic nature of the target receptor; (2) due to the development of insensitivity of the target site to insecticide; (3) due to temperature dependency of the action on the target site.

### A. Intrinsic Nature of Target Receptor

Certain animals possess the nerve intrinsically insensitive to some insecticides. Squid giant axons are not affected by DDT, and this absence of the effect is not due to penetrability of DDT to the nerve membrane. Even when DDT is perfused internally, no effect can be observed. This insensitivity is presumably due to a lack of DDT receptors in the squid axon membrane.

### B. Development of Target Insensitivity

The mechanism of insecticide resistance in some resistant strains of insects cannot totally be accounted for in terms of detoxication and cuticle penetration. The nerve of the resistant strain is generally less sensitive to the insecticide than that of the susceptible strain. A number of such examples are known for DDT-, BHC-, dieldrin-, and organophosphate-resistant strains (Yamasaki and Narahashi, 1958, 1962; Narahashi, 1964; Tsukamoto et al., 1965; Smyth and Roys, 1955). An example of the effect of direct application of DDT to the exposed thoracic ganglia of the housefly is illustrated in Figure 8. DDT induces a burst of impulses in the susceptible (S) strain of the housefly, whereas it has no effect on the resistant (R) strain. Figure 9

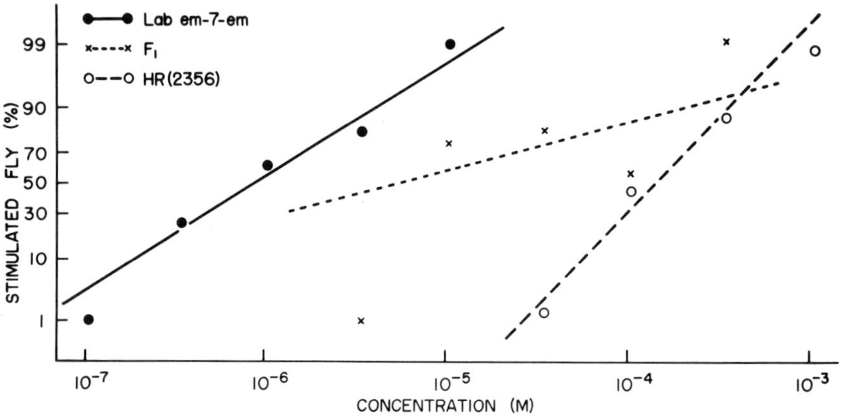


FIGURE 9. Relationship between the profit of percentage of the individual houseflies whose thoracic ganglia are stimulated to discharge impulses by direct application of γ-BHC and the logarithm of γ-BHC concentration in the susceptible strain (Lab em-7-em), in the γ-BHC-resistant strain (HR (2356)), and in their hybrid ($F_1$) (Narahashi, 1964).

illustrates the stimulating effects of direct application of γ-BHC on the exposed thoracic ganglia of the susceptible strain of housefly (Lab em-7-em), the resistance strain of housefly (HR (2356)), and their hybrid ($F_1$). The relative potency of γ-BHC as expressed by the percentage of stimulated houseflies is very high in the susceptible strain, very low in the resistant strain, and intermediate in the hybrid.

The mechanism of such low sensitivity of the nerve remains to be studied, but there is evidence that it is not totally due to a high detoxication in the nervous tissue. In the housefly the lower nerve sensitivity to DDT is exerted by a recessive gene on the second chromosome (Tsukamoto et al., 1965), whereas the higher detoxication of DDT is due to a dominant gene on the fifth chromosome (Tsukamoto and Suzuki, 1964). The receptor macromolecules in the nerve membrane may undergo changes leading to a low sensitivity to insecticide.

## C. Temperature Dependency

Temperature has profound effects on the potency of various insecticides (see Narahashi, 1971a). For example, DDT and pyrethroids are generally more effective in killing insects at low temperature than at high temperature, whereas organophosphates are in many cases more potent at high temperature than at low temperature. In the case of DDT- and pyrethroid-poisoning, the temperature effect cannot solely be accounted for in terms of cuticle penetration and detoxication. The penetrability of insecticide to the cuticle is reduced by lowering the temperature so that this factor plays a role opposite to the negative temperature coefficient of insecticidal activity. Enzymatic detoxication is decreased by lowering the temperature, accounting at least in part for the higher insecticidal activity at low temperature. However, the symptoms of poisoning are totally reversible with respect to the temperature, and this reversibility cannot be explained by detoxication.

The nerve sensitivity to DDT and allethrin is greatly influenced by temperature; lowering the temperature potentiates the effect of DDT in inducing repetitive discharges from sensory cells (Yamasaki and Ishii, 1954a) and the effect of allethrin in blocking nervous conduction (Figure 10) (Narahashi, 1971; Wang et al., 1972). The latter temperature dependency of allethrin blockade can be accounted for in terms of sodium conductance of nerve membranes (Wang et al., 1972). Repetitive responsiveness as induced by allethrin in nerve fibers also is influenced by temperature but in a more complex manner. There is an optimum temperature range where single stimulus easily elicits repetitive discharges in allethrin-poisoned axons. At higher or lower temperatures, repetitive discharges are diminished. This is explained in terms of the temperature dependency of the threshold depolarization and amplitude of the negative after-potential.

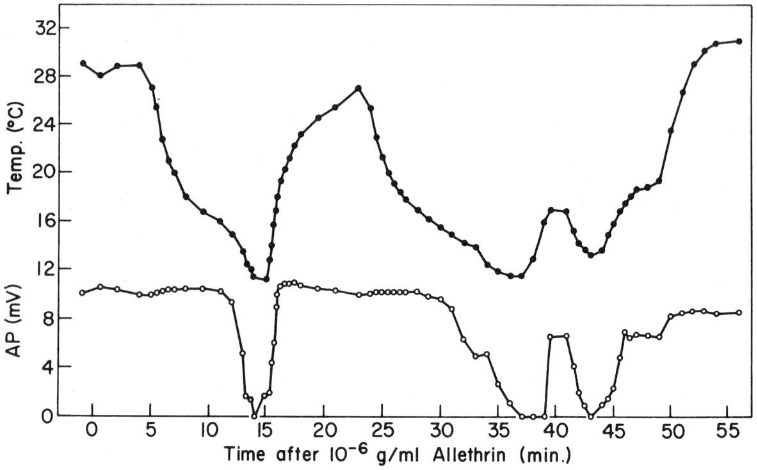


FIGURE 10. Effects of temperature on the action of 1 x $10^{-6}$ g/ml allethrin in suppressing the externally recorded action potential (AP) of the cockroach abdominal nerve cord (Narahashi, 1971a).

The negative temperature coefficient of action of DDT and allethrin on nerve plays an important role in selective toxicity between insects and mammals. Since there is at least a 10°C difference between the body temperatures, these insecticides should be less effective on mammalian nerves than on insect nerves if other conditions are equal.

## VI. CONCLUDING REMARKS

Studies of the target site are essential in elucidating (a) the mechanism of action of insecticides, and (b) the true structure-activity relationship. The target site could be an enzyme or an enzyme system, could be certain ionic conductance components of nerve membranes, or could be the system by which transmitter substance is released from nerve terminals. In any case, the structure-activity relationship as derived from measurements of insecticidal potency is often misleading since a variety of factors other than the primary action on the target site are involved. For many insecticides including DDT, BHC, cyclodienes, and pyrethroids, there are no major metabolic effects that directly cause symptoms of poisoning. Their major target sites are on nerve membranes.

Nerve membrane and postsynaptic membrane have entirely different physiological and pharmacological characteristics and therefore, should be considered separately. Certain insecticides have a special affinity for ionic conductance components of nerve membranes, whereas some other insecticides have a special affinity for ionic conductances of postsynaptic membranes.

## ACKNOWLEDGMENTS

Some of the studies quoted in this paper were supported by NIH Grant NS06855. Unfailing secretarial assistance by Mrs. Frances Bateman and Gillian Cockerill is greatly appreciated.

## REFERENCES

1. Akkermans, L. M. A., van den Bercken, J., van der Zalm, J. M., and van Straaten, H. W. M. 1974. Pestic. Biochem. Physiol. 4, 313.

2. Albuquerque, E. X., Seyama, I., and Narahashi, T. 1973. J. Pharmacol. Exp. Ther. 184, 308.

3. Armstrong, C. M. and Binstock, L. 1965. J. Gen. Physiol. 48, 859.

4. Armstrong, C. M., Bezanilla, F., and Rojas, E. 1973. J. Gen. Physiol. 62, 375.

5. Blaustein, M. P. and Goldman, D. E. 1966. J. Gen. Physiol. 49, 1043.

6. Boistel, J. 1968. The Synaptic Transmission and Related Phenomena in Insects. In Advances in Insect Physiology, Vol. 5, J. W. L. Beament, J. E. Treherne, and V. B. Wigglesworth, Eds., Academic, New York and London, pp. 1-64.

7. Brown, A. W. A. 1964. Animals in Toxic Environments: Resistance of Insects to Insecticides. In Handbook of Physiology, Vol. 4, D. B. Dill, E. F. Adolph, and C. G. Wilber, Eds., Chapter 48, Am. Physiol. Soc., Washington, D. C., pp. 773-793.

8. Deguchi, T. and Narahashi, T. 1971. J. Pharmacol. Exper. Therap. 176, 423.

9. Deguchi, T., Narahashi, T., and Haas, H. G. 1971. Pestic. Biochem. Physiol. 1, 196.

10. Fukami, J. 1961. Bull. Nat. Inst. Agr. Sci., Ser. C. No. 13, 33.

11. Fukami, J. and Tomizawa, C. 1956. Bochu Kagaku 21, 129.

12. Fukami, J. and Tomizawa, C. 1958a. Bochu Kagaku 23, 1.

13. Fukami, J. and Tomizawa, C. 1958b. Bochu Kagaku 23, 205.

14. Furukawa, T., Sasaoka, T., and Hosoya, Y. 1959. Jap. J. Physiol. 9, 143.

15. Gerschenfeld, H. M. 1973. Physiol. Rev. 53, 1.

16. Hille, B. 1967. J. Gen. Physiol. 50, 1287.

17. Hille, B. 1968. J. Gen. Physiol. 51, 199.

18. Hodgkin, A. L. and Keynes, R. D. 1955. J. Physiol. 218, 28.

19. Holmstedt, B. 1959. Pharmacol. Rev. 11, 567.

20. Hubbard, J. I. 1970. Mechanism of Transmitter Release. In Program of Biophysics and Molecular Biology, Vol. 21, J. A. V. Butler and D. Noble, Eds., Pergamon, Oxford and New York, pp. 1-124.

21. Katz, B. 1962. Proc. Roy. Soc., Ser. B, 155, 455.

22. Katz, B. 1966. Nerve, Muscle, and Synapses, McGraw-Hill, New York.

23. Koch, R. B. 1969. J. of Neurochem. 16, 269.

24. Koch, R. B., Cutkomp, L., and Do, F. M. 1969. Life Sci. 8, 289.

25. Koch, R. B., Cutkomp, L. K., and Yap, H. H. 1971. Biochem. Pharmacol. 20, 3234.

26. Koelle, G. B. 1970. Anticholinesterase Agents. In The Pharmacological Basis of Therapeutics, MacMillan, New York, pp. 442-465.

27. Krnjevíc, K. 1974. Physiol. Rev. 54, 418.

28. Lindahl, P. E. and Öberg, K. E. 1961. Exp. Cell Res. 23, 228.

29. MacIntosh, F. C. 1963. Can. J. Biochem. Physiol. 41, 255.

30. Matsumura, F. and Narahashi, T. 1971. Biochem. Pharmacol. 20, 825.

31. Narahashi, T. 1962. J. Cell Comp. Physiol. 59, 61.

32. Narahashi, T. 1964. Jap. J. Med. Sci. Biol. 17, 46.

33. Narahashi, T. 1971a. Effects of Insecticides on Excitable Tissues. In Advances in Insect Physiology, J. W. L. Beament, J. E. Treherne, and V. B. Wigglesworth, Eds., Academic, New York and London, pp. 1-93.

34. Narahashi, T. 1971b. Bull. W.H.O. 44, 337.

35. Narahashi, T. 1972. Fed. Proc. 31, 1124.

36. Narahashi, T. 1974a. Physiol. Rev. 54, 813.

37. Narahashi, T. 1974b. Effects of Insecticides on Nervous Conduction and Synaptic Transmission. In Pesticide Biochemistry and Physiology, C. F. Wilkinson, Ed., Plenum, New York, in press.

38. Narahashi, T. and Anderson, N. C. 1967. Toxicol. Appl. Pharmacol. 10, 529.

39. Narahashi, T. and Haas, H. G. 1967. Science 157, 1438.

40. Narahashi, T. and Haas, H. G. 1968. J. Gen. Physiol. 51, 177.

41. Narahashi, T. and Seyama, I. 1974. J. Physiol. 242, 471.

42. Narahashi, T. and Yamasaki, T. 1960. J. Physiol. 152, 122.

43. Narahashi, T., Deguchi, T., Urakawa, N., and Ohkubo, Y. 1960. Am. J. Physiol. 198, 934.

44. Narahashi, T., Moore, J. W., and Scott, W. R. 1964. J. Gen. Physiol. 47, 965.

45. Narahashi, T., Haas, H. G., and Therrien, E. F. 1967a. Science 157, 1441.

46. Narahashi, T., Anderson, N. C., and Moore, J. W. 1967b. J. Gen. Physiol. 50, 1413.

47. Narahashi, T., Moore, J. W., and Poston, R. N. 1969a. J. Neurobiol. 1, 3.

48. Narahashi, T., Moore, J. W., and Shapiro, B. I. 1969b. Science, 163, 680.

49. Narahashi, T., Albuquerque, E. X., and Deguchi, T. 1971. J. Gen. Physiol. 58, 54.

50. Öberg, K. E. 1961. Exp. Cell Res. 24, 163.

51. O'Brien, R. D. 1960. Toxic Phosphorus Esters, Academic, New York and London.

52. O'Brien, R. D. 1966. Ann. Rev. Entomol. 11, 369.

53. O'Brien, R. D. 1967. Insecticides. Action and Metabolism, Academic, New York and London.

54. Otsuka, M. and Endo, M. 1960. J. Pharmacol.'Exp. Ther. 128, 273.

55. Pitman, R. M. 1970. Comp. Gen. Pharmacol. 2, 347.

56. Sakai, M. 1967. Bochu Kagaku 32, 21.

57. Sakai, M. 1969. Rev. Plant Prot. Res. 2, 17.

58. Seyama, I. and Narahashi, T. 1973. J. Pharmacol. Exp. Ther. 184, 299.

59. Shanes, A. M., Freygang, W. H., Grundfest, H., and Amatniek, E. 1959. J. Gen. Physiol. 42, 793.

60. Shankland, D. L. and Schroeder, M. E. 1973. Pestic. Biochem. Physiol. 3, 77.

61. Simpson, L. L. 1971. The Neuroparalytic and Hemagglutinating Activities of Botulinum Toxin. In Neuropoisons. Their Pathophysiological Actions, L. L. Simpson, Ed., Plenum, New York and London, pp. 303-323.

62. Smyth, T., Jr., and Roys, C. C. 1955. Biol. Bull. 108, 66.

63. Tasaki, I. and Hagiwara, S. 1967. J. Gen. Physiol. 40, 859.

64. Taylor, R. E. 1959. Amer. J. Physiol. 196, 1071.

65. Thesleff, S. 1955. Acta. Physiol. Scand. 34, 218.

66. Treherne, J. E. 1966. The Neurochemistry of Arthropods, Cambridge University Press, New York.

67. Tsukamoto, M. and Suzuki, R. 1964. Bochu Kagaku 29, 76.

68. Tsukamoto, M., Narahashi, T., and Yamasaki, T. 1965. Bochu Kagaku 30, 128.

69. van den Bercken, J. and Narahashi, T. 1974. Europ. J. Pharmacol. 27, 255.

70. Wang, C. M. and Narahashi, T. 1972. J. Pharmacol. Exp. Ther. 182, 427.

71. Wang, C. M., Narahashi, T., and Yamada, M. 1971. Pestic. Biochem. Physiol. 1, 84.

72. Wang, C. M., Narahashi, T., and Scuka, M. 1972. J. Pharmacol. Exp. Ther. 182, 442.

73. Wang, C. M., Narahashi, T., and Fukami, J. 1975. Pestic. Biochem. Physiol. 5, 119.

74. Wilson, I. B. 1959. Fed. Proc. 18, 752.

75. Wu, C. H., van den Bercken, J., and Narahashi, T. 1975. Pestic. Biochem. Physiol. 5, 142.

76. Yamasaki, T. and Ishii,* T. 1954a. Bochu Kagaku 19, 39.

77. Yamasaki, T. and Ishii,* T. 1954b. Bochu Kagaku 19, 106.

78. Yamasaki, T. and Narahashi, T. 1958. Bochu Kagaku 23, 146.

79. Yamasaki, T. and Narahashi, T. 1962. Jap. J. Appl. Ent. Zool. 6, 293.

*Former name of Toshio Narahashi.

## DISCUSSION

Two topics of especial importance were discussed, in other words, the neuromuscular junction and the acetylcholine receptor. Although the neurobiology of vertebrates has been extensively investigated and we understand a good deal about the vertebrate neuromusuclar junction, we know much less about the neuromusuclar junction of insects and other arthropods. There are important differences, and it is well known that the neuromuscular junction of insects is not cholinergic. There has been much speculation about the insect neuromusuclar transmitter. However, Dr. Barton Brown of the Research Institute, London, Ontario, has recently identified a new transmitter substance at the neuromuscular junction of the American cockroach. This was isolated after processing a few million roaches and has been identified as a pentapeptide active at $10^{-10}$ molar concentration. Activity is demonstrated by a roach gut bioassay and also by the increased electrophysiological activity where there is direct correlation between spike height and concentration. The same neuromuscular transmitter has been demonstrated in 8 other species of insects. The pentapeptide is believed to have the amino acid sequence of arginine, tryosine, proline, leucine, and threonine. Incorporation of serine into a similar pentapeptide reduced activity to 1/50th that of the natural substance.

The identification of a pentapeptide neuromuscular transmitter suggests a number of novel thoughts about insect biochemistry. For example, some sort of peptidase or proteinase may be involved in destruction of the transmitter in analogy to acetylcholinesterase. The pentapeptide is inactivated by insect gut extracts, and this suggests that structure optimization to produce a useful insecticide might be challenging. However, small peptide links to aromatic groups have been made by Slama and others, and these apolar molecules were found to have specific juvenile hormone mimic action to Pyrrhocoris. Thus it may be possible to impart appropriate lipophilicity to peptide type molecules. This is an exciting field as there are reports of peptide-like transmitters in mammals such as substance P and one recently isolated by Professor Otsuka and associates in Japan.

There was discussion as to whether the biogenic amines, such as serotonin which has been reported as a neurotransmitter in molluscs, play any role as transmitters in insects. Specific evidence seems to be lacking, and no unique transmitters of this type appear to have been identified in insects. Monoamine oxidase is involved in their inactivation and could play a similar role to acetylcholinesterase.

Postsynaptic receptors in insects are of much interest, and nicotine and nereistoxin are well known to act as competitive synaptic blocking agents of the receptor. In nereistoxin the dimethylamino group and the thiol radicals are presumably necessary for affinity to the receptor since poisoned animals recover after intravenous injection of thiols such as l-cystein, 2-aminoethane thiol, d-penicillamine, or 2,3-dimercaptopropanol (BAL) which serves as a practical antidote. It is now clear from studies with sulfhydryl reagents that sulfhydryl bonds play a very important role in the acetylcholine receptors of the post synaptic membrane, thus explaining the role of thiols in reversing the effects of nereistoxin. However, at present we lack any other clear-cut interpretations of postsynaptic receptor action at the molecular level. Other insecticides such as chlorphenamidine appear to act on synapses, and we know nothing of the detailed mechanisms of action of lindane and the cyclodienes. Aldrin trans-diol has an action similar to tetrodotoxin in blocking nervous conduction through inhibiting the Na-conductance mechanisms without much change in membrane potential. Much more research is needed in this area.

# Specialized Features of the Integument

M. LOCKE
Department of Zoology
University of Western Ontario
London, Ontario, Canada

## I. INTRODUCTION

The integument, that is, the cuticle and the epidermis that secretes it, is the most distinctive feature of the Arthropods. It determines the characteristics of insects in a more profound way than the covering of other groups of animals. It is skin, skeleton, and food reserve (Locke, 1964). If there are uniquely vulnerable targets for our reagents to make them insecticides and not just biocides, we may hope that an understanding of cuticle properties can help us find them.

If I had to make a decision, I should say that it is unlikely that satisfactory chemical control of insects is possible. At best we may alleviate the problem through the sensible use of products and methods similar to those now available. Wigglesworth (1946) reached this conclusion many years ago in a prophetic review on "DDT and the Balance of Nature." This is the nature of things; it is not an admission of our failure as scientists. However, I am not sure and there is still a possibility that a silver bullet exists hidden by our ignorance of the biology of the integument. With relatively little more work on the integument we might know more certainly whether there are some special features that would make insects vulnerable to reagents that we could devise for their control. The principle that we should keep in mind is that insects are relatively small populations of cells with short lives and are most susceptible to acute effects, whereas vertebrates are large-long-lived populations of cells, buffered against acute poisoning but more sensitive to chronic exposure and long-term effects.

This essay outlines the structure and properties of the integument, emphasizing special features that may be important in insect control.

## II. CUTICLE STRUCTURE AND COMPOSITION

Electron microscopy has considerably modified our concepts of cuticle structure based on light microscopy (Locke, 1967a). There are three main regions: the cuticulin or outer epicuticle, a dense homogeneous inner epicuticle below it (the two together make up the epicuticle), and the bulk of the cuticule made from laminae of chitin fibers embedded in protein, the fibers often being oriented to form lamellae. Each of these layers may vary in its properties depending on the type and extent of stabilization (Figure 1).

The cuticulin layer is the primary barrier between an insect and its environment (Filshie, 1970a, 1970b; Locke, 1965a, 1966, 1969b). Typically, it is a membrane about 20 nm thick covering the entire outer surface of insects; it forms the lining of tracheoles, it covers the ducts of glands (Noirot and Quennedey, 1974), the surface of the finest hairs and scales, and all cuticle, including that lining the fore and hind gut (Noirot and Noirot-Timothee, 1969, 1971a, 1971b, 1971c). It is only absent over some sense organs (Lewis, 1970) and the mid gut. In profile (Figure 2) it appears as an electron dense line that may be resolved into a pentalaminar structure with three dense regions (Figure 3). In Calliphorid larvae the outermost lamina may be thick enough to be resolved by light microscopy (Filshie, 1970b). It is an acid-resistant layer that shows only paraffin structures by x-ray and infrared spectroscopy. In some locations, as in tracheoles (Locke, 1966) and the hind gut (Noirot and Noirot-Timothee, 1969, 1971), the cuticulin has a porous appearance as though it is penetrated by channels about 3 nm in diameter. We know nothing about the composition of cuticulin, although we may speculate that it may be a composite of a cross-linked hydrocarbon layer and quinone-tanned protein.

Below the cuticulin lies a homogeneous dense layer forming the inner epicuticle. It varies in thickness from 10 to $10^4$ nm (Figure 2) and constitutes the thin refractile epicuticle seen by light microscopy.

The major part of the cuticle consists of chitin microfibers with varying orientations embedded in a protein matrix stabilized in various ways. The chitin fibers typically align themselves parallel to one another in layers forming laminae (Neville, 1967, 1970; Rudall, 1963). The fibers may keep the same orientation in successive laminae as in structures like bristles and spines where the chitin is predominantly axially oriented (Locke, 1967b); there may be few and abrupt changes in orientation (Neville and Luke, 1969) to make a massive plywood-like structure as in beetle elytra; or most commonly, there is

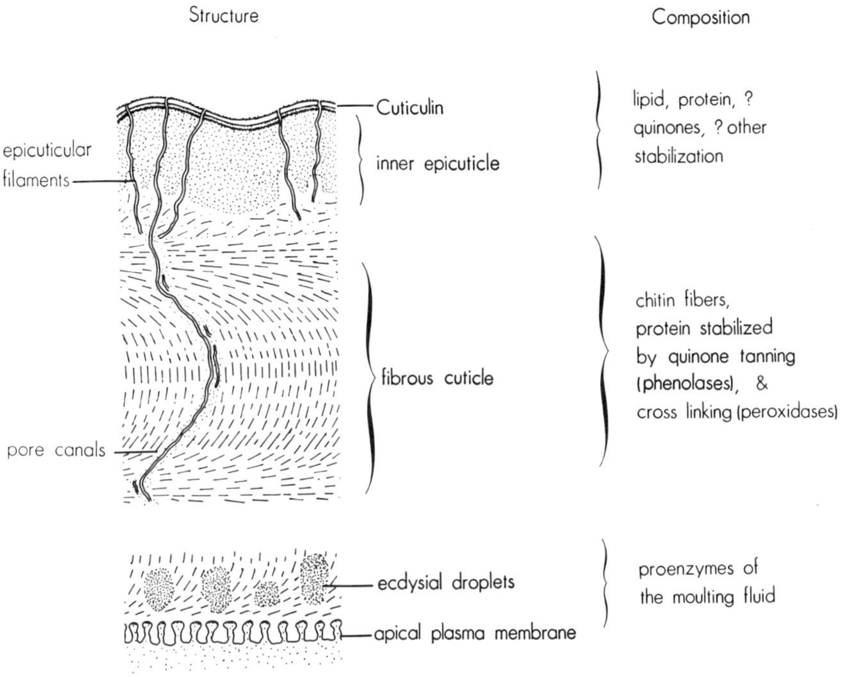


FIGURE 1. An outline of cuticle structure and composition. The cuticle has three regions (cuticulin, inner epicuticle and fibrous cuticle). Filaments provide a route across the epicuticle and pore canals traverse the fibrous cuticle. Chitin fibers in the fibrous cuticle are often helicoidally arranged to form lamellate cuticle which may be hundreds of times thicker than the epicuticle.

a small progessive change of fiber orientation from lamina to lamina giving the helicoidal arrangement that makes up lamellate cuticle (Neville and Caveney, 1969) (see Figure 7).

These three cuticle components are not complete barriers between the epidermis and the environment. The epicuticle is traversed by filaments about 10 nm in diameter. They are the route across the epicuticle for surface waxes and have been called wax canals (Locke, 1960) (Figure 1). Structurally,

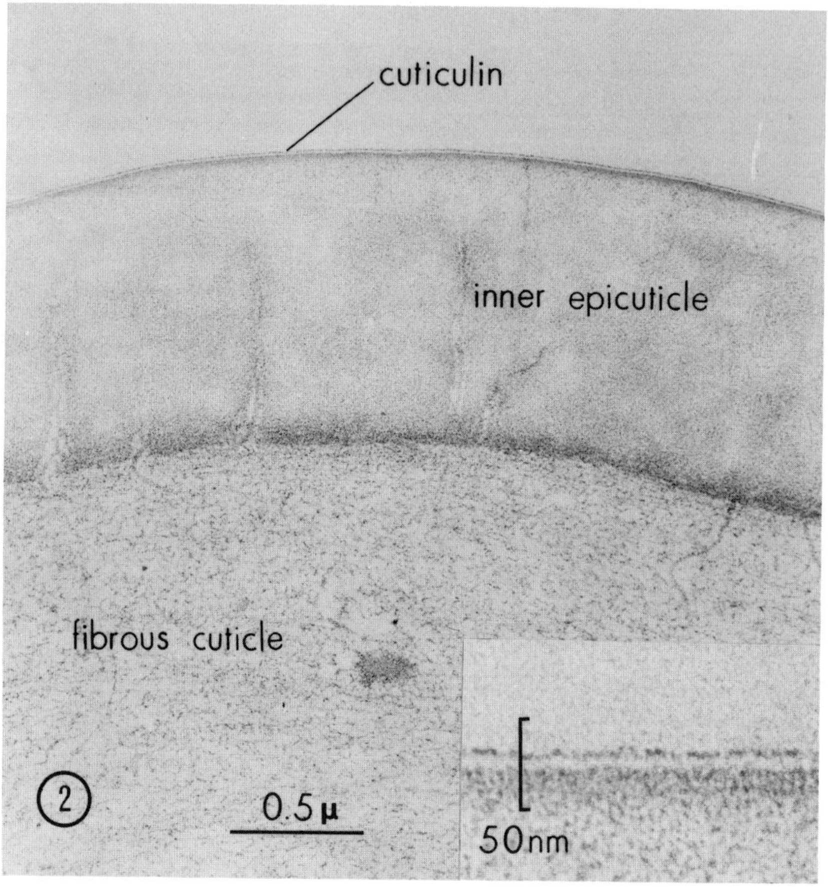


FIGURE 2. The epicuticle. Transverse section of the cuticle of Calpodes larvae showing the cuticulin (insert) and the inner epicuticle overlying fibrous cuticle.

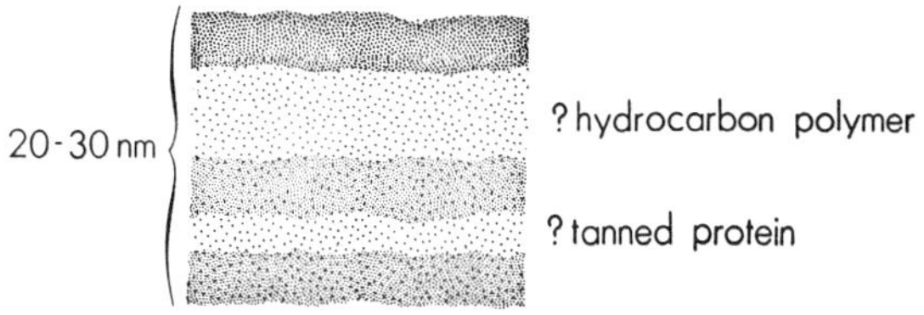


FIGURE 3. The cuticulin. The outer surface of insects is completely covered by this thin membrane of unknown composition.

similar filaments have also been seen in plant cuticles where they have also been called wax canals (Hall, 1967). The epicuticular filaments often form bundles that pass through the fibrous cuticle in pore canals (Delachambre, 1971; Locke, 1961; Neville et al., 1969), channels about 100 nm wide, linking the epidermis with the outermost parts of the cuticle, particularly in strongly stabilized cuticle.

This is the generality of cuticle structure. All cuticles have these components and differ only in variations on this theme. There are three distinct regions: cuticulin, inner epicuticle, and fibrous cuticle; and there are two steps in the route across the cuticle: filaments traverse the epicuticle and pore canals traverse the fibrous cuticle. To this structure we should add one more component that appears at the appropriate stage of development, the moulting gel (Delachambre, 1967; Locke and Krishnan, 1973). Prior to ecdysis the epidermis secretes proenzymes (these are the ecdysial droplets) that, when activated (Katzenellenbogen and Kafatos, 1970, 1971) digest the weakly stabilized components of the old cuticle to allow it to be resorbed.

## III. CUTICLE FORMATION, SYNTHESIS, AND DEPOSITION

Cuticle secretion is the result of an interplay between two main processes (Locke, 1967b): protein synthesis and secretion through the Golgi complex resulting in the discharge of secretory vesicles (Locke, 1969a, 1969b; Locke and Krishnan, 1973) (Figure 4), and transmembrane transport of cuticular components (Locke, 1966), with the final steps of synthesis and orientation

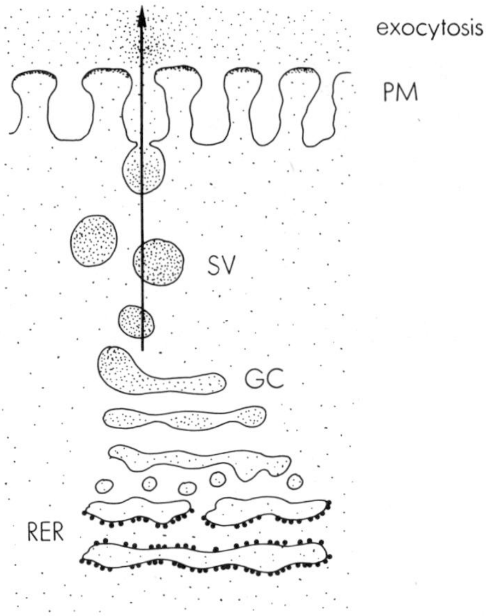


FIGURE 4. The secretion of protein components of the cuticle. Phenolases, peroxidase, and structural components of the cuticle have all been traced through this traditional route for secretion.

being in or at the apical plasma membrane (PM) surface and involving the PM plaques (Figure 5).

Several cuticular proteins have been followed through the traditional route for synthesis and secretion of proteins. The enzymes peroxidase (Locke, 1969a) and phenolase (Locke and Krishnan, 1971), for example, can be detected in the Golgi complex, in secretory vesicles and in the cuticle after exocytosis through the apical plasma membrane. The main part of the epicuticle is secreted in this way (Locke, 1969b) and presumably also the protein matrix for fibrous cuticle. Autoradiographs of the integument after natural pulses of $^3$H amino acids show that the AAs are rapidly cleared from the haemolymph and deposited as layers of protein in the cuticle (Condoulis and Locke, 1966; Locke et al., 1965). The epidermis is not specialized in the way that it secretes proteins, although some of the proteins themselves may be characteristic of cuticle or even unique to it.

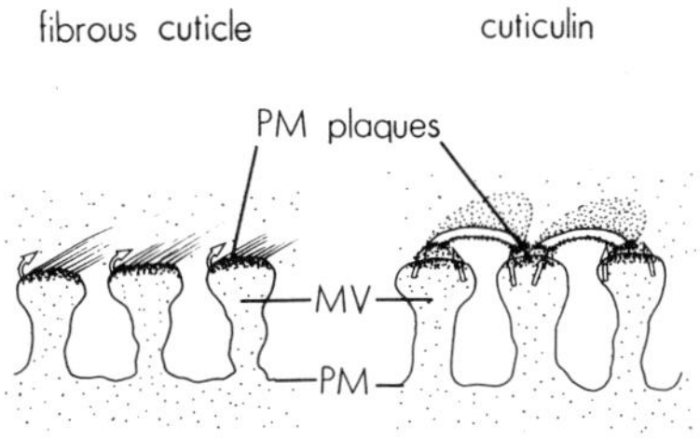


FIGURE 5. The apical plasma membrane in secretion. Chitin fibers and the cuticulin layer both arise at the tips of the microvilli above the plasma membrane plaques. cf Figs. 6 and 7.

On the other hand, the deposition of microfibers of chitin and the membranelike cuticulin layer may be unique to Arthropods. Both involve the plasma-membrane plaques. The cuticulin layer is the first part of a new cuticle to be formed at moulting. It arises at and grows from the plasma membrane plaques at the tips of microvilli (Figures 5 and 6). At a later stage, when fibrous cuticle is being deposited, the microfibers arise at the outer surface of the plaques as though the last step for chitin synthesis takes place at the membrane surface (Figures 6 and 7). This form of deposition at the membrane surface is reminiscent of the way bacterial coats are secreted (Archibald et al., 1973; De Petris, 1967). Although the mechanism of deposition may not be unique to Arthropods, the type of material secreted may be. Unfortunately, although the chitin component has been characterized, the chemical nature of cuticulin (Figure 3) is unknown.

Diagrams 1, 4, and 5 summarize the structure and formation of the cuticle. We may ask what special features of this system may make insects vulnerable, converting the heavy hand of biocidal chemical control into management by insecticides having the chemical precision of dissecting needles.

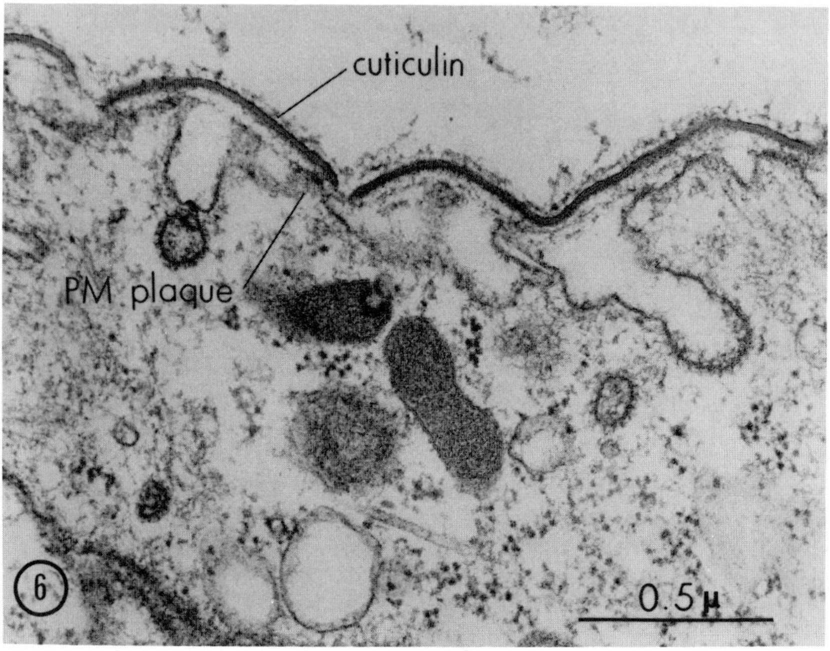


FIGURE 6. The origin of the cuticulin above the plasma membrane plaques. cf. Fig. 5.

## IV. SPECIAL FEATURES

### A. The Cuticulin

The cuticulin layer (Figures 2, 3, 5, 6) looks like a bacterial envelope (De Petris, 1967). It separates the zone influenced by epidermal cells from the environment. Within it the epidermis controls the cuticular milieu, its water content, enzyme reactions, and turnover of components. Its extreme thinness accounts for the effectiveness of abrasive dusts in causing water loss. Part of it is resistant to degradation in a way that suggests that it may be a polythenelike polymer (Filshie, 1970b). During its formation it reacts as though it has a protein component that is being quinone tanned (Locke and

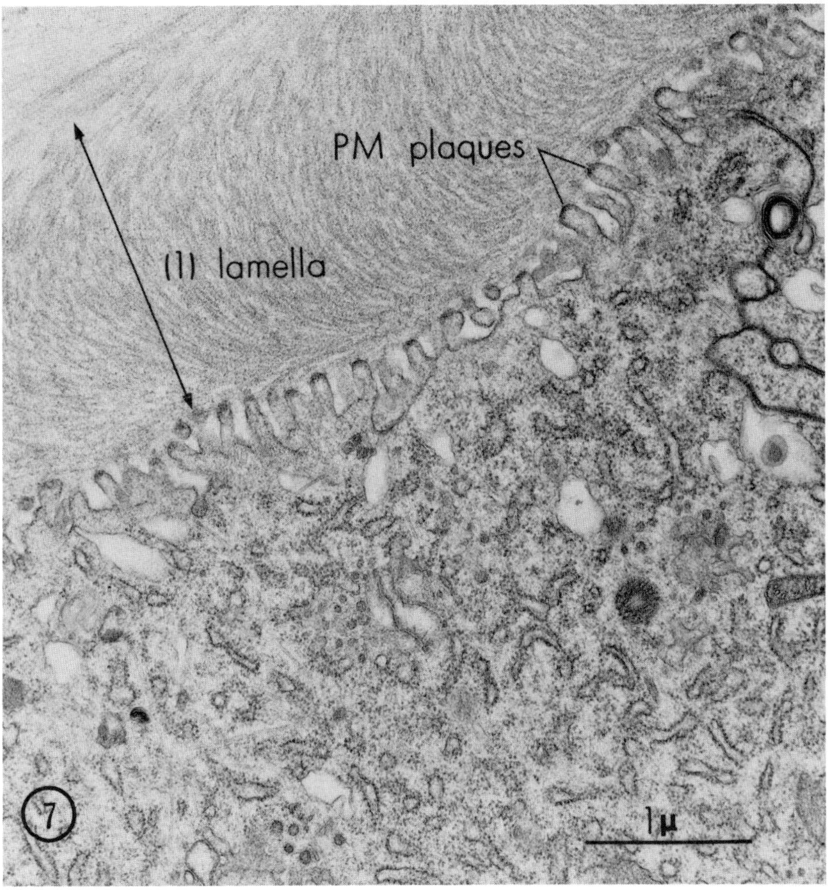


FIGURE 7. The lamellate fibrous cuticle is secreted above the plasma membrane plaques. cf Fig. 5.

Krishnan, 1971). In spite of its importance in the life of an insect, we know very little about its composition. If we knew what its composition was, we might be able to formulate more effective carriers for contact insecticides. If we knew the chemistry of its formation, we might be able to influence survival at moulting in a most dramatic way. I cannot think of a more important neglected aspect of applied biology. The cuticulin layer is all important to an insect: its structure and chemistry may well be unique to Arthropods and yet we know very little about it.

### B. Wax Canals and the Epicuticular Filaments

The epicuticular filaments look like liquid crystals around which the epicuticle forms, thus leaving channels for the transport of lipids to spread from the cuticle to the surface of the cuticulin (Locke, 1965a). They may be lipid-water liquid-crystals since they are similar in appearance to the lipid-water liquid-crystals that are precursors to honey bee wax (Locke, 1961). However, they resist chloroform extraction (Filshie, 1970b) and quinone tan (Locke and Krishnan, 1971) at the same time that the epicuticle is stabilized. The most likely hypothesis is that they are lipoprotein or lipid and protein liquid crystals that are subsequently stabilized to leave permanent transepicuticular channels that can carry both aqueous and lipid solvents. They are the insect's solution to the problem of how to be both permeable and impermeable to water. They account for the observation that insects are resistant to desiccation and yet oils and nonabrasive desiccants can draw water from an insect. They account for the observations that cuticle is more permeable to water going into an insect than leaving it. Any formulation of contact insecticides has to take the wax canals and epicuticular filaments into account. In spite of this, liquid crystals are a neglected field for applied biologists.

### C. Cuticular Hydrocarbons

Cuticulin is important in providing a lipophilic layer upon which waxes can be deposited to play their role in waterproofing and the maintenance of hydrophobic surfaces. The waxes may contain hydrocarbons (Gilbert, 1967) with chain lengths as long as $C_{41}$, the main component of *Calpodes* wax (Locke, 1965b). Hydrocarbons are commonly synthesized by insects but are rare or nonexistent in vertebrates. They make up 75% of the cuticular wax in the cockroach, P. americana, which is mostly $C_{27}$ n-alkene (heptacosa-9, 18-diene) (Gilby and Cox, 1963). The wax is a soft grease on the animal, but it hardens after removal. This poses two general problems for the way that insect cuticles are waterproofed by waxes. How are waxes, and particularly hydrocarbons, transported through the aqueous medium of the cuticle to the surface, and how are the waxes kept liquid until they have covered the cuticulin? Wax filaments, presumably liquid crystals, appear first above the apical plasma membrane. We presume that the final steps in their synthesis occur there and that they traverse the cuticle as lipid-water liquid-crystals, emerging at the surface through the wax canals (Locke, 1960, 1961). There are several possible explanations for the way that the waxes solidify. One possibility supposes that the mixture of hydrocarbons with fatty acids, esters, and aldehydes is liquid until some components are changed chemically, perhaps by the polymerization of aldehydes to resins (Gilby, 1962). Most probably, the change from liquid to solid is due to the physical properties of hydrocarbon, mixed lipid, and water systems about which we know little. If the lipids are transported across the cuticle as liquid crystals in solution, it seems most likely that they continue to cover the surface in this state. The cornicle wax of aphids may be a good model. It is at first liquid but solidifies instantly when touched with a probe. The aphid makes use of this reaction in defense against small insect predators that are instantly cast in hard wax after touching the cornicle (Edwards, 1966).

We know little about the synthesis or precursors of hydrocarbons (Conrad and Jackson, 1971). Most probably the oenocytes are involved (Piek, 1961, 1964; Wigglesworth, 1947, 1948a). Both oenocytes (Locke, 1969c) and the epidermis (Locke, 1964) have a structure appropriate for lipid synthesis. The oenocytes in honey bees lie directly below the wax glands and take up labelled acetate that later appears in wax hydrocarbons (Piek, 1961, 1964). Since hydrocarbons are important and perhaps peculiar to insects, we should do well to find out more about their synthesis, transport, and properties, particularly the properties of liquid crystals.

## D. Phenolases

Although phenolases are widely distributed in animals and plants, particularly in relation to melanin synthesis, they have a special significance for insects because of their role in the stabilization of cuticular proteins (sclerotization) and the way insects respond to wounding (Brunet, 1965; Cottrell, 1964; Hackman and Goldberg, 1971; Lai-Fook, 1966) (Figure 8). Sclero-

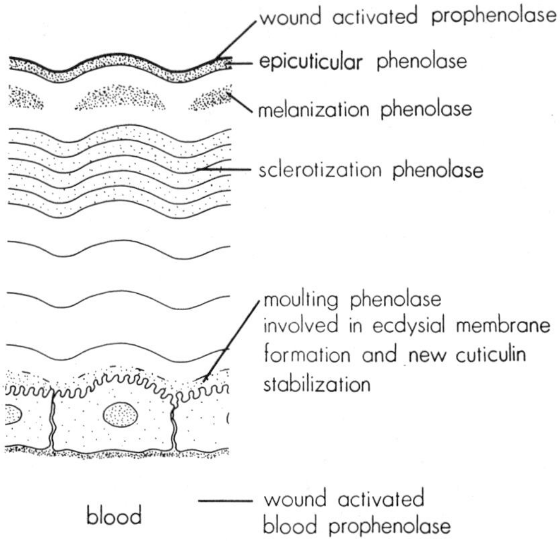


FIGURE 8. The distribution of phenolases.

tization stabilizes proteins by linking their aminogroups through quinones to form dark-colored cuticles. The formation of cross links may also involve oxidation of the β position, and causes cross linking with the formation of stable, light-colored cuticles (Andersen, 1972).

The need to present substrate polyphenols to phenolases at the right time and place also draws attention to another insect speciality, or at least difference from vertebrates. Phenols are, for the most part, present and/or detoxified as β glucosides, whereas mammals detoxify them as glucoside-uronides Mehendale and Dorough, 1972; Winteringham, 1965a, 1965b). Insects use phenolases in two ways. First, phenolases may be secreted as active enzymes. For example, a phenolase has been followed from the Golgi complex through secretory vesicles and exocytosis into the cuticle where it is deposited; it becomes functional at a later time when substrate is made available, presumably by the action of a β glucosidase (Locke and Krishnan, 1971; Shaaya and Sekeris, 1970). Second, they may be present as proenzymes in the epicuticle or blood, where they may be activated by any of the many stimuli occurring during wounding. The prophenolases are delicately balanced, as witnessed by the ease and speed with which dying insects become bags of the blackened products of phenolase reactions and the way in which even the minutest abrasion to soft cuticles may blacken in minutes. There is a big gap in our knowledge of the way insect phenolases are activated and in the way polyphenols are liberated. One would think that insects might be very sensitive to agents that release phenolase substrates or that activate phenolases.

## E. PEROXIDASES

Some cuticular proteins are stabilized through the cross linking of tyrosine residues (Andersen, 1966). The reaction occurs *in vitro* with silk fibroin that forms dityrosine and trityrosine bridges in the presence of peroxidase and hydrogen peroxide. Peroxidases have been detected in some sorts of insect cuticle, where they are presumed to play a role in protein stabilization (Locke, 1969a). The peroxidase arises in the Golgi complex and is discharged from secretory vesicles into the cuticle. Peroxidases have a wide distribution and function in nature, from lignin formation in plants (Harkin and Obst, 1973) to mammalian Kupffer cells (Widmann *et al*., 1972) where they may be bactericidal. The rather precise role that peroxidase plays in insects deserves investigation to see if the inhibition of such oxidases has any profound effect on survival.

## F. Chitin

Polyacetyl glucosamine is the characteristic structural carbohydrate of arthropods, and it is perhaps surprising that it is only recently that reagents whose prime target appears to be chitin deposition have been suggested as insecticides. Insects that have ingested PH60-38 or PH60-40 (1-(4-chlorophenyl)-3-(2,6-dichlorobenzoyl)-urea and 1-(4-chlorophenyl)-3-(2,6-difluorobenzoyl)-urea, two of the many 1-benzoyl-3-phenylureas tested) die in moulting and fail to deposit chitin (Mulder and Gijswijt, 1973). The reagents look promising, but it is premature to speculate until we know their mode of action or the long-term effects on other organisms.

## G. Ecdysis and the Stimulus for Eclosion

The interest in ecdysone and the control of moulting (Willis, 1974) has tended to overshadow an important feature of ecdysis, the control of eclosion from the exuvium (Doane, 1973). It now appears that at the end of the period of preparation for moulting under the influence of ecdysone, the exact moment of ecdysis is determined by the eclosion hormone from the brain (Truman, 1971; Truman and Riddiford, 1970) and that the timing of secondary events after eclosion such as cuticle hardening is determined by the neurohormone, bursicon (Cottrell, 1964; Post and De Jong, 1973; Vincent, 1971, 1972). Without the eclosion hormone, an insect may remain within its old cuticle until it dies from starvation or desiccation. Insects are particularly vulnerable to blocked or mistimed ecdyses. At present, we have too little information to be able to generalize on the mode of action of the eclosion hormone, but it should surely be a worthwhile subject for investigation by those interested in insect control.

The idea that insects may be especially vulnerable during very brief events in their life history has not been exploited. We may even have overlooked potential insecticides because they have only been tested for their effects during relatively resistant stages. Biocides might be 100% effective at moulting but have an unimpressive $LD_{50}$ if tested on insects that are mainly between moults.

## H. The Surface of Insects and the Ingestion of the Exuvium

There is a need for a survey of the surface microflora of insects. The cuticular surface over which there is a constant spread and loss of waterproofing lipids might be expected to supply an environment for a range of bacteria that might periodically infect the gut through the common practice of exuviophagy. This is a completely open field for research. If insect surfaces carry bacteria, one would like to know whether they can become pathogenic; if there is a poor surface flora, one would like to know how the insect keeps it under control.

## I. The Recycling of Cuticle Components

A long while ago, Wigglesworth (1948b) made the point that the cuticle is "alive" up to the epicuticle, even though it is extracellular. The cuticle that is not lost with the exuvium is still within the control of the epidermis, both continually by secretion and resorption and cyclically by digestion and resorption at moulting. The fact that most of the cuticle continues to be within the body metabolic pool may account for the success of contact insecticides. As a structural component the cuticle may be 100 μ thick, but the barrier for insecticides may be only 20 nm or even nonexistent at the entrance through the wax canals.

A characteristic feature of insect cuticle is the efficiency of the recycling of cuticular components. The sloughed exuvium is only a few percent of the functional cuticle, which functions as an intermittant food depot (Condoulis and Locke, 1966). It is added to gradually in the feeding phase and recycled suddenly at each ecdysis. Catabolism and resorption involve very different systems from anabolism and cuticle deposition, and the possibility exists that an insect could deposit products at a low level in the cuticle (either from the gut and through the epicuticle) that could later become harmful when subjected to the sudden lytic action of the moulting fluid. The digestive action of the moulting fluid might itself change the properties of molecules deposited in the cuticle to increase their biocidal effects. Such hypotheses may seem to offer remote possibilities for success in insect control, but they should stimulate us to learn more about the process of cuticle recycling.

### J. Epithelial Transport

Insect epithelia have the peculiar property of transporting information in a polarized way (Locke, 1967b). This fascinates the developmental biologists who study gradient behavior. So far we know that the property is not due to electrical behavior or the flow of very small molecules since insect epithelia are electrically coupled (Caveney, 1974). A potential difference across one epidermal cell plasma membrane is transmitted uniformly throughout the epithelium. This may have no significance for insect toxicology. On the other hand, it is an area of ignorance in our understanding of a very important aspect of integument biology, and it may hide something of importance to us.

## V. CONCLUSION

The preparation for this essay has suggested to me that there may be a misapplication of effort in our attempts at insect control. Anyone who reads the journals reporting applied biological work on insecticides cannot fail to be impressed by the expertise of those working on chemical control. They are skilled and well equipped. However, the work often appears to be specialized and too close to the immediate problems resulting from known insecticides. There is a shortness of vision. There seems to be too little work directed at insects and the generality of their function and too much at the molecules we have invented and their shortcomings with respect to what we should like them to do.

## REFERENCES

1. Andersen, S. O. 1966. Acta Physiol. Scand. 66, (Suppl. 263), 1.

2. Andersen, S. O. 1972. J. Ins. Physiol. 18, 527.

3. Andersen, S. O. and Barrett, F. M. 1971. J. Ins. Physiol. 17, 69.

4. Archibald, A. R., Baddiley, J., and Heckels, J. E. 1973. Nature (London) 241, 29.

5. Brunet, P. C. J. 1965. In Aspects of Insect Biochemistry, T. W. Goodwin, Ed., Academic, New York, pp. 49-77.

6. Caveney, S. 1974. Dev. Biol. 40, 311.

7. Condoulis, W. B. and Locke, M. 1966. J. Ins. Physiol. 12, 311.

8. Conrad, C. W. and Jackson, L. L. 1971. J. Ins. Physiol. 17, 1907.

9. Cottrell, C. B. 1964. In Advances in Insect Physiology, Vol. 2, V. B. Wigglesworth, J. W. L. Beament, and J. E. Treherne, Eds., Academic, London, pp. 175-218.

10. Delachambre, J. 1967. Z. Zell. 81, 114.

11. Delachambre, J. 1971. Tissue Cell 3, 499.

12. De Petris, S. 1967. J. Ultrastruct. Res. 19, 45.

13. Doane, W. W. 1973. In Developmental Systems: Insects, Vol. 2, S. J. Counce and C. H. Waddington, Eds., Academic, London, pp. 291-497.

14. Edwards, J. S. 1966. Nature (London) 211, 73.

15. Filshie, B. K. 1970a. Tissue Cell 2, 181.

16. Filshie, B. K. 1970b. Tissue Cell 2, 479.

17. Gilbert, L. I. 1967. In Advances in Insect Physiology, Vol. 4, J. W. L. Beament, J. E. Treherne, and V. B. Wigglesworth, Eds., Academic, London, pp. 69-211.

18. Gilby, A. R. 1962. Nature (London) 195, 729.

19. Gilby, A. R. and Cox, M. E. 1963. J. Ins. Physiol. 9, 671.

20. Hackman, R. H. and Goldberg, M. 1971. J. Ins. Physiol. 17, 335.

21. Hall, D. M. 1967. Science 158, 505.

22. Harkin, J. M. and Obst, J. R. 1973. Science 180, 296.

23. Katzenellenbogen, B. S. and Kafatos, F. C. 1970. J. Ins. Physiol. 16, 2241.

24. Katzenellenbogen, B. S. and Kafatos, F. C. 1971. J. Ins. Physiol. 17, 823.

25. Lai-Fook, J. 1966. J. Ins. Physiol. 12, 195.

26. Lewis, C. T. 1970. Symp. Roy. Ent. Soc. Lon. 5, 59.

27. Locke, M. 1960. Quart. J. Micro. Sci. 101, 333.

28. Locke, M. 1961. J. Biophys. Biochem. Cytol. 10, 589.

29. Locke, M. 1964. In Physiology of Insects, M. Rockstein, Ed., Academic, New York, pp. 379-470.

30. Locke, M. 1965a. Science 147, 295.

31. Locke, M. 1965b. J. Ins. Physiol. 11, 641.

32. Locke, M. 1966. J. Morph. 118, 461.

33. Locke, M. 1967a. Adv. Morph. 6, 33.

34. Locke, M. 1967b. Essay presented to Sir V. B. Wigglesworth, in Insects and Physiology, J. W. L. Beament and J. E. Treherne, Eds., Oliver and Boyd, Edinburgh, pp. 69-82.

35. Locke, M. 1969a. Tissue Cell 1, 555.

36. Locke, M. 1969b. J. Morph. 127, 7.

37. Locke, M. 1969c. Tissue Cell 1, 103.

38. Locke, M., Condoulis, W. V., and Hurshman, L. F. 1965. Science 149, 437.

39. Locke, M. and Krishnan, N. 1971. Tissue Cell 3, 103.

40. Locke, M. and Krishnan, N. 1973. Tissue Cell 5, 441.

41. Mehendale, H. M. and Dorough, H. W. 1972. J. Ins. Physiol. 18, 981.

42. Mulder, R. and Gijswijt, M. J. 1973. Pestic. Sci. 4, 737.

43. Neville, A. C. 1967. In Advances in Insect Physiology, Vol. 4, J. W. L. Beament, J. E. Treherne, and Wigglesworth, V. B., Eds., Academic, London, pp. 213-286.

44. Neville, A. C. 1970. Symp. Roy. Ent. Soc. Lon. 5, 17.

45. Neville, A. C. and Caveney, S. 1969. Biol. Rev. 44, 531.

46. Neville, A. C. and Luke, B. M. 1969. Tissue Cell 1, 689.

47. Neville, A. C., Thomas, M. G., and Zelazny, B. 1969. Tissue Cell 1, 183.

48. Noirot, C. and Noirot-Timothee, C. 1969. Z. Zellforsch. 101, 477.

49. Noirot, C. and Noirot-Timothee, C. 1971a. Z. Zellforsch. 113, 361.

50. Noirot, C. and Noirot-Timothee, C. 1971b. J. Ultrastruc. Res. 37, 335.

51. Noirot, C. and Noirot-Timothee, C. 1971c. J. Ultrastruc. Res. 37, 119.

52. Noirot, C. and Quennedey, A. 1974. Annual Rev. Entomol. 19, 61.

53. Piek, T. 1961. Proc. K. ned. Akad. Wet. 64c, 648.

54. Piek, T. 1964. J. Ins. Physiol. 10, 563.

55. Post, L. C. and De Jong, B. J. 1973. J. Ins. Physiol. 19, 1541.

56. Rudall, K. M. 1963. In Advances in Insect Physiology, Vol. 1, J. W. L. Beament, J. E. Treherne, and V. B. Wigglesworth, Eds., Academic, London, pp. 257-313.

57. Shaaya, E. and Sekeris, C. E. 1970. J. Ins. Physiol. 16, 323.

58. Truman, J. W. 1971. Proc. Nat. Acad. Sci. 68, 595.

59. Truman, J. W. and Riddiford, L. M. 1970. Science 167, 1624.

60. Vincent, J. F. V. 1971. J. Ins. Physiol. 17, 625.

61. Vincent, J. F. V. 1972. J. Ins. Physiol. 18, 757.

62. Widman, J. J., Cotran, R. S., and Fahimi, H. D. 1972. J. Cell Biol. 52, 159.

63. Wigglesworth, V. B. 1946. Atlantic Monthly, 107.

64. Wigglesworth, V. B. 1947. Proc. R. Soc. B. 134, 163.

65. Wigglesworth, V. B. 1948a. Biol. Rev. 23, 408.

66. Wigglesworth, B. V. 1948b. Discuss. Faraday Soc. 3, 172.

67. Willis, J. H. 1974. Ann. Rev. Entomol. 19, 97.

68. Winteringham, F. P. W. 1965a. In Aspects of Insect Biochemistry, T. W. Goodwin, Ed., Academic, New York, pp. 29-37.

69. Winteringham, F. P. W. 1965b. In Studies in Comparative Biochemistry, K. A. Munday, Ed., Pergamon, Oxford, pp. 107-151.

## DISCUSSION

Despite the fact that it comprises a clearly distinctive arthropod feature, the insect integument appears to have been largely neglected as a potentially selective insecticide target. To a large extent this has been caused by a lack of detailed understanding of the biochemical events occurring in the cuticle and the absence of appropriate techniques and chemical tools (enzyme inhibitors, etc.) with which to study these events.

The recent discovery of the substituted benzoyl ureas is, therefore, of considerable potential importance since these represent the first group of insecticides that appear to exert their action on the insect integument. Several hundred compounds of this type have been evaluated to date and one, PH-6040 [1-(4-chlorophenyl)-3-(2,6-difluorobenzoyl) urea], has been selected for commercial development.

The compounds are particularly effective against mosquito larvae and concentrations as low as 1 ppb will prevent the development of Culex and Anopheles species; this makes them 10 to 50 times more active than the juvenile hormone mimics recently developed. Insecticidal activity is now, however, restricted to mosquito larvae since they are stomach poisons to 90 to 100 other species against which they have been tested. Studies on the mode of action of PH-6040 and related compounds have revealed that treated insects are impaired with respect to their ability to lay down chitin in the integument. It is not yet clear, however, whether this effect results from an interference with chitin synthesis or chitin catabolism. Initial studies established the failure of treated insects to incorporate $^{14}C$-glucose into the endocuticle, and it was assumed that the effect was on chitin synthesis. More recent studies by Ishaaya and Casida, however, have shown that housefly larvae fed on diets containing PH-6040 at concentrations of 1 to 3 ppm exhibited increases of up to 2.5- and 2.0-fold in cuticular chitinase and polyphenol oxidase activities, respectively, just prior to ecdysis. These changes were associated with a four-fold decrease in the level of chitin in the cuticle, but there was no appreciable difference in cuticular protein. These changes presumably cause drastic changes in cuticle flexibility that could prove fatal at ecdysis. These data suggest that the major action of PH-6040 occurs through arrested development in the period just before eclosion at the high chitinase and polyphenoloxidase activities. Consequently, the activity may not occur directly on the enzyme systems but may involve other regulatory mechanisms. Further studies are needed in this area.

It is of interest that molecular models of the substituted benzoylureas show a structural resemblance to griseofulvin. This material is not known to have any action on chitin synthesis or metabolism, but at high concentrations it appears to interfere with ecdysis. In mammals griseofulvin interferes with porphyrin synthesis by inducing the activity of the enzyme δ-aminolevulinic acid (δ-ALA) synthetase. PH-6040 has no effect on δ-ALA synthetase and is not an inducer of microsomal oxidation when fed to armyworm larvae in an artificial diet.

The potentially bright future of these compounds is somewhat marred by their remarkable chemical and environmental stability. They have an extremely low aqueous solubility (probably about 10 ppb) and are somewhat anomalous because they also exhibit a very low lipid solubility. Studies by Metcalf and colleagues have established that the substituted benzoyl ureas are extremely stable in the environment. PH-6040 exhibits some photolability to form products such as 2,6-difluorobenzoic acid, 2,6-difluorobenzamide, *p*-chlorophenylurea, and *p*-chlorophenylamine and little or no biological degradation has been observed. They can pass unchanged through the gut of the salt marsh caterpillar (*Estigmene acrea*) and are not metabolized *in vitro* in an active sheep liver microsomal system. Furthermore, they are not degraded by a strain of *Pseudomonas* that is recognized for its ability to metabolize many compounds in the environment. There are, therefore, serious questions to be raised concerning the release of these materials into the environment where they may come into contact with beneficial insects, lobsters, crabs, crayfish, and shrimps, all of which rely on a similar mechanism for cuticle formation. It is possible that the troublesome environmental stability of PH-6040 could be modified to some extent by incorporating a degradophore into the molecule. Herbicidal compounds of a similar general structure have also been shown to be extremely stable in the environment. They were apparently not adsorbed onto soil particles, nor degraded by microorganisms, and they disappeared from the soil only by the process of leaching. The latter could pose a real problem when applied to these new compounds since if they were used extensively in a river basin (e.g., the Mississippi Valley), they would be leached out into the river and could decimate the shrimp industry in the Gulf. The effects of the materials on shrimps have not yet been evaluated but should receive attention very shortly.

Like most other substituted ureas, these new compounds have an extremely low acute toxicity (around 10,000 mg/kg) possibly because they are not absorbed. As a result of this and their low fat solubility, they do not concentrate in fatty tissues.

In summary, it is clear that the discovery of the substituted benzoylureas is an important step toward the development of a new group of insecticides. If some of the undesirable environmental features of these materials can be overcome, they may be an extremely useful group of materials.

The biochemical pathways involved in dopamine metabolism might also constitute potential insecticide target sites in view of the importance of dopamine in the sclerotization of insect cuticle. In mammals carbon disulfide is known to interfere with catecholamine metabolism and prevents the formation of noradrenaline by blocking the enzyme dopamine β-hydroxylase. Under these conditions, dopamine accumulates. It is of interest that $CS_2$ is toxic to insects, but its biochemical lesion is unknown. Additional research is needed in this area.

Another area of research that might prove worthwhile concerns the nature of the biochemical events involved in the activation of blood prophenolase. This important process occurs in all insect species, and it is probable that interferences with it could have severe consequences on cuticular development.

# Hormone Mimics

WILLIAM S. BOWERS
New York State Agricultural Experiment Station
Geneva, New York

## I. INTRODUCTION

Chemical toxicants have been outstandingly successful in ensuring our agriculture and public health, but most of these toxicants have certain built-in drawbacks such as the development of resistance, hazards of overcontrol, nonselectivity, and nonbiodegradability. The reasons for these drawbacks are multiple, but to a large extent, the method of insecticide selection through random screening of industrial by-products and modification of war gases and so on, is essentially faulty.

The demise of such chemicals through the development of insect resistance or social and environmental concerns leaves an almost nonexistent data base from which to develop safer and/or more efficacious insecticides. Our hope for the future must depend on the development of chemical insecticides that affect some biochemical processes unique to insects.

This is indeed a difficult approach since there is a substantial lack of knowledge of the biochemical and physiological differences between vertebrates and invertebrates. Most of us share an intuitive belief that extensive biochemical innovation has occurred among vertebrates and invertebrates since they diverged from a common ancestry some 600 million years ago. The discovery and delineation of these differences must receive very high priority in our future research if we are to develop a strategy of insect control based on the fundamental differences between vertebrates and invertebrates rather than on the use of indifferent toxicants that affect the common and shared metabolic pathways. Even with this lofty and uncomfortably distant ideal before us, we must be mindful that our best efforts are unlikely to achieve complete control or eradication of any insect pests. We should instead project our efforts toward achieving an acceptable attenuation with those few insects that are actually pestiferous. Although insect control for the present and immediate future will continue to rely heavily on the use of conventional toxicants, our long-range planning must acknowledge the ultimate necessity of developing a variety of chemical, biological, behavioral, and cultural options for insect control that depend on and emphasize the differences between insects and other animals.

Since vertebrates and invertebrates share a common evolutionary ancestry, their basic metabolic pathways remain identical. There are, however, certain adaptive biochemical changes that have arisen through divergent evolution and of these, the most clearly recognized innovations are the control mechanisms that regulate the basic biochemistry (Bowers, 1971; Schneiderman, 1969). Two of these control mechanisms are juvenile hormones (JH) and pheromones.

The discovery and identification of the juvenile hormones that regulate insect development has generated a great deal of effort toward the development of hormonal methods of insect control.

## II. JUVENILE HORMONE (JH)

### A. History

The history of the discovery and identification of these hormones is a very exciting chapter in the progress of insect physiology and biochemistry. Since this history has been reviewed in depth (Bowers, 1971; Wigglesworth, 1964; Berkoff, 1969; Gilbert, 1964), only the highlights will be noted. Wigglesworth (1934) discovered a gland in the insect head that controlled insect differentiation. Williams (1956) subsequently reported the preparation of an active extract of a differentiation-controlling hormone from the adult male cecropia moth. Schmialek (1961) isolated from the feces of the yellow mealworm *Tenebrio molitor*, two compounds with juvenile hormone activity: farnesol (1) and farnesal (2). Although these compounds were active in a variety of assays, they were not chromatographically or biologically equivalent with the JH in the cecropia extract. Through chemical degradation and regeneration of the JH activity in cecropia extract we obtained a profile of the necessary structural moieties (Bowers *et al*., 1965) and incorporated these elements into a sesquiterpenoid framework based on farnesol, resulting in the synthesis of E.E., 10,11-epoxy methyl farnesenate (3). This compound possessed all of the biological activities of the hormone in the cecropia extract. The teamwork of Roller and his colleagues (1967) finally led to the isolation and identification of one of the cecropia hormones (4). This was followed by the isolation and identification of a second JH (5) from the cecropia extract by Meyer *et al*. (1968). Judy and his colleagues (1973) isolated JH from organ-cultured corpora allata of *Manduca sexta* in a brilliant study that authenticated the hormone, previously discovered by Bowers (3), as a natural juvenile hormone; in addition, it confirmed the presence of (5) in this insect. The sesquiterpenoid hormone (3) has been subsequently identified in adult grasshoppers (Judy et al., 1973) and in beetles: Melolantha melolantha by Schweiter (1974) and Tenebrio molitor (Judy, 1974). The identification of (3) as the only JH in larval blood of Manduca by Judy et al. (1973) is the first identification of a JH from a juvenile stage. It is interesting to speculate that the conventional isoprenoid hormone (3) may be the genuine juvenile hormone, whereas the homo- (5) and dihomo- (4) hormones may be the adult gonadotropic hormones. Table 1 summarizes the source and structures of the known natural juvenile hormones.

Table 1. Identify of Juvenile Hormones in Various Insects

I II III

| Species | Stage | Source | Hormone | Reference |
|---|---|---|---|---|
| Hyalophora cecropia | Adult ♂ | Intact insect | I | Roller et al 1967 |
| Hyalophora cecropia | Adult ♂ | Intact insect | II | Meyer et al 1968 |
| Samia cynthia | Adult ♂ | Intact insect | I & II | Roller & Dahm 1974 |
| Manduca sexta | Adult ♀ | Organ culture | II & III | Judy et al 1973 |
| Manduca sexta | Larval | Blood | III | Ibid |
| Schistocerca vaga | Adult ♀ | Organ culture | III | Judy et al 1973 |
| Melolantha melolantha | Adults | Intact insect | III | Schwieter 1974 |
| Tenebrio molitor | Adult ♀ | Organ culture | III | Judy 1974 |

## B. Biological Activities

The enthusiasm for developing hormonal methods of insect control is, to a large extent, rooted in the understanding that JH regulates processes in insects for which there are no physiological counterparts in man and other so-called higher animals. In man adult development is dependent on the secretion of certain hormones, especially hypophyseal gonadotropins; on the other hand, juvenile hormones prevent insects from maturing and consequently, must be absent during certain periods of insect life for adult development to occur. Thus the application of exogenous JH to an insect at a time when it should naturally be absent results in a derangement of adult morphogenesis and the development of intermediates that are unable to feed, mate, or reproduce, and that soon die. Although very little is known of the endocrine events that occur in the insect egg during embryonogenesis, the treatment of eggs with JH interferes with development and is effectively ovicidal (Slama and Williams, 1966; Riddiford and Williams, 1967). These two effects are the principal basis for an insecticide based on JH.

Other processes are regulated by JH. Ovarian development in many insects is dependent on the juvenile hormone (Wigglesworth, 1936; Chen _et al._, 1962; Girardie, 1962) that is gonadotropic in the adult insect. The production of sex attractants (Barth, 1962) in certain insects requires the secretion of JH. Certain insects that enter diapause as adults, when the secretion of JH is stopped (deWilde and deBaer, 1961; Stegwee, 1964), are reawakened when exposed to exogenous JH (Bowers and Blickenstaff, 1966). All these biological activities regulated by JH would be sensitive to an antihormone that could act by inhibiting hormone biosynthesis or interfering with hormone action at a receptor site.

## C. Biosynthesis of Juvenile Hormones

Following identification of the cecropia juvenile hormones, intense interest into the biosynthetic origin of these unusual homoterpenoid compounds developed. Initial investigations (Metzler _et al._, 1971, 1972) yielded essentially negative incorporation of conventional precursors into the JH skeleton, although the ester methyl of both cecropia hormones could be derived from the S-methyl of methionine. Finally, in an elegant and definitive study, Schooley _et al._ (1973) show _in vitro_ incorporation of labeled acetate, propionate, and mevalonate into the juvenile hormones (3) and (5), respectively. By degradation of these hormones and label location, the biosynthesis scheme in Figure 1 can be rationalized. Thus JH biosynthesis proceeds via conventional isoprenoid synthetic pathways with the difference that homoisoprenoid units arise through conversion of propionate into homomevalonate. The unique feature of JH biosynthesis therefore, is that the adult corpora allata possess a mixed functional capacity to convert acetate and/or propionate into normal, and homoisoprenoid, precursors.

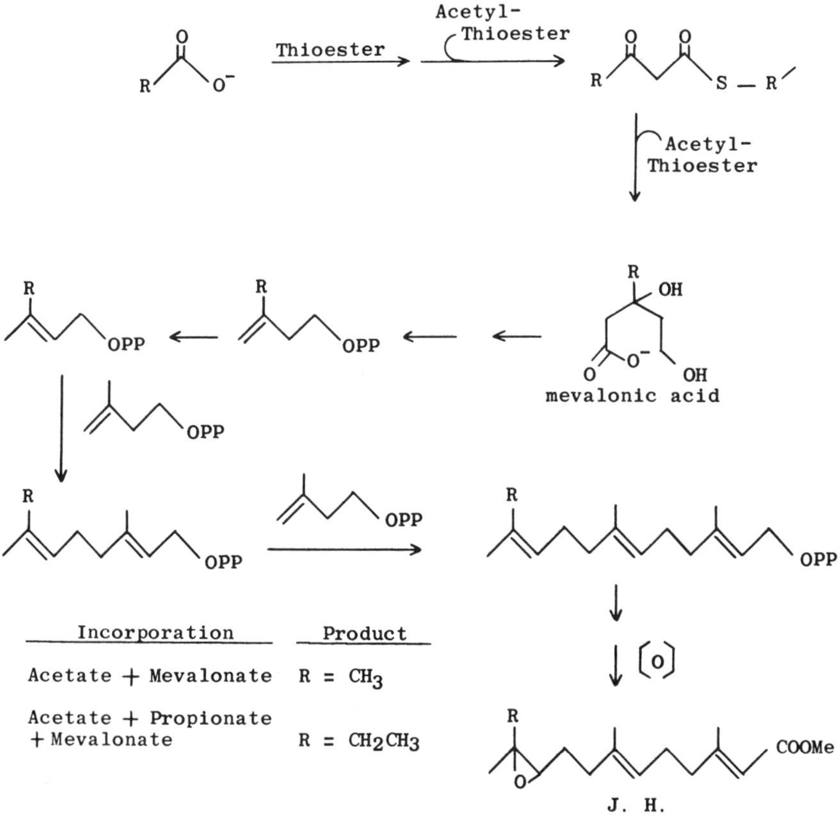


FIGURE 1. Hypothetical Scheme of Biosynthesis of Juvenile Hormones.

## D. Metabolism of Juvenile Hormones

All of the juvenile hormones are bifunctional molecules containing an $\alpha,\beta$-unsaturated methyl ester and a 10,11-epoxide moiety. Carboxyesterases (O'Brien, 1967) and epoxide hydrases (Krieger et al., 1971) are known to exist in insect tissues. Slade and Zibett (1972) have shown that cecropia hormone (4) is readily metabolized by esteratic cleavage and epoxide hydration as shown in Figure 2. Similar metabolites were obtained

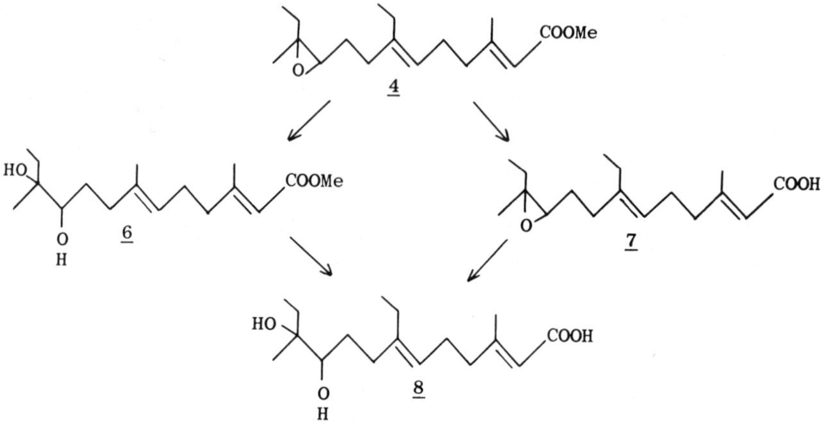

FIGURE 2. Metabolism of Cecropia Juvenile Hormone 4.

with 10,11-epoxy methyl farnesenate (3) by White (1972). Additional metabolites of cecropia hormone (4) were presumptively identified (Figure 3) through cochromatography with synthetic standards by Ajami and Riddiford (1973). It might be pointed out that either ester hydrolysis or epoxide hydration results in the loss of morphogenetic activity.

## III. JUVENILE HORMONE - STRUCTURE OPTIMIZATION

Following Schmialek's discovery that farnesol and farnesal possessed JH activity, many terpenoids were screened for biological activity in the hope of gaining an insight into the structure of the natural juvenile hormone and in an attempt to develop hormonal methods of insect control (Bowers and Thompson, 1963; Schneiderman et al., 1965). Several of the compounds discovered through early synthesis and testing procedures were extraordinarily active morphogenetic agents. One of these (3) later turned out to be a natural hormone (Bowers et al., 1965). Hydrochlorination of farnesenic acid (Law et al., 1966) produced a dichlorofarnesenic ester (12) with very high biological activity (Romanuk, 1967). The discovery of JH

1

2

Schmialek, 1961

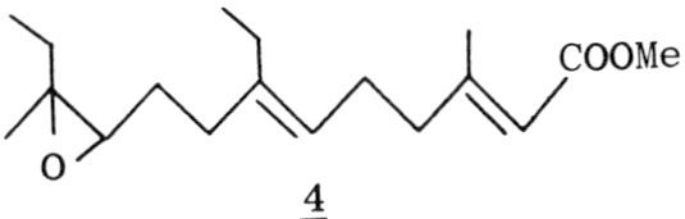

4

Roller, 1967

3

Bowers, 1965

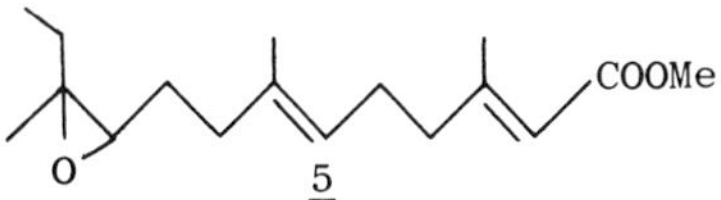

5

Meyer, 1968

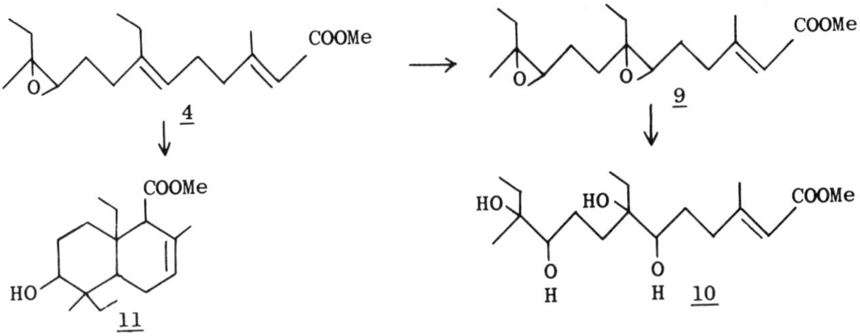

3

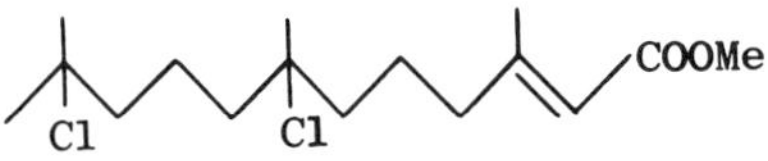

12

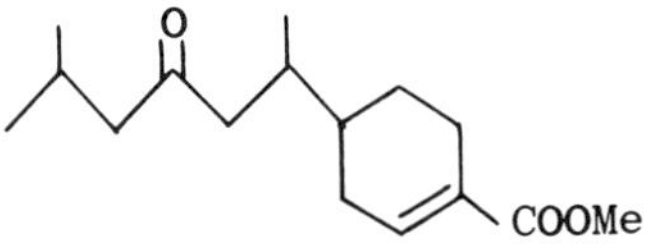

13

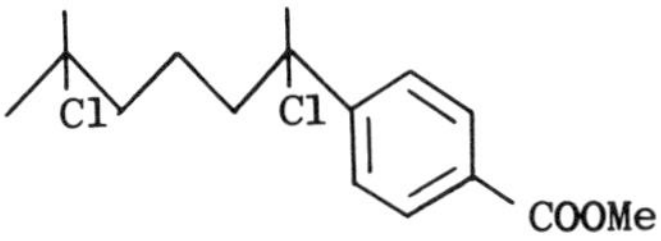

14

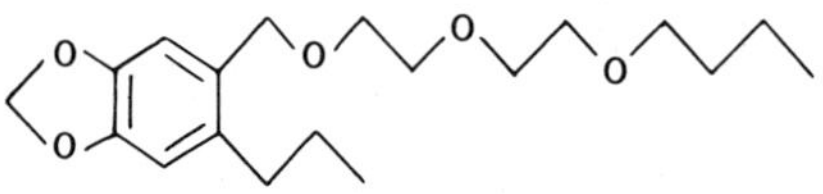

15

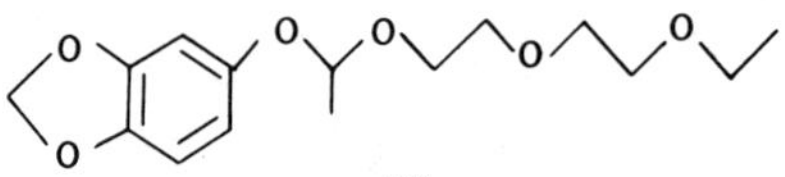

16

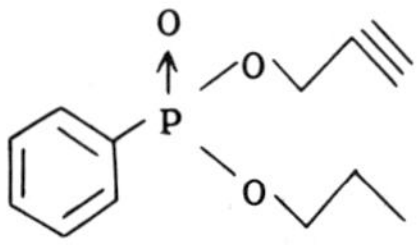

17

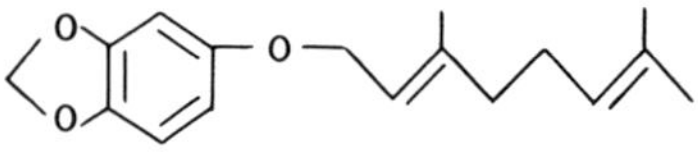

18

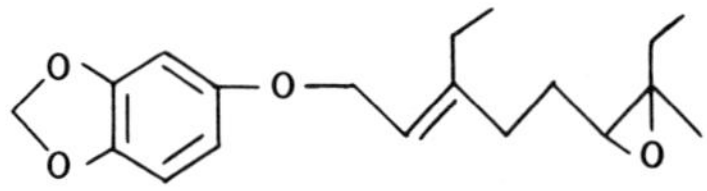

19

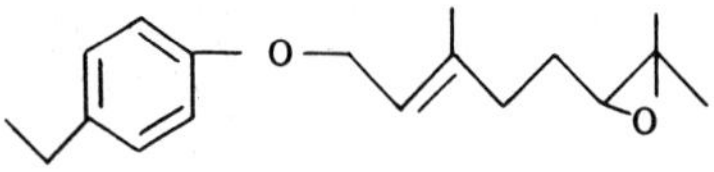

20

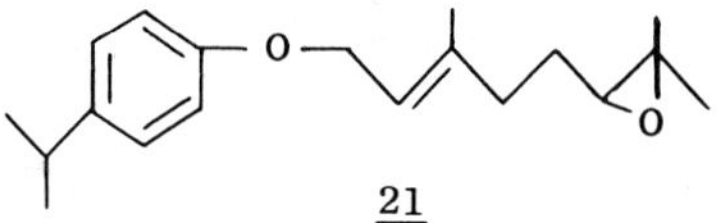

21

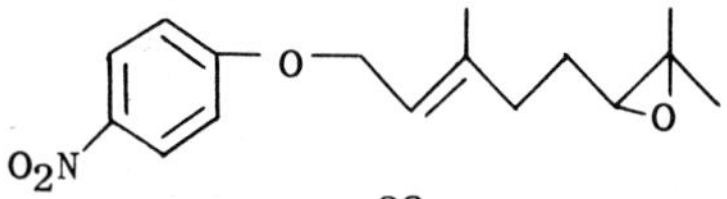

22

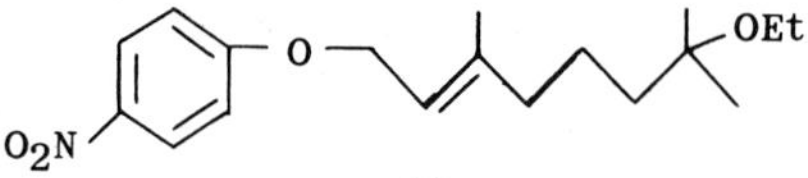

23

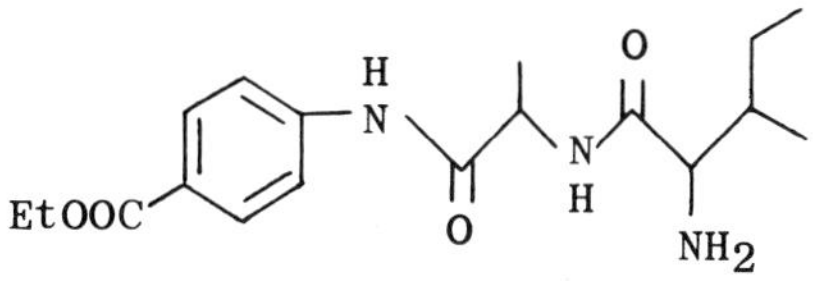


24

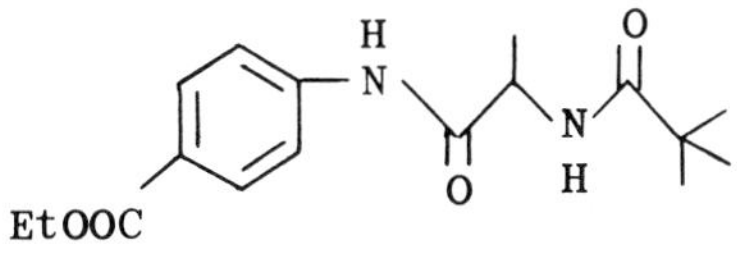


25

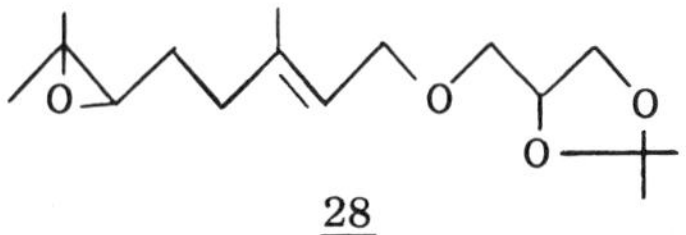

28

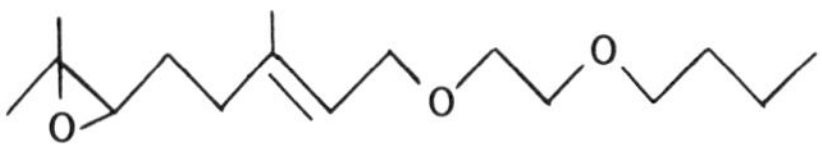

27

26

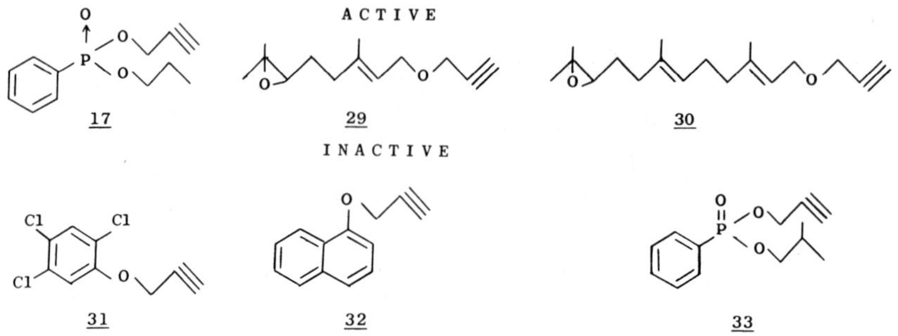


FIGURE 3. Minor Metabolites of Cecropia Juvenile Hormone 4.

activity in wood products (Slama and Williams, 1965) led to the isolation and identification of juvabione (13) (Bowers et al., 1966). One of the unique features of juvabione was its degree of specificity for one family of insects (Pyrrhocoridae). This specificity of action was retained in compounds developed as isosteres of juvabione (14) (Slama et al., 1968). These studies provided the first inclination that morphogenetic agents could be highly selective in action and indicated the possibility for the development of specific hormonal insecticides.

In 1968 we found that certain insecticide synergists possessed JH activity and that certain of these (17) retained essentially ordinal specificity (Bowers, 1968). We were able to show that JH activity depended on the entire synergist molecule (such as 15), since various simple methylenedioxy aromatics and/or several simple polyethoxy ethers were completely inactive. We then found that highly active compounds resulted when the methylenedioxyphenoxy ethers of simple terpenoids were prepared (18) (Bowers, 1969). Synthesis of the homo- and dihomo-analogs (19) resulted in compounds of even greater activity. Subsequent studies (Bowers, 1971; Pallos et al., 1971; Slama, 1971) revealed several additional aromatic-terpenoid ethers (20-23) with excellent JH activity in various assays. Zaoral and Slama (1970) discovered excellent biological activity with a series of aromatic-peptidic analogs (24, 25) that were completely specific for Pyrrhocorid insects.

Acting on information consistent with the composition of the polyether side chains of the methylenedioxyaromatic synergists, we synthesized two terpenoid glycol derivatives and a similar isopropylidene derivative (Bowers, 1971) that turned out to be very active compounds (26, 27, 28). Although it is perfectly obvious, if one sums the carbons and oxygens in the shortest path through each of the nonsynergist compounds shown here, it is clear that there are 12 carbons and 2 ethereal oxygens in each. The point is, that the composition and length of the molecule are highly critical to activity.

I have established this background to explain my interest in the following study that initially seemed to depart from these facts. When we first found that synergists possessed juvenile hormone activity, most of those we studied, and indeed most of those available, were methylenedioxyphenyl derivatives. However, (17) was one nonmethylenedioxy synergist that was active in our tests. This compound (Niagara 16388) was developed as a synergist for carbamate insecticides. It is significantly active in the Tenebrio genitalia test. Even the most liberal interpretation of the structural features necessary for juvenile hormone activity cannot be accommodated to this molecule. Several other synergists of similar structure were found to be entirely inactive (31, 32, 33). The one structural feature that most of these other synergists had in common with NIA-16388 was the propargyl group. In attempting to discover the reason for the activity of this compound, we prepared the expected epoxy terpenoid ethers of propargyl alcohol. The geranyl ether has low grade activity, whereas the farnesyl ether is somewhat more active (in the submicrogram range). The corresponding propyl or allyl ether of epoxy-geraniol is completely inactive, and the same ethers of farnesol have low-grade activity. So the propargyl group coupled to a simple terpenoid is active. The thought, therefore, occurs—does the propargyl group itself possess some unique reason for JH activity? To investigate this possibility, the next step was to devise a way of supplying the propargyl group to the test insect in a form in which the carrier portion of the molecule would fit our general notions of the structural requirements for activity and yet not complicate our interpretation of the contribution of the propargyl group to any observed JH activity. The simplest procedure, therefore, was to prepare saturated, straight-chain alkyl propargyl ethers. We prepared a series of these simple ethers, and quite surprisingly, several were sufficiently active to establish a carbon chain-activity relationship (Figure 4).

This series of such alkyl ethers from $C_{10}$-$C_{18}$ reveals a distribution of activity that is optimal at a chain length of 14 carbons. The activity of these compounds must be due to the

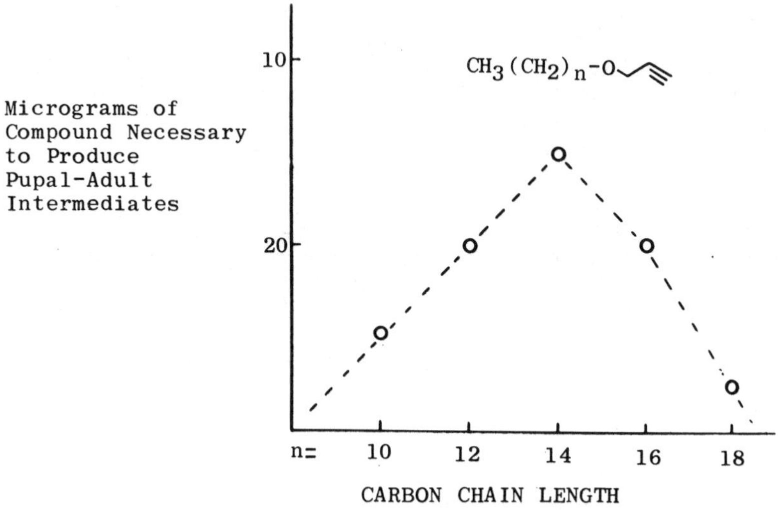


FIGURE 4. JH Activity of Mono-propargyl Ethers.

propargyl group since the corresponding propyl and allyl analogs are inactive. However, I retained a degree of skepticism about whether the propargyl group represented a uniquely active chemical grouping. In seeking a way to intensify the activity relative to the propargyl group without radically changing the molecule (so that any increase in activity could be unambigously attributed to the propargyl grouping), we made the dipropargyl ethers. The most active of these was the compound containing a ten-carbon central chain (Figure 5). It is noteworthy that the increase in the number of propargyl groups increased activity approximately 3 times. It is also apparent that the chain lengths of the most active mono and dipropargyl ethers are identical (34, 35). The most active mono and dipropargyl ethers contain 18 units, counting both carbon and oxygen atoms as one. In view of the foregoing, it appears that the propargyl grouping is genuinely and uniquely a juvenile hormonophoric moiety.

Recently, there have been developed a series of sesquiterpenoid, conjugated dienoates (36, 37, 38) that are being extensively tested in field applications for insect control (Hendrick et al., 1973). Methoprene (36) has been found to be

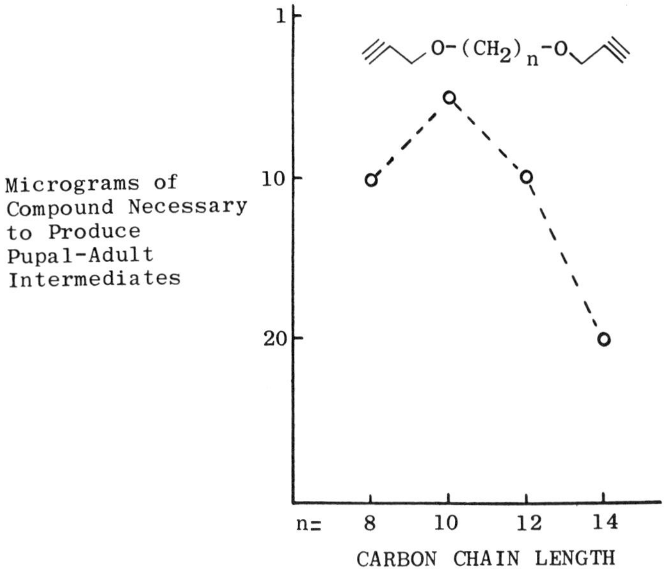
O-(CH2)n-O
Micrograms of Compound Necessary to Produce Pupal-Adult Intermediates
1
10
20
n= 8 10 12 14
CARBON CHAIN LENGTH

34

35

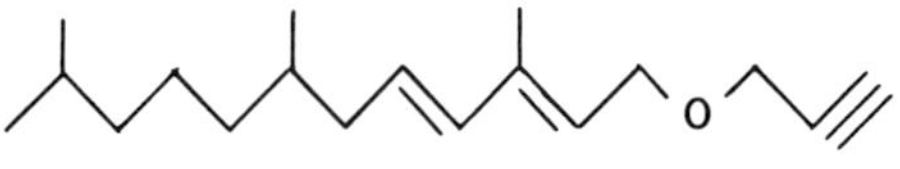

38

37

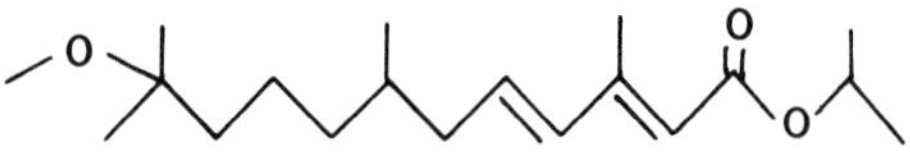

36

FIGURE 5. JH Activity of Di-propargyl Ethers.

highly efficacious as a dipteran larvacide. It is effective against mosquitoes, houseflies, stable flies, horn flies, and face flies. Early in the testing of hormonal compounds for insect control, it was recognized that compounds that were highly active in the laboratory frequently performed poorly in field tests. The presence in these experimental compounds of epoxides, simple esters, and several olefinic double bonds constituted built-in lability. Under field conditions, photochemical-induced polymerization, bisepoxidation, and *cis-trans* isomerization, and so on, occurs (Pawson *et al.*, 1972). In plants and animals rapid epoxide hydration, *cis-trans* isomerization, double bond cleavage, and so on, have been observed (Gill *et al.*, 1972). Even prior to metabolism studies, vigorous synthetic efforts were being directed toward the development of JH-active compounds that lacked the suspected labile moieties. In Figure 6 we recognize efforts to avoid esteratic cleavage by terminal substitution with aromatic isosteres or through the use of conjugated unnatural esters that resist facile hydrolysis (Weirich and Wren, 1973). In the same vein (Figure 7), substi-

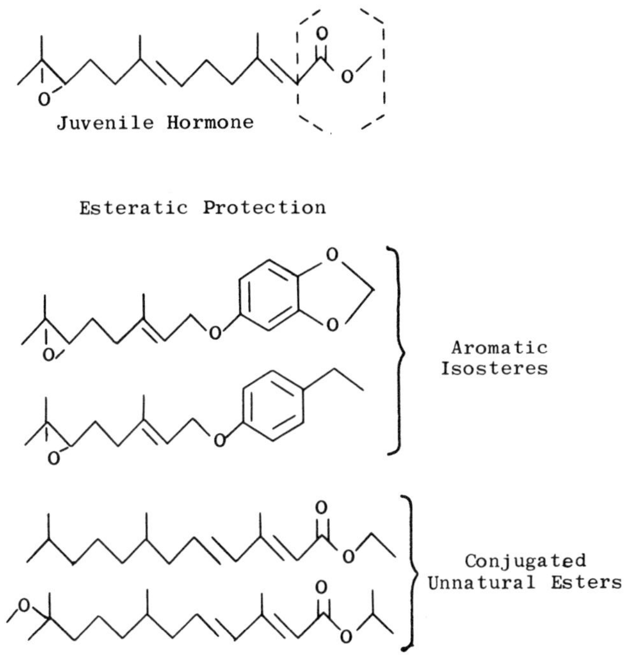


FIGURE 6. Ester Protection/Substitution.

tutions for the labile epoxide with simple and extended alkyls or with simple alkoxy groups result in retention of hormonal activity and obviate the effect of epoxide hydration. It seems clear that the ethyl branching of the cecropia hormones may confer some resistance to epoxide hydration and explain in part the difference in the biological activity among the three natural hormones. However, the work of Mori (1971) and Ozawa (1973) shows clearly that an extension of the alkyl at the 11 position results in compounds as active with or without the epoxide. This fact indicates perhaps some unique function for a bulky substituent at this end of the hormone molecule apart from epoxides or alkoxides.

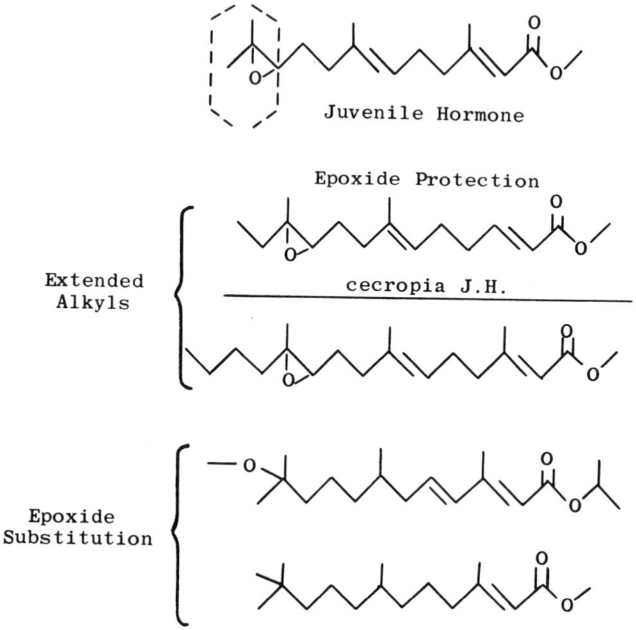


FIGURE 7. Epoxide Protection/Substitution.

## IV. COMMERCIAL DEVELOPMENT

Since early trials indicated excellent although short-term performance of methoprene in flood-water mosquito control, an in-depth program of formulation was undertaken to protect methoprene from rapid environmental degradation. They developed a microencapsulation procedure that protected the methoprene and maintained operational stability under field conditions for about two weeks. Another bonus of their formulation technique was the reduction from one-eighth to one-fiftieth of a pound active material per acre to achieve satisfactory control. The development of such a formulation for methoprene has put the control of insects on an operational basis and methoprene is the first insect morphogenetic agent to receive an experimental label and to be offered for sale under temporary registration as a mosquito larvicide.

The control of livestock pests by feeding hormonal compounds to cattle (Harris et al., 1974) has proceeded to a relatively advanced stage. Incorporation of methoprene (36) into cattle feeds or salt blocks has resulted in excellent control of various dipteran insects that breed in cattle manure. In initial studies, several compounds have shown relatively little decomposition when passed through cattle stomachs, and formulation will probably increase the effectiveness of these compounds.

Laboratory studies with hormone mimics indicate excellent prospects for the control of aphids (Nassar et al., 1973), scale insects, mealybugs (Staal et al., 1973), and numerous stored-product insects (Strong and Dickman, 1973).

Hormone analogs have been successfully applied to the control of social insects. Feeding of the hormone analogs (36) and (37) caused inhibition of metamorphosis of larvae of the imported fire ant, *Solenopsis invicta*, sterilized the queens, and caused caste alterations from workers to alate males (Troisi and Riddiford, 1974). Similar results might be expected with other social insect pests such as termites and perhaps the African hybrid bees.

## V. FUTURE PROSPECTS

In general, hormone analogs are becoming effective insect-control agents, especially in situations where conventional pesticides are ineffective because of the presence of resistant insect populations and also in situations where conventional toxicants were never highly effective anyway, such as the control of biting flies and face flies in cattle. Even the control of plant feeding insects, where environmental instability of hormonal chemicals has been a serious drawback, is giving way to isolated, but multiplying, instances of successful insect control. The potential of employing hormone analogs for the control of social and soil insect pests deserves greater attention.

The facile biodegradability of hormonal pesticides has been the chief drawback to their field efficacy. No problems of toxic or harmful residues have been reported. Perhaps for the first time, the almost mystical art of formulation has been the pivot point for efficacy. This spin-off in genuinely sophisticated formulation technology will benefit not only subsequently developed hormonal pesticides but should improve the performance and decrease the environmental hazards of conventional pesticides as well.

Only the naive can believe that hormones and pheromones will provide final or ultimate answers to insect control. We can only hope to lengthen the interval of success between susceptability and resistance between outbreak and remission while we develop a long-range strategy of insect control to exploit the biochemical and behavioral differences that set insects and man apart. The development of biocides, or general protoplasmic poisons, provides only a temporary respite, and when they are lost through the development of resistance or the recognition of environmental hazard, little is left to build on.

Hormones and pheromones represent only a crude beginning toward the rational development of a cohesive strategy of insect control. No effort of man has ever eradicated an insect species nor will future efforts likely succeed. We must instead realistically develop a continuing program of research into fundamental methods of insect control. Our best hope is to devise chemical, cultural, and biological methods coupled with the development of host-plant resistance that will relegate those few species of insect pests into noneconomic status.

## REFERENCES

1. Ajami, A. M. and Riddiford, L. M. 1973. J. Insect. Physiol. 19, 635.

2. Barth, R. H. 1962. Gen. Comp. Endocrinol. 2, 53.

3. Berkoff, C. E. 1969. Chem. Soc. Quart. Rev. 23, 372.

4. Bowers, W. S. and Thompson, M. J. 1963. Science 142, 1469.

5. Bowers, W. S., Thompson, M. J., and Uebel, E. C. 1965. Life Sci. 4, 2323.

6. Bowers, W. S. and Blickenstaff, C. C. 1966. Science 154, 1673.

7. Bowers, W. S., Fales, H. M., Thompson, M. J., and Uebel, E. C. 1966. Science 154, 1020.

8. Bowers, W. S. 1968. Science 161, 895.

9. Bowers, W. S. 1969. Science 164, 323.

10. Bowers, W. S. 1971. Bull. Soc. Ent. Suisse 44, 115.

11. Bowers, W. S. 1971. Juvenile Hormones. In Naturally Occurring Insecticides, M. Jacobson and D. G. Crosby, Eds., Dekker, New York, pp. 307-332.

12. Chen, D. H., Robbins, W. E., and Monroe, R. E. 1962. Experientia 18, 577.

13. deWilde, J. and deBaer, J. A. 1961. J. Insect Physiol. 6, 152.

14. Gilbert, L. I. 1964. The Physiology of Insects, Vol. I, M. Rockstein, Ed., Academic, New York, pp. 149-225.

15. Gill, S. S., Hammock, B. D., Yamamoto, I., and Casida, J. E. 1972. In Insect Juvenile Hormones, J. J. Menn, Ed., Academic, New York, pp. 177-189.

16. Girardie, A. 1962. J. Insect Physiol. 8, 199.

17. Harris, R. L., Chamberlain, W. F., and Frazer, E. D. 1974. J. Econ. Entomol. 67, 384.

18. Hendrick, C. A., Staal, G. B., and Siddall, J. B. 1973. Agric. Food Chem. 21, 354.

19. Judy, K. J., Schooley, P. A., Dunham, L. L., Hall, M. S., Bergot, B. J., and Siddall, J. B. 1973. Proc. Natl. Acad. Sci. U.S. 70, 1509.

20. Judy, K. J., Schooley, D. A., Hall, M. S., Bergot, B. J., and Siddall, J. B. 1973. Life Sci. 13, 1511.

21. Judy, K. 1974. Zoecon Corp., personal communication.

22. Krieger, R. L., Feeny, P. P., and Wilkenson, C. F. 1971. Science (Wash.) 172, 579.

23. Law, J. H., Yuan, C., and Williams, C. M. 1966. Proc. Natl. Acad. Sci. U.S. 55, 576.

24. Metzler, M., Dahm, K. H., Meyer, D., and Roller, H. 1971. Z. fur Naturforsch 26b, 1270.

25. Metzler, M., Meyer, D., Dahm, K. H., and Roller, H. 1972. Z. fur Naturforsch 27b, 321.

26. Meyer, A. S., Schneiderman, H. A., Hangman, E., and Ko, J. H. 1968. Proc. Natl. Acad. Sci. U.S. 60, 853.

27. Mori, K., Mitsui, T., Fukami, J., and Ohtaki, T. 1971. Agric. Biol. Chem. 35, 1116.

28. Nassar, S. G., Staal, G. B., and Armanious, N. I. 1973. J. Econ. Entomol. 66, 847. Pallos, F. M., Menn, J. J., Letchworth, P. E., and Mianllia, J. B. 1971. Nature 232, 486.

29. Pawson, B. A., Scheidl, F., and Vane, F. 1972. In Insect Juvenile Hormones, J. J. Menn, Ed., Academic, New York, pp. 191-214.

30. Riddiford, L. M. and Williams, C. M. 1967. Proc. Natl. Acad. Sci. U.S. 57, 595.

31. Roller, H., Dahm, K. H., Sweeley, C. C., and Trost, B. M. 1967. Angew. Chem. 79, 190.

32. Roller, H. and Dahm, K. H. 1974. Proc. Conf. Workshop Horm. Heterophyly, in press.

33. Romanuk, M., Slama, K., and Sorm, F. 1967. Proc. Natl. Acad. Sci. U.S. 57, 349.

34. Schooley, D. A., Judy, K. J., Bergot, B. J., Hall, M. S., and Siddall, J. B. 1973. Proc. Natl. Acad. Sci. 70, 153.

35. Schmailek, P. 1961. Z. Naturf. 16b, 461.

36. Schneiderman, H. A., Krishnakumaran, A., Kulkarni, V. G., and Friedman, Lester. 1965. J. Insect Physiol. 11, 1641.

37. Schneiderman, H. A. 1969. Endocrinological and Genetic Strategies in Insect Control, Proceedings of the Symposium on Potentials in Crop Protection.

38. Schwieter, V. 1974. F. Hoffmann-LaRoche, personal communication.

39. Slade, M. and Zibitt, C. H. 1972. In Insect Juvenile Hormones, J. J. Menn, Ed., Academic, New York, pp. 155-177.

40. Slama, K. and Williams, C. M. 1965. Proc. Natl. Acad. Sci. U.S. 54, 411.

41. Slama, K. and Williams, C. M. 1966. Nature 210, 329.

42. Slama, K., Suchy, M., and Sorm, F. 1968. Biol. Bull. 134, 154.

43. Slama, K. 1971. Ann. Rev. Biochem. 40, 1079.

44. Staal, G. B., Nassar, S., and Martin, J. W. 1973. J. Econ. Entomol. 66, 851.

45. Stegwee, D. 1964. J. Insect Physiol. 10, 97.

46. Strong, R. G. and Dickman, J. 1973. J. Econ. Entomol. 66, 1167.

47. Troisi, S. and Riddiford, L. M. 1974. Environ. Entomol. 3, 112.

48. Weirich, G. and Wren, J. 1973. Life Sci. 13, 213.

49. White, A. F. 1972. Life Sci. 11, 201.

50. Wigglesworth, V. B. 1934. Quart. J. Microscop. Sci. 77, 191.

51. Wigglesworth, V. B. 1936. Quart. J. Microscop. Sci. 79, 91.

52. Wigglesworth, V. B. 1964. Adv. Insect Physiol. 2, 243.

53. Williams, C. M. 1956. Nature (London) 121, 572.

54. Zaoral, M. and Slama, K. 1970. Science 170, 92.

## DISCUSSION

The insect hormone field is developing very rapidly, yet it is experiencing many difficulties in reproducing the impressive performances of hormone mimics in the laboratory in practical field tests. The major problems foreseen by the conferees in the practical development of hormone mimics are

(1) Need for better methods of evaluation and assessment of field effectiveness.

(2) Evaluation of the fate of adult insects from treated immatures that emerge and fly away. Are they and their offspring sterile and is this a bonus effect in the $F_1$ and $F_2$ generations?

(3) What are the environmental biological effects of hormone mimics on nontarget organisms?

To date there has been no major success in controlling important phytophagous insects such as boll weevil, cotton bollworms, or codling moth with hormone mimics. It seems unlikely that these compounds will be effective against such important crop pests because the crop damage results from the immature stages that would be prolonged by treatment with juvenilizing agents. Too, these insects live in relatively protected locations and can be affected only by relatively large dosages of the very expensive hormone mimics. The development of such compounds with systemic properties would be potentially useful. The use of methoprene in controlling insects with innocuous immature stages, such as the mosquito, is much more promising, and methoprene is being used in Indonesia in a field trial of 100 km of ditches.

Juvenile hormone mimics are lipophilic, and some may be moderately stable environmentally. The fate of these and their major degradation products is of concern in regard to nontarget organisms. For example, hormone mimics used as cattle feed supplements to control horn flies and other insects breeding in cattle feces obviously pass through the mammalian digestive tract undegraded. How much of such compounds will be found in the milk of treated cows? Invertebrates other than the Insecta, such as crayfish, crabs, shrimp, and lobsters employ the same or related juvenile hormones as the insects. Will the treatment of marshes, swamps, lakes, and streams for mosquito or black-fly control disperse enough of the juvenile hormone mimics to affect the breeding of these arthropods in estuaries and coastal waters? This problem may become more important with the use of more persistent slow release and microencapsulated formulations.

A point of some concern was raised about the epoxide structure common to natural juvenile hormones and many of the more active mimics. Epoxides are known to be biochemically involved in interactions with DNA to produce cancer in mammals, for example, the K-ring epoxide of benzo-(a)-pyrene that is the active chemical carcinogen, and the epoxides dieldrin and heptachlor epoxide that are strongly carcinogenic in mice. Does the environmental use of farnesol epoxide hormone mimics pose a hazard in mammalian carcinogenesis? The answer is apparently unknown, although it was pointed out that the farnesol derivatives are test epoxides and may behave somewhat differently from epoxy-benzo-(a)-pyrene.

The development of insect resistance to hormone mimics has been demonstrated in several laboratories, and recently, wild strains of the housefly and mosquitoes have shown cross resistance to methoprene. These problems can be expected to increase as a result of resistance mechanisms depending on increased mixed-function oxidation (Oppenoorth, Chapter III).

# Pheromones

WENDELL L. ROELOFS
New York State Agricultural Experiment Station
Geneva, New York

## I. INTRODUCTION

A recent report in Science (Müller-Schwarze et al., 1974) describes the response behavior of pronghorn antelope to the secretion of their subauricular glands - one of 4 different scent glands found with pronghorn. The males and females both respond by sniffing, licking, and thrashing the scent marks with their horns. Of particular note is that the bucks apply these odoriferous compounds to vegetation for marking territories, and thus there is recognition of marks of individual animals. Eight compounds have been identified from the subauricular glands, with isovaleric acid accounting for most of the activity; but will it ever be possible to define the complete pheromone with all the various components in their proper ratios when this apparently is different for each individual? Perhaps not, but typical responses can be elicited by releasing the main components in approximate ratios. This is the point where I believe the field of insect sex pheromones has finally arrived.

I am not suggesting that there is individual recognition in insect-mating behaviors where long-range attractancy is involved, although this may be found in some of the more sophisticated social insects, but I would suggest an analogy to species recognition. Discriminating the correct female emanations from those of a number of closely-related sympatric females could be as subtle as individual recognition with the pronghorn.

Previous thinking that each insect species utilized a unique pheromone compound for reproductive isolation was soon to be proved incorrect. It became apparent (some recent reviews of the pheromone literature are given in Birch, 1974; MacConnell and Silverstein, 1973; Evans and Green, 1973; Priesner, 1973; Jacobson, 1972; Law and Regnier, 1971) that many species utilize the same primary pheromone component. In many cases additional components, sometimes obligatory for attractancy, were implicated. These additional components create unique pheromone blends for the species, utilizing a common pheromone compound. Some examples are as follows:

## II. PHEROMONE BLENDS WITH ONE COMPOUND IN COMMON

### A. *cis*-9, *trans*-12-Tetradecadienyl Acetate

This compound has been found to be a pheromone component for a number of pyralid and noctuid species (Table 1). With the noctuid species, both *Spodoptera litura* and *S. littoralis* were found (Tamaki *et al.*, 1973) to use a second component, *cis*-9, *trans*-11-tetradecadienyl acetate, whereas *S. eridania* requires the presence of *cis*-9-tetradecenyl acetate. At least five pyralid species use the common pheromone component, although specificity has again been found with blends. Males of *P. interpunctella*, *Cadra cautella*, and *Ephestia elutella* are preferentially attracted to their own females and female gland extracts. An additional component, *cis*-9-tetradecenyl acetate, was identified (Brady, 1973) for *C. cautella* and evidence presented for one in *E. elutella* (Brady and Nordlund, 1971). A compound, *cis*-9, *trans*-12-tetradecadien-1-ol, was found (Sower *et al.*, 1974) in the female pheromone glands of *P. interpunctella* that does not seem to act as a pheromone component for attractancy, but does decrease *C. cautella* male responses to the common attractant compound. The data also suggest that this alcohol has some input into the *P. interpunctella* CNS because it also reduces the inhibitory effect of *trans*-9, *trans*-12-tetradecadienyl acetate on the pheromone response of male *P. interpunctella*.

### B. Bark Beetle - *Ips* Species

Pheromone specificity has been shown (Wood, 1970) for a number of sympatric and allopatric species, with blends being based on combinations of 4 active $C_{10}$ terpene alcohols released from male frass. The components are as follows (Figure 1):

(1) 2-Methyl-6-methylene-7-octen-4-ol (Ipsenol)

(2) 2-Methyl-6-methylene-2,7-octadien-1-ol (Ipsdienol)

(3) *cis*-Verbenol

(4) *trans*-Verbenol

Reports (Young *et al.*, 1973; Vite *et al.*, 1972) on the various blends indicate that males and females aggregate on host material (a) with *I. grandicollis* in response to 1; (b) with

TABLE 1. Lepidopterous Insects Using *cis*-9, *trans*-12-Tetradecadienyl Acetate (I) as a Pheromone Component

| | Pheromones |
|---|---|
| *Noctuidae* | |
| *Spodoptera exigua* | I + ? |
| *S. eridania* | I + c9-14:Acet. |
| *S. litura* | I + c9,t11-14:Acet. (1:9) |
| *S. littoralis* | I + c9,t11-14:Acet.(1:20) |
| *Pyralidae* | |
| *Plodia interpunctella* | I + ? + (c9,t12-14:OH) |
| *Cadra cautella* | I + ? + c9-14:Acet. |
| *C. figulella* | I |
| *Anagasta Kühniella* | I |
| Ephestia elutella | I + ? |

*I. latidens* in response to 1 and 2; (c) with *I. paraconfusus* in response to 1, 2, and 3 - a mixture that inhibits the response of *I. latidens*; (d) with *I. calligraphus* in response to 3 and 4; and (e) with *I. typographus* in response to 4. Additionally, a variety of mixtures of these 4 components have been reported in the hindgut of 8 other *Ips* species, although the identification of the compounds was based only on gas chromatographic retention times. A more dramatic description of pheromone specificity is the report (Lanier *et al.*, 1972) that populations of *Ips pini* from California, Idaho, and New York have detectable differences in both pheromone production and reception. Geographical variation in pheromones of a species is probably quite common and ranges from subtle differences to differences so great that the two allopatric populations are not cross-attractive. An example of the latter situation is with

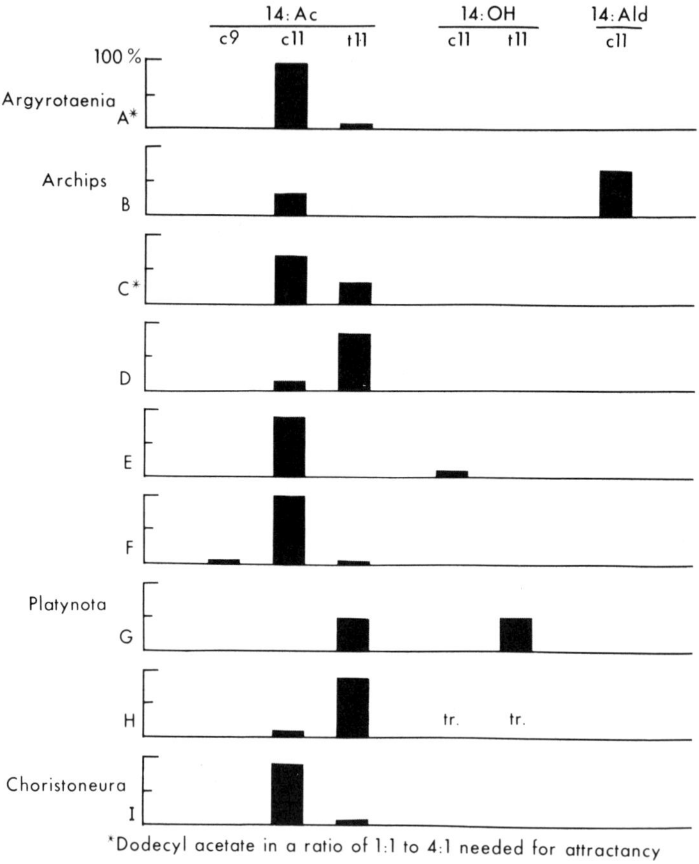


FIGURE 1. Pheromone Compounds Found in Various <u>Ips</u> Species.

European corn-borer moths. In Iowa (Klun *et al.*, 1973) the males are attracted to predominantly the *trans*-isomer - a 50:50 mixture of the two pheromones is not attractive to either population. In Pennsylvania the two populations appear to be sympatric (Carde *et al.*, 1975).

### C. *cis*-11-Tetradecenyl Acetate (c11-14:Ac)

This compound attracts males of species from a number of families (Roelofs and Comeau, 1971), including the European corn borer mentioned above, but has been found most commonly with Tortricinae species (Lepidoptera: Tortricidae) (Roelofs and Carde, 1974). The closely-related alcohols, acetates, and positional and geometrical isomers can be blended together to form specific pheromones for each species. Figure 2 shows examples of some of the tortricid species with which we have been working. It shows that species A and I both use 92:8 *cis*/*trans* blends, but A has a 3rd component, dodecyl acetate. Species C uses a 70:30 blend with dodecyl acetate as a 3rd component, whereas D uses 15:85 and H uses 10:90 with traces of the corresponding alcohols as synergists. Species B uses the corresponding aldehyde in addition to c11-14:Ac, and E uses the corresponding alcohol. Species F adds some *cis*-9 positional isomer and a small amount of *trans*-isomer to c11-14:Ac for attractancy. Species G is somewhat of an exception in that its pheromone does not include c11-14:Ac, but rather is a mixture of t11-14:Ac and t11-14:OH. These mixtures have been defined from the female gland extracts as well as from field attractancy studies. The finding that species H uses an extremely small amount (0.2%) of alcohol to synergize attractancy indicates that it will be extremely difficult to define fully the pheromone system of any of these species.

Even though the complete pheromone system is not defined, it has been found that good attractancy and specificity can be obtained with the main components - sometimes with only one compound. This information is needed on the important pest species for the application of pheromones in pest management programs. The electroantennogram technique (Roelofs *et al.*, 1971) has helped to speed up the identification of the main components of lepidopterous species. It is useful for determining the position and configuration of unsaturation in primary pheromone components and for locating secondary components that do not give behavioral responses by themselves. Chemical and instrumental analysis of these components is conducted to

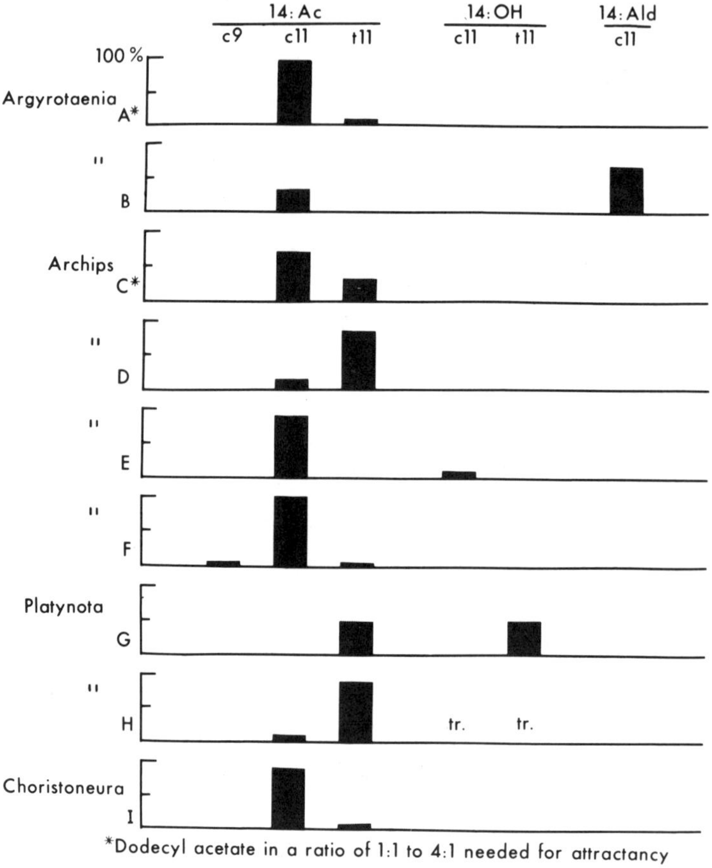


FIGURE 2. Pheromone Blends Found Attractive to Various Tortricid Species.

rigorously prove the chemical structures. Field studies with the various components to optimize release rates, component ratios, trap design, and so on, provide the basic information needed for using the pheromones in insect monitoring and control tests.

## III. PRACTICAL USES OF PHEROMONES

### A. Monitoring

Attractants, such as methyl eugenol, that have been found by screening hundreds of chemicals have been used for surveillance at international points of entry for the Mediterranean fruit fly, the melon fly, and the oriental fruit fly. These attractants presumably are not naturally-occurring pheromones, but showed the benefits of having the availability of a surveillance system. Pheromones now provide a very sensitive surveillance tool for a number of pest species. Thousands of gypsy-moth traps baited with the pheromone, *cis*-7,8-epoxy-2-methyloctadecane (disparlure), are used throughout the states to detect new infestations and to assess moth densities in known infestations.

State and federal agencies also rely on pheromone traps to monitor uninfested and lightly infested cotton areas for the spread of the pink bollworm. Recent identification (Hummel *et al*., 1973) of the pheromone, *cis*-7,*cis*,*trans*-11-hexadecadienyl acetate (50:50) (gossyplure), should provide a much more sensitive tool than the previously used parapheromone, *cis*-7-hexadecenyl acetate (hexalure).

One of the costly aspects of the California red-scale eradication program was manual inspection and searching for incipient infestations. Pheromone traps (virgin females) are now used to detect the red scale at very low densities (Shaw *et al*., 1971).

Pheromone traps are also important tools for insect population surveys. An example of this is the extensive program using codling moth pheromone (*trans*-8, *trans*-10-dodecadien-1-ol) traps to obtain knowledge of codling moth populations throughout western Colorado (Quist, 1973). Through the Tri-River Extension Service, the traps were mailed to fruit growers and returned at weekly intervals throughout the growing season, and the trap counts used to help each grower evaluate his particular problem. The total data were computerized to develop moth population-density maps for each area. The participants in this program concluded that the pheromone trap provides a valuable index of the level of moth activity and that this is needed to get maximum benefit from their pesticide applications. A lack of grower interest was a handicap in some communities.

In Washington (Eves and Chandler, 1973) pheromone traps were used to survey both oriental fruit moth and codling moth populations. The program involved over 1000 traps in 3800 ha of apples and pears and 200 traps in 500 ha of peaches. Trap catches were processed by computer and sent back to the growers to improve spray timing, point out areas of infestation, show migration patterns, and indicate the effectiveness of spray applications.

Another example of using pheromones for survey is in the New York State apple-pest management program (Brann and Tette, 1974) funded by the USDA and operated by Cornell University research and extension personnel. The overall goal is to develop an integrated program for the control of diseases and insects in about 800 hectares of fruit. To obtain base-line information on the insect-pest potential, the entire area was set up on a grid system with monitoring stations located at all the intersections of coordinates forming squares of 4 hectares. Pheromone traps for 6 lepidopterous pest species were placed at each monitoring station and checked each week to give a readout of the activity of each species throughout the area during the entire season. Similarly funded pilot pest-management programs on deciduous tree fruits are also underway in Pennsylvania, Michigan, Washington, and California.

To date, pheromone traps have found their greatest utility in monitoring for the presence and abundance of certain pest populations. Pheromone traps are extremely sensitive for determining the initiation and duration of moth flight. Much use is made of this information for the precise timing of insecticide spray applications. A commercial company* provides monitoring kits throughout the world for this purpose. With some species, the timing for sprays can be obtained directly from the profile of moth catches. An example of this is with the summerfruit tortrix moth on apples in the Netherlands (Minks, 1974). Maximum benefit from insecticidal spray for the destructive _Adoxophyes orana_ larvae is derived in the few days after egg hatch when the larvae migrate to feeding sites. This period is precisely timed by observing the moth emergence profile as given by male catches. Through the extension service of the Ministry of Agriculture, all apple growers are guided in a "supervised control" program. Each grower is provided with two traps per orchard (about 5 ha) and advised when to spray from the resulting trap data. Typically the 5 to 7 preventive sprays used for both generations have been reduced to 3 to 4 precisely timed and thus more effective sprays.

---

*Zoecon Corp., Palo Alto, Calif.

This type of precise spray timing has resulted in insecticide reductions on several crops throughout the world, but sometimes the prediction is not so easy. With codling moth in the States, the early season weather conditions cause problems in timing first-cover sprays. They should be applied at egg hatch, but the interval between the first moth catches and egg hatch is dependent on the variable weather conditions. With codling moth on pears in California, Batiste _et al_. (1973) have supplemented the pheromone trap catches with a system of observing egg hatch from laboratory-reared moths in field-oviposition cages. In Michigan (Riedl _et al_., 1974) the variable duration of the preoviposition period is also a problem. In their initial studies they have used trap catches as a biological fixpoint and then predicted egg hatch after a certain amount of heat or developmental units have accumulated.

Attempting to determine the abundance of moths within a field or orchard and the need for spraying or not is even more difficult. In Washington it was found (Hoyt, 1974) that codling moth abundance was not only affected by male density, but also by weather, trap condition, trap:tree ratio, and the influx of males from outside the orchard. In British Columbia a test (Madsen and Vakenti, 1972) with 14 codling moth pheromone traps in 5 ha gave good results when sprays were applied only when an average of 2 moths per trap per week was reached. In tests (Madsen and Vakenti, 1973) where moths were coming from outside sources, sprays were applied only if border traps captured more than 2 moths per trap per week and infested apples were found.

In New York apple orchards (Trammel, 1973), a system of peripheral and internal orchard monitoring was set up to overcome the problems of attracting moths from outside the orchard. An orchard of 4 ha would have four peripheral and two internal traps, with decisions on potential insect damage being made on the numbers of males trapped and on the peripheral:internal trap count ratio.

In Canada it was found (Miller and McDougall, 1973) that pheromone traps could be used to monitor moth abundance in low population levels of spruce budworm and that the male counts of one generation had a predictive relationship to third instar larval counts of the next generation. This relationship held up reasonably well unless a moth invasion occurred in the area.

## B. Insect Control

1. Mass Trapping. The use of pheromones for the suppression of insect populations has been the goal of many researchers, admin-

istrators, politicians, and so on. Each pseudosuccess is applauded as the forerunner of a powerful insect-control tool. Each pest situation is different, however, and may require the use of pheromones in many different ways. In New York we have shown for 3 years (Trammel et al., 1974) that mass trapping (100 traps/ha) in 25 hectares of apple can control the red-banded-leafroller moth. This method is not economical in New York, however, because of the number of other leafroller species that also must be controlled. Mass trapping could be of value, however, in situations where there are sprays applied for only one major pest. This would be found in the case of the codling moth, but, so far, research in Switzerland, New York, and Washington on the control of codling moths by mass trapping has not shown success. A dramatic success with mass trapping was registered in 1970, however, with the western pine beetle in California (Birch et al., 1974). In a massive cooperative experiment near Bass Lake, California, a test site of about 65 $km^2$ was surveyed with attractant traps on a 0.8-km grid. Two suppression and two check plots (2.56 $km^2$ each) were set up within the surveyed area. Large sticky traps (four vanes, each 0.76 by 2.0 m) were located on 161-m centers and baited with myrcene, brevicomin, and frontalin. The 242 suppression and 99 survey traps caught almost 900,000 beetles. This contributed to a substantial reduction in tree mortality and caused a precipitous decline in the beetle population, which remained very low for the next year as well. Another suppression effort in the next 2 years in northern California gave beetle captures of almost 7 million but no reduction in tree mortality. Apparently there were not enough traps for the larger number of trees and beetles in the area (10 times that occurring at Bass Lake). The tremendous amount of accumulated data from the experiments are being analyzed to develop models that can predict pest abundance and assist in determining the combination of treatment variables that should be applied to reduce a population of known size. The investigators feel that mass trapping has much promise in the overall picture because of their specificity and because "trap-out" methods can decrease aggregation capabilities of the beetles to the point where their damage is tolerable.

Other work (Pittman, personal communication) with bark beetle pheromones has also shown potential for their use in managing the pest populations. In the past two years about 7000 ha of white pine in northern Idaho have been trapped for the mountain pine beetle, *Dendroctonus ponderosae*. In each of the two years, sticky traps (about 0.3 x 1 m) baited with *trans*-verbenol and α-pinene (9:1) attracted enough beetles to cause over 50% reduction in tree mortality from the preceding year. In a different approach, Douglas fir beetles, D. pseudotsugae, were attracted to trap trees with frontalin, α-pinene, and camphene (4:1:1). The baited trees were then harvested along with the resident beetles.

The technique of using a pheromone/host combination was also employed with the cotton boll weevil. Pheromone traps or live males with glued wings placed in aldicarb-treated plots were found (Hardee et al., 1971; Lloyd et al., 1972) to be very effective in suppressing pest populations. The identification of the four component male-produced pheromones (grandlure) in 1969 intensified research efforts on the uses of this pheromone. The traps have been used widely in 13 states and 4 foreign countries for survey and have shown success in suppressing populations. It was found to be particularly effective in attracting both sexes in the spring. Thus the use of grandlure played a key role in the Pilot Boll Weevil Eradication Experiment in 1972 to 1974 (Birch et al., 1974). Fields were trapped with 5 traps/ha, and 3 to 5% of the field was planted early with an early-fruiting variety of cotton that was treated with insecticide and baited with grandlure. Although the overall result of the experiment indicated that the time had not yet come for massive and expensive eradication efforts, the pheromone efforts did look good. It appears that a pest management system utilizing pheromones as one component can be devised to give good control of the boll weevil at costs that are tolerable to society. The development of "in-field trapping" (Hardee, personal communication) throughout the summer adds even more impetus to this approach.

2. Mating Disruption. An entirely different approach to the use of pheromones is that of disrupting mating by permeating the atmosphere with pheromone and/or inhibitor chemicals. The pioneering work (Birch et al., 1974) was done primarily on cabbage looper and pink bollworm. With cabbage looper it was found that widely-spaced (up to 200 m apart) evaporator stations could be used as long as the total amount of pheromone released into the air was greater than 10 mg per hectare per night. It was also found that disruption was apparently density independent. With pink bollworm, hexalure (a moderately active parapheromone) was used in disruption experiments. At least 20 mg per hectare per night were needed to reduce male orientation to traps by 90%. In 1972 an experiment involving 4.8 hectares showed that larval infestation can be drastically reduced by evaporating 750 mg hexalure/ha/night from about 30,000 pieces of string distributed throughout the field. In 1973, stations spaced 20 m apart and releasing 20 mg hexalure/ha/night were effective in

reducing larval-boll infestations to the levels provided by commercial insecticide applications (Shorey *et al*., 1974). The newly identified pheromone (gossyplure) for pink bollworm will be used in 1974 in an expanded test using five evaporating stations/ha throughout 2000 hectares of cotton in California.

The oriental fruit moth was also successfully controlled by disruption in Australia (Rothschild, personal communication) in experiments conducted in 1973 and 1974. High release rates (20 mg per hectare per hour) of pheromone (*cis*-8-dodecenyl acetate) were released from tubes in each tree. Disruption of male orientation and a reduction in crop damage were obtained in New York (Taschenberg *et al*., 1974) with redbanded leaf-roller and grape berry moths by evaporating the pheromones, *cis*-11-tetradecenyl acetate and *cis*-9-dodecenyl acetate, from stations (ca. 80/ha) at rates above 100 mg per hour per station. In Switzerland evidence for disorientation of male codling moths and a reduction of fruit damage was obtained in an experiment (Charmillot *et al*., 1973) using evaporators (5 in a small orchard of 17 trees), each releasing 1.2 mg pheromone/day. Also in Switzerland (Arn, personal communication), evaporation of *cis*-8-dodecenyl acetate from small polyethylene tubes (1 every second tree) gave encouraging results in disorienting male plum fruit moths (*Grapholitha fumeferana*) and giving much reduced oviposition and larval counts compared to a check orchard.

A different approach to atmospheric permeation is that of dispensing the disruptant chemicals by spraying a microencapsulated formulation into the air. The complete coverage of thousands of pinpoint sources in the area could potentially be more effective as well as representing a more feasible method for commercial use. The USDA (Birch *et al*., 1974; Beroza, 1973; Beroza *et al*., 1973) initiated research with microencapsulated formulations on a number of insects. The most encouraging tests are those conducted with the gypsy moth (Beroza and Knipling, 1972; Cameron, 1973; Leonard, 1974). In one experiment (Beroza, personal communication) in Massachusetts (60 $km^2$) the pheromone (disparlure) was distributed in microencapsulated form by aircraft at the rate of 5 g/ha. A check area of similar size was located 10 km from the treated area. The results showed almost complete disorientation of males to lure and live female-monitoring traps, excellent mating suppression at low population levels, and highly significant reductions in egg-mass counts in the treated area. In Pennsylvania tests (Cameron *et al*., 1974) on 16-ha plots sprayed with various rates of microencapsulated disparlure showed that rates of 2.5 to 15.0 g/ha are capable of reducing successful mating of gypsy moths for up to 6 weeks. It was suggested that 15 g/ha might be capable of disrupting mating in incipient populations efficiently enough to keep the population from increasing.

In Washington the USDA (Moffitt, personal communication) conducted tests with the microencapsulated pheromone of codling moth. Aerial application of 7 g/ha in a 0.4-ha block of apples gave a 93% reduction in male response to attractant traps for 6 days. A second test showed over 95% reduction in males trapped for 7 days and a second application made on the eighth day gave 100% reduction in response for 6 more days. Fruit damage and larval counts were substantially lower in the treated plot compared to a check plot. In tests with microencapsulated inhibitor (*trans*-8,*trans*-10-dodecadienyl acetate - inhibits male responses to the pheromone traps), the aerial application of 15 g/ha gave 100% reduction in male response to monitor traps in the first 2 days but only 69% reduction in the third.

The USDA (Klun *et al.*, 1973a; Chapman *et al.*, 1973) also conducted tests with microencapsulated materials on European corn borer and redbanded leafroller moths in Iowa. Two rows of corn were sprayed with microencapsulated *trans*-11-tetradecenyl acetate (part of the pheromone system of both species) or 11-tetradecenyl acetate. Monitoring traps were placed in a row located between the sprayed rows. Male response to the traps in the treated area compared to check traps was reduced by more than 90% with both species with the *trans* compound applied at rates of 25 g/ha and greater, whereas the alkyne gave 95% inhibition at 50 g/ha with redbanded but only 81% inhibition at that rate with European corn borer.

The encouragement given the disruption technique for insect control has underscored the need for more basic research in this area. Much more information must be obtained on formulation, application rates, longevity of effectiveness, as well as on more basic questions of how disruption is actually working. The roles of confusion and habituation are not understood. It is also not known whether the natural pheromone blends are more desirable to use than just one component (either the primary attractant or one of the synergists), an unnatural combination of all components, or an inhibitor. Answers may be different for each species depending on the female's mating behavior and location, the male's behavioral responses to her, and the variables affecting mating behavior in the field (Shorey, 1973).

In summary, the availability of pheromone systems for many insect species has given pest management programs a very sensitive and specific monitoring tool. Preliminary research with pheromones for insect control has been encouraging. Intensive research efforts in the next few years could provide the know-how for using pheromones in certain insect suppression programs. To be used commercially in the states, however, they must be registered as insecticides. To date only one pheromone, Muscamone® (cis-9-tricosene, a housefly pheromone used in sugar-insecticide baits), has been registered by the EPA. Toxicological data has been accumulated on a number of others, although the government may have to play a larger role in pushing registration on pheromones for which there is no patent protection for commercial companies. If the pheromones are used in traps for control, the registration requirements may not be as stringent - Zoecon has already received an experimental EPA permit to use codling moth pheromone traps for control in some experiments, and the American Can Co. was granted permission by New York State to sell gypsy moth traps for control. Other problems, such as chemical expense and availability, should resolve themselves if efficacy is shown and the resulting market becomes profitable.

Research is continuing around the world on sex pheromones of many important species not mentioned in this paper as well as on other types of pheromones, such as alarm and trail pheromones, that have potential practical applications. It appears that we have barely opened the door on this "stimulating" field of insect pheromones.

## REFERENCES


1. Batiste, W. C., Berlowitz, A., Olson, W. H., Detar, J. E., and Joos, J. L. 1973. Environ. Entomol. 2, 387-391.

2. Beroza, M. 1973. Abstr. 3rd Int. Symp. Chem. Toxicol. Aspects Environ. Qual., Tokyo, Japan, pp. 131-132.

3. Beroza, M. and Knipling, E. F. 1972. Science 177, 19.

4. Beroza, M., Stevens, L. J., Bierl, B. A., Phillips, F. M., and Tardif, J. G. R. 1973. Environ. Entomol. 2, 1051.

5. Birch, M. C., Ed. 1974. Pheromones, Elsevier, New York, 495 pp.

6. Birch, M., Trammel, K., Shorey, H., Gaston, L., Hardee, D., Wood, D., Burkholder, W., and Müller-Schwarze, D. 1974. In Phermones, M. Birch, Ed., Elsevier, New York, pp. 411-461.

7. Brady, U. E. 1973. Life Sci. 13, 227.

8. Brady, U. E. and Nordlund, D. 1971. Life Sci. 10 (Part II), 797.

9. Brann, J. L. and Tette, J. P. 1974. New York State Apple Pest Management Project. First Ann. Report.

10. Carde, R. T., Kochansky, J., Stimmel, J. F., Wheeler, A. G., and Roelofs, W. L. 1975. Environ. Entomol. 4, 413-414.

11. Cameron, E. A. 1973. Bull. Entomol. Soc. Amer. 19, 15.

12. Cameron, E. A., Schwalbe, C. P., Beroza, M., and Knipling, E. F. 1974. Science 183, 972.

13. Chapman, D. L., Klun, J. A., Mattes, K. C., and Barry, M. A. 1973. In New Methods in Environmental Chemistry and Toxicology, F. Coulston, F. Korte, and M. Goto, Eds., Intl. Acad. Printing, Totsuka, Tokyo, pp. 163-168.

14. Charmillot, P. J., Baggiolini, M., and Arn, H. 1973. O.I.L.B., Reunion Groupe Carpocapse, Wädenswil, November.

15. Evans, D. A. and Green, C. L. 1973. Chem. Soc. Rev. 2, 75.

16. Eves, J. D. and Chandler, J. D. 1973. Abstr. 48th Ann. Western Coop. Spray Proj., Portland, Oregon, p. 16.

17. Hardee, D. D., Lindig, O. H., and Davich, T. B. 1971. J. Econ. Entomol. 64, 928.

18. Hoyt, S. C. 1974. In Integrated Pest Management Newsletter #4, U. S. International Biological Program, p. 11.

19. Hummel, H. E., Gaston, L. K., Shorey, H. H., Kaae, R. S., Bryne, R. J., and Silverstein, R. M. 1973. Science 181, 873.

20. Jacobson, M. 1972. Insect Sex Pheromones, Academic, New York.

21. Klun, J. A., Beroza, M., Mattes, K. C., Showers, W. B., Berry, E. C., Robinson, J. V., and Barry, M. W. 1973a. Abstr. Natl. Entomol. Soc. Amer. Meeting, Paper #54, Dallas, Texas.

22. Klun, J. A., Chapman, O. L., Mattes, K. C., Wojtkonski, P. W., Beroza, M., and Sonnet, P. E. 1973b. Science 181, 661.

23. Lanier, G. N., Birch, M. C., Schmitz, R. F., and Furniss, M. M. 1972. Canad. Entomol. 104, 1917.

24. Law, J. H. and Regnier, F. E. 1971. Ann. Rev. Biochem. 40, 533.

25. Leonard, D. E. 1974. Ann. Rev. Entomol. 19, 197.

26. Lloyd, E. P., Scott, W. P., Shaunak, K. K., Tingle, F. C., and Davich, T. B. 1972. J. Econ. Entomol. 65, 1144.

27. MacConnell, J. G. and Silverstein, R. M. 1973. Angew. Chem. (Int. Ed.), 12, 644.

28. Madsen, H. F. and Vakenti, J. M. 1972. Environ. Entomol. 1, 554.

29. Madsen, H. F. and Vakenti, J. M. 1973. Environ. Entomol. 2, 677.

30. Miller, C. A. and McDougall, G. A. 1973. Can. J. Zool. 51, 853.

31. Minks, A. 1974. Z. Angew. Entomol., in press.

32. Müller-Schwarze, D., Müller-Schwarze, C., Singer, A. G., and Silverstein, R. M. 1974. Science 183, 860.

33. Priesner, E. 1973. Fortschritte der Zoologie 22, 49.

34. Quist, J. A. 1973. Insect Population Management of Apple Orchards, unpublished report.

35. Riedl, H., Croft, B. A., and Howitt, A. J. 1974. Use of a Synthetic Sex Pheromone for Estimating Seasonal Densities of the Codling Moth, *Laspeyresia pomonella* (Lepidoptera: Tortricidae), in preparation.

36. Roelofs, W. and Carde, R. 1974. In Pheromones, M. Birch, Ed., Elsevier, New York, p. 96.

37. Roelofs, W. L., Carde, R. T., Bartell, R. J. and Tierney, P. G. 1972. Environ. Entomol. 1, 606.

38. Roelofs, W. L. and Comeau, A. 1971. In Chemical Releasers in Insects Pesticide Chemistry, Vol. III, A. S. Tahori, Ed., Gordon and Breach, New York, pp. 91-112.

39. Roelofs, W., Comeau, A., Hill, A., and Milicevic, G. 1971. Science 174, 297.

40. Shaw, J. G., Moreno, D. S., and Furgerlund, J. 1971. J. Econ. Entomol. 64, 1305.

41. Shorey, H. H. 1973. Ann. Rev. Entomol. 18, 349.

42. Shorey, H. H., Kaae, R. S., and Gaston, L. K. 1974. J. Econ. Entomol. 67, 347-350.

43. Sower, L. L., Vick, K. W., and Tumlinson, J. H. 1974. Environ. Entomol. 3, 120.

44. Tamaki, Y., Noguchi, H., and Yushima, T. 1973. Appl. Ent. Zool. 8, 200.

45. Taschenberg, E. F., Carde, R., and Roelofs, W. 1974. Environ. Entomol. 3, 239-242.

46. Trammel, K. 1973. Abstr. 3rd Int. Symp. Chem. Toxicol. Aspects Environ. Qual., Tokyo, Japan, pp. 134-136.

47. Trammel, K., Roelofs, W. L., and Glass, E. H. 1974. J. Econ. Entomol. 67, 159-164.

48. Vite, J. P., Blake, A., and Renwick, J. A. A. 1972. Canad. Entomol. 104, 1967.

49. Wood, D. L. 1970. In Control of Insect Behavior by Natural Products, D. L. Wood, R. M. Silverstein, and M. Nakajima, Eds., Academic, New York, pp. 301-316.

50. Young, J. C., Brownlee, R. G., Rodin, J. O., Hildebrand, D. N., Silverstein, R. M., Wood, D. L., Birch, M. C., and Browne, L. E. 1973. J. Insect Physiol. 19, 1615.

## DISCUSSION

The first question in the discussion of Dr. Roelofs' paper was concerned with patent policies and its relation to the generation of interest among chemical companies to manufacture insect pheromones. Since insect sec pheromones are natural products, it has been stated that they are not patentable. Further, since the principle underlining the mating disruption process by the use of pheromones was published some years ago, this technique for insect control also is considered to be public property. In light of this, how does one persuade a chemical company to develop pheromones on a large scale without patent protection? Dr. Roelofs responded by stating that he did not believe the statements regarding patent policy to be entirely accurate. Natural products have been patented in the past, and there is a good chance that patents also may be secured for pheromones. It was pointed out that pecularities in the identification procedure sometimes may be exploited to secure a patent. For example, the electroantennogram method that Dr. Roelofs uses for the identification of sex pheromones is regarded in some quarters as being inadequate for absolute structure identification. Because of this, it has been possible to secure patents of active pheromones that in all probability are natural products. It was also pointed out that use patents may be obtained even though the disruption method for control may or may not be patentable.

Owing to the demonstrated use with which the electroantennogram method has been used to identify natural sex pheromones, the question was asked if this technique could be applied to identify the American cockroach attractant. Unfortunately, this attractant appears to be significantly different in structure from the compounds that Dr. Roelofs has been working with, and therefore, the electroantennogram method probably would not be useful.

In his position paper, Dr. Roelofs mentioned that certain compounds inhibited pheromone activity, whereas other stimulated it. Does this mean that there are two sensory neurons, one that behaves as an excitatory neuron and the other as an inhibitory neuron, and further, do the inhibitory agents stimulate inhibitory neurons or do the inhibitory agents inhibit the excitatory neurons? The response was that the mechanism is probably much more complex than this. For example, the red-banded leafroller is attracted to a pheromone that consists of 93% cis-isomer and 7% trans-isomer. In this case the trans-isomer appears to be an obligatory component of the attractant. However, when the amount of trans-isomer is raised to 15%, this isomer inhibits attractancy. This means that certain compounds may be both inhibitory and obligatory components of a pheromone, and whether it behaves as an inhibitor depends on its concentration. Dr. Roelofs stated, however, that evidence exists that indicates the presence of single cell response to each component.

Along this same vein the question was raised as to whether laboratory synthesis of a pheromone could present a problem owing to the possibility of the formation of impurities that could act as inhibitors of the attractant. Evidently this is not a problem since <u>cis</u>-isomers or <u>trans</u>-isomers of attractants have been prepared in 99% purity. Although purity is quite important for maximum activity, attainment of pure compounds is not difficult. When pheromones are used in a pest-management program for monitoring or for actual control, it is necessary that the investigator establish prior to this point exactly what blends or ratios of isomers are most efficient and use only those proportions in the program.

Attention was given to the use of attractants as baits, particularly with respect to the methods used in formulating baits. In the monitoring programs conducted by Dr. Roelofs, polyethylene vials are used to hold pheromones that happen to be acetates, and a rubber septum is used for aldehydes and alcohols. Different concentrations and ratios of attractants are incorporated into these cap systems. These caps will be effective for some period, for example, they can be left in the field over the winter and will still be attracting insects the next summer. After this period, however, the caps are somewhat less efficient. Full effectiveness usually is attained for 4 to 6 weeks. Zoecon Corporation has developed a new polymer formulation and extrusion method for incorporation of the codling moth attractant. By this procedure, long extrusion polymers containing the attractant are obtained that may simply be cut into small sections. These pieces have been surprising in that they are much more effective than any other release method that is available. It is active at a very high level for at least 12 weeks, possibly longer.

Discussion followed on the feasibility of using microencapsulated pheromone in the disruption method for insect control. This point was raised because the Riverside pheromone group had been devoting enormous manpower to the fabrication of pheromone dispensing stations for use in the control of the pink bollworm. Evidently, microencapsulation is a relatively new method of formulating pheromones and was first used by the USDA last year. This was accomplished with the aid of the National Cash Register Company, which is one of the two or three companies prepared to offer this kind of service. Since the

Riverside program for the control of the pink bollworm is based almost entirely on dispensing stations as the source of pheromones, it would be unwise at this time to convert to a different method, particularly in view of the magnitude of the project. Preliminary work with microencapsulated pheromone is mandatory before a large-scale field program is undertaken. There is little doubt that the microencapsulation method will work, and the Riverside group is preparing to investigate this possibility in the near future.

The question of resistance development to pheromones was discussed. Is it possible for this to occur? It was agreed that there is always a possibility for resistance development, and investigators working with pheromones are aware of this. However, proof of the existence of resistant insects may be a problem. Mass trapping is probably the place to select for resistant populations, but mass trapping is usually used where low populations are present; therefore, there is not much of a selective material to choose from. On the other hand, if the microencapsulated method is used, that is, where the insect population is bombarded with the attractant, then it is necessary for the insect to change completely to another type of chemical to overcome the effects of habituation, since it is not necessary to use the exact blend or compound to habituate them. If the males do not respond, they should not be able to find the female. It should be added that the problems associated with resistance development to pheromones are matters of considerable debate at the present time.

With respect to the possible use of pheromones to control insect vectors of public health importance, it is unlikely that these materials will become important tools. This is probably true for mosquitoes and the tsetse fly, since most flies and mosquitoes do not use a long-range type of communication system. It may be possible, however, to use a close-range stimulant in the field that may overcome the close-range stimulant actually used by the flies. It was pointed out that the attraction of triatomas to human beings and other mammals is probably not a case of pheromone activity but may be a case of allomone activity. This is an area of considerable research at the present time. For example, mustard oils are currently being used in a variety of ways in the state of New York to determine their usefulness in controlling the cabbage maggot.

In light of the remarkable work done with the tsetse fly with respect to sight and attraction, the question was asked as to why so much effort was being given to the sense of smell rather than to other senses. Dr. Roelofs replied that work with the other senses is being conducted, such as sound and sight. Difference species of mosquitoes, for example, emit different sounds that perhaps can be exploited, although further work is required to define the significance of the specific sounds. Sight has been used in relation to monitoring the apple maggot. This pest does not possess a long-range chemical communication system that can be utilized, but sight has proved useful and yellow panels representing foliage and red spheres giving the image of applies have been used for attraction. They are, however, relatively poor tools for attraction and the results have been less than satisfactory. Additional work in this area is appropriate.

# PERSPECTIVES

# Historical Aspects of Insecticide Development

A. M. BOYCE
University of California
Riverside, California

## I. INTRODUCTION

The major role played by pesticides in the great advances made in agricultural technology and public health is widely recognized. In order to meet the enormous needs for food and fiber in the future, heavy dependence on pesticides will be necessary for more effective control of insects and other pest arthropods, microorganisms, nematodes, weeds, and vertebrate pests. Insecticides will also be very essential for continuing control and/or eradication of the many arthropods of medical importance. In the United States it is estimated that losses resulting from the various pests in farm and forest production and from insects in the home and various structures range from \$10 to \$15 billion per year. Here the estimated annual loss from insects alone ranges between \$4 and \$7 billion. Based on the history of pesticide development during the past 25-odd years, it is confidently expected that compounds--in addition to those available today--will be developed that become a vital part of pest management systems designed to greatly reduce these crop losses without any, or with at least tolerable, adverse effects on environmental quality.

In view of the nature of this conference, this chapter focuses mostly on historical aspects of development of insecticides.*

Extensive personal activity during the period 1926-1950 in field evaluation† on a number of agricultural crops of many of

---

*The Environmental Protection Agency that administers food and drug laws and regulations considers as insecticides those compounds used in the control of insects, spiders, mites, ticks, centipedes, millipedes, and wood lice.

†In the absence of laboratory colonies of the species of insects and mites under study, reasonably precise small-scale spray and dust application methods were developed and used to enable an

the older standard insecticides for potential new uses and also many new compounds, and close administrative association with research of this nature during the remainder of this author's active career as a faculty member of the University of California provided opportunity to participate in, and later to observe, many interesting aspects of insecticide development.

## II. THE FIRST INSECTICIDES AND LATER DEVELOPMENTS*

Prior to the development of Paris green (copper acetoarsenite) in 1867 for the control of the Colorado potato beetle (Leptinotarsa decemlineata) and other chewing insects and the development of kerosene emulsions in 1868 for control of sucking insects, a wide variety of materials were promoted as insecticides. These included soapsuds, turpentine, petroleum oils, fish oils, sulfur, whitewash, lime, lye, brine, wood ashes, hot water, vinegar, herbs, pepper, aloes, tobacco, soot, hellebore, and quassia.

From about 1890 to the 1920s the principal insecticides were sulfur (applied as dust), lime sulfur and wettable sulfur, arsenicals (Paris green, lead and calcium arsenates), plant products, nicotine, pyrethrum, rotenone, petroleum oils of varying composition, whale oil, resins, soaps, carbon disulfide, and hydrocyanic acid as fumigants.

From the 1920s to the early 1940s, entomologists relied principally on lead arsenate (both acid and basic), calcium arsenate, barium fluosilicate, cryolite (sodium fluoaluminate, both natural and synthetic), sulfur dust, wettable sulfur and lime sulfur, a selenium compound (Selocide®, potassium ammonium selenosulfide) for mite control, several thiocyanates, improved preparations of nicotine, pyrethrum, rotenone, petroleum oil, hydrocyanic acid, methyl bromide, ethylene dichloride, and ethylene oxide as fumigants. And, toward the end of that period tartar emetic was used for control of thrips of citrus and

---

†(cont.) approximate evaluation of new compounds on populations in the field. Despite limitations for accurate determinations of dosage-mortality relationships, this field screening procedure did sort out promising compounds for more extensive study; at the same time opportunity was provided to observe injury potential of the compounds to foliage and fruits.

*For a chronologically arranged history of insecticides and some related economic poisons, see Shephard (1951), Martin (1964), and Metcalf (1955).

several other crops; dinitro-o-cresol and dinitro-o-cyclohexylphenol were used in various formulations for dormant spray treatment of deciduous fruit trees for mites, aphids, and certain scale insects. The latter compound was also employed in dust and spray treatments for control of mites on citrus.

The birth of modern insecticides occurred during World War II, beginning with DDT (dichlorodiphenyltrichloroethane) which opened up a whole new vista and philosophy for development of synthetic organic compounds as insecticides. The history of organochlorine, organophosphorus, and carbamate insecticides, and the bridged diphenyl group of acaricides is so well-known and documented that it will not be discussed here.* There is still a great deal of research in progress on these four groups of compounds, with a number of new and promising insecticide candidates being synthesized and evaluated. In addition, there are promising results from research on other groups of compounds such as organotins, pyrethroids, rotenoids, and juvenile hormone mimics. These are discussed elsewhere in this book.

Extensive field experience with DDT and certain other organochlorine compounds strongly emphasized the importance of selectivity and biodegradability in the development of new insecticides, especially from the very important viewpoint of environmental quality, and also in avoiding serious disruptions in field populations of beneficial parasitic and predatory insects.

## III. EARLY PROPRIETARY PRODUCTS AND COMMERCIAL ENTOMOLOGISTS

In the United States until the late 1930s many university, state agricultural experiment station, and federal workers in entomology, as well as persons in other fields, were still skeptical about the ethics of workers in commercial companies and about claims for their proprietary products offered for sale to the farmers and others. The following discussion and excerpt from Howard's History of Applied Entomology reflects the generally extreme view on this matter during the early days, that is, the 1860s (Howard, 1930).

Until the beginning of the use of arsenical poisons for gnawing insects and diluted kerosene emulsions for sucking insects, the effective insecticides used were limited in number, and, after all, were not especially effective. Decoctions of

---

*See footnote on page 470 for references on history of insecticides.

quassia chips, hellebore, limewater, and mixtures of ashes, and several other things, together with what seems to be an early use of nicotine fumes and tobacco water for plant lice, comprised about all. The charlatan was present in the old days, and secret nostrums and utterly ineffective things were recommended and sold. There is no more interesting reading than the fulminations of B. D. Walsh against things of that kind. He had an extraordinary command of language, and poured forth his wrath unstintingly and in wonderful phraseology on charlatanistic claims.

In fact, the advertising of quack remedies went to such an extent that the real entomologists were inclined to frown upon the whole idea of chemical insecticides. So strong was their feeling that it was expressed very forcibly in the opening editorial in the first number of The Practical Entomologist (October 30, 1865). In view of the enormous use of insecticides of great value at the present time, it is worthwhile to quote two sentences from the editorial:

> "The agricultural journals have from year to year, presented through their columns, various recipes, as preventative of the attacks, or destructive to the life, of the "curculio," the "apple-moth," the "squash bug," etc. The proposed decoctions and washes we are well satisfied, in the majority of instances, are as useless in application as they are ridiculous in composition, and if the work of destroying insects is to be accomplished satisfactorily, we feel confident that it will have to be the result of no chemical preparations, but of simple means, directed by a knowledge of the history and habits of the depredators."

Probably the last major vestige of the indicated somewhat critical attitude of "official entomologists," that is, university, state, and federal workers toward "commercial entomologists" vanished in the early 1940s largely as the result of fine working relations and rapport between the two groups at the Federal Orlando (Florida) Laboratory during World War II. All industrial groups were urged to submit their candidate compounds to the laboratory for evaluations as insecticides and repellents for potential use by the Armed Forces. Many thousands of compounds were evaluated, and a number of them were very useful in the war effort, DDT being one prime example.

## IV. RESEARCH AND DEVELOPMENT BY CHEMICAL COMPANIES

During the early 1920s and later, a number of chemical companies showed interest in developing new insecticides and fungicides by the establishment of screening programs to evaluate possibility of pesticidal action of the many new compounds synthesized in their chemical research programs. The increasing difficulties encountered with lead arsenate in control of codling moth (Laspeyresia pomonella) on apples and pears was an important factor influencing the decision to search for new insecticides. With a number of the major chemical companies, pesticide research and development ultimately became a relatively large segment of their total effort, involving highly qualified chemists in the requisite specialties, economic entomologists, and other scientists needed for a full-blown program in agricultural and veterinary medical programs.

Prior to about 1950, the following companies, listed alphabetically, had developed such programs, each of which was very effective: American Cyanamid, California Spray Chemical (later Chevron), Chemagro, Dow, DuPont, Geigy, Monsanto, Naugatuck, Niagara (later FMC), PennSalt, Rohm and Haas, Shell, Sherwin Williams, Stauffer, Union Carbide, and Velsicol. Subsequently, many other companies initiated major research and development programs on pesticides.

Although a few chemical companies abroad were interested in pesticide research and development prior to World War II, during and after the war a number of major companies developed great interest and expertise in this field. Their efforts have been productive, and in many instances useful in meeting pesticide needs in the United States. Within this country the best known foreign companies are Bayer, Ciba-Geigy, Imperial Chemical Industries, Montecatini, Phillips-Duphar, Shell, and Sumitomo.

In general, the industrial development programs on pesticides heavily involved state and federal workers in cooperative work that was often supported by direct grants from the concerned companies to state and federal workers for specific aspects of the program. Much of the field evaluation was conducted in cooperation with commercial farmers. This general procedure was changed somewhat about 15 years ago when Food and Drug Administration (FDA) regulations prevented marketing of any agricultural produce that had been treated experimentally with unregistered pesticides. This FDA policy prompted a number of the major chemical companies to develop their own experimental farms where the needed crops are grown and subsequently destroyed on completion of the pesticide experiments. It is noteworthy, however, that several of the major companies had developed experimental farms prior to the FDA regulation. In some instances, field evaluation is still done by agricultural experiment station personnel on station property and on commercial farms. In the latter case, the usual procedure is for the particular chemical company to fully compensate the farmer for the experimentally treated crops that are destroyed.

It is a fortunate circumstance that chemical industries in this country and abroad have brought to bear their enormous resources on plant and animal protection and public health on a worldwide basis. Estimates indicate that approximately 100,000 synthetic organic compounds are screened for pesticidal activity each year by chemical industries. Probably over 95% of the present pesticidal chemicals in use, or currently under study, were first synthesized by the chemical industries.

## V. RESEARCH AND GRADUATE TRAINING IN INSECTICIDE DEVELOPMENT PRIOR TO WORLD WAR II

Almost from the beginning of the state agricultural experiment stations (Hatch Act in 1887) in the United States, concerned staff members conducted field research to evaluate existing insecticides and to formulate certain modifications of them for control of crop pests. Later, as promising new compounds became available from chemical industries, they evaluated them in the field. However, it was not until the late 1940s that many universities began to undertake research and graduate training in chemical areas oriented around development of insecticides. As expected, most of these institutions are state land grant universities with a college of agriculture and an agricultural experiment station, and in almost all instances the chemical activity is within the existing departments of entomology. The universities that offered these programs prior to World War II are listed alphabetically, without reference to when their program was initiated: Cornell, Illinois, Iowa State, Minnesota, New Hampshire (Crop Protection Institute), New Jersey, Ohio State, Oregon State, Pennsylvania State, and the University of California at Berkeley.

The U. S. Department of Agriculture had chemical research on insecticide development in progress early in the 1900s. Ultimately, this became the largest program in the country; however, emphasis in this particular area was reduced considerably about 10 years ago.

The Boyce Thompson Institute, a private research organization, developed a progressive insecticide research program in the 1920s.

During this same time period, that is, prior to World War II, there was important research and, in some instances, graduate training in insecticide development in Canada and Europe.

In Canada the principal institutions were Science Service Laboratory, Canada Agriculture (now Research) Institute; Macdonald College at McGill University; University of Guelph; and at several of the federal field stations.

In Europe the principal institutions were in England, namely: Rothamsted Experiment Station, Imperial College of Science and Technology, Cambridge University, Wye College of the University of London, and the Agricultural and Horticultural Research Station at the University of Bristol, now the Long Ashton Research Station.

In the Netherlands, the principal institution was the University of Utrecht.

Most major departments of entomology or other administrative units in which entomological work is conducted in the United States, Canada, and other countries, as well as federal agencies in many countries, now have chemical and biological research in progress on insecticide development.

It is gratifying to mention that a number of developing countries are sending graduate students abroad for advanced training in insect toxicology and physiology. A relatively large number of these students are now in the United States.

## VI. RESISTANCE TO INSECTICIDES

The development of resistance by insects and mites to insecticides is perhaps the most serious problem confronting economic entomologists and insect toxicologists. The first record of this phenomenon was the resistance of San José scale (Quadraspidiotus perniciosus) to lime sulfur on apples in 1914, as observed by Melander (1914), followed by discovery by Quayle (1916) that the California red scale (Aonidiella aurantii) and the black scale (Saissetia oleae) insects on citrus in California were resistant to fumigation with hydrocyanic acid gas. Hough (1928) found increasing evidence that codling moth (Laspeyresia pomonella) was developing resistance to lead arsenate. At that time and for years thereafter, there was considerable doubt and actual controversy among entomologists and other specialists in biology about the existence of the resistance phenomenon.

The resistance of citrus thrips (Scirtothrips citri) on citrus in California to tartar emetic was very dramatic, beginning in 1941, about 4 years after the treatment was introduced and its use became widespread (Boyce *et al.*, 1942). Fortunately, DDT, which became available commercially in 1946, was an excellent substitute for tartar emetic. However, citrus thrips developed resistance to DDT within 3 years after its use became widespread (Ewart *et al.*, 1952); it developed resistance to dieldrin after 2 years of general usage (Ewart, 1974).

The resistance of houseflies (*Musca domestica*) and mosquitoes (*Culex pipiens autogenicus*) to DDT was first observed in Italy in 1947 and of houseflies in the United States in 1948 (Babers, 1949).* The classic example to the layman is the resistance of houseflies to DDT. Many people can recall the initial spectacular results obtained by simply applying DDT on the screens of the home in springtime and having protection from flies throughout the entire summer. During the past 25 years of widespread usage of the new synthetic organic insecticides, many species of insects and mites have become resistant to a number of them. At present the total number of resistant species exceeds 225 (March, 1974). Included in this group are a number of species of vectors and other species of public health importance.

This resistance phenomenon involves a number of organochlorine, organophosphorus, and carbamate compounds. There is laboratory evidence now that insects have the capacity to develop resistance to juvenile hormone mimics (Cerf *et al.*, 1972; Georghiou *et al.*, 1974). An added complexity of the resistance matter is cross-resistance and multiple resistance.

The disturbing prospect in regard to resistance in many species of insects and mites is that the development and commercialization of new compounds may not be rapid enough to keep ahead of the resistance phenomenon. The starkest reality of resistance, however, is the effect it can have on negative decisions by management of chemical industries to make the very large investment required to develop a new insecticide. In discussing research and development costs of pesticides Hunter (1974) states, "According to the National Agricultural Chemicals Association in 1970, it took $5.5 million, 6.6 years, and testing of 7430 compounds to get a pesticide on the market." It appears probably that these data are approximate averages involving many chemical companies because Hunter also states "A survey of three companies in 1970 showed that these costs (R & D) were estimated to be from $5 to $10 million for a major product."

---

*For review of resistance in insects of medical importance see Brown and Pal (1971). For review of resistance in insects of agriculture see Anonymous (1970).

## VII. REGULATION OF INSECTICIDES

There has always been concern about safety to the consumer of foods that have been treated with insecticides and other chemicals, either pre- or postharvest, or both. The increasing stringency of laws and regulations governing pesticides, together with increasingly vigorous enforcement, has had a significant impact on insecticide development. The many difficulties encountered in development and use of new insecticides has been a stimulus to search for more effective compounds that produce no problems of safety to the consumer of the treated product, nor adverse effects on environmental quality.

Although many countries now have laws and enforcement mechanisms to regulate pesticides, perhaps the United States has the most elaborate program in this regard.

Since 1906 the United States government by Congressional action has afforded protection of the safety of the country's food supply through the following legislative actions: in 1906 the original Food and Drug Act and the Meat Inspection Act; in 1910, the Insecticide Act; in 1938, the Food, Drug, and Cosmetic Act; in 1947; the Insecticide, Fungicide, and Rodenticide Act; in 1954, the Miller Pesticide Residue Amendment to the Food, Drug, and Cosmetic Act; in 1957, the Poultry Inspection Act; in 1958, the Food Additives Amendment; and a number of subsequent acts and/or amendments.

The Miller Amendment of 1954 is regarded to be the most progressive legislation yet developed to provide for the safe use of pesticide chemicals. The principle of pretesting for safety before a chemical can be marketed is one of the most important aspects of the legislation. In other words, a chemical cannot be used legally in the production of a crop or for public health purposes until its safety has been established. In order to meet these new requirements in the development of pesticides, it was necessary to rapidly develop the relatively new field of residue chemistry. Since the original pesticide chemical usually penetrates plant or animal tissue at least to some degree, migrates, and degrades or changes, it is necessary to detect, characterize, and measure accurately each of the products that result from the compound after its original application. Great progress has been made in residue detection methodology, as evidenced by the fact that with certain compounds their detection is possible in amounts down to 0.001 ppm and, in some instances, even smaller amounts.

Added to the increasing complexity in developing pesticides is the Delaney Cancer Clause in the Food Additives Amendment of 1958. According to this Cancer Clause and subsequent interpretations of it, a compound cannot be used in producing food crops or as a food additive that has been shown to induce cancer in test animals when administered at any dosage.

Some of the history of insecticide use that led to early regulatory action will be mentioned.

As early as the 1880s, several American entomologists expressed concern about the safety of residues of arsenic on apples sprayed with Paris green and London purple. Following the introduction of lead arsenate in 1892, it became widely used for control of codling moth on apples and pears. The arsenic situation reached a very critical stage in the early 1900s when over 6000 people in England suffered arsenic poisoning, with many deaths, from drinking beer that contained traces of arsenic. The source of arsenic was from the acid used to hydrolyze starch. This led to establishment in 1903, by the British Royal Commission on Arsenical Poisoning, of the tolerance of 0.01 grain of arsenic trioxide per pound of solid food (1.43 ppm) (Shepard, 1951a). Later this tolerance was regarded as the unofficial international tolerance. In 1919 a shipment of pears was embargoed by the Boston Health Department because of excessive arsenic residue. There were accounts in the British press in 1925 about illnesses due to eating American apples. British authorities informed United States Government officials that no fruit would be accepted unless it conformed with this official tolerance of 1.43 ppm. The U. S. Bureau of Chemistry, then responsible for enforcing the Federal Insecticide Act of 1910, had already seized a number of shipments of apples in 1925 because the residue exceeded the British tolerance. The situation became chaotic. At the annual meeting of the American Association of Economic Entomologists held at Philadelphia in December, 1926, the lead arsenate residue situation on apples and pears was featured. Many of the leading entomologists vigorously opposed restrictive legislation on arsenical residues on the basis of private opinion that such residues were harmless to the health of consumers. Despite much controversy, the Food and Drug Administration established in 1927 a tolerance of 0.025 gr/lb (3.5 ppm) for arsenic on all fruits and vegetables. This tolerance was modified several times on the basis of new information. Its lowest level was 0.01 gr/1g (1.43 ppm) in 1932, and it was 0.025 gr/lb (3.5 ppm) in 1940 (Shepard, 1951a). The current lead arsenate tolerance on apples and pears is 7 ppm.

Pesticide regulations can serve as a nontariff trade barrier in foreign trade. Since 1959 this particular matter has been of concern to a number of countries, including the United States, that traditionally have had exports of fruits to countries of western Europe. The problem for exports from the United States has been mostly with fresh citrus fruits, dried deciduous fruits, such as apricots and peaches, and others. Initially, the problem began in 1959 in West Germany. It concerned the decay control materials used postharvest on imported citrus fruits, namely, biphenyl and sodium orthophenylphenate and sulfur dioxide residues on dried fruits. Practically all exports of citrus fruits from California contained residues of biphenyl, and many also had residues of orthophenylphenate. All shipments were well within United States tolerances, which were biphenyl, 110 ppm, and sodium orthophenylphenate, 10 ppm. The shipments of dried fruits contained nominal residues of sulfur dioxide used for control of fungi and to preserve natural color of the fruits, which the U. S. Food and Drug Administration has classified as GRAS, that is, generally regarded as safe.

Although a number of examples could be mentioned in which pesticide regulations can serve as nontariff trade barriers, only the biphenyl one will be used to illustrate the point. After extensive negotiations between the United States and West Germany, a tolerance of 70 ppm biphenyl was established (the United States tolerance was 110 ppm) with an imposed labeling requirement that a label must be affixed to cartons, boxes, wrappings, and the like, containing fruits treated with biphenyl, for example, "Treated with biphenyl, peel not suitable for eating." This labeling requirement resulted in substantial sales resistance on the part of many consumers. The most important adverse effect of the required labeling, however, is the use of a voluntary labeling plan by certain competitive shippers of citrus produced in countries of Europe and also in countries close enough to Europe that transit time for shipment is so short that decay control materials are not needed. Examples of this voluntary labeling are" "Not treated with biphenyl or other poisons"; "No chemicals used"; and "Untreated." The German housewife can be greatly influenced by the voluntary label on oranges, grapefruit, and lemons, especially when they are displayed for sale adjoining fruits with the required biphenyl label from the United States, Israel, and South Africa.

For over a decade, each of the countries of the European Economic Community (EEC), (The Common Market), has been developing its own pesticide regulatory program including tolerances for pesticide residues. Most of the pesticides are insecticides. There is substantial variation among the countries in the adopted and/or proposed tolerances for the same pesticides. In general, those tolerances tend to be lower than tolerances in effect in the United States. The goal of the EEC administration is to harmonize pesticide tolerances within the EEC countries. In pesticide tolerance matters now there is cooperation between EEC, FAO, WHO, and Codex Alimentarius.

## VIII. ABOUT SOME OLD INSECTICIDES

In view of great chemical achievements during the past 25-odd years in developing many new insecticides, it is of interest to review briefly the history of each of the relatively few old insecticides that played the major role for many years in control of the most important insect and mite pests of food and fiber crops. Each of these materials still has some use in the United States.

Lead arsenate, calcium arsenate, and cryolite were the most widely used stomach poisons in the United States until the late 1940s. However, the use of these three materials decreased greatly following introduction of DDT and other chlorinated hydrocarbons.

Lead arsenate was first used in 1892. Production of it in 1927 was 21,527,838 pounds which increased to 90,700,000 pounds in 1944 (of which 4,265,513 pounds were exported). Production dropped to 24,600,000 pounds in 1948 (of which 2,039,000 pounds were exported). Of the total of 68,885,021 pounds consumed in this country in 1941, 35,000,000 pounds were used on apples, mostly for control of codling moth (Shepard, 1915b).

The increasing difficulties encountered from the use of lead arsenate during the late 1920s and 1930s such as excessive residues on apples and pears, poorer results in control of codling moth due to the resistance phenomenon, and excessive soil contamination in apple orchards in the Pacific Northwest, stimulated great interest in developing other insecticides to replace it. A number of arsenical and fluorine compounds, fixed nicotine preparations, as well as many synthetic organic compounds were investigated. Cryolite was the only compound to be fairly widely used; however, it was not as effective as lead arsenate in control. Phenothiazine, xanthone, nicotine tannate, and nicotine-bentonite also found substantial use as a substitute for lead arsenate in the late 1930s and early 1940s. However, it was not until DDT came into commercial use in 1947 that spectacular control of codling moth was obtained with as few as three spray treatments per season in areas like the Pacific Northwest where formerly as many as 12 sprays per season with lead arsenate often failed to give good commercial control. Despite the excellent results worldwide in control of codling moth with DDT, its use was phased out because of adverse side effects in which certain noneconomic insect and mite species became major pests, the resistance phenomenon after use for 8 to 10 years, and, more recently, because of excessive pollution of land and water. In fact, the use of DDT along with certain other chlorinated hydrocarbons was virtually banned several years ago in the United States and several European countries on the basis of adverse effects on environmental quality. DDT production exceeded 100 million pounds annually during the middle 1950s.

It is of special interest to note that, recently, lead arsenate now is reappearing in spray schedules for control of codling moth on apples in several states in the United States. The present residue tolerance is 7 ppm.

Calcium arsenate was first used in about 1912. Production of it in 1919 was 1,192,000 pounds which increased to 84,136,000 pounds in 1942 (of which 3,942,000 pounds were exported). Production dropped to 26,000,000 pounds in 1948 (of which 4,569,000 pounds were exported). In 1941, 54,738,856 pounds were used on cotton, and this was 72.8% of the total consumption for that year (Shepard, 1951c). There is still a substantial amount of calcium arsenate used in the United States. The current residue tolerance is 3.5 ppm.

Cryolite, the natural form mined in Greenland, was first used as an insecticide about 1870. The synthetic form of cryolite became available in the late 1920s. Both forms were used in the United States beginning in the late 1920s and their usage on a wide variety of crops increased to about 15,515,000 pounds in 1944 (Shepard, 1951d). Although demand dropped off greatly with the advent of DDT and other new synthetic organic insecticides, there is still a substantial amount of cryolite used on a number of crops. The present residue tolerance is 7 ppm.

The four old standard contact insecticides were nicotine, pyrethrum, rotenone, and petroleum oil; however, for convenience, sulfur and hydrocyanic acid (HCN) are included in this category. Brief comments will be made about each of them.

Tobacco (nicotine) was employed as an insecticide as early as 1690. In 1908 a liquid containing 40% nicotine sulfate was patented. The United States used about 1 million pounds of nicotine in 1938 and about 1.5 million pounds in 1943 (Shepard, 1951e). Following World War II, production of nicotine sulfate declined rapidly; however, there is still very limited use of it.

Pyrethrum, then known as Persian insect powder, was used in the Caucasus and northern Iran prior to 1800. Pyrethrum flowers were grown in Yugoslavia, Japan, Brazil, British East

Africa, the Belgian Congo, and for a while in California and several other states in the United States. There was wide use of pyrethrum as an insecticide for many years in most parts of the world. Millions of pounds of flowers and powder were imported into the United States each year between 1934 and 1948. The two peak years for imports were 1937 with 20,305,000 pounds and 1946 with 20,476,000 pounds. Imports dropped off drastically thereafter (Shepard, 1951f).

Although chemical work on pyrethrum flowers had been in progress since about 1850, it was not until 1924 that Staudinger and Ruzicka determined the composition of pyrethrins I and II, the insecticidal constituent of the flowers. By 1950 synthetic pyrethroids were produced (Shepard, 1951g). As a result of continuous investigation, new and very promising pyrethroids have been synthesized. It is probable that investigations on pyrethrum and pyrethroids have been in progress for a longer period of time than with any other insecticide.

Rotenone, the principal insecticidal constituent in derris and cubé and certain other plants was first reported to be used prior to 1848 as an insecticide in Malaya. *Derris* species are largely grown in Malaya and Indonesia. Cubé refers to *Lonchocarpus* species, grown in South America. Rotenone can act both as a stomach and contact poison. Derris and cubé were under investigation in the United States during the 1920s. However, it was not until the early 1930s when there was adequate chemical information on rotenone and rotenoids to enable standardization, that importation of derris and cubé into the United States became substantial. Imports of rotenone-bearing roots and powder in 1931 amounted to 5000 pounds in 1939; 5,909,000 pounds in 1939; 13,454,000 pounds in 1947; and in 1948 the imports dropped drastically to 4,713,000 pounds (Shepard, 1951h). The rotenoids are still being investigated, and they may again find major usage.

Petroleum oils of varying composition have been used in emulsions as insecticides since the 1860s. Before 1300 mineral oils were known to have insecticidal properties. During the period 1860 to 1900 kerosene emulsion was regarded as practically a universal contact insecticide. The so-called lubricating oils were widely used as emulsions for dormant treatment of deciduous fruit trees and also on citrus from about 1900 to the late 1920s (Shepard, 1951i). Highly refined oils of specific chemical and physical properties were developed in the 1920s and thereafter, and were widely used on fruit trees until about 1950, when many synthetic organic compounds became available to compete with them. There is, however, a large continuing use of petroleum oil. No insect or mite resistance to oil has been evident.

Sulfur is probably the oldest known insecticide and fungicide. About 1000 B.C. Homer referred to the use of sulfur for pest control. Before 1800, mixtures of lime and sulfur were recommended for control of insects; however, boiled lime sulfur was not known until about 1850. The early uses of lime sulfur were mainly for control of mildew diseases on grapes and other crops, but beginning in 1880 it became the standard insecticide for control of the San José scale, a very serious pest of deciduous fruit trees in the United States. In the early 1900s lime sulfur became a standard treatment for control of the apple scab disease, and there is still some use of it for that purpose. The other forms of sulfur that had very extensive usage were as a dust and as a wettable preparation applied as a spray. The widest usage of sulfur as an insecticide was for control of mites, and there is still a substantial amount used for that purpose, despite the large number of very effective acaricides that have been developed during the past quarter of a century. At the peak of sulfur usage as a pesticide in the United States, about 1950, from 75,000 to 90,000 metric tons were used annually. The advent of modern synthetic organic insecticides and fungicides, however, resulted in greatly reduced demand (Shepard, 1951j). Sulfur residues are regarded as safe, and therefore no tolerance has been established.

Hydrocyanic acid (HCN) was the principal early fumigant which had unique usage under tented citrus trees for so many years in California, Spain, South Africa, Australia, and to a lesser extent in a number of other countries (Quayle, 1938). The commercial use of HCN to control scale insects on citrus in California began in 1886. This use terminated in the early 1950s. For many years from 4 to 6 million pounds of liquid HCN were used annually by the citrus industry in the southern portion of the state. There were surprisingly few accidental deaths from the gas. The California red scale, black scale, and citricola scale developed resistance to HCN between 1915 and 1930. Increasingly poorer results in scale insect control, almost prohibitive labor and other costs, competition from new synthetic organic insecticides, and other factors eliminated the use of HCN in citrus insect control.

HCN was also used, beginning about 1900, for the fumigation of flour mills, ships, warehouses, greenhouses, and homes (Shepard, 1951k). These uses, however, have been largely replaced by other materials.

## IX. PUBLIC REACTIONS TO INSECTICIDES

For more than 50 years there has been apprehension by some of the public over the safety of pesticides, mostly insecticides, to human health. Kallet and Schlink's book (1933) produced a big wave of public reaction. Because apples and pears were treated with lead arsenate for control of the codling moth, many people either stopped eating these fruits or peeled them before consumption. The first chapter of the book clearly indicated the focus of the authors as follows: "Chapter I, The Great American Guinea Pig. A hundred million Americans act as unwitting test animals in a gigantic experiment with poisons, conducted by the food, drug, and cosmetic manufacturers." The third chapter is devoted to lead arsenate as follows: "Chapter III, A Steady Diet of Arsenic and Lead. Deaths from minute doses of arsenic--indifferent Government officials permit fruit and vegetables contaminated with deadly insecticide residue to come to our tables--arsenic a cause of baldness and eczema--lead contamination of fruits and vegetables even greater hazard."

Rachel Carson's "Silent Spring" in 1962, undoubtedly did more than any other single publication to inform the public about wideranging adverse effects of pesticides on major aspects of environmental quality including public health (Carson, 1962). Despite certain scientific deficiencies, this well-written book served to stimulate during the following decade the enormous wave of public concern about safety of pesticides, especially DDT and other organochlorine compounds and organophosphorus materials. The carbamates had not come into general use at that time. The book also served to catalyze public opinion on the need for progressive action to prevent the rapid deterioration of environmental quality. During the 1960s more urgently needed legislation at both federal and state levels was developed than ever before dealing with pollution of air, land, and water--and there is pesticide involvement in each of these areas.

Fuller's book in 1972 did not stimulate as much public reaction as the first two books mentioned (Fuller, 1972). Only Chapter 12, "On the Wings of Mercury and Pesticides" is devoted to pesticides.

It has been clearly demonstrated that a sufficiently aroused public can get action. Probably there has been as much sustained adverse criticism of DDT in the public media as with any other single aspect of environmental quality. Therefore, for this and other reasons the public image of insecticides is generally very poor.

## X. PERSONAL REFLECTIONS ON THE USE OF INSECTICIDES DURING THE PAST 50 YEARS

Some of these comments may partially explain why economic entomologists have been in a difficult position during the past decade over use and misuse of insecticides with particular regard to environmental quality.

Until recent years, most entomologists concerned with research on control of insects in the field relied almost entirely on insecticides. Generally, along with control of plant diseases with fungicides, the treatments were essentially regarded as preventative or insurance. The aim of the field entomologist was to obtain the highest possible percentage kill of the pest species. Treatments applied when the population density of the pest species is low, usually provide a longer period of protection before another treatment is needed. In many instances, regular schedules for treatment during the growing season were developed by entomologists and followed by growers. Furthermore, for economic reasons, broad spectrum persistent insecticides were sought and used, because on many crops there are a number of injurious species of insects and mites infesting the crop simultaneously. Because of high application costs, it is often advantageous to combine several insecticides and sometimes a fungicide and/or plant growth regulator in the same treatment mixture.

In retrospect, it is clear that under many conditions there has been an overuse of insecticides. With DDT, for example, repeated applications have caused upsets in the biological balance of the insect and/or mite community which resulted often in one or more species becoming a major pest, whereas previously their populations were too low to cause economic loss. It should be mentioned, however, that on certain crops only one application of DDT produced important adverse side effects.

It is abundantly clear now that only with adequate ecological information can the most intelligent use of insecticides be made. The analysis and application of ecological and other relevant information in regulating populations of insect and mite pests was known earlier as integrated control, but now it is referred to as either pest management or integrated pest management. The full use of adequate pest management information will surely reduce the amounts of insecticides needed for control of many pest species.

Most senior entomologists did not have much formal training in ecology, and the majority of those concerned with field control were too busy "putting out fires" on their respective crops to take the time needed to become knowledgeable in ecological matters and then to develop adequate ecological information.

In reflecting on the past, it is obvious that many of the older entomologists completely underestimated the ultimate importance of the resistance phenomenon in the early days. As mentioned before, this matter was highly controversial among entomologists until about 1948, when the resistance of houseflies to DDT was so dramatic. Even after personal experience with the spectacular resistance of the citrus thrips to tartar emitic in 1941 followed by resistance to DDT in 1949 and to dieldrin in 1952, this author still did not expect the resistance phenomenon would become evident in many groups of compounds. Later, when resistance developed to organophosphorus and carbamate compounds, his thoughts then were that with adequate knowledge on the biochemical, genetical, and physiological basis of resistance it ought to be possible to develop effective compounds to which the insects and mites could not develop resistance, or to develop compounds to use with the insecticide to block the resistance mechanism.

During the earlier years of this author's field work, his feeling was that he was doing some fairly basic research, at least by the standards of those times. Some solutions to pressing entomological problems were developed. In that era, most economic entomologists were doing about the same kind of work, and generally were working up to the capacity of their training. Only very few entomologists had sufficient training in chemistry to extend appreciably the barriers to further progress such as has been done so notably in later years. My aspiration during these early years was to find insecticides that would provide satisfactory control of the various insects and mites on the several major crops with which I was concerned. This information would enable sound development of treatment programs that would be handled by the Agricultural Extension Service. It was hoped that I could develop some expertise in ecology and undertake greatly needed ecological work on these various pests to enable more intelligent use of insecticides. Certainly, I did not expect the virtual explosion in the synthetic organic area with the attendant human safety problems and the necessity for the vitally important emphasis on residue chemistry, nor the resistance problem as well.

So, in conclusion, what does an old timer who has witnessed development of the present troubled insecticide situation think about the future? Well, as indicated earlier, insecticides will be necessary during the foreseeable future in pest management programs, and most likely, insecticides are going to become increasingly controversial and certainly more stringently regulated. Incidentally, a very perplexing aspect of the current insecticide situation is why the chemical industries are still content to market compounds that are so highly toxic to warm-blooded animals when there is evidence that much less toxic and yet effective compounds can be developed. Evidently, performance and cost are the principal factors considered by companies developing new insecticides. More sensitivity to environmental quality is needed, even though the costs may be somewhat higher. It is ironical that the virtual ban on use of DDT in the United States stimulated greatly increased use of parathion, a highly toxic but relatively inexpensive organophosphate.

The continuing necessity for insecticides, together with the needed continuing emphasis on environmental quality, clearly indicates the following musts: (1) make maximum use of all available information in the safe use of existing insecticides; (2) develop needed ecological and other relevant information on the pest species in order to enable most intelligent use of insecticides; and (3) develop much more extensive and imaginative research in the quest for more overall satisfactory chemicals to meet the needs of plant and animal protection and public health.

With the rapid advances in chemistry and biology there is reason to be optimistic about the ultimate development of insecticides to meet satisfactorily the present and future needs.

## REFERENCES

1. Babers, F. H. 1949. USDA Bur. Entomol. No. E-776, 15.

2. Boyce, A. M., Persing, C. O., and Barnhart, C. S. 1942. J. Econ. Entomol. 35, 790.

3. Carson, R. 1962. Silent Spring, Houghton-Mifflin, Boston.

4. Cerf, D. C. and Georghiou, G. P. 1972. Nature 239, 401.

5. Ewart, W. H., Gunther, F. A., Barkley, J. H., and Elmer, H. S. 1952. J. Econ. Entomol. 45, 578.

6. Ewart, W. H. 1974. Personal communication.

7. Fuller, J. G. 1972. 200,000,000 Guinea Pigs, Putnam, New York.

8. Georghiou, G. P., Lin, C. S., and Pasternak, M. E. 1974. Proc. Calif. Mosquito Control Assoc., vol. 42, pp. 117-118.

9. Hough, W. S. 1928. J. Econ. Entomol. 21, 325.

10. Howard, L. O. 1930. "A History of Applied Entomology, 63," Smithsonian Misc. Collections 84, 564 p.

11. Hunter, R. C. 1974. Bull. Entomol. Soc. Amer. 20, 103.

12. Kallet, A. and Schlink, F. J. 1933. 100,000,000 Guinea Pigs, Vanguard Press, New York.

13. March, R. B. 1974. Personal communication.

14. Melander, A. L. 1914. J. Econ. Entomol. 7, 167.

15. Quayle, H. J. 1916. Calif. Univ. Jour. Agr. 3, 333,358.

16. Quayle, H. J. 1938. Insects of Citrus and Other Subtropical Fruits, Comstock Pub. Co., Ithaca, New York.

17. Shepard, H. H. 1951a. The Chemistry and Action of Insecticides, McGraw-Hill, New York, 504 pp.

18. Shepard, H. H. 1951b. ibid pp. 18, 22.

19. Shepard, H. H. 1951c. ibid pp. 23, 26.

20. Shepard, H. H. 1951d. ibid pp. 45, 46.

21. Shepard, H. H. 1951e. ibid pp. 118-119.

22. Shepard, H. H. 1951f. ibid pp. 144-148.

23. Shepard, H. H. 1951g. ibid pp. 149-156.

24. Shepard, H. H. 1951h. ibid pp. 156-163.

25. Shepard, H. H. 1951i. ibid pp. 196-202.

26. Shepard, H. H. 1951j. ibid pp. 54-60.

27. Shepard, H. H. 1951k. ibid pp. 247-248.

# Insecticides and Integrated Pest Management

RAY F. SMITH
Division of Entomology and Parasitology
University of California
Berkeley, California

## I. INTRODUCTION

In the late nineteenth century, when most of the established pest control tactics that we have today had their beginning, the protection of crops from pests was largely achieved by natural and cultural measures. These were sometimes augmented by minimal use of the earliest insecticides and fungicides or by handpicking. Cultural methods of control were very important, because over hundreds of years these were gradually evolved to expose the susceptible stages of the pests to adverse environmental conditions. These methods of control interfered little with natural enemies, and the latter undoubtedly served a very significant role in control. We are now confident that was the prevailing situation, not only from observations made at the time, but also from recent experiments that recreate nineteenth century conditions. Even at that time, approximately a century ago, a few farsighted men warned of the consequences that could result from human modification of the intricate ecology of agro-ecosystems such as by heavy use of pesticides (Forbes, 1880). Forbes, among others, pointed out very clearly the complexities of agro-ecosystems and outlined some of the things which should be preserved in these ecosystems.

In this meeting, Boyce has reviewed some of the historical aspects of the development of insecticides. It is clear from his remarks, especially when coupled with the current situation, that in the future we will not be able to use pesticides with the freedom and abandon that we have in the past. We do not mean by these remarks to suggest that what was done in the use of pesticides 25 years or more ago was wrong; perhaps it was right for that time and environment and it led to better things. Nor do we mean to denigrate pesticides or even to suggest that pesticides will be less important in the future. What we are emphasizing is that we are moving into a new era in pest control. The pesticides we are discussing in this meeting are the pesticides of tomorrow. We should not evaluate them by the criteria of yesterday or today but by the criteria of the future setting of their use. They will be used in a new social, political, and economic environment, and their use must be in harmony with that environment. In general, the pesticides will have to be used more precisely and carefully, in concert with other pest control tactics and with appropriate consideration for environmental quality. This, of course, is the prime objective of integrated pest control.

We would like to digress for a moment to review a bit of the background and meaning of the term integrated pest control. Professional entomologists have long debated the meaning of various terms as they relate to plant protection from pests. Examples are the terms control, regulation, density-dependence (Nicholson, 1933, 1954; Smith, 1935; Birch, 1960; Andrewartha and Birch, 1954; Huffaker and Messenger, 1964). It was the conflict between the groups that wanted to control or kill pests and the group that wanted to regulate them that spawned the first use of the term integrated control (Smith and Allen, 1954). This first use emphasized the integration of the two tactics of biological control and chemical control into a pest management system, as both are essential to efficient pest control. By the early 1960s, there were a number of competing terms meaning essentially the same thing, for example, harmonious control, rational control, and modified spray program. The FAO panel of experts on integrated pest control in its deliberations on this topic emphasized the importance of integrating all tactics in pest control, and the international plant protection community rather uniformly adopted the term integrated pest control (FAO, 1966, 1968, 1970, 1972). One of the factors in the panel's choice of integrated pest control in preference to pest management was the awkwardness of translating the latter into certain languages. Quite recently, in parts of the United States the hybrid term integrated pest management has been offered as a substitute or compromise. We still prefer to use the internationally established term integrated pest control.

Plant protection and pest control are placed in a new context today compared to that prevailing even as short a time ago as 5 years. Our expanding human population requires an increasing supply of food and fiber, and this will require better protection from insects and other pests. Unfortunately, the task of achieving improved pest control has become more difficult in recent years. Great increases in food production have been achieved in recent years through improved varieties, methods of culture, and pest control. However, the protection of this food rests on a shaky foundation. Narrowing the genetic base through planting a narrow spectrum of crop varieties (and in some cases narrowly based varieties) has meant a reduction in genetic factors for resisting pests. Furthermore, the development of resistance to pesticides has greatly weakened the effectiveness of that tactic in certain insect pest populations.

In a similar way the public health sector has increased problems which Jimmy Wright has outlined so well in an earlier chapter. Resistance is one of the problems. In addition, urbanization is exposing people in large numbers to new hazards, and suburbanization, the move to the country, has exposed people in new ways.

Before going on to disucss some details of integrated pest control, it might be well to review briefly the broad philosophy and strategy of integrated pest control. The strategy of integrated pest control relies on an analysis of the biological and economic systems of pest control and crop protection. This embraces all significant factors acting on the complex of real and potential pests and the interactions of such factors among themselves and with other processes in crop production. The integrated control strategy employs the idea of maximizing natural control forces and utilizing any other tactics with a minimum of disturbance and only when losses justifying action are threatened.

Brooks has discussed selectivity of pesticides in an earlier chapter. He mentioned ecological selectivity, but he concentrated his discussion on physiological selectivity--in other words, what happens inside the insect. He talked a lot about "fiddling around with penetration." The integrated control specialist, the pest manager, has to fiddle around with the agro-ecosystem, the world outside the insect. Insecticides must be used in agro-ecosystems; their targets are inside of insects, but we must understand them in the context of agro-ecosystems. This is what integrated pest control attempts to do.

Consequently, the integrated control specialist must be a field-oriented ecologist. He looks at all possible ways of manipulating the agro-ecosystem. What are his tools and tactics? The main ones are

1. Cultural control, with its manipulation of planting dates, crop rotations, plant spacing, strip harvesting, interplants, and trap crops.

2. Host plant resistance. It is especially important to remember that plant resistance need not be complete resistance to be useful in an integrated pest control program.

3. Use of parasites, predators, and pathogens of the pests. This can involve introduction of new spacing, provision of shelter, supplemental food or alternate hosts, and mass release of artificially reared natural enemies.

4. Pesticides.

These tactics are not used each in isolation, but as a combined, integrated system. Sometimes these are very sophisticated systems involving use of computers to handle large amounts of data and systems analysis methodology.

The cadre of integrated control specialists who have been trained in the past 15 to 20 years will be capable of using pesticides in more sophisticated ways. These are the people who will be handling the new pesticides of the future. As they bring these pesticides to the field, they will be asking a new set of questions, particularly as the pesticides relate to the agro-ecosystem, for example, the impact on beneficial organisms and other nontarget organisms. To fit the new environment, the pesticides of the future must be able to stand up satisfactorily to this new set of questions.

In an earlier chapter the question was asked if the use of pesticides could destroy or eliminate a crop. The question was not completely answered, and in fact it is difficult to answer. In most situations, a number of factors are involved in addition to pesticides, but there are examples of pesticide use so aggravating the plant protection situation that it became unprofitable to grow the crop. Perhaps the clearest and most completely documented example is the elimination of cotton production from Northeast Mexico, from Matamoras south to Tampico. The unwise use of pesticides in this area produced huge populations of resistant insects and major secondary outbreaks. The insects could not be controlled by the pesticides available, and their natural enemies had been destroyed. Cotton production ceased in the area, the gins were boarded up, and there was not a good substitute crop (Adkisson, 1973).

The situation in the coastal plain of Central America is very similar but more complex. In that area, growers exceeded 35 applications of pesticides per year. In spite of this heavy use of pesticides, they were not getting satisfactory insect control, and the yields dropped. The economies of several of these countries were seriously threatened because of the large amount of resources expended to import seeds, pesticides, and other inputs required for cotton production. Many farmers fell into financial ruin. There were many other implications of the high pesticide use in cotton, including the developing of resistance in Anopheles albimanus, the vector of malaria, human intoxication, and deaths from pesticides and residues in exported meat. An integrated pest control program for cotton was then introduced with assistance from FAO. The level of pesticide use has been reduced, cotton yields are increasing, and the situation continues to improve.

We have a dilemma in balancing the good and the harm that comes from the use of pesticides. Integrated pest control attempts to solve this dilemma, but it is not a substitute for pesticides. Cotton is a critical export commodity for many developing countries. In almost all situations, pesticides are necessary for satisfactory cotton production. It is essential, however, to use pesticides in a safe and effective manner so that the undesirable consequences of their use are eliminated or at least minimized.

Earlier in the book, the many aspects of physiological selectivity were discussed quite thoroughly. However, very little has been said about ecological selectivity. Unlike physiological selectivity, which stresses activity of the chemical, ecological selectivity emphasizes application and behavior of the insects. Ecological selectivity is achieved by applying compounds, often having rather broad spectra of activity, in such a manner as to ensure contact of the toxic dose with the target species while at the same time avoiding completely, or minimizing greatly, contact of a toxic dose with nontarget species (Unterstenhofer, 1970).

The necessity for achieving selectivity of insecticide action with regard to mammalian toxicity and phytotoxicity has been recognized to be of prime importance for as long as chemicals have been used for insect control. This form of selectivity was comparatively easy to achieve. However, it was not until the synthetic organochlorine insecticides were developed and put to such wide use that the need for much more narrowly differential toxicity within the arthropods was recognized. The broad spectrum of activity of these chemicals, especially their effects on insect predators and parasites and other nontarget species, served to focus attention on the importance of predators and parasites in general, and complexes of native species of these useful insects in particular, for regulation of pest populations. Dramatic examples of resurgence of treated populations and elevation of occasional and secondary species to the status of key pests became commonplace with widescale use of the broad-spectrum synthetic organic insecticides, and entomologists began to plead for the development of narrowly selective chemicals.

For the past 15 to 20 years, the pesticide chemical industry has been increasingly challenged to produce more narrowly selective pesticidal chemicals. From the standpoint of effectiveness in managing pest populations, the availability of chemicals that are toxic to no other species except the boll weevil, Anthonomus grandis, in southern United States, Lygus spp. in western United States, and codling moth, Laspeyresia pomonella, in apple-producing areas, for example, would be of incalculable value. There are enough examples of narrowly selective insecticides to justify the belief that the commercial development of chemicals that are toxic to single families, genera, and even species is not an unreasonable expectation. The discussions in the book, of course, have provided many additional possibilities.

At a time when the need for more narrowly selective compounds becomes increasingly acute, industry appears to have become reluctant to accelerate, indeed even to continue, research having as its objective discovery of more of these desirable chemicals. Unfortunately, in the present system of commercial development of pesticides, the selective material is not favored (von Rümker et al., 1970). Under the current system the commercial companies are forced to develop only those compounds that can be marketed on a very large scale (Persing, 1965). This emphasizes the source of one of our difficulties with modern chemical pesticides, which stems from the manner in which these compounds are developed commercially. In the first place, only limited ecological considerations go into the search for new compounds. The condidate materials are screened in the laboratory on the basis of high percentage kill to a small select group of pest species. A few companies may have more than 20 insects in their screen, but the average is about 5 or 6 (Persing, 1965; von Rümker et al., 1970). Those chemicals showing promise are taken into the field for smallscale trials against a wide variety of important crop arthropods. If a given pesticide candidate is effective against major pests on important crops, it may be able to compete against the current established pesticides. Once the compound is developed for major markets, it is relatively easy to expand those markets by including minor ones. The resultant pesticide material registered for use for a variety of crops has a broad toxicity spectrum and is precisely the type of compound that is so disruptive ecologically. Von Rümker et al. (1970) in their survey of pesticide manufacturers concluded that the following were the most important obstacles to the development of more selective pesticides: (1) high cost of research and development; (2) lack of knowledge of basic plant and animal biochemistry; (3) competition from existing, nonselective, relatively inexpensive pesticides; (4) lack of grower interest resulting from high cost of selective products which the grower is unwilling to pay; (5) cumbersome government registration procedures; (6) fear of consumer complaints and litigation; and (7) lack of interest, support, and experience on the part of agricultural workers. They also point out that the research and development costs of bringing a single compound to market has risen from $1,196,000 in 1956 to $4,060,436 in 1969, while the chance of success for an experimental product has dropped from 1 in 1800 to 1 in 5040.

Perhaps more than any other factor, the most important deterrent to proper emphasis on selective pesticides is the difficulty being experienced by the pesticide industry, growers, and even many entomologists in substituting a philosophy of pest control based on regulation of pest populations at economic injury thresholds for that of maximum kill of pests based on repetitive applications of broad-spectrum, highly effective, relatively cheap insecticides without regard to pest population assessment. It has not proved easy to change quickly a philosophy that has prevailed for a quarter of a century. A generation of entomologists has developed under the strong influence of a philosophy that advocated the complete elimination of every individual of a pest population wherever possible. Such approaches make clearly evident high external costs, that is, external to the decision-making practice within the commercial pesticide manufacturing companies or the pest control decision-making process. Perhaps, if a broader base of those interested in crop production were involved in these decisions, some of the decisions now being made might be reversed in some cases. In the past, other external costs (e.g., hazards to humans) have been brought into the internal decisionmaking process for the production of pesticides. This has been largely through regulations in the use of pesticides and the pesticide registration system. Perhaps similar use of regulations will be necessary to account for other external costs involved in pesticide use.

The ideal selective material is not one that eliminates all individuals of the pest species while leaving all of the natural enemies. The maintenance of low populations of pest species is, of course, essential to the continuity of predators or parasites. In Washington apple orchards, the apple rust mite, *Aculus schlechtendali*, can be tolerated at moderate populations without economic loss. The populations of apple rust mite are a good food source for mite predators which not only prevent eruptions of apple rust mite but also regulate

McDaniel mite, *Tetranychus mcdanieli*, at low densities and help to control the European red mite, *Panonychus ulmi* (Hoyt and Caltagirone, 1971).

After a shift to pesticide chemicals that are more selective in their action, it may be some time before balance is restored. The effects of the previous treatments may last several years. In some instances, effective biological control no longer exists and would have to be reestablished. This may be a slow process (DeBach, 1951; Stern *et al*., 1959). It took four years to reestablish normal predator-prey relationships for mites in apple orchards in the state of Washington after mite control sprays were modified (Hoyt and Caltagirone, 1971).

Clearly, the amount of research required for the development of selective chemicals will not be achieved in the near future without substantial financial support from government agencies. Industry remains unconvinced that concentrating their research toward the discovery and development of narrowly selective (monotoxic or oligotoxic) compounds is the proper step to take. Moore (1970) has fairly stated the attitude of the majority of industry as follows: "For example, it does not pay industry to look for and market specific pesticides, or to study new methods of integrated pest control, when these reduce the amounts of pesticides sold, and therefore reduce profits. So, only governments can sponsor adequate research on specific pesticides and integrated control effectively." Few, if any, universities or governmental agencies have adequate personnel or facilities required to mount research programs of the magnitude that would be required to offer a reasonable chance of significant success during the foreseeable future. Therefore, it appears that the most appropriate subsidies from government to industry to help finance the required research and development costs.

## II. ECOLOGICAL SELECTIVITY

Fortunately, the development of effective and economical systems of pest management for the control of many major pest species is not dependent on physiological selectivity provided by the availability of a large number of narrowly selective pesticidal chemicals.

Pesticides having the broadest spectrum of activity may be used in an ecologically selective manner. As far as pest management is concerned, it appears that the selective use of insecticides (ecological selectivity) will continue to be far more important than use of selective insecticides (physiological selectivity).

Selective action of nonselective chemicals can be obtained by manipulating doses, formulations, timing of applications, method of application, and localization of area to be treated (Ripper, 1944; Ripper *et al*., 1948; Ripper *et al*., 1951; Ripper, 1956; Bartlett, 1964; Stern *et al*., 1959; Lean, 1965).

## A. Dosage

From the time of Paracelsus it has been known that dose alone determines whether a chemical is poisonous or not. The remarkable success of chemotherapy in human and veterinary medicine is based on differences in species-specific response to dosage (Albert, 1965). One of the most elegant examples to illustrate this principle is a recent one from veterinary medicine provided by Thompson *et al*. (1972) in their description of a method for control of the vampire bat, *Desmodus rotundus*, on cattle in Mexico. In agricultural entomology, Ripper (1956) demonstrated the value of reduced dosage rates of the systemic organic insecticide schradan for control of the cabbage aphid, *Brevicoryne brassicae*. Similarly, integrated control of the spotted alfalfa aphid, *Therioaphis maculata*, in California has as one of its key elements the application of demeton at a discriminating dosage rate that gives adequate kill of the aphid while exerting minimal toxic effects upon its predators and parasites (Stern *et al*., 1959).

The philosophy of overkill that has been so widely prevalent in applied entomology for the last 25 years has resulted in far more use of chemical pesticides than required to obtain optimum control of insect pests. Not only has this philosophy resulted in excessive pollution of the environment, rapid development of insecticide-resistant populations, rapid change in status of secondary and occasional pests, and unnecessarily severe effects on nontarget species, it has also been economically unsound. Experimental establishment of minimum dosage rates required to hold pest populations just below economic injury levels for the minimum period of time required for optimum production of the crop being protected and acceptance of these levels in grower practice should be an objective of highest priority in pest management.

## B. Confining Insecticides to Restricted Areas

There are many opportunities for restricting the total amount of pesticides used for control of pest species. Thorough knowledge of the biology, ecology, and behavior of a pest often makes possible its control by application of insecticides to very restricted areas. The effectiveness of such a technique for control of a major pest was demonstrated by Isely (1926) almost 50 years ago by his "spot-dusting" method for control of the boll weevil.

The same type of selective action on a pest-parasite complex can be obtained by treating only those areas where the pest-parasite ratio is unfavorable. On an areawide basis, the balance in favor of the hymenopterous parasite, *Apanteles medicaginis*, as compared to its host caterpillar was shifted in the supervised control program in California, even though many parasites were destroyed in the treated areas.

Other behavioral traits have been demonstrated to occur in various pest species which allow relatively high percentages of the total population in an area to be concentrated in restricted sites or localities. Nishida and Bess (1950) and Nishida (1954) showed that the oriental fruit fly, *Dacus cucurbitae*, in Hawaii spent a relatively small percentage of each day in cultivated crops. Most of its time was spent in vegetation on the field margins and adjoining areas, with the females moving into fields to oviposit and returning to areas outside of the fields for the major part of each day. They found that this pest was controlled more effectively by treating areas adjoining the fields rather than the fields themselves.

Thus, by taking advantage of an understanding of insect behavior, broad-spectrum insecticides may be used in such a way as to attain high degrees of selectivity. Ecological selectivity obtained in this manner deserves much more attention in developing systems of pest management than it has received in the past. Its obvious advantages in reducing the amount of pesticides introduced into the environment, restricting the amount of acreage treated, thereby lowering costs, conserving populations of beneficial insects, minimizing the selection pressure towards pesticide resistance, and reducing the hazard to nontarget species, are far greater than the inconveniences that may be caused by planting earlier than at the optimum time, using less desirable varieties, or similar less-than-optimum cultural practices.

### C. Formulation and Method of Application

Methods of formulating and applying broad spectrum insecticides can be varied in such ways as to obtain, in many cases, a high degree of selectivity. The use of granular formulations in the central whorl of maize plants is an example of this differential impact on lepidopterous pests of maize and pests and their enemies on other parts of the plant (Ortega, unpublished data). It has long been known that the type of carrier and residue deposit may produce differential effects on the insect complex associated with a crop (Flanders, 1941; Holloway and Young, 1943).

## D. Baits

Formulation of insecticides into poisoned baits has provided one of the most effective and long-used means of obtaining ecological selectivity of broad-spectrum insecticides. Arsenical based baits have a long history of effective use for control of such pests as various species of grasshoppers, locusts, Mormon crickets, armyworms, and cutworms. Such baits provided the only effective and economical means of control for many of these pests prior to the synthetic organic insecticide era. With the introduction of these highly effective and relatively cheap insecticides, the arsenical insecticides were generally replaced by them. In fact, the principle of using poison baits for insect pest control was essentially abandoned for several years. Recently, it appears to be making a comeback with the development of serious problems involved in the use of the synthetic organic insecticides and as the value of selectivity in pest management programs is becoming more widely recognized.

Several species of fruit flies have been controlled very effectively by the use of poisoned baits. One of the most striking examples is that of the eradication of the Oriental fruit fly, *Dacus dorsalis*, from the entire island of Rota by male annihilation by using the highly effective male attractant, methyl eugenol. The bait consisted of small cane fiberboard wafers, saturated with methyl eugenol and naled (95:5) and distributed over the island by airplanes (Steiner *et al*., 1964, 1965, 1970). Generally, such baits require the application of much-reduced amounts of insecticide per unit area treated and, frequently, substantial reductions in the total amount of area required to be treated.

Seed treatments provide another means of application of pesticides in a highly selective manner. The synthetic organic insecticides, for the first time, made this method of pesticide use available for the control of a large number of pests, including seed maggots, wireworms, rootworms, fruitflies, and thrips especially. Often, seed dressings at substantially reduced amounts of insecticide on a per-acre basis have given comparable levels of control to those obtained by broadcast or band applications of the same chemical at much higher rates of application. For example, a complex of seedling pests composed of bean maggot, *Hylema platura*, the sand wireworm, *Horistonotus uhleri*, and the southern corn rootworm, *Diabrotica undecimpunctata howardi*, has been controlled by use of seed dressings as effectively as by broadcast or band applications of the chemical to the soil.

## E. Proper Timing of Insecticide Application

The proper timing of insecticide applications is often the most effective and economical way of producing differential selectivity of insecticide action on the pest-natural enemy complex. Detailed knowledge of the biology and ecology of the species involved and of the economic injury threshold is required for the successful application of the timing principle. Much too often sufficient data are not available on these points. The enormous initial successes with the use of synthetic organic insecticide compared with use of insecticides previously available imposed a severe handicap to research on these basic principles of applied ecology. Essentially, work in these areas has been largely ignored by a generation of entomologists. The initial successes with the synthetic organic insecticides have been slowly eclipsed by failures resulting from the well-known symdrome of unwanted side effects, and there is now an awareness that pest management must be based on more and better information on biology, ecology, and economic injury thresholds. Remarkably good results have been obtained in many instances by simply delaying the initiation of insecticide applications until pest populations have reached economic injury thresholds and discontinuing further applications when the cost of further protection equals or exceeds the value of crop harvested.

## F. Systemic Insecticides

Development of systemic insecticides (Ripper, 1956) made possible a previously unknown technique of using insecticides in a selective way. Schradan was one of the first of these new materials to be developed and used widely. Discovery of its unique and useful properties stimulated a great deal of research, and many organophosphorus and carbamate insecticides were quickly made available for evaluation. Many of these chemicals proved to have useful properties for obtaining selective action. When applied to foliage, they are absorbed rapidly, thus becoming essentially unavailable directly to all except those species feeding on the treated plants. They may also be applied to the soil or to seeds in seed dressings, enter the plants through the roots, and be translocated to aerial portions in amounts that are selectively toxic to some species that feed on treated plants.

Such qualities have proved to be so useful that they have resulted in a substantial amount of overoptimism in evaluating the role of systemics in pest management. The relative safety of systemic insecticides to predators and parasites when they are applied in such a manner as not to have contact effects has usually been overestimated (Ripper, 1956; Bartlett, 1964; O'Brien, 1961; Metcalf, 1964; Ridgway, 1969; Cate *et al.*, 1972).

In spite of the fact that systemic insecticides may not be as specific in their effects as is often claimed, they are nevertheless very useful. They have proved to be especially effective in situations in which aphids and spider mites are members of a pest complex. In such cases, they often may be applied at dosage rates so low that aphid and spider mite populations can be regulated at subeconomic infestation levels with little adverse effect on predators and parasites.

Selective pesticides and the selective use of pesticides are now, and must continue to be for the foreseeable future, one of the foundations on which pest management programs are constructed. The use of pesticides is the only method available that can so quickly be brought to bear for regulating populations of a wide spectrum of pests and pest complexes with such predictably reliable results. Indeed, the intelligent use of conventional chemical pesticides provides the only currently known effective means for controlling a large number of the world's most important pests.

Narrowly selective (monotoxic) chemicals appear to offer an almost ideal means of pest control. However, only a very few such chemicals have been discovered, and future prospects for additional discoveries have become very dim. The chemical industry for the most part has historically had little interest in finding and developing such useful compounds.

Problems involved with development of resistance to insecticides in many pest species with resultant rapid obsolescence of the chemical, unwanted side effects, high costs of securing tolerances and registrations for use, and a society that has become increasingly critical of chemical pesticide use, make it unlikely that industry will be willing to make any substantial effort to discover new selective compounds. In fact, the prospects are so unattractive that it is unlikely that industry will attempt to develop compounds previously synthesized and known to possess interesting selective properties which are now sitting on the shelves of chemical laboratories.

A serious limiting factor is a shortage of properly trained, imaginative, and capable applied entomologists dedicated to the development of pest management systems based on the principles of integrated control. Nevertheless, there are many encouraging examples of progress in pest management. Recognition of the need for improved pest management has reached a stage that has resulted in substantial financial support being allocated to a large number of institutions for work in both research and extension in this field. The most encouraging development of all, however, is the accelerating change taking place in the entomological profession that is orienting applied entomologists toward the philosophy of integrated control in pest management.

## III. CONCLUSIONS

In our concluding remarks, we would like to pass on to you a few reflections on the discussions of insecticides in this conference, particularly as they are significant to integrated pest control and plant protection in general.

First, it is not enough to elucidate the physical-biochemical characteristics of pesticides, their mode of action, their differential toxicities to mice and flies, and their toxicity to certain pests. They must also be fitted into a pest management system and a crop production system. They must be carried to the target species in an effective dose with minimal side effects. They must be produced, packaged, distributed, marketed, and otherwise handled in safe, effective, and economic ways, that is, the pesticide management system. In other words, our goal is not simply the discovery and making available of new pesticides. The goal is, in reality, the reduction of important pests to nonpest status. All these steps are a part of achieving that goal.

In this book, there has been great frustration expressed, even depression, over the tremendous barriers in our pesticide production system to obtaining effective and practical utilization of what we might call "good compounds" that are developed by the laboratory and field by research scientists. There are many things involved in creating these barriers in the system. A lot centers around the problem of compounds with very small markets. In fact, even if we have the compound and its potential end use identified, we may not be able to get registration of developed compounds for new use for minor crops like rutabagas or rhubarb. The so-called "free" compounds (i.e., those with patents expired) are also a problem in that no one seems to want to do the necessary costly research on these pesticides, because once researched, anyone can produce and market them. This frustration is not tempered by the suggestion that we will have lots of new compounds identified.

Resistance continues to be a serious problem. There does not seem to be much that we can do about it except find new compounds. Integrated control will help somewhat by lowering the selection pressure and hence delaying development of resistance, but if there is selection pressure, the same result is inevitable. Floris Oppenoorth's concept of "negative resistance" should be followed up to determine explanations as to why some pests have not developed resistance.

Several times it has been pointed out (and we agree) that we are not going to make progress in solving our critical, high-priority pest problems unless we give a sharp focus to our total attack on them. We cannot deal simply with generalities about pest control. The specific characteristics of these major priority problems, their ecological, biological, social, and economical aspects, should be investigated intensively and advantage taken of these special characteristics in designing a specific effective pest control or management program.

## REFERENCES


1. Adkisson, P. L. 1973. The Principles, Strategies, and Tactics of Pest Control in Cotton. In Insects: Studies in Population Management, (P. W. Geier, L. R. Clark, D. J. Anderson, and H. A. Nix, Eds.), Ecol. Soc. Aust. (memoir 1), Canberra, pp. 274-283.

2. Albert, A. 1965. Selective Toxicity, Methuen, London.

3. Andrewartha, H. G. and Birch, L. C. 1954. The Distribution and Abundance of Animals, University of Chicago Press, Chicago, 782 pp.

4. Bartlett, B. R. 1964. Integration of Chemical and Biological Control. In. Biological Control of Insect Pests and Weeds, (P. DeBach, Ed.), Reinhold, New York and Chapman and Hall, London.

5. Birch, L. C. 1960. Stability and Instability in Natural Populations, New Zealand Sci. Rev. 20, 9.

6. Cate, J. R., Jr., Ridgway, R. L., and Lingren, P. D. 1972. J. Econ. Entomol. 65, 484.

7. DeBach, P. 1951. J. Econ. Entomol. 44, 443.

8. FAO. 1966. Report of the First Session of the FAO Panel of Experts on Integrated Pest Control.

9. FAO. 1968. Report of the Second Session of the FAO Panel of Experts on Integrated Pest Control.

10. FAO. 1970. Report of the Third Session of the FAO Panel of Experts on Integrated Pest Control.

11. FAO. 1972. Report of the Fourth Session of the FAO Panel of Experts on Integrated Pest Control.

12. Flanders, S. E. 1941. J. Econ. Entomol. 34, 453.

13. Forbes, S. A. 1880. Bull. Ill. Nat. Hist. Sur. 1(3), 3.

14. Holloway, J. K. and Young, Roy T. 1943. J. Econ. Entomol. 36, 453.

15. Hoyt, S. C. and Caltagirone, L. E. 1971. The Developing Programs of Integrated Control of Pests of Apples in Washington and Peaches in California. In. Biological Control, (C. B. Huffaker, Ed.), Plenum Press, New York, pp. 395-421.

16. Huffaker, C. B. and Messenger, P. S. 1964. The Concept and Significance of Natural Control (Chapter 4). In: Biological Control of Insect Pests and Weeds, (P. DeBach, Ed.), Reinhold, New York, 844 pp.

17. Isely, D. 1926. Ark. Agric. Exp. Sta. Bul. 204, 17 pp.

18. Lean, O. B. FAO's Contribution to the Evolution of International Control of the Desert Locust 1951-1963. Desert Locust Newsletter Special Issue 1965. Food and Agriculture Organization of the United Nations, Rome, 1965.

19. Nicholson, A. J. 1933. J. Anim. Ecol. Suppl., vol. 2, 132.

20. Nicholson, A. J. 1954. Austr. J. Zool. 2,9.

21. Metcalf, R. L. 1964. World Rev. Pest Control 3, 28.

22. Metcalf, R. L. 1971. Bull. W.H.O. 44, 43.

23. Moore, N. W. 1970. Ceres 3 (May-June 1970), 26.

24. Nishida, T. 1954. J. Econ. Entomol. 47, 226.

25. Nishida, T. and Bess, H. A. 1950. J. Econ. Entomol. 43, 877.

26. O'Brien, R. D. 1961. Selective Toxicity of Insecticides, Advan. Pest Control Res. 4, 75.

27. Persing, C. O. 1965. Bull. Entomol. Soc. Amer. 11(2), 72.

28. Ridgway, R. L. Control of the Bollworm and Tobacco Budworm through Conservation and Augmentation of Predaceous Insects, pp. 127-144. In Proc. Tall Timbers Conf. on Ecological Animal Control by Habitat Management, No. 1, February 27-28, 1969, Tallahassee, Florida.

29. Ripper, W. E. 1944. Nature 153, 448.

30. Ripper, W. E. 1956. Ann. Rev. Ent. 1, 403.

31. Ripper, W. E., Greenslade, R. M., and Hartley, G. S. 1951. J. Econ. Entomol. 44, 448.

32. Ripper, W. E., Greenslade, R. M., Heath, J., and Barker, K. 1948. Nature 161, 484.

33. Smith, H. S. 1935. J. Econ. Entomol. 28, 873.

34. Smith, R. F., and Allen, W. W. 1954. Insect Control and the Balance of Nature, Sci. Amer. 190(6), 38-42.

35. Stern, V. W., Smith, R. F., van den Bosch, R., and Hagen, K. S. 1959. Hilgardia 29, 81.

36. Thompson, R. D., Mitchell, G. C., and Burns, R. J. 1972. Science 177, 806.

37. Unterstenhofer, G. 1970. Pflanzenschutz-Nachrichten 23, 264. (English).

38. Von Rümker, R., Guest, H. R., and Upholt, W. M. 1970. Bio. Sci. 20, 1004.

39. Winteringham, F. P. W. 1969. Ann. Rev. Ent. 14, 409.

# STRATEGY FOR ACTION

# Summary and Recommendations

ROBERT L. METCALF AND JOHN J. MCKELVEY, JR.

The pesticides that have come into use over the past 30 years have had a major impact on reducing certain vector-borne diseases and in increasing world capability to produce adequate quantities of food and fiber. However, their continued use presents to the public and to the profession of entomology increasingly serious problems. Insect resistance, secondary effects of pesticides on the environment, and related shortcomings of pesticide chemicals constitute several of these problems.

New information on pesticides alternative to those that have lost their efficacy or that have deleterious side effects in man's environment and on ecological requirements for using new compounds is not being adequately utilized. In large part, inadequate training of staff to implement research and control operations and severe financial and regulatory constraints on the development and use of new materials and related technology impede the application of this new information.

The participants of the Bellagio Conference believe that a comprehensive and coordinated effort constituting an international program is needed to mobilize and implement new initiatives to solve the pest problems of agriculture and public health. Such a program should link together all the organizations and institutional agencies that deal with pest control in its broadest sense. They recommend that a formal international program be established with the following objectives:

1. To marshall the best worldwide talent in the specialties related to control of pests of agriculture and public health in order to build sound programs of research, to develop appropriate training procedures, and to initiate the actions required to solve pressing pest problems on an international scale.

2. To evaluate and to establish the parameters required for safe, effective, and environ-

mentally sound pest control systems through surveys and other monitoring systems.

3. To provide the means for discovery and development to the use stage of new chemicals and devices that would fit the requirements of control operations either directly through this program or in cooperation with world industries.

4. To provide the mechanism for licensing or assisting industries on a mutually advantageous basis to introduce useful products that normally would not be developed by unassisted commercial endeavors.

5. To provide mechanisms to ensure rapid exchange of information and personal discussion among the various groups dealing with or dependent on pest control.

6. To mobilize the financial resources required to carry on a sustained effort in the entire program.

The international program will require an international secretariat operating through a governing board to develop, with assistance from advisory committees, several subordinate programs. The subordinate programs will be implemented through research contracts or other procedures and should involve toxicology, chemistry, environmental impact, biology, international registration standards, and other fields as needed to meet the requirements for different types of basic pest control programs.

A secretariat of the highest quality must be engaged and provided with an adequate and sustained level of financial support. Some potential sources of funding include: (1) The Consultative Group on International Agricultural Research; (2) governments and other sources of bilateral support; (3) multilateral agencies including FAO, WHO, UNEP, UNIDO, UNESCO, and IAEA; (4) patent licenses, industrial fees, and contracts.

The secretariat for this program might be located with FAO in Rome. After detailed consultation with all major potential participants in the program, the governing board would establish the policies under which the secretariat would operate and the program would develop.

The first phase in implementation will require agreement on the procedure for selecting the governing board and determining its composition. This step will depend on the assurance of initial funding to establish the secretariat and to elaborate the specific policies and operating procedures for the program.

The second phase would be the selection of the secretary general of the secretariat. The identification of initial programs and the naming of area coordinators to develop those programs and budgets would follow.

The third phase would be the development of funding agreements. These would determine the extent and phasing of program implementation.

# Index

Acaricides, 358, 359
Acephate, 302
Acethion, 107, 116, 117
Acetylcholinesterase, 42, 55, 62, 105, 113, 142, 305, 360, 380
Acetylthiocholines, 43
*Aculus schlechtendali*, 495
Acylcyanocarbonylphenylhydrazones, 358
Acyl N-methylcarbamates, 119, 121-122
Acyloxyarylindones, 358
*Adoxophyes orana*, 452
Adult morphogenesis, 424
*Aedes aegypti*, 21-23, 28, 29, 43, 54, 95, 237
*Aedes nigromaculis*, 238
*A*-esterases, 198
Affinin, 352
African trypanosomiasis, 24-25
*Alabama argillacea*, 237
Alanosine, 354
Aldicarb, 119, 313, 323, 343, 455
Aldrin, 50, 78, 80, 126, 228, 250
Aldrin
  degradation of, 251
  toxicity of, 152
Aldrin *trans*-diol, 252, 380
Alkaloids, 352
Alkylation of DNA, 148, 150
Allethrin, 128, 153, 164
Alodan, 109, 123, 128, 228
Alternative insecticides, 31-33
American cockroach, 129
γ-Aminobutyric acid (GABA), 379
Amitraz, 358
*Anas platrhynchos*, 226
Anonaine, 352
*Anopheles albimanus*, 28, 61, 493
*Anopheles gambiae*, 28, 232, 237
*Anopheles labranchiae*, 231
*Anopheles sacharovi*, 231
*Anopheles* spp, 20
*Anopheles stephensi*, 29
Antelope, 445
*Anthonomus grandis*, 234, 237, 494
Anticholinesterases, 297
Antifeeding compounds, 351, 358
Antimycin A, 354
*Aonidiella aurantii*, 475
*Apanteles medicaginis*, 498
*Aphis gossypii*, 321
*Apis mellifera*, 105
Apple pest management, 452-454
*Argyrotaenia velutinana*, 103
Arsenical compounds, 143
Arsenical poisoning, 478
Arsenic trioxide, 478
Arthropod-borne diseases, 17
Arthropod venoms, 355
Aryl propynyl ethers, 203
Aspochracin, 354
ATP'ase, 260, 360
Aureothin, 354
Auricular glands, 445
Autosynergism, 56, 118, 203-204

*Bacillus sphaericus*, 30
*Bacillus subtilis*, 20
*Bacillus thuringiensis*, 30, 353, 368
Barium fluosilicate, 470

Bark beetle pheromones, 446
Barthrin, 203
Batrachotoxin, 376
Baygon, 120
Beauveria, 354
Beauvericin, 354
Benzene hexachloride, 22, 26, 227, 232, 234, 380
Benzodioxoles, 205-209, 214, 220, 351, 431
Benzomate, 359
Benzopyrene, 213, 444
Benzothiadiazoles, 201, 208-211, 213
Benzylrethrin, 170
Benzyl northrin, 170
B-esterases, 198
Bidrin, 118
Bifluorylidene, 358
Bioaccumulation of organochlorines, 247-249
Bioallethrin, 180
Biodegradability, 24, 37, 124, 276
Biodegradable DDT analogues, 124, 258-275
Biodegradable lindane analogues, 128, 275, 287-293
Biological control, 30-31
Bioresmethrin, 111, 128, 165, 170, 180, 181
Biphenyl, 479
Biscarbamyl sulfides, 319-323
Bis-chloromethyl ether, 148
Bis-(pentachlorocyclopentadiene-2,4-yl), 228
Blaberus discoidalis, 103
Blackfly, 23, 31
Blattella germanica, 237, 287, 291
Boophilus annulatus, 234
Boophilus microplus, 42, 113
Botanical insecticides, 350-353
Brevicoryne brassicae, 497
Bromodan, 109, 123, 128
Bucculatrix thurberiella, 321
Bufo americanus, 248
Burimamide, 213
Butacarb, 122, 343
Butethrin, 171
Bux, 343

Cadra cautella, 446
Calanus finmarchicus, 88
Calcium arsenate, 470, 480
California red scale, 451
Camphene, 455
Cancellation of organochlorines, 257
Cantharidin, 355
Carbamate insecticides, 313-341
Carbamate insecticides
  anticholinesterase activity of, 151, 316, 318, 322, 328, 330
  derivatized, 314-337, 344
  detoxication of, 118
  human toxicity of, 343
  metabolism of, 315
  selectophores in, 337-339
  synergism of, 101
  toxicological properties of, 316, 322, 324, 329
Carbaryl, 50, 103, 115, 119, 314, 343
  synergism of, 202
Carbofuran, 313, 323, 330, 333, 339, 343, 344
Carbofuran derivatives, 302
Carbonic anhydrase, 152
Carboxylesterase, 43, 46, 117, 298, 314, 425
Carcinogenesis, 145, 148, 444
Carcinogenesis by organochlorines, 255-257
Cartap, 112, 130, 355, 367
Catecholamines, 152
Cattle tick, 42, 45, 53
Chagas' disease, 26-27, 232
Chaoborus astictopus, 247

Chemical carcinogens, 148, 255, 257, 275
Chemical contamination, 69-70
Chemical persistence, 71
Chemical production, 66-68
Chemosterilization, 29, 359
Chickens, dieldrin in, 94
*Chilo suppressalis*, 237
Chitin, 409
Chitin synthesis, 361
Chlorbenside, 229
Chlordane, 108, 123, 227, 233, 234, 253
Chlordecone, 228
Chlordimephorm, 381
Chlorfenthol, 228
Chlorfenvinphos, 107, 301, 303
Chlorobenzilate, 229
Chloropropylate, 229
Chlorphenamidine, 132, 358
Chlorthion, 106
*Choristoneura fumiferana*, 237
*Chrysanthemum cinerariaefolium*, 164
*Cimex hemipterus*, 237
*Cimex lectularius*, 237
Circulation of insecticides, 104
Cismethrin, 165, 173, 182
Clenpyrin, 358
Clerodendrins, 350
Cocarcinogenesis, 213
Cocculolidine, 352
Codling moth, 451, 457
*Coelomomyces* spp, 30
Conjugation, 196
Coroxon, 45, 113
*Corvus frugilegus*, 127
Cotton boll weevil, 455
Coumaphos, 45
Counter, 303, 304
Cross resistance, 238
Cryolite, 470, 480
*Culex fatigans*, 20, 29
*Culex pipiens*, 237, 261-264, 321, 476
*Culex pipiens pallens*, 287, 291
*Culex tarsalis*, 29, 237
*Culex tritaeniorhynchus*, 28
Cultural control, 491
Cuticle composition, 398-401
  deposition, 401-403
  formation, 401-403
  structure, 398-401
  synthesis, 401-403
Cuticulin, 398, 403-409
Cyclodienes, 227
Cyolane, 304
Cytochrome P-450, 50, 198, 205-209
Cytoplasmic incompatibility, 29

2,4-D, 71
*Dacus cucurbitae*, 498
*Dacus dorsalis*, 499
DDA, 250
DDCN, 248
DDD, 108, 122, 224, 248
DDE, 238
DDT, 19, 22, 25, 26, 43, 74,86, 103, 223, 229-232, 256, 485
  analogues, 124
  degradation of, 239
  formulations, 141
  receptor fit, 260
  resistance, 43, 47, 50
  safety of, 146
  site of action, 258-260
  theories of action, 259-260
  toxicity of, 151
DDT analogues
  altered aliphatic substituents, 268-269
  asymmetrical, 263-264, 267-268
  configuration of, 265-266
  degradation of, 269
  insect toxicity of, 260-269
  silicon containing, 265

DDT analogues (cont.)
  symmetrical, 261-262
DDT'ase, 43, 46, 47, 239
O-Dealkylation, 116
DEF, 129, 203
Defense secretions, 355, 368
Degradaphores, 272-275
Delaney Clause, 477
Delayed neurotoxicity, 150, 309
Demeton, 150
Dendroctonus ponderosae, 454
Dendroctonus pseudotsugae, 455
Dendrolasin, 355
Dengue haemorrhagic fever, 21-23
Dermestes lardarius, 117
Dermestes maculatus, 54, 117
Derris spp, 482
Desmodus rotundus, 497
Destruxin A, 354
Desulfuration, 115
Deuterium isotope effects, 240
Deutero-DDT, 43, 54, 240
Developmental inhibitors, 359
Diabrotica spp, 237
Diabrotica undecimpunctata howardi, 500
Diacetoxyscirpenol, 354
Diallyl disulfide, 352
Dianisylneopentane, 108, 122, 269
Diatraea saccharalis, 237
Diazinon, 45, 50
Diazoxon, 116
Dichlorobenzoyl-phenyl ureas, 132
Dichlorodihydrochlordene, 108
Dichlorofarnesenic ester, 426
Dichlorvos, 73, 105, 107, 150, 154
Dicofol, 228, 238, 248
Dicyanovinylanilines, 358
Dicyanovinylphenols, 358
Dieldrin, 22, 25, 26, 78, 84, 103, 126, 228, 232, 234, 250, 256, 380
Dieldrin (cont.)
  hazards of, 146
  toxicity of, 152, 232
O,O-Diethyl, O-phenyl phosphorothionate, 49, 51
Dihydroaldrin, 109, 126
Dihydrochlordene dicarboxylic acid, 83, 89, 252
Dihydroheptachlor, 109, 275
Dihydroisodrin, 109, 126
Dihydronereistoxin, 130, 355
Dimethoate, 45, 103, 107, 116, 117
Dimilin, 112
Dinitrophenols, 358, 471
Dinobuton, 358
Dioxathion, 45
Dipofene, 358
Disparlure, 451, 456
Distantiella theobroma, 237
DMC, 228
DNA, 148
Dopamine, 379
Drosophila melanogaster, 248
D-tubocurarine, 380
Dysdercus peruvianus, 237

Ecdysis, 410
Ecological selectivity, 97, 496
EDTA, 206
Eggshell thinning, 255
Electroantennogram, 463
Endosulfan, 25, 109, 126, 228, 234, 253, 275, 276
Endrin, 80, 85, 126, 228, 253
  hazards of, 146
Environmental chemicals, 65
Environmental chemistry of organochlorines, 248-254
Environmental pollution by organochlorines, 243-248
Environmental properties of organochlorines, 244-245
Environmental quality, 65
Ephestia elutella, 446
Epicuticle, 398
Epilachna varivestis, 54, 119

Epithelial transport, 411
EPN, 117, 203
Epoxide hydrase, 127, 131, 425
Epoxides, 444
Erias insulana, 237
Estigmene acrea, 418
Ethylene dibromide, 470
Ethylene oxide, 470
Ethyl tetrachlorvinphos, 107
European corn borer, 457
European Economic Community, 479

Fenitrothion, 22, 95, 141, 301, 310
Fenozaflor, 358
Fenson, 229
Fensulfothion, 304
Fenthion, 25, 310
Fenton's reagent, 206
Fibrous cuticle, 399
Filariasis, 19-21
Fish, 31
Fluoroacetamide, 352
Fluoroacetic acid, 352
Fluoro-DDT, 240
Fluorthion, 106
Food reserves, 7
Food web transport of organochlorines, 247-248
Foramidines, 358, 360
Formaldehyde, 149
Frontalin, 454

GABA, 379
Galecron, 132
Gambusia, 31
Gambusia affinis, 248
Gammarus lacustris, 226
Gardona, 105, 107, 118
Gastroenteric disease, 233
Genetic control, 28-29
Genite, 229
Glossina austeni, 127
Glossina spp, 24-25, 127
Glucuronyl transferases, 114
l-Glutamate, 379
Glutathione, 116, 254, 300
Glycine, 379
Gonadotropic hormones, 423
Gossyplure, 451
Grandlure, 455
Granulosis viruses, 30
Grapholitha fumeferana, 456
Grayanotoxins, 376
Gypsy moth, 451, 456

HCE, 110, 126
Heliothis spp, 104, 119
Heliothis virescens, 50, 237
Heliothis zea, 50, 237
Hellebore, 352
Hemicholinium, 3, 379
HEOM, 110, 127
Heptachlor, 126, 228, 256
Heptachlor epoxide, 244, 245, 249, 252
Hexachloronorbornene, 128
Hexalure, 451, 455
Hexobarbital sleeping time, 213
Highly hazardous compounds, 141
Histamine receptors, 213
Horistonotus uhleri, 500
Hormone mimics, 443
Host plant resistance, 491
Housefly, 44, 236, 458
Human health hazards of pesticides, 145
Human sleeping sickness, 24
Hydrocyanic acid, 483
Hydroxychlordene, 252
Hydroxychlordene epoxide, 252
Hydroxydieldrin, 252
Hydroxy-endrin, 253
Hydroxyrotenone, 129
5-Hydroxytryptamine, 379
Hylema platura, 500
Hylemya spp, 237
Hymexazole, 367
Hypera postica, 237

Ibotenic acid, 354, 367
Imidazoles, 201, 208-210, 213, 368
Induction of microsomal oxidases, 254-255
Insect growth regulators, 112, 359
Insect hormones, 421-438
Insecticides
  action on nerve, 375
  baits, 499
  cost of, 10
  development of, 8
  dosage, 497
  formulation, 498
  history of, 469-487
  incentives to develop, 12
  in public health, 17-33, 229-233
  intoxication, 27
  laws, 35
  metabolism, 196-199
  methods of application, 498
  patent conventions, 13
  poisonings, 258
  production, 1, 4, 223-228, 480-483
  public, and the, 484
  registration, 63
  registration standards, 13
  regulation, 477
  research costs, 12
  residues in food, 145
  resistance, 18, 27, 28, 41-63, 475-477, 486
  resistance, genetics of, 238
  safety of, 145-154, 157-159
  selective toxicity, 384
  systemic, 500
  target sites, 371, 385
  temperature dependency, 387
  timing of application, 500
Insensitivity index, 43, 45
International agencies, 36-37, 63, 510
International Agriculture Trends, 6
*Ips calligraphus*, 447
Ipsdienol, 446
Ipsenol, 446
*Ips grandicollis*, 446
*Ips latidens*, 447
*Ips paraconfusus*, 447
*Ips pini*, 447
*Ips typographus*, 447
Isobenzan, 228
Isoboldine, 352
Isodrin, 126
Isopropyl parathion, 105, 312
Isopropylphenyl N-methylcarbamate, 318
Isovaleric acid, 445
Isoxazoles, 367

Juvabione, 430
Juvenile hormones, 112, 130-131, 220, 422
Juvenile hormone biosynthesis, 424-425
Juvenile hormone metabolism, 131, 425
Juvenile hormone mimics, 51, 130, 219, 476

Kadethrin, 165, 170, 181
KDR gene, 46
Kelthane, 228
Kepone, 228
Kerosene, 470
Keto-dieldrin, 252
Keto-endrin, 253
Keto-photodieldrin, 252
Knockdown, 163, 179, 180
K-Othrin, 165, 180

*Lagenidium* spp, 30
Landrin, 314, 343
Lannate, 119
Larviciding, 20, 22, 24
*Laspeyresia pomonella*, 233, 237, 473, 475, 494
$LD_{50}$ values, 100, 143, 166-167, 225, 343

Lead arsenate, 470, 480
Lepomis macrochirus, 226
Leptinotarsa decemlineata, 237, 470
Leptophos, 303, 309
Lethane, 384, 359
L-glutamate, 361-367
Lindane, 128, 227, 232, 234, 254, 287-291
Lipid-water partition coefficient, 248
Lipophilic compounds, 196-197
Lipophilicity, 178
Liver tumors, 148, 256
Livestock pests, 437
Lonchocarpus spp, 482
London purple, 478
Lumbricus terrestris, 303
Lumbrineris heteropoda, 130
Lygus spp, 494

Malaoxon, 44, 49, 50, 56, 314
Malaria, 17-19, 231-232
Malaria eradication, 232
Malathion, 22, 25, 56, 43, 44, 103, 107, 116, 198, 310
Malathion monoacid, 116
Malathion resistance, 314
Mammalian selectivity ratio, 113, 115, 298, 299
Mammalian toxicity ratios, 343
Manduca sexta, 103, 131, 422
Manpower needs, 36
Marine toxins, 368
Matacil, 343
Melanoplus differentialis, 103
Melanoplus femur-rubrum, 103
Melolantha melolantha, 131, 423
Membrane potential, change, 376
Mermithid nematodes, 30
Mesurol, 120, 343
Metabolism and selectivity, 114
Methamidophos, 302
Methomyl, 119, 343
Methoprene, 238, 432-433, 436
Methoxychlor, 108, 122, 234, 276
Methyl bromide, 470
Methylenedioxynaphthalene, 119
Methylenedioxyphenyl N-methylcarbamate, 118
Methyl parathion, 106, 113
MGK 264, 210, 212, 214, 219
Microbial insecticides, 353
Microbial toxins, 368
Microencapsulation, 221, 304, 436, 456, 464
Microsomal difference spectra, 208
Microsomal electron transport, 198
Microsomal metabolism, 293-294
Microsomal oxidase inhibitors, 205-211
Microsomal oxidases, 46, 114, 127, 368
Microsomal oxidations, 273-274
Miller Amendment, 477
Mink, PCBs and, 94
Mirex, 228, 234, 254
Mixed function oxidases, 46-51, 56, 129, 197-198, 301
  induction of, 152
  inhibitors, 119
Mon-0585, 359
Monitor, 302
Monoamine oxidase, 132
Mosquito coils, 164
Mothproofing, 236
Moulting gel, 401
Multiple resistance, 476
Musca domestica, 42, 119, 169, 261-264, 287, 291, 476
Muscamone, 458
Mutagenesis, 145, 149
Myrcene, 454
Myristicin, 351

N-acetyl Zectran, 314
NADH oxidase, 360
Naphthodioxoles, 203

Naphthylamine, 148
Naphthyl propynyl ether, 119
Naphthyl thiourea, 148
Natural products as insecticides, 349-362
N-dealkylation, 300
NDRC 143, 192
Negative after-potential, 259
Negatively correlated insecticides, 55
Neopyramin, 164
Nephotettix cincticeps, 42
Nereistoxin, 111, 130, 355, 367, 381
Nerve action potentials, 48
Nerve axon, 259
Nerve excitation, mechanism, 373
Nervous system, 104
Neuromuscular junction, 361, 367
NIA 16824, 201
Nicotine, 352, 381, 383, 470, 481
Nicotine bentonite, 480
Nicotine tannate, 480
Nitrites, 151
p-Nitroanisole, 52
N-nitrosocarbaryl, 151
Non-tariff trade barriers, 479
Nontoxicity, 158
Norepinephrine, 379
Nosema spp, 30
NRDC 108, 182
NRDC 119, 182
NRDC 161, 174, 192
Nuclear polyhedrosis viruses, 30

Octanol-water partition coefficients, 178
Oil-water partition coefficients, 102
Omite, 359
Onchocerca volvulus, 23
Onchocerciasis, 23-24
Organochlorine insecticides, 223-276
  animal toxicity of, 225
  detoxication of, 122-128
  in agriculture, 233-235
  in households, 234
  residues, 257
  resistance to, 236-240
Organometallic compounds, 369
Organophosphorus insecticides, 295-305
  activity, 299
  detoxication, 115
  persistence, 310
  toxicity of, 149
Organothiocyanates, 359
Oriental fruit moth, 452, 456
Orthene, 302
Ortho-chloro-DDT, 43, 54, 242
Ortho-phenyl phenate, 479
Ovarian development, 424
Ovex, 229, 359
Oxadihydroaldrin, 109
Oxazoles, 368
Oxetanes, 269
Oxidative phosphorylation, 360
Oxychlordane, 253
Oxythioquinox, 359

Panonychus ulmi, 496
Panstrongylus spp, 26
Paraoxon, 52, 105, 106, 296
Parapheromone, 455
Parathion, 76, 106, 115, 296, 304
Paris green, 478
PCBs, 94
Pectinophora gossypiella, 233, 237
Pediculus humanus, 229, 233, 237
Penetration of insecticides, 102
Pentac, 53
Pentachlorocyclohexene, 254
Pentachlorophenol, 71

_Periplaneta_ _americana_, 250, 407
Permethrin, 176, 192
Peroxidases, 409
Perthane, 224
Pest management, 345
Petroleum oils, 470, 482
PH 6040, 359, 361, 417
_Phaedon_ _cochleariae_, 169
_Phalocrocoras_ _aristotelis_, 127
Pharmacokinetics, 98-99, 114
_Phasianus_ _colchicus_ _torquatus_, 226
Phenethyl isothiocyanate, 352
Phenolases 407-409
Phenothiazine, 480
Phenothrin, 176
Phenyl barbitone, 148
Phenylthiourea, 358
Pheromones, 445-458
  for mass trapping, 453-455
  for mating disruption, 455-457
  monitoring with, 451-453
  resistance to, 465
  specificity, 447
  traps, 452
Phorate, 303, 304
_Phormia_ _regina_, 234, 261-264
Phosphamidon, 303
Phosphorthionate activation, 300
Phosvel, 309
Photoaldrin, 87
Photodieldrin, 87, 250
Photolysis, 87-89
Photostability of pyrethroids, 175
Phoxim, 303
Physiological selectivity, 97
Physostigmine, 350
Phytoplankton, 90
Piericidin A, 354
_Pieris_ _brassicae_, 126
_Pieris_ _rapae_, 237
_Pieris_ spp, 119
Pinene, 454
Pink bollworm, 451, 455
Piperonyl butoxide, 115, 163, 200, 212, 213, 214, 219, 261-264
Pitressin, 152
Plague, 27-28
_Plasmodium_ spp, 231
Plictran, 358
_Plodia_ _interpunctella_, 446
_Plutella_ _maculipennis_, 237
_Poecilia_, 30
_Poecilus_, 248
Polychlorophenols, 254
Potassium permeability, 376
Prasinons, 354
_Prodenia_ spp, 119
Prolan, 54, 108, 122
Prolan analogues, 268
Propargyl ethers, 431
Proparthrin, 165, 181
Propoxur, 314, 323, 343
Propyl isome, 200, 212, 214, 219
Prothrin, 165
_Pseudomonas_ spp, 418
_Pteronarcys_ _californicus_, 226
Pyresmethrin, 179, 181
Pyrethrins, 25, 110, 163, 350, 470, 481
  safety of, 147
Pyrethroids, 33, 128, 163-183
  acid side chains, 172-175
  alcohol components, 170-172
  detoxication of, 128
  mammalian toxicity of, 181-182
  production of, 192
  properties of, 175-182
  resistance to, 191
  structural features of, 168-175
  toxicity of, 153

_Quadraspidiotus_ _perniciosus_, 475
Quassin, 351

Random screening for insecticides, 356
Rat flea, 28
Red banded leafroller, 463
Red spider mite, 42
Reproduction inhibitors, 359
Reproductive isolation, 445
Residual house spraying, 17, 26, 146, 231-232
Residue tolerances, 478
Resistance, counter measures for, 54-56
Resistance development, 52-54
Resistance-proof DDT derivatives, 240-243
Resmethrin, 165
*Rhodnius prolixus*, 26, 27
RNA polymerase, 354
Rodenticides, 28
Ronnel, 73
Rotenolone, 129
Rotenone, 129, 350, 470, 482
Ruelene, 105
Ryanodine, 352

Sabadilla, 352
Safety practices with insecticides, 153
Safety of insecticides, 145-154, 157-159
*Saissetia olease*, 475
Salicylanilides, 358
*Salmo gairdneri*, 226
*Salvelinus namaycush*, 246, 247
*Sarcoptes scabiei*, 233
Saxitoxin, 376
S-biollethrin, 171, 181
Schradan, 296
*Scirtothrips citri*, 475
Screening for insecticides, 356, 359
Screening for synergists, 142
Seawater, 90
*m*-*Sec*-butylphenyl *N*-methyl *N*-thiophenyl carbamate, 238
Selective toxicants, 360
Selective toxicity, 97
Selectophores, 337-339
*Selenopsis invicta*, 437
Selocide, 470
Sesamex, 49, 126, 131, 169, 174
Sesamin, 200
Sesamolin, 200
Sesquiterpenoid dienoates, 432
S-ethyl parathion, 301
Sex-linked translocations, 29
*Similium* spp, 23
*Sitophilus* spp, 237
SKF 525-A, 201
Sodium gate, 259
Sodium permeability, inhibition, 376
Soil insecticides, persistence of, 304
Soil persistence of organochlorines, 246
*Solenopsis richteri*, 234
Solvents, 141
*Spodoptera eridania*, 446
*Spodoptera exigua*, 237
*Spodoptera littoralis*, 237, 446
*Spodoptera litura*, 446
Sterile male technique, 28-29
Steroid metabolism, 255
*Stomoxys calcitrans*, 127
*Streptomyces prasinus*, 354
Strontium 90, 94
Structural optimization, 356-359, 367-368, 426
Substance P, 379
Sulfenylated carbamates, 323-336
Sulfoxide, 163, 200, 212, 214, 219
Sulfur, 470, 483
Sulfur dioxide, 479
Sulphenone, 229
Sumioxon, 113
Sumithion, 106, 113, 116
Surangin B, 351, 360

Synaptic transmission
  effects of drugs, 379
  mechanism, 378
Synergism, 101, 115, 195-216
Synergistic ratio, 115, 201, 287, 291
Synergists, 55, 61, 163, 199-205, 430-431
  commercial use of, 214-216
  mammalian toxicity of, 211-214
Synergophores, 221
Synthetic organic insecticides, 474
Synthetic pyrethroids, 214
Systemic insecticides, 296

2,4,5-T, 71
Tartar emetic, 470, 475
Telanomid wasps, 31
Temephos, 22, 24, 33, 95
Temik, 119
*Tenebrio molitor*, 131, 422
Tenuazonic acid, 354
Teratogenesis, 145, 147
Tetrachlorvinphos, 107, 118, 303
Tetradecadienyl acetate, 446
Tetradecenyl acetate, 446, 449
Tetradifon, 229
Tetraethyl pyrophosphate, 296
Tetramethrin, 164
Tetranactin, 354
*Tetranychus cinnabarinus*, 321
*Tetranychus mcdanieli*, 496
*Tetranychus urticae*, 42, 43
Tetrasul, 229
Tetrodotoxin, 376
Thalidomide, 157
*Thamnophis sirtalis*, 248
*Therioaphis maculata*, 497
Thiadiazoles, 368
Thin eggshells, 152, 255
TOCP, 203, 301
Toxaphene, 75, 128, 227, 234, 254, 275
Toxic effects of insecticides, 146
Toxicological judgements, 158
*Toxorhynchites* spp, 31
*Triatoma* spp, 26
Triatominae bugs, 26-27
*Tribolium castaneum*, 117
Trichloromethylbenzylanilines, 108, 123, 266
Trichloromethylbenzyl phenyl ethers, 108, 266
Tricholomic acid, 354
Trichomalic acid, 367
*Trichoplusia ni*, 50, 237
Trimethyl phosphate, 150
Triorganotins, 358
Triphenyl phosphate, 49, 117
Triphenyltin acetate, 358
Tropital, 163, 200, 212, 214, 219
*Trypanosoma cruzi*, 26
Tse-tse flies, 24-25, 31
Tsumacide, 343
Typhus, 229-231
Typhus fever, 28

ULV spraying, 22, 95, 164, 192
Uridine diphosphate glucuronic acid, 114

*Vanessa io*, 131
Veratridine, 352
Verbenol, 446, 454
Versimide, 354
Vinyl chloride, 148, 153
Vitamin D, 152
Volta river basin, 24

WARF-antiresistant, 56
Waterbuffalo toxicity, 309
Wax canals, 406
Western pine beetle, 454
Wildlife toxicity, 339
*Wuchereria bancrofti*, 19

Xanthone, 480
*Xenopsylla cheopis*, 237
Zectran, 120, 343
Zineb, 77

*PRACTICAL LEGAL ADVICE*
*FOR*
*BUILDERS AND CONTRACTORS*

# PRACTICAL LEGAL ADVICE FOR BUILDERS AND CONTRACTORS

EDWARD E. COLBY
*Attorney at Law*

PRENTICE-HALL, INC., ENGLEWOOD CLIFFS, NEW JERSEY

10 9 8 7 6 5 4 3 2 1

ISBN: 0-13-692079-9

Library of Congress Catalog Card Number: 79-170888

Printed in the United States of America

**PRENTICE-HALL INTERNATIONAL, INC.,** *London*
**PRENTICE-HALL OF AUSTRALIA, PTY. LTD.,** *Sydney*
**PRENTICE-HALL OF CANADA, LTD.,** *Toronto*
**PRENTICE-HALL OF INDIA PRIVATE LIMITED,** *New Delhi*
**PRENTICE-HALL OF JAPAN, INC.,** *Tokyo*

*To my sons Ray and Robert*

" . . . a house built with love."

# CONTENTS


1 *STARTING YOUR OWN CONSTRUCTION BUSINESS* 1

- 1-1 *Name of New Construction Business* 2
- 1-2 *Location of New Business* 4
- 1-3 *Financial Needs of New Business* 5
- 1-4 *Financial Statements–Accounting* 6
- 1-5 *Land Development* 14
- 1-6 *Laws–Rules and Regulations* 18

2 *BUILDING AND CONSTRUCTION CONTRACTS* 19

- 2-1 *Construction and Effect* 19
- 2-2 *Compensation* 22
- 2-3 *The Architect and the Contractor* 27
- 2-4 *Substantial Performance* 29
- 2-5 *Excuses For Nonperformance* 32
- 2-6 *Accidental Destruction of the Building* 33
- 2-7 *Damages* 35

2-8 Assignment 37
2-9 Termination 39
2-10 Remedies 42
2-11 Minors and Contracts 42

3 PARTNERSHIPS 45

3-1 Formation of Partnership 46
3-2 Dissolution of a Partnership 49
3-3 Limited Partnership 49
3-4 Joint Business Ventures 50

4 SUBCONTRACTORS 51

4-1 General Information 51
4-2 Subcontractor and General Contractor 52
4-3 Subcontractor and Owner 55
4-4 Regulations Pertaining to Subcontractors 58
4-5 Liability of Subcontractors for Injuries 59

5 CORPORATIONS 63

5-1 Types of Corporations 64
5-2 Incorporation and Organization 66
5-3 De Facto and De Jure Corporations 68
5-4 Corporate Name, Seal, and Domicil 69
5-5 Bylaws 71
5-6 Records and Reports 71
5-7 Capital and Capital Stock 72
5-8 Stock–Common, Preferred, and Miscellaneous 73
5-9 Value of Stock 75
5-10 Transfer of Stock 77
5-11 Stockholders 77
5-12 Stockholders' Meetings and Elections 78
5-13 Liability of Stockholders 79
5-14 Dividends and Stock Splits 79
5-15 Bonds and Debentures 80

6 INSURANCE–WORKMEN'S COMPENSATION AND PUBLIC LIABILITY 85

6-1 Workmen's Compensation 85
6-2 Public Liability 89
6-3 Automobile Insurance 92
6-4 Miscellaneous Insurance 93

7 NEGOTIABLE INSTRUMENTS 99

7-1 History 99
7-2 Uniform Commercial Code and NIL 102
7-3 Negotiable Requirements 103
7-4 Checks 105
7-5 Bond for Employees 109
7-6 Accommodation Paper 109
7-7 Nonnegotiable and Negotiable Instruments 110
7-8 Usury and Conflicts 111

8 LIENS 115

8-1 Origin and Background 115
8-2 Types of Liens 117
8-3 Mechanics' Liens 118
8-4 Right to Lien 119
8-5 Procedure for Obtaining Mechanics' Liens 124
8-6 Operation and Effect of Lien 125
8-7 Enforcement of Lien 126

9 CONDOMINIUMS AND COOPERATIVE APARTMENTS 127

9-1 Definitions and Background Information 127
9-2 Condominiums 128
9-3 Statutory Provisions for Condominiums 127
9-4 Cooperative Apartments 129
9-5 Condominiums vs. Cooperative Apartments 130

10 FORECLOSURES 133

10-1 Mortgage 134
10-2 Interest 134
10-3 Parties 139
10-4 Mortgage Notes 141
10-5 Mortgage Foreclosure 143
10-6 Deficiency Claims 145
10-7 Equity of Redemption 145
10-8 Trust Deeds 146
10-9 Trust Deed Foreclosure 147

11 BANKRUPTCY 149

11-1 Introduction 149
11-2 Filing of Bankruptcy Petition 151
11-3 Referee in Bankruptcy 152

11-4 Involuntary Bankruptcy Proceedings 154
11-5 Claims of Subcontractors, Laborers and Materialmen 155
11-6 Discharge in Bankruptcy 156
11-7 Exemptions 157
11-8 Debtor Relief Provisions 158
11-9 Wage Earners' Plans 160

12 ARBITRATION AND AWARD 163

12-1 Explanation 163
12-2 Arbitration vs. Appraisement 165
12-3 Arbitration Statutes 166
12-4 Arbitration Agreements 167
12-5 Right to Lien 170
12-6 Pending Lawsuits and Revocation 171
12-7 Who and What May Be Arbitrated 172
12-8 The Award 174

13 SURETY BONDS 177

13-1 Definitions 177
13-2 Performance Bonds 182
13-3 Who Is Covered 183
13-4 Discharge of Surety 186
13-5 Public Construction Work 189
13-6 Who Is a Subcontractor under the Miller Act 191
13-7 Rights and Remedies of Sureties 192

14 ARCHITECTS 195

14-1 Who and What Is an Architect 195
14-2 Employment and Duties of Architects 196
14-3 Payment 197

15 ODDS AND ENDS 201

15-1 Easements 201
15-2 Escrow 202
15-3 Fixtures 204

GLOSSARY 205

INDEX 227

# PREFACE

This book was written for practical use by builders and contractors who are just beginning or are about to begin their new business career in the category of being in business for themselves for the first time. In other words, they are entering a brand new realm of experience, a thrill that comes only once in a businessman's lifetime.

The book explains some of the do's and don't's, the potential rewards and the possible pitfalls that face our ambitious adventurer who is about to start his own business for the first time in his life. Therefore, the material has been streamlined with an attempted deletion of all ambiguous, repetitious, and verbose material. It is written in simple and plain language so as to be easily understood by all. Even technical, legal terms are stated in simplified form.

There has been no attempt to educate the reader on the basis of a short academic course in real estate or construction law. Rather, the emphasis is to educate the reader on the general types of problems commonly encountered in such business situations and to encourage the builder to seek competent advice, be it in the field of law, accounting, or financing, *before it is too late.* Doctors and dentists have advocated preventive medicine and dental care as a means towards health; similarly, preventive law can be of tremendous help to the brand new aspiring building and construction business, possibly to the extent of saving its young commercial life.

Acknowledgment is made for research material emanating from American Jurisprudence 2d.

Acknowledgment is also made for forms furnished through the courtesy of The American Institute Of Architects, 1735 New York Avenue, N.W., Washington, D.C. 20006.

Acknowledgment is also made for forms furnished through the courtesy of The Associated General Contractors Of America, 1957 E. Street, N.W., Washington, D.C. 20006.

Finally, acknowledgment is also made for the encouragement, understanding advice and constructive criticism of Alonzo Wass, Author and Professor, and my Wife, Anita Rhee Colby, who made this book possible. Their cooperative assistance has been indispensable to my literary efforts.

Edward E. Colby

*PRACTICAL LEGAL ADVICE*
*FOR*
*BUILDERS AND CONTRACTORS*

# 1

# STARTING YOUR OWN CONSTRUCTION BUSINESS

When an individual, who has previously been employed for a number of years by a large construction firm, decides he wants to go into business for himself on a small scale, it is recommended that he seek competent legal advice *before* becoming involved in legal and business ramifications rather than *after*.

An attorney is best equipped to advise the prospective businessman about the considerations he must take in starting a new business venture. For instance: the amount of money he has available to put into his new business; his ability to support his family during the early period when the money is going out and nothing is coming in; his financing needs; his bank or other source of borrowing construction money; his financial statement; his choice of an accountant; whether he wants to use his own name, a trade name, or possibly form a new corporation; and the various licenses and forms that are required by the city, county, state, and federal authorities.

Modern business life is complicated. Certain laws, rules, and regulations must be complied with, and the best time for you, the prospective businessman, to learn about them and their requirements is at the beginning. The Internal Revenue Service, usually located in the Federal building, will furnish the necessary forms for registering as an employer and the information needed for your bookkeeper to make the required payroll deductions.

Appreciating that the laws of different states will vary on almost every branch of commercial endeavor, the basic requirements follow the same general pattern. Most states require that a builder and contractor take and pass an examination given at regular intervals before he may engage in the construction business. Usually, the contractor's licensing board or similar agency has a handbook that sets forth the requirements of a particular state. Frequently, licensing requirements are stricter for plumbers and electricians than for other crafts because of the element of public safety involved.

Many states require that either a sales or a school tax must be paid by all businesses including construction firms. If this applies in your state, the Bureau of Revenue or comparable agency will provide the necessary forms for your bookkeeper.

Let us not overlook the usual requirements of most cities or municipalities. At city hall the city treasurer will advise the requirements for an occupational license and the building inspector will state the requirements and fees that must be paid in connection with approval of plans and specifications before a building permit may be issued.

Another aspect to consider in the construction business involves the purchase of land or lots. Here, too, an attorney can best advise the prospective businessman about the zoning requirements for residences as well as how to handle the matter of title insurance to insure a valid, legal title for the land.

Furthermore, the attorney will discuss the advisability of insurance coverage for the new business, including, but not limited to, tools and equipment; public liability insurance for all motor vehicles used by the employer or his employees; workmen's compensation insurance to cover his employees; and performance or surety bonds that are required by customers on certain jobs.

As you can see, starting a new business is a chore but not an impossible task. The important requirements for the young builder and contractor are that he knows his business, is well organized, is well advised; is adequately financed, and willing to work much harder than he ever did as an employee of a large construction firm. Working for oneself is entirely different from working for someone else; but there is no joy to compare with that of succeeding in one's own business venture, in spite of headaches and heartaches that it may include.

The purpose of this preliminary discussion on how to venture forth and start a brand new construction business is not to discourage the novice builder and contractor; rather, it is to enlighten him. In the following chapters, each and every step necessary for the builder and contractor will be covered in detail.

## 1-1 NAME OF NEW CONSTRUCTION BUSINESS

Just as young parents are faced with the thought provoking chore of naming their new baby, the same problem faces the new businessman in selecting a name for his construction business.

There are a number of choices and caution should be used, together with due deliberation, before deciding on your new firm's name. The answer may not be as simple as it seems.

Legally, you can use your own name for a new firm if it does not conflict with an existing construction company. The old legal adage of "prior in time is prior in right" holds true in protecting an existing name. You are not privileged to use your own name when it would create confusion and possibly hurt the existing firm with the same or similar name.

A new name and two alternatives should be selected as possible choices. A check should then be made at the state office of the secretary of state or comparable agency to find out if any or all three choices for a proposed new name are already being used by someone else. If this is the case, it will be necessary to select another name. If not, you can then have the new business name registered to protect yourself for the future.

Whether you use, for example, the name of Jones Construction Company or a trade name, such as Pacific Construction Company, it has no legal significance. If Jones ever became involved in a lawsuit, he would be referred to as Jones, doing business as (DBA) Pacific Construction Company.

There have been many lawsuits involving business names that supposedly conflict with other names that are either identical or similar. The courts have ruled that if a new name causes confusion among clients who think they are dealing with the original firm, then the proposed new name must be changed to avoid that confusion. An outstanding example is the Coca Cola Bottling Company, who has spent millions of dollars over the years to protect its trade name as well as its distinctive bottle so that they may not be used by anyone else.

If you select the trade name of Pacific Construction Company and discover there is a construction firm already registered under the trade name of Pacific Coast Construction Company, this could represent a borderline case. In order to avoid a potential suit, it is recommended that another trade name be selected to avoid any possible claim of confusion.

The best legal advice that any attorney can offer his businessman client is to stay out of court whenever possible. Legal proceedings are very expensive and time consuming to the extent that the average small businessman cannot afford the luxury of litigating his problems. Whenever possible, business agreements should be clear and concise so that disputes will not arise and problems will not have to be presented to the court for a ruling. The cost of consulting with your attorney before signing any document is minimal and inexpensive insurance against a misunderstanding that may later lead to a dispute in court. The few hours spent in your attorney's office are nothing compared with the days required for preparation and actual time spent in court.

All clients are unhappy when they learn from their attorney that court proceedings are not only expensive and time consuming, but it takes many months and sometimes years before the dispute is finally ruled on by the court. The reason why reference is made to the court instead of a judge is that sometimes the business dispute is presented to a jury of twelve people or less for a verdict, rather than a ruling by a judge. In smaller communities it frequently takes one year from the time the lawsuit is filed until the matter comes to trial before a judge. In larger cities, the trial waiting period of one year can stretch out to two or three years.

Jury trials usually take longer because there are more cases waiting. Corporate litigants, including, but not limited to, bonding and insurance companies, are not in any rush to settle litigation pending as they prefer the longer waiting periods afforded by the jury docket. It takes longer for any matter to be presented to a jury than to a judge and the more lawyers involved, the more time will be consumed in court proceedings. Most jury trials will take a minimum of two days in court and will sometimes drag out for long periods of time when the problems are complicated.

In larger cities, the court calendar is usually clogged and it may take years before the pending lawsuit awaiting a jury trial actually comes to trial. Large corporate litigants can better afford costly and time consuming appeals than can the small businessman.

When you talk about delays before your dispute comes to trial in the lower court, the real delay can occur when your matter is appealed and many months and even years can drag along before the appellate court rules on the case.

It may sound trite to quote the old saying, "A poor settlement is better than the best lawsuit" but it is still true. The attorney who successfully keeps his client away from the courthouse is doing a better job for his client than the attorney who encourages litigation and is anxious to make new law. The client who tells his attorney that he is willing to spend $5,000 to prove his point to win a $1,000 dispute is just not being practical. "Discretion is the better part of valor" and frequently, a courthouse winner later discovers that he was actually a loser. Spend your time and energy in pursuing your construction business instead of trying to prove points of law in the courtroom.

My efforts will have been successful if the reader only learns one thing: *consult your attorney before signing any papers.*

A consoling factor about attorneys' fees is that they are deductible as a business expense for income tax purposes.

## *1-2 LOCATION OF NEW BUSINESS*

Mature thought should be given to the location of the new construction business. Before you think about the need of a small office and a construction yard for equipment and materials, you should think about *overhead.*

Overhead has been defined as business expenses not chargeable to a particular part of the work. The concept sounds very simple but is really complex. Some of the main items of overhead are rent, office and labor payroll, payroll deduction expenses, insurance, cost of equipment, motor vehicles, taxes and license fees, depreciation, legal and accounting fees, utilities, cost of borrowed money, and other hidden items. More new businesses have failed because of high overhead than any other factor. The expenses of a new business must be watched very carefully to avoid that pitfall. At the beginning of the new business venture, all money is going out and, temporarily, no money is coming in. If a new business can keep its head above water the first year by breaking even, it is doing exceptionally well. If the business prospers as time goes by, it is very easy to expand the overhead by moving into larger offices.

A fancy office is not a requirement nor is it recommended for your new construction business. The odds are favorable that you will be calling on your prospective customers at their homes or offices, rather than their calling on you at your office. Whether you combine your office location with your construction yard is a matter of personal choice. Although a central location is desirable, the cost is the most important factor. If cheaper rent is available farther out, then that is where you belong.

Equipment for your small office should also be held to a minimum. Gadgets, such as dictating machines, are luxuries that should wait until your new business is established and prospering. It is so easy to buy things, but paying for them later is not so easy for a new business. Even items sold by the telephone company, such as fancy new dials or telephone instruments, sound like small items, but when you figure the annual cost it becomes another item of expense that can be eliminated at the start of the new business.

Many builders and contractors have discovered that they are better off to rent construction equipment when needed for a job rather than tie their capital up in a purchase of the same equipment. The same principal of being frugal would apply.

The actual location of the new construction business should be dictated by the rental cost more than any other factor. Most communities have zoning law requirements that restrict the construction office and yard to certain limited areas. When you are dealing with your prospective landlord, you have to inquire about the zoning regulations and make sure that the zoning is proper for your particular business. On the question of renting by the month or arranging a lease for a fixed period of time, individual desires would control. A lease for a reasonable period of time offers protection to the tenant as well as the landlord. A lease for a long period of time should be avoided by the tenant because your construction business is new and your needs may change in the future, and you certainly don't want to be tied up by a long term lease. From the tenant's point of view, the most desirable leasing arrangement would be to start with a short period up to three years and then have an option for renewal for a longer period at an agreed rental. The option for renewal gives the tenant a free choice to renew the lease or permit it to expire at its original date. Most landlords want their tenants to stay put for a long period of time. Leases that provide that the tenant has the option to renew for a given period of time at a rental to be agreed upon by the parties is a meaningless situation because the tenant may say $200 per month and the landlord may say $500 per month. The option must name the agreed rental in order to protect the tenant.

Many people have a misconception on the question of whether a tenant may break a lease. A lease is a written agreement or contract between the landlord and the tenant. Like all other contracts, it may be broken by either party for only valid and justifiable reasons. A recent example involved a tenant leasing a business location for the painting and body repair of motor vehicles. The city condemned the building because the electrical wiring was defective and the landlord refused to make repairs to comply with the requirements of the building inspector and fire department. As a result, the tenant broke his lease. We would all agree that the tenant was justified in breaking his lease. A good way to answer the question, "May the tenant break his lease?" is to ask the tenant what his answer would be if the landlord said, "I will cancel your lease effective the first of the month." If the tenant says to the landlord, "I have a valid lease and you are not privileged to cancel it," then the tenant has answered his own question. If it is a valid lease on one side, it is a valid lease on the other.

Another example of a similar situation arises when people sign a contract to purchase land, pay a small deposit and later change their mind and ask if they can break the contract. When you ask the buyer what his reaction would be if the seller said, "Buyer, I have another party who wants to buy and pay me more money than you agreed to pay and I therefore cancel your contract," the buyer will complain bitterly that he has a valid binding contract and that the seller cannot cancel. If the land purchase contract is valid and binding on the seller, then it is valid and binding on the buyer.

Over the years, comments have been made about the location of Woolworth stores on the main streets of cities throughout the country and the high rents that are involved. The Woolworth company has been very successful, and the high rents that it pays are a small part of its overhead as compared with the volume of business. In other words, if you are doing a high volume of business and are successful, then the overhead is of small concern and presents no problem. But when you are just starting out in your new construction business, overhead expense is of vital concern and could mean the success or failure of your business.

## 1-3 FINANCIAL NEEDS OF NEW CONSTRUCTION BUSINESS

How much money does it take to start a new construction business? There is no logical answer to this question. If you ask the cost of a new car, you must state whether you want a Ford or a Cadillac. Similarly, in starting a new construction business, you can start off with a flourish of "don't spare the horses" or go to the other extreme of "poor boying" it. A happy medium would probably be the most desirable result.

The first year of a new business is comparable to the first year of a marriage in that it is a critical testing period. If the first year is reasonably successful, then half of the battle is won. More new businesses fail in the first year than in all other years combined because of inadequate financing.

Until the first construction job begins or the sale of the speculative home develops, the money is going out and nothing is coming in. Even when the first construction job is started and a profit is in the offing, it still takes time before payday comes around. Meanwhile, the usual items of overhead continue. Landlords complain about the long wait for rent day to arrive, while tenants complain about the days flying by.

It therefore behooves the new builder and contractor to sit down and figure his regular overhead in order to determine his financial needs. This would be similar to a personal budget that some people try to live on. When you finish listing the various items of overhead, you add on at least 10 to 15 percent for miscellaneous items that always show up. No matter how hard you try to be accurate, some hidden expense item will rear its ugly head.

Now that you have an idea of your monthly overhead, you are in a better position to answer the original question, "How much money does it take to start a new construction business?" You know that the first income will take some months and your personal opinion controls as to how long you can operate with deficit spending. The government has good borrowing power and can go into the hole for long periods of time. How good is your borrowing power and how long can you continue with deficit spending?

In addition to the regular expenses of your new construction business, don't overlook the financial needs of your family and household. Your former weekly paychecks when you were an employee are now a thing of the past. In addition to knowing the overhead budget of your new business, you obviously should know what is needed to maintain your family. The financial needs of your new business should be kept separate from your personal expenses. Some builders and contractors borrow money from their banks for construction or business purposes and then make the fatal mistake of living from the borrowed money. As you can readily see, this practice is soon going to create a shortage for the needs of the business. When the time to repay the loan arrives and the money is not available, the new business is going to suffer, and your banker will be unimpressed with your explanation that you needed some of the money to live on. It is important to the success of your business that your bank credit be maintained on a prompt pay and good credit basis. If your conduct shakes the bank's faith in you, the chances of your being successful are greatly reduced. If you try a second or third bank, each will want to know why you are leaving the first bank, and the facts will speak for themselves.

Always shoot straight with your bank and it will pay dividends in the long run. Anticipate your money needs and don't wait until the last minute and then run to your bank in a panic. Bankers are calm and cool in making loans. If you are able to make your bank payment ahead of schedule, it creates a favorable impression and becomes a part of your prompt payment record. If you are unable to make your bank payment when it is due, discuss the problem with your banker in advance of the due date. He will be impressed with your businesslike efficiency even though you are unable to pay promptly.

## *1-4 FINANCIAL STATEMENTS—ACCOUNTING*

We are now living in a computerized world and things are different in the field of finance. The buyer as well as the seller has learned that you can't fool the computer. If an error has been made, it takes time for the computer to make the correction. The use of credit cards as a substitute for money has grown in popularity throughout the world. It has been suggested by various banking officials that the need for the use of money is decreasing, while the use of computers is increasing.

With the passing of time, banks, savings and loan associations, insurance companies, and other financial institutions have profited from their experiences in making loans to builders and contractors. Whenever a builder or contractor is unable to pay his bank loan, the bank suffers a loss. The bank then analyzes the cause of the financial failure. When a building project results in a loss, it is very important that the builder or contractor understand why he lost money. One builder, who built and sold homes for $35,000 each, never did learn that his cost was greater than his sales price; consequently, he went bankrupt. It is a tragedy that he did not have any idea about his costs, and his entire building operation was doomed to failure. No business can operate very long unless it earns a profit. And cost accounting is a must in order to know the selling price if the business is going to earn a profit.

A financial statement is a detailed list of your assets and liabilities—a list of what you own and what you owe. When you subtract your liabilities from your assets, your financial statement will show your net worth.

Knowing that many clients are too optimistic in their estimates of the value of assets, the banks automatically reduce the figures. Banks and other financial institutions have learned that a financial statement presented by the client who wants to borrow money must be certified by his accountant to really mean anything. The financial statement forms used by most banks and financial institutions include detailed questions about income taxes paid last year and the estimated income tax for this year. This information not only helps the bank to determine the client's net worth but also helps the loan officer to set a credit limit or line of credit on borrowings. (Additional information on financial statements will be found in Chapter 10, section 10-3.)

There is no substitute for taking care of one's business. The old saying, "Who is taking care of the store?" was never more important in the success of a new, struggling business. Whenever you see a new successful business, you can be assured that the owner is properly supervising the business and it is not an accident or a lucky break.

Selecting your accountant is nearly or equally important as selecting your attorney. Frequently, both will work together to help solve your financial or tax problems, and it is desirable that they work as a team for your benefit. The requirements of various states for the accounting profession will vary. Some states permit bookkeepers to call themselves accountants; others have regulations providing for registered accountants. However, it is best to employ a certified public accountant (generally referred to as a CPA) because no reputable CPA will certify an audit or financial statement, unless he himself knows that the information contained therein, is accurate. Some builders and contractors like to estimate the value of materials on hand and they usually guess high for purposes of building up their assets. The CPA will not estimate values but will insist on an inventory being taken so that accurate figures can be established.

Most professional people charge for their services on a time basis. Members of the accounting profession are no exception and operate on a time basis for computing their charges. Even though the fees of a certified public accountant may be slightly higher than other accountants, it is worth the difference from your point of view. Your bank will be impressed by the signature of your CPA, as will the Internal Revenue Service when they see your CPA's signature on your tax returns.

The time to hire your certified public accountant is at the beginning when you are ready to start your new construction business. Just as you do with your attorney, you

# *FINANCIAL STATEMENT*

*Condition at close of business................................................ 19......*

| **ASSETS** | **Dollars** | **Cts.** |
|---|---|---|
| **1. Cash:** (a) On hand $................, (b) In bank $................, (c) Elsewhere $........ | | |
| **2. Notes receivable** (a) Due within 90 days........ | | |
| (b) Due after 90 days........ | | |
| (c) Past due ........ | | |
| **3. Accounts receivable from completed contracts, exclusive of claims not approved for payment** ........ | | |
| **4. Sums earned on uncompleted contracts as shown by Engineer's or Architect's estimate** ........ | | |
| (a) Amount receivable after deducting retainage........ | | |
| (b) Retainage to date, due upon completion of contracts........ | | |
| **5. Accounts receivable from sources other than construction contracts**........ | | |
| **6. Deposits for bids or other guarantees:** (a) Recoverable within 90 days........ | | |
| (b) Recoverable after 90 days........ | | |
| **7. Interest accrued on loans, securities, etc.** ........ | | |
| **8. Real estate:** (a) Used for business purposes........ | | |
| (b) Not used for business purposes........ | | |
| **9. Stocks and bonds:** (a) Listed—present market value........ | | |
| (b) Unlisted—present value ........ | | |
| **10. Materials in stock not included in Item 4** (a) For uncompleted contracts (present val.) | | |
| (b) Other materials (present value)........ | | |
| **11. Equipment,** book value........ | | |
| **12. Furniture and fixtures,** book value ........ | | |
| **13. Other assets** ........ | | |
| Total assets ........ | | |
| **LIABILITIES** | | |
| **1. Notes payable:** (a) To banks regular........ | | |
| (b) To banks for certified checks........ | | |
| (c) To others for equipment obligations........ | | |
| (d) To others exclusive of equipment obligations........ | | |
| **2. Accounts Payable:** (a) Not past due........ | | |
| (b) Past due ........ | | |
| **3. Real estate encumbrances** ........ | | |
| **4. Other liabilities** ........ | | |
| **5. Reserves** ........ | | |
| **6. Capital stock paid up:** ........ | | |
| (a) Common ........ | | |
| (b) Common ........ | | |
| (c) Preferred ........ | | |
| (d) Preferred ........ | | |
| **7. Surplus** (net worth) Earned $................ Unearned $........ | | |
| Total liabilities ........ | | |
| **CONTINGENT LIABILITIES*** | | |
| **1. Liability on notes receivable, discounted or sold** ........ | | |
| **2. Liability on accounts receivable, pledged, assigned or sold** ........ | | |
| **3. Liability as bondsman** ........ | | |
| **4. Liability as guarantor on contracts or on accounts of others** ........ | | |
| **5. Other contingent liabilities** ........ | | |
| Total contingent liabilities........ | | |

* For co-partnerships a separate list of contingent liabilities of individual members is required.

## *DETAILS RELATIVE TO ASSETS*

**1** **Cash**
- (a) on hand ............ $..........
- (b) deposited in banks named below ............ ..........
- (c) elsewhere—(state where) ............ ..........

| NAME OF BANK | LOCATION | DEPOSIT IN NAME OF | AMOUNT |
|---|---|---|---|
| | | | |
| | | | |
| | | | |
| | | | |

**2★** **Notes receivable**
- (a) due within 90 days ............ $..........
- (b) due after 90 days ............ ..........
- (c) past due ............ ..........

| RECEIVABLE FROM: NAME AND ADDRESS | FOR WHAT | DATE OF MATURITY | HOW SECURED | AMOUNT |
|---|---|---|---|---|
| | | | | |
| | | | | |
| | | | | |
| | | | | |
| | | | | |
| | | | | |

Have any of the above been discounted or sold? ............ If so, state amount, to whom, and reason ............

..........

**3★** **Accounts receivable from completed contracts exclusive of claims not approved for payment** ............ $..........

| NAME AND ADDRESS OF OWNER | NATURE OF CONTRACT | AMOUNT OF CONTRACT | AMOUNT RECEIVABLE |
|---|---|---|---|
| | | | |
| | | | |
| | | | |
| | | | |
| | | | |
| | | | |

Have any of the above been assigned, sold, or pledged? ............ If so, state amount, to whom, and reason ............

..........

**4★** **Sums earned on uncompleted contracts, as shown by engineer's or architect's estimate:**
- (a) Amount receivable after deducting retainage ............ $..........
- (b) Retainage to date due upon completion of contract ............ ..........

| DESIGNATION OF CONTRACT AND NAME AND ADDRESS OF OWNER | AMOUNT OF CONTRACT | AMOUNT EARNED | AMOUNT RECEIVED | RETAINAGE | | AMOUNT EXCLUSIVE OF RETAINAGE |
|---|---|---|---|---|---|---|
| | | | | WHEN DUE | AMOUNT | |
| | | | | | | |
| | | | | | | |
| | | | | | | |
| | | | | | | |
| | | | | | | |
| | | | | | | |
| | | | | | | |

Have any of the above been sold, assigned, or pledged? ............ If so, state amount, to whom, and reason ............

..........

## *DETAILS RELATIVE TO ASSETS (continued)*

**5*** **Accounts receivable not from construction contracts** ................................ $..............

| RECEIVABLE FROM: NAME AND ADDRESS | FOR WHAT | WHEN DUE | AMOUNT |
|---|---|---|---|
| | | | |
| | | | |
| | | | |
| | | | |
| | | | |
| | | | |

What amount, if any, is past due ................................ $..............

**6** **Deposits with bids or otherwise as guarantees** ................................ $..............

| DEPOSITED WITH: NAME AND ADDRESS | FOR WHAT | WHEN RECOVERABLE | AMOUNT |
|---|---|---|---|
| | | | |
| | | | |
| | | | |
| | | | |
| | | | |

What amount, if any, has been assigned, sold or pledged ................................ $..............

**7** **Interest accrued on loans, securities, etc.** ................................ $..............

| ON WHAT ACCRUED | TO BE PAID WHEN | AMOUNT |
|---|---|---|
| | | |
| | | |
| | | |

What amount of the above, if any, has been assigned or pledged ................................ $..............

**8** **Real Estate** { (a) Used for business purposes ................................ $..............
**Book value** { (b) Not used for business purposes ................................ ..............

| DESCRIPTION OF PROPERTY | IMPROVEMENTS | | TOTAL BOOK VALUE |
|---|---|---|---|
| | NATURE OF IMPROVEMENTS | BOOK VALUE | |
| 1. | | | |
| 2. | | | |
| 3. | | | |
| 4. | | | |
| 5. | | | |
| 6. | | | |
| 7. | | | |

| LOCATION | HELD IN WHOSE NAME | ASSESSED VALUE | AMOUNT OF ENCUMBRANCES |
|---|---|---|---|
| 1. | | | |
| 2. | | | |
| 3. | | | |
| 4. | | | |
| 5. | | | |
| 6. | | | |
| 7. | | | |

* List separately each item amounting to 10 per cent or more of the total and combine the remainder.

**9** **Stocks and Bonds**
(a) Listed—present market value .................... $..........
(b) Unlisted—present value .................... ..........

| DESCRIPTION | ISSUING COMPANY | LAST INT. OR DIV. PAID | | PAR VALUE | PRESENT MARKET VALUE | QUAN-TITY | AMOUNT |
|---|---|---|---|---|---|---|---|
| | | DATE | % | | | | |
| 1. | | | | | | | |
| 2. | | | | | | | |
| 3. | | | | | | | |
| 4. | | | | | | | |
| 5. | | | | | | | |
| 6. | | | | | | | |
| 7. | | | | | | | |

| WHO HAS POSSESSION | IF ANY ARE PLEDGED OR IN ESCROW, STATE FOR WHOM AND REASON | AMOUNT PLEDGED OR IN ESCROW |
|---|---|---|
| 1. | | |
| 2. | | |
| 3. | | |
| 4. | | |
| 5. | | |
| 6. | | |
| 7. | | |

**10** **Materials in stock and not included in Item 4, Assets:**
(a) For use on uncompleted contracts (present value) .................... $..........
(b) Other materials (present value) .................... ..........

| DESCRIPTION OF MATERIAL | QUANTITY | PRESENT VALUE | |
|---|---|---|---|
| | | FOR UNCOMPLETED CONTRACTS | OTHER MATERIALS |
| | | | |
| | | | |
| | | | |
| | | | |
| | | | |

**11*** **Equipment at book value** .................... $..........

| QUAN-TITY | DESCRIPTION AND CAPACITY OF ITEMS | AGE OF ITEMS | PURCHASE PRICE | DEPRECIATION CHARGED OFF | BOOK VALUE |
|---|---|---|---|---|---|
| | | | | | |
| | | | | | |
| | | | | | |
| | | | | | |
| | | | | | |
| | | | | | |
| | | | | | |
| | | | | | |
| | | | | | |
| | | | | | |
| | | | | | |
| | | | | | |
| | | | | | |
| | | | | | |
| | | | | | |
| | | | | | |

Are there any liens against the above? .......... If so, state total amount .................... $..........

* If two or more items are lumped above, give the sum of their ages.

## *DETAILS RELATIVE TO ASSETS (continued)*

**12** **Furniture and fixtures at book value** .................... $..........

**13** **Other assets** .................... $..........

| DESCRIPTION | AMOUNT |
|---|---|
| | |
| | |
| | |
| | |

TOTAL ASSETS $..........

## *DETAILS RELATIVE TO LIABILITIES*

**1** **Notes payable**
- (a) To banks, regular.................... $..........
- (b) To banks for certified checks.................... ..........
- (c) To others for equipment obligations.................... ..........
- (d) To others exclusive of equipment obligations .................... ..........

| TO WHOM: NAME AND ADDRESS | WHAT SECURITY | WHEN DUE | AMOUNT |
|---|---|---|---|
| | | | |
| | | | |
| | | | |
| | | | |
| | | | |
| | | | |
| | | | |

**2** **Accounts payable**
- (a) Not past due.................... $..........
- (b) Past due .................... ..........

| TO WHOM: NAME AND ADDRESS | FOR WHAT | DATE PAYABLE | AMOUNT |
|---|---|---|---|
| | | | |
| | | | |
| | | | |
| | | | |
| | | | |
| | | | |

**3** **Real estate encumbrances** (See Item 8, Assets).................... $..........

**4** **Other liabilities** .................... $..........

| DESCRIPTION | AMOUNT |
|---|---|
| | |
| | |
| | |
| | |

**5** **Reserves** .................... $..........

| INTEREST | INSURANCE | BLDGS. AND FIXT. | PLANT DEPR. | TAXES | BAD DEBTS | | |
|---|---|---|---|---|---|---|---|
| $.......... | $.......... | $.......... | $.......... | $.......... | $.......... | $.......... | $.......... |

**6** **Capital stock paid up**
- (a) Common .................... $..........
- (b) Preferred .................... ..........

**7** **Surplus: $** .................... $..........

## Condensed Operating Statement

For period beginning ........................ 19...... and ending ........................ 19......

**Income**

| | |
|---|---|
| Gross receipts from contracts | |
| Value of work performed on contracts but not yet paid for | |
| Return from investments | |
| From other sources (specify) | |
| | |
| Total income | $ |

**Expense**

| | | |
|---|---|---|
| Costs of construction, labor, material, etc., not listed below | | |
| Salary of officials, partners or proprietor | | |
| Depreciation: Plant $........ Buildings $........ | | |
| Interest paid | | |
| Federal taxes paid during fiscal period | | |
| Reserved for interest, taxes, etc., excepting depreciation | | |
| Travel, estimating, rent and other expense of doing business | | |
| Other expenditures (specify) | | |
| Total Expense | | |
| Net Profit or Loss | | $ |

## Reconcilement or Surplus

Undivided surplus (net worth) at close of previous fiscal year ........ $........

Items not applicable to current year:

Add ........ $........

Add ........ ........

Deduct ........ ........

Deduct ........ ........

Net addition or reduction ........ ........

Balance ........ ........

Net profit or loss as above ........ ........

$........

Less dividends or withdrawals by partners or proprietor, except as salary above ........

Undivided surplus (See fifth page, item 7, liabilities) ........ $........

## Contracts on Hand (1)

| | Name and Address of Owner | Name and Address of Architect or Engineer | Name of Surety |
|---|---|---|---|
| 1 | | | |
| 2 | | | |
| 3 | | | |
| 4 | | | |
| 5 | | | |
| 6 | | | |
| 7 | | | |

| | Character of Work | Probable Date of Completion | Amount of Contract | Amount Receivable on Contract | Amt. Disbursed on Contract | Amt. Owning on Contracts (2) | Estimate of Cost to complete work |
|---|---|---|---|---|---|---|---|
| 1 | | | | | | | |
| 2 | | | | | | | |
| 3 | | | | | | | |
| 4 | | | | | | | |
| 5 | | | | | | | |
| 6 | | | | | | | |
| 7 | | | | | | | |
| 8 | Total of all contracts not listed above | | | | | | |

Total for all contracts on hand ........ $........

should sit down with your CPA to discuss your accounting setup in detail. His advice on how to handle matters from an accounting point of view can and will be of material help. With the tax situation becoming more complex with the passing of time, it is vital that you obtain your tax information from your CPA or attorney or both before you sign your name to a land contract or other important financial transaction. You will soon learn that a capital gain tax is only 25 percent, whereas ordinary tax liability may be considerably higher. A capital gain is the profit from an increase in value of an investment, held more than six months. It could be shares of stock you bought on the New York Stock Exchange or land under certain conditions. Tax laws, generally speaking, have become so complicated in recent years that an expert is needed to properly advise taxpayers in complicated business transactions. Not all attorneys are qualified as tax experts and it is unfortunate that some clients suffer because the attorney will not admit that he is not familiar with current tax laws.

Whether it is better to operate your construction business under your sole ownership or to form a new corporation presents a serious tax problem. Your attorney and CPA, working together as a team, should be able to properly advise you. Usually, however, the small business is better off under individual ownership. When the business has grown substantially, it may be more advantageous to incorporate. Since each case must be analyzed on its own facts and figures and no hard and fast rule can be applied, it is best to patronize your attorney and certified public accountant.

Most CPAs advise their small businessman clients to use a bookkeeper for the daily book entries, payroll records, accounts payable and accounts receivable, and other records that are maintained on a daily basis. The cost of your bookkeeper's services are obviously much lower than those of your CPA. Your CPA will assist your bookkeeper in setting up the necessary books and records that are to be maintained on a daily basis. The bookkeeper acts under the instruction and supervision of your CPA, and your CPA may come in once a month to work up the important figures for the previous months' business.

Holding your overhead expenses to a minimum is most important. If your wife is interested, she could help out with the bookkeeping chores until your business is better established. She would work under the supervision of your CPA and would be able to see the problems involved in starting a new business and the importance of being thrifty during the struggling period.

A few builders and contractors have run into serious problems because of inefficient business practices. Whether it is the fault of the bookkeeper or the boss is not important. When a check is received from a client, the first thing to do is place your bank endorsement stamp on the back side of the check. This stamp usually reads, "For deposit only to the account of Jones Construction Company" or comparable. This procedure is important because in the event the check is lost or stolen, the thief will not be able to cash the check, even though he may attempt to forge your name. All checks should be deposited in your business account and should be deposited promptly, preferably daily. The quicker your client's check arrives at his bank for payment, the better the chance that it will be honored and paid. Many things can happen during the time the check is sitting in your bookkeeper's desk drawer.

## *1-5 LAND DEVELOPMENT*

It is of great interest to all builders and contractors to have some knowledge of "zoning" and "planning," legal title, and general acquisition of land. However, the matter of land acquisition or development is of greater interest to the home builder, rather than the commercial contractor.

Zoning is basically the control of the use of land by city or county authorities, who are granted the power to limit certain parcels of land to specific uses. The laws of many states permit this on the theory that it is for the promotion of the public welfare. This is a great power and must be used justly and wisely.

"Eminent domain" is the right of government to take private property for public use, provided fair compensation is paid to the owner of the land. No money is paid to the owner of the land for zoning restrictions because this is not a taking of his property.

The orderly development of a city and county depends a great deal upon how its growth is guided. Haphazard growth can make a city or county unsightly, transportation difficult, and property ownership more risky. Evidence of this may be seen in some old cities where millions of dollars are spent in an attempt to correct deficient street systems and blighted areas caused by lack of foresight and planning. Other old cities which were systematically laid out are now noted for their beauty and accessability.

To guide orderly growth, progressive communities have adopted planning and zoning laws. A planning commission is appointed to study city growth, recommend policies to the governing body, and also recommend which areas of the community should be used for certain purposes. A "master plan" is usually drafted by the commission. This plan is a guide for the development of the land and specifies street alignment, improvements, size and shape of lots, curbs and gutters, as well as street paving. The commission reviews any new subdivision development to determine whether it fits in with the master plan.

Zoning and planning laws usually divide the community into residential, business, and unrestricted districts. Residential zones are sometimes classified as R1 for single family; R2 for duplex or up to four family units; R3 for up to eight units; and R4 for larger apartment units. C1 is for retail business; C2 and upwards are for other types of commercial business; M1 and upwards are for light manufacturing; and similar designations are used for industrial districts.

"Spot zoning" is a term used for creating a small area within the limits of a zone inconsistent with those permitted in the larger area, and it is usually frowned upon by planning commissions. An example would be the zoning of a prominent corner in a residential district for a needed service station.

Zoning and planning laws also regulate the height of buildings, setback requirements from the street or curb line, minimum floor areas for dwellings, and the height of walls surrounding the home.

Before entering into any binder arrangement to purchase land, the builder and contractor should verify the planning and zoning requirements to make sure that he will be able to use the land for his purposes. If the builder wants to construct a four-unit apartment building and the zoning on the land that he is considering is R1, or single family residence, he is making a serious mistake. Most communities have a planning and zoning book that is sold at city hall and it is recommended that the requirements be studied.

It is also imperative that the buyer first learn about the title to the land. There are cases where people buy and pay for land without checking the ownership or title and, as you can well imagine, serious problems result. Buying land in the dark is comparable to playing "Russian roulette" and the odds are against you.

Checking title to land is the proper function of your attorney and not the real estate broker trying to sell you the land. Title to land is usually evidenced by an "abstract of title" or "title insurance." An abstract of title is a digest or summary of documents or records affecting title to property. The records of the county clerk are checked and searched by trained employees of the title company. It is important to learn of anything that would affect the title to your particular parcel of land; for example, a lien field against the property; unpaid taxes; a mortgage of record; unpaid assessments for improvements; or a judgment against the seller. When the abstract has been brought down to the current date, it is delivered to your attorney for examination. He will then give you his written opinion of the status of the title and whether it is merchantable and clear of defects. The seller must give you a clear title before you can proceed with the purchase of the land.

Title insurance policies are issued by title companies and offer protection to the buyer of real estate against loss because of a defective title. The cost of a title insurance policy or the cost of bringing the abstract of title current falls on the seller. But the cost of having an attorney examine the abstract of title falls on the buyer. More and more land throughout the country is being sold with title insurance rather than with abstracts of title because the insurance companies and banks who lend money on land or homes are demanding title insurance policies.

Whether you buy land with title insurance or an abstract of title is unimportant, but it is important and vital that you have the title to the land checked before committing yourself by any written agreement.

In the past it was comparatively simple to conclude a sale of land after the buyer and seller agreed on the price and terms. Modern tax problems, however, have complicated matters materially. Now, after the buyer and seller make an agreement, the seller says, "I can go no further in our negotiations until I clear your offer with my attorney and tax advisor to find out how I work out tax-wise."

Whether a builder and contractor should acquire land for future construction is a personal choice. If you are building homes for speculative sale, arranging for your future land needs is mandatory.

In many boom areas of the country, home builders have done very well financially by arranging for their future real estate or residential lot needs in advance of actual construction. The average home builder cannot afford to tie up large sums of money in unimproved lots, waiting for the day when he is ready to start his construction program. Frequently, the home builder is able to enter into an agreement with the owner of the land, providing that the entire acreage will be used for the construction of new homes within a given period of time and that each lot will be paid for when construction begins and the deed is issued in favor of the home builder. In this manner, the builder is able to use his money for his construction program and the land owner receives his money as each lot is being used.

Frequently in the development of large residential areas, involving a substantial number of acres, the planning of homesites results in a portion of the acreage usable for commercial purposes. The value of commercial land is higher than that used for building homes, and some astute builders have done very well financially by selling off the commercial land and building on the residential portion. Other builders have discovered that they are better off confining their efforts to construction and letting the land owner be the real-estate speculator. It is difficult to become a specialist in more than one business.

A home builder who finds it necessary to anticipate his land needs for future home building will usually purchase his raw acreage and then subdivide it into attractive home sites. The subdivision plat must be approved by the Planning Commission or comparable governmental agency.

Planning commissions throughout the United States have adopted a new policy requiring that a certain portion of newly subdivided land be dedicated to the community as a park. "Dedication" is a gift of land by the home builder to the community for public use.

## WARRANTY DEED (Joint Tenants)

JOHN DOE and MARY DOE, his wife

..., for consideration paid, grant...

to RICHARD ROE

and JANE ROE, his wife

as joint tenants the following described real estate in Santa Fe County, New Mexico:

Lot numbered Ten (10) in Block numbered Sixteen (16) of Tijeras Place, an Addition to the City of Santa Fe, New Mexico, as the same are shown and designated on the Plat of said Addition, filed in the office of the County Clerk of Santa Fe, New Mexico, on the 24th day of August, 1923.

SUBJECT TO: Restrictions, Easements and U.S. Land Patent Reservations of record, if any.

with warranty covenants.

WITNESS our hands and seals this ... day of December, 19...

...(Seal) /s/ JOHN DOE (Seal)
JOHN DOE

...(Seal) /s/ MARY DOE (Seal)
MARY DOE

### ACKNOWLEDGMENT FOR NATURAL PERSONS

STATE OF NEW MEXICO } ss.
COUNTY OF Santa Fe }

The foregoing instrument was acknowledged before me this ... day of December, 19...,

by John Doe and Mary Doe, his wife.
(Name or Names of Person or Persons Acknowledging)

My commission expires:
(Seal)

Notary Public

FOR RECORDER'S USE ONLY

### ACKNOWLEDGMENT FOR CORPORATION

STATE OF NEW MEXICO } ss.
COUNTY OF ... }

The foregoing instrument was acknowledged before me this ...

day of ..., 19...,

by ...,
(Name of Officer)

... of ...,
(Title of Officer) (Name of Corporation Acknowledging)

a ... corporation, on behalf of said corporation.
(State of Incorporation)

My commission expires:

Notary Public

If the community has more land dedicated for park purposes than it can hope to use, the builder is given the option of paying money as a substitute for dedicating land. Sometimes the formula for paying money instead of donating land is to require money or land on the basis of five acres for every one-hundred acres of subdivided land. If the home builder refuses to comply with the new requirement, the commission can refuse to approve his plot plan and he may not build homes on his land. There is no point debating the legality of the planning commission's demands because the courts have generally approved this new substitute for taxes.

There is no such thing as a dedication between an owner and an individual. The public must be a party to every dedication. This would apply in cases where the dedication is involuntary, in the sense that it is required in order to obtain approval of a platting of land.

Another type of dedication occurs when it is entirely voluntary. This happens when an individual makes a dedication of land to his church or other charitable organization. Note the distinction between a dedication to the church and a gift of land to the church. If it is an outright gift of land to the church, the person makes out a deed and title to the land passes with no strings attached.

The essence of dedication is that it shall be for the use of the public at large. There may be a dedication for special uses, but it must be for the benefit of the public, not for private uses.

Dedication is applicable not only to highways, alleyways, and land, but to recreation areas and bridges as well. Sometimes, a community will want to widen a highway but does not have the money to acquire the additional land needed by purchase or condemnation. Because the widened highway will increase the value of the land involved, the property owners will often get together and dedicate the needed land to the community.

## *1-6 LAWS – RULES AND REGULATIONS*

The courts have ruled that the business of building or construction contractors is one that the state legislature may regulate in the interest of public welfare. The state legislature, however, may delegate this power to the legislative body of a city or county.

The city or county has the power and authorization to enact legislation for the regulation of the construction business within the jurisdiction of the municipality. The regulation of such business by statute or ordinance must, however, be reasonable and not discriminatory.

The courts have said that the purpose of a contractor's licensing act is for the protection of the public against the consequences of incompetent workmanship and deception. A legislature may provide that building permits be issued only to duly licensed contractors or it may make it unlawful for any unlicensed person to act in the capacity of contractor for the construction or alteration of any building or structure.

Failure to comply with a licensing statute has been held to preclude a contractor from asserting a mechanic's lien for work performed. A license may be revoked for violation of a particular provision of a statute or law, such as abandonment of a construction project that the licensee has undertaken or failure to complete a project for the agreed price.

Many state laws provide that an unlicensed contractor may not prevail in a lawsuit to collect monies due him; others provide that an unlicensed contractor may not collect on the signed contract but may collect for the reasonable value of the labor, services, and materials furnished by him. The moral involved is obvious, namely, be sure you are properly licensed to do business in any state where you plan to do construction work.

# 2

# BUILDING AND CONSTRUCTION CONTRACTS

Building and construction contracts could mean the success or failure of your new business. It is most important that you get competent legal advice *before* you sign any contract. The examples cited in this chapter are actual court cases where the provisions of the construction contract were not clear, and the disputes between the parties were settled by a lawsuit.

The law of contracts is complicated. Law students spend a good part of their first year in school studying this subject. No attempt is being made to make you an expert on the law of building and construction contracts; however, if you understand a few of the important points, it could prove to be very rewarding and helpful in the success of your construction business.

## 2-1 CONSTRUCTION AND EFFECT

The rules of law pertaining to building and construction contracts are the same as applied to all other contracts.

The legal definition of a contract is an agreement by two or more competent persons to do or not to do some lawful act. There must be valuable consideration and the subject matter of the contract must be legal. In order to have a binding contract, there must be an offer and an acceptance. For example, if *A* makes an offer to *B* in writing and *B* accepts the offer in writing, you would have a binding contract. Unfortunately, in most instances *B* never does what he is supposed to do and problems result. What appears to be a simple problem often turns out to be very complicated.

A building or construction contract should be in writing. Even though the builder might be protected under a verbal contract, it would be poor business not to have the construction agreement in writing. Too often, when a dispute arises between the builder and the client, the parties wind up in court because the agreement between them was verbal. Remember, if the construction services are important enough to be performed, the agreement should be in writing. It will save many headaches as well as time and money if the written agreement covers the matter in dispute. This is where the services of an experienced and competent attorney can be invaluable. Don't forget the admonition of Samuel Goldwyn, the motion picture producer who supposedly said, "Verbal contracts aren't worth the paper they are written on."

The parties to a building and construction contract may change the contract or vary its terms. A new written agreement should be prepared covering the changes as well as the additional charges that must include the cost of labor and materials, overhead expenses and a reasonable profit whenever any changes or alterations are ordered while the building is under construction. It is important that the new written agreement be signed before the work involved is started. Don't permit your client to demand that you start the new work under the change order before the new written contract is properly drawn and signed by both sides.

Contractors shudder when the client watches the daily progress of construction and requests wholesale changes. However, as long as the written contract provides that all changes made by the client be charged for by the contractor at a fair price, the contractor has protection. Under the written change order, the client agrees in writing to pay *X* dollars for each change ordered and is privileged to make as many changes as he wishes. Thus, the usually unpleasant subject of client changes is now handled in a pleasant, businesslike, and profitable manner. To digress slightly but to emphasize how an annoying task can be handled in an efficient and businesslike manner, I must tell you about a client, who built tract homes in large quantities. He found from experience that as soon as his clients moved into the new tract homes they would complain about such minor defects as a door that fit imperfectly or a leaking faucet. He suggested to each client that they prepare a list of the little things that needed correction and advised them when his cleanup crew would take care of each problem. His handling of this unpleasant part of home construction pleased everyone and his cost was minimal compared to his cleanup crew running to each home on a daily basis.

The general rule that the grammatical and ordinary sense of the words in a contract are to be followed applies to building and construction contracts.

*Examples*

1. If the specifications require a cellar in certain depth, the contractor must sink the walls sufficiently to get a sure foundation.
2. A contract to erect a building implies an obligation to pay for all of the work and materials that enter into the structure and to defend against all mechanic liens.
3. A contract to repair an existing structure means to restore or supply in the original structure that which is lost, destroyed, or missing.
4. A contract to install certain equipment binds the contractor not only to set up and connect the parts but also to furnish suitable and adequate material.

5. A contract to furnish all the dimension stone that may be required in the construction of a building does not include the dimension stone used in the approaches or steps leading up to the building.

Even a simple word like similar can create problems. The word "similar" does not mean that the work is to be an exact duplicate of that in the building named, but as nearly like it as the provisions of the contract and the general conditions will permit. Under a building contract calling for material of a certain make "or equal" the contractor is not bound to furnish the material specified even if available and he may substitute other materials if they are equal. If the same contract had specified a certain make and then said "or equal" could be substituted if the named make were not available, we would then have the opposite result. These few examples demonstrate how technical and important the language contained in a building contract can be and how important it is to the builder and contractor to know his business or to get help from a qualified attorney.

A provision in a contract for the construction of a cellar according to specifications that "the whole to be perfectly watertight and guaranteed" binds the contractor only so far as his own work specifications will produce a watertight cellar. After all, the plans and specifications are drawn by an architect, and whether they produce the desired result as advertised, is the responsibility of the architect and not the contractor.

The only thing the builder and contractor should guarantee is that each will perform his construction work in a workmanlike manner and follow the plans and specifications furnished by the architect. Of course the contractor is responsible for any acts of negligence that he may have committed or for work that was not done in a workmanlike manner. Negligence is defined as doing something wrong or improperly or it can involve failing to do something which should have been done.

A construction contractor who follows plans and specifications furnished by the owner (or his architect or engineer) which prove to be defective will not be responsible to the owner for loss or damage which results. However, if the contractor fails to follow the plans and specifications of the contract, he is liable to the owner for damages caused by any weaknesses in the building.

Building and construction contracts frequently provide that the work be done in a first-class or workmanlike manner. Even if the contract is silent in this regard, the law will assume that the builder is qualified and therefore, will hold him responsible for the knowledge, skill, and judgment of builders in that particular community.

The term "workmanlike manner" sounds vague and indefinite but it really is not. Whether or not certain work is done in a workmanlike manner will be determined by other qualified and experienced contractors in that area. The client is not qualified to testify in court as to the quality of the work performed. He may only state the facts and may not offer his opinion; the judge will determine whether the builder performed in a workmanlike manner.

The qualified experts will examine the disputed work and testify as to how the work was done. The true test is not how it would have been done in New York or Chicago or Los Angeles, but only in the area in dispute. Each side usually arranges to hire its own experts and, as you can well imagine, there will be differences of opinion. (We have all heard or read about expert medical testimony and how the opinions of different doctors will widely vary.) Sometimes, if a judge is hearing a case without a jury, the judge has the right to consult with an experienced contractor of his choice and then decide the case with the help of the court appointed expert witness.

The extensive use of testimony by expert witnesses on the part of some litigants has caused criticism by judges as well as the legal profession. A trial is not supposed to be a contest to see how many expert witnesses one side can hire. We all know that the testimony of the expert witnesses that you hire will support your position or you would not have them testify in your behalf. In criminal trials, each side brings forth many expert witnesses, whose testimony will favor the side that hired them. As a result, the jury frequently ignores all of the expert testimony and substitutes its own judgment. In workmen's compensation lawsuits, one doctor testifying in behalf of the claimant will swear that he is 100 percent disabled and the insurance company doctor will testify that the claimant is only 50 percent disabled. As a result of this wide variance, the jury will usually compromise the medical testimony and strike an average by awarding the claimant compensation based on 75 percent disability.

The term "turnkey job" refers to an agreement by the contractor to complete a certain home or building in accordance with the plans and specifications for an agreed sum of money. If the contractor runs into problems along the line of increased cost of materials, he may not bill his client for these items because he is committed under the turnkey job agreement. Other types of contracts that give the contractor greater flexibility and protection against the unusual are obviously more desirable.

Turnkey job contracts should be avoided whenever possible. Even if the parties enter into such a contract, it should give the contractor some protection against changes made by the client. All changes are costly to the contractor and the agreement should provide for compensation to protect the contractor. Let us assume that the change made by the client is estimated to cost the contractor \$200, and if the agreement merely provides that the contractor shall be reimbursed for his cost of the change, the contractor is still the loser in this transaction. It costs money to operate a business and if provision is not made for the builder's overhead and profit, he is going to lose money on the transaction. If you lose money on too many transactions, you are not going to stay in business very long.

It is important and vital to bear in mind that overhead is simply the cost of doing business and does not take into consideration the element of profit. No business can operate very long without earning a profit. Don't overlook a salary, drawing account, or some compensation for the builder or boss. The contractor is contributing his time, attention, and knowledge to the job and must be paid for his efforts. Now, it is easier to appreciate that the \$200 change ordered by the client is not a good deal for the builder who suffered a loss due to no fault on his part. Clients who order changes are obligated to pay the reasonable value of the change and this has to include not only the cost but a reasonable profit as well.

Reference has been made previously to plans and specifications. They are commonly called "plans and specs." The specifications are the particulars or details of the plans. They are usually prepared by an architect and are under his supervision and control.

When a building contract refers to the plans and specifications and makes them a part of the contract, they are to be considered together. However, in cases in which the plans and specifications are referred to in the building contract for a particular specified purpose, then they are foreign to the contract for all other purposes. When there is a conflict between the building contract and the provisions of the plans and specifications, the positive language of the contract will prevail.

## 2-2 COMPENSATION

The right to compensation under a building or construction contract is determined from the terms of the contract. When the amount of compensation is specified, it is controlling. It is hard to imagine an agreement that does not specify an amount of money, but it does happen. In such cases the law dictates a promise to pay the reasonable value of such services. In other words, the law assumes that the parties agreed that the contractor would be paid for his labor, materials, overhead, and profit and adding them all together, the law determines what would amount to reasonable value of his services. Failure to have the contract specify the amount due the contractor is a poor business arrangement that may wind up in court, with the contractor a loser regardless of the outcome of the lawsuit. If the building contract does not mention time of payment for work, no payment can be demanded by the contractor until the work is substantially performed. In short, loose business practices on the part of the contractor in not specifying when and how the work progress is to be paid for can only result in trouble and should be avoided.

A contractor may recover reasonable value of additional work required by a material change in the specifications. If the cost of labor or materials provided for in a building contract has been increased due to the fault of the owner, an increase in the contract price is warranted.

A "cost plus" building contract is the opposite of a turnkey job in that the exact amount is not determined until the construction has been completed. Under a cost plus contract, the contractor is entitled to recover all of his costs, plus the agreed percentage of profit. Like all other contracts, if the details are spelled out and covered there is no problem or disagreement.

Many lawsuits are filed when the parties do not specify what is included under costs. In some cases, the courts have ruled that the contractor is not entitled, in addition to the percentage called for in the contract, to charge for his general or overhead expenses such as salaries, telephone service, office supplies, time spent in superintending the work, carfare for laborers, cost of extra work not called for by the original contract, or cost of doing over the work which was not properly done. In other cases, the courts have ruled that the contractor is entitled, in addition to the percentage called for in the contract, to charge for materials and supplies furnished, wages of workmen, salaries of superintendents, and accident and indemnity insurance. As you can readily see, these problems were present in court because the building contracts were poorly drawn and the necessary details not spelled out. In such cases, the judge is called upon to practically draft a new contract for the parties. If you are involved in a cost plus building contract, your contract should spell out in great detail exactly which items are to be included under costs.

Opinion is divided as to whether or not a general contractor under a cost plus contract is entitled to a percentage on the profits of subcontractors. Some courts hold that the contractor's percentage can be charged only on the actual material and labor furnished by subcontractors and not upon its profits.

"Retainage" is the agreed percentage of withholding of money due the contractor when he presents his monthly bill for work completed during the preceding month. Although the amount of retainage will vary from contract to contract, it is more often approximately 10 percent. When the job is completed, the contractor will normally present his bill for the final month's work as well as the withheld 10 percent of all previous monthly bills.

Provisions in construction contracts permitting the owner to retain a portion of the contractor's current billings until the completion of the work are usually included for the protection and benefit of the owner or of the owner and the contractor's surety or guarantor. The reason for retaining a certain percentage of the installments paid on a job during the course of construction is to create a fund to insure the payment of claims against the job, and to provide greater protection for the owner or the contractor's surety company

## 8800 JOB OVERHEAD SUMMARY

Building ______ Listed by ______ Sheet No. ______

Location ______ Checked by ______ Estimate No. ______

Stories ______ Size ______ Cube ______ Floor Area ______ Date ______

| Item | Class of Expense | Amount | Item | Class of Expense | Amount |
|---|---|---|---|---|---|
| 8801 | **Equipment, Tools, Accessories** | | | Brought Forward | |
| .1 | Rental (See A. G. C. Rental Schedule) | | 8818 | **Telephone and Telegraph** | |
| .2 | Freight | | | (Regular, Long Distance: Connect. Fees) | |
| .3 | Hauling | | 8819 | **Rents** | |
| .4 | Loading, unloading, erecting, dismantling | | .1 | Offices | |
| 8802 | **Job Organization** | | .2 | Land, Unloading and Storage Facilities | |
| .1 | Superintendence—(supt.; assistant) | | 8820 | **Permits**—Building, street, sidewalk, water, | |
| .2 | Time and Material—(timekeeper; mat'l clerk) | | | sewer, hauling over boulevards, use of equip- | |
| .3 | Accounting—(bookkeeper; accounting) | | | ment, etc. | |
| .4 | Clerical—(stenographer, clerk, office boy) | | 8821 | **Insurance**—Miscellaneous— | |
| .5 | Shops—(blacksmith, machinist, tool man) | | | Fire, tornado, earthquake, riot, theft, boiler, | |
| .6 | Safety—watchman; safety foreman) | | | plate-glass, automobile, payroll, etc. | |
| .7 | Miscel.—(job chauffeurs; teamsters; waterboy) | | 8822 | **Petty Cash Items** | |
| 8803 | **Light, Power, Water: Connections** | | 8823 | **Interest**—On deposits and job funds | |
| .1 | Electricity—(light and power) | | 8824 | **Cutting and Patching for Trades** | |
| .2 | Carbide, Gas—(lights, cutting, welding) | | 8825 | **Contingencies**—Guarantees; (strikes, wages, | |
| .3 | Gasoline, Heating and Illuminating Oils | | | labor, output, rains, freezing, floods, cyclones, | |
| .4 | Coal and Coke—(power, heating, thawing) | | | earthquakes, material shortage and price, trans- | |
| .5 | Water—(boilers, sprinkling, mixing, etc.) | | | portation, sub-soil conditions, finance, drastic | |
| 8804 | **Supplies** | | | and ambiguous contract provisions, supervising | |
| .1 | Office—(stationery, time-books, forms, etc.) | | | personnel) | |
| .2 | Job Shop—(steel, smiths coal, etc.) | | 8826 | **Cold Weather Expense** | |
| .3 | Equipment—(oils, waste, boiler comp., etc.) | | .1 | Thawing Materials, (Plant installation and | |
| 8805 | **Traveling and Hotel Expenses** | | | operation—See Item 8803.4) | |
| .1 | Material—(expediting) | | .2 | Weather Protection—(window and door clo- | |
| .2 | Labor—(procurement and transportation) | | | sures, temp. walls, canvas, etc.) | |
| .3 | Officials and members of permanent force | | .3 | Temporary Heat—(installing, maintaining), | |
| 8806 | **Express and Miscellaneous Freight** | | | See Item 8803.4 | |
| 8807 | **Demurrage Allowance** | | 8827 | **Repairs**—Streets, sidewalks, property | |
| 8808 | **Hauling**—hired for odd jobs | | 8828 | **Pumping**—De-watering (If preferred, cover un- | |
| 8809 | **Advertising**—labor, material, equip. | | | der Excavation or Sheeting of Main Summary) | |
| 8810 | **Signs**—company, warning, notice, etc. | | 8829 | **Final Clean Up**—Windows, walls, floors, ceil- | |
| 8811 | **Engineering, Surveys and Inspection** | | | ings, fixtures, and premises | |
| .1 | Layout—(lines, levels, batters boards, etc.) | | 8830 | **Association Dues**—Job share | |
| .2 | Public Inspectors—(boilers, wiring, etc.) | | 8831 | **Share of General Company Overhead** | |
| .3 | Inspection of sub-contract work | | | Expense of home office, shops and yard, | |
| .4 | Lot survey | | | quantity survey, estimating, investigations, | |
| 8812 | **Tests** | | | dead time of permanent field force, taxes and | |
| .1 | Soil—(test, pits, borings, bearing power) | | | all other company expense, not chargeable | |
| .2 | Material—(cement, steel, aggregates, etc.) | | | to the specific job. | |
| .3 | Structure—(floor loading) | | | **SPECIAL ITEMS NOT LISTED ABOVE** | |
| .4 | | | | | |
| 8813 | **Drawings**—shop and setting | | | | |
| .1 | Drafting | | | | |
| .2 | Extra Prints | | | | |
| 8814 | **Photos** | | | | |
| 8815 | **Patents and Royalties** | | | | |
| 8816 | **Legal**—Attorney and Notary Fees | | | | |
| 8817 | **Medical and Hospital Expense** | | | Total Job Overhead | |
| | Carried Forward | | | Equals ______ % of direct cost | |

## 8800 JOB OVERHEAD SUMMARY

Building........ Listed by........ Sheet No........
Location........ Checked by........ Estimate No........
Stories........ Size........ Cube........ Floor Area........ Date........

| Item | Class of Expense | Quantity | Labor Cost Unit | Labor Cost Amount | Material Cost Unit | Material Cost Amount | Total Cost Unit | Total Cost Amount |
|---|---|---|---|---|---|---|---|---|

in the event the job is not completed by the contractor and it becomes necessary to hire a second contractor or to pay all unpaid bills against the job.

It should be understood that when you talk about the contractor's surety company, it means that the construction job has been bonded. As businessman owner, you may decide that you cannot afford to gamble on the contractor's ability to pay the various bills incurred during the course of construction. In order to protect yourself, the building contract should provide that the contractor agrees to furnish a performance bond for the amount of the contract. The cost of the performance bond falls on the owner, even though it may be billed to the contractor who passes the charge along.

Some contractors are not bondable which hurts their chances of getting certain building contracts. The usual reasons for a contractor not being bondable are failure to finish a previous job which cost the bonding company money, inability to pay bills promptly when due, liens filed against the job which means not paying certain labor or materials bills, a poor credit record, and a poor financial statement. To emphasize the importance of ability to furnish a performance bond, all government construction jobs can only be awarded to bondable contractors. A contractor who is unable to furnish a performance bond is advertising the fact that he is inexperienced or not in very good financial condition.

The performance bond in favor of the owner or customer, says, in effect, that if the contractor fails to perform as per the conditions set forth in the building contract and fails to pay the various construction bills incurred, the bonding company will agree to complete the construction job by hiring a second contractor and paying all of the bills. The retainage money withheld from the original contractor, who has now defaulted, is turned over to the bonding company by the owner, to reduce their liability under the terms of the performance bond.

Bonding companies have suffered some severe losses and, as a result, have become quite conservative and cautious in issuing performance bonds to contractors. Many contractors are not strongly financed or adequately experienced to satisfy the requirements of the bonding company. Refusal to issue a performance bond does not automatically mean that the contractor is dishonest or incompetent. The young builder or contractor should start off with small construction jobs and leave the big jobs to larger firms that are already established. Building up a good reputation is very important and the small builder has a better chance of obtaining a performance bond for nominal amounts. When he completes the job, the bonding company will be favorably impressed with his performance, and it will be easier to be bonded in the future for a larger job.

Just as contractors try to protect themselves by providing that all changes made by the owner will be charged for, owners try to protect themselves by providing that no recovery may be had for changes, alterations, or extra work by the contractor without a written order approved by the owner or his designated representative. The written order required by the contract must be obtained before the work is done. The temptation to do the work first and obtain written approval later is great, but once the contractor gets burnt he will learn the hard way to conform to the terms of the contract. The contractor's liability is fixed by the terms of his contract, and he is obligated to perform according to those terms.

When the construction contract is clear and absolute, or there is an express guaranty without qualification or exception, the contractor must abide by his contract no matter the cost. A contractor accidentally involved in a bad deal, whereby he will lose money on the contract, will get no sympathy in court. He is committed by the contract and the law presumes that he knew what he was doing when he signed the contract.

The advantages of a written contract over a verbal contract are numerous. In a written agreement, whether it be a building or construction contract or any other type of contract, we turn to the written instrument to learn what the parties agreed upon. If the written agreement is thorough and complete, it will cover practically all situations that might arise.

Even though a verbal contract is binding, the biggest problem is the lack of understanding and agreement between the parties. Human nature being what it is, the client is going to remember the facts in a manner favorable to himself and the same applies to the contractor. How very difficult for a judge to settle a dispute where each side claims to be right!

The law assumes certain facts to be true in a verbal construction contract. For example, it is assumed that the contractor in a verbal agreement warrants that the building will be erected in a workmanlike manner and in accordance with usage and accepted practices in the community in which the work is done. Furthermore, it is assumed that the home builder warrants that the house when completed will be fit for human habitation. These assumptions by the law are called implied warranties on the part of the builder and contractor, and they are his obligation and responsibility. On the other hand, it is assumed that the client agrees to pay the reasonable value of services rendered by the builder, including the cost of labor and materials, overhead expenses, and a fair profit.

In effect, the judge has to provide a contract between parties that do not enter into a written agreement, spelling out exactly what the parties mutually agree upon. Some judges have said they will not write a new contract for the parties, where the terms of the agreement are not clear, however, in the case of a verbal contract, the judge is forced to do a certain amount of assuming in his effort to do justice.

In short, *do not enter into any verbal construction agreements.* They are unsatisfactory and dangerous and usually wind up in court. A well-drawn written contract has a fighting chance of avoiding problems.

## 2-3 THE ARCHITECT AND THE CONTRACTOR

What part does the architect play in the building and construction contract between the owner and contractor?

First of all, the architect is responsible for and controls the plans and specifications that are eventually approved by the owner. Next, the job is put out to bids by contractors who are selected or invited by the owner or his architect. The owner is not obligated to award the contract to the low bidder but is privileged to select any bidder of his choice. The cost of figuring a bid is a normal hazard taken by the contractor because there is no assurance that he will receive the contract for the job. On occasion, the owner will ignore the low bidder and award the construction contract to a high bidder. Even though the owner is trying to save money, he may be concerned about the low bidder's ability to perform or the contractor's ability to earn a profit on the job. If the job is bonded, the owner will be better off awarding the bid to the next higher bidder who is financially sound and will be able to complete the job without defaulting. A defaulting contractor who is unable to complete the construction job, creates many problems for the owner, in spite of the fact that the job is protected by a performance bond. It takes time for the bonding company representative to take over and hire a second contractor to finally complete the job. Owners who have gone through this experience find it very time consuming and will avoid this unpleasant procedure whenever possible.

Governmental agencies, whether they are federal, state, or municipality, are required by law to award the construction job to the low bidder, providing he is qualified by the bid requirements. If the bid of the lowest bidder is a great deal lower than the next lowest bidder, the difference between the two bids is referred to as "money left on the table."

When the owner selects his contractor and all the necessary details are completed, including the furnishing of a performance bond, construction is ready to begin. The owner may or may not decide to hire the architect as his personal representative to oversee the construction. If the architect is retained by the owner, whatever duties and functions are assigned to the architect are incorporated into the construction contract. Three parties are now involved, namely, the owner, the contractor, and the architect.

The architect's primary duty is overseeing the construction to insure that materials called for in the plans and specifications are properly used by the contractor. In addition, the owner provides in his agreement with the contractor that his architect is given the authority to comment on the quality and amount of work completed and to certify when the job is finished.

Frequently, the architect is designated by the owner to approve the monthly statements of the contractor as to the amount of work completed and how much money shall be paid to the contractor. Some contracts provide that the work shall be done to the satisfaction of the architect, and in such a case, the decision of the architect is binding on both the owner and the contractor. Even though the architect is employed and paid by the owner, it is taken for granted that his opinion will be fair and honest and not tainted by any plan or scheme.

A stipulation in a building contract that the work shall be satisfactory to the owner is valid and enforceable. An honest dissatisfaction with the work can prevent a recovery on the contract. In one case, whereby the contractor's agreement offered to satisfy the architect and the owner, the architect was satisfied but the owner was unhappy and the court ruled against recovery by the contractor. No builder or contractor should enter into a contract that subjects him to the whim of the owner. Trying to please some people in the construction business is like trying to please others with a portrait of themselves. Even though some courts have ruled that the owner may not unreasonably withhold his approval, if it takes a lawsuit to collect your money, it is not a desirable business transaction and should be avoided.

In order to learn the authority granted the architect, you must examine the contract between the owner, architect, and contractor. When the contract provides that the contractor's monthly draw or progress payments must be approved by the architect, his decision will be binding upon the owner as well as the builder. Usually, the architect issues his certificate authorizing these payments to the contractor. A fraudulent scheme on the part of the architect and builder, based on a phony certificate by the architect authorizing payment by the owner, can result in a lawsuit for conspiracy against both the architect and the contractor. When the building is to be constructed to the satisfaction of the architect, his final certificate of completion is binding on the owner as to visible defects but not latent defects. A latent defect is present but not visible or active.

If the architect exceeds the authority granted him under the contract, his opinion will become invalid. Even in cases where the architect is given the right to make final decisions regarding disputes between the owner and builder, the architect's rights are limited because he does not have the power to construe a contract, that being a legal function. An example would be a contract where the question is whether a building contractor is required to do certain work, or is required to use certain materials, and the answer involves the construction of the contract which is beyond the scope of the architect's powers. If the architect accidentally performed a legal function, his decision would not be binding on either party. The court would then have to rule on the dispute.

If the building contract requires the architect's certificate to be in writing, his oral approval would not be in compliance with the contract and would be invalid. The architect named in the building contract has no right to delegate his authority to another architect, unless the owner and builder agree to such a substitution. If an architect accepts the completion of a building upon condition that the contractor is to correct certain defects in workmanship or materials used, his final architect's certificate may not be legally issued until all work has actually been completed.

A qualified architect can make a valuable contribution to the success of a construction job. He is a member of the construction team and can render a valuable professional service to both the owner and the contractor. Many contractors actually prefer to work with the architect, rather than the owner. The knowledge and ability possessed by the architect is usually lacking on the part of the owner. (The duties and functions of the architect are discussed in greater detail in Chapter 14.)

## *2-4 SUBSTANTIAL PERFORMANCE*

Translated into simple language, substantial performance occurs when the contractor completes the structure and it is defective or when he fails to complete the structure. Substantial performance by a contractor under the terms of a building contract is a question of fact that varies with each case.

Courts have universally ruled that a substantial performance of a building contract will enable the contractor to collect his money, either on the contract or for the reasonable value of the labor and material furnished. Three reasons are given for this decision. First, the owner of the land receives the benefit of the builder's labor and materials, and it is only fair and equitable to require the owner to pay the reasonable value for benefits that he received. Second, it is sometimes next to impossible for a builder to comply with all of the minute specifications in a building contract. And third, the law presumes that the parties agreed to do what is fair and reasonable under all the circumstances with regards to performance. Even in cases where the contract provides that the work is to be performed to the satisfaction of the owner or his architect, their judgment has to be exercised reasonably and not arbitrarily and the courts will permit the contractor to recover his money in case of substantial performance.

There are certain obligations that fall on the contractor to justify his failure to perform in accordance with the terms of the contract. The contractor must have acted in good faith. The contractor cannot recover for substantial performance in cases where his failure to perform is intentional or is due to his carelessness.

The following are actual court cases that rule both ways on the question of substantial performance, where the contractor did not comply with the terms of the building contract and filed a lawsuit for the reasonable value of his labor and materials.

*Examples*

1. The erection of homes which are not fit for human habitation even when completed cannot be regarded as even being close to substantial performance of an agreement to build homes.
2. A slight deviation in color of brick siding on a house did not show a failure to comply with the substantial provisions of a contract to provide the siding.
3. Where the contractor has honestly performed the contract in all its substantial and material particulars, he will not be held to have forfeited his right to collect by reason of technical, inadvertent, or unimportant omissions.

4. A contractor would not be required to demolish a residence and install a different brand of pipe, in order to collect his money for building the residence, where the contractor had inadvertently installed pipe of a different but equal grade with that specified in the contract. Instead, an allowance could be made for the difference in value of the two brands of pipe, which might prove to be nominal or nothing.

Although it is generally held by courts that proof of substantial performance of a building contract will permit a contractor to collect his money on an action based on the contract, it is obvious that the contractor should not be permitted to recover the full contract price as though he had exactly performed. The true measure of recovery by the contractor would be the sum stipulated in the contract, less the damage sustained by the owner, because of the contractor's failure to perform.

The doctrine of substantial performance is based on the good faith of the contractor and does not ordinarily extend to a wilful or intentional failure to perform in accordance with the contract requirements.

*Example*

1. A contractor who substituted iron pipe for lead pipe was required to pay the owner the cost of laying a lead pipe as provided for in the contract, rather than the difference in cost between the iron and lead pipe.

Obviously, if the contractor has committed some mistake that can be corrected reasonably, he is obligated to correct his faulty construction. Whether or not a building contract has been substantially performed is not decided by the courts on a percentage basis. Each situation is weighed on its own particular facts and problems involved. Usually, the award to a contractor for substantial performance will be reduced by the cost to the owner of hiring a second contractor to complete the building.

Many building contracts contain the phrase "time is of the essence." This means that failure to comply with time schedules listed in the construction agreement are binding upon the parties involved and they will be held responsible for being late in accordance with the provisions of the contract. Usually, the contract will provide for payment of money for failure to complete certain work within the time prescribed. The mere statement of a date in such a contract does not make time of the essence. Failure to complete the work within the specified time does not terminate the contract, but only subjects the contractor to damages because of the delay. A party to a building contract who has performed part of the work but is prevented by the other party from completing the contract, may recover his money for the work performed and the materials furnished.

When the parties to a building contract have agreed that the contractor shall pay a liquidated or agreed sum for each day's delay, the amount recoverable by the owner is to be computed by the contractor's total delay, minus the time chargeable to the owner that caused part of the contractor's delay. The contractor is legally bound to complete the work within the time period specified in the contract. Because the courts dislike penalties, most building contracts provide for a substitute known as "liquidated damages." The parties agree that if the contractor does not perform his work within the specified time, the contractor has to pay a certain sum of money to the owner as liquidated damages.

A provision commonly found in construction contracts is that the owner shall not be liable to the contractor for damages caused by delay on the part of the owner or other contractors employed by him. In cases where delays are caused by the owner, the contractor would be entitled to this additional time for completion of the work without penalty to the contractor. Delays by other contractors are matters that should be anticipated and if a contractor wishes to relieve himself for his own late performance because of such delays, he should see to it that the building contract so provides.

The acceptance by the owner of a building which has been defectively constructed, the defects being unknown and not discoverable by inspection, does not prevent the owner from making claim against the contractor for damages for a breach of the contract. This is known as a latent defect because it is not visible but is hidden and not reasonably discoverable until after the contractor has been paid and the building accepted.

The fact that the owner of a building goes into possession with knowledge that the new building contains defects in its construction as well as inferior material will not prevent his claiming damages for such defects as an offset against the contractor's claim for the contract price. The owner's occupancy and use of the building do not constitute an acceptance of the work as complying with the contract, or amount to a waiver of the defects.

In cases where the owner is occupying the home or building when he hires the contractor to repair or remodel the structure, you run into a small problem. In most instances, the contractor either does not finish the job or if he does, the job is not done in a workmanlike manner. The owner may then refuse to pay the contractor for what he feels is a below par job and the contractor may file a lawsuit for his money under the contract.

The contractor will argue in court that the owner continued to occupy the home or building and that this occupancy is the same as an acceptance of the job. The owner will reply that he was already occupying the structure when he hired the contractor and that the law does not require the owner to move out of the home or building, in order to claim that the job was not done in a workmanlike manner and is loaded with defects. The owner can file a counterclaim against the contractor for damages that he suffered and it is up to the judge to rule on the dispute. Based upon the facts recited, the judge will probably rule in favor of the owner because the law does permit the owner to occupy the premises without losing his claim against the contractor.

*Examples*

1. No inference of acceptance of a painting contractor's work may be drawn from the owner's continuing to use the house upon which the contractor had spread his paint.
2. A contractor for the construction of a roof who does not substantially perform cannot put upon the owner the alternative of abandoning the house or removing the roof in order to avoid acceptance.
3. Home owners who lived in the house during remodeling paid $5,600 of the $6,200 contract price for remodeling and observed the work from day to day, thereby waived any claim for damages resulting from one of the bedrooms being made too small to be used any longer as a bedroom.
4. Where a person agreed in the construction of a mill with different dimensions from those agreed to in an agreement, and afterwards accepted it, and also accepted policies of insurance upon it, he is estopped from afterwards complaining that the dimensions had been varied from those contained in the contract.

Even though the owner pays for a building constructed for him, this fact does not prevent his recovering damages for defects that are discovered later. If his late discovery was reasonable, he still has not lost his right to file a claim against the contractor for defects that are chargeable to the contractor.

Usually, the fact of partial payment by the owner does not constitute an acceptance or waiver of defective construction by the contractor. The most important requirement is the owner's knowledge at the time of making the partial payment. If the owner had knowledge of the defects and ignored them and made the partial payment, he may have condoned the situation and lost his right to claim damages against the contractor for the defects.

## 2-5 EXCUSES FOR NONPERFORMANCE

When you talk about legal excuses for nonperformance of a contract, the same rules of law that apply to contracts generally also apply to construction contracts. If a contractor quits the job without justification, the owner is not required to perform his part of the contract, before filing a lawsuit against the contractor to recover damages for his breach of contract.

Earthquakes, unusual floods, tornadoes, lightning, and hurricanes are some of the more common disasters inflicted by nature that are classified as "acts of God." In order to protect himself, the contractor should see to it that the contract specifies certain itemized acts of God as legal excuses for nonperformance of the building contract. When the building contract is silent in this area, the court rulings are mixed, some courts holding that it is an excuse.

Most written construction contracts contain provisions protecting the contractor in case he is late justifiably by virtue of a labor strike or some other condition beyond his control. Modern labor unions are sometimes responsible for secondary labor strikes. This means that one union is fighting another union and the poor contractor is an innocent victim who is caught in the middle of the union struggle. Whether a new construction business should employ union or nonunion labor is a matter of opinion to be decided on an individual basis. Many builders and contractors feel that they are just as well off to pay union scale or better but not to operate a closed or union operation. Governmental agencies deal with union contractors only. However, the new contractor does not figure to get any government contracts so he is really not affected. The size of your community and local custom and ground rules should help you in deciding how to handle the union situation.

What are the rights of the contractor with regards to prices of labor and materials that he is obligated to furnish under the provisions of the construction contract? If the contract is silent about an unexpected increase in cost of labor and materials, the contractor is committed and the increased cost is his problem. The time for the contractor to think about protecting himself is while the contract is being prepared and more important before he signs it. From past experience, many established contractors have learned to obtain cost plus contracts which automatically include protection against increased prices for labor and materials that are beyond his control.

If the owner wrongfully withholds payment of money due under the terms of a building contract, the contractor must check his written contract to learn what his rights are. If the point is not covered, he has a problem and complaining to his attorney, who overlooked the point, will not alter the situation. Although the owner wrongfully withheld payment of the money, the test is whether or not the owner's wrongful conduct prevents performance by the contractor. If this is the case, he is legally excused. The attorney will advise that the contractor is legally excused from performance of the contract because of the wrongful conduct on the part of the owner.

The death of a party to a construction or building contract does not terminate the contract. If a contract is of a personal nature, such as a portrait painter, then death would terminate the contract. A building or construction contract ordinarily does not involve the personal skill or taste of the contractor, who follows the plans and specifications of the architect. In actuality, most building contracts are obtained by low bid.

## 2-6 ACCIDENTAL DESTRUCTION OF THE BUILDING

During the course of construction, the owner and contractor should make provisions in their building contract placing the responsibility for losses due to accidental destruction. For example, provisions should be made for a building's accidental destruction by fire. This problem has been presented in court many times and in those cases where the contract was silent, neither party was protected. The responsibility for this type of loss can be placed on the owner or the contractor.

The party responsible under the terms of the building contract is able to protect himself against an unusual loss by appropriate insurance. If the contractor is to be held responsible under the terms of the contract, he cannot afford to gamble and must protect himself by adequate insurance coverage. The cost of this insurance, like performance bonds and other insurance coverage, is a normal part of the overhead and has to be included when figuring how much to bid for a job. Rather than get into a hassle with the owner about who is going to be responsible for accidental destruction of the building while under construction, it is good business for the contractor to be responsible and add the cost of necessary insurance coverage to the owner's bill. If the contractor is lax about including *all* items of expense in figuring his overhead, he will later discover that instead of making a profit on the job, he suffered a loss.

In cases where the building contract is silent, you are faced with the problem of fixing the responsibility for the loss. You cannot point the finger at the contractor and hold him responsible for the loss without determining whether the contract is entire and indivisible or divisible and severable. If the building contract is entire and indivisible, the loss falls on the contractor. If the building contract is divisible and severable, the loss falls on the owner.

Previous discussion covered the point of liability if the building under construction was destroyed by some act which was not the fault of the contractor or the owner. The answer to the question of liability is in the contract. Now the question concerns liability in case of the destruction of a building after construction has been completed. The cause of the destruction is not the fault of the contractor or the owner. The rule of law that applies here is different because the contractor's obligation is ended upon the completion of the structure, in accordance with the terms of the contract. If the contract says the builder is responsible, your insurance agent's advice is needed for adequate coverage. Once the building has been completed and accepted by the owner, the responsibility of appropriate insurance coverage is transferred to the owner.

You have a different situation in a case where the contractor agrees to contribute only a portion of the work and materials in the erection of a building. This building contract further provides that the owner and other independent contractors are to do part of the work and furnish part of the materials. Now comes the accidental loss of the building before completion, where the contractor is not responsible to the owner for the loss. The contract involved here is divisible and severable and the contractor is discharged from his obligation by the destruction of the building before completion. In this case, the contractor can collect for the reasonable value of labor and materials furnished on the basis of work that he has already done.

A similar result was reached in a case where the contract provided that the contractor was to furnish the work and the owner of the land was to furnish the materials. This contract was clearly divisible and severable. The accidental loss of the building before completion fell on the owner and the contractor could collect for his labor performed.

When a contractor completely installs his part of the work, he may collect his money, despite the fact that his work is damaged by other persons who are not his employees, but who are doing other construction work. Another exception involves a building contract where the contractor agrees to furnish labor and materials to do regular work on the owner's existing building and the building is destroyed without fault of either party. Of course, the contractor is not held responsible for the loss of the owner's building and he may collect the reasonable value for the work already done. The reasoning of the law in this case is that the destruction of the building without fault of either party, excuses performance of the contract by the contractor. In this case, if the owner had paid the contractor the full amount for his repair work in advance, he is entitled to recover that portion that would represent an overpayment.

You must examine the building contract carefully to learn whether the contractor agreed absolutely and unqualifiedly to erect a building for a stipulated price. If this is the case, then it appears that the contract is entire and indivisible and the contractor is stuck for the loss of the accidental destruction of the building before completion. The reason reference is made to accidental destruction of the building before completion is that a different rule of law applies to accidental destruction of the building after it is completed. It has been decided that the contractor agreed to build a building for a stipulated price; that the loss occurred without the fault of either party; and that the owner is entitled to receive what the contract called for, namely, a completed building at an agreed price.

The theory of the law is that the destruction of the building is no legal justification for nonperformance of the contract. The contractor can protect himself by having the contract provide that he will not be responsible for loss of the building while under construction. The law makes no distinction between accidents that can or cannot be foreseen when the contract is entered into. If the building contract is clear, unqualified, and absolute, it must be performed no matter the cost.

If the contractor refuses to rebuild or carry out his contract, the owner of the building may sue on the contract for damages for nonperformance. The measure of damages is the difference between what it would have cost the owner to construct the building provided for in the contract, and the amount he would have to pay a second contractor to complete it. When a contractor fails to comply with the duty imposed by the terms of the contract, a breach or violation results. In this case the owner may file a lawsuit to recover damages suffered by the owner due to the contractor's breach of contract.

If the owner does something to prevent the contractor from performing the contract, this would be a breach or violation on the part of the owner, and the contractor could file a lawsuit for any damages that he may have suffered.

Where a contractor announces his intention not to perform under a building contract, the owner may treat the contract as broken and file a lawsuit for breach of contract against the contractor immediately, without waiting for the time fixed for performance. The above situation is called an "anticipatory breach" of contract. An actual example involved the refusal of a contractor to build more than twenty of the thirty-five homes which he had contracted to build, unless he was paid more money than was originally agreed upon. The owner was not obligated to wait and was able to file his lawsuit promptly for damages for breach of contract.

## 2-7 DAMAGES

The ordinary rule of damages, which applies to building and construction contracts, says that an injured party is entitled to compensation for the loss sustained by the other party's breach of contract. Anyone who is a party to the contract can recover damages if there is a breach of building contract. Normally, it would be the contractor or the owner, or possibly the architect.

It is a simple matter for someone to file a lawsuit for damages for breach of contract. However, proving the allegations of your complaint in court is something else, and proving your alleged claim for damages is often impossible.

The fact that an owner or contractor breached their building contract does not mean that the other party has an automatic valid claim for damages. He may not be able to prove his damages in court or his claim for damages may be too remote or speculative, and therefore denied. A good example of how difficult it sometimes is to prove actual damages occurs in a slander lawsuit. Smith sues White for one million dollars, claiming that White called him a crook in the presence of other persons. Smith contends he is not a crook and this slanderous statement embarrassed him in front of other people. Proving any damages, let alone one million dollars damage, is nearly impossible.

A jury could award punitive or punishment damages in a slander lawsuit that would not apply to breach of a building contract.

The owner or contractor may claim such damages as may be reasonably considered as arising from the breach of contract itself. This includes damages for a breach of the contract by delay in performance. One case held that if a contractor defaults in the performance of his contract, he is liable in damages to the owner. The owner is entitled to recover that compensation which will leave him as well off as he would have been had there been full performance. In other words, the owner is entitled to be made whole. If damages are too remote or too speculative, the law refuses to rule in favor of the alleged victim.

*Example*

1. One failing to perform his contract to erect in an office building an elevator which will work at its rated capacity is not liable to the owner of the building for the loss of rents during the time the unsatisfactory service continues; for even if such damages could be regarded as having been within the contemplation of the parties, they are too uncertain and remote to form a basis for recovery.

The law requires a party injured by the breach of a contract to make reasonable efforts to reduce or minimize the resulting damages as much as possible. If a tenant breaks a lease without legal justification, the landlord is still obligated to try to rent the premises, to reduce the tenant's liability for damages. If the second tenant pays equal or more rent than did the first tenant, the landlord has suffered no loss and therefore cannot recover any damages.

When the owner breaches a building contract, the contractor is obligated to try to obtain other construction work, in an effort to minimize the damages. An example would be the rental of equipment by the contractor, who is a victim of a breach of contract by the owner.

Where there is a breach of contract on the part of the contractor and no construction at all, the measure of damages is the cost of completing the contract; that is, the cost in excess of the contract price which would be incurred by the owner. In other words, the measure of damages is the amount above the contract price that it costs the owner to complete the structure in accordance with the terms of the contract. If the bid of the second contractor equals that of the first, the owner will have no claim for damages because he suffered no loss.

A contract provided that the contractor is to receive as his compensation a stated percentage of the cost of construction. The contract is terminated before completion without a breach or fault of the builder. The contractor may recover the agreed percentage of the cost of materials and labor supplied or furnished within the terms of the contract.

In dealing with the measure of damages for defects or omissions in the performance of a building contract, the basic principle of law which controls is that a person is entitled to have what he contracts for or its equivalent. Trying to decide what is the equivalent depends upon the circumstances of each case. Whether or not the defect or omission can be remedied without tearing down the building is sometimes a difficult decision.

Generally, the measure of damages is the cost of correcting the defects or completing the omissions. In some cases, however, the measure of damages can be the difference in value between what ought to have been done in the performance of the contract and what has been done. Where there is a mere defect in a building which can be remedied by repair, the measure of damages is the cost of such repair, rather than the difference between the value of the building as constructed and that contracted for. On the other hand, if a contractor has not fully performed the terms of the contract, and to repair the defects would require a tearing down and rebuilding of the structure, the measure of damages is the difference in value of the defective structure and that of the structure if properly completed.

Some courts have ruled that a residence is different than a commercial structure because a house has an esthetic value and must be constructed as the owner wants it, even though the finished dwelling may be just as good. Therefore, in the case of a home, where it is not built according to plans and specifications because of the contractor's negligence, the damages would be the amount required to reconstruct it to make it conform to the plans and specifications, rather than the difference in market value.

*Examples*

1. The difference in value is not the appropriate measure where foundations and footings were not in accordance with the plans, and attempts to correct the situation might have caused disastrous results.

2. Where the entire house was placed 10 feet forward on the lot, with the result that support was insufficient and the garage rendered inaccessible, and the defects were not reasonably remediable.

3. The particulars in which the dwelling was defective were that the porch and carport ceiling were finished in plywood instead of pine, the hall walls were sheetrock instead of panel pine, other material was substituted for formica counter tops over fir plywood, a rafter in the roof had been spliced and not braced so that there was a sag in the roof, walls were not plumb, the tile in the kitchen was bottom side up, ceramic tile in the bathroom was improperly installed, and a long board which had been nailed outside the diagonal sheeting had not been removed before the asbestos siding was installed, thus creating an unsightly ripple.

In the case of decorating work, such as painting or paperhanging, where the work is worthless and has to be redone, the owner is entitled to be made whole and to claim the expense of redecorating.

The parties to a building contract may agree upon the sum of money to be paid the other in case of violation of the contract, provided the amount stipulated is reasonable. Where the damages are uncertain and not readily capable of ascertainment, the sum of money agreed upon is called liquidated damages. Courts dislike penalties and forfeitures, and contracts therefore use the words liquidated damages instead. Some judges have ruled that the term is really a subterfuge or a trick and have denied recovery to the claimant.

Clauses in construction contracts inflicting severe and drastic penalties for nonperformance are not favored by the courts. A building contract invariably provides for a completion date, and further states that the contractor agrees to pay the owner a certain sum of money for each day that the contractor is late beyond the agreed date. The contract calls this daily payment liquidated damages and ordinarily it would be enforced by the courts. The facts of each case stand on their own and will be interpreted by the court. Where the contractor fails to finish the job, his nonperformance can be rectified by hiring a second contractor and holding the original contractor responsible for any losses suffered as a direct result of his nonperformance.

It is difficult to lay down a general rule as to when a stipulation in a construction contract is to be regarded as a penalty or as liquidated damages. If the amount agreed upon is out of proportion to the actual damages, some courts hold that the parties could not, in fact, have intended liquidated damages but actually intended a penalty.

*Example*

1. A clause in a building contract providing for the payment of 10 percent of the total contract price, for failure of the contractor to complete the work by a given date, could not be properly regarded as an agreement or a settlement of liquidated damages. It was actually a penalty and the court refuses to enforce that provision of the building contract.

As a general rule, damages by way of lost interest are allowable in damage actions based on building contracts. The controlling date for computing interest would be the date the lawsuit is filed and not the date of the alleged breach of the contract. Where the owner as well as the contractor is responsible for substantial delays in the performance of the construction contract, the court will not keep books as to whose delay was greater.

## 2-8 ASSIGNMENT

Assignment means to transfer or sell. Building and construction contracts are assignable because a contractor frequently needs money to operate his business and large sums of money are owed him. By assigning some of the monies due him on jobs that are nearly completed, to his bank or other creditors, he is able to borrow money and pay his bills promptly. When he completes the job involved in the assignment, the owner pays the amount due to the bank or whoever is named in the assignment as assignee. It is most important for the contractor to maintain a high credit rating. When he needs a performance bond for a construction job, the bonding company orders a credit report to find out about his record for payment of bills.

Contractors, as well as all other businessmen, have found that they can make money by paying their bills promptly and taking advantage of discounts that are offered, over and above the cost of interest that they pay the bank for the borrowed money. In other words, the discounts for prompt payment of the bills are greater than the cost of the borrowed money. In addition and equally important, the contractor is able to maintain a high credit rating by prompt payment of his bills.

The assignment of a contract operates not only as an assignment of the moneys to be earned later, but of the whole contract with its obligations and burdens. No set form is required in making an assignment of the proceeds of a building contract. The intent to assign or transfer, or sell on the one side and agree to receive on the other, is the essential. The one who makes the assignment is called the assignor and the one on the receiving end is called the assignee. If you were assigning a note, you could write on the backside, "For value received, I hereby assign all of my right, title and interest in and to the note on the reverse side to the Second National Bank" and sign your name. You are the assignor and the Second National Bank would be the assignee.

It is not necessary that an assignment of money under a building contract be recorded with the county clerk. It is recommended as good business practice for the person receiving the assignment to notify the debtor, who in the case of a building contract would be the owner. Send him a copy of the assignment so that when the debtor is ready to pay, the money will go to the assignee and not to the assignor, who was the original payee but who sold or transferred his right to the money. In the example cited, the Second National Bank would be entitled to the money, they being the assignee.

If the obligation is past due, inquiry should be made as to when the owner debtor intends to pay his obligation. Remember, the assignment of the proceeds of the building contract does not become effective until you give notice to the one who owes the money, namely, the owner. The importance of prompt notice of the assignment by the incoming assignee to the owner debtor cannot be emphasized too strongly. One sound reason for the prompt notice is that after the owner debtor receives notice of an assignment of the proceeds of the building contract, he may not pay the original contractor, who was the assignor. If the owner ignores the assignment and pays the contractor after receiving notice, he is paying the wrong party and such payments are at his own peril. If the owner debtor made such an improper payment after notice of the assignment, the assignee may recover the assigned amount, even though it amounted to a second payment.

As between more than one assignee who comes first? This does not mean that the contractor did anything wrong by issuing two or more assignments. The contractor could have $10,000 due him under a building contract; the first assignment was for $5,000; the second for $3,000; the third for $2,000. Again, the importance of giving notice of the assignment to the owner debtor must be realized by the assignee. Courts have ruled that the assignee who first gives notice to the owner debtor of his assigned claim is entitled to be paid first, even though the assignment to him is later in date than those to other assignees. This is another important lesson to be learned from the old legal adage, "prior in time is prior in right." If you are ever placed in the position of being an assignee, don't sleep on your rights by failing to notify the debtor promptly of the assignment.

A contractor may make an assignment of the percentage retained by the owner until the completion of the building. If the total building price was $50,000 and the contract provided that the owner could withhold 10 percent of the monthly construction statements, the contractor has a $5,000 asset available to assign to another.

For years the courts have ruled universally that the parties to a building contract had a right to provide that the contractor may not assign any funds payable to him. However, the Uniform Commercial Code saw fit to ignore the established law in this situation. Now it will be up to the courts of the various states to decide whether or not to follow the established law of permitting the parties to a contract to agree not to assign or follow the new doctrine. The Uniform Commercial Code doctrine provides that assignments of money can be made regardless of the contractual provisions.

If you were the owner and this problem arose, it would be smart to deposit the money you owed with the clerk of the court and let the court decide whether you pay the money to the contractor assignor or to the receiving assignee. This type of lawsuit filed by the owner is called an action in "interpleader," because all of the parties claiming they are entitled to all or part of this money are named as defendants. Sometimes, a similar disagreement arises between the contractor and lien claimants and others in regard to funds due under a construction contract and still in the hands of the owner, who is willing and anxious to pay his debt but does not know to whom. There is an alternative to the expense of filing an interpleader lawsuit. The owner can instruct all parties that make some claim to the money that he owes to file their own joint lawsuit, and let the court order the owner as to how the money should be paid and to whom.

It is unfortunate that the law is not always clear and concise. The law is a changing process and modern life with its new inventions has caused many transformations. Years ago, the ownership of land meant that you owned the parcel of land from the bottom of the earth to the sky. Then came aviation and the realization that aircraft, by necessity, had to fly through airspace that you thought you already owned. We now all agree that we cannot prevent aircraft from flying through airspace, even though they are flying over our land. No one has been able to figure out a way to handle lawsuits more efficiently and quicker than they are now being handled. Actually, inflation has affected lawsuits and litigation, just as it has affected many other facets of modern life. However, the legal profession and the judges are trying to modernize some of the archiac procedures.

The oldtime attorney used to start his lawsuit by saying, "Comes now the Plaintiff for his cause of action against the Defendant and alleges for his first cause of action." The modern version of the same thing is two words "Plaintiff states." We cannot change the law or procedures overnight, but progress is being made.

## 2-9 TERMINATION

A building contract may be discharged in the same manner as any other contract. The more common methods of discharging a contract are (1) agreement of the parties; (2) performance; (3) breach of violation; (4) impossibility; (5) operation of law; and (6) rescission because of fraud, mistake, duress, or undue influence.

"Rescission" involves the act of cancellation, revocation, or termination. The effect of rescinding the contract is to nullify it as if it never existed. A building contract may be rescinded upon a number of grounds including by agreement of the parties. If the parties to a contract agree to rescind, the agreement of all of the parties involved is necessary. For example, in a valid building contract, the owner and contractor could not agree to rescind and ignore the architect, who is also a party to the contract.

Many construction contracts contain provisions for the right to rescind. A common provision requires the contractor to supply a given number of workmen to perform the job, and his failure to do so entitles the owner, after giving notice to the contractor, the right to terminate the contract and make other arrangements. Another common provision authorizes the architect to certify that the contractor is in default and the architect's certificate is sufficient grounds to justify the owner in terminating the contract.

Usually, the contractor must be notified of intention to terminate, and he must be given an opportunity to correct the situation complained about. Even if the building contract is silent about rights of rescission, any of the parties involved may rescind the contract upon proper grounds. The right of a party to the contract to rescind must be exercised promptly or within a reasonable period of time, otherwise it may be ruled that he waived his rights. In legal language, one who had rights but failed to act within a reasonable period of time is said to be guilty of "laches," which means sleeping on his rights.

A building contract may be rescinded for default in performance by the contractor where the owner defaults, or by the owner when the contractor defaults. The party rescinding cannot be in default himself. There must be a substantial default that goes to the roots of the contract in order to justify rescission. Minor omissions on the part of the contractor will not justify cancellation by the owner.

*Examples*

1. One who has contracted to construct a tunnel may treat the contract as terminated upon refusal of the owner to furnish the necessary timber for shoring up the work as the contract requires him to do.
2. A contractor employed to take down portions of a building, whose employees became demoralized and refused to work in the building on account of its weakened and dangerous condition, resulting from the negligence of the owner and his architect in stripping off the sheathing, rafter, etc., so that spars fell and killed some of the contractor's employees, may abandon his contract and recover damages from the owner.
3. Continued refusal of the contractor to remove work not done in accordance with a building contract, where the contract authorized the architect to order said removal and further authorized the owner to rescind the contract upon notice to the contractor.
4. The contractor's failure to build cupboards in the kitchen did not justify the owner in rescinding a contract to build a house, where the house had been constructed in substantial conformity to the contract.

If the parties enter into a new or second construction contract and proceed under it, no claim can later be made under the first contract that was breached. In this situation, the first contract was replaced by the second contract. In adjustment of the contractor's breach of the first contract, the parties enter into a supplemental agreement imposing additional duties upon the contractor. If the contractor fails to perform under the supplemental agreement, the owner may then recover his damages for the contractor's breach of the first contract.

If the owner discovers reasons for declaring a default against the contractor, entitling him to rescind a building contract, but he leads the contractor to believe that it is still in effect, the owner has waived his right to rescind. By permitting the contractor to continue work on a building contract after the expiration of the time specified for completion, the owner waives the right to rescind on that ground. Likewise, an owner cannot rescind a building contract because of variations from the specifications for the foundation, after he permits the contractor to nearly complete the building.

The owner's failure to pay an installment of the contract, as provided in a building contract, is a substantial breach of the contract. It gives the contractor the right to consider the contract at an end, to cease work, and to recover the value of the work already performed. The default of the owner in making payments not only justifies a building contractor in abandoning the work but entitles him to recover as damages his profits on the uncompleted portion of the work. If the contractor is in default and has not fulfilled his obligations under the contract according to its terms at the time he demands payment, he is not justified in leaving the job.

The mere fact that there is a delay in the performance of a building contract will not terminate or justify rescinding the contract. Rescission of a contract for delay will not be permitted unless time has been made the essence of the contract. Where time has been made the essence of a building contract by its terms, late performance does not fulfill the requirements of the contract. In such cases, the innocent party is given a chance to rescind the contract.

Each case goes on its own facts and there is no hard and fast rule that will fit all situations. For example, it has been held that an owner cannot rescind a contract merely because the contractor runs a few days over the time set for the completion of a home. The law takes knowledge of the fact that in construction work minor details, which can be the subject of controversy, are bound to arise. The law requires only substantial performance within the time set. If there is a substantial performance by the contractor within the time fixed for completion of a building contract, the owner cannot rescind. But if the contractor's performance is in fact, defective, negligent, and worthless, it can be treated as no performance.

A refusal by the contractor to fulfill a building contract must be absolute in order to authorize the owner to rescind. A building contractor who is wrongfully forbidden to complete his work may treat the contract as rescinded and he may maintain a claim for the reasonable value of his services and materials furnished in the construction of the building. The contractor may also rescind for acts of the owner in preventing performance.

It is a general rule of law that a contract induced by fraud may be terminated by the defrauded party. When material misrepresentations are made by the owner and relied on by the contractor, pertaining to the cost or expense of a building contract, the contractor is entitled to rescind on ascertaining the facts, providing he acts promptly. A building contract may also be rescinded where the architect designated by the contract to approve the work is later discovered to have a financial interest in keeping construction costs down.

In order to justify the rescission of a building contract on grounds of fraud or misrepresentation, it must be shown that the misrepresentations amount to a statement of fact, rather than the expressions of opinion. A contractor seeking to establish fraud must show not only that he relied on the false representations, but also that he had a right to rely on them. A contractor cannot rely on the owner's representation of the probable cost of a building and base his price on it. The law says that the contractor is presumed to be the expert and has no right to rely on the opinions of the owner.

Sometimes building contracts provide that the contractor agrees that he is relying solely on his own investigation, or that the owner is released from any responsibility for statements made by the owner. If actual fraud on the part of the owner can be shown, the courts will ignore statements in the building contract, attempting to release the owner from liability.

A "mutual mistake of fact" by both parties to a building contract may be a ground or reason for rescinding the contract. By mutual we mean the mistake was made by both parties to the contract. By "unilateral" we mean that the mistake was made by only one of the parties. Under some circumtances a unilateral mistake of fact may constitute a ground for rescission of a bid. For example, an experienced contractor, who accepts a bid for material for two roofs with sufficient notice to put him on inquiry as to whether or not the bidder was mistakenly bidding on materials for one roof, cannot hold the bidder to his contract. He knew or should have known that the price was too far off not to involve a mistake on the part of the bidder. The law says that one may not knowingly take advantage of another's mistake.

The courts are not agreed on the right of the bidder to rescind a contract or bid solely on the ground that he made a mistake in computing the amount of his bid. Some cases lay down the rule that when there is some reasonable excuse for an error made in calculating the bid, and the party receiving the bid knows or should have known of the mistake at the time the bid is accepted, the bidder has the right to rescind. A bid based on a unilateral mistake that is so great that it must be considered fundamental, may be voided when the mistake is honestly made, without negligence, and the acceptor will not be prejudiced by the cancellation. The bidder cannot be relieved when the mistake is caused by his own carelessness and it is not so gross as would justify the court in saying as a matter of law, that the acceptor of the bid is put upon notice that a mistake has been made.

## 2-10 REMEDIES

Generally speaking, any material breach or violation of a building contract gives the innocent party a right to recover any damages sustained. However, only one action can be maintained for several breaches of an entire building contract. Installment payments are the exception to this rule because the contractor can file a claim for each installment payment due and not paid by the owner. The contractor is not required to wait until the building is completed before filing his lawsuit.

The parties to a building contract may stipulate as to the remedies that shall be available for a breach of the contract. The remedies agreed upon are legal and binding upon all parties to the contract. A contractor, who is entitled under the building contract to receive monthly payments which have not been paid by the owner, has a number of choices as to what action to take: The contractor may treat the contract as rescinded by so notifying the owner; he may file a lawsuit for the work performed; he may continue the work and file his claim for each installment when it falls due; he may complete the entire contract and then file his claim or lawsuit for the balance of the contract price; or he may elect to suspend work until the past due installments are paid, and later sue for special damages that he suffered caused by the owner's failure to pay as agreed and the resulting delay.

*Example*

1. A wrongfully discharged contractor was not limited to sue for the reasonable value of his services in supervising the construction of a twenty-unit addition to a motel; but when he was hired for 10 percent of the total cost of construction, he could elect to stand upon the contract and collect his 10 percent agreed fee.

In summary, always consult your attorney before you sign a contract or enter into any important business transaction. The same admonition applies to matters concerning your accountant and insurance agent.

Know your business and its problems. Be honest with your banker; don't go off on tangents and have your business suffer as a result.

Take care of your customers; they are your most valuable asset. A satisfied customer can do you a tremendous amount of good by recommendations to friends and relatives. A disgruntled customer can do considerable harm and should be satisfied, if at all possible.

## 2-11 MINORS AND CONTRACTS

One who deals with a minor does so at his own peril. The general rule of law involved is that an infant or minor has the choice when he reaches his majority, which is usually twenty-one years of age, of either affirming or disaffirming his contract. If the minor disaffirms his contract when he attains his twenty-first birthday, he is obligated and must pay the reasonable value of the subject matter involved. For example, if he buys a car, he can return the car and pay a reasonable rental sum for the use of the car. In any event, it is undesirable and unprofitable for an adult businessman to attempt to do business with a minor. If a very young customer comes along, it is important that you determine his age. The law places the burden on the adult businessman, however, the minor who looks older may not perpetrate a fraud by lying about his age. Remember, if it takes a lawsuit to prove your point, it is not good business.

Many mortgage lending institutions are reluctant to deal with young people, even married couples in their early twenties, because of their lack of financial stability. The important thing is whether or not the mortgage company will lend mortgage money to young people.

Some states have a procedure called "emancipation of a minor," wherein a young married man under the age of 21 may obtain a court order declaring him emancipated from his parents. The young married man can then legally deal with the business world as an adult. The emancipation procedure, however, will ordinarily not solve the problem facing the young couple in obtaining their mortgage money when they desire to buy their first home. The only practical solution is to have the parents guaranty the mortgage obligation.

# 3

# PARTNERSHIPS

A partnership is an association between two or more persons or entities to unite their property, labor, or skill in prosecution of some joint or lawful business, and to share profits and losses in certain proportions. When forming a partnership, the parties should discuss division of the profits as well as division of the losses since there is no guarantee that the partnership venture will prove to be successful. There have been cases where two men, who have been close friends for years, decide to form a partnership and discover that there is a personality clash between them as business partners. Consequently, they are ready to dissolve the partnership.

If the partners are able to get along and business is good, it can be a pleasant and successful relationship. On the other hand, if there is discord, bickering, and a feeling of distrust between the partners, the business is bound to suffer. In this case, something should be done about dissolving the partnership. (The procedure for dissolution of a partnership will be discussed later in this chapter.)

As mentioned above, a partnership involves two or more persons or "entities." An entity is a legal status or classification. Examples would be a person, one or more; a partnership; a corporation; or an association. Thus, a partnership could be formed between an individual, a partnership, and a corporation. The fact that three different types of entities are involved in the example above does not create problems. With the current trend towards big business and international commercial activity, combines of various types are formed between corporations in the United States and foreign corporations.

The discovery of oil deposits in the state of Alaska and the sale of oil and gas leases have resulted in a number of oil companies, American and foreign, forming a combine and jointly bidding millions of dollars for some of the oil and gas leases that were offered for sale. The tremendous sums of money involved and the element of gamble as to the quantity of petroleum available were factors that caused these large corporations to form partnerships and combines of varied interests. On a \$30 million bid, X Oil Company may have provided \$20 million, whereas Y Oil Company and Z Oil Company may have provided \$5 million each. Now, their partnership or joint venture agreement would provide for X Oil Company having a two-third interest, and the remaining one-third interest is divided equally between Y and Z.

Sometimes, professional people like doctors, lawyers, and accountants decide to form partnership relationships within their own profession. Some work out well, while others work out poorly. Newly formed partnerships in which the individuals have not known each other for a very long period of time create an additional handicap and suffer as a result. Two lawyers form a partnership and soon discover a serious problem in that one produces most of the legal business, while the second partner produces very little, causing friction between them.

## *3-1 FORMATION OF PARTNERSHIP*

The procedure in forming a partnership contract or agreement is strictly legal, and all details should be handled by an attorney. The partnership agreement or contract is called "articles of copartnership." Even though it is possible to establish a partnership relationship verbally, this would not be a very businesslike arrangement and is strongly discouraged. Law books are filled with disputes where one party testifies in court that the parties agreed to white, and the other party testifies that the parties agreed to black, and the judge is called upon to be a Solomon. Written articles of copartnership will eliminate most of the doubt as to what each party agreed to do.

In due course, the partnership agreement between the parties is reduced to writing and signed by all of the partners involved; sometimes, the provisions of the partnership agreement are approved by the wives of the married partners involved, because in some jurisdictions the wife has a community interest in the marital assets. It solves possible future problems when the wife has knowledge of the partnership arrangements and cannot later complain that her husband acted without her knowledge or approval. (As a practical matter, most wives are delighted that they are being advised about the details of their husbands' business affairs, and it can lead to happier marriages.) Even though the wives are not active in the business affairs of the partnership, it is still desirable that they know and understand the important details involved. The reader will later appreciate the advisability of keeping wives knowledgeable about the business of the partnership, when the articles of copartnership provide for buy and sell agreements between the partners, what happens in event of extended illness or death, "keyman insurance," and other provisions.

When the attorney is instructed to prepare the necessary papers to form the partnership, he then arranges a meeting for all of the persons involved. At this time, he advises the partners that it will be necessary to check the records of the Secretary of State, or a comparable public official, to ascertain whether the names they have chosen have already been registered and are being used by others. If so, the partners will have to select another name. Obviously, you cannot open up a new bottling business and call your soft drink Coca-Cola. In the case of common names, the same principle applies; for example,

Jones and Brown will have to select another trade name if there is already a Jones & Brown Construction Company in business whose name was registered with resulting priority.

Your attorney will want to know the address or addresses where the partnership business will be conducted. The designation of contractor and builder indicates that you are going to be engaged in the construction business. The attorney will question you as to whether you are licensed to do business as general contractors by the state or local authorities. This is a vital requirement and cannot be ignored. There have been many cases where the contractor performed his services and built the structure agreed upon and then, when his customer refused to pay and the contractor filed a lawsuit for his money, lost his case because the judge ruled that he had no standing in court, being an unlicensed contractor.

What is the length of time your partnership business is scheduled to exist? If it is only for the period of time required to construct a certain project, then your attorney will so provide in the partnership agreement. If, on the other hand, Jones and Brown are thinking about a permanent partnership, the attorney will state that the partnership shall continue for twenty-five years or forty-nine years, and for so long thereafter as the partners may agree. *Death of any of the partners automatically terminates and dissolves the partnership.*

The attorney will also inquire about what contribution each of the partners is making to the partnership and whether the contribution is represented by cash, materials, tools, or knowledge and experience. It is not necessary that each partner contribute an equal sum of money to the partnership. Particularly in the construction business, one partner may have the construction know-how while the other partner will contribute the money. While most two man partnerships are 50-50, the percentage interest of each partner may vary as agreed. There is no requirement that each partner hold an equal interest.

Assuming that each of the partners is going to devote all of his time and attention to the partnership business, it is important to discuss the amount of salary or drawing account to be permitted each partner. It is desirable to draw as little as possible, so that the new partnership business will endure growing pains and be strong financially. Any additional or special draws of money should be discussed by both partners and whatever amount is agreed upon for one should also apply to the other. If one partner draws partnership monies without the knowledge or approval of the other, discord and mistrust are bound to result.

The duties to be performed by each of the partners call for serious discussion. Once agreed upon, the duties should be spelled out in detail; there should be no overlapping. In a small construction firm, an ideal arrangement would be for one partner to be in charge of the construction and the other to be in charge of the office. The construction partner is the outside man and advises the inside partner where and when to purchase materials and supplies. The inside partner then handles the paper work and bookkeeping involved. The books and records are kept by the inside partner, under the supervision and control of an accountant selected by both of the partners, and the outside partner is privileged to examine the books and records at any reasonable time. Both partners must be honest and open with each other at all times. The monies of the partnership are banked in a bank that has been approved by both partners, and they agree as to who is to sign partnership checks. Sometimes they agree that either partner may sign checks, while others may prefer to have each check signed by both partners. If the partnership wants to purchase land for future development, this would be a major undertaking in which the partners ought to work together as a team and consult each other.

The partnership agreement should state how much time is to be contributed by each partner. If each of the partners is obligated to devote his entire time and attention to the partnership business, there is no problem. Sometimes, one of the partners has outside business interests that are not involved in the partnership. This can lead to serious misunderstandings, unless the details are spelled out as regards the amount of time for outside interests and the amount of time for partnership business.

It has already been stated that death of any of the partners automatically terminates and dissolves the partnership. Similarly, in the event of serious illness or disability of either partner, both partners may agree that any illness or disability of six months duration will result in the termination of the partnership. Furthermore, the well partner is privileged to buy out the interest of the ill or disabled partner on a formula to be worked out by the partnership accountant and approved by both parties. Usually, the partnership business is not strong enough financially to provide for payment in a lump sum and, as a result, a schedule is arranged for installment payments over a period of time, so that the financial structure of the partnership will not be impaired.

If the partners are unable to agree on any buy or sell arrangement, the remaining choice would be to agree to voluntarily dissolve or terminate the partnership. Under this choice, the assets of the partnership would be placed on the market and sold. After all debts of the partnership were paid, the remaining proceeds would be divided between the partners. Whenever partners get stubborn and refuse to work out a buy and sell arrangement that is fair and equitable, both partners will be hurt financially because the partnership business has a greater value to the inside partners than to any outsider.

In the event of an unexpected or unusual happening, an arbitration clause is usually included in most partnership agreements. An arbitrator is a neutral referee or informal judge. The articles of copartnership provide that if the partners are unable to agree and settle their differences, each partner shall select an arbitrator, and the two arbitrators selected shall select a third arbitrator. The dispute shall be submitted to the three arbitrators, and their decision shall be final and binding upon each of the partners.

The procedure of submitting disputes to arbitration, rather than to the courts, is quick and inexpensive. Recently, the author was involved in a dispute between an architect and the owner of an apartment building. Their agreement provided for arbitration, and three attorneys were involved. The apartment building owner contended that the architect failed to render qualified services in that the air-conditioning system was not functional. The architect rebutted this by arguing that the malfunctioning of the air-conditioning system was the responsibility of the subcontractor and not the fault or responsibility of the architect. The arbitration board sat for four days, heard all of the testimony, examined the exhibits, and issued their decision ten days later. It was mutually agreed between the attorneys that the same dispute presented in court before a judge and jury would have consumed three weeks of time instead of four days.

Reference was made earlier to "community property." Community property is property that is acquired by a husband and wife from the time of their marriage onward, as a result of their joint efforts. The eight western states of the United States operate under the community property concept. This means that the wife has a greater voice and control over property acquired by the married couple. The earnings of a working wife are community property, the same as the earnings of the husband. Property that the husband or wife brought into the marriage would be their separate property. Property acquired by gift or inheritance and the income from separate property are also classified as separate property.

"Key man insurance" is a modern concept that is recommended in many partnerships. As the name indicates, life insurance is taken out on the lives of key men in the partnership, including the partners, who make a sizable contribution to the success of the partnership. The premiums are paid by the partnership, and the proceeds of the life insurance policy go to the partnership. Some partnership agreements that provide for the surviving partner to buy out the interest of the deceased partner's widow, use the proceeds of key man life insurance to provide the money to accomplish the buy out mission. In other key man cases, the life insurance proceeds are used to obtain a replacement for the deceased key man.

In spite of all of our efforts, the written partnership agreement cannot cover every conceivable situation that may arise during the life of the partnership. The main purpose of the written partnership contract between the partners is to set up guidelines in the event of problems or difficulties. Each partner is supposed to be candid and forthright with his partner. If the time ever comes when one partner attempts to take unfair advantage of his partner, and a personality clash results, the partnership should be terminated or dissolved.

## 3-2 DISSOLUTION OF A PARTNERSHIP

A partnership may be dissolved by voluntary or involuntary acts. As previously stated, death of any one of the partners automatically terminates and dissolves the partnership as of the date of death. This would be a prime example of dissolution of a partnership by an involuntary act. The partners could agree to go out of business, thereby dissolving the partnership by their voluntary actions. In this case, creditors would be notified as to the closing of the partnership business, all bills would be paid, and then the partnership assets would be sold and the proceeds distributed.

More often than not, a partnership is dissolved by mutual agreement of the partners, based on one partner buying out the other partner. The terms and conditions of the buy and sell agreement is personal and of no interest to outsiders. However, creditors are vitally concerned about who is going to continue operating the business, and, in particular, who is going to be responsible for the partnership bills. Creditors are also concerned with the financial stability of the partner who is going to continue operating the business, as regards future credit. Most jurisdictions require that a voluntary dissolution of a partnership be advertised in legal publications, so that creditors will know who is retiring, who is operating the business, and who is paying the bills.

The reader may wonder how to solve the problem whereby death of any one of the partners terminates the partnership. This particular problem can be solved by the principal parties, two or more, forming a corporation. Corporations are a complex subject that will be discussed in detail in a later chapter. Suffice it to say that death of any one of the principal parties involved in a corporate entity does not terminate or dissolve their business relationship as it does in partnerships.

## 3-3 LIMITED PARTNERSHIP

As the name implies, the term limited refers to the limitation of the liability of a partner, one or more, who is making only a capital investment in the partnership business. Frequently, a "limited partner" is referred to as a silent partner for the reason that he has invested a certain sum of money in the business, yet limits his liability if the partnership business should fail. If the business goes broke, the silent or limited partner loses the money he has invested but is not liable for any additional sums. The limited or silent partner does not advertise the fact that he has made a capital investment, and creditors are not advised of his inactive participation. Creditors cannot complain because they did not extend credit on the basis of the silent partner.

The general partners, one or more, actively operate the partnership business, and credit is obtained on the financial stability of the general partners. If the business should go broke, the general partners are personally liable for partnership obligations and are not protected by the limited liability of the silent partner.

## 3-4 JOINT BUSINESS VENTURES

A "joint business venture" is a hybrid relationship that is neither a partnership nor a corporation. It usually occurs when two or more companies are interested in a certain project and they think more can be accomplished by acting jointly, instead of in their individual capacities. The parties to a joint business venture agreement could be an entity, persons, partnerships, corporations, or any combination of same. For example, two or three large oil companies might hold oil leases that are adjacent to each other in a foreign country. The management of the oil companies decide that a joint exploration program, wherein all of the involved companies participate, would be more advantageous and less expensive than each of the three companies drilling independently. They would enter into a joint business venture agreement that would set forth the terms, conditions, and interest of each of the three companies involved. The joint venture agreement would further provide that the relationship applies only to the one transaction and would terminate upon its completion.

# 4

# SUBCONTRACTORS

## 4-1 GENERAL INFORMATION

When the architect has completed the plans and specifications for a construction job, contractors are then invited to bid on the job. A copy of the plans and specifications are then furnished to each contractor who is invited or interested in bidding. In order to bid intelligently, the contractor or his estimator must first obtain bids from the various subcontractors involved. If the construction job is to be bonded, then a performance bond must be available, in connection with each subcontractor's bid. The cost of preparing and submitting a bid by each subcontractor or even the general or prime contractor is a hazard of the business. Usually the low bidder gets the job contract and all of the others get the experience but no compensation for their efforts in submitting the bid.

This discussion about general contractors, subcontractors, and submitting bids pertains to large and small construction jobs. For example, a prospective customer wants your price to build his new home. He and his wife have been dreaming about their new home for a number of years and finally hired an architect to draw the plans. You examine the plans and specifications and then decide you need bids from the following subcontractors: plumber; electrician; roofer; heating and air-conditioning; painter and mason. After you have received all of the bids from the subcontractors, you are then in a better position to figure your own costs on labor, material, and overhead expenses, and finally add your anticipated profit. Small builders who feature price only in offering bargain basement deals to their prospective clients, overlook charging for their own supervisory time and many other necessary items of overhead expenses, and wind up going broke. They are no longer with us as contractors but are now employed as carpenters and various other craftsmen within the construction industry. If you fail to include all items of cost as overhead expenses and charge for them, you are doomed to failure because your business can continue only if you earn a profit.

Experience is a great teacher. In spite of all precautions, situations will arise where things go wrong and after the home or building is completed, you discover that your cost was higher than anticipated and that you have lost money on the job. If you learn what happened and avoid this pitfall in the future, the experience was worth the monetary loss. The sad cases are those contractors who do not know that they lost money on the job and those contractors who never find out why they lost money. They are doomed to failure. If being successful in the construction business would be a simple matter, every carpenter would aspire to become a contractor. It does take ability, hard work, and knowhow to be successful in any business endeavor.

The temptation in starting a new business is to feature price. However, it is far better and often easier to sell your client on the desirability of quality rather than a bargain basement price. Over a long period of time, quality construction will prove to be better value than cheap construction. Pride in workmanship goes along with quality materials, rather than the philosophy of "we can make it a little cheaper for a little less money."

Some years ago during a home building boom period, a young, inexperienced builder tried a novel approach to building tract homes in wholesale quantities at a bargain basement price. He decided to eliminate the element of risk and obtained bids from subcontractors for every conceivable part of the home to be built. He subbed out the foundation, the framing, the insulating, the flooring, the plastering, and every other part of the house. He then totalled the various subcontracts and added his cost of the land and thought he was playing a winning game. The plan sounded good on paper but did not work out well in theory and he is no longer in the construction business. In his eagerness to attain quick success through volume sales of inexpensive homes, he overlooked two important items of overhead: the cost of insurance which he mistakenly thought would be paid by the subcontractors, and compensation for himself for salary, sales, and supervision.

Going to the other extreme, some people have wondered why the home builder does not do all of his own subcontracting work and thus eliminate the subcontractor entirely. The answer is that most states have laws prohibiting this practice. Particularly in the field of plumbing and electrical work, the welfare of the general public is involved and a novice has no business dealing with electricity or plumbing if he does not know what he is doing: it would be dangerous for the workman as well as the occupants of the structure. Most state laws require rigid examinations before a plumber or electrician is licensed to perform his work.

Successful home builders who feature quality rather than price deal with subcontractors who also feature quality over price. This type of home builder has dealt with the same plumber, electrician, roofer, and painter for a number of years and knows that a cheap price is a poor substitute for quality and experience.

## *4-2 SUBCONTRACTOR AND GENERAL CONTRACTOR*

The relationship between the general contractor and his subcontractors should be a very congenial one. If they do not get along or if there is a personality clash, the quicker they part company the better it will be for all parties involved. Although they may not be able to part company until the present contract has been completed, they should ignore each other for future construction contracts. It is important that the two get along well together for they actually form a team, and teamwork is the key to successful construction. Many home builders who have built up a favorable reputation for building quality homes at a fair price, realize the importance of using the same subcontractors over a period of years. They have found it more desirable to deal with the same subs, rather than switch to a new subcontractor for benefit of a few dollars cheaper bid. The old adage of "you get what you pay for" still holds true in the construction business.

The parties to a building or construction contract may modify it or waive their rights under it and arrange new terms in the same manner as may be done with any other contract. It is also the rule that an agreement by a prime contractor to pay more money for work to be done by a subcontractor, which the latter was already bound to perform at a previously agreed price, is unenforceable because it is without consideration.

*Examples*

1. An agreement between a builder and a plasterer to pay more for the work than the price specified in the original contract is unenforceable.
2. An oral agreement between a subcontractor and a building contractor abandoning a prior written contract is without consideration where it obligates the contractor to do no more than he was legally bound to do.
3. A post-contract agreement to pay the contractor for extra dirt used in grading was binding on theory of unforeseen difficulty.

A subcontractor or contractor may not refuse to perform the conditions of his written contract and seek additional compensation because he has made a bad bargain. It is this type of transaction that is best charged off to experience. Experience is a valuable asset and it is worth a sizable sum of money to acquire it. If you profit by your unfortunate experiences, then it was well worth while.

Clauses in construction contracts involving subcontractors that inflict severe and drastic penalties for nonperformance are not favored by the courts. They will be strictly construed and the courts will circumvent them whenever possible.

Building and construction contracts frequently provide that the work shall be done in a workmanlike manner. A guaranty by a contractor or subcontractor that the workmanship in a building shall be of the first class and satisfactory in all respects protects the owner against the use of bad or unsuitable materials.

*Example*

1. Where a contractor for a building guaranteed that the workmanship should be first class and satisfactory in every respect, and the plaintiff sued for damages caused by the failure of a subcontractor to provide proper materials for the construction of a granolithic floor, a request to charge that the defendant contractor did not guarantee that the floor would be perfect, but only agreed to use his best knowledge, skill, judgment, and energy in the business, and if he did that, and kept all the other parts of the contract with reference to workmanship, the jury should find for him, is properly refused, as contrary to the terms of the contract.

Where there is a conflict between, or an inconsistency in, the provisions of a building contract and the provisions of the plans and specifications, the positive language of the contract should prevail.

*Example*

1. A reference in a subcontract to the provisions, plans, and specifications of the general contract imports them into the subcontract, but if the subcontract contains words of definite limitation, they will be given effect and the difference will be limited accordingly; hence, where a subcontract for electrical work extended only to inside wiring, the fact that the contractor was also required to do outside electrical work did not affect the terms of the subcontract.

A subcontractor is liable to the prime contractor for the amount which the latter has been required to pay as workmen's compensation to an injured employee of the subcontractor where the agreement between them provides that the subcontractor will indemnify the principal contractor for any such loss. On the other hand, the failure of the general contractor to furnish a foundation upon which a subcontractor's work could be superimposed is a failure to provide labor and materials not included within the subcontract. The general contractor's failure to provide such labor and materials binds him to reimburse the subcontractor for any loss that he suffered. The contractor's liability is fixed by the terms of his contract, and he is obligated to perform according to those terms.

The subcontractor as well as the general contractor does not guaranty the sufficiency of the specification, but only the skill with which he performs his work and the soundness of the materials that he uses.

*Example*

1. A defect in bricks, caused by unfit clay, and not discoverable by the exercise of care and skill on the part of the subcontractor after they are manufactured, is not answerable to the owner for damage to the building caused by such defect.

However, one constructing a building impliedly warrants that the building will be erected in a workmanlike manner and in accordance with good usage and accepted practices in the community in which the work is done. The contractor is liable as a matter of law if the foundation of a building cracks so as to leak and crumble immediately after its completion, whereas if it had been properly constructed it would have done neither.

If a subcontractor follows the plans furnished by the owner or his architect and he is not guilty of any negligence, the subcontractor is not responsible for any loss or damage which results if the structure proves to be defective or insufficient. If the subcontractor departs from the specifications, he is liable to the owner for an inherent weakness in the building. If the subcontractor is foolish enough to give an express guaranty, he must abide by his contract and perform his undertaking no matter what the cost. Under his guaranty he cannot recover compensation for extra work made necessary by a fault in the specifications.

*Example*

1. A contractor is ordinarily not liable for consequences resulting from changes made in plans and specifications prepared by an architect, where the changes are made with the full knowledge and consent of the owner, unless the consequences of the changes are so obvious or well known to the contracting trade that the contractor should know of them from his experience or the nature of his undertaking.

A subcontractor is not bound, by a stipulation of his general contractor, to submit his work to the decision of the architect, engineer, or others.

Three reasons have been given for the rule that a substantial performance of a subcontractor's building contract will support recovery. First, even if rejected, the owner of the land receives the benefit of the subcontractor's labor and materials, which is not the case where a chattel is constructed, because the chattel may be returned. The law deems it equitable for the owner to pay for benefits that he gets. Second, it is next to impossible for a builder to comply literally with all the minute specifications in a building contract. And third, the law presumes that the parties have impliedly agreed to do what is reasonable under all of the circumstances.

As a general rule, time is not of the essence in a building contract, in the absence of a provision in the contract making it such. The mere statement of a date in such a contract does not make time of the essence. A failure to complete the work within the specified time does not automatically terminate the contract, but only subjects the contractor to possible damages for the delay.

*Examples*

1. A provision that time was of the essence "on the part of the subcontractor" did not enable the subcontractor to recover damages for delay occasioned by the contractor's failure to complete more promptly work which was necessary to the doing of the subcontractor's work, because if the parties had intended that the provision should apply to both parties, they would have stated only that time was of the essence.
2. Failure of a subcontractor for a heating and water system in buildings under construction, who by the terms of the subcontract had sixteen months in which to complete the work, to comply with the principal contractor's demand that he have such system ready in seven months for temporary use during the winter season, constitutes a default in respect of which damages are recoverable, where the subcontract provided that the work contracted for should be done under the jurisdiction and control of the principal contractor, and was entered into with specific reference to the building specifications, one of which was that the contractor might operate the heating plant, and the subcontractor did not question the right of the contractor to make such demand, and the demand itself might have been complied with by the exercise of reasonable diligence.

The subcontractor is bound to complete the work within the time specified in the subcontract, plus such additional time as has been lost by the fault of the other party. The amount recoverable for delay is to be computed after deducting from the total delay, the number of days delay caused by the acts of the employer. Some courts have held that if the subcontractor is delayed by the acts of the other party, the stipulation as to the time for completion becomes ineffective and instead a reasonable time period is substituted.

## 4-3 SUBCONTRACTOR AND OWNER

The relationship between the subcontractor and the owner is most unusual in that there is no contract between them, yet the subcontractor plays a vital part in the construction project that is taking place on the owner's land. Normally, the important middle man between the owner and the subcontractor is the general or prime contractor. If the owner is of the opinion that a subcontractor is doing something wrong or not doing his work in a workmanlike manner, the owner is obligated to report the incident and work through the general contractor, who then will deal with the subcontractor.

The construction contract is between the owner and the general contractor. The contract provides that the general contractor shall be authorized to be the representative of the owner and that he is directed to enter into separate contracts with each subcontractor involved. As a result of the contractual relationship between the general contractor and the subcontractor, the subcontractor is not obligated to deal with the owner and he is answerable only to the general contractor. Under normal construction conditions, the subcontractor would probably not even know the owner, let alone have any contact with him.

In spite of the lack of contact between the subcontractor and the owner, peculiar situations have arisen which have resulted in dealings between them. In one case, the subcontractor gave up his right to file a lien because the owner promised to pay the subcontractor his money that should have been paid by the general contractor. The court ruled that the owner was bound by his promise to the subcontractor. In another case, the legal question presented was whether or not a promise by an owner to the subcontractor, to pay the subcontractor additional compensation to perform his contract with the general contractor is enforceable.

*Example*

1. The court held there was consideration for an agreement between a brick subcontractor and the owner, by which the owner offered to pay a bonus of $350, in addition to the contract price to be paid by the general contractor, if the subcontractor would pay his striking bricklayers the increased rate of wages demanded by them and have them start work immediately. The judge said the subcontractor was not bound to perform for the owner, but only to perform his subcontract with the general contractor.

The important legal question involved here is whether or not, at the time of the promise from the owner to the subcontractor, the general contractor has breached his contract. If the general contractor has breached his contract with the owner, then the subcontractor is relieved from further performance of his original contract. In this situation, a promise by the owner to the subcontractor would be binding on the owner. On the other hand, if the general contractor did not breach his contract with the owner, a promise of additional compensation by the owner to the subcontractor, is not enforceable. In the latter case, the subcontractor was already legally obligated to perform under his contract with the general contractor.

Subcontractors who agree with the owner to move and fit up a building in a workmanlike manner are liable to the owner for negligent injury to the building in doing the work, even though there is no privity of contract between them. By "privity of contract" is meant there is no agreement, understanding, or connection between them. The only contract or agreement is between the subcontractor and the general contractor. The gist of this type of claim would be the breach of duty owed by the subcontractor to the owner not to injure his property negligently, and this duty does not depend on any contract between them.

As in all other claims based upon negligence, the owner in order to recover against a subcontractor for personal injuries must establish not only that the subcontractor was negligent but that such negligence was the direct cause of the injuries. The subcontractor might be able to defeat the owner's claim if he can prove that the owner's own negligence contributed proximately to the injuries, or that the owner assumed the risk of such injuries.

*Examples*

1. The act of the owner of a building in leasing it with knowledge that it is structurally defective and does not comply with the specifications and the municipal ordinances, which facts could not be discovered by any inspection which the intending tenants could make, operates, whether the owner discloses the facts to the tenant or not, to break the casual connection between the contractor's wrongful act in failing to construct a proper building and the injury to the tenant through the fall of the building due to such defects, and prevents his being liable to the tenant for such injury.

2. One entering a dark building under construction may be guilty of contributory negligence as a matter of law, thus barring any recovery.

Unusual accidents often occur when a person is walking along the sidewalk. A few years back, a tourist visiting the city of New York was killed by a falling flowerpot while walking along the sidewalk. A maid on the twenty-fourth floor had accidentally caused the flowerpot to fall while she was attempting to clean it. The important legal question arises as to how to prove the facts that caused the accident that is referred to as negligence.

Proving negligence in court is most difficult when the facts are unknown to the victim or his surviving family. In order to circumvent this unfair handicap, the law has adopted the doctrine known as *res ipsa loquitur* (the thing speaks for itself). The innocent claimant under this doctrine merely has to prove that he was walking along the sidewalk and was struck by some object that was under the supervision and control of the defendant. Once the doctrine of *res ipsa loquitur* has been applied, the vital requirement of proof of negligence has been solved.

Getting back to the construction industry, many persons have been injured from falling tools, building materials, or temporary scaffolds.

Normally, the plaintiff or claimant has the burden of establishing the negligence of the defendant and that such negligence was the proximate cause of the plaintiff's injuries. This burden is eased considerably when the facts warrant the application of the *res ipsa loquitur* doctrine. Each case goes on its own facts and sometimes, the results will vary in different states.

*Examples*

1. In an action against a contractor and an owner for injuries suffered by one of the contractor's employees as a result of an insufficient foundation, evidence that the contractor's foreman informed the owner that the support was not strong enough is admissible to show that both the contractor and the owner had notice of its insufficiency.

2. Where a four-year-old child was found on the floor of a newly constructed concrete foundation which had been unprotected and the only evidence regarding the happening of the accident consisted of testimony of the child's babysitter, who had not seen the incident, a verdict for the defendant was properly directed, since the question of whether the child fell, was pushed, or in some other manner was precipitated to the floor of the cellar would require conjecture or guesswork on the part of the jury.

When a person about to enter a store was injured by the sudden collapse of a scaffold erected at the store entrance by a contractor engaged in remodeling the entrance, the doctrine of *res ipsa loquitur* was held properly invoked in an action against the contractor.

The doctrine is not available, however, when it is not established that the defendant contractor or subcontractor had exclusive control over the instrumentality which caused the injuries. Before the claimant can establish the doctrine of *res ipsa loquitur* in his favor, he must prove that the defendant had the exclusive control over the flowerpot or ceiling, etc.

*Example*

1. In an action by a restaurant proprietor's employee against a contractor who had constructed a metal ceiling in a restaurant that fell upon and injured the plaintiff, in which action the plaintiff relies upon the doctrine of *res ipsa loquitur* and succeeds in making a "prima facie" case, the defendant has the burden of offering evidence to rebut the inference that it was negligent in the construction of the ceiling.

## 4-4 REGULATIONS PERTAINING TO SUBCONTRACTORS

The designation of contractor and subcontractor can change on different jobs. Persons or concerns that engage in such operations as roofing, stonework, brickwork, and iron and steel work can be contractors or subcontractors depending on the contract and the nature of the work to be performed. For example, if you hire a roofer to reroof your home, he is a contractor and not a subcontractor. If the same roofer is awarded a contract by a general contractor to place a roof on a building, in this instance the roofer becomes a subcontractor. Usually, the contractor hires subcontractors, while the owner hires contractors.

The business of building or construction contractors is one that the legislature may regulate in the interest of public welfare. Many courts have held that the primary purpose of a contractor's licensing act is the protection of the public against the consequences of incompetemt workmanship, imposition, and deception.

The power of the legislature to regulate the business of construction contracting may be delegated to the legislative body of a municipality, and that body may regulate the construction business within its jurisdiction. The regulation of such business by statute or ordinance must, however, be reasonable and not discriminatory. A legislature may provide that building permits be issued only to duly licensed general contractors, or may make it unlawful for any unlicensed person to act in the capacity of contractor for the construction or alteration of any building or other structure.

Failure to comply with a licensing statute has been held proper grounds to prevent a contractor from asserting a mechanic's lien for work performed. In some states, licenses may be revoked for violation of particular provisions of a statute, such as abandonment of a construction project that the licensee has undertaken, or failure to complete a project for the agreed price.

Some examples of where laws of different states went too far and were held to be unconstitutional are mentioned below.

*Examples*

1. A statute discriminating in taxation between foreign and domestic companies by imposing a $25 tax on domestic and a $100 tax on foreign contractors was held to be in violation of the federal constitution, which provides that the citizens of each state shall be entitled to all privileges and immunities of citizens in the several states.
2. Holding unconstitutional an ordinance requiring heating contractors who installed heating equipment in dwelling houses to give a bond to indemnify the purchaser.

3. Holding that the police power does not extend to require one contracting to lay concrete pavements to give bonds for the faithful performance of the work, and to protect customers against defects for a term of years.

Whether one comes within a statute or ordinance regulating the business of contractors depends on the language of the particular statute or ordinance in question. A statute is a state law whereas an ordinance is a law passed by a city or county. It has been held that a bricklayer who contracts with builders or general contractors to lay part or all of the bricks required in any work, and who sometimes has one or two helpers to do the necessary work, is not a building contractor within the meaning of a statute requiring that such contractors be licensed. Another case held that one who erects buildings as a carpenter by employing a large number of persons to do the work, and only occasionally does some work himself, is not exempt from procuring a contractor's license.

*Example*

1. Persons rendering supervisory services in connection with the construction of a drive-in theater were contractors within the meaning of a state business and professional code requiring that contractors be licensed.

## 4-5 LIABILITY OF SUBCONTRACTORS FOR INJURIES

Just as public liability insurance coverage is a *must* for the prime contractor, the same applies to the subcontractor. The chances of accidents resulting in serious personal injuries or even death are so great, that no subcontractor can afford to gamble by not being adequately covered by way of insurance protection. The legal profession has made lawsuits more popular with new philosophies of liability and defending lawsuits is an expensive matter for a subcontractor, even if he is lucky enough to prevail in the litigation.

The cost of necessary insurance protection is another item of overhead that must be included with the cost of doing business, when the subcontractor submits his bid to the general contractor.

Whenever a construction employee, be he employed by the general contractor or some subcontractor, suffers an injury on the job, liability may fall on the contractor, the subcontractor, and sometimes even the owner. The construction contract between the owner and prime contractor covers the question of insurance coverage, cost and liability between them. Similarly, the same questions pertaining to public liability and workmen's compensation insurance coverage, as between the prime contractor and each subcontractor is or should be covered in great detail in the written contract between the parties. Contrary to popular opinion, the prime contractor does not deal with his various subcontractors on a verbal basis. The agreements between them are in writing and necessarily, in great detail. If the agreement between the general contractor and his various subcontractors are not in writing, they are going to be very good clients for the legal profession when misunderstandings and disagreements arise. Both parties are doing business in a very unbusinesslike manner and will suffer severe penalties of loss of time and money for their poor business tactics. In many cases, where the litigants deal with each other under a verbal arrangement and the testimony of each side is diametrically opposed to one another, the decision of the judge will make both sides unhappy.

The question of liability for injuries on the job and who carries insurance coverage is usually covered in the various written agreements. The contract between the owner and the general contractor would automatically go into detail as regards liability and insurance coverage. The same arrangement would hold true in the separate written contracts between the prime contractor and each subcontractor. The important thing to remember is not who is responsible for injuries or death to construction employees, regardless of whether they are employees of the general contractor or a subcontractor, but which one is going to provide the necessary insurance protection.

The matter of tort or accident liability of contractors and subcontractors should not be neglected. Under certain circumstances, a subcontractor can become liable to the owner, even though there is no contractual relationship between them. The gist of the action is the breach of duty owed by the subcontractor to the owner not to injure his property negligently, and such duty has nothing to do with any contract between the parties. As in other negligence cases, the claimant must prove not only that the subcontractor was negligent but that this negligence was the proximate cause of the injuries. The claim may be defeated if it is established that the plaintiff's own negligence contributed proximately to the injuries or that the claimant assumed the risk of such injuries.

*Example*

1. Action against house movers for injuries sustained by Plaintiff who fell into hole while attempting at night to return to house on premises involved; held that finding that negligence of movers was sole proximate cause of accident was justified.

Injuries to children always presents a greater problem than if the injuries were suffered by an adult. Attorneys for the child and parents will always attempt to apply the "attractive nuisance" doctrine. This doctrine when it is applicable, means that the subcontractor is under a greater duty to a child because of the attraction of construction to the infant.

The courts of many states have ruled that buildings under construction, or unfinished buildings, are not attractive nuisances. One court held that there was nothing unusual or unique in the construction of a basement of a building that would be attractive to children in a large city where there is much construction. The opinions of the courts, in general, vary materially, and many courts have said that the doctrine of attractive nuisance will depend upon the facts of the particular case.

*Examples*

1. The fact that a construction job is involved does not put a case in a special category. It would be just as wrong to exempt all construction projects from the attractive nuisance doctrine as it would be to say that builders are insurers against all accidents involving children on premises under construction. Whether or not the situation is one that presents a jury question must depend on the facts of the particular case.

2. Where a seven-year-old boy was injured when falling through an open stairwell of a house which the defendant was constructing on his property, and the defendant did not and could not know that children were likely to trespass, and had no reason to anticipate that harm of the general nature of that suffered was likely to result, the evidence was insufficient to take the case against the defendant to the jury.

3. To permit recovery by a ten-year-old boy, who knew he should not play in a construction area and who fell into a pool of water between the foundation walls of a building under construction while throwing rocks at frogs in the water, without a further showing of negligence on the builder's part, would be unfair and unjust and would carry the doctrine of attractive nuisance to extremes.

4. The question of whether or not a contractor was liable under the attractive nuisance doctrine for the loss of an eye and other injuries sustained by a twelve-year-old boy struck in

the face with plaster thrown by another boy while playing on the lot on which the defendant contractor was building a home was for the jury, where there was evidence from which the jury could have found that the defendant knew or should have known that neighborhood children were likely to use his construction site as a playground, and that his knowledge was sufficient to show the reasonable foreseeability of injury.

5. Where a nine-year-old boy was injured while climbing a concrete block column which constituted part of a garage under construction, whether or not danger of the column falling from its use by children was foreseeable, and whether the contractor and subcontractor exercised reasonable care to take precautions to prevent such a structure from being blown down is a jury question. The court held that a directed verdict for the defendants would be reversed.

We next come to the matter of injuries to employees of the owner. In order to create liability on the part of the subcontractor there must be negligence. A general contractor in control of a structure or premises owes to the employees of any other contractor rightfully there a duty to exercise ordinary care to keep the premises in a safe condition for their use.

*Examples*

1. A deliveryman was rightfully on premises controlled by builders when delivering quantities of stone for building purposes under a contract between his employer and the builders.
2. One employed by a general contractor in the construction of a building, and injured by the servants of a subcontractor, who, after calling to him to remove his wheelbarrow, let a plank fall upon him, was rightfully, as a licensee with interest, at the place where he was injured.
3. The duty owed by a subcontractor on a construction contract to the employees of other contractors on the job is similar to, and no greater than, that owed by an employer to an employee or by the owner of real property to an invitee on the premises. Furthermore, the law holds no one to a higher responsibility than the fair and reasonable standard of his trade or undertaking. In an action by subcontractor's employee against another subcontractor on construction project, the court held there was no primary negligence and that, even if there had been, plaintiff was fully aware of protruding bolts on which he had tripped, and had assumed the risk of injury.

Where a general contractor is in control of the construction premises, he is liable to employees of other contractors rightfully on the project for injuries sustained by them because of his negligent supervision. However, if the negligence which causes the injury is that of the injured person's own employer, then the general contractor would not be responsible.

A subcontractor whose equipment is used by an employee of another subcontractor is not liable for his ordinary negligence in maintaining the equipment. The reason for this result is that the user of the equipment is a mere licensee of the owner of the equipment. If the injured employee was using someone else's equipment without permission, it becomes his own responsibility. A different result would be reached if the subcontractor receives compensation for the use of his equipment by employees of other subcontractors. If it is the practice of subcontractors to use each other's equipment, then there is a duty to see to it that the equipment is in reasonably safe condition for those that he knows will use it.

*Examples*

1. The plaintiff, an employee of the prime contractor, who knew that a subcontractor's lift operator was not authorized to allow him to use the lift, could not recover against the subcontractor for injuries sustained on the lift.

2. The plaintiff, using a ladder belonging to another subcontractor, was not an invitee or licensee, and the other subcontractor was not liable for the injury sustained by the plaintiff while using the ladder.
3. The subcontractor-owner of a scaffold who left it at the site of construction work where another subcontractor's employee used it was not liable for injury to the employee, due to the defective condition of the scaffold, where the use was without the knowledge or permission of the subcontractor-owner.

Even after the construction work has been completed and accepted by the owner, situations have arisen where the contractor and subcontractors have been sued and held responsible for negligence on their part; for example, where the articles constructed or installed are inherently dangerous, such as a gas heater installed without a vent; a furnace installation; a defective chimney; and a faulty elevator.

*Examples*

1. A gas heater installed without a vent to carry away carbon monoxide gas created by it was a nuisance, and the contractor who installed it was liable to one whose daughter died as a result of inhaling the gas while rightfully on the premises.
2. An instrumentality need not cause damage on first use in order to be imminently dangerous, and therefore it could not be said as a matter of law that a furnace installation was not imminently dangerous because it was completed before November 3, 1952, and the Plaintiff's house did not burn until January 4, 1953.
3. "Inherently dangerous" means a type of danger inhering in an instrumentality or condition itself at all times, requiring special precautions to be taken to prevent injury, and not a danger arising from mere casual or collateral negligence of others under particular circumstances. This court held that a glass door was not inherently dangerous.

# 5

# CORPORATIONS

A corporation has been defined by the United States Supreme Court as "an artificial being, invisible, intangible, and existing only in contemplation of law." This is the definition most frequently accepted.

Another definition of a corporation is "an artificial being existing only in contemplation of law; a legal entity; a fictitious person, vested by law with the capacity of taking and granting property and transacting business as an individual; it is composed of any number of individuals authorized to act as if they were one person. The individual stockholders and the board of directors are the constituent or component parts, through whose intelligence, judgment, and discretion the corporation acts."

It is within the power of the legislative department of a state to define a corporation. In some states, corporations are defined to include associations and joint-stock companies, that have powers or privileges not possessed by individuals or partnerships. The statutes or laws of some states expressly provide that the word "company" shall include corporations.

The right to act as a corporation is a special privilege conferred by the sovereign power of each state by its laws. Until there is a grant of such right, either by special charter or under general corporate law, there can be no corporation. The laws of each state are supreme for forming a new corporation in that particular state. The fact that the requirements for a new corporation vary from state to state does not alter the necessity of complying with the ground rules of the state in question.

The builder or contractor starting his new construction business will want to know if he should form a corporation or operate under his personal or trade name. In order to answer this question, thought and deliberation has to be given to many factors. The builder should sit down with his attorney and his accountant for a detailed discussion of the pros and cons of forming a new corporation. Tax considerations and consequences are the most important items to be considered. There are situations where it is possible to incur income tax liability on a personal basis as well as corporate, with the result of double taxation or at least increased taxes.

Taxation is a special field of law and is very complicated. The contractor's attorney may or may not be qualified to offer competent tax advice. If not, he should be candid and so advise his client. Several millionaires pay little or no income taxes on substantial incomes due to competent tax advice that enables them to take advantage of tax loopholes. Courts have universally held that it is against the law to evade taxes but it is perfectly legal to avoid taxes.

How much does it cost to incorporate or form a new corporation? This is another thing to consider when trying to decide about the advisability of doing business under a corporate status. The filing fee and other charges for starting a new corporation will vary with each state. As a rough average, the cost of filing articles of incorporation; recording; required legal advertising of the new corporation; stock certificates; corporate minute book; corporate seal; and other miscellaneous charges, will run about $150. Similarly, the attorney fees will vary from state to state, however, an average fee would be about $350, making the total cost of forming a new corporation roughly around $500. If you are starting a small construction business, the $500 expense to form a corporation may not be practical or necessary, and your attorney and accountant may jointly advise that you forget about doing business under a corporate status until your business has increased in volume.

There are certain advantages to a builder or contractor doing business as a corporation rather than under his personal or trade name. Aside from any income tax implications, the contractor can limit his personal liability to the assets of his corporation. Of course, the contractor waives this benefit in cases where he guarantees payment of an obligation of his corporation. If a contractor does business under his personal or trade name, or under a partnership, and the contractor or a partner dies, then the business entity terminates. Death of a principal does not terminate the business of a corporation.

## 5-1 TYPES OF CORPORATIONS

Corporations fall into various broad classifications such as public and private; profit and nonprofit; quasi-public; and foreign and domestic.

The powers of a corporation are determined by its articles of incorporation, sometimes called the corporate charter. The provisions of the corporate charter or articles of incorporation are usually made very broad so that the new corporation will have the power to do many things that may come up in later years. There is no harm in granting many powers to the corporation even though they do not seem pertinent at the present time. For example, the corporate charter may provide that the new corporation is granted the power to buy and sell real estate; borrow and lend money; build homes, buildings, roads, and bridges; and many other privileges that may not ever arise. After all, the corporate charter may provide that the corporation's existence is for fifty or ninety-nine years and may be renewed. Many activities may come up in the future and it is a good idea to anticipate and provide for the possibility of unusual events.

*Example*

1. In determining the character of a corporation, the court will look to its articles of incorporation and not to any specific acts or results flowing from an exercise of its corporate powers.

A corporation may not engage in activities that are beyond those provided in its charter. If it does, these acts are called "ultra vires" by law, which means beyond its legal powers that are provided in its charter or articles of incorporation.

Corporations may be classified as public or private. A public corporation is created for public purposes only, usually connected with the administration of the government. Private corporations are created for private purposes, as distinguished from purely public purposes.

A quasi-public corporation may be said to be a private corporation which has been given certain powers of a public nature. For example, the power of eminent domain, in order to enable it to discharge its duties for the public benefit. Eminent domain is the right to take private property for public use, provided it serves a necessary public use and fair compensation is paid to the owner. A quasi-public corporation is entirely different than a public corporation. A good example would be the local gas or electric company. A public utility or quasi-public corporation is one private in its ownership but holding an appropriate franchise from the state to provide for a necessity or convenience of the general public. A public utility corporation, be it the gas or electric or telephone company, has a monopoly by virtue of the franchise granted them.

Private corporations may be further classified, according to whether or not they are incorporated for the purpose of earning profits for their stockholders. A corporation for profit is primarily a business corporation.

A holding company has been defined as a super corporation that owns or controls a dominant interest in one or more corporations, enabling them to dictate their policies through voting power. Although the popularity of holding companies has materially increased in recent years, it really involves big business and would not be of particular interest to the small construction firm.

A corporation is an entity entirely separate and apart from its individual members or stockholders. Stockholders as natural persons are merged into the corporate identity. Even in cases where a corporation such as a mutual fund corporation buys stock of another corporation, the entity of the stock selling corporation is not affected. By the very nature of a corporation, its property is vested in the corporation entity itself, and not in the stockholders.

The stockholders do not have the power to represent the corporation or act for it in relation to its ordinary business. Nor are they ordinarily personally liable for the acts and obligations of the corporation.

All laws pertaining to corporations vary from state to state. It is necessary and vital that the laws of your particular state be understood so as to govern the conduct of your particular corporation. On the question of possible liability of stockholders for obligations owed by the corporation, some states have enacted statutes that provide for "no stockholders liability" by adding the letters NSL after the corporate name.

The possibility of stockholder's liability arises only when the corporation is in financial trouble and is unable to pay its debts. There is no hard and fast rule of law because of the variance of the statutes in the different states. Some states provide for protection of the original stockholder who organized the corporation by letting the creditor know that liability for the corporate debt is limited to assets of the corporation: this is the reason for the use of the letters NSL immediately following the corporate name. Other states provide that if the original stockholder has not paid all of the money that he owes the corporation for his original stock issue and the corporation goes broke, then the creditors of the corporation can force the stockholder to pay his balance to apply on corporate debts.

The corporate entity is distinct even though all or a majority of its stock is owned by a single individual. The same rule applies when the corporation is a "family" or "close" corporation. Usually, the contractor who is ready to convert his business into a corporation will operate as a family corporation by having his wife join him as an officer and stockholder of the corporation. In those states where three incorporators are required, the attorney or his secretary will act as the third incorporator, which has no legal significance. If one share of stock must be owned by each of the three incorporators, then one share is issued to the attorney, who immediately endorses it back to his client; thus complying with the technical requirement of the laws of that particular state. Obviously, the attorney does not acquire any interest in his client's new corporation. Invariably, family or close corporations are closely knit private businesses where stock is not available to be purchased by outsiders.

When you hear that a certain firm or corporation is "going public" it means that the stock of their corporation, which had previously been held by their own group, is now going to be offered to the public. The price of the newly issued stock is always determined by the management of the corporation. A word of caution about public stocks might be appropriate at this point. Many contractors have found that they can use any accumulated profits that might become available in their own business, rather than buying stock in someone else's company. Several smart contractors have said, "I have devoted years of time and study to my business and think I know a great deal about it. Why should I gamble my money on someone else's gambling operation." The doctrine that a corporation is a legal entity, existing separate and apart from the persons composing it, is a legal theory introduced for purposes of convenience and to serve the ends of justice.

The corporation entity is disregarded where it is used as a cloak to cover for fraud or illegality, or to work an injustice, or where necessary to achieve equity.

*Example*

1. While, for all practical purposes, a corporation is regarded as a legal entity separate and apart from its stockholders, the corporation will be regarded as an association of persons where the notion of legal entity is used to defeat public convenience, justify wrong, protect fraud, or defend crime.

The mere fact that a corporation is organized to take over a business formerly operated by a firm or individual is not of itself sufficient to render it liable for a debt, or be responsible for an obligation incurred by such firm or individual in conducting such business. The new corporation is separate and distinct from its predecessors. A corporation that lawfully acquires the property of a partnership does not become liable for the partnership debts. There have been cases where the liability of the new corporation for partnership debts is based upon the theory that a transfer of assets to the corporation is fraudulent as to creditors of the partnership.

## 5-2 INCORPORATION AND ORGANIZATION

No corporation can be formed or exist without the consent of the state or federal government. Until there is a grant of such right, there can be no corporation. A federal savings and loan association is not chartered by any particular state, but is granted a charter by the federal home loan bank, an agency of the United States government. A similar situation would apply to a national bank that receives its charter from the federal reserve bank. On the other hand, a state chartered bank operates by virtue of being incorporated under the laws of a particular state.

The general corporation laws of most of the states provide for the formation of corporations for any lawful business purpose, other than the practice of a profession. A group of doctors, lawyers, or accountants can associate themselves in a firm for the practice of their profession, but they are not permitted to form a corporation. A corporation may not practice law or dentistry or any other profession.

In spite of the publicity that was given many years ago to a criminal group in New York, known to the police authorities as "Murder Incorporated," a corporation cannot be organized for any illegal purpose. However, a corporation may be organized for the purpose of escaping or limiting personal liability on the part of the stockholders. The organization of a holding corporation for the express purpose of gaining control by another corporation is perfectly legal.

Some years ago, the states of Delaware and Nevada enacted corporation laws that made it simple and attractive to incorporate under the laws of those two particular states. Hundreds of thousands of new corporations took advantage of the liberal laws and incorporated in Delaware or Nevada, even though the new corporation was actually going to have their main place of business in some other state. However, with the passing of time, the former advantages offered by those two states have disappeared and most new corporations are formed in the particular state where the corporation has its main office and intends to do business. There is no prohibition against incorporating in one state and actually doing business in another. Many large corporations are authorized to do business in all of the states, yet are incorporated in only one state. This is done to protect their rights, just as in the case of a contractor's doing business in a state where he is not originally licensed.

The number and qualifications of incorporators depend on the terms of the state law involved. Most states require at least two or three incorporators. The modern trend is to permit dummy parties to act as incorporators so that the true identity of those persons actually interested in the new corporation can be hidden. By dummy incorporators, the practice is to use the names of legal secretaries and young attorneys in the office so that the general public will not read the names of the actual interested parties. For example, if you read in the local newspaper that Mary Swift, Joan Long, and Nancy White were the incorporators of Southwest Construction Company, you would not know at that point that John Jones, the contractor, was the real party of interest.

The corporation laws of many states require that the articles of incorporation state the purpose of the corporation; the place of business; the names and addresses of the original incorporators; the amount of authorized capital stock; the amount of paid-in capital with which the corporation will start business; the name and addresses of the directors for the first year, and similar information.

After the articles of incorporation have been prepared by the attorney, they are ready for the signature of the prescribed incorporators. The articles of incorporation are then filed with the county clerk or state corporation commission or similar public officials. In most states, it is the duty of the proper official to determine whether or not the articles comply with the laws of the state and are allowed to be filed. The official will also determine if the name of the new proposed corporation is in conflict with an existing corporate name. If so, the articles of incorporation for the new proposed corporation will be returned to the attorney involved.

In connection with forming a new corporation, the word organization means the election of officers; the subscription and payment for the capital stock; the adoption of bylaws; and other steps that are necessary to endow the new legal entity with the capacity to transact the business for which is was created.

The corporate charter can be accepted and the first meeting held only within the particular state that created the new corporation. The notice of the original organizational meeting may be waived by the subscribers to the original stock who intend to start the corporate life. In closely held or family corporations, such meetings are usually a mere formality.

*Example*

1. As soon as the statutory requirements for the organization of a corporation are complied with by obtaining subscriptions to the requisite amount of stock and by the election of officers and directors, the corporation comes into full and complete existence, with power to sue and to be sued, to enter into contracts, and to begin prosecution of the business for which it was organized.

### *5-3 DE FACTO AND DE JURE CORPORATIONS*

De Facto and de jure corporations are technical terms that you should be familiar with. Problems can occur if your attorney fails to comply with all of the requirements necessary to form a new corporation and the corporation starts to do business.

A corporation de facto is, in short, a corporation in fact. It means a bona fide attempt was made to organize a corporation under a valid law but somehow it was irregularly formed. In spite of the irregularity and the fact that the newly formed corporation had no right to exercise corporate franchises or rights, its formation may be ratified either directly or impliedly. If the state corporation commission or comparable government agency ratifies the acts of the invalid corporation, by so doing it legalizes the existence of the corporation and authorizes it to act in a corporate capacity.

A de facto corporation is an apparent corporate organization that acts as a legally organized corporation even though it has failed to comply with all the requirements of the law. It exercises corporate rights and franchises under color of law and can be challenged only by the state involved.

A de jure corporation is one that has been regularly created in compliance with all legal requirements. Being legally organized, it has the right to exercise its corporate franchises and cannot be challenged or attacked by the state involved.

If a group gets together and combines their capital in a business venture it is important that they be protected as to their personal liability in the event the venture fails. As you know, they would automatically lose their capital investment. Now a very important question arises, would the investor be liable for any additional sums of money? If they formed a partnership, this would not be good because all of the partners are liable for bills owed by the partnership. The recommended procedure would be to protect themselves by forming a valid corporation that would limit their loss to the investment only and nothing beyond that amount.

How long may a corporation be authorized to exist? The period of corporate existence is a matter which concerns the state of sovereignty of incorporation only. The initial period of time provided for corporate existence will vary from fifty to ninety-nine years in different states. All states make provisions for renewal of corporate charters so that for all practical purposes a corporation may continue to operate for an indefinite period of time. This is one of the many advantages of operating a business under a corporate status.

In order to avoid any confusion, the word franchise when used in connection with a right granted by a state to a newly formed corporation has nothing to do with food franchises that are sold. Kentucky Fried Chicken Corporation may sell a franchise authorizing a purchaser to operate a chicken food establishment, using the name and recipe of the parent organization. When discussing corporation law, franchises are rights or privileges that are granted or given to newly organized corporations.

A corporate charter is the authority from the state bestowing rights or privileges. After the attorney for the new corporation-to-be has filed the articles of incorporation, the state corporation commission or comparable agency will issue its certificate of incorporation which is also known as the corporate charter. It is comparable to a license issued to the new corporation to start operations.

A corporate charter has been described as a three-way contract. It represents a contract between the state and the newly formed corporation; it also becomes a contract between the corporation and its stockholders; and finally, it becomes a contract between the stockholders themselves. Each of the three entities, namely, the state, the new corporation, and its stockholders, have certain rights and obligations owed to each other.

*Example*

1. In accepting stock in the newly formed corporation, the stockholder agrees to be bound by all of the terms, conditions, and limitations of the corporation's certificate of incorporation, and all amendments thereto.

The legislature of each of the states has the power to authorize a corporation to amend its corporate charter. However, after a corporation has been in business for a period of time, the board of directors of the corporation may decide that a corporate change of name would be desirable. If, for example, the original corporate charter does not authorize the corporation to lend money, or drill for minerals, or open branch offices in foreign countries, or other similar unusual situations, the board of directors may decide to amend the charter. The attorney for the corporation will then prepare an amendment to the corporate charter, authorizing the omitted items and the change of name. Mergers of one corporation into another frequently result in a change of name or some other corporate power change, requiring an amendment to the original corporate charter. There is no limit as to how many times a corporate charter may be amended.

## 5-4 CORPORATE NAME, SEAL, AND DOMICIL

The presence or absence in a trade name of the word "company" has no bearing on whether or not the concern is a corporation or a copartnership or even an individual. There is nothing to prohibit your use of the word "company" in your trade name, even though it is a one-man business. The same rule applies to a copartnership as well as a corporation. Some firms use the word "corporation" and some use the word "incorporated" as a part of their name. None of these designations are required by the state that issued the corporate charter. There have been cases where a theatrical person has formed a corporation using his name only, such as Jack Dempsey or Jack Benny. This is usually done for tax reasons where the individual derives some benefits from operating under a corporate status.

A corporation may adopt any name it desires, except it may not use an existing name or one close enough to cause confusion. Charlie Chaplin successfully instituted a lawsuit to enjoin another actor from using the professional name of "Charlie Aplin." To make matters worse, Charlie Aplin attempted to duplicate the peculiar garb made famous by Charlie Chaplin, including, moustache, derby hat, cane, and a shuffle manner of walking. The general moviegoing public could be confused by the similarity of names and thereby cause irreparable damage to Charlie Chaplin.

A corporate seal is a small circular metal contraption that contains the name of the corporation as well as the state of incorporation and sometimes the date of incorporation. By squeezing the two pieces of metal, the printed matter is impressed upon any paper that is inserted. It is very similar to the type of seal that is used by a notary public, or other public officials.

The laws of each state provide for the details that must be contained in the corporate seal. The right to adopt and use a corporate seal is an incidental corporate power. Many state laws provide that the corporate seal must be affixed to any contract, deed, or document, in order to make it legal. Regardless of legal requirements, the presence of the seal establishes the fact that the instrument is the act of the corporation.

Banks have learned to protect themselves when dealing with corporations by requiring a copy of a resolution of the board of directors along with the corporate seal. This will insure the bank that what the individual customer wants done has been approved by the corporation. Another example of this technical requirement on the part of corporations occurs when a corporation is in financial trouble and ready to file a voluntary petition in bankruptcy in the United States Federal District Court. The clerk of the court will not accept the corporate bankruptcy petition unless it is accompanied by a certified copy of a resolution of the board of directors of the corporation, authorizing the filing of the bankruptcy petition, together with the affixing of the corporate seal. Remember, even though you may operate a one-man type of corporation, the corporation is a separate entity created by law, and the technical requirements are necessary and must be complied with.

The domicil of a person is the place where he lives. Important lawsuits have been won and lost on the question of domicil, both as to persons as well as corporations. The rule of law as applied to a person is that domicil is a matter of intent. The rule of domicil as applied to a corporation belongs exclusively to the state or sovereignty under the laws of which it was created. If a charter is issued by the United States of America through one of its governmental agencies, the principal place of business of its main office would be its domicil, but it is still subject to control by the federal authorities.

The state of incorporation is the domicil of the state chartered corporation. Any corporation has the power to contract or do business in any state of its choice. There is no restriction to doing business only in the state of corporate domicil. However, if your corporation does do business in other states, it is good business judgment to comply with the rules and regulations of those other states. This does not mean that you are required to incorporate in every state that you do business. Many states have requirements regarding who may be served with a copy of the legal papers in the event your corporation is sued, as a prerequisite to your corporation doing business in that state. Also, many states have taxation rules and regulations that must be complied with, in order to legally do business in that state. The old adage, "If you can afford to drive a motor vehicle, you can afford adequate insurance coverage" would apply equally here. If you can afford to do business in states other than your home state, you cannot afford not to comply with their rules and regulations and laws.

## 5-5 BYLAWS

A bylaw of a corporation is a self-imposed rule of law by which it is governed. The bylaws prescribe the rights and duties of the members with reference to the internal government of the corporation; this includes the management of the corporation's affairs, as well as the rights and duties existing between the members themselves.

A resolution applies to a single act of the corporation while a bylaw is a continuing rule to be applied on all future occasions. A resolution is a written agreement by the directors of the corporation to do or not to do a certain thing. A good example of a corporate resolution would authorize, for example, the bookkeeper to sign corporate checks up to the amount of $1,000.

Bylaws must be reasonable and for a corporate purpose, and must always be within charter limits.

*Example*

1. The validity of a contract consisting of a corporate bylaw for disposition of the shares of a deceased stockholder, unanimously adopted and agreed to by the stockholders, has been held intended to be governed by the laws of the state of incorporation and not of the states where the stockholders purchased their shares.

Corporate bylaws should be construed according to the general rules governing the construction of contracts.

*Example*

1. A bank is not chargeable with notice of a bylaw of a corporate depositor, requiring two signatures to check upon the deposit account, if it has not been the custom to comply with the bylaw.

## 5-6 RECORDS AND REPORTS

The records of a corporation should include the transcript of its charter and bylaws. These could be called the constitution of the corporation. If any question ever arises as to what the corporation may or may not do, the charter and bylaws should furnish the answer. If the charter and the bylaws do not permit what the board of directors want to do, an amendment to the charter might be in order.

In addition, formal records should be kept of all important activities of the corporation. A written record of all meetings of the board of directors and stockholders would be mandatory as well as desirable. Even the smallest of corporations should not be forced to guess as to what was said and done at any official meeting of the board of directors or even of the stockholders. In family corporations, the minutes may be on an informal basis, but they should be transcribed promptly and kept with the permanent records.

It is good business practice for every corporation, large or small, to keep books containing the accounts of its official doings as well as the written evidence of its contracts and business transactions. These permanent records are the property of the corporation and do not belong to the officers or employees.

Practically all of the states require that corporations make and file with the state corporation commission or comparable official, annual reports of the financial condition of the corporation. This gives the appropriate state officials information needed for purposes of taxation. There is invariably a filing fee charged for the annual report.

A stockholder has a right to inspect the books and records of the corporation at reasonable times and for a proper purpose. Larger corporations have solved inspection problems by publishing and mailing quarterly financial reports to all stockholders of record. In recent years, many courts have been quite liberal in granting rights to stockholders who complain about poor management or other transgressions of the officials of a corporation. Not only have stockholders prevailed in their lawsuits to inspect certain records of the corporation, but they have also been granted the right to the list of names and addresses of all the stockholders so that they can be canvassed by mail. If the stockholder prevails in his lawsuit against the corporation and he is able to prove some type of wrongdoing on the part of the corporate management, courts have been ordering the corporation to pay for the attorney's fee incurred by the stockholder. (The matter of proxy fights will be discussed later in this chapter.)

## 5-7 CAPITAL AND CAPITAL STOCK

There are material differences between the terms "capital" and "capital stock." The capital stock of a corporation is the amount of money or property, paid by the shareholders as the financial basis for the prosecution of the business of the corporation. Once this amount of money or valuation is determined, it is a fixed amount and does not vary. If property were involved, the money value of the property would be used as a substitute for actual money.

Although the word "stockholder" is used more often than the word "shareholder" they mean the same, namely, the owner of shares of stock in a corporation.

The term capital is used broadly to indicate all the assets of the corporation. It is said that the capital belongs to the corporation, and capital stock when issued belongs to the stockholders. Capital may be either real or personal property, but capital stock is always personal property.

A share of stock is a unit of interest in a corporation. You can own a share in a partnership business but it would not be a share of stock because it is not a corporation. Each share of stock represents a distinct and undivided share or interest in the common property of the corporation.

### *Examples*

1. A share of the capital stock of a corporation is an entity, although intangible and incapable of manual possession and delivery. Both the title and possession thereof, insofar as it is susceptible of possession, are in the stockholder. His share is an interest in the corporation itself, not something held by the corporation for him.

2. The capital stock of a corporation is the capital upon that which the business is to be undertaken and is represented by the property of every kind acquired by the company, while shares are the mere certificates that represent a subscriber's contribution to the capital stock and measure his interest in the corporation.

## 5-8 STOCK—COMMON, PREFERRED, AND MISCELLANEOUS

Corporations are frequently authorized to create and issue two or more classes of stock. The two most popular types of stock are common and preferred.

The basic class of stock issued by corporations is common stock. This is the kind of stock ordinarily issued, and in the absence of other classes of stock having superior rights, represents the complete ownership and interest in the corporation.

It is fundamental that the owners of the common stock in a corporation are entitled to a pro rata share in the profits of the corporation and in its assets upon dissolution.

Preferred stock has different characteristics and is entitled to certain preferences over common stock. In one sense, preferred stock is better than common stock and in another sense, common stock can be better for the stockholder.

The peculiar characteristic of preferred stock is that it is entitled to priority over other stocks in the distribution of profits. The reason for the name "preferred stock" is that it is entitled to a certain percentage or amount as a dividend and this payment is ahead of all other stocks. However, the amount is limited and if the corporation does well financially, the holders of common stock can do better than the holders of preferred stock. Preferred stock is safer and more cautious, whereas common stock is more of a gamble with a bigger payoff if the corporation does well. If the corporation goes broke, the holders of preferred stock would be first in line and the holders of common stock would receive the leftovers, if any were available.

Some corporations that issue preferred and other types of stock place certain restrictions, preferences, voting powers, and similar qualifications in the stock. These limitations are fixed by the corporation's articles of incorporation or bylaws, but these records are rarely checked by the small investor.

On occasion, preferred stock has been given other names, such as "guaranteed stock" and "interest-bearing stock." Stock that is to be sold to the general public comes under supervision of either state or federal regulatory agencies. If the stock is to be sold only to residents of the state where the corporation is located, then the state regulatory agencies control and act as policemen over the issuing corporation. This is a desirable policy to protect the unsuspecting investing public from unscrupulous promoters, who would otherwise cheat the ill-advised investor.

If the stock is to be sold to residents outside the home state of the corporation, it requires the approval of the Securities Exchange Commission, commonly known as the SEC, a federal agency. The SEC requires that the organizers and promoters of the corporation issuing the new stock make a full disclosure of many pertinent and embarrassing details. They must swear under oath as to the amount of stock they are receiving for their promotional efforts; how much money has been invested and by whom; the salaries of the leading officials of the corporation; how much is to be paid the stock brokerage firm that is going to try to sell X numbers of shares; what happens if X number of shares are not sold; exactly what business the corporation is going to engage in; if oil or mineral exploration is involved, the chances of success; if it is speculative, a complete disclosure of the speculation involved; and many more incriminating but important disclosures. At least the general public and the uninformed investor is given some protection by these stiff requirements.

"Unissued stock," as the name implies, is stock that has been authorized to be issued, however, for a variety of reasons has never been issued.

"Treasury stock" is entirely different than unissued stock. Treasury stock is corporate stock which has been issued, subscribed, and paid for, but has later been reacquired by the corporation by purchase, donation, forfeiture, or other means. Sometimes, when the price of a well-known stock is down, the officials of the same corporation decide that the corporation is in a very good cash position and purchase their own stock on the New York Stock Exchange. This stock that is purchased is then known as treasury stock. When the price of the same stock goes up and if the corporation is in need of money, the directors may decide to sell this treasury stock. This procedure is perfectly legitimate and is not underhanded or tricky.

Stock which is merely to be held as unissued or unsubscribed for, is not regarded as treasury stock. Treasury stock acquired by the corporation becomes part of its assets. It may be sold by the corporation at any time upon such terms as it desires.

"Convertible stock" is the designation where stock is changed into another class or into other obligations of the corporation.

The largest and most prestigious stock exchange is the New York Stock Exchange, located in New York City on Wall Street. Before any stock may be listed on the New York Stock Exchange, certain rigid requirements must be met. For example, the company must be in business for a number of years; their assets must exceed millions of dollars; and they must have paid dividends for a given number of years. As a result, only the large and well-established companies are able to comply with the requirements to be listed on the New York Stock Exchange. Presently, the various companies that are listed total about 1,400.

The American Stock Exchange also located in New York City on Wall Street is the next largest stock exchange. Their requirements are not nearly as rigid as those of the New York Stock Exchange. The American Stock Exchange is not as old or as large or as prestigious as their big brother.

Most of the new corporations just starting out are unable to qualify for any large stock exchange and are therefore sold "over the counter." This means that their stock is not listed on any stock exchange and stockbrokers sell this stock at the best available price. Information as to the best available price is not always readily available. If you own shares of stock in a large company that is listed on the New York or American Stock Exchange and want to sell, your stockbroker is able to find a ready buyer within a matter of minutes. Over the counter stock is not always that easy to sell and there may not be any market or buyers interested in buying.

Some thirty-one million Americans own stock in various companies throughout the country; most of this stock is common stock. Some of the companies represent a good investment, while others are very speculative and are a gambling investment. Many small investors are of the opinion that the value of their stock holdings can only go up in price. This is a false premise; if stocks can go up, they can also go down. Therefore, the stock market is not a guaranteed profit investment program. Playing the market by buying and selling various stocks is actually a legalized form of gambling.

When starting a new construction business, gambling is not recommended. The new business represents a big enough gamble for the builder or contractor. The contractor has the advantage of gambling in his own business where he has certain knowledge and experience and a reasonable chance to succeed. He is a novice when he gambles on the other person's business.

## 5-9 VALUE OF STOCK

The value of corporate stock, as well as the methods of computing their value, may vary for different purposes. Generally speaking, the valuation of stock is determined largely by the capital and surplus of the corporation, including the value of the franchise as a part of its property. The franchise may be very valuable, while the visible capital is of little value.

There is no presumption whatever as to the value of corporate stock. Nor is there a controlling presumption that the value of stock of a corporation is its book value, without taking into consideration the earning capacity of the corporation.

To emphasize how complicated it is to determine values of stock in many corporations, a recent news article discussing merger of two corporations is quoted. "A couple of years back, the big conglomerate corporations were riding the merger trail on the backs of inflated stocks, accounting gimmicks, and a handsoff government antitrust policy. The Accounting Principles Board of the American Institute of Certified Public Accountants killed off the accounting gimmicks by curtailing the use of warrants, convertible debentures, and other "funny money" used by the conglomerates in making acquisitions."

The book value of corporate stock is the figure obtained upon dividing the amount of the difference between assets and liabilities by the total number of outstanding shares.

*Example*

1. Book value normally means the value of the corporation as shown on the books of account of that corporation, after subtracting liabilities.

The courts have recognized a difference between "book value" and "market value," that is, the price obtainable for the stock if sold on the market.

The subject of good will and trying to place a realistic value on good will is a serious problem. Some companies in an attempt to balloon the value of their assets, place an arbitrary high value on the intangible known as good will. Some prominent accounting firms argue that good will should be excluded from the calculation of book value. In the case of your newly formed construction firm, doubt has been eliminated and you know that the value of good will is negligible if anything.

The courts have frequently approved book value findings on the reasoning that they were shown to have been compiled in accordance with accounting methods customarily used by the corporation.

A recent case involving a foreign mutual fund corporation is of interest on this point. The mutual fund bought eight million acres of land at $1 per acre and later sold one million acres for $12 per acre. They then listed the remaining seven million acres on their books as a $84 million dollar asset and charged fees based on that fabulous sum of money. Consequently, they were criticized for this unusual type of accounting.

The subject of "par value" of stock presents a most interesting situation. Whenever a new corporation is formed, the question arises as to what value to place on the common stock as par value. This question is determined by the decision of the people in control. When promoters are involved, they are more concerned with the future sale of the stock to investors than with any other factor. That is why in times of prosperity and inflation, when the gullible public will buy anything, uranium stocks were offered for 1 cent per share as par value. The small investor thinks that the inexpensive price of the stock will enable him to sell it shortly at a profit. Sometimes he does make a profit and on other occasions he is stuck with a worthless stock that has no salability or value. Some of the uranium claims that were supposedly going to make everyone a millionaire, turned out to be valueless.

Par value stock has imprinted upon its face a dollar value. Some stocks actually have a par value of $1 per share of common stock and this amount is arbitrarily fixed as the nominal value of interest so specified.

Par value and actual value are not synonymous and there is often a wide disparity between them. Some people are shocked when they buy stock in a large company for $90 per share and later discover when they see the stock certificate that the par value when the stock was originally issued was $1. There is nothing basically wrong with par value having been set at $1 per share and, with the passing of years, increasing in value to $90. Every business day millions of shares of stock are sold on the various stock exchanges, and only the passing of time will determine whether or not the buyer or the seller made a good deal.

Some corporations have seen fit to issue stock without any nominal or par value and this stock is called "nonpar value."

Many state corporation commissions charge a certain fee based on the valuation of stock issued. Thus, a higher fee is charged when one million dollars of common stock is issued when compared to $100,000. Many state corporation laws authorize a corporation to change the par value of its outstanding shares by amending its corporate charter.

The amount of capital stock that a corporation may issue is ordinarily fixed by the articles of incorporation. It is within the discretion of the incorporators to determine the amount of stock to be issued and the amount of money with which the corporation is going to start its business. Most states will permit a new corporation to incorporate for one million dollars, even though the corporation is going to start off with $10,000.

The issuance of stock by a corporation is regulated and controlled exclusively by the laws of the state of its incorporation. Consequently, the validity of an issue of corporate stock is to be determined solely by that law. The authority to issue corporate stock is initially vested in the stockholders of the corporation. After the corporation has been organized, this power is then delegated to the board of directors by the stockholders.

Shares and their ownership are created by the payment or agreement to pay for stock, accepted by the corporation. The capital stock of a corporation which has been subscribed for is actual and valid stock even though payment for it may not have been made. When certificates of stock are officially executed and delivered by a corporation to its stockholders, they are issued in the ordinary sense.

A "certificate of stock" is a written instrument signed by the proper officers of the corporation, stating that the person named therein is the owner of a designated number of shares of its stock.

In recent years, there have been stock-option plans available to key employees and officials of certain corporations. Most of these stock-option plans accomplish a dual purpose. First, key employees and officials in the employment of their company are tempted to stay and not join a competitor. Second, there are tax benefits to be gained if the price of the stock rises over the figure named in the stock-option plan. As the name implies, the employee has the option or choice of taking advantage of the offer. If the stock price goes down, he will naturally ignore the stock-option.

Stock-option plans would not be feasible in the case of your new construction firm, because your corporate stock is not being sold on any stock exchange and has no appeal to the general public. Some contractors have adopted a plan of rewarding the superintendent or other key employees by permitting them to share in the profits earned by the new construction firm or even buying a limited number of shares of stock in the construction corporation. Key employees who share in the success of the construction firm are usually outstanding in their performance and interest in the welfare of the company.

## 5-10 TRANSFER OF STOCK

Shares of stock are personal property. They are evidenced by a certificate indicating the number of shares held.

The transfer of shares of corporate stocks is governed by express provisions of the Uniform Stock Transfer Act and the Uniform Commercial Code, one or the other of which is now in force in all the states and in the District of Columbia.

Certificates of stock are expressly declared to be negotiable instruments by the Uniform Commercial Code. Certificates of stock may be transferred by indorsement and delivery in the same manner as bills of exchange and promissory notes.

## 5-11 STOCKHOLDERS

A "stockholder" is the owner of one or more shares of stock in a corporation that has a capital stock. If a corporation has no capital stock, the incorporators and their successors are called "members."

Ordinarily, nonprofit corporations have no stockholders. The most common nonprofit corporations are fraternal orders and benefit societies, clubs, and religious groups. Wealthy families have established foundations, the primary purpose being charitable and educational. Some of the better known and older established Foundations are the Mellon Foundation, the Ford Foundation, and the Rockefeller Foundation. These foundations have been ruled to be nonprofit corporations and therefore tax exempt by the Internal Revenue Service. The extent of wealth of these large foundations is so great that they continue to grow in size, in spite of the many millions of dollars that they contribute and spend.

In recent years, smaller family foundations have been established and their motives are being questioned by the taxing authorities. There is a serious question as to whether they were established primarily for tax advantages, rather than charitable or educational advancement.

All churches have enjoyed tax exempt status for many years. Presumably, they are nonprofit organizations or corporations and thus not subject to various kinds of taxes.

Different church denominations now own and operate various types of business enterprises. At first the churches automatically claimed tax exempt status on the business ventures, however, modern tax rulings have changed and now churches are being taxed on business activities that have nothing to do with religious functions. If a church member donates valuable business land to his church and the church operates a parking lot open to the public, then real estate taxes and income taxes on the proceeds of the parking lot business are properly removed from the tax exempt status. Churches that own and operate apartment houses are properly ruled not to be tax exempt for these business activities.

A corporation is a legal entity distinct from the body of its stockholders. The stockholders are represented by the corporation in all matters within the scope of its powers. The relation between a corporation and its stockholders is contractual. This contractual relationship embodies the charter, articles of incorporation, bylaws, provisions of the stock certificates, and the pertinent statutes of the state of incorporation. The rights, interests, and obligations of the stockholder arise out of such contract. A stockholder is not by reason of his ownership of stock a creditor of the corporation.

A stockholder cannot resign from a corporation. He remains a stockholder until he sells or transfers his shares to another. A stockholder's status can be terminated by the forfeiture of his stock under certain conditions.

## 5-12 STOCKHOLDERS' MEETINGS AND ELECTIONS

Stockholders' meetings are called for the purpose of electing directors and transacting other business requiring the consent of the stockholders. Frequently, the stockholders pass on the amendment of the articles of incorporation, sale or mortgage of corporate assets, consolidation or merger of the corporation, or any other business that may properly come before the meeting. When the stockholders act on corporate matters, they must act as a body and the majority vote prevails. Members of a partnership may informally agree among themselves but stockholders actions are handled on a more formal basis. All stockholders are bound by the action of the majority at corporate meetings of which required notice is given.

Most corporate bylaws provide for at least one stockholders' meeting each year. The number of meetings is discretionary with each corporation. The authority to call special meetings for the stockholders rests with the president of the board of directors of the corporation.

The corporation laws of most states require that stockholders' meetings be held at a specified place, usually the corporation's principal or main office. Some large corporations arrange to hold their stockholders' meetings at different cities for the convenience of their many stockholders. You must realize that different ground rules apply to large corporations with hundreds of thousands of stockholders than to small family controlled corporations where no outside stockholders are included.

It is the required duty of a corporation to keep a detailed record of the meetings of its stockholders. Usually the secretary will keep a record of the stockholders who were present, the date of the meeting, and the details of all business that was discussed and transacted. These records are called corporate minutes and they are later transcribed by the secretary and signed by the appropriate officials of the corporation. The minutes are then filed and kept with the permanent records of the corporation.

A "quorum" must be present before any business can be transacted at a stockholders' meeting. A quorum means a majority of the voting stock issued and not a majority of the actual bodies of the stockholders. In small family type of corporations, practically all the stock is owned by the husband and wife, and the stockholders' meeting is conducted with only two or three persons present. In cases of extremely large public corporations, large auditorium halls are rented as the stockholders' meeting place to take care of the hundreds and thousands of stockholders that are present and eligible to attend the meeting. In other types of organizations, a quorum could mean a majority of the members.

The right to vote stock at a stockholders' meeting is a privilege of stock ownership. This is an essential attribute of ownership of stock that enables the stockholder to protect his investment.

A "proxy" to vote on shares of stock is a written document empowering a person to act for another. Frequently, proxy fights are organized by unhappy stockholders who want changes in the corporate management, or they are sponsored by individuals or groups who are attempting to gain control of a corporation and have bought a percentage of its stock.

Such individuals or groups will solicit help from other stockholders by advising of their plan to improve the corporation's business activities and asking for the proxies of other stockholders to enable them to gain control of the corporation and carry out their plan. If you as a stockholder approve of the proposed new plan by those not in power, you are privileged to give them your proxy. If you attend the stockholders' meeting in person, there would be no need for you to give your proxy, because you would vote at the meeting in accordance with your desires. You are also privileged to ignore the proxy fight that usually goes on between the outside faction trying to gain control and the present officials of the corporation who are already in control and want to continue in office.

## *5-13 LIABILITY OF STOCKHOLDERS*

Stockholders are not liable for any of the obligations of a corporation. It does not matter about the character of the debt or the manner in which it was incurred. The mere fact that a corporation has done some business outside that authorized by its articles of incorporation does not render the stockholders, as such, liable for its corporate debts.

One of the prime advantages of operating your business under a corporate status is that, if your corporation should fail, your personal assets are not responsible for corporate debts. Consequently, when a corporation goes broke it is still possible for the principal stockholder to remain a wealthy man and not be held responsible for corporate debts. Obviously, if there is any wrongdoing on the part of the corporate officials, the law will not protect the wrongdoer.

The reasoning of the law is that the corporation is an entity entirely separate and apart from its stockholders or members, and corporate debts are not the responsibility of its members. However, creditors of a corporation may bring suit against stockholders to recover the corporate assets wrongfully distributed.

Stockholders who have been guilty of fraudulent misrepresentations in procuring credit for the corporation may be held liable as individuals to any creditor injured. Any person guilty of fraud would always be held responsible for his wrongful conduct. Fraud is the loss of property due to use of deceit, cheating, false promises, and similar misconduct.

Ownership of the majority or all of the stock of a corporation by one person does not in itself impose any additional or different liability upon the single stockholder. The identical ground rules apply as to situations where the corporate stock is owned by a number of persons.

## *5-14 DIVIDENDS AND STOCK SPLITS*

A payment to the stockholders of a corporation as a return upon their investment is called a dividend. The meaning of the term "dividend" has been regarded as indicating that there must be surplus or profits to be divided. If there are no profits there can be no dividends. Profits are not dividends until so declared or set aside by the corporation.

Even in cases where the corporation has earned a substantial profit, it is not mandatory upon the corporate officials that a dividend be declared. The board of directors could decide that it is to the corporation's benefit to use the profits accumulated for expanding the corporation's commercial activities or buying new equipment or some other expansion program. Even in the case of a small construction corporation, the profits earned are retained to build up the business, in preference to declaring a dividend. Sometimes, tax advantages can be gained by keeping the earnings in the corporation. The advice of your accountant and attorney can prove helpful in deciding this question.

A "stock split" is merely dividing up the outstanding shares of a corporation into a greater number of units. There are various reasons for a stock split. Some corporations whose stock is sold on the larger stock exchanges find that the cost per share has increased so much that people with limited funds to invest cannot buy the stock. In order for their stock to appeal to a larger segment of the investing public, this stock is split three for one, which means if one share was priced at \$300, the new split share which is really one-third of the former share is now priced at \$100. Other corporations offer bonuses to their stockholders in the form of a stock split. In this case, the stock bonus is used as a substitute for a money dividend.

A corporation owes its existence to the laws and will of the state or government. A corporation has only such powers as are expressly granted in its charter or in the statutes under which it is created.

*Example*

1. A corporation has no natural rights or capacities such as an individual or a partnership has, and if a power is claimed for it, the words giving the power from which it is necessarily implied must be found in the charter, or it does not exist.

Donations for political purposes are beyond the power of the corporation. Federal laws as well as the laws of some states expressly prohibit political contributions by business corporations. Some corporations have been caught making political contributions in violation of federal law and attempting to hide their unlawful act by calling it advertising or legal expense or some other false designation.

Unless a corporation is especially authorized to do so, the execution or indorsement of accommodation paper for the benefit of third persons is an act beyond the scope of its corporate authority.

A corporation may sue and be sued. A corporation may employ an attorney to act in its behalf in the same manner as an individual.

## 5-15 BONDS AND DEBENTURES

Corporations have the power to issue bonds or debentures as evidence of their indebtedness. Usually, a "bond" secures only the general assets of the issuing corporation. A "debenture" is a certificate of indebtedness. Frequently, when a corporation is in need of additional capital, instead of borrowing money from their bank on short term payback, they will sell bonds or debentures on a long term payback. You will probably not be faced with this situation in your small construction corporation because outsiders are not interested in investing in a small private corporation, and your bank will be your best and only source of money needed for your business. Whether money is borrowed by use of bonds or debentures or from the bank, it has to be repaid on due date.

The president of the corporation is the general manager of its corporate affairs, with automatic authority to act for the corporation. The vice-president has limited power other than substituting for the president and presiding over meetings of the board of directors, in the absence of the president.

The board of directors are the executive representatives of the corporation and as such they have very broad powers. Many articles have been written about the duties and functions of members of the board of directors of a corporation. However, the fact of the matter is that their performance varies from "do nothing" to "do everything." Certain board members sit in on all of the meetings of the board of directors and never participate in discussions, or ever make any comment or contribution to the success of the corporation. Directors are supposed to be knowledgeable about what is going on within the corporation and make some worthwhile contribution towards its success. If the actions of certain corporate officials are objectionable or not in the best interests of the corporation, a good director will do something about calling the weaknesses to the attention of the entire board of directors for appropriate board action.

The following example will emphasize the point that a good director is required to act in good faith for the best interests of the corporation and its stockholders. A certain large mutual fund corporation, handling millions of dollars of other people's money, saw fit to make extremely large personal loans to certain high ranking officials of the corporation. A member of the board of directors discovered this improper procedure and reported the facts to the entire board. The unsecured loans were then immediately repaid and this objectionable handling of corporate funds was stopped.

It is elementary that a director or officer of a corporation is not permitted to make a private or secret profit out of his official position. A bank official who was also a member of the board of directors of a large corporation was severely criticized for instructing the trust department of his bank to purchase shares of stock in the same corporation, after he learned of favorable mineral developments at the board meeting. To make matters worse, the instructions were given via long distance telephone immediately upon his leaving the meeting of the board of directors.

The authority to fix the salaries and other compensation of the executive officers of a corporation, such as president, treasurer, and secretary, is usually vested in the board of directors.

A "dummy director" as indicated by the name, is one who is a mere figurehead and in fact discharges no duties. This occurs sometimes in a family corporation when the required third director is an outsider and owns no stock in the corporation.

The well-established general rule is that a corporation is liable for the torts and wrongful acts or omissions of its officers, agents, or employees acting within the scope of their authority or in the course of their employment.

*ARTICLES OF INCORPORATION*
*OF*
*J & B CORPORATION*
*(No Stockholders' Liability)*

We, the undersigned, all citizens of the United States of America, and of the State of New Mexico, do hereby associate ourselves for the purpose of forming a Corporation under the Laws of the State of New Mexico, United States of America, and do hereby certify:

*FIRST*

That the corporate name is J & B CORPORATION, No Stockholders' Liability.

*SECOND*

That the registered office of the Corporation is located at 701 Main Street, N.E., Albuquerque, New Mexico; and John Doe is designated as the Statutory Agent located there and upon whom process against the Corporation may be served.

### *THIRD*

The objects for which this Corporation is established are:

To take, purchase, lease, exchange, hire or otherwise acquire lands or any interest therein, whatsoever and wheresoever situated; to erect, construct, rebuild, enlarge, alter, improve, maintain, manage and operate houses, buildings, or other works of any description on any lands owned or leased by the Corporation, or upon any other lands; to sell, lease, sublet, mortgage, exchange, or otherwise dispose of any of the lands or any interest therein, or any houses, buildings, or other works owned by the Corporation; to engage generally in the real estate business as principal, agent, broker, or otherwise, and generally to buy, sell, lease, mortgage, exchange, manage, operate, and deal in lands or interests in lands, houses, buildings or other works; and to purchase, acquire, hold, exchange, pledge, hypothecate, sell, deal in, and dispose of tax liens and transfers of tax liens on real estate.

To make, enter into, perform, and carry out contracts for research, designing, constructing, building, altering, improving, repairing, decorating, maintaining, furnishing and fitting up buildings, tenements and structures of any description. To advance money to and enter into agreements of any kind with architects, decorators, builders, contractors, property owners and others for certain purposes.

To carry on and conduct any and every kind of business of general contractors, builders, architects and engineers, including the erection, alteration, repair, demolition, or any other work in connection with any and all classes of buildings and improvements of any kind and nature whatsoever. To design, draw and prepare plans, specifications and estimates for and to supervise, bid upon, and enter into and execute contracts for the construction of buildings, structures, houses, piers, wharves, canals, docks, slips, dams, bridges, viaducts, railroads, railways, rights of way, cuts, fills, roads, avenues, streets, highways, fortifications, conduits, pipe lines, electric or other transmission lines, subways, tunnel foundations, mines, shafts, wells, water works, manufactories, plants, machinery, drainage, irrigation and sewage disposal systems, and any other engineering or construction project or enterprise of any nature whatsoever.

To purchase, hold, cancel, reissue, sell, exchange, transfer or otherwise deal in its own securities from time to time to such terms as the Board of Directors of the Corporation shall determine; provided that the Corporation shall not use its funds or property for the purpose of its own shares of capital stock when such use would cause any impairment of its capital, except to the extent permitted by law; and provided further that shares of its own capital stock belonging to the Corporation shall not be voted upon directly or indirectly.

To promote, organize, manage, aid or assist, financially or otherwise, persons, firms, associations or corporations engaged in any business whatsoever, to such extent as a corporation organized under the Laws of the State of New Mexico, may now or hereafter lawfully do.

To purchase or otherwise acquire, own and hold unlimitedly such real and personal property of every kind and description within and without the State of New Mexico and in any part of the World, suitable, necessary, useful or advisable in connection with any or all of the objects hereinbefore or hereinafter set forth, and to convey, sell, assign, transfer, lease, mortgage, pledge, exchange or otherwise dispose of any of such property.

To acquire and carry on all or any part of the business or property of any company engaged in a business similar to that authorized to be conducted by this company, or with which this company is authorized under the laws of this State to consolidate, or whose stock the company under the State laws of this State and the provisions of this certificate is authorized to purchase, and to undertake in conjunction therewith, any liabilities of any person, firm, association or company described as aforesaid possessing property suitable for any of the purposes of this company, or for carrying on any business which this company is authorized to conduct, and as for the consideration for the same, to pay cash or to issue shares, stocks, or obligations of this company.

To purchase or otherwise acquire any and all letters patent and similar rights granted by the United States or any other Country or Government, licenses and like, or any other interests therein, or any inventions which may seem capable of being used for or in connection with any of the objects or purposes of said corporation, and to use, develop, sell and grant licenses in respect to or other interests in the same, and otherwise to turn the same to account.

In general to carry on any other business, in connection therewith, not forbidden by the Laws of the State of New Mexico, and with all the powers conferred upon corporations by the Laws of the State of New Mexico.

To lend money to customers and others having dealings with the corporation and to guarantee the performance of contracts by any such persons.

To enter into, make and perform and carry out contracts of every sort and kind with any person, firm, association or corporation, municipality, body politic, county, territory, state, government or colony or dependency thereof, and without limit as to amount to draw, make, accept, endorse, discount, execute and issue promissory notes, drafts, bills of exchange, warrants, bonds, debentures and other negotiable or transferable instruments, and evidences of indebtedness whether accrued by mortgage or otherwise, as well as to secure the same by mortgage or otherwise, so far as may be permitted by the Laws of the State of New Mexico.

To enter into contracts or arrangements with any government or authority national, state, municipal, local or otherwise conducive to any of the purposes of this Corporation, and to obtain from such government or authority any and all rights, easements, privileges and subsidies.

To purchase, acquire, apply for, register, secure, hold, own or sell, or otherwise dispose of, any and all copyrights, trademarks, trade names and distinctive marks.

In the purchase or acquisition of property, business rights, or franchises, or for additional working capital, or for any other object or in or about its business or affairs, and without limit as to amount, to incur debt and to raise, borrow and secure the payment of any lawful manner including the issue and sale or other disposition of bonds, warrants, debentures, obligations, negotiable and transferable instruments and evidences of indebtedness of all kinds, whether secured by mortgage, pledge, deed of trust or otherwise.

To do any and all things necessary, suitable and proper for the accomplishment of any of the purposes or for the attainment of any of the objects or for the exercises of any of the powers herein set forth, whether herein specified or not, either alone or in connection with other firms, individuals or corporations, either in this State or throughout the United States and elsewhere, and to do any other act or acts, thing or things incidental or pertinent to, or connected with, the business hereinbefore described, or any part or parts thereof, if not inconsistent with the Laws under which this Corporation is organized.

The Corporation may use and apply its surplus earnings or accumulated profits authorized by law to be reserved to the purchase or acquisition of property, and to purchase or acquisition of its own capital stock from time to time and to such extent and in such manner, and upon such terms, as its Board of Directors shall determine; and neither such property nor the capital stock taken in payment or satisfaction of any debt due to the Corporation shall be regarded as profits for the purpose of declaration or payment of dividends, unless otherwise determined by a majority of the Board of Directors, or by a majority of the Stockholders.

### *FOURTH*

The total authorized stock of this Corporation shall be One Hundred Thousand ($100,000.00) Dollars, divided into One Thousand (1000) Shares of the par value of One Hundred ($100.00) Dollars each. The amount of Capital Stock with which this Corporation shall commence business is One Hundred and Thirty (130) Shares of the value of One Hundred ($100.00) Dollars each, said Thirteen Thousand ($13,000.00) Dollars being assets contributed to the Corporation by the original Subscribers hereinafter named. All stock shall be common stock, fully paid and nonassessable.

*FIFTH*

The full name and postoffice addresses of the Incorporators, and the number of Shares of this Corporation subscribed for by each, respectively, are as follows:

| | | |
|---|---|---|
| John Doe | 701 Main Street NE<br>Albuquerque, New Mexico | 34 shares |
| Jane Doe | 701 Main Street NE<br>Albuquerque, New Mexico | 1 share |
| Robert Roe | 204 Broadway NE<br>Albuquerque, New Mexico | 94 shares |
| Mary Roe | 204 Broadway NE<br>Albuquerque, New Mexico | 1 share |

*SIXTH*

The term of existence of this Corporation shall be for a period of Fifty (50) years from the date hereof.

*SEVENTH*

The Board of Directors of this Corporation shall have power to make, alter, amend and repeal By-Laws for the government of this Corporation and fix the amount to be reserved as working capital, and, at their option, to select and appoint an Executive Committee which shall have and exercise all of the powers of the Board of Directors in the management of the business affairs of said Corporation when the Board of Directors is not in session.

/s/ John Doe (seal)

/s/ Jane Doe (seal)

/s/ Robert Roe (seal)

/s/ Mary Roe (seal)

# 6

# INSURANCE—WORKMEN'S COMPENSATION AND PUBLIC LIABILITY

## 6-1 WORKMEN'S COMPENSATION

Workmen's compensation insurance protects the builder and contractor against claims resulting from the injury or death of his workmen through industrial accident. In most states, the law provides that the employer must provide protection for the employee who is injured or killed while on the job.

In the case of an automobile accident, the law says that before you can be held responsible for an injury to the other party involved, you have to be guilty of some type of negligence. Some automobile accidents are referred to as "unavoidable accidents" meaning that neither party did anything wrong, and therefore no negligence was involved and no liability on either party. A tort is a civil wrong other than a crime. Thus, an automobile accident would be a tort. In the law of torts, negligence is a required factor to establish liability.

Negligence is not a factor in any industrial accident when the workman is injured or killed while on the job. The mere fact that the workman is injured during the course of his employment creates the liability on the part of the employer builder and contractor and his insurance company. Even in cases where two workmen are fooling around and one of them is injured, the injured party is entitled to collect compensation under the provisions of the laws of most states. This might impress you as being unfair to the employer but there are other factors involved.

Protection afforded the employee who is injured on the job has advantages and disadvantages. The example of the employee who was injured while engaged in horseplay is on the employee's plus side, because without workmen's compensation insurance he would not be able to recover due to his own negligence. The big disadvantage to the employee, which means a big advantage to the builder, is that liability for injury or death to the employee is limited by state law. Appreciating that the laws pertaining to compensation to injured employees or victims of a fatal accident will vary with different states, nevertheless, the liability of the builder is limited and he is fully protected by his insurance coverage.

Whenever reference is made to the word "statutes" in this book, it has the same meaning as the word "laws." The right of the injured employee to workmen's compensation is created by the statutes of the states, and you must turn to these statutes to measure the rights. Whenever reference is made to "Examples" it means that some high court has ruled on a disputed point of law, and the words used belong to the judge who wrote the opinion.

*Example*

1. Workmen's compensation is not a charity, but the recognition of a moral duty and the erection of it into a legal obligation of the public, not of the mere employer, to compensate reasonably those who are injured while in the employment of others, as a part of the natural, necessary cost of production.

The principle of workmen's compensation has had rapid development, and has been adopted in practically all of the states. The name or procedure may vary from state to state but the net result is the same. In some states, workmen's compensation claims are handled by the courts, sometimes by the judge alone and in other cases by a jury. Some of the insurance companies feel that juries are too sympathetic and liberal in favor of the poor injured workingman against the supposedly rich insurance company. By a slight increase in the amount of compensation, the insurance companies have succeeded in convincing the state legislature that the lawsuit should be heard by a judge without a jury trial.

In most lawsuits involving any wrongful conduct or negligence, when insurance protection is involved, the fact that any award by a jury will be paid for by the insurance company is withheld. For example, in an automobile accident lawsuit, the mere mention of the words "insurance company" in the presence of the jury is grounds for a mistrial. Again, insurance companies feel that juries are sometimes inclined to be too liberal in favor of the claimant and prejudiced against the insurance company.

Workmen's compensation lawsuits are different because they are controlled by the statutes of each state and the provisions of said laws. Usually, the title of the lawsuit would be Bill Smith vs. John Jones, doing business as Jones Construction Company and Blank Insurance Company. The fact that an insurance carrier is involved is no longer a mystery, and everyone really knows that employers are required to carry protection for their employees in the form of workmen's compensation insurance.

Whenever an ordinary lawsuit is filed in the local courthouse, a filing fee must be paid by the plaintiff or claimant. The amount will vary from about $15 to $30 for each case filed. When a workmen's compensation lawsuit claim is filed, most state's laws provide that no filing fee is paid at the time of filing, but is due when the case is settled. The filing fee is then paid by the insurance company. The general reasoning is that the workman is definitely not a wealthy man and his claim should be heard by the courts without placing any handicaps on him. The laws of many states also provide that a workmen's compensation claim shall be given time priority over other lawsuits and that they shall be placed at the head of the court docket.

Attorney's fees are also different in connection with a workmen's compensation lawsuit than with most other accident lawsuits. In the average automobile accident lawsuit and other similar types where negligence is a prime factor, most attorneys will handle the lawsuit on a "contingent fee" basis, which means a percentage of the recovery, gambling that the lawsuit has a favorable result for the claimant. If the claim is denied in court, then the attorney's contingent fee claim goes down the drain and he receives no compensation. In another type of lawsuit, many attorneys will charge a retainer fee, that is a down payment, with additional compensation agreed to be paid later. The laws of most states provide that attorneys fees in workmen's compensation claims shall be set by the court and are usually paid by the insurance company and not by the workman client. The attorney knows there is no gamble involved in nearly all workmen's compensation lawsuits, because he does not have to prove that anyone was negligent, merely that his client was injured on the job. A workmen's compensation lawsuit is practically a sure winner for the claimant, because the only element of uncertainty is the amount of the award.

A very high percentage of workmen's compensation claims are honored and paid by the insurance company directly to the injured workmen with no need for an attorney or a lawsuit. There are, however, unusual situations when the insurance company and the claimant are not in agreement as to the correct or fair amount to be paid, consequently, the dispute winds up in court.

A recent example involved a client who was seriously injured by a falling tree while working on clearing the way for utility poles to be installed by his employer. This is a typical industrial accident covered by the statute for an injury occurring during the course of his employment and no question about his right to compensation. The insurance company complied with state law and paid all of his hospital and medical expenses; they also paid the $60 per week that was provided for based on his regular pay which had been considerably higher. The doctors said the injured employee was unable to work because of his serious injuries. However, about a year later the doctors said the injured employee could not perform his regular work of felling trees in the mountains but he was able to perform some other type of light work. The insurance company wanted to pay him a lump sum settlement in full payment of his claim, rather than continue with paying $60 per week for ten years. The dispute between the parties was the amount offered by the insurance company as a lump sum settlement. The only serious question involved the medical opinion as to the percentage of disability which was later established as 75 percent. The dispute was settled out of court on the basis of 75 percent of the total award, minus the payments already paid to the claimant and a discount for the lump sum settlement.

In some states, workmen's compensation claims are handled by the Industrial Accident Commission instead of by the courts. The procedures may vary but the basic results are the same.

Although the employer builder and contractor pays the bill, namely, the premiums to the insurance company for workmen's compensation insurance coverage, his responsibility ends with reporting the employee's injuries to the insurance company. All of the other details about the workmen's compensation claim are handled by the insurance company through their claim agent. If a lawsuit should result, the attorneys are hired and paid for by the insurance company, as well as court costs and all other incidental expenses. In most lawsuits, "depositions" are taken of the principal parties involved, which would include the employer and the charges could be substantial. A deposition is the taking of sworn testimony in question and answer form, by a court reporter after the lawsuit has been filed but before the trial. It is used by most attorneys as a legal means of discovery; consequently, the attorney knows in advance what the testimony of each important witness will be before he takes the witness stand in court. Deposition is a recommended procedure, but it means that court litigation is becoming more costly.

Although the workmen's compensation claim including all medical expenses, court expenses, and attorney's fees are paid by the insurance company, there are still two undesirable factors from the viewpoint of the builder and contractor. If the insurance company does not settle the workmen's compensation claim directly with the claimant and the claim winds up as a lawsuit, then comes the employer's time in having his deposition taken and more time spent in court in connection with the trial. In addition, the insurance rate of the employer is affected by the amount of money paid out by the insurance company on claims filed by his employees and the more money paid out, the higher the premiums paid by the employer.

In order to learn the amount due an employee injured or even killed on the job one has to examine the laws of the state involved. Most of the states have a regular schedule set up as a part of the workmen's compensation law. The schedule will remind you of insurance policies that pay you *X* dollars if you lose one eye, *Y* dollars if you lose two legs, and *Z* dollars if you lose one eye, one arm, and one leg. The subject matter is rather gruesome, but it cannot be avoided. Basically, the amount of compensation is determined by referring to the schedule which is invariably based upon the loss of earning power. It will usually provide for a minimum and maximum amount per week. In between amounts are calculated on a percentage of the actual earnings of the injured employee. The important factor is that the amount paid to the injured employee for workmen's compensation is never equal or even close to his normal earning power when he was able to work. For example, if a skilled employee earning $200 per week is injured due to no fault of his own, his state law may provide for a maximum of $60 per week as workmen's compensation. Obviously, the employee and his family cannot live on $60 per week; consequently, the employee will want to return to work at his earliest opportunity.

Most states' workmen's compensation statutes provide for payments to be made every fourteen days or comparable for whatever period of time is involved. The injured employee is privileged to use a doctor of his own choice and he is not obligated to use the company's doctor or a doctor named by the insurance company. If the insurance company is not satisfied with the medical report given to them by the employee's doctor, they may legally insist that he be checked by a doctor of their choice in view of the fact that all medical expenses are paid for by the insurance company. The compensation paid to the injured employee is not affected by the insurance company paying all medical, surgical, hospital, nursing, and even burial expenses.

Most states do not require every employer to have workmen's compensation insurance coverage automatically. Some states exempt certain employers on the basis of the nature of the employment; for example, a law office would not ordinarily be required to provide such coverage for their secretaries. Other states provide that workmen's compensation insurance coverage is mandatory when three or more employees are involved.

Many states' laws provide for large employers to have the choice of providing normal workmen's compensation insurance coverage for their employees or to be self-insured. This would usually be restricted to the telephone company, gas and light company, and certain other large employers who qualify by posting a bond as protection for the payment of claims filed by their employees for workmen's compensation claims. From the employee's point of view, there is no difference in the protection offered or in the procedure to be followed.

The injured employee's right to compensation is not based upon the idea of damages for a wrong. Rather, it is liability for compensation that is imposed as an incident of the employment relationship, as a cost to be borne by the business enterprise, which means the employer.

In many of the states, the compensation acts are elective, and the employer and employee are privileged to jointly accept the provisions of the law or ignore them. In other states, no choice is granted, and both the employer and employee are bound by the provisions of the workmen's compensation law. The claimant may file an ordinary type of lawsuit against the employer and his insurance company based on negligence.

*Example*

1. A recent front page newspaper article advised of a jury award of $3,600,000 for personal injuries suffered in an industrial accident based on negligence. The claimant was left blind, speechless, and paralyzed from the neck down when a control cabinet being hoisted into a truck being assembled broke loose and fell on him. An eyebolt on the cabinet broke and the jury found that the cabinet was defective and the judgment was against the maker of the cabinet and the manufacturer of the truck.

Some workmen's compensation statutes provide that if the employer has failed to provide a safe place for the employee to work and then the employee is injured, the employee may claim an extra 50 percent over and above the regular award.

Many of the workmen's compensation acts contain provisions preserving the ordinary remedies available at law for injuries resulting to employees from the employer's willful act or misconduct. The willfulness or misconduct is generally held to mean something more than mere negligence or carelessness.

*Example*

1. An employer who intentionally and maliciously assaults and beats an employee while engaged in the employment, inflicting injuries which disable, is not, in an action to recover damages therefor, entitled to a directed verdict on the ground that the only redress the employee had is under the workmen's compensation act.

Aside from the fact that the builder and contractor usually has no choice, but must provide workmen's compensation insurance coverage for his employees, he cannot afford to gamble without adequate insurance coverage. Many people pay for fire insurance on their homes for a lifetime and never collect on their policy. Yet, one cannot afford to gamble without the protection.

## 6-2 PUBLIC LIABILITY

The dictionary defines insurance as "coverage by contract whereby the company agrees to indemnify or guarantee the customer against loss by specified contingent event or peril."

Public liability insurance as the name implies, is protection against liability for the injury or death of a person. Although the term public liability insurance is used most often in connection with automobile insurance, it is not restricted to persons injured by motor vehicles. It is vital that the builder and contractor be protected against the claim of any person injured or killed because the amount of the claim for damages can be a very substantial sum of money. Some business people have gone bankrupt only because they failed to have adequate insurance coverage to pay for unforeseen accidents.

As mentioned earlier, workmen's compensation insurance offers protection against the claim of an employee who is injured or killed on the job. On the other hand, public liability insurance offers protection against the claims of all other persons who may be injured or killed, for which they claim the builder and contractor is responsible. Even though the litigant's claim does not prevail, the cost of defending a lawsuit is expensive. American law provides that anyone can file a lawsuit against another and even if the defender is the winner, he still must pay for substantial attorney's fees. Canada and other countries require the claimant to post a bond in favor of the defendant's attorney's fees and legal expense, should the defendant prevail. This seems to be a more equitable procedure but, unfortunately, has not been adopted in the United States. The defendant who prevails in a lawsuit may recover his court costs but not his attorney fees.

Most home and car owners carry fire insurance on their home and public liability and other insurance on their car, and some are fortunate enough never to have a loss. Many people have a misconception about insurance coverage in general. They think that if they suffer a loss they will profit on the transaction; such is not the fact of life as regards insurance claims. A good example would be the case of the client who overinsures his home and then the home is destroyed by fire. In spite of the insurance contract for the higher amount, the insurance company will only pay the fair value as compensation for the loss. In short, do not overinsure, paying higher premiums will not be profitable.

The selection of your insurance agent is most important. A competent and qualified insurance agent, who is interested in your welfare, as well as your insurance business and the resulting profit, can contribute towards your success. Contrary to general opinion, the cost of various insurance coverage will vary materially, and there is nothing wrong with shopping around for the best price and insurance coverage available. Although complete and adequate insurance coverage is desirable, the cost is a part of the overhead expense that must be watched. Overinsuring is nearly as bad as underinsuring. A conscientious insurance agent should be your insurance advisor and not just a salesman interested in making a sale and earning a commission.

A builder and contractor is liable in money damages for his wrongful or negligent conduct that causes injury to persons rightfully on the building site; or occupants of adjacent property; or persons lawfully using the streets on which the construction adjoins.

*Examples*

1. Claim by pedestrian against a general contractor for injuries sustained when pedestrian fell on accumulation of ice caused by the contractor.

2. A contractor is liable for negligently leaving an excavation unguarded and unprotected at night so that one rightfully on the premises falls into it.

3. A contractor is liable for negligence in excavating the foundation for a new building so as to cause the collapse of a building on adjoining land.

4. A building contractor leaving accumulations on premises where he was working has been held liable to the tenant of the next door property for damages caused by flooding.

5. A passenger in an automobile which struck a pile of building materials left in the street at night, without lamps being placed as a warning, was entitled to recover from the contractor for violating a municipal ordinance requiring the placing of warning lights near such materials.

6. A pedestrian struck by a falling crowbar was entitled to recover damages from the contractor on the project next to the public sidewalk, for his failure to erect a roof over the sidewalk as required by city ordinance.

Negligence invariably involves some type of wrongdoing. A claimant other than an employee, must not only prove that the contractor was negligent but also that this negligence was the direct cause of the injuries. The contractor can defeat the lawsuit if he can prove that the claimant's own negligence contributed to the accident. A person entering a dark building under construction may be guilty of contributory negligence as a matter of law and therefore cannot collect on his claim for injuries.

Injuries to young children often involve the "attractive nuisance" doctrine of law. Years ago, the doctrine started with certain situations that would automatically attract children of a tender age and the law said that the adults knew or should have known that young children would be attracted and precautionary steps should have been taken to prevent the youngsters from being injured. This used to include the ice delivery wagon; today's ice cream vendor; a house or building under construction; mounds of dirt that have been excavated by contractors; and similar cases that would automatically attract small children.

A "trespasser" is one who unlawfully enters the land of another. Technically, the young child that enters the contractor's construction area is a trespasser in the eyes of the law. Whether or not this area would be an attractive nuisance is a legal question. If it is, the contractor owes a greater duty to protect the child than if it is ruled not to be an attractive nuisance in which case the child is actually a trespasser.

Some courts have ruled that buildings under construction or unfinished buildings are not attractive nuisances. One ruling held that there is nothing unusual or unique in the construction of a basement of a building in a large city where there is much construction. Other courts have ruled just the opposite, namely, that the attractive nuisance doctrine is applicable to construction sites, which means the claimant does not have to prove any negligence on the part of the defending contractor, in order to prevail on his claim.

*Examples*

1. Injuries resulting to children from playing with mortar from an unguarded mortar box set up near a path known to be used by children.
2. Falling through the floor or off a ladder.
3. Catching a finger in a hoist; all of the above accidents happening in unlocked and unbarricaded buildings under construction.

A general contractor in control of a structure owes to the employees of any other contractor or subcontractor working there, a duty to exercise ordinary care to keep the premises in a safe condition for their use.

After a detailed talk with your newly selected insurance agent, you will then order adequate protection by way of public liability insurance. It is important that the amount of coverage be sufficient to protect you against all reasonable claims; if your coverage is for \$10,000 and you are sued for \$100,000, the insurance company attorneys are only interested in the insurance company's maximum \$10,000 liability. There is no point in having inadequate insurance coverage and then winding up by hiring an attorney to handle the lawsuit in your behalf. A good example is public liability insurance in connection with a motor vehicle. Some builders and contractors have been ill-advised by carrying \$10,000 and \$20,000 coverage, which is definitely inadequate. The cost of \$100,000 and \$200,000 coverage is just slightly higher and is needed for protection against current lawsuits for high amounts. Some people are confused by the meaning of \$10,000 and \$20,000 public liability insurance coverage or \$100,000 and \$200,000. The smaller amount is the insurance coverage for the injury or death of one person, and the larger amount is the insurance coverage for two or more persons being injured or killed. You will be pleasantly surprised when you check with your insurance agent and learn the slight increase in cost for the higher and necessary coverage.

## 6-3 AUTOMOBILE INSURANCE

Adequate insurance coverage for all motor vehicles owned or operated by the builder and contractor as well as his employees is an absolute must. No businessman can afford to gamble on driving a motor vehicle, or permit an employee to drive a company motor vehicle, unless it is covered with reasonable insurance. Public liability insurance coverage is mandatory because of the possibility of large claims against you. Fire and theft insurance coverage is not very expensive and should be carried. Property damage insurance coverage which protects against damage done to the other car or even running into a building is also inexpensive and should be carried. Reasonable medical protection for the driver and up to five passengers is also inexpensive and is therefore recommended coverage.

However, the most expensive item on your car insurance bill is the collision portion that partially protects against the cost of repairing the damage to your vehicle. Some people carry $100 deductible, which means in the event of an accident with property damage to your car, you pay the first $100 and your insurance company pays all over that amount. Others carry $50 deductible which is obviously much more expensive insurance. The cost of this protection is nearly prohibitive and is subject to question. If the vehicle covered is a half ton pickup truck or an older passenger car, you might be better off acting as your own insurance carrier for this one item of coverage. After all, the amount of property damage to your motor vehicle cannot amount to a tremendous sum of money. Another reason for not carrying this item of insurance coverage is the possibility that the other car is responsible for the accident and that you can be reimbursed by the insurance company covering the other car. If you or your employee is responsible for the accident, you are covered for any property damage caused to the other car.

Your insurance agent will distribute printed instruction cards listing things to say and do when an accident occurs, and your employees should be cautioned about following the instructions issued by your insurance company. All accidents should be reported to your insurance agent promptly.

Recently, a businessman was involved in a one car accident that caused extensive property damage to his own motor vehicle and a property owner's wall. He was also charged by the police with driving while intoxicated and promptly notified his insurance agent of the accident. The insurance agent suggested that he not file his claim with his own insurance company immediately because of the pending charges of driving while intoxicated. In due course of time, the trial was held and the defendant was found not guilty by virtue of the fact that he was ill and not intoxicated. The insurance agent continued to recommend that he not file a claim with his own insurance company; on the other hand, his attorney recommended that he file his claim after he was vindicated by the court's ruling. Automobile insurance companies have been complaining that it cost them more to process a small claim than the amount of money involved. In short, it is good business not to affect your insurance claim record by filing a number of small claims; for example, if your hubcap is stolen you are better off to pay for the loss yourself. However, in this case, where about $700 was involved, there is no reason for paying insurance premiums and then not filing a legitimate claim for reimbursement.

The question of whether a builder and contractor is better off to lease or buy his motor vehicles is worthy of discussion. There are arguments pro and con. The one big advantage of leasing motor vehicles is that it does not tie up money that can be used to better advantage in your new business. The counter reasoning is that the cost of leasing motor vehicles is higher than purchasing your own because the leasing dealer must make his profit out of the transaction. There is no tax question involved because the expenses are deductible either way.

The cost of the operation of a motor vehicle is necessary to figure as part of your overhead. For example, a builder buys a new half ton pickup at a cost of $2,500. The builder who thinks that his cost of operation of this pickup is limited to the gas and oil for the first year because the pickup is brand-new and needs no repairs or tires is fooling himself.

The proper way to figure the cost of operating the pickup is to use the figures charged by automobile rental agencies for a similar type vehicle. The cost of repairs, tires, batteries, and similar items should be averaged out over a period of time. Depreciation of a motor vehicle cannot be overlooked when the builder learns that the $2,500 pickup is valued at about $1,300 one year later. Depreciation of motor vehicles as well as all equipment and tools is a necessary part of overhead expense.

## *6-4 MISCELLANEOUS INSURANCE*

The desirable type of insurance agent is not only knowledgeable in his profession but is also interested in the welfare of his client. During the period of time that you pay your insurance premiums and have no claims, all insurance agents are "nice guys." The acid test of your choice of insurance agent is when you are involved in a claim of loss and you need help in the presentation of your claim. A high percentage of clients do not know the name of their automobile insurance company and naturally feel that they do know their insurance agent and that they are dealing with him on a personal basis. The true worth of your insurance agent will be tested by the manner in which he handles your claim with your insurance company. If he is aggressive and fair and obtains satisfactory results, you will be pleased and continue to patronize him. If he does not render efficient service in your behalf, you obviously will change insurance agents.

Unfortunately, some insurance companies have adopted an unfair attitude in regard to honoring claims with the result that they have many alibis why you are not covered under the provisions of your insurance contract. You expect such unfair technical treatment from an opposition insurance company but not from your own insurance company. A famous story involves the contractor who telephoned his insurance agent to report that his office safe was broken into during the preceding night. The insurance agent replied that his policy only protected him against a lefthanded burglary of the office safe and therefore denied the claim.

In addition to workmen's compensation insurance coverage, public liability, and automobile insurance the builder and contractor still needs insurance coverage for fire loss or theft of his tools and other personal property. Later, as the need develops there is a possibility of additional insurance coverage for materials on the job, possible liability to subcontractors, and other coverage tied in with actual construction.

Lloyds of London have received international publicity regarding their insurance coverage for the unusual, such as rain insurance for an athletic contest. The new builder and contractor need not have insurance coverage for every conceivable eventuality, mainly because it is expensive and really not necessary. Listen to the advice offered by your insurance agent and after detailed discussion of your insurance needs, do your own deciding.

# INSURANCE CHECK LIST

## *Part I: Liability Forms Of Insurance*

☐ A. **WORKMEN'S COMPENSATION INSURANCE** protects employers from claims arising from various state laws relating to injuries sustained by employees in the course of employment. In some states, this coverage is underwritten by private companies; in others there is a monopolistic state fund. State law may also require coverage for subcontractor employees if they are not otherwise covered. The following endorsements may be added to Workmen's Compensation coverage:

☐ 1. Longshoremen's and Harbor Workers' Compensation Act Liability—This endorsement should be added if there are operations on boats or docks.

☐ 2. Extra-Legal or Additional Medical—If state provisions pertaining to medical expenses are not broad enough, extra voluntary coverage for employees may be provided by adding this endorsement.

☐ 3. Voluntary Compensation—Although the contractor may not be liable for accidents while his employees are not in the course of their employment, he may wish to cover them voluntarily, for example, when they are participating in company-sponsored athletics. State law may also permit this coverage for employees not otherwise protected under workmen's compensation law.

☐ 4. Universal (All States Endorsement)—If a contractor's employees travel away from their home state, they may become subject to the Act or Acts of other states. This coverage automatically covers contractor liability in such locations, and furnishes coverage as well should liability arise under the Longshoremen's and Harbor Workers' Compensation Act.

☐ 5. Special maritime endorsement—Covers work in connection with navigable waters.

*Ins. Co.* ______________________________ *Policy No.* ________ *Limit: $*______________

☐ B. **COMPREHENSIVE GENERAL (PUBLIC) LIABILITY INSURANCE** protects against legal liability to the the public. There are many forms of liability insurance, but the one usually recommended is the Broad Form Comprehensive Liability Policy (Automobile included). All forms of liability insurance are combined in one contract. Physical damage may also be included on all owned automobiles. The following forms of liability insurance may or may not be included in the Comprehensive form. If not, separate policies may be arranged.

☐ 1. Premises-Operations—This coverage protects against the legal liability for bodily injury to persons other than employees, and damage to property of others which is not in the contractor's care, custody, or control.
Exclusions should be checked carefully; for example, Explosion, Collapse, and Underground Damage are normally excluded from coverage under the basic policy for most types of work. (These are usually designated exclusions "x", "c", and "u" respectively.) While in some cases these exclusions cannot be removed, each project should be carefully examined for exposure to these hazards, and coverage secured where possible and necessary.

☐ 2. Personal Injury—This protects against legal liability for claims arising from false arrest, libel, wrongful entry or eviction, and other related wrongs against a person. A check of the policy will reveal exclusions.

☐ 3. Independent Contractors—Contractors Protective Liability—Coverage provides for the insured's legal liability which may arise from acts or operations of a subcontractor or his employees, and damage to property of others if that property is not in the care, custody or control of the contractor. Certificates of insurance should be secured from subcontractors and the scope of their insurance verified.

☐ 4. Elevator—This is for permanent elevators such as in a contractor's office. Temporary material hoists are covered under premises-operations.

☐ 5. Contractural or Assumed Liability—Many construction contracts include a clause in which the contractor assumes the liability of someone else toward third parties. Such clauses can usually be recognized by the words "hold harmless" or "indemnify". It is recommended that insurance policies of contractors and subcontractors provide "blanket contractual liability". Assumed liabilities are then covered automatically.

☐ 6. Completed Operations (Products Liability)—This is an optional coverage which, subject to exclusions, protects against liability to persons or property of others which may arise after a project is completed; for example, from an accident due to faulty workmanship or materials. Actual replacement of faulty work cannot be covered.

☐ 7. Umbrella Excess Liability—This insurance provides catastrophe coverage for claims in excess of the limits of liability afforded by other policies, and also for some hazards normally excluded in underlying liability policies. It may be subject to a large deductible feature—$10,000 to $25,000, and is a reasonably inexpensive way to protect a business from claims which could arise from a disastrous occurrence.

☐ 8. Automobile—All liability from existence or operation of any owned, hired or non-owned vehicle may be included in this provision. This insurance should be on an "automatic" basis to provide coverage of newly added equipment. A special endorsement may be needed if employees or their families use company cars. A "Use of Other Car" endorsement, naming each person so protected, would provide coverage for individuals using cars not owned by them or their employer.

☐ 9. Automobile Medical Payments—This may be added to a policy. In some states it is required.

☐ 10. Automobile Physical Damage—This covers damage to property of others and also may be endorsed to include Comprehensive and Collision coverage on owned vehicles. A high deductible—$250 or $500—can result in a considerable savings in premiums.

*Ins. Co.* ______________________________ *Policy No.* ________ *Limit: $*____________

# *Part II: Property Forms Of Insurance*

☐ **A. STANDARD BUILDER'S RISK INSURANCE** protects against physical damage to the insured property during the construction period resulting from any of the perils named in the policy. This coverage provides reimbursement based upon actual loss or damage rather than any legal liability which may be incurred. There are four principal methods used to establish amounts of coverage and to determine the premium:

☐ 1. Completed Value. This method is based upon the assumption that the value of a project increases at a constant rate during the course of construction. While the policy is written for the value of the completed project, the premium is based upon a reduced or average value. The dollar coverage provided is the actual value of completed work and stored materials at any given time. This form of Builder's Risk must be taken out at the start of construction. It is recommended that this method be used for the typical building project.

☐ 2. Reporting Basis. The contractor must periodically notify the insurance carrier during construction of the increase in value of a project. Coverage and premiums are based upon this reported value. This method is advantageous where completed value is low during most of the construction period, but increases very rapidly toward the end. However, failure to report an increase in value may result in lack of proper coverage.

☐ 3. Automatic Builder's Risk. This policy form gives a contractor temporary protection automatically, pending the issuance of a specific policy for each project.

☐ 4. Ordinary Builder's Risk. This seldom-used type of policy form is written for a fixed value. Coverage may be increased by endorsement at the request of the insured.

Within the framework of the above methods of writing a Builder's Risk policy, the following perils may be covered:

☐ 1. Fire and Lightning

☐ 2. Extended Coverage—covers windstorm, hail, riot, civil commotion, non-owned aircraft, smoke, and explosion (other than from boilers, machinery, or piping).

☐ 3. Vandalism and Malicious Mischief—excludes pilferage, burglary, larceny, theft, and damage to glass (other than glass block).

☐ 4. Additional Perils—The standard Builder's Risk policy may be endorsed to provide for specific additional perils. These may include collapse (not caused by design error, faulty material or workmanship), landslide, ground water, surface water (other than flood), sprinkler leakage, explosion or rupture of boilers, machinery or piping, breakage of glass, pilferage, and theft.

It may be possible to obtain endorsements or separate policies to cover perils of flood and earthquake; however, such coverages can be difficult to obtain. For these perils, it is recommended that the contractor request the project owner to secure this coverage in conjunction with his permanent insurance for the completed structure.

Some of these additional perils can be covered by adding an "all risk" type endorsement, or a Multiple Perils Builder's Risk policy (described below) may be preferred. These additional coverages generally require a deductible clause.

*Ins. Co.* ________________________ *Policy No.* ________ *Limit: $*____________

☐ **B. MULTIPLE PERIL (ALL RISK) BUILDER'S RISK INSURANCE** is a non-standard type of policy which provides similar but broader coverage than the Standard Builder's Risk policy. While the name "all risk" is widely used, all perils are *not* covered—it is a relative term denoting broader-than-usual coverage. Generally each insurance carrier writes its own form of multiple peril.

Rather than naming the perils insured, this type of policy insures against all risks of direct physical loss or damage to property from any external cause *except* those specifically excluded in the policy. Thus, coverage is determined by what is excluded, not included, and the policy must be checked closely to determine the coverage provided. Some forms will require a deductible clause for some of the perils covered.

This type of policy (written on a Completed Value basis) is recommended over a Standard Builder's Risk policy *if* sufficient care is taken at the outset to make sure that all desirable coverage is included. Protection can be tailored to an individual contractor's needs.

*Ins. Co.* ________________________ *Policy No.* ________ *Limit: $*____________

☐ **C. BOILER, MACHINERY, AND POWER PLANT INSURANCE** insures against damage caused by explosion or rupture of steam boilers, turbines, piping, or associated machinery. It is written as a separate policy covering specific equipment, and is usually a combined property insurance and public liability insurance policy. Temporary equipment of this nature should be insured by the contractor. Where a project contains permanent machinery of this type, it is recommended that the owner purchase a policy as part of his permanent coverage *prior* to start-up and testing of the equipment.

*Ins. Co.* ________________________ *Policy No.* ________ *Limit: $*____________

☐ **D. FLOATER POLICIES.** Some portable machinery, tools, equipment, and materials are not covered by Builder's Risk policies. The term "floater" is used in the sense that the insurance follows the property wherever it might be located (possibly subject to territorial limits). In addition to special floater policy forms which may be available to meet special requirements, the following are examples of the more common types:

☐ 1. Contractor's Equipment Floater—This insurance provides coverage on any or all equipment of the contractor (and optionally that of other's in his possession), other than that moved regularly under its own power over highways (licensed or unlicensed). This coverage is usually written on a "named peril" basis, but may also be included under a multiple peril plan.

*Ins. Co.* ________________________ *Policy No.* ________ *Limit: $*____________

☐ 2. Transportation Floater—This provides coverage against damage to property (the contractor's or other's) while being transported. It may be obtained on a per trip, project, or annual basis.

*Ins. Co.* ________________________ *Policy No.* ________ *Limit: $*____________

☐ 3. Installation Floater—This insures the contractor on a named peril or all risk basis while he is installing or moving valuable equipment, materials, or machinery. Premiums can be reduced by using a deductible clause. (Steam Boiler and Machinery coverage only applies after machinery is installed and ready to use; hence the need for this type of coverage.)

*Ins. Co.* ________________________ *Policy No.* ________ *Limit: $*____________

# 7

# NEGOTIABLE INSTRUMENTS

## 7-1 HISTORY

The dictionary defines "law" as a rule of conduct or action established by custom or laid down and enforced by a governing authority. The "lex mercatoria" is the law merchant or the custom of merchants that developed out of international trade, and is a combination of the rules of law and business conduct of many nations.

Since the law of negotiable paper is designed to be an aid to trade and commerce, its rules should be in harmony with general business practices. Negotiable paper is more commonly referred to as negotiable instruments and is known in simpler terms as bills and notes. The word "bills" is the abbreviation for bills of exchange and the word "notes" is the abbreviation for promissory notes. A promissory note is a written agreement for the payment of money. Failure to pay the promissory note or any installment when due would be a breach of contract.

Negotiable instruments include checks, notes, drafts, bonds, bills of lading, warehouse receipts, and many other forms that are used in every day commercial life.

The origin of bills of exchange is shrouded in some obscurity. It is very questionable whether or not they were known to the countries of ancient times. They were in use among the maritime and commercial communities located on the shores of the Mediterranean as early as the fourteenth century. From this region, it is most probable, they were introduced into England about the year 1381. The facilities and ease which they afforded for the safe transmission of money or values from one country to another soon brought them into general use among merchants.

Of the various negotiable instruments, the main one of interest to the contractor is checks. A builder would have more to do with checks than with any other type of negotiable instrument.

In the law of bills and notes a note is a promise. The essential requirement of a promissory note is a written unconditional promise to pay another person a certain sum of money at a certain time. The fact that a note is payable at a bank has no legal significance. There are many different types of notes. They actually run the gauntlet from a simple I.O.U. (I owe you) to a more complicated note that is used in connection with a real estate mortgage loan. It is important that the reader know about the various types of notes and their distinguishing features.

An I.O.U. is a written acknowledgement that one person owes another a certain sum of money. Usually, the written acknowledgement of a certain sum of money does not provide the date when the money is to be repaid; however, it is a legal and enforceable instrument. If the debtor does not pay the obligation within a reasonable period of time, the creditor could file a lawsuit for the amount named in the I.O.U., and the written instrument would be presented to the judge as evidence of the debt. Although written evidence of a transaction is easier to prove in a lawsuit, it is not impossible to prove your case with verbal evidence and testimony. All evidence is a matter of proof.

There are several different types of promissory notes. Various types of notes are reproduced on the following pages so that you will have the opportunity of examining them and learning the different features of each note.

An installment note is a promissory note that calls for specified payments to be paid at regular intervals, usually monthly. Installment payment notes are used frequently by retail stores that sell items of furniture, jewelry, clothes, appliances, and similar merchandise that is to be paid for over a period of months. Most installment notes state that if payments are not made when due and it becomes necessary for the creditor to turn the note over to an attorney for collection, that the defaulting debtor agrees to pay a named amount as attorneys' fees. Other notes state that a reasonable attorneys' fee will be charged if the note is referred to an attorney for collection. A number of years ago, a few credit jewelry stores attempted to take advantage of the attorneys' fees provision by trying to charge the debtor $100 collection expense on a $25 or $50 balance. The courts stopped this unfair business tactic by substituting reasonable collection expense and the judges ruled that $5 was a reasonable collection expense on a $50 balance. If the judge is of the opinion that provisions of a written instrument are unconscionable, he has the legal power and right to substitute terms that will be conscionable.

Another type of promissory note is called a "lump sum note" because it calls for one lump sum payment at a specified date. The laws of most states provide that collection expense or attorneys' fees can only be collected if the written note so states. If the written note is silent in that regard, no collection expense or attorneys' fees can be charged to the delinquent debtor.

Real estate mortgage notes are another form of promissory notes. They are usually quite lengthy because of the larger sum of money involved and because of the security that is included. The security is a mortgage which is the collateral for the mortgage loan. The note is the evidence of the obligation. An important feature to look for in a real estate mortgage note is whether the note gives the borrower the right to pay ahead of schedule or even to pay the entire balance at anytime *without penalty.* The words that give this protection are "or more." If you borrowed $25,000 on your home and your mortgage note called for monthly payments of $300 or more, then you are privileged to pay ahead or pay in full at anytime without any penalty being charged.

Other printed real estate mortgage notes do not contain the words, or more, and the borrower may not pay ahead of schedule or pay the balance in the future in a lump sum. It is important that you read and understand the written provisions of your real estate mortgage note to know what your rights are, and if you are getting reasonable protection.

Even in cases where the note says *X* dollars per month, or more, serious disagreements have resulted as to just what that means. Some people think that if they pay a triple monthly payment today, they are excused from paying for the next three months. This is not the case, the triple payment is not an advance payment and the note specifically calls for monthly payments. This means that one payment as a minimum must be paid each month. The extra money paid will reduce the principal balance of the mortgage loan but will not substitute for monthly payments.

In order to keep payments to an even amount, some mortgage lending institutions use printed notes that give the borrower the right to pay two or more payments at anytime.

Some insurance companies that make home mortgage loans have printed notes that give the borrower the right to pay 20 percent of his remaining balance in any one calendar year. This means a mortgage could be paid off ahead of schedule in five years, without any penalty being charged.

$30,000.00 Albuquerque , New Mexico

day of December, 19

**REAL ESTATE MORTGAGE NOTE**

(For individual borrower. Installments include interest.)

After date, as hereinafter set forth, for value received, I, we, or either of us, promise to pay to the order of JOHN JONES and /or MARY JONES, his wife

at Albuquerque National Bank, Albuquerque, New Mexico

the sum of Thirty Thousand and No/100 - - - - - - - - - - - - Dollars ($30,000.00)
in manner following, that is to say:

Two Hundred Thirty-Three and No/100, or more Dollars ($233.00 or more

on the ........ day of January , 19...., and Two Hundred Thirty-Three and No/100, or more Dollars ($233.00 or/ more) on the ........ day of each and every month thereafter until the entire balance hereof with the interest thereon, as hereinafter set forth, shall have been fully paid.

The said monthly installments of $233.00 or more shall include interest on said principal amount and/or on the unpaid balance thereof at the rate of Seven per centum (7%) per annum, and when said installments are paid, they shall be apportioned between interest and principal, and applied first to the payment of all interest due at date of payment, and the balance applied on the principal amount.

And I, we, or either of us, agree, in case of default of the payment of any of said installments, when, by the terms hereof, the same shall fall due, that such installments shall bear interest from the date of their respective maturities until paid at the rate of ten per cent (10%) per annum; and that, if any one of said installments or any interest due thereon is not paid within ten days after the same becomes due and payable, the whole of the principal sum then remaining unpaid, together with the interest that shall have accrued thereon, shall forthwith become due and payable without notice or demand, at the option of the holder of this note, with ten per cent (10%) additional on amount unpaid should this note be placed in the hands of an attorney for collection. The makers, endorsers, and sureties hereof hereby severally waive protest, demand, presentment, notice of dishonor, and notice of protest in case this note, or any installment due thereunder, is not paid at maturity, and agree that after maturity of this obligation, or any installment thereof, the time of making payment of the same may be extended without prejudice to the holder and without releasing any maker, endorser, or surety hereof.

This note is secured by a ~~Real Estate Deed of Trust~~ Real Estate Mortgage bearing even date herewith.

The makerS reserve the right to pay two or more installments at any time.

/s/ RICHARD ROE

/s/ JANE ROE

Form 113 (Rev. 4/64) The Valliant Co. Albuquerque, N. M.

For value received, the undersigned, jointly and severally, endorse, guarantee and promise to pay the note on the reverse hereof and all extensions and renewals thereof, and hereby waive (a) presentment, demand, protest, notice of protest, notice of dishonor, and notice of non-payment; (b) the right, if any, to the benefit of, or to direct the application of, any security hypothecated to the holder, until all indebtedness of the maker to the holder, howsoever arising, shall have been paid; (c) the right to require the holder to proceed against the maker, or to pursue any other remedy in the holder's power; and agree that after maturity of this obligation, or any installment thereof, the time of making payment of the same may be extended without prejudice to the holder and without releasing any endorser, surety, or guarantor; and agree that the holder may proceed against the undersigned directly and independently of the maker, and that the cessation of the liability of the maker for any reason other than full payment, or any extension, forbearance, change of rate of interest, or acceptance, release or substitution of security, or any impairment or suspension of the holder's remedies or rights against the maker, shall not in anywise affect the liability of the undersigned hereunder. In case this note is placed with an attorney for collection, the undersigned, jointly and severally, promise to pay an additional 10% of the total amount unpaid hereon as attorney's fees.

**PAYMENTS**

| DATE PAID | REC'D BY | INTEREST PAID | | PRINCIPAL PAID | BALANCE OF PRINCIPAL |
|---|---|---|---|---|---|
| | | AMOUNT | PAID TO | | |
| | | | | | |

Other lending situations use printed mortgage notes that specify certain penalties or charges for the privilege of paying off your mortgage loan ahead of schedule, or even paying the entire balance in a lump sum. The amount of the penalty for prepayment may vary from 1 percent to 5 percent. Therefore, the time to discover the provisions of your mortgage note is before you sign and not afterwards.

## 7-2 UNIFORM COMMERCIAL CODE AND NIL

Many of the rules of the law merchant have been enacted into laws by the various state legislatures and placed in a book called a code. There are federal codes for federal laws and each state has its own code for the laws of its state. Unfortunately, there are so many laws, both federal and state, that too many volumes are required to contain them.

Each state took the law merchant and placed it in their code under the provisions of the Uniform Negotiable Instruments Act. Each state further provided that if any situation was not covered by their new law, that the law merchant would govern. The Uniform

Negotiable Instruments Act, referred to as "NIL," was adopted over a period of thirty years by all of the states.

In the years since it has been adopted, the NIL has become obsolete due to great changes in our economy and business methods. Amendments were added in many states and different interpretations given to its provisions by the courts of the various states.

Studies on the Uniform Commercial Code started in 1942 and the first draft was finished ten years later. It was the culmination of years of study and research by experienced organizations with no self-interest to be served, the sole objective being improvement on the law.

The Uniform Commercial Code is arranged in ten subdivisions designated as Articles, and each Article is concerned with a particular type of commercial activity. Treating commercial transactions as a single subject of the law, the Code includes the law of:

(a) Sales
(b) Commercial Paper
(c) Bank Deposits and Collections
(d) Letters of Credit
(e) Bulk Transfers
(f) Documents of Title
(g) Investment Securities
(h) Chattel Secured Transactions

The Uniform Commercial Code represents an ambitious and commendable effort to put together most of the phases of the law governing commercial transactions. These transactions were formerly treated separately and differently by various states. Now there is greater uniformity and certainty in our commercial law.

## *7-3 NEGOTIABLE REQUIREMENTS*

Negotiable means an instrument or note can be transferred by a sale and indorsement or delivery. It can also be transferred by assignment. When a negotiable instrument is sold or assigned it means you are transferring the debt or obligation from the original payee to the incoming buyer or assignee. The new incoming party is called a "holder in due course" which means if the original debtor does not pay as provided for in the agreement, the new holder can sue in his own name. Equally important, the holder is not subject to any defenses that were available against the original payee. As a result, this commercial paper may pass from party to party and be accepted as a substitute for money.

An example would be where Bill Smith issues his promissory note to Tom White for \$1,000, payable in 180 days, plus interest at 7 percent per annum. White is the payee and 30 days after he receives the note, White assigns the note to you in payment of \$1,000 White owes you. By virtue of this assignment, White was the assignor and you are the assignee. You are now an innocent purchaser of the \$1,000 note for value and you become a holder in due course. If Smith does not pay the note when it becomes due, you can sue him in your name and any defenses that Smith might have had against White are not applicable against you. The big advantage of negotiable paper is that it is a substitute for money and can be passed from one person to another and the receiver acquires protection. There is no obligation on the part of Smith, the maker of the note, to do anything or pay any money before the note matures or becomes due. Upon maturity, Smith is obligated to pay the \$1,000 plus interest accumulated to the holder in due course, regardless of who that might be.

For an instrument to be negotiable it must conform to the following requirements:

(a) It must be in writing and signed by the maker or drawer

(b) It must contain an unconditional promise or order to pay a sum certain in money

(c) It must be payable on demand or at a definite time

(d) It must be payable to order or to bearer

What is money? Is there a substitute for money? How does it work out? Money represents personal property and passes as such currently from hand to hand without any necessity for inquiry as to title.

Paper money is a negotiable instrument. Negotiable instruments approach money in character and use and are a temporary substitute. Negotiable commercial paper are representatives of money. They circulate in the commercial world as substitutes for money, and for all practical purposes pass by delivery as money. This refers specifically to bills of exchange, promissory notes, bank drafts, and checks. Frequently, checks are passed from the original maker to the payee, and then endorsed and given to a third person, who finally deposits the check in the bank for collection. There is no limit to the number of times a check can and does pass hands.

The earliest type of commercial paper still in use is the bill of exchange. A bill of exchange requires three parties, the drawer, the drawee, and the payee. The prime function of the bill of exchange is to collect for the drawer from the drawee, residing in another place, money to which the drawer is entitled. A bill of exchange has been defined as a written order of one person who is called the drawee, upon another who is called the payee and is usually a bank for the payment of money to a third person who is called the drawer.

The term draft is synonymous with bill of exchange. Actually, a draft is a bill of exchange. Banks are the greatest users of drafts and they sell them to persons who desire to transmit funds. Thus, a draft has been defined as a check drawn by a bank. A draft in the law of bills and notes is a "drawing" and has been defined as an order by one person on another to pay a sum of money to a third person on demand or at a future specified time. The only distinguishing feature between a draft and an ordinary check is the character of the drawer. A cashier's check is a draft. A bank draft is a check drawn by one bank upon another bank in which it has deposits, much the same as the ordinary depositor draws his check upon his bank.

A trade acceptance is a draft or bill of exchange drawn by the seller on the purchaser of goods sold, and accepted by such purchaser. Its purpose is to make the book account liquid and permit the seller to raise money on it before it is due under the terms of the sale. The trade acceptance is ordinarily turned into cash by the seller through indorsement and discount. The situation is substantially the same as if the buyer had given the seller his promissory note.

The word discount used in the previous paragraph is different from the discount offered by a retail store. In order to sell commercial paper before its maturity date, the buyer wants a better deal than just collecting the interest called for in the note or draft or trade acceptance. The seller is forced to offer the buyer a discount which means selling the $1,000 obligation for $950 or some other reduced amount. Discounting of commercial paper is not unusual and occurs frequently. In the case of a land contract where the payout is scheduled to take a number of years, the seller or owner is required to pay discounts of up to 25 percent or even 50 percent to a buyer of the paper, in return for a lump sum payment.

## 7-4 CHECKS

A check is a written order drawn upon a bank for the payment of a specified sum of money to the order of cash or to a named person upon demand. The sole function of a check is to transfer money. Legally, a check is more than an order for payment of money because it is also a contractual obligation of the drawer. Although a check is usually drawn on check forms provided by your bank, they can be written on any piece of paper or even a substitute material. On rare occasions, and as a novelty, checks have been written on a piece of wood. It is difficult for the banks to handle the cumbersome piece of wood, when their computers and machines are designed for pieces of paper, but it has happened.

A check is a bill of exchange or draft drawn on a bank and payable on demand. A check is a negotiable instrument even though it is possible for the drawer to "stop payment" on the check.

The procedure of stopping payment on a check is quite simple. You go to your bank and advise them of the details in connection with the reasons why you want to stop payment on a check that you issued and your bank will take appropriate measures to refuse to honor this check if and when it should ever be presented for payment. You must act promptly in having your bank issue the stop payment order; if the holder of your check has already cashed it your stop payment order becomes a nullity. The most common reasons for stopping payment of a check are the following: loss of your check; your payee did not receive it or he lost your check; your check was stolen from you or from your payee; an error was made in the amount of your check and it represents an overpayment; and many other legitimate business reasons.

A "cashier's check" is different than a personal check in that it is drawn by a bank upon itself and accepted by the act of issuance. Cashier's checks are usually considered to be "as good as gold," however, there have been cases where the signature of the bank official is forged and then problems occur. When large sums of money are involved, cautious and prudent financial institutions will get on the long distance telephone and verify the signature by calling the issuing bank. A casher's check is also a bill of exchange.

A "certified check" is a personal check issued by a depositor on which payment is guaranteed by the depositor's bank. If you receive a check from Tom White for $1,000 and you want to make sure that his check will be honored when presented for payment, you take his $1,000 check to his bank and ask that they certify White's $1,000 check. White's bank will verify his bank balance and assuming White does have the money in his account, they certify his check by adding their stamp of certification. They charge the $1,000 against White's account immediately and guarantee that you will receive your $1,000 whenever you present or deposit White's $1,000 check. You can only get White's check certified at his bank and not your bank. You may be wondering why White's $1,000 check is not being cashed instead of having it certified. The answer is that there may be many valid reasons for not cashing White's check at that particular time.

*Examples*

1. Examples could be where the bookkeeper was sent to White's bank to have his $1,000 check certified and the boss does not want her to carry $1,000 around in cash.
2. Another reason could be where the $1,000 is not to be paid until certain work is performed, however, you want to be certain that the $1,000 will be available when the work is completed.

A certified check is an accepted bill of exchange.

"Travelers checks" have special attributes and functions. They differ from ordinary cashier's checks in that they are sold by banks, express companies, and other financial institutions that require both signature and counter-signature of the purchaser. They constitute a complete purchase and sale of credit because the traveling customer pays the money for the value of the travelers checks in advance and avoids carrying money on his trip. Travelers checks have the characteristics of a cashier's check where issued by a bank and are "foreign bills of exchange." A foreign bill of exchange involves persons living in different states or countries whereas an "inland bill of exchange" concerns persons living within the same state.

Checks that are payable in the future are commonly referred to as "postdated checks." A postdated check is, in effect, a representation by the drawer that he expects to have funds in his bank account with which to pay the check on the date specified. In this respect there is no essential difference between a postdated check and a promissory note. Some people are born optimists and issue postdated checks regularly. They are hopeful that they will have the money in their bank account when due date of their postdated check arrives. This is a very bad business habit and can only lead to problems such as having your postdated check later returned for insufficient funds. Issue your check only when you have adequate funds in your account to cover and don't guess about future deposits. Bankers as a group are cautious and conservative and for valid reasons. When you deposit a check, you assume that the check is good and will clear, however, the banker does no assuming and he waits for the check to clear before he will permit you to draw against uncollected funds.

Most states have laws making it a criminal offense to issue a check when there are insufficient funds to cover the check. In order to successfully prosecute the offender, the District Attorney is usually required to prove intent to defraud on the part of the issuer of the check and this is not always easy to prove. Checks that are returned to the depositor because of insufficient funds on the part of the maker have become so voluminous that it is not practical for every violator to be prosecuted. If you visited your bank in the morning and observed a bank official going through a pile of hundreds of checks that do not have adequate balances in their accounts to cover the check, you would realize how serious this problem has become. In cases of good customers, the bank official may approve the check being cleared even though the balance is insufficient, which results in an overdraft in the customer's account. Banks don't like to lend money on an involuntary basis and you can hurt your bank credit by becoming a problem customer.

The fact that most District Attorneys will refuse to initiate a criminal prosecution on postdated checks that are returned because of insufficient funds is little consolation. This refusal to prosecute criminally is based on the theory that a postdated check is merely a promissory note and not in violation of the law pertaining to insufficient funds.

Some contractors are always wondering about their bank balance. As a result, they are not sure if they have sufficient funds in their account to send checks to material firms, other suppliers, and meet their payroll. The easy way to know your bank balance at all times is to keep a running balance. The procedure is very simple. Each time you issue a check, deduct the amount of the check from your present balance. Each time you make a deposit, add the amount of your deposit to your present balance. The cardinal rule is that if you don't have sufficient funds in your account now, don't issue a check in anticipation that additional checks from customers will arrive later and be deposited.

If checks that you have issued are returned by your bank because of insufficient funds, you are hurting your credit with your bank as well as your material firm or other supplier.

Aside from the fact that your bank will charge your account with a service charge for the returned check, your material firm or other recipient of your check will also question your financial stability and business ability. Your own banker will wonder about the advisability of lending you money if you cannot keep your check records straight. In short, *never issue a check unless you have adequate funds in your account to cover.* Your creditor will think more of you even if you are late in paying your account, than if you issue your check and it is returned for insufficient funds. Having a check returned for insufficient funds is comparable to advertising the fact that you are financially in bad shape.

"Kiting" of checks is an old trick that is not very popular with the banking profession. Merely for educational purposes, the procedure for kiting of checks is as follows: I issue my check to you for $1,000 and you deposit my check in the ordinary course of business. My check is good and clears, and you immediately issue your check to me for $1,000. As a result of your depositing my $1,000 check, your $1,000 check to me is now cleared. This unethical procedure can be kept up for an indefinite period of time. The punchline to this constant exchange of checks between two persons is that one of the two is broke and is taking advantage of the time that it takes for his check to clear from his bank to the other bank, and to that extent he is using the bank's credit. It might help to explain this procedure by relating the procedure of what happens to my check drawn on my bank and your check that you draw on a different bank. All checks are sent to a central clearing house and then mailed to the bank on which the check is drawn to be charged against the maker of the check. This all takes time and time is the primary desire on the part of the check kiter. Eventually, my bank or your bank or both will discover the game we are playing, namely, using the bank's money for our benefit. Consequently, we will be notified by both banks that they do not appreciate our game, and we will probably be invited to take our banking business elsewhere.

In this modern day of computers, it is important that your monthly bank statement containing your cancelled checks, deposit slips, and bank balance be checked promptly. Computers have been known to make errors and the quicker any mistake is discovered the better. If your procedure of keeping a running balance has been maintained, it should be comparatively simple to go through your cancelled checks and deduct those checks that have not cleared as yet, and compare your balance with that listed by your bank statement. If they do not jibe, something is wrong and it is necessary and important that the mistake be discovered because the two balances must agree.

Although banks hate to admit it, sometimes they have been known to make mistakes. If you discover that your bank has made an error in your favor, it is advisable that you notify your bank promptly. Be just as prompt in notifying your bank of the error as if you were the victim instead of the benefactor. It is not a good idea to play possum after you learn of the bank's error in your favor, because the mistake will be discovered in due course of time. Your banker will appreciate your honesty in telling him of the bank's error in your favor and you can also build up goodwill at the same time.

If there is any doubt in your mind as to your own bank balance, you are privileged to ask your bank teller for this information. It is better to be safe than to gamble on issuing an insufficient check.

The importance of the use of checks in the conduct of your new construction business cannot be over emphasized. First of all, your bills should be paid by check so that you will have a permanent record of all transactions. The use of a petty cash fund for small amounts should be held to a minimum because it is more difficult to maintain records of payment. In addition to cancelled checks serving a worthwhile purpose of permanent records, many builders use their check book as an important source for obtaining income tax information. If all monies received are run through your business bank account, your bank records will be very helpful to your accountant.

Should and if your cancelled checks ever be lost or stolen, your bank maintains excellent records and microfilmed copies of your cancelled checks are available.

An assignment of a bill or note means the transfer of ownership which is called title. It involves two persons, the assignor and the assignee. The assignor is the person selling the title and the assignee is the person buying or receiving the title. The assignee takes only such title as his assignor had, subject to all of the defenses available against his assignor. When you receive a check and indorse your name on the backside, and pass it along to another, you are really assigning the check to the other person.

In addition to checks which you will use to pay your employees and bills, clients will pay you by check. The first thing that should be done with your client's check when it is received, is to place your rubber stamp bank endorsement on the back side. The reason for this is that if the check is lost, stolen, or the subject of a robbery, it is of no value to the finder or thief. He cannot obtain the money from any attempt to cash the check because of your bank endorsement on the back side. The important part of your bank endorsement stamp reads, "For deposit only to the account of, etc."

In the event of loss of this check, it is fairly simple to contact your client and have him stop payment on the lost or stolen check and issue you a new check as a replacement. You can stop payment only on checks that you issue and your client would have to stop payment on his check to you.

Now your client's check has been deposited to your bank account and your records reflect that it is worth the bit of extra trouble to make their deposit slips in duplicate. The original goes to your bank with the check to be deposited and the duplicate bank deposit slip is stamped by the bank teller and becomes a part of your permanent records. On your duplicate deposit slip, add a brief explanation identifying the transaction, such as the name of your client and possibly the address of the building project. Years later, some Internal Revenue Service representative may be making a routine check of your income tax returns and he will want to know the source of this deposit for $5,000 and your duplicate deposit slip will automatically furnish the needed information. Otherwise, hours of time can be wasted attempting to obtain this information regarding a transaction that occurred many months or even years ago. Even if the audit by the Internal Revenue Service agent discloses no errors in your income tax returns, it still costs you money because of the time consumed in conducting the audit. You cannot prevent the audit but if your records are in clean shape, even the tax agent will be favorably impressed.

All checks that are received from your clients should be deposited promptly. It is just not good business to deposit checks every once in a while or weekly or at other irregular intervals. Your client's check may be good today and not good a few days from now. Other checks that he issued may have cleared his account ahead of your check. The first check that is presented has the better chance of going through and being honored. Even Internal Revenue Service has learned about the advantages of depositing taxpayer's checks promptly. Years ago, it took Internal Revenue Service a number of weeks to get around to depositing taxpayer's checks and many checks were returned for insufficient funds. Now, a photostatic copy is made of the taxpayer's check and it is deposited within a day or so after it is received and the possibility of returned checks has been greatly reduced. You should profit by the experience of Internal Revenue Service and deposit incoming checks promptly.

## 7-5 BOND FOR EMPLOYEES

All employees that handle your company's money should be covered by a fidelity bond. A fidelity bond protects the employer against embezzlement or any other unlawful taking of money or property by an employee. The bonding of an employee is not a reflection on his character or honesty. It is just good business and an honest employee should have no objection to being bonded. No bank or other type of financial institution would think of hiring an employee that was not bonded. The cost of the fidelity bond obviously falls on the employer and not on the employee. Many bank tellers as well as other bank employees are bonded for one million dollars and more and the cost of bonding is actually quite low. Your insurance agent will advise you that there is no need for a one million dollar fidelity bond coverage for your new construction business employee.

In addition to the protection afforded to the employer by the fidelity bond on the employee, there are other factors involved. Sometimes, bonding companies find out background information about employees that is of interest to the employer. If the employee has ever been involved in any criminal matters or any other type of difficulty, the bonding company's investigation will bring these facts to light. If a bookkeeper is not bondable, it is important that this fact be known before large sums of money are placed in the care and trust of this employee. Not being bondable means that the bonding company has had an unfortunate experience with this person and they are of the opinion that the person is a bad risk. If the bonding company won't pass your bookkeeper, you cannot afford to hire that person in your business. You don't have to tell the bookkeeper that she was turned down by the bonding company but you can be diplomatic and advise that you are unable to hire any employee unless they are bondable.

The New York Stock Exchange recently ran a check on all employees of member stock brokerage firms, including fingerprinting, and the results were surprising. Some did not pass the investigation with flying colors and their services were necessarily dispensed with.

The law books are filled with cases where the old time bookkeeper, who had been with the firm for many years, suddenly was discovered to have had "sticky fingers" and embezzled large sums of money over a period of years. Many banks, financial institutions and other large corporations now insist that employees who handle and are responsible for their employer's money, take annual vacations to avoid the opportunity of hiding any possible manipulations of their records.

Now that money-handling employees are bonded, it is possible that they can be authorized to sign company checks for limited amounts. When you start your new construction business, you will probably be the only signature on your firm's checks. However, as your business continues to grow, you may find it very convenient and time-saving to have your bookkeeper authorized to sign company checks up to a certain reasonable limit. You can afford this luxury, only if your bookkeeper is bonded.

## 7-6 ACCOMMODATION PAPER

Under the broad topic of negotiable instruments and bills and notes we come to "accommodation paper." As the name indicates, accommodation paper occurs when a friend comes to you to either borrow money or ask that you sign his note at the bank or finance company. A word of caution is in order to the young contractor just starting up in business for himself.

In trying to help your friend by signing his note, you are affecting and impairing your credit without benefit to yourself. Reference to this age old problem is even made in the bible, with the admonition, "Neither a borrower nor a lender be." If your friend meets his obligation and pays the note that you guaranteed to pay if he failed, it is a break even business transaction. If your friend does not pay as he agreed, the bank or finance company will call on you for payment. You discover that you inherited someone else's problem and you can ill afford to be paying bills incurred by others. In addition to losing your money, you have probably lost a friend for life.

Experienced bankers have stated that seven out of ten accommodation endorsers, who want to help a friend out, wind up as losers by having to pay the obligation that belonged to their former friend. Many attorneys have come across this situation on numerous occasions, and invariably the debtor who requested the help is now unhappy with the creditor who got stuck in the deal. It is truly amazing when you listen to the weird excuses and alibis on the part of the original borrower and sometimes even the animosity he now bears to the friend who tried to be of help. Then comes the lawsuit and a beautiful friendship has been suddenly terminated and the former friends are now bitter enemies. There is no logical answer to this peculiar result.

Turning down your friend when he asks you to cosign his note is advisable. It is recommended that you tell your friend, "My banker instructed me not to sign any accommodation paper because it will affect my line of credit at the bank and I am sorry to have to turn you down." You will actually be doing your friend a favor by turning him down, even though he may not appreciate it at the time.

## *7-7 NONNEGOTIABLE AND NEGOTIABLE INSTRUMENTS*

The various types of documents or instruments listed are the most common used in the ordinary course of business. There are others that are unique or unusual, however, they would not be of interest to a contractor and thus, they are being ignored.

A "CD" is the abbreviated name for a "Certificate of Deposit." It is a written acknowledgment by a bank of the receipt of a sum of money, left on deposit for a given period of time, for which the bank promises to pay to the depositor or to his order, plus an agreed rate of interest. It could be called a time deposit for the money is left with the bank for an agreed period of time and may not be withdrawn before that date. An ordinary savings account is different in that the depositor may withdraw all or part of his savings deposit at any time. Technically, the fine print of the bank savings passbook states that the bank is entitled to demand thirty days notice before any withdrawal is made, however, banks generally do not take advantage of this provision. If the depositor withdraws his savings deposit before interest paying time, usually semi-annually, the bank pays nothing for the use of the depositor's money. A Certificate of Deposit creates the relationship of debtor and creditor between the bank and the depositor. Both the Certificate of Deposit and the savings passbook account are protected up to a limited amount by the guaranty of the Federal Deposit Insurance Corporation, an agency of the government.

"Bonds" represent an obligation by way of a written promise to pay money at a definite time of maturity. To that extent it is similar to a promissory note. Bonds and notes have been used interchangeably in the business world. No distinction exists as a matter of law for the reason that the essence of each is to pay a certain sum of money to the bearer.

"Debentures" are promissory notes, that is, promises to pay, which basically are not secured. A debenture is a debt and may be either a simple acknowledgment or a promise to pay. The biggest difference between a corporate bond and a corporate debenture, is that the bond is usually secured by a mortgage while the debenture is an unsecured obligation.

"Stock certificates" are evidence of title to a given share in the assets of a corporation rather than an unconditional promise to pay a certain sum of money. Stock certificates are negotiable under the provisions of the Uniform Commercial Code dealing with investment securities.

An "Interim Certificate" is a receipt issued on a subscription for stock or a purchase of bonds or other securities, to be held until the securities sold are ready for delivery. The interim certificate is the promise or obligation of the issuer to deliver securities when they are later issued. An interim certificate is not a negotiable instrument.

"Bills of lading" are evidence of title and represent the goods that are described therein. The function of bills of lading are entirely different from that of a bill or note. They are not negotiable instruments.

A "warehouse receipt" is a bilateral contract. It represents that goods are in the hands of a warehouseman and is a symbolical representation of the property itself. It is not a negotiable instrument.

A "post-office money order" is an order for the payment of money to the payee named, drawn by one postoffice upon another, under authority of an act of Congress and departmental regulations. It is governed by the postal laws and is not a negotiable instrument. Bank and express company money orders are the same as the post office variety.

A "letter of credit" is a written instrument issued by your bank, addressed to any other bank in the world, authorizing that bank to pay you a sum of money within the limits stated on a guaranteed basis. It is usually used by most people while they are traveling. It eliminates the necessity of carrying large sums of money on your person. When you need money, you take your letter of credit into the strange bank and they will advance any sum of money within the limits stated in your letter of credit. Your cash withdrawal is then sent back to your bank through channels and the money advanced to you is repaid. Your bank has guaranteed to pay your withdrawals that are advanced by the foreign bank. It is not a practical substitute for travelers checks because you already have the travelers checks in your possession and can cash them anywhere. Your letter of credit is best used for larger transactions and withdrawals. When you return home, your bank will have charged your account with your withdrawals or present you with a statement, according to your prior arrangements.

## *7-8 USURY AND CONFLICTS*

Interest charges are a necessary phase of bills and notes. Sometimes interest rates are in violation of law and this is called usury. Usury is the lending of money at interest rates in excess of that permitted by law. What is usury in one state may not be usury in another. The statutes of the various states limit the amount of interest that may be charged in connection with the lending of money. Most states permit a higher rate of interest to be charged on loans that are not secured as compared with loans that are secured. Secured means the lender has a mortgage on land or a chattel mortgage on a car or furniture as collateral for his loan.

Any interest charge in excess of the legal rate is in violation of law and is labelled usurious. The penalties provided by law as against the lender for charging illegal rates of interest will obviously vary from state to state. The laws of some states provide that the usurious lender may not collect his original loan. Other states provide that the borrower may collect a penalty in the form of treble damages, meaning a recovery of three times the illegal rate of interest charged. The legal reasoning behind these various penalties against the lender is that the consideration for the loan fails because the entire lending transaction has been declared to be illegal.

Some people feel that the entire matter of interest charged by a lender is a personal matter between the borrower and the lender. However, the public interest is definitely involved and the rate of interest that can be charged should be controlled by the legislature of each state. Some borrowers are not concerned with interest charges when they borrow the money, but there is a day of reckoning and the borrowed money must be repaid and the interest charges should not be unreasonable.

In this connection, even the federal government through the United States Congress enacted legislation referred to as "Truth in Lending." This law compels the seller to make a complete and honest disclosure of all of the various charges that are included when time payments are provided. Even mortgage companies are now required to advise the borrower how much his interest will be and how much it will total for the life of the mortgage, whether it be twenty years or any other period of time. The car dealer advertises his car as only costing *X* dollars per month, but formerly did not have to tell the customer how many months you have to pay *X* dollars. A good portion of the public is gullible and should be protected from being taken advantage of, even if they are not concerned about their own welfare. At least the new law provides for the pertinent information to be made available to the customer and it is up to the customer to protect himself.

Some financial institutions including banks have attempted to circumvent the usury laws by using different names for charges in connection with making a loan. The poor borrower has to pay the bill whether you call the charge interest or credit report charge; or appraisal charge; or investigation charge. The courts of many states have ruled against this questionable procedure which attempts to avoid the purpose of the usury law.

Finance companies and pawn brokers usually come under a special category. Most state laws provide for a definite schedule of charges for interest that may be made by a pawnbroker and finance companies. Most people who are forced to borrow money from a pawnbroker by pledging some item of personal property have reached the end of the line as regards obtaining credit. They have no bank credit, nor do they have the type of collateral that would be of interest to the average bank. They probably already owe one or more finance companies and have no further credit available from that source. They know the pawnbroker's charges are high and that they probably will not redeem the item pawned as a pledge, but their need for the money is so great that they do not care about the charges.

The charges of most finance companies are not inexpensive. They are usually licensed and regulated by each state. The invitation issued by the finance company to come in and borrow their money is very tempting but it behooves the borrower to investigate the interest charges involved. More often than not, the borrower is already in financial difficulty and the new payments that he obligates himself to pay to the finance company are most difficult if not impossible.

Finance companies serve a worthwhile purpose for those who do not have and cannot obtain bank loans at more reasonable interest rates. The builder should obtain his money from his bank and not from a finance company because his margin of profit is not great enough to pay the interest charges of the finance company.

The laws of most states provide that payments or notes or other obligations that fall due or become payable on Sunday or a holiday, are payable on the next business day. Some states have a similar law for instruments that fall due on Saturday.

When there is a conflict between the written and printed provisions of an instrument, the general rule of law provides that the written portion prevails over the printed part. A good example that occurs frequently is an ordinary check. A check is typed out for $100 and the maker discovers he made a mistake and he wants the amount changed to $105. It is very simple, he crosses out the $100 and writes in $105 and initials the change. A typewritten instrument is considered the same as a printed instrument.

The Uniform Commercial Code declares that handwritten terms control over typewritten and printed terms. Further, that typewritten terms prevail over printed terms. If you are ever faced with the problem of making a change in a printed or typewritten agreement, cross out the objectionable portion and write the new provisions and then have all parties involved sign their initials next to the new portion. If it is too complicated it should be passed on to your attorney for proper correction. It has been said that "one who acts as his own attorney, has a fool for a client."

Bills and notes are like other contracts as regards illegal consideration or those that grow out of illegal transactions. Examples would be transactions that violate the law, public policy, or good morals.

In order to prevent recovery on an instrument upon the grounds of illegality, the connection of the claimant with the illegal transaction must be direct and not remote. Anti-gambling statutes declare that bills and notes issued with the background of gambling consideration are void whether based on law or public policy.

Even in the state of Nevada where gambling has been ruled legal, problems have arisen in attempting to collect checks issued in connection with gambling activities. Gambling casinos in Nevada have partially solved the problem by having cashier's located in the gambling area. The patron goes to the cashier to cash his check, and he is then privileged to spend the proceeds of the check as he pleases. He may want to purchase food, drink, clothes, or lodging if he so desires. However, the fact of the matter is that the patron uses the money to return to the gaming table to continue with his gambling.

The courts have ruled that the check was not issued for gambling purposes and the maker is responsible for payment. Sometimes, when the credit of the patron is established at the gambling casino, credit is given to him at the gaming tables and later he issues his check for the amount of the value of the chips that were issued to him. Under these facts, the courts have ruled that the check was issued in a gambling transaction and is therefore unenforceable.

The rights and duties of the United States on commercial paper which it issues cannot be governed by local law, because the jurisdiction falls on the federal courts only. State courts do not have any rights over the federal government or any of its agencies.

# 8

# LIENS

## 8-1 ORIGIN AND BACKGROUND

A lien is a legal claim against the property of another for the satisfaction of a debt. The reasoning behind lien laws are based upon the fairness and equity of the property owner paying for work done or materials delivered to his property, in connection with improving his property.

All lien laws are statutory which means we must look to the law books to ascertain the rights and liabilities of the contractor, subcontractor, workmen or materialmen. The legislatures of each of our states have enacted laws that permit the four categories listed to file a lien against the construction property as an important and invaluable aid to collect the money due them. The laws or statutes of each state are different and it is important that the contractor learn the provisions of the lien laws of his particular state.

The examples that are listed in the following pages of this chapter are used to indicate the requirements of one particular state and probably do not apply to your state.

Most legal problems arise when someone fails to do something that they are supposed to do under the provisions of the pertinent contract. If the property owner pays his construction bills promptly when they are due, there is no problem and obviously no need for the filing of a lien. Liens are only filed against the construction property owned by the property owner when the contractor, subcontractor, workmen or material supply firms are not paid and it becomes necessary for them to take appropriate legal measures to enforce payment. In other words, the filing of a lien is a powerful weapon that helps the contractor, subcontractor, laborer or materialmen to collect their past due bill.

The property owner would never have occasion to file a lien because he is the debtor that owes the money and not the creditor that has money due him. The property owner does have certain rights, even though filing a lien is not one of them.

If the contractor does something wrong and is guilty of negligence, the property owner could file a lawsuit for his damages. The property owner has a similar remedy against the subcontractor or workmen for their negligence or failure to do their work in a workmanlike manner. Likewise, if the material supply firm delivers defective materials, the property owner can return them for full credit or hold the supply firm responsible for any damages attributable to the materials that were not up to par. This type of lawsuit for damages that could be filed by the property owner has nothing to do with filing a lien. For example, the cement subcontractor did his work in a sloppy, unworkmanlike manner and the concrete is loaded with big cracks from one end to the other. The result is the same whether the materials he used were defective or his men just did a sloppy job. The property owner has a valid lawsuit based on the negligence of the cement subcontractor for his failure to do his job in a workmanlike manner. This lawsuit is on the theory that the property owner had a legal right to assume that all construction work would have been done in a workmanlike manner.

A lien filed by a contractor, subcontractor, workmen or material firm in order to collect money due them does not mean that the property owner is defenseless in a proper case. If the property owner has a good reason for his refusal to pay a bill, as in the case of the cement subcontractor referred to above, he can defeat any lien that may be filed by the cement subcontractor. If, for example, the cement subcontractor refused to be fair and redo the job, the property owner could file a lawsuit for his damages. Or, the property owner could defend the lien claim filed by the cement subcontractor and in due course of time, the matter would be heard by a judge. Like all other trials, testimony would be given by the various witnesses, expert or otherwise. Based on our assumed facts, the judge rules in favor of the property owner and denies the lien claim filed by the cement subcontractor. As a result of the court's ruling, the lien claim is removed from the public records and is no longer a cloud on the property owner's title.

Some of the lien laws enacted by many legislatures throughout the country were the result of pressure exerted by large construction and materials firms who demanded special treatment. As a result of this pressure, some of these lien laws were really unfair to property owners and the general public. Situations arose with the passing of time that were inequitable and very bad. A case that comes to mind involved a contractor who was building a large volume of homes and awarded the plumbing subcontract to a licensed plumber. The home builder paid the plumber as his work progressed in accordance with the terms of their contract. In due course of time, the plumber completed his work and was paid the balance due him in full. Some months later, the wholesale plumbing firm advised the home builder that they had not been paid in full by the plumber subcontractor and demanded payment of their bill. The dispute was brought into court and after all testimony was concluded, the court ruled in favor of the wholesale plumbing firm. The judge said that the payment to the plumber contractor did not protect the home builder and that he was ordered to pay the wholesale plumbing firm for their materials.

This double payment result proved the unfairness of this law and it was later changed. As a result, the wholesale plumbing firm would now be required to notify the home builder that they are furnishing plumbing materials and supplies to the plumbing subcontractor so that he has the opportunity of protecting himself from double payment. Consequently, the contractor is now able to protect himself by making his checks payable to both the plumber subcontractor and the wholesale plumbing firm, to insure that the money will reach the proper parties.

Another undesirable type of lien law permitted a tenant to order new electrical fixtures, or other items, without the knowledge or approval of the landlord, with possible responsibility for the bill falling on the unsuspecting landlord. Under the old law, if the tenant did not pay the electrician's bill which he incurred, the electrician was privileged to file a lien against the property with the result that the landlord was stuck for the bill. Many states have changed this unfair lien law and the electrician is now required to obtain the approval of the landlord, if he intends to look to the landlord for payment of his bill. Otherwise, his claim is against the tenant only, he being the person who incurred the obligation.

## 8-2 TYPES OF LIENS

There are many types of liens, some of which involve personal property and others that involve real property. Examples of liens involving personal property could be a garageman's lien for possession of the automobile on which he did mechanical work; a pawnbroker who retains possession of your pledged ring until you pay him the $100 that you borrowed, plus his interest charges; and the moving and storage firm who refuse to release your personal belongings until you pay the storage bill.

Liens involving real property, which means land and the improvements or buildings that are placed on the land, are also numerous and varied. An example of a voluntary lien involving real property is a mortgage. A home owner voluntarily places a mortgage lien on his property by borrowing money and issuing his note and mortgage as security for the loan.

Real property taxes constitute an involuntary lien, as they automatically become a charge against the particular parcel of real estate involved until they are paid.

A tax lien in favor of the government falls into a peculiar category, different than any other type of lien. When income taxes or other taxes due the federal government are due and remain unpaid for a certain period of time, the government files a tax lien against the delinquent taxpayer. The tax lien is recorded with the county clerk's office and becomes a matter of public record. More important, the tax lien attaches to and becomes a lien against *all* real estate in the name of the delinquent taxpayer in that county. If the government has information about real property owned by the taxpayer in other counties or even other states, identical tax liens are filed in the other communities. Only the federal government has the right to the advantages of such a powerful and all-inclusive lien against real estate. It should be obvious that taxes due the government, federal, state, or local should be paid promptly when they are due. If they are not paid when due, interest and other penalties run the tax balance up. Tax obligations are not dischargeable debts even if the contractor goes broke and winds up in bankruptcy.

A judgment lien is another example of a general lien. If you obtain a court judgment against your debtor and he fails to pay the judgment, your judgment can become a lien against any and all real estate that you debtor may own in that particular county. Your attorney who handled your lawsuit and was successful in obtaining the judgment will file a transcript of your judgment with the county clerk and this will then become a lien against your debtor's real estate. If your debtor owns real estate in more than one county, your attorney can file transcripts in more than one county. The laws of some states provide that the obtaining of a judgment does not automatically constitute a lien unless the transcript of the judgment is recorded. Your judgment lien is effective only if the debtor tries to sell his real estate or refinance his mortgage loan and then your judgment becomes a stumbling block. Otherwise, it is necessary for your attorney to file a new lawsuit against your debtor, asking the court to sell his real estate and apply the proceeds towards your judgment. If he owes a mortgage balance on his real estate, the obligation would take priority over your judgment lien. However, the pressure of the second lawsuit and the possibility of losing his property will often get the judgment debtor started in making payments to apply on his judgment obligation.

## 8-3 MECHANICS' LIENS

Liens pertaining to the construction industry are called "Mechanics' Liens." They involve a particular class of creditors called contractors, subcontractors, laborers or workmen, and materialmen or material supply firms.

The various lien laws are based on the fairness of paying for materials furnished and work done. The general effect of mechanic's lien laws is to make the building and the land on which it is located, responsible for the payment of debts due for work done or materials furnished. The purpose of mechanics' lien laws is to permit a lien to be filed against the construction property when bills have not been paid for labor or materials.

Even though the material firm or subcontractor does not know the property owner and may never have dealt with him, the law says that they may protect themselves, by filing a lien against the construction property for their unpaid bill. This right comes from the contractor who ordered the labor or materials and the contractor is the agent of the property owner by virtue of their written contract. In turn, the law reasons, this labor and material has benefited the owner's property and he is obligated to pay the bills.

The basic philosophy of the lien law is fair and sound. One who has received benefits should pay for them. If we tried to change the procedure, it would be necessary for the subcontractor or the workmen or the material supply firms to have the property owner approve each order for labor or materials. This procedure would be cumbersome and unnecessary inasmuch as the contractor is the property owner's agent for this purpose.

Please do not confuse the example given on an earlier page of the garageman who did mechanical work on the automobile with our Mechanics' Lien title above. In spite of the established legal name of Mechanics' Lien it actually involves construction workmen and laborers as well as materialmen. Examples of materialmen are the lumber firm who furnished the lumber; the hardware store that furnished the roofing tar paper and nails; and the cement firm that furnished the concrete.

Even the word laborer does not necessarily refer to unskilled or the lowest rung of the construction work ladder. It could refer to the plumber, who was the subcontractor, who files his Mechanics' Lien claim for his labor and materials. It could also refer to the electrician, the bricklayer, or the floorlayer.

A mechanics' lien is a special lien or claim that the law permits to be filed against the real estate involved in construction, when the workmen or persons furnishing materials are not paid for their labor or materials. The filing of a lien claim with the county clerk where the construction land is located constitutes a powerful factor in aiding collection in behalf of the labor or material claimant involved.

You might wonder why, for example, the roofer, who is owed a balance of $800 does not file an ordinary lawsuit against the property owner. While the ordinary type of lawsuit might be pending for a year or so, the property owner could sell the house in question and leave the country. On the other hand, the Mechanics' Lien filed by the roofer, immediately affects the title to the real estate and the property owner would not be able to sell his newly constructed home with a pending lien claim filed on record. As you will see, the filing of the mechanics' lien is only a preliminary step and that other legal measures are taken to enforce collection of the lien claim.

Each state has its own special laws pertaining to the filing of Mechanics' Liens and you are urged to consult with your attorney to ascertain the laws of your particular state and how they affect you. Your attorney will advise you as to when you are required to file your lien claim; he will assist you in the preparation of your lien claim and the itemized statements and affidavit that are usually required to be attached to your lien claim; he will file your lien claim with the county clerk where the construction work is located; and he will also write to the property owner and advise him of the filing of your lien claim and probably enclose a copy of your itemized statement and lien claim. Some state laws provide that a lawsuit must be filed to enforce the lien claim within one year from the time the lien is recorded. In due course of time and within the time limits allowed by law, your attorney advises, for example, that Tom Brown, the property owner is at odds with the contractor and refuses to pay your bill. He recommends that a lawsuit be filed against Brown to foreclose your lien claim, hoping this procedure will result in the collection of your money. Frequently, this lawsuit will make Brown realize that you mean business and force him to pay your long overdue bill. By now, the amount of your bill has increased, and Brown has to pay court costs that you have expended; recording fees for filing your lien claim; as well as attorney's fees that may be allowed by your local statute or law.

The attorney's fee award may or may not be adequate to pay your attorney in full. Definite fee arrangements should be made in advance so that there will be no misunderstanding on the question of attorney's fees or collection costs. Many attorneys handle lien claims on the same basis as other collection items, that is, the attorney's fee is a certain percentage of the money collected on a contingent fee basis. If no money is collected, you have to suffer your loss and the attorney lost his fee gamble and is paid nothing for his time. Other attorneys charge for lien claims on a time basis, regardless of the success or failure in collecting your money. This contingent fee arrangement is used frequently in automobile accident lawsuits for personal injuries where the attorney is willing to gamble on the chances of recovery of money. If the lawsuit is lost, the attorney receives no compensation for his time. If he wins and collects the money, his fee will vary from 25 percent to 40 percent of the recovery. Who pays the court costs and expenses is subject to agreement between the client and the attorney.

Some builders and others involved in filing mechanics' liens think that the attorney is being overly technical in his handling of the lien claim. However, most lien laws are quite technical and the attorney has no choice but to follow the procedures laid down by the statutes of the state involved. After all, that is the law of your state and it must be followed in order for you to collect your money.

## *8-4 RIGHT TO LIEN*

What is the general thinking and philosophy behind the law that provides for the right to file a mechanics' lien by the workmen or materialmen? It is based upon a contract, either verbal or in writing, between the owner and the contractor. Under the provisions of the contract, the builder is authorized to employ subcontractors and others to assist in constructing the home or building ordered by the owner. The contractor becomes the authorized agent of the owner and he orders labor and materials to be furnished in the construction of the home or building. It is the use of the materials furnished and the work and labor by the subcontractors and the others, for the benefit of the owner's home or building, that gives the laborers and materialmen their lien rights under the law.

Many problems arise, for example, if the contractor quits the job in the middle of construction without justification. May the contractor file his own lien for the value of his services rendered? Based upon the above statement, the contractor would not prevail in his lien claim, because of his failure to perform without any legal justification.

If there was some valid reason for quitting the job in the middle of construction, the contractor might have a good claim in an ordinary lawsuit for the reasonable value of his services rendered. Like any other complicated lawsuit, the owner could have a counterclaim for damages that he sustained by virtue of the contractor's action in quitting the job.

Cases have happened where the material firm furnished only a portion of the materials ordered and refused to complete the order for no valid reason. To make matters worse, the material firm would like to file a mechanics' lien for the materials that they did furnish. Anyone can file a lien but the big problem is to have the lien stand up and be approved by the court when the right to file the lien is disputed. Under the facts presented, most likely, the material firm could not prevail in filing a lien for partial delivery of materials but could file an ordinary lawsuit for the reasonable value. Then the owner could file his counterclaim for any damages that he may have suffered because of the failure of the material firm to deliver all of the goods as agreed. The proof of damages is always a necessary element in any lawsuit. If the missing materials could be obtained elsewhere at the same or lower price, then there was no loss and no damages resulted. Generally speaking, the law attempts to compensate the person who has been damaged by suffering a loss by making him whole, but will not permit anyone to make a profit at the expense of the wrongdoer.

**CLAIM OF LIEN**

SMITH SHEET METAL COMPANY, INC.,
a corporation
Claimant,

vs.

RICHARD ROE, Owner and
JOHN JONES, d/b/a MODERN BUILDERS

STATE OF NEW MEXICO, } ss.
County of Bernalillo

I hereby certify that this instrument was filed for record on the ........ day of ........, A. D. 19...., at ........ o'clock ........ M., and was duly recorded in Book ........ of Records of Mechanic's Liens, page ........ on this ........ day of ........ A. D. 19........

........
Clerk and Ex-Officio Recorder.

By........
Deputy.

PRINTED AND FOR SALE BY VALLIANT PRINTING CO., ALBUQUERQUE

# CLAIM OF LIEN

To RICHARD ROE owner or reputed owner of the premises hereinafter described, and

To JONES JONES, d/b/a MODERN BUILDERS contractor, and to all who may be concerned:

You and each of you are hereby notified that SMITH SHEET METAL COMPANY, INC.,

hereinafter called the claimant, hereby claims a lien in the sum of $ 452.00 upon the following described

property, to-wit: 110 South Main Street, also known as

Lot numbered Ten (10) in Block numbered Six (6)

of the Park Addition to the City of Albuquerque,

Bernalillo County, New Mexico, as the same is

shown and designated on the plat thereof.

together with all improvements thereon, for materials furnished by claimant to be used in the construction, alteration and repair, ~~(labor done by the claimant in the construction, alteration and repair)~~ of a certain

home situated thereon.

That a statement of claimant's demands after deducting all just credits and offsets is hereunto annexed and marked Exhibit "I" and made a part of this claim as if the same were herein set out

in full

That the name of the owner or reputed owner of said property is RICHARD ROE

and the name of the person by whom claimant was employed (to whom claimant furnished the said materials) is

JONES JONES, d/b/a MODERN BUILDERS

That the terms, time given and condition of claimant's contract are as follows:

Payable in full on the following 10th of the month.

Wherefore, claimant claims a lien upon said property and improvements thereon in the amount aforesaid.

SMITH SHEET METAL COMPANY, INC. Claimant

By GEORGE SMITH,
President

STATE OF NEW MEXICO, }
County of Bernalillo } ss.

On this ........ day of December 19...., before me the undersigned authority came

George Smith, President of Smith Sheet Metal Company, Inc.

to me personally known to be the same person who signed the foregoing statement of lien claim, and on oath did state that he has read the foregoing statement of lien, and that the matters of fact set forth therein are true.

Witness my hand and seal on this the day and year last above written.

.................................. Notary Public.

Bernalillo County, New Mexico

My commission expires ..................................

NOTE. 1. If you do not know the owner's name, find it out from the county records; be careful about this.
2. Describe the property carefully. Don't stop with giving the street number; put the lot and block numbers, or legal description, into your proof. State whether the work or material was for a store building, dwelling house, barn, or other building.
3. Attach a statement of your account to the proof marked "Exhibit I." Specify hours of labor, or materials furnished, prices, and all payments.
4. Sign and swear to the claim, and have it recorded by the County Recorder of the county where the work was done. Original contractor has 120 days and all others 90 days from the completion of the job. File your lien in time.

Another problem that frequently occurs is when the prime contractor and the owner are fighting each other and the innocent materialmen and subcontractors are not being paid. If the subcontractors and materialmen are without fault, they are entitled to protect their valid claims by filing a mechanics' lien. The owner cannot use his dispute with the general contractor as an alibi for not paying the valid claims of the subcontractors and materialmen.

In connection with filing mechanics' liens, it is sometimes important to know whether the claimant is a contractor or subcontractor or materialman because the time for filing liens will vary with each category. Usually, the general or prime contractor is granted sixty days after completion of the home or building to file his lien,whereas the subcontractors, workmen, and material firms may only have thirty days.

Appreciating that the lien laws of each state vary, there is a time limit for filing mechanics' liens; the time being different under particular circumstances. The laws of one state provide, "but in no case less than thirty days or more than ninety days after completion of a building or completion of a repair job." The time limit for filing these liens depends upon whether or not a "notice of completion" is filed. A notice of completion is a notice filed by the owner in the county clerk's office, giving public notice that the construction job has been finished. The completion notice must be filed within ten days after all work is done. A "notice of abandonment" has the same effect, and is filed when the work has stopped on an uncompleted building. When these notices are filed, the original contractor has sixty days in which to file a lien, and all subcontractors, workmen, and material firms have thirty days. If no notices are filed, then anyone furnishing labor or material to the job has ninety days to file their lien. Not all states have laws providing for notices of completion or abandonment.

Generally speaking, most state laws require a mechanics' lien claim to name the property owner; a legal description of the premises; a detailed itemized statement of the work performed or the materials furnished; the balance due showing what payments have been made, if any; from whom and for what it is due; and an affidavit signed by an official of the lien claimant stating that the information contained in the lien claim is true and correct.

The right to file a mechanics' lien is not restricted to new construction. Under modern state laws, a mechanics' lien may be filed for repairs, alterations, and additions. Some state laws provide that if the plumber who installs a hot water heater in your home is not paid, he has lien rights for the value of the heater as well as his labor for installation.

Sometimes, the tenant or renter orders the plumbing work and then fails to pay and the landlord or property owner winds up getting stuck for the plumber's bill. The agreement between the landlord and tenant would determine which one was obligated to pay the plumber's bill, whether it was for repairs or even a new hot water heater.

The right of a material supply firm to file a mechanics' lien is contingent upon proof that the materials involved were delivered to the job-site. It is a very simple matter for the bookkeeper of the material firm to prepare itemized statements of the materials involved and list the balances due and unpaid. However, when this problem is presented in court, usually a year or so later, the technical attorney for the property owner denies that the materials were delivered to this particular job-site. If the testimony reveals that the driver of the material supply firm dumped his entire load at one address and the materials were used for five separate homes, the lien claim will probably be denied by the judge. The fact that the itemized statement bills all of the materials to one address is further proof that the law requiring delivery to the job-site was not complied with and that the lien claim has to be denied.

There have been many cases where the contractor's driver goes to the material supply firm and picks up miscellaneous materials and the material firm bookkeeper charges all of the items to one job, whereas the materials are actually going to be used on more than one job. If the bill is paid when due there is never any problem. The problem always arises when the bill is not paid and then comes the lien claim and the embarrassing question is brought up in court about failure to comply with the law requiring delivery to the job-site. Even though the lien claim is denied in court because of some technicality, it does not mean that the property owner is excused from paying his bill. It only means that the lien claim is denied because the technical lien law was not followed. The property owner or in given cases the contractor still owes the bill for the materials that his driver picked up and signed for. The material supply firm can file an ordinary type of lawsuit against the contractor or property owner if he is involved, and can collect their just debt.

You may wonder why the lien claim is so important if the material supply firm still has their claim against the contractor and can file a regular lawsuit against him. The answer is that usually the failure of the lien claim is bad news because the contractor has gone broke and the only hope of payment for the material supply firm is to collect from the property owner, via the lien route. Denial of the lien claim by the court means the property owner is excused from payment and the contractor is unable to pay. Now the material supply firm has the opportunity of writing off the uncollectable account as a bad debt. By writing off is meant the procedure whereby an account on your books that indicate a balance due is determined to be a bad or uncollectable debt and is written off. Consequently, taxwise you are able to claim a credit for the bad debt that proved to be uncollectable. Some tax agencies require that a lawsuit be filed against the debtor and a judgment obtained, even though it is impossible to collect the judgment, in order to claim the bad debt as a tax write off. If you ever collected the money at a later date, taxes would be due on the amount collected as business income. Under existing tax laws, writing off a bad debt is some consolation in case you are unable to collect the money due you.

The laws of most states provide that every building or improvement or remodeling constructed upon any land, with the knowledge of the owner, shall be subject to a lien, unless the property owner shall give notice that he will not be responsible for payment of the bills. The "notice of nonresponsibility" of the property owner is accomplished by posting a notice on the land itself in writing to that effect.

In this manner, the contractor has notice that he must look for payment of his money only to the person who he is dealing with, and he will not have the protection or security of having lien rights against the land and collecting from the owner of the property. This third person may be a tenant; or he may be buying the land under a real estate contract or land contract; or he may hold a life estate. But the net result is that the contractor may not look to the land as security, or the property owner for protection under the lien laws.

In some states, the property owner's notice of nonresponsibility is filed with the county clerk, in addition to being posted on the land itself. Obviously, it behooves the contractor to learn how to protect himself by either obtaining the approval of the property owner or making sure that the credit of the third party is good enough to warrant ignoring the property owner and giving up the protection of lien rights.

If you are dealing with a partnership firm that owns land who want you to construct a building for the firm, it is not necessary that each individual partner sign the construction contract. The signature of any one of the partners is sufficient to bind the partnership firm, and if problems later develop your lien rights are protected.

If you are dealing with a corporation the procedure is more formal and quite different.

The corporation owns the land and the president wants you to construct a building for his firm. Aside from the fact that the construction contract should be prepared or examined by your attorney, the question concerns the necessary corporation signatures on the contract. The building contract should be signed by the president and secretary of the corporation and their signatures should be properly notarized. In addition, the corporate seal should be affixed to the contract, close to their handwritten signatures. As discussed in an earlier chapter, a corporation may only enter into contracts by an appropriate act of its Board of Directors. In addition to the corporate signatures referred to above, you should be given a copy of a resolution of the Board of Directors authorizing the president to enter into the construction contract. This procedure may sound technical and unnecessary to you, but the law books are filled with cases where shortcuts were attempted and the contractor was hurt in the process.

## 8-5 PROCEDURE FOR OBTAINING MECHANICS' LIEN

Most state laws provide that in order to obtain a mechanics' lien, the lien claimant must give notice to the property owner and file a claim or statement in the manner and form provided for by the laws of his particular state. The courts of most of the states have ruled that strict compliance with the lien laws must be shown. A good example would be in the address of the property to be liened. Your bookkeeper shows the address as 110 S. Main Street on your itemized statement and your state law provides for the address to indicate the legal description. Therefore, the address should be shown as 110 S. Main Street, also known as lots numbered nine (9) and ten (10) in block numbered sixteen (16) of Westside Addition, an Addition to the City of San Diego, California. The required information as to the legal description of the land involved can be obtained from any title company or the county clerk's office.

The laws of most states provide that the claim of lien must be filed within a certain time limit in order to preserve the validity of the lien. The law usually provides that the lien filing shall be within a certain time from the completion of the building or improvement, as well as the last labor performed. This requirement may sound simple but actual cases in court have proven otherwise.

The pertinent question is, "When is a building completed?" Some states provide for a notice of completion to be filed with the county clerk and that date would control. In some court cases, the contractor testifies that a door knob was added two weeks later and he claims the later date as being the completion date. On the point of when the last labor was performed, the contractor testifies the property owner called him to take care of a sticky door and that chore changed the date of when the last labor was performed. The judge will usually substitute substantial completion as the controlling date. Judges will not permit the contractor to take advantage of some technicality, such as the door knob or the sticky door or some comparable trivial repair.

The reason for this discussion is that if the lien is not timely filed, the lien claimant is thrown out of the court and probably will not be able to collect his bill. The lien claimant still has a valid claim for the money due him and he is privileged to file an ordinary lawsuit for his money, but the security of the land and improvements is gone and his chances of collecting from a contractor who is broke are not very good. All lien claimants are therefore urged to file their mechanics' lien claims ahead of schedule and not lose out on their lien rights due to some technicality. Some contractors refuse to be practical and hope that their bill will be paid before the lien filing time expires, in order to circumvent paying any attorney's fee. I would certainly not criticize any contractor or other businessman who tries to avoid paying any legal expense if it is unnecessary. However, there is no point in hurting yourself by waiting too long and losing your lien rights entirely, while trying to save a part of your past due bill.

## 8-6 OPERATION AND EFFECT OF LIEN

The operation and effect of a mechanics' lien is dependent upon a number of circumstances. Among these are the amount of the lien and the time of its accrual. The amount of a mechanics' lien secured to a contractor depends upon the contract price. Another important circumstance is its priority over other liens and mortgages. Subcontractors, workmen or material supply firms have the right to file their own mechanics' lien against the property involved in the construction, independent of the contractor's agreement with the property owner.

Many courts have ruled that interest is allowable on a mechanics' lien from the date of the filing of the lien. The legal rate of interest used to be six percent but is subject to change by each state's legislature.

The question of priority of a mechanics' lien presents unique problems and, as between different mechanics' liens, depends primarily on the language of the local law. Some state statutes provide that they take precedence over each other in the order in which they were filed. In other words, prior in time is prior in right. Other state laws place all mechanics' liens on the same footing. This means that all lien claimants share pro rata in the proceeds of the sale of the property if there is not enough money to satisfy all claims. If there is sufficient money to pay all claims, then the order of priority becomes unimportant. Some state laws give priority to the lien claims of laborers and materialmen over those of subcontractors.

What is the order of priority between a real estate mortgage and a mechanics' lien claimant? The answer would depend upon the time when each lien attaches to the land involved, and the method of computing the time is purely statutory, in accordance with the laws of each state.

Most state laws provide that the real estate mortgage lien is first *if* the mortgage is recorded before any construction work is performed or even started. That is why all financial institutions, including banks, savings and loan, and insurance companies send a man out to the proposed job-site to insure that no work has been accomplished, and then have him sign an affidavit to that effect before recording the mortgage. If one spade of dirt has been excavated or construction tools have been placed upon the job-site, the financial institution would fear that their proposed mortgage would be second to that of subcontractors, laborers, or materialmen and they would refuse to go through with the proposed mortgage loan. In short, the recorded mortgage lien would have priority over lien claimants because it was recorded first and no construction had been started prior to such recordation. There have been cases where no investigation as to prior start of construction was made and as a result, all of the construction people, subcontractors, laborers, and materialmen had priority over the mortgage lien holder. If the mortgage company was unable to protect the priority of its mortgage, after the usual investigation, it would be impossible for the average person to arrange a mortgage loan for their home or any other building.

A mechanic's lien, like all other liens, may be satisfied by payment of the obligation, and this would discharge the lien. In a given case, mechanic's liens could have been filed by the contractor, subcontractor, laborers, and materialmen and payment to any one of the group would ordinarily have no effect over the lien claims filed by the others.

### 8-7 ENFORCEMENT OF LIEN

A mechanic's lien, being entirely statutory, that is, governed by the laws of each state, gives the procedure to enforce our lien. The laws of each state relates the ground rules and they are similar to foreclosure proceedings of a mortgage. If all of our normal attempts to collect the money due from the property owner have failed, the final step is to file a foreclosure lawsuit against the property owner. The purpose of the foreclosure lawsuit is to foreclose our lien. By this procedure the judge will rule on the priority of our lien and order the property sold to satisfy or pay our lien.

All lawsuits take time, frequently too much time, and it is unfortunate that the dockets of many courts in the country result in long delays from filing time until trial time. The one advantage of your lawsuit to foreclose your lien over other types of lawsuits is that title to the property involved in the lawsuit is tied up, and that the property owner has been unable to sell or dispose of it until the clouded title caused by your lien has been disposed of or ruled on by the judge. Thus, the long court delay is hurting the property owner just as much as it is hurting you.

Finally, your lawsuit comes to trial and the judge rules in your favor. Your lien is established as number one in priority and the property is ordered sold to satisfy your lien claim and possibly other lien claims. If the proceeds of the court ordered lien foreclosure sale are adequate to pay your claim in full, there is no problem. If the proceeds are not adequate, the judgment awarded you is only partially satisfied and the court will grant you a deficiency judgment against the property owner for the difference between your original judgment and the balance still due.

Under the deficiency judgment, you now have the same rights of collection as you would have under any other judgment. The property owner has been given credit for the money that you received from the sale of his foreclosed property, and he still owes the balance indicated by the deficiency judgment. If the property owner has any other property or bank accounts or automobiles that you can locate, your are privileged to try and collect your deficiency judgment. Bear in mind, however, that deficiency judgments are usually difficult to collect because the judgment debtor, the property owner, is not in good financial shape. At least your claim against the property owner is now in judgment form, and most states provide that a judgment is valid for six years and that it can be renewed for an indefinite period.

# 9

# CONDOMINIUMS AND COOPERATIVE APARTMENTS

## 9-1 DEFINITIONS AND BACKGROUND INFORMATION

The odds are favorable that the newly established builder or contractor will not be called upon to construct a large condominium or cooperative apartment complex. However, it is still desirable that you familiarize yourself with some of the basic differences between the two types of apartment projects.

One never knows what might favorably impress your client, resulting in a recommendation or possibly a future construction job. The construction business is no exception to recommended standards or guides of other businesses and professions. It pays dividends to be cordial and to appear to be knowledgeable in your chosen business or profession. You may receive an inquiry regarding a small repair job or other type of minor construction that seems to be too small to even bother with, yet it could and frequently does lead to bigger and better jobs in the future.

As discussed earlier, goodwill of a business and a satisfied client is a very important factor. Every job that you complete should add another satisfied client who will tell his friends about the builder that he discovered. This type of advertising cannot be purchased, it can only be earned.

Both standard and legal dictionaries recognize that the term "condominium" refers to joint ownership of real property. In recent years, however, the meaning has changed slightly. Condominium is now defined as the individual ownership of a unit in a multi-unit structure such as an apartment building. Under this more modern definition, a condominium refers to an apartment building, each of whose residents are called "unit owners." Although a condominium could be an office building, custom provides for its use in connection with apartment structures.

Each unit owner enjoys exclusive ownership of his individual apartment or unit, holding a fee simple title to his portion of the entire building. In addition, each unit owner retains an undivided interest, as a tenant in common with all of the other unit owners, in the common facilities and areas of the building and grounds that are used by all of the residents of the condominium. In order to eliminate any doubt, a brief explanation of fee simple title and tenants in common title is in order. A "fee simple title" to land is the highest and best title with nothing withheld. A purchaser of land who pays cash normally will receive a warranty deed granting fee simple title to the land which is the absolute title, there being no higher title. As the name implies, "tenants in common" means ownership of land along with others. Visualize a pie representing a parcel of land and it is owned by six people as tenants in common. Each owner has a one-sixth interest in the entire parcel of land but does not own a particular part of the land.

This chapter deals with two types of community apartment buildings.

(a) Condominiums in which each unit owner has a fee title to his apartment while being a tenant in common with all of the other unit owners of general facilities such as land, hallways, stairs, and elevators.

(b) Cooperative apartment buildings in which each resident is a stockholder of the corporation owning the building and at the same time he is a tenant of such lessor-corporation.

## 9-2 CONDOMINIUMS

The major characteristics of a condominium are as follows:

(a) Individual ownership of an apartment or unit

(b) An undivided interest in certain designated common elements that serve all the unit owners in the condominium

(c) A written agreement among the unit owners regulating the administration and maintenance of the condominium property

We again come to the usual problem of variance in the laws of the different states. The tendency is growing for uniformity of laws throughout the United States. Consequently, there is a lack of uniformity in defining the basic terms and laws that are involved in this area. Generally speaking, the term condominium or condominium project is used to refer to the entire apartment building and grounds.

The person who buys an apartment in a condominium is generally referred to as a unit owner. He owns and receives a warranty deed to the interior surfaces of the perimeter walls, floors, ceilings, windows, and doors. He does not receive a warranty deed to the bearing walls, columns, floors, roofs, foundations, elevator equipment and shafts, central heating system, refrigeration and air conditioning equipment, pipes, conduits, wires, and other utility installations.

The common elements to which the unit owner receives a deed as tenants in common with all of the other unit owners, covers all portions of the condominium property except the units. An example of some of the main items that are owned by the unit owners as tenants in common are the land, foundations, main walls, roofs, halls, stairs, janitor's lodgings, swimming pool, elevators, and garbage and trash incinerators.

## 9-3 STATUTORY PROVISIONS FOR CONDOMINIUMS

The legality of fee ownership of an apartment or of part of a building is now recognized in the great majority of states that have specifically enacted legislation regarding the condominium form of ownership.

The laws of most states provide that in order to establish a condominium it is generally necessary to use three basic documents: the declaration; the individual apartment deed; and the bylaws.

The declaration is the primary instrument by which the property is committed to a condominium plan of ownership. The declaration should contain a legal description of the entire property involved; detailed floor plans of the buildings; a description of the facilities to be owned in common by all of the unit owners; a legal description of each apartment unit; the percentage of ownership interest in the common elements allocated to each unit; and a certificate consenting to the recordation of the declaration signed by the owner of such property and all record holders of interest in the units. The declaration may also provide for an association of owners; a method of sharing the common expenses; and the percentage of votes assigned to each unit in the management of the building.

Ownership of a unit in a condominium is evidenced by individual apartment deeds. These deeds should contain a direct reference to the declaration that has already been recorded, and should contain an accurate description of the property and of the apartment number as set out in the declaration.

There is considerable variety in the various statutes of the different states as to what the bylaws, by which the condominium will be governed, must contain. Usually, the bylaws provide for the management of the project by a board of governors who are elected by the unit owners; and for voting majorities and rules that act as guidelines for the board of governors. Other provisions of the bylaws provide for the power to maintain fire, casualty, liability, workmen's compensation, and other insurance coverage; for employment of personnel necessary for operating the building; and for payment of maintenance, utilities, and other services benefiting the common areas.

## 9-4 COOPERATIVE APARTMENTS

A cooperative apartment house is a multi-unit building in which each owner has the following:

(a) An interest in the entity owning the building

(b) A lease entitling the owner to occupy a particular apartment within the building

During and after World War II, it became common to have cooperative apartment leases running from twenty to fifty years. Some leases were drawn for one year, renewable indefinitely at the option of the owner-lessee unless terminated by the lessor-corporation for cause. In typical American style, the term cooperative apartments has been shortened to coop apartments and even referred to as coops.

Legally speaking, the cooperative apartment project has been characterized as a "legal hybrid." It is a peculiar relationship in that the stockholder-tenant possesses both stock in the corporation as well as a lease with the corporation, acting in its capacity as a landlord. The rights of the unit owner may very well depend on whether or not he is considered to be a lessee or a shareholder.

The purchaser of a cooperative apartment becomes the owner of stock in the cooperative association as well as a tenant of the same corporation.

A cooperative apartment association is ordinarily originated by a promoter who either has a building he wishes to sell or who plans to build or acquire an apartment building. After acquisition, the apartment building is then sold to a cooperative association that issues stock of a total par value equal to the purchase price. Such stock is then allocated among the various apartments according to their estimated relative value.

The purchaser of such shares is entitled to a proprietary lease, or an occupancy agreement, that carries with it the right to occupy a specified apartment.

Among the essentials of a proprietary lease are provisions limiting the maintenance charges per share unless the charge is approved at a stockholders' meeting. Also, provisions to the effect that the lease may not be assigned nor the apartment sublet except with the prior approval of the board of directors.

Ordinarily, nonprofit corporations may be used as the vehicle for an association owning a cooperative apartment house, since such an association makes no pecuniary profit.

The tenants in a cooperative apartment project are the owners of stock but are not entitled to receive a deed for the fractional part of the real estate they occupy. As stockholders they have no more legal or equitable right to the real estate owned by the corporation than stockholders in any other type of corporation owning real estate.

The relationship between the various tenant-stockholders in a cooperative apartment project has been compared with a partnership, even though the legal setup is expressed in corporate terms.

In order to give tenant-stockholders a voice in the selection of other tenants, most cooperative apartment leases provide that the lease and stock of a member cannot be transferred without the approval of the incoming buyer by elected representatives of the remaining members. A tenant-stockholder wishing to dispose of his interest in the cooperative apartment may sell his stock and assign his lease at the best price he can get, *unless* the lease and stock certificate contain restrictions dealing with the right of the association to repurchase the stock, or requiring the approval of the association prior to any transfer.

The relationship between the cooperative apartment association and the tenant-stockholder is that of landlord and tenant to the extent that the association may maintain a summary proceeding for the failure to pay as rent, a proportionate share of the operating expenses of the apartment building. Objectionable conduct on the part of the tenant may be restrained by court injunction proceedings.

Promoters of the cooperative apartment project are entitled to make a legitimate profit on the transaction but are answerable for any fraudulent activity or scheme or misrepresentation.

## 9-5 CONDOMINIUMS VS. COOPERATIVE APARTMENTS

The basic differences between condominiums and cooperative apartment houses are as follows:

(a) In condominiums, individuals take title to their units in the form of a fee title deed, and become tenants in common with the other unit owners on the other areas and facilities that are used by all of the residents.

(b) In cooperatives, individuals buying apartments have stock ownership in the corporation or association, and the right of occupancy of a specific unit.

(c) In condominiums, unit owners vote on a proportionate basis, while in cooperatives, each apartment buyer has one vote regardless of the size of his unit.

(d) In condominiums, unit owners are taxed separately on their apartments, while in cooperatives, apartment buyers pay their share of taxes on the entire project in their monthly carrying charges.

(e) In condominiums, each unit owner is responsible for taxes and mortgage indebtedness on his own apartment, while in cooperatives, the taxes and mortgage indebtedness are billed to the corporation and each apartment buyer is dependent upon the solvency of the entire project.

Accordingly, the unit owner in a condominium has an interest in real property, which upon his death, passes to his heirs as any other realty would. By way of contrast, the tenant stockholder in a cooperative association is the owner of shares of stock, which upon his death, passes as personalty to his representatives. In other words, stock is personal property whereas an interest in real estate as evidenced by a deed is real property.

Another important difference between condominiums and cooperative apartments concerns the unit owners right to sell his interest and to whom and at what price. In the case of a condominium, the other unit owners are frequently given the right of first refusal. This means a unit owner wants to sell his unit and under the provisions of the written bylaws or agreement, the other unit owners have the first opportunity of buying out the selling unit owner's interest. If the other unit owners do not exercise this right, then the selling unit owner is free to sell his interest to any outsider who wants to buy. In the case of a condominium, the selling unit owner is privileged to sell his interest at the market price, which could mean making a profit.

However, in the case of a cooperative apartment, the written bylaws or agreement usually binds the tenant-stockholder to sell his interest back to the corporation for the same amount that he paid. Thus, he is unable to realize a profit even though the value of his interest may have increased. Most country clubs and other private clubs usually operate the same way, where the member receives a certificate of stock in the nonprofit organization. When the member wants to drop out of the country club, he turns in his certificate of stock representing his membership; and when the club is able to find a replacement, the member dropping out then receives the amount he paid to become a member, minus federal tax charges. A recent actual example resulted in a $650 repayment for a country club membership that cost $1,000 originally, and the repayment took about a year.

Frequently, a cooperative apartment lease terminates on death of the tenant-stockholder and the written provisions of the bylaws gives the surviving family a right to occupy the leased apartment for only a limited period of time. Things work out differently in a condominium and the unit owner knows that his family will have a place to live, even after his death. The reason for this protection is that the unit owner's widow or family or heirs will inherit his real estate and they will become the new owners of the unit. Remember, the condominium uses two deeds, one fee simple warranty deed giving total and complete interest to the inner portion of the apartment proper, and a tenancy in common deed for the fringe areas, that is owned together with all of the other unit owners. Whereas, the cooperative apartment arrangement results in a share of stock in the corporation and the buyer now becomes a tenant of the corporation. Legally, there is considerable difference between the two interests.

When you consider the differences, you can readily appreciate that the advantages of the condominium apartment far outweigh its disadvantages as well as the advantages of the cooperative apartment. In the cooperative apartment arrangement, although each tenant-stockholder is supposed to pay his proportionate share of the monthly mortgage payment in the name of the corporation as well as his share of the real estate taxes, his failure to pay creates serious problems. The failure of one tenant-stockholder to pay his share results in a default that must be cured by the other tenant-stockholders if foreclosure on the corporation's property is to be avoided. The financial institution that made the mortgage loan wants their mortgage money every month when it is due, and they are not interested in the sad story that one or more tenant-stockholders had personal problems and were unable to pay their share. The corporation made the mortgage loan in the name of the corporation and the monthly payments become the problem of all of the tenant-stockholders.

In the case of the condominium apartment, no corporation is involved and each unit owner arranges his own mortgage in his own name, and he is not affected by the failure of his neighbor to pay his monthly mortgage payment. The individual unit owner acts on his own and arranges for his own mortgage with payments and schedules to suit his own convenience. If he wants to pay ahead of schedule, he arranges his own terms and conditions.

The conclusions to be drawn are quite obvious. From the buyers point of view, he is better off buying into a condominium apartment arrangement in preference to a cooperative apartment. The condominium apartment buyer has greater protection and flexibility and controls his own fate, much more so than if he bought the same apartment from a cooperative apartment corporation.

# 10

# FORECLOSURES

Foreclosure results in the sale of pledged property to collect a debt that is delinquent. It is the nightmare of all builders and mortgage lenders. When and if the borrower repays his obligation, there is no problem. The problem arises when the borrower fails to pay; certain rights then accrue to the lender.

The mechanics or paper work involved to protect the lender in giving him collateral or security for his loan in the event of the borrower's failure to repay is covered by an instrument known as a mortgage or deed of trust or trust deed. There is no difference between a deed of trust and a trust deed, and they use the identical instrument. There are many important differences between a mortgage and a deed of trust; they will be explained in detail later in this chapter. Both mortgages and deeds of trust have the same purpose, in that they make real estate security for money owed by the owner of the property. Whenever a loan is made with real estate as security, the instrument would be either a mortgage or deed of trust. If personal property were involved, such as an automobile or furniture to be used as collateral for a loan, then the instrument would be a chattel mortgage. Chattel mortgages are for personal property, while a mortgage or deed of trust is for real property which is real estate. As you will see, the methods by which mortgages and deeds of trust accomplish their mission are different.

## 10-1 MORTGAGE

A mortgage is the time-honored method of securing real estate loans. It is a contract in writing whereby real estate is hypothecated to secure a debt or obligation. To hypothecate is to give a thing as security without giving up the right to use it. An example is the home owner who mortgages his home yet retains the use and benefit of it as long as he keeps his part of the bargain by making the monthly payments as provided in the mortgage and real estate mortgage note that goes along with the mortgage instrument. Thus, whenever you refer to a mortgage, there is automatically a second instrument which is a real estate mortgage note, and the note spells out the terms of the real estate loan and the provisions for repayment as well as the rate of interest. By its terms, the real estate mortgage note states whether the note may be paid ahead of schedule without penalty. This is accomplished by the note providing that the borrower may pay two or more payments at any time. Usually, mortgage loans on homes are made for a long period of time, from ten to forty years, and the borrower is a wage earner. Thus, the lender prefers to have the mortgage and note provide for regular monthly payments in preference to quarterly, semi-annual or annual payments. These would create too much of a burden for the borrower. On the other hand, farm loans are tied in with agricultural crops as the prime source of income and frequently provide for annual mortgage payments. In conclusion, a mortgage is a written contract which makes a piece of real property the legal security or guarantee that a loan will be repaid.

## 10-2 INTEREST

Interest is the price paid for the use of money.

A food store sells groceries and earns a profit by buying at wholesale prices and selling at retail prices. Out of the difference or gross profit comes the varied items of expense or overhead, including but not limited to rent, wages, insurance, utilities, and losses caused by shoplifting. After all expenses are paid including the cost of the groceries, the remaining amount is called the net profit.

The builder and contractor needs money in the operation of his business, and he obtains the money from a bank, savings and loan association, insurance company or similar financial institution. Their charge for the use of money is known as interest. In addition, many lending institutions add to the interest charge an "origination fee" for the origination of a new loan; this extra charge is called "points." One point is the same as 1 percent and a charge of two points on a $10,000 loan would amount to $200. This charge is in addition to the interest charge, and must be included in the cost of borrowing money. Many builders have learned by bitter experience that borrowing money is one thing and repaying the borrowed money, plus the interest and point charge is another.

What is a fair rate of interest? How much can a builder or contractor afford to pay for the use of money that he requires in his building operation? One of the more serious pitfalls leading to the downfall of builders is caused by high rates of interest and other charges that he just cannot afford to pay. If the builder is being charged a high rate of interest and points, while his net profit in selling the completed home is limited and does not include the cost of his borrowed money, then he is doomed to failure. The more houses he builds, the more money he will lose.

Frequently, the novice builder and contractor is overawed by the fact that some

**This Indenture,** made this ........ day of ........ December ........, 19........,

by and between ........ JOHN JONES and MARY JONES, his wife ........

........,

hereinafter, whether singular or plural, masculine, feminine, or neuter, designated as "Mortgagor," which expression shall include mortgagors, jointly and severally, and shall include Mortgagor's heirs, executors, administrators, assigns, and successors in interest, and ........ RICHARD ROE and/or JANE ROE, his wife ........

........,

hereinafter, whether singular or plural, masculine, feminine, or neuter, designated as "Mortgagee," which expression shall include Mortgagee's heirs, executors, administrators, assigns, and successors in interest, WITNESSETH:

WHEREAS, the Mortgagor is indebted to the Mortgagee and as evidence thereof has made, executed, and delivered to the Mortgagee s ........ certain note ........ of even date herewith, described as follows, to-wit:

WHEREAS, the Mortgagor desires to secure the payment of said indebtedness;

NOW, THEREFORE, in consideration of the premises and of the sum of One Dollar ($1.00) to it in hand paid, the receipt whereof is hereby acknowledged, the Mortgagor has granted, bargained, sold, and conveyed, and does hereby grant, bargain, sell, and convey, unto the Mortgagee forever the following real estate, situate, lying, and being in the County of ........ Bernalillo ........, State of New Mexico, to-wit:

Lot numbered Ten (10) in Block numbered Eight (8) of TIJERAS PLACE, an addition to the City of Albuquerque, New Mexico, as the same are shown and designated on the Plat of said Addition filed in the office of the County Clerk of Bernalillo County, New Mexico, on the 24th day of August, 1923.

Together with all and singular the lands, tenements, privileges, water rights, hereditaments and appurtenances thereto belonging or in anywise appertaining, and the reversion and reversions, remainder and remainders, rents, issues, and profits thereof, and all the estate, rights, title, interest, claim, and demand whatsoever of the Mortgagor, either in law or equity, of, in, and to the above bargained premises; TO HAVE AND TO HOLD said premises above bargained and described, together with all and singular the lands, tenements, privileges, water rights, hereditaments and appurtenances thereto belonging or in anywise appertaining, and the reversion and reversions, remainder and remainders, rents, issues, and profits thereof, and all the estate, right, title, interest, claim, and demand whosoever of the Mortgagor, either in law or in equity, of, in, and to the above bargained premises, to the Mortgagee forever as security for the faithful performance of the aforesaid promissory note, and each and all of them if there be more than one, and as security for the faithful performance of each and all of the covenants, agreements, terms, and conditions of this Mortgage.

The Mortgagor hereby covenants and agrees with the Mortgagee as follows:

1. That, at the time of the ensealing and delivery of these presents, the Mortgagor is well seized of the premises above conveyed in fee simple, and has good right, full power, and lawful authority to grant, bargain, sell, convey, and mortgage the same in manner and form aforesaid, and that the same are free and clear from all former and other grants, bargains, sales, liens, taxes, assessments, and encumbrances of whatsoever kind and nature, except

and that the Mortgagor shall and will forever warrant and defend the Mortgagee's quiet and peaceable possession of the same against the lawful claims and demands of all persons, except as in this paragraph stated;

2. That the Mortgagor shall promptly pay and otherwise perform all amounts and obligations, as provided in the obligation secured hereby, or any renewal or extension thereof, and in the manner, form, and at the time or times provided in said obligation or in any renewal or extension thereof; and that the Mortgagor shall promptly pay all such additional sums as may hereafter be advanced to the Mortgagor or expended by the Mortgagee on behalf of the Mortgagor for any purpose whatsoever and evidenced by notes, drafts, open account, or otherwise, together with interest thereon at rates to be fixed at the time of advancing or expending such additional sums; provided, however, that the making of such advances or expenditures shall be optional with the Mortgagee; that this mortgage shall secure the payment and performance of all renewals or extension of the obligation secured hereby and shall secure the payment and performance of all such additional sums as may hereafter be advanced to the Mortgagor or expended by the Mortgagee on behalf of the Mortgagor for any purpose whatsoever and evidenced by notes, drafts, open account, or otherwise together with interest thereon, and for all of which this mortgage shall stand as continuing security until fully paid; that the Mortgagee may apply any payments made on any indebtedness secured hereby, at its option, on any such indebtedness;

3. That the Mortgagor shall perform the conditions of any prior mortgage, encumbrance, condition, or covenant;

4. That the Mortgagor shall pay when due and payable all taxes, charges, and assessments to whomsoever and whenever laid or assessed upon the mortgaged premises or on any interest therein; that the Mortgagor shall pay when due and payable all rent, charges for electrical, gas, sewage, water, and all other utility and other charges, fines, or impositions, and all laborers', mechanics', or materialmen's or other liens of whatsoever kind that may be laid or assessed upon the mortgaged premises or on any interest therein;

5. That the Mortgagor shall, during the continuance of any of the indebtedness secured hereby, keep all buildings and other destructible improvements now existing or hereafter erected on the mortgaged premises in good order, condition, and repair at Mortgagor's own expense and shall not commit or suffer any strip or waste of the mortgaged premises;

6. That the Mortgagor shall keep the buildings and all other destructible improvements now existing or hereafter erected on the mortgaged premises insured in the sum specified by the Mortgagee from time to time but in no event less than the total amount of the indebtedness secured hereby and against the hazards of fire and those covered by extended coverage insurance and against such additional hazards as the Mortgagee may from time to time specify, for the benefit of the Mortgagee, its executors, administrators, successors, and assigns, such insurance to be in such form and in such insurance companies as the Mortgagee shall approve; that the Mortgagor shall pay when due and payable all premiums for such insurance, shall deliver such policy or policies of insurance to the Mortgagee, and shall deliver to the Mortgagee at least two days prior to the expiration of any policy on such premises a new and sufficient policy to take the place of the one so expiring; that in the event of loss or damage, Mortgagor shall give immediate notice by mail to the Mortgagee, who may make proof of loss or damage, if not made promptly by the Mortgagor; and that each such insurance company concerned is hereby authorized and directed to make payment for such loss or damage directly to the Mortgagee alone, and not to the Mortgagor and Mortgagee jointly, and the insurance proceeds, or any part thereof, may be applied by the Mortgagee at its option to the reduction of the indebtedness secured hereby or to the restoration and repair of the property damaged or lost;

7. That in the event of the failure or refusal of the Mortgagor to keep in good order, condition, and repair all buildings and other destructible improvements now existing or hereafter erected on the mortgaged premises at Mortgagor's own expense, as in Paragraph 5 above provided; or to keep the premises insured or to deliver the policies of insurance, as in Paragraph 6 above provided; or to pay all taxes, charges, and assessments, and all rent, charges for electrical, gas, sewage, water and all other utility and other charges, fines, or impositions, and all laborers', mechanics', or materialmen's or other liens of whatsoever kind that may be laid or assessed upon the mortgaged premises, as in Paragraph 4 above provided; or to perform the conditions of any prior mortgage, encumbrance, condition, or covenant, as in Paragraph 3 above provided, the Mortgagee, and its executors, administrators, successors, or assigns may, at its option, make, do, perform, and/or pay all such things, and all moneys thus paid and/or all such expenses and costs thus incurred shall bear interest at the rate of ten per centum (10%) per annum and shall be payable by the Mortgagor on demand and shall be so much additional indebtedness secured by this mortgage;

8. That the Mortgagee shall have the right at any time by it deemed necessary to incur the expense of procuring and/or continuing an abstract of title to the said mortgaged premises, or other showing of title, and any expense so incurred shall bear interest at the rate of ten per centum (10%) per annum and shall be payable by the Mortgagor on demand and shall be so much additional indebtedness secured by this mortgage;

9. That, if there is any default in or breach of any covenant, term, condition, or agreement of this mortgage or of any indebtedness secured hereby, all indebtedness secured by this mortgage, whether the same shall be due and payable according to the tenor and effect thereof or not, and anything herein to the contrary notwithstanding, shall, at the option of the Mortgagee, immediately become due and payable without notice to the Mortgagor of the exercise of such option. Upon the happening of such event, this mortgage shall be subject to foreclosure at the option of the Mortgagee, and the premises may be sold in the manner and form prescribed by law; that, in the event of any sale hereunder, the Mortgagee may become the purchaser of the mortgaged premises or any part thereof and shall be entitled to a credit on the purchase price in the amount of its interest in the proceeds of such sale.

10. That, if there is any default in or breach of any covenant, term, condition, or agreement of this mortgage or of any indebtedness secured hereby, the Mortgagee shall have the right, and it is hereby so authorized, to take possession of the mortgaged premises and collect and receive the rents, issues, and profits thereof, and the Mortgagee shall have the right to apply the residue of the same, after deducting all charges and expenses of collection, to the payment of all sums due hereunder or due upon any indebtedness secured hereby; that, whenever any application to any court or referee shall be made to compel the payment of any indebtedness secured hereby or to foreclose this morgage, the Mortgagee shall have the right and is hereby entitled to the immediate appointment of a receiver, without notice to the Mortgagor, to take possession of the mortgaged premises and to collect and receive the rents, issues, and profits thereof and to apply the residue of the same, after deducting all charges and expenses of collection to the payment of all sums due hereunder or due upon any indebtedness secured hereby, under the direction of the court or referee appointing such receiver; and that the right to the appointment of such a receiver shall not be dependent upon the solvency or insolvency of the Mortgagor or upon the value of the mortgaged premises.

11. That the Mortgagor will pay to the Mortgagee in addition to all of the other indebtedness secured hereby ten percent (10%) of the total amount due hereunder, including all indebtedness secured hereby, as attorney's fees whenever any applications to any court or referee shall be made to compel the payment of any indebtedness secured hereby or to foreclose this mortgage, and the amount of such attorney's fees shall be so much additional indebtedness secured hereby.

12. That no waiver of any obligation hereunder or of the obligation secured hereby shall at any time be held to be a waiver of the terms hereof or of the indebtedness secured hereby. That no sale of the mortgaged premises and no forbearance on the part of the Mortgagee and no extension of time for the payment of the indebtedness secured hereby given by the Mortgagee shall operate to release, discharge, modify, change, or affect in any way the original liability of the Mortgagor.

13. That all of the grants, covenants, terms, conditions, and agreements hereof shall be binding upon and inure to the benefit of all of the heirs, executors, administrators, assigns, and successors in interest of the parties hereto.

14. If this mortgage is foreclosed the redemption period after judicial sale shall be one month in lieu of three lonths.

IN WITNESS WHEREOF, the Mortgagor has executed this indenture the day and year first above written.

JOHN JONES

MARY JONES

ACKNOWLEDGMENT FOR NATURAL PERSONS

STATE OF NEW MEXICO } ss.
COUNTY OF BERNALILLO

The foregoing instrument was acknowledged before me this ........ day of December, 19......,

by John Jones and Mary Jones, his wife
(Name or Names of Person or Persons Acknowledging)

My commission expires:
(Seal)
Notary Public

ACKNOWLEDGMENT FOR CORPORATION

STATE OF NEW MEXICO } ss.
COUNTY OF..........

The foregoing instrument was acknowledged before me this ........ day of ........, 19......,

by ........ (Name and Title of Officer), ........ of ........ (Corporation Acknowledging)

........, a corporation, on behalf of said corporation.

My commission expires:
(Seal)
Notary Public

financial institution is willing to lend him a sum of money for construction purposes that impresses him as being staggering. To a large financial institution establishing a maximal credit limit which is called a "line of credit," the sum of $100,000 or even $250,000 is not a spectacular sum of money. A line of credit signifies that the builder may use this agreed sum of money in connection with his building program, and interest is charged from the time the money is used until it is repaid. Yet the inexperienced builder has often not yet learned that the mere ability to borrow sums as large as our $250,000 example does not insure his success. To some builders, this sum of money gives them ideas of grandeur and results in unnecessary purchases.

Needless to say, there comes the day when borrowed money has to be repaid, and the day of reckoning must be planned long in advance. The excessive spending of money that was borrowed for construction purposes on personal items that were nonessential has caused serious financial problems for many inexperienced builders and contractors. In many cases, if the interest and point rate is too high, the small builder is doomed to failure before he gets started, even though it takes some time until he discovers his plight. In other words, his profit in the sale of the completed home is insufficient to cover the high cost of his borrowed money.

Other financial problems for the contractor or builder who is either seeking to borrow money from a financial institution or attempting to repay money already loaned to him by such an institution may be caused not by his own individual business practices but by the condition of the general national economy. One possible example is that a so-called "tight money" situation may prevail from time to time. In this case, it may be difficult for a beginning contractor to negotiate a loan of sufficient size to meet his initial expenses.

During such periods financial institutions become less willing to make large loans to business enterprises that are not already well established. An even more common problem is that inflationary trends may reach the money market, drastically reducing the actual buying power of money already borrowed under more stable conditions. A perceptive understanding of the state of the national economy can aid the beginning contractor or builder in deciding on the proper timing for a certain business venture; many potential crises can thereby be avoided or minimized.

Accountant's fees, like those of the attorney, are a normal item of overhead to the builder and contractor. All items of overhead must be taken into consideration in order to compute the cost and proposed selling price of the completed home. The professional services of a certified public accountant are vital to the builder and contractor of limited experience in determining cost and selling prices. More small builders have become involved in financial difficulties and even foreclosures because they did not realize that they simply could not afford the high rate of the cost of their borrowed money, and they failed as a direct result.

## 10-3 PARTIES

Just two parties are involved in a mortgage, the property owner who borrows, and the lender. The borrower is called the mortgagor, and the lender is called the mortgagee. The borrower or mortgagor is not limited to a single entity or quantity. The borrower may be one person, a married couple, a group of individuals, a partnership, a corporation or other entities. What happens if the mortgagor or borrower dies or absconds? The mortgagee or lender has the choice of filing his claim against the estate of the deceased mortgagor or of looking to the mortgaged property to secure his loan balance. If the mortgagor absconds, the mortgagee would look to the mortgaged property as his protection, and, if necessary, would foreclose.

Many corporations are formed by builders and contractors to cover specific building projects for tax benefits. Thus, the builder and contractor would be well-advised to retain his attorney and accountant early in his business career. *The best time to seek advice is before the problem has arisen rather than afterwards.* The attorney representing the builder or contractor can do a far more efficient job if he is consulted regularly on daily problems in advance of the final decision. In particular, the best time to consult with your attorney is before you sign a contract rather than afterwards.

Likewise, tax implications are becoming more important than ever before, and your accountant can do a better job in your behalf if you consult him before you make your land purchase or enter into a binding contract for the sale of land.

Many financial institutions have found from experience that they are better off when making loans to your newly formed building corporation, to require that you and your wife also sign the note and mortgage, or both, in your personal capacity, creating personal liability as well as corporate liability. If, for example, the note is signed John Jones Corporation by John Jones, president, and Mary Jones, secretary, and the corporation seal is added, this creates corporate liability only, and the money lender can only look to the assets of the corporation for his money. If the note is signed John Jones Corporation by John Jones, president, and Mary Jones, secretary, the corporate seal is added, and *then* the note is signed by John Jones, individually, and Mary Jones, individually, the money lender can now look to the personal assets of John Jones and Mary Jones as well as to the corporate assets.

When dealing with a bank or similar type of financial institution, it behooves the contractor or builder to become familiar with their lending policies and requirements before applying for a loan to be used for construction. Many financial institutions require a certified financial statement, that is, one that has been prepared by a Certified Public Accountant, who certifies as to the accuracy of the statements and figures contained in the financial statement. The reason for this requirement is to avoid the situation where an overly ambitious or optimistic builder decides to place a valuation on some acreage that he owns that is not realistic with the current market value of his land. In years past, when financial institutions accepted financial statements that were not certified and represented puffing values by the builder, they would compensate for his optimism by reducing his values by 25 percent to 50 percent.

If you walk into your bank and inquire about a loan, your banker will immediately ask for a current and detailed financial statement. The banker will even supply you with a blank form designed by his bank, asking not only for detailed information as to your assets and liabilities, but also how much federal income tax you paid last year and your estimated tax liability for the current year. The field of taxes and tax liability has become very important in connection with the net worth of the builder and contractor, and your accountant's assistance is truly mandatory.

Thus, you should enter your bank prepared, and have your current certified financial statement available to leave with your banker. He is bound to be impressed with your efficiency and knowledge of establishing your credit limits with your bank. The information contained in your financial statement and the source of the information are equally important. Your statement that you purchased five acres of land that is valued at $10,000 per acre is neither accurate, nor acceptable to your bank. The bank will want to know when you purchased the five acres of land; how much you paid for it; whether the title is clear; do you owe money on your land purchase; has it been appraised and, if so, who was the appraiser; do you have a copy of the appraiser's report; what is the zoning regulation covering your five acres; and other detailed information.

The process by which anyone arrives at a judgment of value is called an "appraisal." The figure arrived at is called the appraised value. The person who renders or performs this technical service is called an appraiser. A city or municipality does not employ its own official appraiser as it would for most other public officials. Whenever there is need for the city to determine value of land, it would hire an appraiser to examine the land involved, ascertain recent sales of adjacent land of comparable value, determine the replacement cost of building improvement, minus their depreciation, and finally issue his written report as to his professional opinion as to the current market value of the land involved in the appraisal. The art or science of appraising is not exact and represents one person's opinion, even though he is classified as an expert appraiser. Frequently, there is considerable difference of opinion between qualified appraisers, as to the value of a given parcel of land. The builder and contractor would hire an appraiser in his own behalf, just as does the city, county, state, federal government, or other entity. Appraisers are usually qualified real estate brokers who have had years of experience and have studied modern procedures and techniques, and have qualified as certified appraisers of real estate.

Most urban communities have found a need for all land within its confines to be zoned. This means that the city or county has control of the use of land within its boundaries, and also the power to limit property to specific use. Good zoning regulations improve over-all values in the area. R-1 is usually the designation for single family dwellings. R-2 could be the zoning designation for up to four family units, and R-3 could be multiple or apartment units. Obviously, if our five acres of land was purchased in a nice residential area, we would not want an apartment house or junk yard to be built adjoining our acreage. Commercial zoning usually starts with the letter C and C-1 is more restrictive than would be C-3. Zoning regulations are an important factor to take into consideration. One must ascertain the exact zoning of a parcel of land before entering into any agreement to purchase same.

Most municipalities have a building code which provides for laws or ordinances requiring sound construction for protection of health and safety. Building restrictions and local zoning laws may affect property values favorably or unfavorably. Nevertheless, these requirements are a necessary evil and eventually are beneficial to all who are subjected to them. Many building codes require that a residence not be built on a lot smaller than 5,000 square feet. If we did not have this requirement, some small lot owner would build a small home on a 2,500 square foot lot, and the appearance of the neighborhood would be adversely affected. Under the category of health and safety, a usual requirement makes it mandatory for all electrical and plumbing work to be accomplished by a licensed electrician as well as a licensed plumber.

The mortgagor or borrower does not transfer the title of his property to anyone. The debt is evidenced by a real estate mortgage note signed by the mortgagor, and this is merely a promissory note wherein the borrower promises to pay the borrowed money back at so much per month with interest at so much. The note refers to the mortgage into which the parties have entered. The mortgage contract makes the property the security for the payment of the note. The real estate mortgage note is not recorded; the mortgage instrument or contract is recorded with the county clerk or comparable public official, who officially stamps it showing the date, time of day, and the book and page in which it will be entered. It is important that the mortgage be recorded promptly, because the recording will constitute notice to the world that the lender has a claim against the borrower's home or property as evidenced by the mortgage contract. The law books are filled with cases where the lender was negligent in failing to promptly record his mortgage and meanwhile the borrower went to a second lender who had no knowledge of the first transaction, loaned money and took a mortgage on the same land, and the second lender promptly recorded his mortgage. This resulted in the second mortgage acquiring priority over the first mortgage. If the first mortgage holder or lender had recorded his mortgage promptly, the recordation or placing of record constitutes notice to the world of his transaction. Had he done so, the second mortgage holder or lender, in checking the records of the county clerk, would have been able to ascertain knowledge of the first mortgage. When the first mortgage holder failed to act or record promptly, and the second mortgage holder did act and record promptly, the second mortgage holder acquired first position, ahead of the first mortgage holder. An old legal saying of the advantage of recording any instrument is that "prior in time is prior in right." A person that fails to record his mortgage promptly is guilty of a cardinal sin and would lose out to a secondary obligation, lien or mortgage that was recorded first. The subject of liens is covered in detail in Chapter 8.

When the mortgagor or borrower repays his obligation in full, he naturally wants his recorded mortgage released on the records of the county clerk. Most jurisdictions require the mortgagee or lender to file a Satisfaction or Release of Mortgage with the same county clerk or comparable public official that handled the recording of the original mortgage. If the mortgagee failed to release the recorded mortgage promptly after being paid in full, a reminder from the mortgagor would ordinarily accomplish the mission.

## 10-4 MORTGAGE NOTES

Real estate mortgage notes are barred or outlawed by the Statute of limitations four years after they are due and payable. The Statute of limitations is a state law limiting the time in which certain court actions may be brought. If a person expects to secure relief by court action, he cannot wait an unreasonable time to start his suit. The law sets certain time limits. In the case of a note, including our real estate mortgage note, the time limit set is four years from the date of last payment or transaction. If a lender delays in taking action to recover his money until after that time, he cannot collect by court action. It should be born in mind that most mortgage notes provide for monthly payments and are payable over a period of many years, usually from ten to forty years. Thus, the four year period as provided by the Statute of limitations only begins to run from the date of last payment or transaction and not from the original date of the note. For example, if the mortgagor falls eight months behind in his monthly payments and the mortgagee wants to help him, then the parties might agree that the mortgagor will pay back interest only for the eight month's delinquency and regular monthly payments thereafter as provided in the real estate mortgage note. The four-year Statute of limitations period would then begin to run with the date of the interest payment and not from the date of the note. The fact that the original note is ten years old is of no consequence.

What happens to physical possession of the original note itself? We have already ascertained that the mortgage was recorded and is now a matter of public record. If the mortgage instrument itself is now lost, it is not very serious because the records of the county clerk are available to obtain a certified copy. The usual charge for a certified copy of any recorded instrument is $1.00 per page and our four-page mortgage would cost $4.00 and be available in a day or so. Loss of the original real estate mortgage note could create serious problems; it is therefore kept in a safe place. Accordingly, the original note is usually left with your bank or similar financial institution for collection as well as safekeeping. Another advantage of leaving the original mortgage note for collection with your bank is that the bank keeps your records for you for a nominal charge. Each month that the mortgagor makes his monthly mortgage payment to your bank, you will promptly receive a small printed notice advising the date and amount of the payment; the portion that was applied to interest, the portion that was applied to principal, and the resulting new balance. At the end of the year, the twelve monthly payment slips will furnish the necessary information needed for your income tax return. If your mortgagor fails to pay promptly, you automatically know this by the absence of the monthly payment slip from your bank. It is now your responsibility to contact your mortgagor and dun him for his overdue payment because your bank is only acting as an "escrow agent" and not as a collection agent. In other words, your bank does not send out delinquent payment notices to your mortgagor when it is acting as escrow agent and charges about 75 cents per month for the bookkeeping chore of handling the monthly payment. An escrow agent is a neutral party, in this case selected by the mortgagee to collect and disburse the monthly payments as they are made by the mortgagor. When and if the mortgagor pays, the bank acting as escrow agent, follows its instructions and disburses the money, usually to the credit of the mortgagee. However, if the mortgagor does not pay as he agreed, this is the problem of the mortgagee and not the escrow agent.

A mortgagee or lender may sell his real estate mortgage note and mortgage to another without first obtaining the permission of the borrower. The mortgagee then assigns the obligation of the mortgagor by endorsing the real estate mortgage note over to the new owner or purchaser. This happens frequently and is referred to as the sale of negotiable paper.

What about the mortgagor? May he sell or assign his equity in the real estate mortgage note? In other words, may the mortgagor sell his home while it is still mortgaged? To obtain the answer we must turn to the mortgage contract to ascertain its provisions. Some mortgage contracts provide that sale of the home by the mortgagor may only be accomplished by paying the mortgage balance in full. This would force the incoming buyer to refinance the mortgage obligation, either with the same mortgagee or a new lender. Most mortgage contracts provide that the mortgagor may sell his home but that his personal liability still remains under his real estate mortgage note. In this situation, if the incoming buyer failed to pay the monthly mortgage payments as agreed, the mortgagee would be privileged to demand payment from the original mortgagor and in turn, the original mortgagor would have his legal claim against the incoming purchaser.

If the mortgagee thinks that the incoming purchaser is as financially strong or stronger than the original mortgagor, he will agree to an assignment from the original mortgagor in favor of the incoming purchaser. In this case, the mortgagee is agreeing to substitute the parties, and the original mortgagor is released from any liability on his real estate mortgage note and the incoming purchaser now assumes the liability. Obviously, the parties can always change the terms of the original mortgage contract if they both agree.

What about paying ahead of schedule? May the mortgagor pay off the entire remaining mortgage balance at his pleasure? To determine the answer, we must turn to the real estate mortgage note. If the note provides for monthly payments of $100, or more, then the mortgagor is given the choice of paying his obligation ahead of schedule. If the note is silent in this regard, and does not provide "or more" or a similar provision, then the mortgagor may not pay his obligation ahead of schedule.

In some cases, the note provides for a penalty for the privilege of lump sum payment ahead of schedule. The note would be silent as to lump sum payment, and the mortgage contract provides for a penalty of 2 percent of the remaining balance if the mortgagor desires to pay the remaining balance in a lump sum. Thus, if the mortgage balance was $25,000 the mortgagor would have to pay $500.00 for the privilege of paying the remaining mortgage balance in a lump sum. In other cases, the mortgage contract provides that the mortgagor may pay ahead of schedule only to the extent of 20 percent per year of the remaining balance. This would result in a five year payoff period. Sometimes, private mortgagees provide that the mortgagor may not pay his mortgage obligation ahead of schedule because the mortgagee has spent considerable time and effort in investigating the mortgage loan, and he wants his money repaid only as provided in the mortgage contract and payment schedule. From the viewpoint of the mortgagor, it is to his advantage to have the "or more" payment privilege even if he does not use it. This choice may be of interest to an incoming purchaser, or refinancing may offer better terms than those provided in the original mortgage.

If all mortgages were paid according to terms, we could drop the subject. Unfortunately, mortgagors do not always do what they are supposed to do and problems result. Mortgagors are human and meet with various types of misfortune and illness. If they are unable to pay as agreed, the mortgagee or lender is required to foreclose against the secured property in order to collect his debt.

## *10-5 MORTGAGE FORECLOSURE*

To foreclose means to file a lawsuit in court to sell the mortgaged home or property to pay off the loan, interest, attorney's fees and costs of the lawsuit. This is done under a special section of the statute or law which sets forth the exact procedure required. Each state or jurisdiction has its own statutes; however, generally speaking, the provisions would be similar. If the mortgage note has been sold or assigned to someone else, the purchaser can institute the foreclosure lawsuit.

Your mortgagor is now five months in arrears on his monthly payments, and you have been unsuccessful in your efforts to get him to pay. You now need help, and your attorney is the proper person to file your foreclosure lawsuit. He will need detailed information regarding your entire transaction with your mortgagor, including the current balance; the date of last payment; your recorded mortgage which will show the date, book and page, and other detailed information in connection with the recording of your mortgage. Your attorney will also need the original note which you will have to remove from your bank's collection department. Now that your attorney has this necessary information, he now needs to know the status of the mortgaged home or property from the time your mortgage was recorded onward. He notifies the title company and requests a title search of the legal description that covers the home or property involved in your mortgage loan. The title company searches the records and issues its written report to your attorney. The title search report indicates that after your mortgagor borrowed money from you, he went elsewhere and created additional liability. For example, he ordered a new hot water heater and other plumbing services for a total of $500 and failed to pay his obligation, and the plumber filed a mechanics lien against the mortgaged property. He also failed to pay last year's taxes in the sum of $500. He paved a portion of the backyard area and failed to pay the mason the $300 agreed price, so the mason is threatening to file a lawsuit for payment of his bill. Now your attorney has the necessary information, he is ready to file the foreclosure lawsuit, and he will name the mortgagor and all of the other claimants that make some claim adverse to your claim to clear title. The mortgagor and all of the other defendants are served with copies of the foreclosure complaint and eventually the defendants involved in your lawsuit either default or contest the lawsuit, causing the case to come to trial. The court rules in your favor as against your mortgagor by finding that your mortgage is valid, that he failed to pay as agreed, and that you are entitled to foreclose against his interest in the home or property involved in the mortgage. The court further finds that the mechanics lien filed by the plumber for his $500 bill is secondary to your mortgage because the mortgage was recorded properly and is prior to the plumber's claim. The court further finds that the plumber's $500 claim is a valid one and awards the plumber a $500 judgment against the mortgagor personally and not against the mortgaged property. The court further finds that the mason's $300 bill falls in the same category as the plumber's with the same result. The real estate taxes in the sum of $500 represents an exception to the rule, even though the tax liability was created after your mortgage was properly recorded. The $500 tax claim is not against your mortgagor or even you but is against the mortgaged property and may not be defeated by a prior mortgage because it attaches to the land. Thus, even though you appeared to be declared a winner in your foreclosure lawsuit against your mortgagor and the plumber and mason, you are a loser as to the $500 tax liability. You or your successor in interest will have to pay the $500 tax bill, plus the usual penalties and interest for late payment.

Many financial institutions that make mortgage loans in large quantities have learned how to solve the problem of unpaid taxes. Instead of gambling that the mortgagor will pay his annual tax bill and also pay his annual premium for fire insurance on the mortgaged house, the mortgage contract provides for payment of monthly "impounds" to cover 1/12th of the tax bill and insurance premium. For example, the annual taxes are $600, and the annual fire insurance premium is $180. By dividing these amounts by twelve, we arrive at $65 which is one twelfth of the total. At the end of the year, the mortgagor has accumulated the $780 needed to pay the $600 tax bill and the $180 fire insurance premium. This $65 monthly impound payment is added on to the regular payment provided for in the mortgage to pay the interest and apply on the principal. The mortgagor is pleased with this arrangement because when he receives his tax and insurance bill at the end of twelve months, he gives them to his mortgagee who pays them with the monies accumulated by the impound. The mortgagee is willing to take care of these details in return for the assurance that no problems will result on taxes and insurance.

Your attorney has now completed the judgment in connection with your pending foreclosure lawsuit, and he now obtains a court order for the sale of the property involved. The sale is a public sale and is conducted by the sheriff or "special master" appointed by the court. Anyone may attend the sale and bid; however, the court order provides that all bidders other than the mortgagee who filed the lawsuit must cover their bid with a "cashier's check" for the amount of their bid. A cashier's check is issued by a bank and represents a bank check rather than a check issued by an individual. The reason the mortgagee is excepted from posting a cashier's check for his bid is that he already has been awarded a judgment against the mortgagor for the mortgage balance, plus interest, attorney's fees and costs, and the court order provides that the mortgagee is entitled to credit for his judgment award as against his bid. An example may prove helpful to the reader. The court judgment awarded the Mortgagee $25,000 principal balance; plus $1,000 interest; plus $2,500 attorney's fees; plus $500 costs, for a total of $29,000. The mortgagee is present at the public sale and bids $29,000, the full amount of his judgment. He need not post any cashier's check because he is the creditor, and the judgment provided that he is entitled to credit on his bid to the extent of his judgment award. If any outside bidder bid $29,500, he would be required to put up a cashier's check for the $29,500. What would happen to the $500 in excess of the judgment? It would go to the mortgagor. This happy result is a rarity similar to the finance company selling the repossessed car for a sum greater than the balance due the finance company.

## *10-6 DEFICIENCY CLAIMS*

The public sale of the mortgaged property resulted in the mortgagee being the only and high bidder in the sum of $25,000 which is $4,000 less than the judgment award. May the mortgagee obtain a $4,000 deficiency judgment against the mortgagor, in addition to recovering the mortgaged property? Yes, under some circumstances but not if the mortgage was a "purchase money mortgage." A purchase money mortgage is one given as part of the purchase price when buying property. In the latter situation, the mortgagee must look to the foreclosed property only to recover his mortgage loan balance. As a practical matter, mortgagors who are foreclosed against are not in good financial condition, and it would therefore be difficult to collect a deficiency judgment.

## *10-7 EQUITY OF REDEMPTION*

The public sale resulted in the house or property being sold to the highest bidder, and usually this is the mortgagee. Even though the mortgagee was the high bidder at the public sale, the mortgagor is still in possession of the house and has certain rights and obligations. The mortgagor has the right to retain possession of the house and reside in it, however, he is obliged to pay a reasonable value for rent to the high bidder, be it the mortgagee or other person. The mortgagor may legally sell his "equity of redemption" to another. Even though the mortgagor did not pay as agreed, and his actions forced the mortgagee to file a foreclosure lawsuit and obtain a court judgment of foreclosure, the mortgagor is granted a right by law of redeeming his house or property by paying the mortgagee's judgment in full. This right is known as the mortgagor's "equity of redemption." The period of time during which the mortgagor may redeem varies with different jurisdictions. California law provides for a one-year period of redemption from the date of the public sale. This is the main reason why California real estate lenders dislike mortgages and much prefer to use a deed of trust also known as trust deed. Under the provisions of a deed of trust, the borrower who is known as the trustor has no right of redemption after the sale.

Many institutional lenders such as insurance companies, savings and loan associations, banks, and so forth refused to make real estate loans where the local law provided for redemption rights to the mortgagor covering a long period of time. As a result of this pressure and the need for mortgage loan money to be made available, many States enacted legislation providing that the period of redemption for the defaulting mortgagor could be reduced by written agreement of the parties from one year to a lesser period of time; in one case, down to one month. Thus, one can see the conflict between the parties; the borrower wants the best terms available, and the lender wants the greatest amount of protection in the event of default on the part of the mortgagor. As you can well imagine, the struggle is uneven, the lender usually prevails because he is in the driver's seat.

We have observed that the rights of redemption of the mortgagor are contingent upon his tendering the entire remaining balance to the mortgagee or other high bidder at the foreclosure sale. There is nothing to prevent the parties from making a new contract, and the mortgagee may permit the mortgagor to pay the delinquent payments and expenses of foreclosure and resume their former relationship.

## 10-8 TRUST DEEDS

We had previously learned that a mortgage and a trust deed basically accomplish the same mission, wherein both instruments are used as collateral or security of property for the real estate loan, but differ materially otherwise. In a mortgage, the mortgagor or borrower makes the property security for the loan; whereas, in the case of a trust deed the borrower or owner actually deeds title to his land to a third party for him to hold until the loan is paid. The third party, who is in effect a stake holder, is called the trustee.

The borrower, who under a mortgage was called the mortgagor, is now called the trustor under a deed of trust. The lender who was called the mortgagee under the mortgage contract, becomes the beneficiary under the deed of trust. The mortgage contract involved two parties, namely the borrower or mortgagor and the lender or mortgagee whereas now, under the deed of trust, there are three parties. The borrower becomes the trustor; the so-called stake holder becomes the trustee; and the lender becomes the beneficiary.

Unlike the mortgage, the law does not specifically provide for the creation of trust deeds. They came about through use. Trust deeds originated because lenders sought a method whereby they could foreclose more quickly without court action and consequent delays with expense involved. They also did not like that one-year equity of redemption that was granted to the mortgagor under the mortgage contract. Lenders were always unhappy about the foreclosed property being tied up and not saleable during the one-year redemption period. Not only was their loan money tied up, but their prospect of selling the property during the redemption period was practically nil.

In defense of the use of trust deeds over mortgages, lenders argue that they can risk more liberal loans if they can recover their loans more quickly in case of default. The great preponderance of real estate loans and financing in the State of California is secured by trust deeds.

The trustee or stakeholder is able to foreclose in case the borrower or trustor defaults because the borrower has given him the "power of sale." The trust deed he signed provides for it. Power of sale, as the name indicates, is a power or right given to the trustee by the trustor to sell the property in the event the borrower or trustor fails to pay as agreed and defaults.

In the case of a mortgage, the loan is evidenced by a note. Similarly, in the case of a trust deed, the loan is evidenced by a note. The trust deed provides that the property shall be the security for payment of the note. A note is a written promise to pay a certain sum of money under a certain payment schedule, at a certain rate of interest and with other similar provisions.

If the borrower or trustor under a trust deed pays off the loan, the lender or beneficiary requests the trustee or stakeholder to deed back the property to the borrower. The trustee draws up and signs a deed of reconveyance that clears the trust deed from the records when it is recorded.

## *10-9 TRUST DEED FORECLOSURE*

Should the borrower or trustor default, that is, not pay as provided in the note, then and in that event, the lender or beneficiary gives the trustee or stakeholder a written notice stating that the borrower has defaulted.

The trustee then notifies the borrower that he is in default in certain respects, and that the beneficiary chooses to sell the property or house to recover the debt. A copy of this notice is then recorded with the county clerk of the county where the property is situated, or a comparable public official. If the holder of a junior lien has filed with the county recorder a request for notices of default on the part of the borrower or trustor, the holder will be so advised.

What is the significance of a junior lien holder? Just as in the case of a mortgage, you can have a first mortgage, a second mortgage and even a third or more mortgages, and they take priority by virtue of their recording date. Trust deeds work out identically and you can have a first trust deed, a second trust deed and even a third or more trust deeds. A junior lien holder is one who holds a deed of trust that is junior to the first deed of trust. Let us assume that the junior lien holder represents the second deed of trust, and he is concerned as to whether the borrower or trustor is paying the first trust deed holder promptly each month when the payment is due. It would be difficult to check the records each month at the county recorder's office to ascertain if a default has occurred, so instead, the junior lien holder has filed a written request with the county clerk to be notified in the event the borrower or trustor has defaulted and is under notice by the trustee in regards to said default. This procedure enables the holder of the junior lien to decide what course of action to take. If he feels that property is worth considerably more than the balance due under the first deed of trust, he will want to notify the borrower or trustor to cure the default by prompt payment to the first trust deed holder or advise the borrower that he, the second trust deed holder, will advance the delinquent payments to the holder of the first deed of trust and institute proceedings in his own behalf against the borrower. If the junior lien holder is of the opinion that the value of the property is not worth more than the amount involved in the first deed of trust, he would probably ignore the matter and realize that his junior lien will soon lose its security.

After notifying the borrower as described heretofore, the trustee must then wait three months, after which he publishes a notice of the sale in a newspaper of general circulation, giving the time and place of the sale and a description of the property. The sale must be set for at least twenty-one days from the day the advertisement first appears. At least twenty days before the day of the sale, a copy of the notice of sale must be posted prominently on the property to be sold.

The property is then sold and the lender paid from the proceeds. This would occur only if someone other than the lender proved to be the high bidder. If there is any excess over the loan balance and costs, it goes to the borrower. If the sale does not bring enough to satisfy the loan, the lender must be content as he can only look to the property to recover his loan.

No deficiency judgment may be obtained after such a sale. Only when liens are foreclosed by court action, as described when discussing mortgages, may a deficiency judgment be secured. A deficiency judgment may not be secured on a purchase money deed of trust. To repeat for the purpose of emphasis, the lender would prefer to foreclose under a deed of trust which takes about four months even though no deficiency judgment is involved rather than foreclose under a mortgage and obtain a deficiency judgment because of the redemption rights of the borrower under a mortgage that he does not have under a deed of trust.

After the sale under foreclosure of a trust deed, the trustee issues a "trustee's deed." As the name implies, this is a deed that passes the title held by the trustee to the successful bidder at the trustee's sale, whether it be the lender or beneficiary or someone else. This is a valid deed, and now the borrower has lost his title to the property.

In the foreclosure of a trust deed, and during the three months waiting period before the trustee starts to advertise, the delinquent borrower may reinstate his loan by paying up any delinquent payments, interest and so forth. After advertising starts, the borrower must then pay up the entire loan plus costs in order to save his property.

# 11

# BANKRUPTCY

## 11-1 INTRODUCTION

The dictionary says that a bankrupt is a person reduced to financial ruin. The average man on the street would say that a bankrupt is a person who is broke. Both answers are about the same. A person who went into bankruptcy used to carry a certain stigma of disgrace, however, times have changed. From the time that former President Harry Truman was forced into filing a petition in bankruptcy when his haberdashery business failed after World War I, the element of dishonor has been reduced if not eliminated. Mr. Truman later repaid all of his creditors that were included in his bankruptcy petition, on a voluntary basis and he was applauded for his display of honor.

The first Bankruptcy Act in the United States was passed by the Congress in 1800. Since that time, each amendment to our Federal Bankruptcy Acts has followed a major or minor business depression. All bankruptcy laws are enacted by the Congress of the United States and are therefore federal laws and they apply equally in all of the states. Even though the individual states are permitted to enact laws for the benefit of citizens of that state, bankruptcy laws remain within the sole province of the federal government.

There are two main objectives involved in bankruptcy proceedings. The first is to enable the bankrupt debtor to start his financial life all over when he is overburdened with debts. The second is to provide for the conservation of the bankrupt's assets, if any there be, for the protection of his creditors. More often than not, when a man has failed in his business endeavors, he is so hopelessly in debt that there is no possible way for him to work his way out of debt. For example, a home builder has met with financial reverses due to conditions beyond his control and he owes $100,000 and his assets are negligible. Every creditor naturally wants to be paid and he is unable to continue operating his construction business. Liens and lawsuits are being filed against him in wholesale quantities and the harassment of his creditors makes life dismal. Meanwhile, the broke contractor has a wife and children to support and he obtains a job as a carpenter. Certain tough creditors are threatening to garnishee his wages and in order to protect himself, he files a voluntary petition in bankruptcy. For the moment, the bankruptcy petition takes precedence over the state court garnishment and he is able to provide the necessities of life for his family. When you stop to think about it, his modest earnings as a carpenter make it impossible for the debtor to ever even hope to pay off the $100,000 indebtedness. By the time his weekly paycheck has been reduced by the usual deductions for state and federal income tax, social security and withholding, his take-home pay is barely enough to support his family.

To digress for a moment, the famous case involving Joe Louis, the former heavyweight champion prizefighter, and his financial and tax problems are in point. Even though his earnings during the height of his pugilistic career ran into millions of dollars, Joe Louis was the victim of bad financial advice and when his big-earning career ended, he owed back income taxes totalling about one million dollars. It was a physical impossibility for Joe Louis to ever repay one million dollars in back taxes, even if he earned one hundred million dollars because of the high tax liability on his future earnings. There was no practical solution to his tax liability problem, so the Internal Revenue Service finally relented and permitted Joe Louis to continue his work as a wrestling referee and paying a small portion of his income as a settlement on his one million dollar tax debt.

The problem of conserving the bankrupt's assets for benefit of his creditors has never been adequately solved. The objectives sound wonderful on paper but many creditors who have been victims of bankruptcy petitions will agree that most bankruptcy cases wind up with the same result, namely nothing left for the creditors. By the time the bankrupt's assets are converted into money and the expenses of the referee in bankruptcy, the trustee for the bankrupt, the attorney for the trustee are paid, very little if anything is left over for benefit of the creditors. Most business firms have learned from bitter experience involving many debtors, that as soon as they receive notice of the filing of a bankruptcy petition, their debt is down the drain and gone. They might just as well salvage a part of their loss by charging the debt off as a tax write-off. Under Internal Revenue Service regulations, a bad debt can properly be charged off against taxable income. Thus, if you are in the 40 percent tax bracket, a bad debt loss of $400 could save $160 on your income tax bill. If you reverse the situation and collect the $400 debt and if it represents taxable income, then $160 goes to pay the tax and you net $240 out of the transaction. The moral is that there is no way to beat high taxes.

The problem of conserving the bankrupt's assets for benefit of his creditors has never been satisfactorily solved. Lack of interest on the part of the bankrupt's creditors is partially responsible for the poor result of converting the bankrupt's assets into money. The reason for the lack of interest on the part of the creditors is that they have learned over a period of time, that bankruptcies rarely produce any money for the creditors involved. Too many people have their fingers in the pie and as a result, the cost of administration eats up the money in the bankrupt's estate.

Two examples based on actual personal experiences of bankruptcies handled in my law office will emphasize the fallacy of the present system. In one case, the bankrupt who was my client, had an equity in his home of $7,000 but this asset was abandoned and released by the trustee in bankruptcy. It is true that the house had a substantial mortgage which constituted a first lien but the fact still remains that the home was later sold by my client for $7,000 over and above the mortgage balance. The trustee in bankruptcy took the position that in order to sell the house, the trustee would have to hire a real estate appraiser and then pay a brokerage fee of 6 percent to a real estate broker for selling the property. You and I would both agree that the $7,000 equity was too large a sum for the trustee in bankruptcy to ignore and he was guilty of poor judgment.

The second actual example involved a large department store corporation that had a number of retail stores in two different states. The amount of liabilities amounted to about $3 million dollars and various assets were sold for $200,000. Because of the large sum of money involved, and the different locations of the various stores, there was considerable litigation involved. By the time the smoke died down and the various officials and their attorneys were paid, the entire $200,000 was eaten up by expenses and nothing was left for the creditors.

Some years ago, during a depression when bankruptcies were very prevalent, it was discovered that some Referees in bankruptcy were earning commissions legally and as a result, the Referees' income in certain large cities was greater than the salary paid to federal judges. The Congress of the United States soon corrected this evil and all Referees in bankruptcy are paid a salary and no commissions are allowed. Some enterprising creditor or a professional investigator like Ralph Nader will expose the poor results attained by trustees in bankruptcy in converting the bankrupt's assets into money for benefit of his creditors and the Congress will enact new legislation to correct the present weaknesses in the system.

You may wonder how the bankruptcy laws can affect you in your new construction business. There is a definite connection and it is quite important. Every person starting off in a new business venture naturally thinks his efforts will be successful and negative thinking is not very popular. Partners in a new business always talk about dividing the profits and no one wants to talk about how to divide losses, if they should occur. The number of new businesses that fail every year and wind up in bankruptcy court are mute testimony to the facts of commercial life, namely there is never any guaranty that a new business will succeed.

Therefore, if your new construction business is unsuccessful, you could become financially involved and wind up as a statistic in a bankruptcy court. Some business people say that they will never fail in business but never is a long, long time. Officials of the large Pennsylvania Railroad did not anticipate that financial problems could arise and become so serious that a voluntary petition in bankruptcy became necessary. Although the assets of the Pennsylvania Railroad amount to many millions of dollars, their liabilities are also in the millions and more important, they were not able to pay their payroll and other necessary expenses and payments. When things reach that stage, the corporation or individual debtor becomes insolvent and the only remaining choice is bankruptcy.

Whether we like it or not, knowledge of bankruptcy proceedings is a necessary part of business education for the contractor.

## *11-2 FILING OF BANKRUPTCY PETITION*

Petitions for adjudication of bankruptcy are filed with the clerk of the United States District Court. This is the lowest of the three federal courts and is usually located in the federal building. The forms for the bankruptcy petition are printed and usually the details will be handled by an attorney. It is not mandatory that the bankrupt hire an attorney, but most clients prefer to be represented by an attorney. They are handicapped by lack of knowledge and are in fear of court proceedings in general.

The bankruptcy forms contain many questions that must be answered by the petitioner, that is the person who is seeking to be adjudicated a bankrupt, and he will be required to sign his name many times. The petition is filed in triplicate and the attorney needs one for his file and the client will think he is signing his name a thousand times but it really is not that bad. Like all federal forms, the information required is very detailed and pertinent as to the financial plight of the individual concerned. The bankruptcy form even inquires as to the amount of money paid as attorney's fees for services rendered and to be rendered in connection with the bankruptcy proceedings. Appreciating that legal fees will vary in different sections of the country, an ordinary bankruptcy petition filed for an individual that is not complicated would be billed at the $250 to $300 range, plus the filing fees. Even though the person is broke, it still costs money to go into bankruptcy.

The bankruptcy petition that we have been discussing is a voluntary petition. More details will follow later on the filing of an involuntary petition by the creditors of the alleged bankrupt.

The filing fee in federal court has been caught in the inflationary trend like everything else and is currently $50 for each name that is involved. If a petition is filed for husband and wife, the filing fee would then be $100. Other than the increased filing fee, there is no additional work for the attorney in preparing the bankruptcy petition for both husband and wife. If the wife is an ordinary housewife and is not gainfully employed, there is little to be gained by including the wife's name in the bankruptcy petition. If creditors are threatening to pester the wife by garnisheeing her wages, then it would be best to include her name in the bankruptcy petition and pay the additional filing fee.

The information required by the bankruptcy forms must be completed in great detail or the petition will not be accepted by the federal court clerk for filing. For example, the form provides for the name, address, date of transaction and balance due each unsecured creditor. Some bankruptcy petitions include a long list of creditors and they must be listed in alphabetical order. By unsecured creditors, we mean those that do not hold any security. The gas station, grocery store, shoe store, dry goods store, public utilities, would be good examples of the average unsecured creditor.

Some examples of secured creditors that must be listed separately would be the mortgage holder of your house mortgage; the finance company that holds a chattel mortgage on your household furniture; the furniture store that holds a conditional sales contract for the title to the furniture that you are paying on; and any other creditor that holds some type of security as protection for your obligation.

Now your bankruptcy petition has been filed in the federal district court and the clerk refers your petition to one of the federal judges available who automatically and routinely adjudges the petitioner to be a bankrupt and the petition is then referred to a Referee in Bankruptcy. Bankruptcy courts have limited jurisdiction in that they only handle bankruptcy matters and nothing else. A federal judge is not limited in his jurisdiction and he handles many different types of matters that are filed in his court.

## *11-3 REFEREE IN BANKRUPTCY*

A Referee in Bankruptcy is an attorney who is appointed by the federal judges of his district, as an official similar to a judge with limited power and jurisdiction. In certain smaller districts, the Referee in Bankruptcy is a part-time job and he is permitted to maintain his private practice of law without any conflict with his official duties. In larger areas, the Referee in Bankruptcy is a full-time job and he handles bankruptcy matters only all day long. To this extent, he is an assistant to the federal judges and works under their control.

The Referee in Bankruptcy maintains his office and official files in the federal building and also presides over a small courtroom where official bankruptcy matters are conducted. The Referee in Bankruptcy is required to keep a docket of all bankruptcy matters handled by his office and his clerk's docket is open at all times to public inspection. Any and all bankruptcy papers would automatically be filed in the office of the Referee in Bankruptcy and not with the clerk of the federal district court.

The delays in court proceedings that have been the subject of widespread complaint in recent years, do not apply to federal courts in general and to bankruptcy courts in particular. Within a matter of a few weeks after the bankruptcy petition has been filed in the federal district court, all of the creditors listed in the petition are notified by mail, of the First Meeting of Creditors to be held in the courtroom of the Referee in Bankruptcy. Personal appearance of the bankrupt at the first meeting of creditors is mandatory but is optional for the creditors. Bankruptcy courts, like all other courts are open to the public and if you are curious, stop in and watch and listen to the proceedings. If you walk in on an average bankruptcy first meeting of creditors hearing, you will be shocked by the fact that the only persons present in the bankruptcy courtroom will usually be the referee sitting as a judge and the bankrupt and his attorney. The reason that the creditors are conspicuous by their absence, is that they have learned from experience that it is a waste of time to attend the first meeting of creditors hearing.

While it is true that all creditors and their attorneys are given the opportunity of questioning the bankrupt debtor regarding his assets and liabilities, this procedure will not produce any money for payment of their bill. More often than not, the debtor has no assets but a long list of liabilities and prospects for collecting their bill are not very good. That is why most creditors realize the hopelessness of collecting their money when they receive the news about their debtor having filed a voluntary petition in bankruptcy.

Upon the conclusion of the First Meeting of Creditors, which sometimes takes as little as about fifteen minutes, the Referee in Bankruptcy then appoints a Trustee in Bankruptcy to handle all of the business details. Usually, the trustee appointed is an attorney and he has the chore of selling any available assets of the debtor and converting them into money and he later renders his report to the Referee in Bankruptcy. If, as frequently happens, there are no assets, the debtor's bankruptcy is then scheduled for discharge within a few months. During this waiting period, a creditor is privileged to file his objections to a discharge in bankruptcy being granted to the debtor based upon proper grounds, such as fraud or withholding assets from the bankruptcy court, etc. If objections are filed, the matter is set down for a hearing before the Referee in Bankruptcy. If not, the discharge is issued by the Referee in Bankruptcy and mailed in duplicate to the bankrupt's attorney.

If creditors are of the opinion that there may be assets available in the bankrupt's estate to pay all or part of their bill, they are privileged to file a "Proof of Claim" to substantiate their bill.

A proof of claim should be filed if the debtor's bankruptcy schedule indicates there are assets that might result in a dividend payment to unsecured creditors. There is no fee charged for filing the proof of claim and it is a simple form that the bookkeeper can fill out and attach an itemized statement of the account.

Under existing federal bankruptcy laws, a discharge in bankruptcy has certain limitations. The debtor is discharged from certain obligations that he listed in his bankruptcy petition but he is not discharged from obligations that were not listed or that are classified as nondischargeable. Divorced men have found out to their chagrin that judgments for alimony due or to become due are not dischargeable obligations. The same rule applies to obligations for the support or maintenance of a minor child. All tax claims in favor of the federal, state, county or municipal government are not dischargeable in bankruptcy. A claim for money or property obtained by false representations or credit obtained by false written financial statements is not dischargeable in bankruptcy. A judgment for wilful and malicious injuries is not dischargeable. The word wilful means intentional or deliberate; however, injuries caused by mere negligence are dischargeable in bankruptcy.

In cases where the bankruptcy proceedings have been complicated, the Referee in Bankruptcy, acting under the provisions of the Federal Bankruptcy Act, orders a final meeting of creditors.

All orders issued by the Referee in Bankruptcy are subject to review by the federal judge involved.

## *11-4 INVOLUNTARY BANKRUPTCY PROCEEDINGS*

Previous discussion covered the filing of a voluntary petition in bankruptcy. An involuntary petition in bankruptcy may be filed by three or more creditors against the alleged bankrupt, regardless of whether he is an individual, partnership, or corporation. Their petition alleges that the person, firm or corporation is insolvent in that it is unable to pay its current debts and that its liabilities are greater than its assets.

A copy of the involuntary petition is served on the alleged bankrupt and if the facts are disputed, an answer is filed. In due course of time, the disputed petition is set down for a hearing before the Referee in Bankruptcy. After all of the testimony and evidence is submitted, the Referee in Bankruptcy rules on the involuntary petition. If he finds in favor of the petitioning creditors, the matter then proceeds like any other voluntary bankruptcy petition. The main reason for three or more creditors filing the involuntary petition in bankruptcy, is to prevent the person, firm or corporation involved from continuing to do business to the detriment of existing creditors.

On rare occasions, it has happened that two or more involuntary petitions have been filed in the same federal court by different creditors, against the identical alleged bankrupt. Usually this unusual event would result in the federal judge ordering all of the petitions to be consolidated into one hearing.

Bankruptcy courts have been referred to as courts of equity.

*Example*

1. The equitable powers of the bankruptcy court have been exercised in passing on a wide range of questions arising out of the administration of bankrupt estates. Generally speaking, they have been invoked to the end that fraud will not prevail, that substances will not give way to form, and that technical considerations will not prevent substantial justice from being done.

One of the many advantages of having bankruptcy under federal control rather than state control, is that the bankruptcy court now has jurisdiction over property anywhere in the United States.

There are certain groups who are excepted from having involuntary bankruptcy petitions filed against them. They are farmers, wage earners, railroads, insurance companies, banking corporations and building and loan associations.

The Federal Bankruptcy Act defines six separate "acts of bankruptcy," any one of which is ground for an adjudication in bankruptcy.

(a) To conceal or remove any part of his property with intent to defraud his creditors or make a fraudulent transfer of any of his property.

(b) Preferential transfer or payment while insolvent with intent to prefer one creditor over his other creditors

(c) By permitting a lien or attachment to be made against his property

(d) General assignment for benefit of creditors

(e) Permit the appointment of a receiver to take charge of his property

(f) Written admission of inability to pay debts and willingness to be adjudged a bankrupt

Any transfer of property or payment of money to creditors within four months of the filing of the voluntary bankruptcy petition is under a cloud of suspicion. If they are later found to be preferential payments to one creditor over another, the bankruptcy court through the referee in bankruptcy, may order the creditor so preferred to disgorge or return the money. Then this money goes into the bankrupt's estate for benefit of all creditors.

*Example*

1. Payments in ordinary course of business were not considered to be made with intent to prefer.

An outstanding example the other way, would be a case where the debtor owes $1,000 to each of ten creditors and he has a total of $1,000 available in money. He attempts to favor one friendly creditor by paying him his full amount of $1,000, thus leaving the other nine creditors out in the cold. The creditor receiving the $1,000 would obviously be ordered to disgorge.

The final order in a proceeding on an involuntary petition is a dismissal of the petition or an order to adjudication in bankruptcy.

## 11-5 CLAIMS OF SUBCONTRACTORS, LABORERS AND MATERIALMEN

Even though we say that bankruptcy laws are controlled by federal laws enacted by the Congress, in certain situations state law actually prevails. For example, when a mechanics' or materialmen's lien is duly filed and recorded in accordance with the laws of a particular state, it takes precedence over a trustee in bankruptcy, who takes the bankrupt's property subject to the existing lien. Similarly, a lien given to a landlord by a state law, or one created in good faith by contract between a landlord and a tenant, is good against the trustee in bankruptcy of the tenant.

If a construction contract provides that funds retained by the owner of the construction property are held as security for indebtedness of the general contractor to subcontractors, laborers, and materialmen, they are generally considered to be the property of the trustee in bankruptcy of the prime contractor. This particular court ruling sounds unusual where the contract said the withheld money was to be security for money due the subcontractor, laborer and materialman. It emphasizes the fact that the law is not always perfect or certain. More often than not, you can find a court ruling one way and another court seems to be ruling exactly the opposite.

*Example*

1. The jurisdiction of the bankruptcy court is exclusive with reference to the determination of laborers' or materialmen's liens upon a fund due to the contractor by the owner under a contract for a public improvement governed by the New York Lien Law, not filed until after the petition in bankruptcy of the contractor.

In many cases, the state courts have assumed jurisdiction to enforce the liens of subcontractors, laborers and materialmen, after the bankruptcy of the principal contractor, without any discussion in the opinions rendered respecting the jurisdiction of the state courts of such action.

*Example*

1. Where the trustee in bankruptcy of a head contractor has, of his own motion, become a party to an action in a state court to enforce mechanics' liens against a fund held by the public and owing to the head contractor, the state court has the right to declare the rights and interests of the parties, although it has not the power to take the fund out of the court of bankruptcy into which it has been paid by the public fundholder.

## 11-6 DISCHARGE IN BANKRUPTCY

A discharge in bankruptcy is the release of a bankrupt from all of his debts which are proveable in bankruptcy, other than those that are excepted by the Bankruptcy Act. A discharge is a legal right to be granted by the bankruptcy court, unless good cause for refusal is shown. One of the primary purposes of the bankruptcy laws is to discharge a bankrupt from his indebtedness and afford him a fresh start in life.

A discharge in bankruptcy does not mean that the debtor has paid the obligations that he listed in his bankruptcy schedule. It means that the creditors so listed cannot legally enforce their claims to collect monies due them. Technically, a creditor of a discharged bankrupt could sue him for his debt, however, the lawsuit could be defeated by the debtor raising the valid defense of his discharge in bankruptcy. In this connection, a state court cannot make an order directing that a judgment debtor make payments out of his income on a judgment as to which the debtor has a valid discharge in bankruptcy, properly pleaded and brought to the attention of the court in the proceedings in which the order is sought.

Even after receiving a discharge, the bankrupt can revive the obligation by signing a new promise in writing agreeing to pay the same obligation from which he was discharged by the bankruptcy court. A finance company whose claim was discharged by the bankruptcy court was successful in hounding a poor client of mine who signed a new note in order to "get rid of the pest from the finance company." Small wonder this client is still poor and it is most difficult to help a client who refuses to heed the advice of his attorney.

*Examples*

1. A promise sufficient to render enforceable a debt discharged in bankruptcy is made by a letter, stating: "In regards to your claim against me, you will be paid every dollar of it, with interest, as soon as I sell the mill."
2. An example of a sufficient promise exists in the case of a bankrupt's declaration that if he gets his discharge he will be in shape to pay and he is going to pay.

It is not mandatory that a debtor list all of his creditors. The debtor may omit one or more of his creditors if he so desires. Sometimes, the relationship between the debtor and certain creditors is so friendly and cordial that the debtor knows that his friendly creditor will never harass him for the money that he owes and intends to pay even after being discharged in bankruptcy. Creditors whose names have been omitted by the debtor are not affected by the discharge in bankruptcy and legally could pursue their legal remedies to enforce payment.

A novel and interesting point of bankruptcy law came up recently when another client asked if it was possible to pay all creditors he had listed in a bankruptcy twenty years earlier and have the bankruptcy proceeding removed from the public records. My curbstone opinion was that you can always pay creditors but I did not know about any procedure whereby the earlier bankruptcy could be wiped off the bankruptcy court records. In due course of time I learned the answer and discovered that it was legally possible to accomplish this unusual mission. All of the creditors involved in the original bankruptcy were contacted and they were not only delighted to be paid off in full but they added the comment that they always knew that my client was a true gentleman and the victim of unfortunate circumstances. After proof of the payment in full to all of the creditors involved, the federal judge in charge of the district issued his order wiping the original bankruptcy off of the records. This accomplished the desired result and my client was able to truthfully say: "I have never gone into bankruptcy."

The Bankruptcy Act provides that a bankrupt may not receive a discharge within six years from having received a prior discharge in bankruptcy. On one occasion I had the dubious distinction of putting a client into bankruptcy for his second time. He is now convinced that he is not suited to be in business for himself and enjoys being an employee instead of a boss.

## *11-7 EXEMPTIONS*

In addition to the list of secured and unsecured creditors that are included in the debtor's voluntary petition in bankruptcy, the forms also provide for a list of the debtor's assets. The Bankruptcy Act provides that the petitioner shall make oath to and file with his petition, a schedule of his property showing the amount and kind of property as well as the location and its monetary value. As you can well imagine, questions asked pertaining to the debtor's property are very detailed and pertinent.

Some people are inclined to treat this requirement very lightly and think that it is unimportant if they "forget" to list some of the property that they own. Most attorneys, including myself, advise the debtor-client that the federal authorities are the wrong people to "play games with" and accurate information is strongly recommended.

We have all heard statements made pertaining to some person of prominence filing a voluntary petition in bankruptcy and then shortly after starting up in business a second time and doing well financially. The whispering campaign intimates that the bankrupt-debtor withheld a lot of his money or property from the bankruptcy court. What does actually happen on occasion is that the bankrupt-debtor has friends or relatives who have faith in his business ability in spite of his bankruptcy and they lend him money to start up in business anew. Although a few people might get away with withholding their assets from the bankruptcy court, others have been caught up with and were prosecuted by the United States Attorney for their criminal offense.

The bankruptcy petition consists of about twenty-five pages and the first portion is called "Schedule A" which covers the liabilities of the debtor. Then comes the second portion, called "Schedule B" which covers the debtor's assets, regardless of whether he has any property or not. Then, bringing up the rear of the bankruptcy petition is the debtor's claim for certain items of his property to be exempted. The Bankruptcy Act provides that the bankrupt shall claim his exemptions in the schedule of property which he is required to file.

In order to learn just what property is exempt from the bankrupt-debtor's estate, we must turn to the laws of the particular state in which the bankrupt-debtor resides. Even though the Bankruptcy Act is federal, the laws pertaining to exempt property are state laws and the federal judges and referees in bankruptcy are obligated to honor the state law involved.

Because there is considerable variance between the various state exemption statutes, the awards of exempt property in bankruptcy court are not uniform throughout the United States. The bankruptcy court is bound to follow the construction placed by the highest court of a state upon a state exemption law, even though a narrower construction appears equitable. In certain situations, state law is superior to federal law and applying the exemption laws in favor of a bankrupt-debtor is one of them.

Property which, because of its exempt character, is not subject to levy and sale under the state law, cannot be made to respond for the debts of the bankrupt under the Bankruptcy Act. The exemption laws of some states are more liberal than other states and this is purely a matter of local state law.

A good example involves life insurance policies, particularly those that have a cash value. At first blush, you might guess that the cash value of a life insurance policy represents an asset just as does money. However, if the state law involved says that the life insurance policies and their cash surrender values are exempt from garnishment, attachment or from creditors in general, then the trustee in bankruptcy must honor the state law and this asset is exempt in favor of the debtor. If there is dissatisfaction with this provision of the law, then it is up to the legislature of that state to change its law.

All states have exemption statutes that cover items of necessity such as clothes, tools used in the debtor's work, one motor vehicle of limited value and similar items. The purpose of the Bankruptcy Act is to distribute the assets of the debtor among all of his creditors but there are reasonable limitations under exemption laws that give the debtor a bit of dignity and the opportunity of supporting his family while trying to get started financially.

A later discussion under the Wage Earners Plan will bring out the point that frequently the bankrupt-debtors financial plight has been aided and abetted by "easy credit" that is made available to all that are interested. Advertisements that are read in newspapers and heard on radio and television, suggesting the purchase of merchandise on the installment plan, commonly referred to as $1 down and $1 per week, contribute to the downfall of the weak debtor. He buys a color television set when he can ill afford it and he is concerned only with the amount of his payments and not the total cost of his purchase. Finally, comes the embarrassing day of reckoning, when he discovers for the first time that his payments exceed his income and he is a prime candidate for bankruptcy court. You can point the finger of blame at the debtor but I place the blame equally on the creditor who contributes to this sad situation that permits a champagne taste with a beer pocketbook.

## 11-8 DEBTOR RELIEF PROVISIONS

Debtor relief provisions are the outgrowth of the concept that bankruptcy procedures may be invoked successfully without a liquidation or sale of assets. The general idea is that one of the essentials of the successful reorganization or composition is to save a going business. Every businessman knows that as soon as the doors of a going business are closed, even though only for a short period of time, the business has been materially hurt and is usually doomed to failure.

The Congress in its wisdom and in an attempt to help save going businesses from closing their doors and filing a voluntary petition in bankruptcy, added Chapter XI to the Bankruptcy Act, providing for arrangement proceedings. This procedure is fairly new and it enables the debtor to be given a reasonable opportunity to rehabilitate itself despite the fact that some losses may be sustained in the transitional period. Every creditor has learned that bankruptcy proceedings for his debtor is invariably bad news and the usual final results are very little if any money available to pay his debts. The proceedings under Chapter XI may result in a loss but there is some hope that the debtor's business will be saved and at least part of his bill paid, sometimes in a lump sum and other times on the installment plan. On rare occasions, there is a possibility under Chapter XI that the creditor's obligation will eventually be paid in full.

*Example*

1. Proceedings for a composition or extension are not the equivalent of equity receivership proceedings; they contemplate a feasible plan, promptly presented, whereby the over-burdened debtor may, through the creditor's cooperation secure a scaling of debt or interest, or an extension of the due date. If none of these objects can be accomplished, the case is then one for liquidation through bankruptcy proceedings.

Under normal conditions, we say that a person or a company is insolvent when their liabilities are greater than their assets. Under the provisions of Chapter XI which are available to corporations as well as individuals, the debtor says that he is insolvent or unable to pay his debts when they mature.

This difference is very important because there are cases where the debtor's assets are greater than his liabilities, yet he is unable to pay his debts when they mature. Now Chapter XI comes to the rescue by providing for an extension of time to pay indebtedness, as provided for in the plan of arrangement.

Instead of filing a voluntary petition in bankruptcy, the debtor files a plan of arrangement for an extension of time or compromise of the amount due to each creditor, as approved by the creditor's committee. The plan of arrangement must be approved by 50 percent of the dollar amount as well as 50 percent of the numerical count of the creditors. If this provision is not met, the plan fails and the Chapter XI proceeding is then converted into an ordinary bankruptcy proceeding. The obvious advantage of the plan being approved by the majority of the creditors as required, is that the business will continue to operate as a going business by the debtor. If the plan fails, the debtor is through and a trustee in bankruptcy takes over. The trustee is not permitted to continue to operate the business, so the doors are closed and then the assets are converted into money. Usually, the value of the bankrupt's assets shrink horribly and there is little or no money left over for benefit of the unsecured creditors.

If the plan of arrangement is approved by the creditor's committee and confirmed by the federal bankruptcy court, then it becomes binding on all of the unsecured creditors who now have a choice of the alternatives included in the plan of arrangements.

The $3 million dollar department store bankruptcy previously referred to, started off under a Chapter XI plan of arrangement.

Every unsecured creditor was given three choices: (1) 40 percent of your bill if you want cash now; (2) 75 percent of your bill, payable on the installment plan over a three year period; and (3) 100 percent of your bill, payable over a five year period, with no interest. Of the hundreds of creditors involved, the three choices were divided with the larger creditors selecting the 40 percent cash payment plan. In this particular case, the passing of time proved the 40 percent cash payment the wiser choice because the business continued in operation for a limited time and later wound up in ordinary bankruptcy proceedings.

At least, Chapter XI proceedings under the plan of arrangements, gives the creditors a voice in the matter and the debtor an opportunity to work himself out of his financial difficulties over a period of time, if his plan is approved by a majority of the creditors. Any plan, if it has any reasonable hope of success, is better for the creditors than ordinary bankruptcy.

In those cases, where the plan of arrangement is approved by the creditors and the debtor is permitted to continue operating his business, it is vital that credit be extended to the financially involved debtor. Naturally, the existing unsecured creditors don't want to give further credit, so the plan provides that all new credit extended is given priority protection. If the existing unsecured creditors do not desire to deal with the debtor that is naturally their choice. Then, new creditors who extend credit after the plan of arrangement has been approved, are offered the priority protection.

If the plan of arrangement is successful, then we have a happy ending and the creditors did much better than they would have under an ordinary bankruptcy proceeding. If the plan of arrangement fails and the debtor is unable to pay as agreed, then the pending Chapter XI proceedings are by court order, converted into an ordinary bankruptcy proceeding.

It is my considered opinion that the Chapter XI plan of arrangement has merit, even though it seems to fail more often than it succeeds. The procedure may not be perfect but it is certainly an improvement over the ordinary bankruptcy petition. At least, the unsecured creditors have reasonable hope of partial recovery of their loss under the plan of arrangement, and a faint possibility of 100 percent recovery over a period of time. From the creditor's point of view, any degree of success is an improvement over the ordinary bankruptcy result.

## 11-9 WAGE EARNERS' PLANS

Just as Congress enacted Chapter XI of the Bankruptcy Act for benefit of big business, either in personal or corporate form, there was need for similar legislation for benefit of the small man, the wage earner.

Chapter XIII of the Bankruptcy Act on "Wage Earners' Plans" was added to give the financially involved debtor more time to pay his creditors in full, even though it is on the installment plan and thus avoid going into bankruptcy.

Under the wage earners' plan, he is granted an extension of time in which to pay his debts in full, rather than to discharge his obligations via bankruptcy. Under the provisions of Chapter XIII, the wage earner is not adjudicated a bankrupt but is known as the "debtor." This removes the stigma of being called a bankrupt and is an honorable opportunity for the debtor to work his way out of his financial dilemma. From the creditor's point of view, he has reasonable hopes of collecting all or at least a portion of his bill. Again, any plan is better for all parties concerned than the ordinary bankruptcy route.

If a wage earners' plan under Chapter XIII is accepted by a majority of his creditors, the court appointed trustee comes into the picture. The wage earner ordinarily is granted three years to pay his creditors in full and this time may be extended by the court. It would only be extended for good cause.

The court appointed trustee is usually an attorney and he handles all of the money involved that is paid in by the debtor. The trustee then makes partial payments to the creditors whose claims have been approved by the Referee in Bankruptcy. Some creditors have complained about the small payments and the length of time it takes the debtor to pay his debts in full. The money is prorated and every creditor receives fair treatment.

Even the attorney for the debtor and the trustee appointed are required to wait for their money on the installment plan, the same as all other creditors.

If the wage earner or debtor fails to pay as he agreed, then his Chapter XIII petition is converted by the Referee in Bankruptcy into an ordinary bankruptcy petition.

Chapter XIII of the Bankruptcy Act on Wage Earners' Plans is of more significance from the social aspect than any other part of the Bankruptcy Act. It is intended to assist the financial rehabilitation of the debtor in mind as well as in finances by providing a practical method for the workingman to keep his job, maintain his family, and pay his debts in full. The wage earners' plan may not be perfect but it is the best arrangement presented so far for joint benefit of the creditor as well as the debtor. It may help to prevent the creditor from offering easy credit where the debtor cannot afford it and it may help the debtor from abusing his credit and buying things that he does not need and cannot afford.

# 12

# ARBITRATION AND AWARD

## 12-1 EXPLANATION

Arbitration is a substitute for using the courts to settle disputes between parties who are unable to agree on a settlement of their differences. Arbitration is a method of settling disputes through the investigation and determination, by one or more arbitrators who have been selected and agreed to by the parties to the contract.

The object of arbitration is the final disposition of differences between the parties in a faster, less expensive, and less formal manner than is available in an ordinary lawsuit in court. Usually the arbitrators are selected by the agreement of the parties that provides for each side to select one arbitrator and then the two arbitrators so selected, are charged with the responsibility of choosing the third arbitrator. It is presumed that if you only had two arbitrators, your arbitrator would automatically favor your position in the dispute and my arbitrator would favor my position and the result would be a tie vote. The selection of the third arbitrator makes it possible for any two of the arbitrators to arrive at a majority decision which is binding on both parties.

For all practical purposes, the arbitrators have the function of a judge. The investigation and determination of matters of difference are placed in the hands of the arbitrators, except that the proceedings are less formal and not subject to the usual legal technicalities, such as rules of evidence, swearing in of witnesses, and having a court reporter take all testimony down in shorthand and later transcribing same.

*Example*

1. An agreement to arbitrate under a modern statute is a method for settling disputes in which the parties create their own forum, pick their own judges, waive all but limited rights of review and appeal, dispense with the rules of evidence, and leave the issues to be determined in accordance with the sense of justice and equity they believe their self-chosen judges possess.

The decision of the arbitrators is called the "award." It is just a variation of the judgment of the court, which occurs when the judge has completed the case and is ready to rule. The award is binding on the parties involved in the arbitration proceedings, as to all matters properly submitted and investigated by the arbitrators under the authority of the arbitration submission agreement. There can be no arbitration proceedings unless it has been agreed to by the parties involved in the dispute. If the written contract does not provide for arbitration, no one can be coerced to submit a dispute to arbitration. To do so would be unconstitutional as depriving the party of liberty and property without due process of law.

The entire arbitration procedure is comparatively new, however, it is rapidly gaining in popularity. As court proceedings take longer and become more expensive, more and more disputants unable to settle their differences, turn to arbitration as a better and faster solution. The first Uniform Arbitration Act, which proposed to bring about a uniform system of arbitrating controversies, was not generally adopted by the various states. But a second Uniform Arbitration Act, based on the New York arbitration act and similar acts of fifteen other states, has met with wider acceptance.

The popularity of arbitration proceedings as a substitute for court proceedings has even reached the field of professional sports. Formerly, by the time the judge got around to ruling on the dispute between the professional athlete and his team owner, a good portion of the sport's season had gone by. Now, the modern trend permits the professional athlete to practice and play with his team while his wage contract dispute is being determined by a board of arbitrators. The speedy award of the arbitrators is good for the athlete as well as his employer.

The change from a judge in a court of law to an arbitration panel meeting away from the courthouse, may make a radical difference in the final result because the procedures are entirely different. The tremendous popularity of arbitration over formal court proceedings tells its own story. This popularity continues even though arbitration proceedings carry no right to trial by jury that is guaranteed in court proceedings, by both the federal and most state constitutions. Arbitrators do not have the benefit of judicial instruction on the law, but sometimes "horsesense" is more desirable than technical legal knowledge. Arbitrators are not required to give reasons for their award but the same applies to judges in court lawsuits. The formal record of arbitration proceedings is not as complete as it is in a court of law and this may be a good thing. The hours and hours spent by court reporters in taking down testimony of all of the witnesses in a trial is frequently time wasted if the judge's decision is not appealed by one of the litigants; in which case, the notes of the court reporter are thrown away. The judicial review of an arbitration award is more limited than a judicial review of a court trial with its automatic right of appeal. Litigants in court proceedings become very discouraged when they learn that the ruling of the trial judge is a mere preliminary to the long delay that follows when the trial court's ruling is appealed and sometimes there is more than one appeal. The long wait that sometimes takes years and the increasing cost of litigation is forcing more and more parties to a contract to decide on arbitration as a more desirable method of settling disputes. Peculiar as it may seem, even the judges are encouraging arbitration as a means of settling disputes.

## *12-2 ARBITRATION VS. APPRAISEMENT*

Arbitration and appraisement are not the same. Because certain of the rules of law that apply to arbitration also apply to appraisement, the two are sometimes confused. Certain construction contracts that provide for the determination by the architect of amounts to be paid to the general contractor based upon progress of construction, are often called arbitration agreements. Actually, this designation is incorrect and they should more properly be designated as agreements for appraisal.

Arbitration should not be confused with what takes place in any case where the parties to a contract, refer to some persons to be selected to perform some ministerial duty or some matter that involves only the ascertainment of facts. This procedure does not require a hearing and is thus not an arbitration proceeding.

*Examples*

1. The determination of a fire loss is not an arbitration proceeding.
2. Where parties to a contract, before a dispute arises and in order to avoid one, provide for a method of ascertaining the value of something related to their dealings, the provision is one for an appraisement and not for an arbitration.

An agreement for arbitration ordinarily covers the disposition of the entire controversy between the parties upon which award a judgment may be entered. Whereas an agreement for appraisal extends merely to the resolution of the specific issues of actual cash value and the amount of monetary loss.

Appraisers are generally expected to act on their own skill and knowledge and do not hear evidence or testimony of witnesses. Arbitrators, on the other hand, must meet together at all hearings in their capacity as junior judges. Arbitrators may receive evidence or hear testimony of a witness or party to the dispute only in the presence of the other side. Attorneys usually represent the disputants and to this extent, the arbitration hearing is an adversary proceeding.

*Example*

1. The valuation of property or estimate of damages is frequently confined to the personal skill, knowledge or experience, or even acquired information of appraisers.

In cases where the agreement so specifies, the results of an appraisal may be just as binding as an award by arbitrators.

*Example*

1. Where appraisers proceed on the mistaken belief that they are acting as arbitrators, and file an award on which the lower court enters a judgment, if substantial justice has been done thereby, the appellate court may act under the harmless error rule and affirm the judgment.

The "harmless error rule" arises when the ruling of a lower court has been appealed to a higher court. The appellate court decides that the judge of the lower court committed an error but states that his error was of such a minor nature that it was harmless and therefore does not warrant a new trial or a reversal of his decision.

An arbitration is distinguishable from a reference by a court, which occurs when the judge appoints a referee to obtain facts and make recommendations to the judge. An arbitration rests on mutual agreement of the parties to submit their matters of difference to selected persons, to be accepted as a substitute for the judgment of a court. Whereas, a reference is made in a pending court lawsuit to obtain facts and rests on the constitutional and statutory power of the court.

*Example*

1. Where an agreement to arbitrate is waived and the parties stipulate for a reference of the controversy by the court to referees of the judge's own choosing, the stipulation gives the court jurisdiction to fully settle the matter without further arbitration proceedings.

To further emphasize the limitations of a referee, it is important to remember that nothing can originate before a referee and nothing can terminate by his recommendation, because the referee must report back to the appointing judge, who may modify, correct, reject, or set aside his report as the law and the facts may require. The referee acts only in an advisory capacity. A submission of a dispute to arbitration of a pending matter presumes that there will be a final determination of the entire controversy by arbitrators, and that the court will thereupon enter the award as a judgment.

Originally, some of the courts leaned strongly against enforcing arbitration agreements as tending to oust the courts from their lawful jurisdiction by private agreement. Since the enactment of arbitration statutes by most of the states, beginning with the New York act in 1921, the courts have generally looked with favor on arbitration as a shortcut to substantial justice.

*Examples*

1. It should be the policy of the law to encourage the settlement of business disputes by arbitration, and agreements which provide for arbitration should be given a liberal construction in furtherance of such policy.
2. Courts, in all proper cases, should be reluctant to limit the scope of such useful remedy as arbitration.
3. The adoption of arbitration statutes indicates a strong public policy favoring this method of settling disputes.

## *12-3 ARBITRATION STATUTES*

The method of settling disputes by arbitration is of common law origin, but in many jurisdictions special statutes relating to arbitration have been enacted.

*Examples*

1. It is reasonable that parties who voluntarily agree in writing to arbitrate should be bound by the statute and should not as an afterthought be permitted to escape from their contract through the portals of the common law.
2. In the state of Washington, common law arbitration is not recognized or permitted; arbitration is wholly statutory, and rights of the parties are governed and controlled by statutory provisions. In the states that still retain common law arbitration alongside the statutory method, the parties may choose which method they will follow.
3. The rights and liabilities of the parties are governed in the state of Illinois by common law and by a statute entitled "Arbitration and Awards" and the agreement and the statute should be read in harmony, if possible.

Diversity of citizenship alone is not enough to empower federal courts to compel arbitration under the Federal Arbitration Act, for the contract in which the arbitration clause is found must involve interstate commerce. By diversity of citizenship we mean a citizen of one state against a citizen of another state. The citizen need not be a person, it can be a partnership, company or corporation.

Arbitration statutes that change the characteristics and procedural rights of the common law have been upheld against numerous constitutional objections. It has been held that arbitration statutes do not deprive parties of property without due process of law. That they do not unconstitutionally confer judicial powers on private individuals. That they do not violate constitutional provisions vesting the judicial power in constituted courts. That they do not impair but rather strengthen the obligations of contracts. An arbitration statute does not violate a constitutional guaranty of jury trial, for such a right may be waived by the parties.

Statutes in some jurisdictions provide for the arbitration of disputes in certain classes of cases, and such statutes have been upheld against various objections on constitutional grounds.

*Examples*

1. A provision for arbitration in a standard insurance policy, the form of which has been approved by the legislature, cannot be declared void as being against public policy.
2. Coverage under a contract of automobile insurance is one of fact to be determined by the court, notwithstanding a contractual provision for arbitration.

## *12-4 ARBITRATION AGREEMENTS*

Except where a compulsory arbitration is provided by statute, the first step toward the settlement of a dispute by arbitration is the entry of the parties into a valid agreement to arbitrate. An agreement to arbitrate is a contract and the relation of the parties is contractual. The rights and liabilities of the parties are controlled by the law of contracts.

*Example*

1. A common law arbitration agreement consists of two elements:
   (a) a contract between the parties whereby they mutually agree to submit their controversy to named arbitrators
   (b) a grant of power by each party authorizing the arbitrators to act in his behalf and to settle and determine the matter in controversy

If the contract between the parties was found to be an illegal contract, then a provision for arbitration of disputes would automatically become unenforceable. An example would be if the original contract was found to be void because of usury or against public policy.

Arbitration agreements fall into two principal classifications:

(a) agreements to submit a present controversy to arbitration, including all or a portion of the issues in actions then pending before the court

(b) agreements to arbitrate future controversies

An agreement to arbitrate is not required to be started in any particular form or wording, and the use of technical or formal words is not required.

A common law agreement to arbitrate or a submission of a dispute to arbitration with an agreement to abide by the award may be either oral or written. A parol or oral submission to arbitration is valid in any case where an oral agreement of the parties with respect to the matters submitted would be valid and enforceable. Arbitration agreements may be subject to the statute of frauds.

*Example*

1. An oral submission and award of the amount due for rent for past occupation of real estate under a parol lease is not within the statute of frauds, and hence is valid and binding so far as it fixes the sum payable.

Statutes that make arbitration agreements valid, enforceable, and irrevocable, almost universally provide for a written agreement, and the requirement as to a writing is strictly enforced.

The courts seek to uphold arbitration agreements even where they are somewhat uncertain and indefinite, provided the deficiency can easily and certainly be supplied. It has been held that subjects not within the strict letter of the agreement, but plainly and necessarily within the spirit, are included.

*Examples*

1. Ambiguity or uncertainty in a written arbitration agreement may justify admission of parol evidence to remove the ambiguity. (Parol evidence means oral or verbal.)
2. The federal policy is to construe arbitration clauses liberally, to find that they cover disputes reasonably contemplated by the language, and to resolve doubts in favor of arbitration.
3. Courts are not overtechnical in their interpretations and will give effect to the spirit, as well as the letter of the agreement.

When the language of the arbitration clause indicates a desire on the part of the parties to include all controversies that may arise under the principal agreement, the clause is generally given effect and no more specific designation of arbitrable matters is needed. Parties to a contract may agree to submit disputes over that contract to arbitration and may also agree upon a particular tribunal for reviewing the arbitration award. An agreement that an arbitration award shall itself be final and binding upon the parties will generally preclude a court review.

The question of the validity of the basic contract is essentially a legal question for determination by the court. It may be presented in a proceeding for an order directing arbitration under the statute, or in a proceeding to confirm or vacate the award under the statute.

*Example*

1. Where a corporation official denies the making of a contract to arbitrate, and the purported contract is contained in an order form that is unsigned by the corporation or its officers, there is an issue of fact that must be tried by the court before arbitration may proceed.

Arbitration which dealt only with value of services performed in construction work and failed to deal with the matter of the value of materials used was invalid since it served no useful purpose and did not avoid a lawsuit which was the purpose of the submission. The determination whether a particular dispute is arbitrable under an arbitration contract is a legal question for the court rather than for the arbitrators.

*Example*

1. To permit the arbitrators to decide whether a dispute is subject to arbitration is outside the usual understanding of the relations between the court and arbitrators. Generally, the court will determine the arbitrability of a matter unless the parties to the contract have clearly stated otherwise.

It is for the court, not the arbitrators, to decide whether or not the latter have exceeded their powers under the submission. But if an entire contract is submitted to arbitration, the submission includes all issues of law or fact, including the interpretation of the terms of the contract. Arbitration of disputes arising under a contract may be provided for by reference to outside documents.

*Example*

1. Where a pamphlet prepared by the American Institute of Architects, and containing a requirement for arbitration in certain cases, is included by reference in an agreement between two parties, it becomes part of the contract and the parties are obligated to arbitrate as therein specified, provided the terms of the arbitration agreement comply with the statutes of the state.

With the passing of time, the scope of arbitration agreements have spread to varied fields that even involve the public interest. For example, arbitration provisions concerning alimony or wife support have usually been held to be valid. The same is true of arbitration proceedings dealing with payments for the support of a child or a wife and child. The cases are divided as to the validity of arbitration provisions concerning child custody or rights of visitation. Even attorneys have not been exempted in connection with disputes with their clients. If the parties have agreed to settle their differences by the use of arbitrators, the courts will approve of arbitration and will uphold the arbitration award in the absence of good reason to reject it. Some clients feel that the judge is a former attorney and will be partial to their former attorney and they prefer to take their chances of fair treatment by using arbitration as a substitute for a court hearing.

An agreement to submit a controversy to arbitration does not operate to stop the running of the statute of limitations. This is clearly so where the agreement to arbitrate is not acted upon. There is general agreement that a demand for arbitration must be made within a reasonable time, but opinion differs as to what is a reasonable time. Usually, a reasonable time is the statutory period and a party may not prolong the statute by failing or neglecting to take the necessary steps to perfect his claim for arbitration. There is contrary authority to the effect that where there is no showing of injury as a result of the delay, the right to enforce an arbitration clause accrues at the time the demand was made for arbitration and not at the time of the breach of the principal obligation.

Cases have arisen involving third persons who are not parties to the original contract which contained the arbitration agreement between the parties. Courts usually hold that such third persons are not bound by the arbitration agreement because they are not claiming rights under or through the parties to the contract.

The American Arbitration Association has adopted construction industry arbitration rules. This organization, also known as AAA, includes the American Institute of Architects;

Associated General Contractors; Consulting Engineers Council; Council of Mechanical Specialty Contracting Industries; and the National Society of Professional Engineers.

The Association does not act as arbitrator but does maintain throughout the United States a national panel of arbitrators consisting of experts in all trades and professions. They contend that when an agreement to arbitrate is written into a construction contract, it may expedite peaceful settlement without the necessity of going to arbitration at all. Thus, the arbitration clause is a form of insurance against loss of good will.

Some owners and contractors use this organization instead of selecting their own arbitrators and having their own two selected arbitrators, select the third arbitrator. In such cases, the AAA suggests that the contract provides as follows:

"Any controversy or claim arising out of or relating to this contract, or the breach thereof, shall be settled by arbitration in accordance with the Construction Industry Arbitration Rules of the American Arbitration Association, and judgment upon the award rendered by the Arbitrators may be entered in any Court having jurisdiction thereof."

## *12-5 RIGHT TO LIEN*

The mere fact that a construction contract contains an arbitration clause has been held not to be a waiver of lien rights. But there is a difference of judicial opinion as to whether the actual submission of the dispute to arbitration, in accordance with the agreement of the parties, constitutes a waiver of the right of one of the arbitrating parties to a mechanic's lien. To add to the confusion of rulings of different courts, one case held that the filing of an action to foreclose a mechanic's lien was not consistent with the agreement to submit any dispute to arbitration.

A submission to arbitration is sometimes referred to as an agreement for submission. It is a contract between two or more parties who, for settlement of their respective legal rights and duties, thereby refer disputed matters to the decision of arbitrators, by which decision they agree to be bound. The submission agreement must embrace everything necessary to give the arbitrators jurisdiction over the parties and subject matter in dispute, and to enable them finally to determine the controversy.

*Example*

1. An agreement for the submission of issues to arbitrators defines and limits the issues to be decided and constitutes the charter of the entire arbitration proceedings. The submission agreement determines what has been submitted to the arbitrators and, since the award is limited by the submission, the submission agreement should show clearly what disputes are to be arbitrated.

A submission to arbitration under a statute must conform to the statute in every essential particular, and a submission that departs substantially from statutory requirements is void. Parties to a contract may agree that any and all disputes growing out of it in any way shall be submitted to arbitration, and where a controversy under such an agreement is submitted to arbitration, the arbitrators are empowered to make such award as will fully settle the controversy. In such a submission the law, facts, and rules of evidence are submitted unreservedly to the arbitrators. After an award has been made, neither party can have the matter heard by either a court or another board of arbitrators. Arbitrators are not bound by strict adherence to legal procedures and to the rules of the admission of evidence that are applicable in court trials. Similarly, the arbitrators are not required to hire a court reporter to take down the testimony of witnesses nor to spell out the reasons for their decision or award.

The legal question of whether or not the parties have committed a particular dispute to arbitration is up to the court to decide. If the court's ruling is that the submission agreement withdraws a given controversy from arbitration, then the judge will issue his order cancelling or staying the arbitration.

The submission is the commission of the arbitrator. By force of it, the arbitrator becomes a judge, with absolute power over the things submitted to his judgment, both as to subject matter as well as procedural rules. It therefore follows that the submission agreement must be valid, otherwise the arbitrator has no jurisdiction.

## 12-6 PENDING LAWSUITS AND REVOCATION

The subject matter of a pending lawsuit may be submitted to arbitration, both at common law and under statutes. It can be accomplished by private agreement of the parties or by order of the court.

*Example*

1. Where a state recognizes arbitration as a desirable method of settling differences between parties, and the disputed matters are not contrary to public policy, the fact that the controversy is in court does not limit the right of the parties to agree either upon arbitration or the kind of arbitration they desire.

However, after a valid and final award has been made by the arbitrators, revocation is impossible because the agreement to arbitrate is no longer executory. The laws of many states now provide rules governing the procedure required for the revocation of an arbitration agreement. A common provision of these statutes is that an arbitration agreement is valid and enforceable and irrevocable, except upon such grounds as exist at law or in equity for the revocation of any contract.

*Example*

1. The rule of the common law that a submission may be revoked at any time prior to the award has been so far modified by statutes that courts are directed to enter summary orders to enforce arbitration contracts pertaining to either present or future differences arising out of the contract.

The United States Arbitration Act makes provisions for arbitration in contracts involving maritime transactions as well as transactions involving interstate commerce. Maritime relates to ships and commerce on the high seas. Interstate commerce refers to commerce transacted between the borders of more than one state, while intrastate commerce is limited to commerce within the borders of one state. The federal courts have ruled that the arbitration agreements authorized by the United States Arbitration Act are valid and enforceable and irrevocable, and that the federal courts will stay pending lawsuits on issues referable to arbitration under such contracts.

Whenever the revocation of an arbitration agreement has been accomplished, the authority and power of the arbitrators to determine submitted differences is terminated. Once revocation has taken place, the parties are now restored to their respective rights against each other as they existed before the arbitration agreement was made. They may resort to the courts for the settlement of their controversy with the same effect as if there had never been an arbitration agreement.

Generally, death of a party to an arbitration proceeding before an award has been rendered therein effects a revocation of the submission and terminates the authority of the arbitrators to act further. However, this result can be changed if the parties by a clear and express stipulation in the arbitration agreement, provide that the submission should not be revoked by the death of either party.

In those states in which an arbitration agreement is irrevocable, such an agreement may be enforced against an assignee as well as against the original contractor. Arbitration contracts would be of no value if either party could escape the effect of the arbitration clause by assigning to a third party a claim subject to arbitration between the original parties. By way of variation, a contractor's assignment of his contract to another general contractor was not an enlargement of his rights. A subcontractor whose subcontract did not provide for arbitration was not bound by the provisions of an arbitration clause in an assigned contract. A party to an arbitration agreement cannot unilaterally extend the scope of the submission agreement because the modification requires the consent of all of the parties.

## *12-7 WHO AND WHAT MAY BE ARBITRATED*

A novel situation arose when one party to an arbitration dispute said he did not deny his liability under the contract and had no defense to the disputed claim. He did contend that he was entitled to an arbitration hearing based on his refusal to pay the claim because of his counterclaim against the claimant, arising out of an unrelated contract. The court ruled that his argument was valid and that he was entitled to have the counterclaim dispute heard by a board of arbitration.

Where a party to an arbitration agreement appears before the arbitrators without objection to any provisions of the agreement, he ratifies such agreement and may not later object to the provisions the agreement contains.

The right to arbitrate given by a contract may be waived, even in those jurisdictions where a contract arbitration is irrevocable. Thus, where one party files a lawsuit, he waives his right to arbitration, because his conduct is clearly inconsistent with a claim that the parties were obligated to settle their differences by arbitration.

*Example*

1. The waiver of the right to arbitrate need not be expressed in terms; it may be implied from the acts, omissions or conduct of a party. This is especially true of stipulations in contracts for appraisement of damages or arbitration of amounts, made operative only by some affirmative action of the parties, as, for example, a request or demand in writing.

A dispute, controversy, or honest difference of opinion between parties concerning any subject in which both are interested is sufficient and a matter that is simply in doubt may be submitted to arbitration. In any case it is generally held that there must be a controversy of some type between the parties. Until there is a controversy there is nothing to arbitrate.

All of the legal authorities agree that any matters concerning a dead person's estate which all parties concerned could settle by agreement are proper subjects for a valid arbitration.

*Example*

1. An agreement in writing by a widow and the heirs of a landed proprietor who dies intestate to submit the division of the land to three disinterested persons is a valid common law submission and the award of the arbitrators is binding.

Intestate means to die without leaving a will. If you leave a will you die testate.

The arbitration agreement signed by one partner only may be validated by ratification of the remaining partners. To play it safe, it is recommended that the arbitration agreement should be signed by all of the partners. Where all of the partners sign an agreement for arbitration of future disputes that may arise under a contract, a valid submission or demand for submission may thereafter be made by one of the partners. Partners may and frequently do, agree to submit disputes between themselves to arbitration.

Corporations have the same power as individuals to submit their controversies to arbitration.

Arbitrators may be defined as private, extraordinary judges of a domestic tribunal, chosen by parties by whose agreement they are invested with quasi-judicial power to decide, finally and without appeal, matters in dispute between the parties. They are considered sometimes as the substitutes, and sometimes as the judges, of the parties to the submission and it has been said they are both judges and jury.

An "umpire" is a person selected by the arbitrators, according to the authority of the submission, to decide the matter in controversy when the arbitrators are unable to agree. The umpire stands in the same position as a sole arbitrator of the issue originally submitted to the arbitrators.

There is a clear distinction between an umpire and an additional arbitrator. The umpire has the power to settle the dispute by his sole award. The additional arbitrator acts only in conjunction with the other arbitrators. Whether a given person is an umpire or an additional arbitrator depends not at all on what he is called by the parties, but on his power under the agreement of submission to make a sole award. If one party to an arbitration agreement fails or refuses to appoint an arbitrator in the manner agreed, the courts will take a hand and make the appointment.

*Examples*

1. While it is true that settlement of a controversy by arbitration is favored by the courts, nevertheless, the authority of the arbitrators is derived from the mutual assent of the parties to the terms of the submission.
2. A submission agreement is as binding on the arbitrator who accepts his duties as on the parties to the agreement.
3. The submission is the commission of the arbitrators.
4. The arbitration agreement is the law of the case in an arbitration matter.

Arbitration agreements very frequently provide that the decision of the issues may be made by a majority of the board of arbitrators. That is why most arbitration agreements provide that each side will designate one arbitrator and the two arbitrators so designated will select a third arbitrator. Ordinarily, any competent, disinterested person may be an arbitrator no matter what his or her legal status may be. A trial court judge has been held not to be eligible to serve as an arbitrator.

The arbitrators that are usually designated by the parties are experts on the subject in controversy and bring to their task a wealth of specialized knowledge.

*Example*

1. Arbitrators did not improperly delegate their responsibilities where the arbitration agreement in connection with a contract for the exploration and production of uranium ore contemplated the employment of expert personnel by the arbitrators and the determination of the dispute in question was independently made by the arbitrators based on their own skill and a review of the work of an expert employed by them.

If parties are to be encouraged to arbitrate their disputes, arbitration proceedings must be conducted with the same degree of impartiality as the courts afford. Public policy requires that arbitrators not only be completely impartial, but also that they have no connections with the parties or the dispute involved which might give the appearance of their being otherwise. Obviously, a person is disqualified to act as an arbitrator if he himself is a party to the dispute. Mere personal friendship with one of the parties does not mean an automatic disqualification. The mere fact that there is some business relationship between the arbitrator and one of the parties to the arbitration is not in and of itself sufficient to disqualify the arbitrator.

The fact that an arbitrator, after lengthy proceedings, indicated that a settlement might be advantageous to the parties, and expressed his view of the case, did not indicate bias on his part. Bias or prejudice on the part of an arbitrator, or undue interest in the dispute to be arbitrated, may be sufficient reason to warrant removal of an arbitrator, prior to the award.

There is an implied agreement to pay such sums as will fairly and reasonably compensate arbitrators for expenses incurred and services rendered. What is reasonable is governed by the nature of the case, the time consumed, the amount involved, and other particular circumstances. Arbitrators have a lien for their fee on the award they make and may withhold it from the parties until payment is made. In addition to their remedy by way of lien on the award, arbitrators may maintain a lawsuit for their compensation. The costs of arbitration are usually divided equally between the parties and are deductible as business expenses for federal income tax purposes.

The essence of arbitration is its freedom from the formality of ordinary legal procedures. Although informal in nature, an arbitration proceeding is a quasi-legal procedure.

*Example*

1. Arbitration may or may not be a desirable substitute for trials in court, but when the parties adopt it they must be content with its informalities. They may not hedge it about with those procedural limitations which it is precisely its purpose to avoid.

It is the duty of arbitrators to hear the evidence adduced by the parties and to make a fair and impartial award. Each party must be allowed an opportunity to present the claims he has in full. Ordinarily, arbitrators have no power to compel a witness to attend a hearing and testify. Each party has a right to cross-examine a witness produced by the adversary.

## 12-8 THE AWARD

The decision of the arbitrators determining the disputed matters submitted to them is the award. The award is final as to all matters submitted. An award is ordinarily sufficient if the decision represents the honest judgment of the arbitrators, acting reasonably and fairly. An award is not required to be stated in technical language.

The Uniform Arbitration Act and the statutes in some of the more populous states on which the uniform act is based, require that the award be in writing and signed by the arbitrators joining in making it. Arbitrators need not disclose the facts or reasons behind their award unless the arbitration agreement or submission, or an applicable statute requires them to do so. Otherwise, they are no more bound to go into particulars, or to give reasons for their award, than a jury for its verdict.

A final determination of the submitted disputes exhausts the power of the arbitrators and thereafter the matter is beyond their control. The board of arbitrators have no power to recall, reconsider, amend, or otherwise alter the award, except they may correct clerical mistakes or any error appearing on the face of the award.

The courts favor arbitration proceedings as being a quick, amicable and inexpensive method of settling private disputes and will make all fair presumptions in order to sustain the award. A court will enforce an arbitration award if it is valid and will set it aside if it is not, but the merits of a controversy submitted to arbitration on which an award is made are not subject to judicial review. The award of the arbitrators acting within the scope of their authority determines the rights of the parties as effectually as a court judgment, and is as binding as a court judgment until set aside or its validity questioned in a proper manner.

Once a valid award is made it is the only basis for determining rights as to any demands embraced in the submission because the claims are merged in the award. The award is regarded as the judgment of the court of last resort.

When the arbitration agreement provides that disputes arising thereunder may be arbitrated in another state, or in another country according to the laws obtaining in that country, the agreement is binding even though neither of the parties live there.

The investigation of arbitrators is in the nature of a judicial inquiry and ordinarily involves a hearing and all that is thereby implied. Arbitrators should not proceed independently, and without notice to the parties, to make personal inquiries or investigations on which they intend to base their awards. It follows that an ex parte investigation by arbitrators may constitute misconduct such as will warrant setting aside the award.

*Example*

1. Choosing arbitrators wholly disinterested is an admirable standard to aspire to, but the parties seldom do that, and if all awards were set aside in which it was not done, few awards would stand.

# 13

# SURETY BONDS

## 13-1 DEFINITIONS

A surety is one who becomes a guarantor for another person. Many people have had the experience of helping out a friend by cosigning his note at a bank or finance company. When you cosign the note, you become a guarantor and in effect are saying, "If my friend does not pay as agreed, I agree to be responsible and pay his obligation." If your friend pays as agreed, then your obligation as guarantor terminates by virtue of the contract having been completed. If your friend does not pay as agreed, the creditor calls on you for payment on the remaining balance still due on the obligation.

When talking about surety bonds, you are talking about a different type of guarantor and the subject matter has been changed materially. In the construction business, the amounts involved can be very sizeable sums of money and even friends of the contractor are not interested in "sticking their neck out" by becoming his guarantor for thousands of dollars. With the passing of time, the business of becoming a professional guarantor was taken over by surety or bonding companies, who charge a premium and issue their bond. There is no difference between the terms surety company and bonding company and they are used interchangeably.

Just as life insurance companies charge premiums for life insurance policies that are based on the longevity or life of the insured, bonding companies charge so much per thousand dollars of the bond that they issue. The surety bond is ordered by the general contractor or by the subcontractor and the person that usually benefits under the terms of the bond is the owner.

The owner of the property and the building to be constructed is called the "obligee" under the bond issued by the bonding company. The bonding company that issues the bond is called the "obligor." Any other persons that claim benefits under the bond are called "third party beneficiaries." Third party beneficiaries could be laborers, materialmen, or subcontractors, all of whom have a claim for something they furnished or did that benefited the owner's construction project, that is covered under the surety bond issued by the bonding company. The law does not require that the third party beneficiary and the owner know each other or even require that the materialman knew that the job was covered by a surety bond, in order to create liability for the bonding company.

Before the advent of bonding companies, the courts used to construe the rules of suretyship strictly in favor of the surety. However, this has changed since the bonding companies are professionals in the business of acting as surety for compensation. For all practical purposes, bonding companies are essentially insurers and the rules are now interpreted by the courts strictly against the surety company and in favor of obligees and the third party beneficiaries under the bond. No hardship is being imposed upon the surety company by this rule of construction, since this factor, as well as all others, is known by the bonding company when they determine their risk and in turn, when they fix their premium fees for executing the bond.

Arranging for the issuance of surety bonds is a proper function of your insurance broker. Some bonding companies contend that they suffered substantial losses on surety bonds and as a result they raised the premiums and became more stringent in their requirements. In rare instances, the requirements reached the ridiculous stage where the bonding company wanted assets of the contractor to be about ten to one over the amount of the surety bond. This unreasonable requirement is nearly as bad as a few ultra conservative and greedy banks who ask that the borrower leave 50 percent of his borrowings with the bank, at no interest, as collateral for the bank loan. If you stopped to figure out the rate of interest you are paying for the 50 percent of the money you are permitted to use, it is shocking and unconscionable.

Your insurance broker can make a valuable contribution to the success of your construction business, by really pitching in and lending a helpful hand, to convince the bonding company of your ability and integrity and the advisability of their issuing the required surety bond. If one bonding company turns down your application for a surety bond, your insurance representative should select another bonding company and sell them on your financial responsibility. The fact that your insurance broker will earn a brokerage fee for writing the surety bond is aside from the more important fact that he has rendered an aggressive and helpful service. There are two occasions when your insurance broker can prove his true worth as your insurance representative. One takes place in helping you obtain a necessary and important surety bond and the other occurs when you need help in settling a claim against your own insurance company and they are dragging their feet. If your insurance man passes those two tests with flying colors, then he is truly a valuable asset to your construction business. If he fails in either test, you might decide to look elsewhere for a replacement.

The smiling insurance broker who thanks his customer for his insurance business and the monthly check for premiums has not had the acid test of doing something for his customer. The best time to discuss your requirements in connection with the handling of your insurance claims and surety bonds is before you make the selection of your insurance agent.

INDEMNITY BOND

Know all men by these presents, that we, JOHN JONES, as principal, and abc bonding company, as surety, are held and firmly bound unto RICHARD ROE, in the sum of $25,000 dollars, to be paid to the said RICHARD ROE, his heirs, executors, administrators, or assigns, for which payment, well and truly to be made, we bind ourselves, our heirs, executors and administrators, jointly and severally, firmly by these presents.

Dated and signed this _________ day of December, 197_.

Now, therefore, the condition of this obligation is such, that if said JOHN JONES, his heirs, executors, administrators, successors and assigns, shall well and truly save harmless and keep indemnified the said RICHARD ROE, his heirs, executors, administrators, successors and assigns, from and against all claims and demands of any person or persons whomsoever as well as all costs, expenses, damages, or attorney's fees for which said RICHARD ROE may become liable or answerable by reason of any claim or demand of any such person or persons by reason of the condition of this obligation, then this obligation shall be null and void; otherwise to remain in full force and effect.

/s/ JOHN JONES (L.S.)
John Jones

/s/ ABC BONDING COMPANY (L.S.)
By /s/ William Wright
Statutory Agent

# SURETY BOND CHECK LIST

## *Part I: Types of Bonds*

☐ **A. BID OR PROPOSAL BONDS**—These bonds provide that if the contract is awarded the contractor will execute the contract documents and furnish the required Performance and Payment Bonds, if requested. The penalties assumed by the Surety under Bid Bonds vary from as little as 2% of the total amount of the bid to as much as 20% on certain Federal Government projects.

*Surety Co.* ______________________ *Form No.* __________ *Am't.* ____________

☐ **B. PERFORMANCE BONDS**—In short, these bonds provide for performance of the agreement by the contractor and payment by the Surety of the owner's loss up to the bond penalty if the contractor defaults.

*Surety Co.* ______________________ *Form No.* __________ *Am't.* ____________

☐ **C. LABOR AND MATERIAL PAYMENT BONDS**—These bonds provide that the furnishers of labor and materials going into the project shall be promptly paid. (See "Double Jeopardy," page 10.)

*Surety Co.* ______________________ *Form No.* __________ *Am't.* ____________

☐ **D. MAINTENANCE BONDS**—Maintenance bonds protect the owner, usually for a period of one year, against defects in workmanship and materials. In many states, if the contract provisions are adequate, separate maintenance bonds are not necessary because the Performance Bond automatically applies to such maintenance guarantees.

*Surety Co.* ______________________ *Form No.* __________ *Am't.* ____________

☐ **E. SPECIAL INDEMNITY BONDS**—In some instances, a special indemnification is required against injury to persons or property. Such provisions usually involve "Contractual Liability" from an insurance standpoint and the contractor should always have Contractual Liability insurance to protect himself against the liability assumed under such Indemnity Bonds or under the contract. (See Part I, paragraph B-5, Insurance Check List.)

*Surety Co.* ______________________ *Form No.* __________ *Am't.* ____________

☐ **F. RELEASE OF LIEN BONDS**—These bonds are given to obtain payment from the owner when a mechanic's or other lien has been filed against the premises or the unpaid contract price. In substance, they assure the owner that he will not suffer loss or damage from the lien claimed, if, notwithstanding this claim of lien, payments continue to be made to the contractor.

*Surety Co.* ______________________ *Form No.* __________ *Am't.* ____________

# *Part II: Factors in Bonding*

**A SURETY BOND IS NOT INSURANCE**—It will not insure the owner or contractor against loss or guarantee against financial failure. Basically it is comparable to a conditional promissory note guaranteeing protection to the owner or obligee.

**CONSIDERATIONS WHEN BONDING**—Surety companies consider many factors prior to underwriting a contract bond.

When requesting bonds a contractor should be prepared to present factual information regarding his firm or company including: a general historical background of the firm, as well as information on:

(a) recent projects constructed;

(b) financial status, including bank credit and secondary assets;

(c) equipment and machinery, including amount immediately available;

(d) organization, including experience, skill and responsibility of key personnel.

**SELECT SURETY CAREFULLY**—To facilitate the handling of bonds, contractors should select a thoroughly qualified and experienced agent who has bona fide surety company affiliations to enable him to speak for and bind the company he represents and to offer promptly the bonding service needed by the contractor. Obtain from the bonding agent a certified copy of the power of attorney which clearly sets out the extent of his authority to act for his surety company.

**FORCED PLACEMENT OF INSURANCE AND BONDS**—Any action or contractual requirement depriving the general contractor from selecting the reputable surety companies of his choosing is a departure from traditional practice and is contrary to the best interest of the owner and contractor.

**DOUBLE JEOPARDY**—Many labor and material payment bonds used today stipulate that the general contractor (or principal) shall promptly make payment for all labor and material used or required in the performance of a contract. This is particularly true of the Labor and Material Payment Bond Form A311 prepared and published by The American Institute of Architects.

This type of clause has been interpreted to mean that the general contractor can become liable for any unpaid bills of a subcontractor who has defaulted, even though the general contractor has paid the subcontractor for his just proportion of service or materials installed.

For protection against double payments of this nature the following factors should be considered:

(a) select all subcontractors with care:

(b) require all subcontractors to furnish evidence (affidavits, receipts, payroll vouchers, etc.) to verify all material bills paid; or

(c) require all subcontractors to furnish a performance and labor and material payment bond.

*NOTE: Every state has some form of mechanics and materials lien law. Since the requirements of these laws affecting private work vary from state to state (and since the provisions contained in these laws may impose additional liabilities) it is important that these laws be reviewed for that state or local political subdivision where the project is located.*

### 13-2 PERFORMANCE BONDS

There are two types of contractors' bonds and sometimes they are combined in a single bond. They are performance bonds and labor and material bonds. A performance bond guarantees that the contractor will perform the contract. It usually provides that if the contractor defaults and fails to complete the contract, the surety company can itself complete the contract or pay damages up to the limit of the bond. The liability of the bonding company is limited to the amount of the bond and under no circumstances could its liability exceed the named amount. If the amount of damages are greater than the amount of the bond, it can lead to problems by the bonding company taking advantage of its position by offering a lesser amount to the owner, he being the loser and the beneficiary. It is sad but true that some insurance companies will offer a lesser amount than is due under the provisions of the bond, knowing that a lawsuit by the owner is costly to him and he can only recover the amount named in the bond. This procedure is really chiseling; however, some unscrupulous insurance companies will take advantage of this type of situation.

The second type of contractors' bond is a labor and material bond that guarantees the owner that all bills for labor and materials contracted for and used by the contractor will be paid by the surety company if the contractor defaults. For the protection of the owner, it is desirable that both types of bond be combined in a single bond. Even though the general contractor arranges for the surety bond and pays for it, in the final analysis, all costs of construction including bond premiums are paid for by the owner. If you are going to pay insurance premiums, you should have adequate coverage to protect the beneficiary. If the contractor goes broke and is unable to complete the construction job and the bonding company suffers a loss in completing the project, they have a claim against the contractor for their loss, regardless of their chances of collecting from him.

With regard to construction contractors bonds, performance bonds are to be distinguished from indemnity bonds. Under the usual form of a performance bond, the surety company guarantees not only that the building will be completed within the contract price without extra cost to the owner, but also that payment will be made by the contractor to the subcontractors and to those who furnished labor and materials.

On the other hand, the purpose of the indemnity bond is for the sole benefit of the owner, to secure for him the completion of the building within the terms of the contract and without extra cost to him because of liens for unpaid subcontractors and labor and material, for which he might otherwise be liable. As a rule, an owner of premises on which construction work is being done may require the contractor to give a bond for the payment of claims for material purchased by him on the credit of the building and land.

In some states, laws have been enacted permitting or requiring an owner to exact from the contractor, a surety bond protecting subcontractors, materialmen and laborers. Some state laws provide that the owner shall obtain a surety bond from the contractor, and that on his failure to do so the owner shall be liable to any person furnishing labor or materials used in the building, for the amount due him.

Now the owner has complied with the state law and obtained a surety bond from the contractor protecting subcontractors, materialmen and laborers and the contractor went broke and failed to pay the bills. The liability of the surety company is the same as that of the owner, provided the contractor's default falls within the provisions of the surety bond. Don't forget, the surety bond is a contract just as is the building contract between the owner and the contractor. Whenever you are dealing with building contracts that are usually technical and complicated, and a default occurs, then the bonding company does not automatically pay out money like a slot machine. If there are any legal loopholes, the attorneys for the bonding company will argue the liability of their company with your insurance agent and your attorney. If the insurance agent of the contractor does not also represent the owner, he really is not interested in what position the bonding company takes in dishonoring a claim.

Where a construction contract required the contractor to furnish all of the materials, then the surety company who issued their performance bond is liable for claims for materials which the contractor failed to pay.

*Example*

1. Under a contract providing for wallpapering a house to the owner's taste, with an allowance of a stated amount to cover the cost of materials used in wallpapering, the performance bond includes the price of wallpaper sold to the owner.

A question frequently arises regarding whether the furnishing of certain labor, services, or materials falls within the terms of a contractor's performance bond, apart from the question whether the person furnishing them is entitled to the protection of the bond. Some courts have held that the labor and materials, as specified in a performance bond, must be necessary to construct the work in accordance with the contract.

The following examples may help to clarify the legal ruling as set down by the courts. Lumber used in the making of forms for concrete in a structure is material within the meaning of a performance bond to pay all indebtedness incurred for labor and materials used in the contract work. But merchandise and provisions furnished to the employees of a subcontractor, were not included within the language of a bond for the payment of all laborers and materialmen who supply the contractor with provisions or goods of any kind. Furthermore, freight charges were held by the court not to fall within the terms of a contractor's performance bond to secure all claims for labor or material which might be the basis of mechanic's liens. The surety company is liable for gas and oil and for necessary, ordinary repairs to the equipment used in the performance of the contract, but is not liable for equipment rental or transportation of equipment.

## *13-3 WHO IS COVERED*

The coverage of any type of bond depends upon its terms and the laws of the applicable state under which the contract is written. Frequently, there is no clear cut answer to the problem as there is in answering a mathematical problem. Some courts hold one way under the provisions of their own state law and another court holds differently under its state law that varies from the first and the final result is a hodge-podge of rules and exceptions to the rule. When discussing various legal principles and rules, textbook writers and legal scholars will give a rule of law as representing the majority opinion and then list the many and varied opinions that veer off on different tangents. The old phrase, "there is an exception to every rule" probably applied to this situation.

Even the highest court in the land, the United States Supreme Court with its nine justices, seem to wind up with more five to four decisions than any other numerical count. Whenever a complicated legal problem is decided by five justices voting yes and four justices voting no, it is easy to appreciate the fact that there is merit to both sides of the legal question. That is why the client cannot always appreciate a vague answer from his attorney when the client asks, "What is the law on my particular problem?" and the attorney is forced to hedge.

A contractor's performance bond provides that the owner-obligee shall be held harmless for any breach of contract on the part of the contractor. A mechanics' lien was filed against the owner's property and the owner paid the lien claim and then claimed protection for reimbursement under the contractor's bond. The court held that the owner was protected under the provisions of the bond for his payment of the mechanics' lien. Another case decided that the surety company was not responsible under its performance bond issued to the contractor, for payments made by the owner to materialmen where no valid lien could have been asserted against the property. This is a good example of a rule and then comes an exception to the rule.

A number of ticklish situations have occurred when someone lends money to the contractor who is in a financial bind, to help him pay for labor and material furnished on the project and then the contractor goes broke and a claim is made against the bonding company on its performance bond. Generally speaking, a lender of money to the contractor cannot recover his loan from the surety company, even though the borrowed money was used to pay the cost of labor and material used in the project. Under a performance bond that is conditioned on the satisfaction of all claims for labor, materials, or otherwise, the bonding company was held not liable for money loaned to the contractor expressly for paying the cost of labor and material. However, where a construction contract provided that the contractor would pay all just debts, due and demands incurred in the performance of the work, and the performance bond was conditioned on faithful performance of the construction contract, it was held that the performance bond did cover money loaned to the contractor and used to pay for labor and material. The court rulings indicate clearly that if the money loaned to the contractor could be used by him for any purpose he chose, then the lender of money would not be protected under the contractor's performance bond. Even though it may sound repetitious, money lending and borrowing should be handled by banks and other financial institutions, who are professional money lenders and are in a better position to stand an occasional loss. Plus the fact that personal friendship will not be lost because of financial dealings.

Courts have uniformly held that even though money loaned to a building contractor has been used to pay laborers or materialmen on the project, the lender does not become subrogated to the claims and rights of such laborers and materialmen against the contractor's performance bond.

To subrogate means to stand in the shoes of or in the position of another as regards his legal rights. When your car is involved in an accident resulting in \$300 property damage and your insurance company pays \$250 and you pay the other \$50, your insurance company becomes subrogated to your claim against the driver of the other car for the \$250 that it paid on your claim. If the \$300 claim is recovered from the other party, then \$250 goes to your insurance company and the other \$50 belongs to you.

A money lender cannot, merely by agreement with the contractor, become subrogated to the rights of laborers or materialmen by furnishing money for their payment. An entirely different situation results where the money lender pays the laborers directly and takes a written assignment from them of their claim against the contractor. In addition, the money lender obtains a written acknowledgment from the contractor recognizing his claim as valid. Under these conditions, the money lender can recover from the surety company on its contractor's bond that it issued, conditioned on the payment of laborers and all furnishers of materials and supplies

Premiums for insurance, such as workmen's compensation insurance, are not covered by the contractor's bond conditioned upon faithful performance of the principal contractor and/or the payment of labor and materials, which contains no general obligation to pay all claims incurred. Even in cases where the contractor is required by law or by contract to procure or maintain the insurance, his surety bond does not cover the premium cost.

*Example*

1. Where the contract obligates the contractor to maintain insurance, and he fails to do so, the owner may not recover upon a bond for insurance premiums he has paid, even though the bond is conditioned in general terms upon faithful performance of the contractor's agreement.

The contract or suretyship under a contractor's performance bond imports entire good faith and confidence between the parties in regards to the whole transaction. The slightest fraud or alteration of the contract cancels it. Thus, a surety to a building contractor's performance bond is not bound if there is concealment or nondisclosure of information by the obligee. Moreover, if the principal construction contract is invalid, no action can be maintained against the bonding company. The obligee, or owner of the premises on which the construction work is being performed, is of course entitled to the benefit and protection of a contractor's performance bond. Where the contractor's performance bond recited that the bond accrued to the benefit of all who might become entitled to the liens under the building contract, the owners were entitled to recover on the performance bond for the amount of liens paid after the contractor had abandoned the work.

An important and well recognized rule of law provides, "When a promise is made to one, upon a sufficient consideration, for the benefit of another, the latter may sue the promisor for a breach of his promise." This rule is also known as and applies to the third party beneficiary type of contract, which means that frequently, the parties do not even know each other. It is unnecessary, in order for a laborer or materialman to recover on a performance bond in favor of the owner, that the former be known at the time the bond is made. Nor is it necessary that there be any privity or dealings between the owner-obligee on the bond and the third party beneficiary, in order for the latter to recover. Most of the courts agree and hold that anyone furnishing materials or labor may recover on a building contractor's performance bond to the owner of the property where the bond contains a condition for the benefit, and is intended for the protection of laborers and materialmen. This is true even though the owner is the only obligee named in the bond and there is no express provision that it shall accrue to the benefit of laborers or materialmen or that they may avail themselves of the security offered by the performance bond. This rule of law pertaining to third party beneficiaries is sound law and has been expanded over a period of years.

*Examples*

1. The fact that a materialman was not entitled to a lien against the property of the owner-obligee did not prevent him from maintaining an action on the bond, where it was conditioned on the payment of such claims, since it is not essential that there be privity of contract between the obligee and the third party beneficiary, or that the obligee should owe any duty to the beneficiary.

2. The action upon the bond is not in any sense based upon the personal liability of the contractor, but is based upon the obligation of the bond, since the bond provides a separate and distinct and statutory remedy. The obligation of the bond, therefore, is enforceable without reference to any contract between the materialman and the contractor.

The mere fact that laborers and materialmen did not know of the existence of a contractor's performance bond to the owner, conditioned for their benefit, at the time they furnished the materials or labor, does not prevent them from availing themselves of the protection of the bond. Moreover, a laborer's or materialman's right to recover on the building contractor's performance bond to the owner is not dependent upon any law giving such a remedy or requiring such a bond, or upon the right of the laborer or materialman to a lien on the owner's property. Since the rights of laborers and materialmen are independent of the rights of the obligee in the bond, their right to recover against a surety company on such a bond cannot be defeated by any act of the owner-obligee, even though it could affect the owner's right of recovery.

The intention of the parties to a contractor's performance bond is the controlling factor in determining the rights of laborers and materialmen to recover on the bond. However, this intention is to be determined by the terms of the performance bond construed in the light of the construction contract in connection with which it is executed. The intention must be gathered by reading the construction contract and the performance bond in the light of the circumstances under which they are executed.

*Example*

1. The owner's motive is not important, the real question being whether or not a promise has been exacted that laborers and materialmen shall be paid, rather than a mere promise to protect the owners if they are not paid.

## 13-4 DISCHARGE OF SURETY

A surety company or any individual surety on the performance bond of a building contractor may be discharged from his obligation in the same manner as any other surety may be discharged, such as, by a change in the terms of the contract, or by a change in principals.

*Example*

1. Where a partnership of three persons contracted to construct a building, and a surety guaranteed the faithful performance of the contract, and later one of the partners withdrew from the partnership, assigned his interest to one of the remaining partners, and was released from his obligations under the contract, the surety was also released from liability for the default of the new partnership.

If the obligee of a bond violates any provision of his contract made for the benefit of his principal or his sureties and injury results therefrom, the sureties are discharged. When the owner of property fails to insure it as required by a building contract and a loss results that injuriously affects the rights of the contractor's sureties, the latter are discharged from their liability.

It is sometimes provided in a contractor's bond that the surety shall not be liable for damages resulting from strikes or labor difficulties. A strike in the common understanding of the term must occur. The mere act of union employees in leaving the contractor because he loses his standing in the employees' association does not come within the limits of an exception releasing the sureties in case of a strike or labor difficulties.

A material alteration made without the consent of the surety on the contractor's bond will discharge such surety. If the contract between the owner and contractor permits alterations to be made in the work to be done, or if the contractor's bond itself permits the alterations, the surety on the bond will not be discharged, where such alteration is one contemplated by the terms of the stipulation. This rule is based upon the doctrine of prior consent by the surety.

As a general rule, the acceptance by the owner of work performed under a construction contract prevents him from later recovering for a breach of the construction contract on a bond conditioned on the performance of the contract or guaranteeing the completed work. This is clearly so, where the contractor's bond expressly provides that it shall be annulled by the acceptance of the work by the owner. An owner's failure to give notice to the surety of the principal's default does not discharge the surety unless the bond specifically provides for such notice.

A provision in a performance bond contract to give notice of default is valid providing it is reasonable and is a condition precedent to an action so based.

*Example*

1. A surety on a contractor's bond is not entitled to notice of the progress of the work, or of the death of the contractor, and of the steps taken thereafter to complete the work, in the absence of an express agreement. Where no time is specified as to the giving of notice of default, it must be given within a reasonable time under all of the circumstances. If the surety already has notice or is chargeable with knowledge of the default, a failure to give the notice required by the bond does not relieve it from liability.

In the absence of any express provision to the contrary in the bond, payment, in accordance with the terms of the construction contract, made by the owner to the contractor, does not release the surety on the contractor's bond from liability to the owner. However, where a construction contract makes express provisions with regard to the times or amounts of payments made to the contractor, the retention of percentages, or the production of certain certificates, estimates or receipted bills and a material departure is made by the owner without the surety's consent, a different result is reached. Under these facts, a material departure by the owner from the requirements of the contract, operates to discharge the surety from liability to the owner on the contractor's bond, to the extent that such unauthorized payments result in injury or prejudice to the surety. This rule has been justified on the grounds that such provisions in construction contracts are not only for the protection of the owner, but also for the benefit of the surety on the contractor's bond, and that they afford a security in the hands of the owner for the protection of the surety.

The above rule calling for the discharge of the surety by reason of improper payments to the contractor is subject to certain qualifications. In order to have the effect of discharging the surety on the contractor's bond from liability to the owner, departures from the terms of the contract with respect to payments made by the owner to the contractor must be material or substantial.

*Example*

1. Where a contract required that payments be made on the fifteenth day of each month, the making of payments on other dates, not earlier or before they become due, was merely a technical departure from the contract and did not release the surety from liability to the owner.

When payments made by an owner do not injure or prejudice the surety on the contractor's bond, the surety is not discharged even though the payments do not comply with the provisions of the building contract. A premature payment to the contractor will release the surety from liability to the extent of such payment.

*Example*

1. Where the contractor abandons the work long before completion and the owner finishes the work himself, the surety's liability is not affected by the fact that the owner did not reserve the stipulated percentage of the contract price, such provision being predicated on the contractor's performance of the contract.

It is sometimes provided by the building contract or the performance bond that an owner shall withhold payments from the contractor in order to meet claims or liens of third persons against the owner or his property. The effect on the surety's liability of an owner's failure to withhold such payments depends on the language of the contract itself as well as the performance bond in each particular case and also upon the circumstances under which the payments are made to the contractor.

When a bond provides not merely for the performance of the contract, but also for the payment of materialmen and laborers, their rights cannot be affected by violations of the contract by the owner, and as to them the surety is not released by the owner's violation of a provision in the contract for retained percentages. Although where laborers and materialmen seek no relief from the surety, but file liens on the premises, the owner cannot recover against the surety.

*Example*

1. The theory is that the liability created by the bond in favor of materialmen is independent of that assumed as to the obligee and cannot be affected by the subsequent acts of the obligee.

An owner's failure to require, as a condition of payment to the contractor, a certificate from his architect, or receipts showing payment of claims for labor and materials, does not release the surety from liability on the contractor's bond to the owner, since such payments cannot be said to have injured or prejudiced the surety.

*Example*

1. A surety on a building contractor's bond cannot claim exoneration from liability because payments were made on the contract price without a certificate from an architect or other representative of the owner employed for the purpose of supervising the work where the contract, as interpreted by the provisions of the bond, leaves the employment of an architect or representative optional with the owner, although the contract provides that payment shall be made "only upon certificates of the architect."

The surety on the contractor's bond is not discharged from liability to the owner by reason of payments made in good faith in accordance with overestimates or erroneous payments, although such payments exceed in fact the amount due under the contract. When payment is made by the owner on the basis of receipts forged by the contractor, the surety on the contractor's performance bond is not discharged.

One of several sureties on a contractor's bond, who, upon the default of the contractor, finishes the work, cannot compel the co-sureties to contribute to the expense without permitting them to share in the amount received by him under the contract. A surety on a contractor's bond is entitled to be subrogated to the rights of the obligee or of laborers and materialmen against the principal. The surety is also entitled to any and all funds still in the hands of the owner-obligee, after the surety has completed and paid for the building project.

## *13-5 PUBLIC CONSTRUCTION WORK*

Usually, the giving of a performance bond by the successful bidder for a public construction contract is prescribed by statute or by ordinance. A contractor for public building or construction work is ordinarily required to give a performance bond conditioned on the faithful performance of the contract or indemnification of the obligee government or public body, and on the payment of the claims of laborers, materialmen and subcontractors in the prosecution of the work.

Under the provisions of the Miller Act of 1935, before any contract exceeding $2,000 in amount, for the construction, alteration or repair of any public building or public work of the United States is awarded to any person, such person shall furnish to the United States a performance bond for its protection and a payment bond for the protection of all persons supplying labor and material in the prosecution of the work provided for in the contract.

The Miller Act provides that the contracting officer is authorized to waive the requirement of a performance bond and payment bond when the work is to be performed in a foreign country, if he finds that it is impracticable for the contractor to furnish such bonds. It is also provided in the Miller Act that the requirements of performance and payment bonds may be waived by the Secretary of the Army, Navy, Air Force, or Treasury, with respect to cost-plus-a-fixed fee and other cost-type contracts.

Public bodies are under a moral obligation or public duty to protect persons furnishing materials and labor to their contractors, against the default of such contractors. Without such bonds, laborers and materialmen may suffer much injustice and loss by reason of the failure of irresponsible contractors to pay for the labor and materials used in making public improvements, in view of the fact that such laborers and materialmen have no right to file a mechanics' lien upon public property. The Miller Act enacted by Congress is designed to give those who supply labor or materials for use in federal construction the same protection that state lien laws give to persons furnishing labor or materials for use in private construction. The property of the United States is not subject to liens or to state lien laws and thus, the bond protection of the Miller Act is in lieu of liens provided by state laws.

*Example*

1. The Miller Act, being highly remedial in nature, is entitled to a liberal construction and application in order to properly effectuate the congressional intent to protect those whose labor and material go into federal projects.

The legal theory that entitles a laborer, materialman or subcontractor to file a claim against the surety company based upon a bond issued by a surety company to a city, county, state or the United States, is that a third person may, in his own right and name, enforce a promise made for his benefit even though he is a stranger both to the contract and to the consideration. This is called a third party beneficiary contract.

*Examples*

1. Where one person for a valuable consideration engages with another to do some act for the benefit of a third person, the latter may maintain an action against the promisor for the breach of the agreement.

2. An agreement by a contractor with an owner of a building to perform remodeling of the lobby of the building "in such a way as to cause a minimum of disturbance to the daytime operations of the building" is made directly for the benefit of tenants in the building, so as to raise a cause of action for breach thereof under the third-party beneficiary statute in favor of a tenant-furrier whose merchandise, displayed as to be viewed by customers, is damaged by dust arising from the contractor's operations.

The fact that laborers and materialmen did not know of the existence of a contractor's performance and payment bond, conditioned for their benefit, at the time they furnished the materials or labor, or that they did not act upon the faith of or in reliance upon a public contractor's bond, does not prevent them from availing themselves of the benefit of the bond. Furthermore, the fact that the public body is not liable to laborers, materialmen, or subcontractors does not defeat the right of the latter to sue on the contractor's bond.

An action upon the payment bond of a contractor for federal work, given pursuant to the Miller Act, must be brought in the United States Federal District Court for any district in which the contract was to be performed and executed. Under the Miller Act, the federal court has original jurisdiction of the action regardless of the amount in controversy. The rights and liabilities of the parties to a bond given to the United States under the Miller Act are governed by federal and not by state law.

A bond given for the faithful performance of a public contract, in pursuance of a local law, is a contract governed by the law of the state where the work is to be performed.

Under the federal Miller Act, as amended in 1959, there are certain time limitations before and after a lawsuit may be filed in federal court. A subcontractor, laborer, or materialman who furnished labor or material to a contractor for a federal public building or works, in respect of which the contractor has furnished a payment bond, cannot file a lawsuit before ninety days has expired after the day when the last work was done. The time limitation the other way is one year after the day on which the last of the labor was performed or material was supplied.

The courts have taken the position that the total liability of a surety on a contractor's bond to the several persons protected by the obligation of the bond is the penal sum named, plus interest. Consequently, when this sum is exhausted by the payment of judgments on claims within the protection of the bond, that is the end of the bonding company's liability. The mere fact that the amount of the bond was not adequate to pay all claims is just too bad. The bond specifies the amount of liability on the bonding company and once that sum is paid, the bonding company has fulfilled its contract of liability. The same situation would apply in the case of public contractors as well as private contractors.

Statutes and ordinances requiring the furnishing of bonds by contractors for public work, sometimes raise the question of what constitutes public work. Highway construction and improvements have been held to be public work. On the other hand, work done in preparation for a map and plat book system and delinquent tax list, to be delivered to a county, has been held not to be public work. The decisions disclose a definite tendency on the part of the courts to sustain the validity of performance bonds for contractors for public work and improvements, if that holding is reasonably possible.

Public bodies, such as a school district or a municipal corporation are liable to laborers and materialmen where there is a failure to exact the required statutory bond for the protection of such persons. The theory of this rule is that the statutory bond is intended for the protection of laborers and materialmen by giving them a substitute for unavailable lien rights. Further, that if such protection is lost through the negligence of the public authorities in failing to exact the required bond, the laborers and materialmen should recover from the public body or from its agents individually.

The obligation of the surety company in connection with a performance bond issued to the contractor for public work is measured by that of the contractor. If the contractor is not liable to a claimant, then the surety may not be held liable.

Most of the statutes governing bonds of public contractors contemplate a single bond protecting both the public body and laborers and materialmen. Yet the two undertakings are as distinct as if they were contained in separate instruments. The rights of the laborers and materialmen are independent of the rights of the public body under the provisions of the performance bond.

When a contractor for public work defaults and the surety company on his bond undertakes to complete the building, the surety stands for all practical purposes in the shoes of the original contractor. Now the surety company becomes the contractor, subject to all of the obligations and entitled to all of the rights and benefits of the original contractor. The surety company now assumes the obligation under the building contract to complete the building and to pay all subcontractors, materialmen and laborers who contribute to the construction of the building, regardless of whether the services or materials are furnished while the original contractor is in charge, or afterwards.

The principal contractor may look to the surety of the defaulting subcontractor for indemnification or reimbursement.

*Example*

1. Under a subcontract requiring that the subcontractor furnish all materials necessary for doing the work covered by the subcontract, the subcontractor's surety was liable to the prime contractor for sums which the contractor was required to pay for supplies furnished to the subcontractor where the subcontractor's bond guaranteed the performance of the subcontract.

## *13-6 WHO IS A SUBCONTRACTOR UNDER THE MILLER ACT*

Under state law a subcontractor has been defined as one to whom the principal contractor sublets a portion or all of the contract itself, or as one who performs or contracts to perform part of the principal contractor's building contract.

*Example*

1. One who furnishes gravel to a contractor for a state highway, but does not contract to do any work embraced in the original contract, is not a subcontractor. Similarly, one who furnishes curbstone, sand and gravel, crushed stone, or marl for highway construction is not a subcontractor. The same is true of one furnishing brick or ready-cut stone for school construction, or furnishing reinforcing steel from stock for a county courthouse.

The word subcontractor as used in the Miller Act, covering all federal construction, has been held to exclude ordinary laborers and materialmen, and to mean one who performs for and takes from the prime contractor a specific part of the labor or material requirements of the original construction contract.

The question frequently arises, in determining the liability of a surety on a contractor's performance bond, whether a certain person, firm or corporation is a subcontractor within the meaning of the bond or statute which requires it. This question arises both where one seeks to recover as a subcontractor on the prime contractor's bond, as well as where recovery is sought against a prime contractor's bond for labor or materials furnished to a subcontractor.

Under state law it has been held that a supplier of material is a materialman rather than a subcontractor, if he has not agreed that the original contract should be the standard by which the performance of his contract should be judged. This would be the case even though he agrees to furnish material in accordance with certain measurements, which in fact comply with the terms of the original construction contract. However, one furnishing materials in strict accordance with plans and specifications, and delivering them as directed, is a subcontractor and not simply a materialman. Thus, one furnishing interior trim for a building, in conforming with plans and specifications, is a subcontractor as distinguished from a materialman.

Where a subcontractor furnishing work and materials gave the general contractor a false receipt in order that the contractor might effect a final settlement with a school board in connection with the construction of a high school, the subcontractor was not entitled to recover against the contractor's surety on any claim for which it had acknowledged payment. The old rule of equity might well apply to this situation where a claimant has by his own act of chicanery prevented himself from coming into court with clean hands. Many judges will refuse to aid a claimant who has "dirty hands" by virtue of his voluntary actions and leave the parties as is, which is called the status quo.

## 13-7 RIGHTS AND REMEDIES OF SURETIES

The rule is well settled that the surety on the performance bond of a public contractor who is required to make good a default of the contractor becomes subrogated to rights which the owner-obligee had. This means that money held back by the owner, under the terms of his construction contract with the contractor, are now made available to the surety company. This money was earned by the contractor before he defaulted and the surety company is required to have completed the building and paid all of the bills. This right of subrogation in favor of the surety company is applicable to such as by the terms of the building contract are reserved and retained by the owner until complete performance and acceptance of the work.

*Example*

1. Under a public contractor's bond conditioned for faithful performance of the contract and the payment of laborers and materialmen, where the contractor completed the contract but did not pay certain materialmen and laborers, and the surety company paid them and claimed reimbursement out of the moneys retained by the public body from the contractor, but the United States Government intervened and claimed the fund by reason of a tax lien against the contractor after completion of the contract for back income taxes, the surety company had the prior right, inasmuch as the surety's equitable lien in the retained moneys arose at the time of the giving of the bond, and thus was superior to the later tax lien filed by the government.

The stipulation in the construction contract for the retention of a certain percentage of the contract price until the completion of the contract, although principally for the protection of the public authorities against any breach of contract on the part of the contractor, also has other protective features. It protects the laborers and materialmen for moneys due them. It also acts as indemnity for the surety company who guarantees the performance of the contract, and the surety is deemed to have an equity in the fund that takes precedence over later assignments of the funds by the contractor. The surety's right to this fund attaches at the time the contract of suretyship is entered into, and the surety's right to subrogation dates as of that time.

*Example*

1. An order by the contractor upon the public authorities to pay the amount of the percentage of the contract price retained, to a bank from which he has borrowed money to pay for labor and materials after the contract of suretyship was entered into, although an equitable assignment, does not preclude subrogation of a paid surety to the public authorities' right to the fund to the extent he has paid claims for labor and material. It is not important that the surety's subrogation rights will exhaust the fund and leave nothing for the bank, because from the date of the contract of suretyship the bank was bound to know that the surety had an equity in the funds to be reserved. Further, that when the bank loaned its money to the contractor, it did something that it was not obliged to do, and it must be deemed to have acted with a full knowledge of the rights of the surety.

# 14

# ARCHITECTS

## *14-1 WHO AND WHAT IS AN ARCHITECT*

An architect is a person who plans buildings and oversees their construction. The two functions are entirely separate and apart. On many occasions a property owner will hire an architect to draw plans and specifications only, with no arrangements for superintending the construction program. In this case the services of the architect are completed and terminate when he delivers the plans and specifications to the owner. The superintendence of the construction alone, without having drawn the plans and specifications, is not the practice of architecture. Very often the architect is named in the building contract between the owner and contractor, to act as a referee for the determination of disputed questions. More often than not, these disputes involve the amount of work done in connection with payments to the contractor and the quality of materials used in the construction of the building. Usually, the building contract obligates the contractor to build the structure in accordance with plans and specifications and under the superintendence of the architect. The architect is then granted the power of acting as referee and approving or disapproving of the work done and the materials supplied.

It may sound as if there is a conflict of duties when the architect acts in a dual capacity but it does not work out that way. In the first instance, the architect has been retained by the owner to prepare plans and specifications. In the second case, the architect has been included in the construction contract between the owner and the builder, to oversee the construction and to act as referee between the parties. When acting in the latter capacity, the architect is the representative of the owner but he is still obligated in his capacity as referee to be fair to both parties. Most contractors would prefer to deal with the architect rather than the owner because the architect is a professional man who is familiar with the construction business, whereas the owner is handicapped by lack of knowledge.

Most states have laws which regulate architects in the practice of their profession. It is well settled that a state may, in the exercise of its police power, regulate the profession of architecture. Provisions of a city building code that provide that only architects licensed by the state, can prepare plans and specifications for buildings of a public nature, have been held to be a reasonable exercise of the police power. Generally, the laws of each state provide for an examination to be given to those intending to engage in the profession of architecture, before they are given a license or certificate to practice. In order to be eligible to take an examination to be licensed as an architect, the applicant must have graduated from a recognized school of architecture and received his degree.

A draftsman is a person who draws plans for a home or building, yet he is not an architect. Because he is not an architect, the draftsman is not licensed by the state and he does not enjoy the professional status of the architect. Normally, a draftsman works with blueprints and is able to draw plans for a home or a building that is not too complicated. Some home builders have found that the normal charges of an architect are too expensive for a modest type of home and have used draftsmen as a substitute. The cost of plans from a draftsman are obviously much less than those furnished by an architect. To this extent, the profession of architecture is regulated in the same manner that the medical, legal and other professions are regulated and controlled. The architectural board which possesses the right to grant licenses to architects may be empowered to revoke such licenses for proper cause.

It has been generally held by the courts that designing a building for another or the furnishing of plans and specifications for such a building, constitutes architectural services. However, the making of blueprints for an architect does not constitute the performance of services as an architect. The courts have recognized that there may be an overlapping of services rendered by an architect and an engineer. The two professions have a tendency to overlap in given situations. The same services may in one instance constitute architectural and in another engineering.

If an architect is not licensed in the state where he performed his professional services, he is usually barred from recovery of payment for his work. One case held that the architect's employment contract was not rendered enforceable by virtue of the fact that his employer was informed that he did not have a license. In another lawsuit, a draftsman and construction engineer, who did not claim to be an architect, was held not precluded from recovering for his services rendered.

## 14-2 EMPLOYMENT AND DUTIES OF ARCHITECTS

The employment of an architect is ordinarily a matter of contract between the property owner and the architect. If the architect's duties include supervision of construction or approval of materials and the determination of the percentage of work completed, these provisions would be a part of the construction contract between the owner and the builder. In any event, the terms of the architect's employment can only be determined by the contract between the owner and the architect, or the contract of construction between the owner, the contractor and the architect. Normally, the contract will determine the amount of compensation that the architect is to receive. Naturally, an additional charge has to be made by the architect where he is charged with superintending

# CERTIFICATE OF SUBSTANTIAL COMPLETION

*AIA DOCUMENT G704*

OWNER ☐
ARCHITECT ☐
CONTRACTOR ☐
FIELD ☐
OTHER

PROJECT:
(name, address)

CONTRACTOR:

TO (Owner)

ARCHITECT'S PROJECT NO:
CONTRACT FOR:

CONTRACT DATE:

DATE OF ISSUANCE:

PROJECT OR DESIGNATED AREA SHALL INCLUDE:

The Work performed under this Contract has been reviewed and found to be substantially complete. The Date of Substantial Completion is hereby established as
which is also the date of commencement of all warranties and guarantees required by the Contract Documents.

**DEFINITION OF DATE OF SUBSTANTIAL COMPLETION**

The Date of Substantial Completion of the Work or designated portion thereof is the Date certified by the Architect when construction is sufficiently complete, in accordance with the Contract Documents, so the Owner may occupy the Work or designated portion thereof for the use for which it is intended.

A list of items to be completed or corrected, prepared by the Contractor and verified and amended by the Architect, is appended hereto. The failure to include any items on such list does not alter the responsibility of the Contractor to complete all Work in accordance with the Contract Documents.

ARCHITECT ______ BY ______ DATE ______

The Contractor will complete or correct the Work on the list of items appended hereto within ______ days from the above Date of Substantial Completion.

CONTRACTOR ______ BY ______ DATE ______

The Owner accepts the Work or designated portion thereof as substantially complete and will assume full possession thereof at ______ (time) on ______ (date).

OWNER ______ BY ______ DATE ______

The responsibilities of the Owner and the Contractor for maintenance, heat, utilities and insurance shall be as follows:
(NOTE — Owner's and Contractor's legal and insurance counsel should determine and review insurance requirements and coverage)

duties or furnishing certificates of progress and generally acting as a referee. The compensation due the architect for his professional services is usually based on a percentage of the actual cost of the completed structure.

As a general rule, the architect who prepares the plans and specifications is acting in the capacity of an independent contractor. He has been hired by the owner to perform certain professional services and he is neither an agent or employee of the owner. However, when the architect is also retained by the owner to perform other services that are specified in the construction contract, his status changes from independent operator to that of agent or representative of the owner for whom the construction work is being done. When the architect is engaged to supervise the construction of a building as an agent of the owner, his authority is limited by the provisions of the construction contract.

*Examples*

1. The architect's duties are limited to the supervision and direction of work to be done by the contractor, or those acting under him, and to see that the materials and workmanship are in accordance with the specifications.

2. An architect, as the agent of both builder and owner to construe plans for a structure and settle disputes in that regard, has no authority to change the plans.

The relationship between an architect, employed to furnish plans and specifications and superintend construction, and his employer, which is frequently characterized as an agency, is one of trust and confidence. Good faith and loyalty to his employer constitute a primary duty of the architect. The architect is duty bound to make a full disclosure of all matters, of which he has knowledge, which is desirable or important that his employer learn.

Like all other professional persons, the architect represents himself as being qualified in his particular line of work and is employed because he is believed to be such. An architect, in contracting for his services as such, implies that he possesses skill and ability, including taste, sufficient to enable him to perform the required services at least ordinarily and reasonably well.

Over a period of years, many architects have taken the position that the preparation of plans are similar to an artistic creation and that even after the owner has paid for them and the building has been completed, that the plans belong to the architect. This position has not been upheld by the law. In the absence of any agreement to the contrary, where the architect has prepared the plans and superintended the construction and has been paid for his work, the owner of the building is entitled to possession of the plans.

*Examples*

1. An alleged custom among architects to retain the plans in such case is unreasonable and will not be given effect.

2. An employer is not liable for the services rendered by an architect in preparing plans for work which has been abandoned by him, unless the plans were delivered to him, in the absence of knowledge on his part of a rule of the American Institute of Architects that drawings and specifications remain the property of the architect.

## 14-3 PAYMENT

Generally, when plans are submitted in competition in response to a notice that offers a reward for the best plan, or for a number of plans to be selected, or for each plan entered, an architect who is granted an award has received full compensation for his services in preparing the plans and has no further property right in them.

When the contract between the owner and the architect contains an express stipulation as to the amount of compensation for architectural services, or the manner of determining such amount, that stipulation furnishes the standard for measuring the amount of recovery for the performance of such services. If there has been no agreement respecting compensation, the architect is entitled to be paid for the reasonable value of his services. The reasonable value of an architect's services may be shown by the customary charges for similar services by other architects.

Arguments have arisen as to when the architect is to be paid by the owner. If the contract for architectural services stipulates the time at which payments to the architect are to be made, there is no problem as the architect would be paid accordingly. If the contract is silent as regards time of payment, the architect would be entitled to be paid only when the construction was completed.

An architect may be entitled to additional compensation, upon the happening of certain contingencies, if such compensation is expressly provided for in his contract with his employer.

*Examples*

1. If the employer abandons construction of a building after the architect furnishes, and his employer accepts, a set of plans and specifications pursuant to the contract to prepare plans and specifications and superintend construction of a building for an agreed compensation, and the owner orders the architect to prepare plans and specifications for an entirely different kind of structure, there is an implied promise on the part of the employer to pay for the extra and independent work in drawing the second set of plans.
2. An architect cannot recover extra compensation for services as an arbitrator between the contractor and the owner, where, in the building contract prepared by him, he inserted the usual clause that such disputes should be referred to him for final decision, since such services fall within that part of the contract in which he agreed to perform the "usual and customary services of an architect."

If the contract of employment is breached by the employer by way of wrongful discharge of the architect, his rights are the same as anyone else who is the victim of a breached contract. If such wrongful termination is made before the plans are completed, the architect is entitled to recover for work already done on them.

So far the discussion has centered on the rights of the architect with no mention made as to his obligations. The architect's right to compensation for his services may be lost by reason of his unskillfulness or negligence in preparing and furnishing defective plans and specifications. The owner could assert this defense in an action by the architect for compensation for his services. If the architect was guilty of negligence and the owner suffered damages as a result of the architect's negligence, the owner could recover damages for his loss. Just as the builder is legally obligated to construct the building in a workmanlike manner, the architect is legally obligated to prepare his plans and specifications in a professional manner.

With the never ending increase in the cost of construction, due to constant increases in the cost of labor and materials, some architects have seen fit to "stick their neck out" by guaranteeing to their owner-customer that his building will not cost over a certain sum of money. The owner agrees to retain the architect on his representation that the cost of the building will not exceed a named amount. The architect completes the plans and specifications and they are let out to various contractors for bids. All of the bids materially exceed the amount that the architect assured his owner customer was the high water mark and you can see the problem coming. The architect says, "I did my work and I ought to be paid even though you have decided not to build the building" and the owner replies that, "I would not have ordered the plans and specifications but for your representation as to the cost of the building and now I have no need for your plans and specifications." The key to the legal problem is that all of the bids materially exceeded the amount predicted by the architect and the courts would deny the architect's claim for compensation. After all, the assured price was a part of the contract that was entered into by the parties and the architect saw fit to practically become a guarantor of the price-limit.

Some state's laws provide that an architect has lien rights for money due him that is unpaid. However, in most of the states mechanics' lien laws, architects are not expressly mentioned. Legal opinion is divided as to whether the architect has lien rights for his services in preparing plans and specifications. Some cases hold that the architect is entitled to lien rights if his contract calls for him to supervise the job and not otherwise.

On the question of whether or not an architect is entitled to a lien for services rendered in merely drawing plans and specifications, there is a sharp conflict of legal opinions. This conflict exists both in cases in which the plans were actually used as well as in cases where they were not used. Even in those states where the architect does not have lien rights, he is still entitled to file an ordinary lawsuit against the owner for compensation which he claims is due him. The owner is privileged to set forth his defense and from there on out, the dispute is the same as any other litigated matter.

An architect may be held responsible for his negligence in failing to exercise the ordinary skill of his profession, which results in the erection of an unsafe structure whereby anyone lawfully on the premises is injured. An architect's liability for negligence resulting in personal injury or death may be based upon his supervisory activities or upon defects in the plans and specifications.

# 15

# ODDS AND ENDS

## *15-1 EASEMENTS*

An "easement" is the right to use another persons land for a particular purpose. A farmer may give an easement to a neighbor to drive his cattle across a certain field to reach a pasture. An easement cannot be a verbal arrangement, it must be in writing. If the above cattle arrangement were verbal it would not be an easement but would be a license.

A "license" is a personal privilege to commit some act upon the land of another without having any formal rights in the matter. Usually, it is some minor act which is agreed upon verbally or negative approval is accomplished by the owner ignoring the act. The most common example of a license occurs when people walk over another's vacant land by way of a shortcut. The pedestrian travel across the grass or weeds results in a pathway that is obvious to the naked eye. The owner of the land observes that his vacant land is being used as a shortcut by many persons, however, he is not concerned and he ignores the entire matter. By his actions of negative approval, the law says he has granted a license to the shortcut users. They have no formal rights under their license and the owner can close the pathway anytime he sees fit.

Easements are a vital necessity for the telephone, gas and electric, and all other public utilities. In order to have access to utility poles to provide services to the property owner, an easement is mandatory for the use of the rear five feet of the lot. If the owner refused to give the five foot easement to the public utility, these vital services would not be available to him. Thus, easements are always given to the public utility companies and they do not pay for the easement. The easement has nothing to do with the ownership of the five feet or payment of real estate taxes. The recipient of the easement has the right to use the rear five feet for necessary poles and to walk over the land to service same. Sometimes, easements are given to permit a gas company to run a pipeline across someone's land.

Actual cases have occurred where the owner of a large parcel of land sells a portion of inner land and the buyer discovers that his land is hemmed in with no access to a road. The right to cross over the owner's land or use his road to get to your newly purchased land is granted through an easement for purposes of "ingress" and "egress." Ingress is the right to reach your land by passing over another's land without trespassing. Egress is leaving your land under identical circumstances. In the example cited above, if the seller refused to grant an easement to his buyer, the courts would force him to be fair by giving the buyer some reasonable means of ingress and egress.

A practical type of easement occurs frequently when adjoining home owners get together for a common driveway, leading to their respective garages. Actual ownership of each side of the joint driveway is unchanged; however, each party has the right to use the entire driveway by virtue of the reciprocal easements.

A "party wall" with your next door or rear neighbor falls into a different category. If the wall is built on your neighbor's property and you agree to share in the cost, it is a party wall but your neighbor is the legal owner of the wall. On the other hand, if the party wall is constructed in such a manner that it is occupying one half of each owner's land, then reciprocal easements should be used, giving each party equal rights and equal ownership.

What happens if a builder constructs a home on another's land by mistake? An old legal principle held that it becomes the property of the landowner, and the latter is said to have acquired "title by accession." This concept of law has changed materially and one is not permitted to take advantage of another's mistake. This situation actually occurred when a builder client thought he bought all twenty lots in a block on which he built tract type of homes. When his construction program was about two thirds completed he discovered that he only purchased nineteen and did not own lot number twenty. The lot owner was contacted and the problem was solved by purchasing their lot for a higher price than prevailed at the time. If the problem had been presented to the court, most probably, the result would have been the same.

Many other common type of business errors have been corrected by the courts, who insist that equity and fair play prevail. We have all read about cases where a bank teller made an honest mistake and gave the customer more money than he actually had on deposit. When the bank records clearly show the true figures and the customer is unable to dispute that fact, the bank's claim will prevail in court. Whether the customer has spent the money or is financially responsible to pay the court judgment is another matter, because we are only concerned with the legal principles. Honesty is still the best policy in the event of any business mistake.

Acquiring title to land by prescription presents an entirely different situation. If a person openly occupies another's land for a period of five years continuously, claiming it as his own in defiance of the legal owner's rights, and pays all taxes on the land during that time, he may acquire "title by prescription." Occupying property and claiming title in that manner is called "adverse possession."

## *15-2 ESCROW*

An "escrow" is the depositing of papers and money with a third neutral party along with instructions to carry out an agreement. Escrows may be opened for any lawful purpose to carry out the wishes of two or more parties to a contract. They may involve the completion of a lease, the transfer of personal property, placing of a mortgage, or more frequently, a real estate transaction.

The handling of all papers and money in connection with business transactions placed in escrow is involved, and needs someone skilled in this work to handle the matters promptly and accurately. For this reason the custom has developed to entrust some able firm or person with these important details. He (or they) carries out written instructions from both parties to a transaction and is called an "escrow holder." The escrow holder may be an attorney, a bank or other financial institution, a title company or more common now in larger cities, a professional escrow company.

While all of the documents, money, and written instructions are in the hands of the escrow holder, they are said to be in escrow. All escrow money is placed in a trust account, waiting for the transaction to be completed so that the money can be distributed to those involved in the escrow.

More often than not, escrows are used in connection with the closing of real estate transactions. Buyer and seller are together on the terms and conditions of their home or land purchase. The seller wants his money and the buyer wants his Warranty Deed or similar evidence of title. If the escrow instructions so indicate, the buyer is also to receive a title insurance policy indicating that the legal title to the land involved is clear and free of any liens or defects. The escrow agent is really a stake holder.

Acting under written instructions furnished by the parties, the escrow agent is told to deliver the pertinent papers to the buyer when the buyer has paid the agreed sum of money for credit to the seller. When all of the escrow papers are properly signed, delivered, and recorded, the escrow agent is ready to account for the money left in trust with him. At this point, the escrow agent then makes up a written statement and a copy is given to the buyer, and another version of a copy is given to the seller, together with a distribution of the money due the seller and others. Under the escrow instructions, the escrow agent is usually instructed to pay a certain amount of money to the real estate broker that handled the transaction; possible mortgage balances; certain public utilities bills; and any other claims that may be involved. The reason for the variation between the copy of a statement to the buyer and the seller is that the buyer's statement will show the disposition of his money paid to the escrow holder, whereas the seller's copy will show how the money was spent by the escrow agent. Insurance premiums and taxes are usually pro-rated between the parties.

Just who pays the escrow, title, and recording fees in connection with an escrow is a matter of agreement between the parties to the escrow. The escrow fee is usually divided equally between buyer and seller. The cost of recording a deed, about a $3 item, is usually charged to the buyer. The cost of the title insurance policy is usually the responsibility of the seller. The cost of preparing the deeds, contracts and other legal papers are usually split between the parties. Sometimes, the real estate broker will pick up the tab for the preparation of the necessary papers. The legal expense for having the abstract of title or title insurance policy checked, is usually the responsibility of the buyer. Customs vary in different states and areas as regards who pays escrow charges. The parties are privileged to make their own arrangements for payment of the various escrow items. In some cases the seller will agree to sell his land at a certain price—"net to me," which means that all charges are to be paid by the buyer.

On rare occasions, an attorney will act as escrow agent. The main objection to any attorney being escrow agent is that he is usually involved in the transaction and is already representing one of the parties, whereas the escrow agent should be entirely neutral. An attorney's time is normally too expensive to handle minor details that are connected with simple escrows. Some escrows are quite simple, while others can be very complicated. In short, the professional escrow companies or banks or title companies are able to render a more efficient service to the parties. Attorneys are trained and better equipped to give legal advice and handle complicated legal matters and leave the specialties of escrows to those that are trained in this field.

As indicated earlier, escrows can and are used for many different types of business transactions, in addition to real estate matters. Sometimes stock transactions are handled by use of an escrow. The main theme being that the seller does not want to deliver the stock or deed until he has received his money and the buyer does not want to pay the money until he has received the stock or deed. With this type of honor system, someone has to give. Thus, the escrow agent is the neutral third party who in effect is representing the seller and the buyer and he follows the written instructions of both sides.

## *15-3 FIXTURES*

"Fixtures" are things that are attached to property and cannot be removed as ordinary personal property because they have become a part of the realty. A fixture is an object that was originally a personal chattel and has been affixed to the soil itself or to some structure legally a part of the soil.

Unless otherwise stated in the contract, personal property annexed or attached to real property becomes real property. The mortgagor and mortgagee of realty are free to determine by their contract what are and what are not fixtures within the meaning and security of the mortgage.

Trade fixtures are in a special category by themselves. Almost without exception, the tenant is accorded the right to remove his trade fixtures when the lease expires. The removal of the trade fixtures must not substantially injure the landlord's property. It is still advisable in order to avoid any disputes, to have the lease provide for the removal of shelves, showcases and other trade fixtures, even though they have been nailed to the building proper.

Factory equipment and machinery in a leased building are usually declared to be removable as personalty. Particularly, when it can be removed without injury to the property, and may readily be used elsewhere.

Originally, the law of fixtures said they became attached to the realty and became a part of the realty. With the passing of time, custom and by contract, trade fixtures are exempted.

# GLOSSARY

ABANDONMENT. To release claim or forfeit rights—as in the case of a homestead.

ABEYANCE. Pending or temporarily suspended. Such as an action held in abeyance.

ABSTRACT OF JUDGMENT. Record of a court's judgment. Creates lien when recorded.

ABSTRACT OF TITLE. A digest or summary of documents or records affecting title to property.

ACCELERATION CLAUSE. Provision sometimes inserted in a mortgage or trust deed note causing it to become payable at once, in a lump sum under certain conditions—such as in event property is sold, leased or payments not paid promptly.

ACCEPTANCE. Giving consent to an offer—as when seller signs an offer to purchase.

ACCESSION. Acquiring title to unauthorized improvements to your land.

ACCOMMODATION PAPER. An act of trying to help a friend by cosigning his note at a finance company or bank.

ACCRETION. Addition to your land as by deposits from a stream or lake.

ACKNOWLEDGMENT. A formal declaration before a notary public or other qualified officer, in signing a document, that it is your voluntary act.

ACQUISITION. The process by which property is procured—through purchase, inheritance, gift, foreclosure, etc.

ACRE. An area of land containing 43,560 feet.

ACT OF GOD. A disaster inflicted by nature such as an earthquake, unusual flood, tornado or hurricane. Sometimes a valid legal excuse for not performing a building contract.

ACTION TO QUIET TITLE. A lawsuit to determine status of title to land. Often to remove a defect or cloud on the title.

ACTUAL NOTICE. Notice given by open possession and occupancy of property.

ADMINISTRATOR. A man appointed by a court to take charge of an estate of a deceased person who left no will. If a lady were appointed, she would be called the administratrix.

ADULT. A person who has reached an age established by law to attain certain privileges, such as the right to vote, to enter into binding contracts, etc. Those persons of lessor age are minors.

AD VALOREM. Based upon the value—property taxes for example.

ADVERSE POSSESSION. Openly holding possession of land under some claim of right which is opposed to the claim of another.

AFFIANT. One who makes a sworn statement, such as an affidavit.

AFFIDAVIT. A sworn statement in writing before an officer authorized to administer oaths.

AFFIRMATION. A solemn declaration—usually by one opposed to oaths on religious grounds.

AGENCY. Act of representing a principal in the capacity of an agent.

AGENT. One who is authorized to represent another person, as in a real estate transaction.

AGREEMENT OF SALE. A written contract whereby buyer and seller agree on terms of the sale.

ALIAS. An assumed name.

ALIEN. A resident who is a citizen of a foreign country.

ALIENATE. Act of transferring title to property.

ALIENATION CLAUSE. Provision in a mortgage or trust deed providing for full payment if the property is sold.

ALLUVIUM. Deposit of soil on or adjoining property, as by flow of a river or stream, or by tides.

AMERICAN STOCK EXCHANGE. The second largest stock exchange in the United States. Their requirements are not as rigid as the New York Stock Exchange.

AMORTIZE. Pay off debt in installments; or gradual recovery of an investment.

ANNEXATION. Adding land to another unit. Such as bringing a tract within the limits of a city.

ANNUITY. Money paid annually or in other agreed periods.

ANTICIPATORY BREACH. An announced violation of contract that permits a lawsuit before the completion date of the contract.

APPRAISAL. An opinion as to value based upon facts and experience, by one skilled in such work.

APPURTENANCE. A thing or right which attaches to or becomes incident to the land, so as to become a part of the realty. A house, fence, etc. which, when the land is conveyed, goes with it without special mention in the deed.

ARBITRATION. A substitute for court proceedings to settle disputes between parties to a contract.

ARCHITECT. A professionally trained person who plans buildings and sometimes is employed to oversee their construction.

ASSESSED VALUE. Value placed on property by the assessor as basis for the levy of taxes.

ASSESSMENT (SPECIAL). Levy against particular properties for cost of improvements which particularly benefit them. (Sewers, sidewalks, drains, etc.)

ASSESSOR. An official, usually county or city, who determines the value of property for tax purposes.

ASSIGN. To endorse over to another, such as a promissory note or lease.

ASSIGNEE. One to whom property or a right is transferred.

ASSIGNOR. One who assigns or transfers a property or right to another.

ASSUMPTION OF MORTGAGE OR TRUST DEED. Taking title to property and assuming personal liability for payment of existing notes for which the property is security.

ATTACHMENT. Seizure of property by court order in connection with a pending lawsuit.

ATTORNEY IN FACT. A person to whom a power of attorney is given authorizing him to do all or specific acts for another.

ATTRACTIVE NUISANCE DOCTRINE. A theory of law that anything that attracts children and the child is injured does not require proof of negligence. It used to be the ice man and now could be a construction project.

AUTHORIZATION TO SELL. Commonly called a listing by real estate brokers.

AUTHORIZED CAPITAL STOCK. The amount of common stock specified in the articles of incorporation.

AVULSION. Sudden removal of land by flowing water.

AWARD. The decision of the arbitrators. The same as a judge's ruling.

BALLOON PAYMENT. Usually an extra large payment on an installment note at the time it is payable in full.

BANKRUPTCY. A federal court legal proceeding to aid those that are unable to pay their obligations when they fall due.

BANKRUPTCY PETITION. The required form to be filled out and filed in federal bankruptcy court by those who seek relief from financial problems.

BASE AND MERIDIAN. Principal survey lines in an area from which townships are numbered.

BENCH MARK. Permanent marker placed by surveyors at an important point, upon which local surveys are based.

BENEFICIARY. One who is a recipient of benefits from a trust; such as a lender of money secured by a trust deed.

BEQUEATH. To leave property by will.

BEQUEST. What is bequeathed by will—an inheritance.

BID. An offer by a contractor to build a certain structure for a fixed price.

BILATERAL CONTRACT. A contract by which both parties agree to perform certain acts.

BILL OF SALE. Signed document which transfers ownership of personal property.

BILLS OF LADING. Paper evidence of title that represents the goods that are described in it.

BILLS AND NOTES. Abbreviation for a bill of exchange and a promissory note. A check is one of many examples of a bill of exchange that are used in daily business transactions.

BINDER. A preliminary agreement for sale of real estate requiring a deposit, and providing for a formal deed or contract at some future date.

BLANKET MORTGAGE. A single lien covering two or more lots or parcels of land.

BOARD OF DIRECTORS. A group of persons that manage a corporation for all important matters.

BONA FIDE. In good faith—honest.

BOND (SURETY). A pledge to pay a sum of money in case of failure to fulfill obligations, inflicting damage, or mishandling funds. Usually written by a bonding company for a fee.

BONDABLE. An employee who is able to obtain a bond to protect the employer. Also a contractor who is able to do the same for his customer.

BONDS (CORPORATE). Evidence of indebtedness of a corporation which is secured by its general assets.

BOOK VALUE. The value of a property as carried on the owner's accounts.

BOOT. A profit gained in exchange of properties, not reflected by cash, upon which income tax is not deferred.

BREACH. Failure to perform a duty or fulfill an obligation.

BROKER LOAN STATEMENT. A statement of charges to be made in connection with a loan, for information of borrower.

BUILDING RESTRICTIONS. Laws or ordinances requiring sound construction for protection of health and safety.

BULK SALES LAW. State law requiring that the sale of a business be advertised beforehand, for protection of creditors.

BUSINESS OPPORTUNITY. A going business, including physical assets, good will and perhaps a property lease.

BYLAWS. Rules for internal management of a corporation within its corporate charter limits.

CAPITAL GAIN. Profit from increase in value of an investment. If held more than six months, it becomes taxable at a lower rate.

CAPITALIZATION. In appraising, using a predetermined interest rate and the net earnings of a property as a basis of computing value.

CASHIER'S CHECK. A check drawn by a bank on its own account and signed by the bank's official.

CAVEAT EMPTOR. Means "Let the buyer beware." Dealing "at arms' length." Used car dealer selling a jalopy "as is."

CD (CERTIFICATE OF DEPOSIT). A receipt for a time deposit of money left with a bank at an agreed rate of interest.

CERTIFIED CHECK. A customer's check that is guaranteed by his bank to be good when presented for payment.

CERTIFIED FINANCIAL STATEMENT. A statement of assets and liabilities that is confirmed by the person's or firm's certified public accountant.

CHAIN OF TITLE. Detailed account of all actions and events affecting a title to land as far back as the original government patent, if possible.

CHANGE ORDERS. A variation in the plans and specifications of the architect that require a written memo.

CHATTEL. Personal property; moveable property.

CHATTEL MORTGAGE. A mortgage on personal property.

CLOSING STATEMENT. A final accounting in closing a transaction. Mandatory for real estate brokers.

CLOUD ON THE TITLE. Anything affecting clear title to property. Term usually used in connection with minor nuisance items that must be eliminated by quitclaim deed or quiet title lawsuit.

COLLATERAL SECURITY. Additional sums or things of value posted to guarantee fulfillment of a principal contract.

COLLUSION. A secret arrangement to defraud someone.

COLOR OF TITLE. A title which appears to be good on the surface, but actually is not good.

COMMERCIAL ACRE. What remains of an acre after allowing for deductions for streets, alleys, etc. Something less than an acre.

COMMERCIAL PAPER. Notes assigned in the course of trade, bills of exchange, etc.

COMMINGLING. Situation where husband or wife has confused separate property with community property to the extent it cannot be separated.

COMMON STOCK. A printed certificate that represents the complete ownership of a corporation.

COMMUNITY PROPERTY. That property acquired by husband and wife from time of marriage onward as a result of their joint efforts.

COMPACTION. Tamping of filled ground to make it more suitable for building. Done extensively in subdividing of hilly ground.

COMPARATIVE ANALYSIS. Appraising a home or lot by comparing it with others of similar qualities with known recent sales prices.

COMPETENT PARTIES. Persons mentally fit; legally capable of entering into a contract.

COMPOUND INTEREST. Earnings on the original investment and the accumulated interest therefrom.

CONDEMNATION. Ruling by a public agency that property is not fit for use. Also refers to taking of private property for public use by right of eminent domain, by paying the fair value.

CONDITIONAL SALES CONTRACT. The purchase of personal property on contract, usually on the installment plan. Buyer does not receive a deed or title until he has made all payments called for by the contract.

CONDITIONS. Limitations imposed in a deed.

CONDOMINIUMS. Apartments or other types of properties in which the owner has fee title to the part actually occupied, with an undivided interest in areas used by all occupants.

CONGLOMERATE CORPORATIONS. A group of companies that have been merged into a single ownership and are usually engaged in non-related businesses.

CONSIDERATION. Something of value to induce a person to enter into a valid contract. The consideration for a gift between husband and wife or parents and children can be love and affection.

CONSPIRACY. An agreement between two or more persons to commit an unlawful act.

CONSTRUCTIVE NOTICE. Notice given by the public records, as opposed to actual notice.

CONTIGUOUS. Adjoining, touching. As two contiguous parcels of land.

CONTINGENT FEE. A gambling fee by an attorney based upon favorable result of collecting money.

CONTRACT. Agreement to do certain things, or not to do them.

CONTRACTOR. One who is licensed to build or erect a home, building, or other structure for another.

CONTRIBUTORY NEGLIGENCE. The fault or negligence of the claimant that weakens or defeats his claim against the alleged wrongdoer. One who enters a dark building under construction may be guilty of contributory negligence.

CONVENTIONAL LOAN. A loan not guaranteed or insured by a governmental agency. Usually made by banks, insurance companies and savings and loan associations for home mortgages.

CONVERTIBLE STOCK. Designation where stock is changed into another class or into other obligations of the corporation.

CONVEYANCE. Transfer of title from one person to another. This is accomplished by the use of a deed.

COOPERATIVE APARTMENTS. Each occupant owns his apartment by receiving a share of stock and a lease from the cooperative apartment corporation.

CORNER INFLUENCE. In appraising, the additional value given to a corner lot due to its advantages, especially in business property.

CORPORATE CHARTER. Articles of incorporation that give the corporation the right and power to do many things over a period of many years.

CORPORATE MERGERS. Usually, the larger corporation buys all or the controlling interest in a smaller corporation and controls its fate.

CORPORATE MINUTES. A detailed written record of what went on at meetings of the stockholders and also of the board of directors of a corporation.

CORPORATION. A legal creation authorized to act with the rights and liabilities of a person.

COSIGNER. One who guarantees that if his friend who borrowed money does not pay as agreed that the good guy will get stuck.

COST PLUS CONTRACT. An agreement that the owner will pay the cost of all labor and materials plus a certain profit to the contractor.

COUNTY RECORDS. A recording system for documents maintained by each county as provided by state law.

COVENANT. Agreement in a deed to control the use and acts of future owners. Used also in other instruments such as leases and conditional sales contracts.

CPA (CERTIFIED PUBLIC ACCOUNTANT). A skilled professional accountant who has been licensed by his state after passing a difficult examination. He is to the accounting profession what the MD is to the medical profession.

CREDITOR'S COMMITTEE. A group of creditors who decide if the debtor's petition for an extension of time or compromise of his obligations shall be permitted. This is handled under Chapter XI of the federal Bankruptcy Act.

DAMAGES. Compensation the court may award to a person who has been injured physically or financially by another.

D/B/A. Abbreviation for "DOING BUSINESS AS", used in lawsuits to identify a trade name. John Jones, d/b/a Highway Motors.

DEBENTURES. An unsecured note given by a corporation for money it has borrowed on a long term payback.

DECLARATION OF HOMESTEAD. Document recorded to declare a homestead under state law.

DECLARATION OF RESTRICTIONS. A list of restrictions to a tract imposed by a subdivider and recorded.

DECREE. A decision by a court or others authorized to make decisions. Frequently mispronounced, as if spelled degree.

DEDICATION. Acceptance of land from an owner by a city or county for particular use by the public.

DEED. A written instrument which conveys title to real estate.

DEED OF RECONVEYANCE. Deed given by a trustee under deed of trust when loan is paid.

DE FACTO AND DE JURE CORPORATIONS. A de facto corporation although irregularly formed, exercises corporate rights under color of law. A de jure corporation is one which has been created in compliance with all legal requirements.

DEFAULT. Failure to perform a duty or keep a promise, such as to make payments on a note.

DEFENDANT. Persons who are being sued in a civil lawsuit.

DEFICIENCY JUDGMENT. A judgment awarded by a court against a person when after foreclosure the security for the loan does not realize enough money at a sale to pay the balance of the loan.

DELIVERY. Formal transfer of a deed to the new owner, without the right to recall it. Essential to a valid transfer of title.

DEPOSITION. Sworn testimony by way of questions and answers given outside of the courtroom in a pending lawsuit. Preliminary to the court trial.

DEPRECIATION. Loss of value to property from any cause.

DEVISE. Gift of real estate by will.

DEVISEE. One who inherits property by will.

DISCHARGE IN BANKRUPTCY. A document that legally excuses one from being forced to pay certain obligations listed in his bankruptcy schedule that are legally dischargeable.

DISCOUNTING BILLS. Paying bills promptly when due and thereby earning a discount. Builds up good credit.

DISGORGE. A creditor who has been favored by a bankrupt debtor at the expense of all other creditors and he is ordered to return the tainted money to the trustee in bankruptcy.

DIVERSITY OF CITIZENSHIP. A citizen of one state becoming legally involved with a citizen of another state.

DIVIDENDS. A payment of a portion of corporate profits to its stockholders. Also payment by a bank to its depositors as interest payment for use of their money.

DOMICILE. Place of residence. In court proceedings, residence is a matter of intent.

DOWER. Interest of wife in her husband's estate after his death. Term not used much in community property states.

DRAFTSMAN. One who is trained in the skill of drawing plans. Frequently employed by architects for drawing plans and specifications.

DUMMY DIRECTOR. A figurehead in the formation of a new corporation who has no duties to perform.

DURESS. Unlawfully forcing someone to do an act against his will by use of force.

DWI (DRIVING WHILE INTOXICATED). Commonly called "drunk driving." One who is charged with the criminal offense of driving a motor vehicle while under the influence of alcohol to an extent prohibited by law.

EARNEST MONEY. A deposit of money given to bind an agreement or an offer.

EASEMENT. The right or interest of one person in another's property.

ECONOMIC LIFE. Life of a building during which it earns enough to justify maintaining it.

EGRESS. A means of leaving property without trespassing.

EMANCIPATION OF A MINOR. A legal proceeding to permit one under age to transact business as an adult.

EMINENT DOMAIN. The right of government to take private property for public use, provided it serves a necessary public use and fair compensation is paid to the owner.

ENCROACHMENT. Building in whole or in part on another's property.

ENCUMBRANCE. A debt on property. Anything that burdens the title to property.

ENDORSEMENT. The signing on the back of a check or note for the purpose of transfer.

ENDORSEMENT IN BLANK. Signing to transfer rights to a check or note without qualification, making endorser equally responsible for payment.

ENDORSEMENT WITHOUT RECOURSE. Signing to transfer a check or note in this manner makes no guarantee to future holders.

EQUITABLE OWNER. One who has hypothecated his property. He has conveyed title in trust perhaps, but retains the right to use and enjoy the property.

EQUITY (OWNER'S). Value of owner's interest in property in excess of the lien's against it.

EQUITY OF REDEMPTION. Owner's right to redeem property after foreclosure sale for a period provided by law.

ESCALATOR CLAUSE. Provision in a lease whereby the rents increase under certain conditions, such as every year, or based upon a periodical appraisal of the property.

ESCHEAT. Process by which property reverts to the state for lack of private ownership. Bank accounts that lie dormant for a number of years, with no apparent owner alive, are subject to escheat laws.

ESCROW. The depositing of papers and money with a third neutral party along with instructions to carry out an agreement. Such as the transfer of title to a home.

ESCROW HOLDER. One who undertakes to carry out escrow instructions.

ESTATE. The interest of a person in property; as to real property, the degree, quantity and extent of his interest.

ESTATE FOR LIFE. Use of property only during the life of the person given the interest; after which it reverts to the original estate or others designated.

ESTATE TAX (FEDERAL). A tax on estates of deceased persons in excess of an amount specified by law, with exemptions.

ESTATE AT WILL. A lease which may be terminated at will by either party.

ESTATE FOR YEARS. Another term for a lease or leasehold estate.

ESTIMATOR. A person who is skilled in the cost of construction to determine the amount the contractor should offer to build in his bid to the owner.

ETHICS. A standard of moral practice and fair play.

EXCHANGE AGREEMENT. A contract for the exchange of properties.

EXCLUSIVE LISTING. An authorization to sell which gives sole agency or right to sell to one real estate broker. He is entitled to his brokerage fee even if the owner finds his own buyer.

EXECUTE. To sign and consent to carry out an agreement to completion.

EXECUTOR. Man named in a will to handle and dispose of an estate. If a woman is named for the same purpose, she is called EXECUTRIX.

EXEMPTION STATUTES. Laws that save property from creditors in connection with attachment or bankruptcy or other legal proceedings.

EXPERT WITNESS. One who testifies in a lawsuit who is not directly involved in the dispute but is qualified by reason of experience or educational background.

EXTRAS. Anything that is ordered by the owner after construction has started that was not included in the original contract.

FAMILY CORPORATION. One whose stock is owned and controlled by immediate members of a family with no stock available to the general public.

FEDERAL HOUSING ADMINISTRATION (FHA). Federal government agency, which insures loans on residential property.

FEDERAL SAVINGS AND LOAN ASSOCIATION. A financial institution that is chartered by the federal home loan bank board in Washington, D.C. and whose accounts are insured by an agency of the government.

FEE SIMPLE ESTATE. Highest and best estate possible.

FELONY. A serious crime punishable by a sentence of more than one year in a state penitentiary.

FICTITIOUS NAME. A name which does not identify the person.

FIDELITY BOND. The dictionary says, "the quality or state of being faithful." A bond to protect the employer against employees who embezzle money.

FIDUCIARY RELATIONSHIP. A position of trust and confidence requiring loyalty.

FINANCING STATEMENT. A list of assets and liabilities which means a list of what you own and what you owe. Required by banks and other financial institutions before they will lend money.

FINDER'S FEE. Money paid to a person who furnishes information helpful to arranging a loan or completing a deal.

FIRST MEETING OF CREDITORS. A bankruptcy court proceeding held in the courtroom of the Referee in Bankruptcy to question the bankrupt debtor.

FIXTURES. Things that are attached to property and cannot be removed as ordinary personal property, as they become a part of the realty.

FORECLOSURE. The sale of pledged property to cover a defaulted debt.

FORFEITURE. Loss of a deposit or earnest money for failure to perform.

FOUNDATIONS. Nonprofit corporations that are formed for charitable or educational purposes.

FRANCHISES. A right granted by a state to a newly formed corporation. Sometimes used to denote a food type of franchise sold by a national company, authorizing a purchaser to use their brand name and recipe.

FRAUD. Causing loss of property due to use of deceit, cheating, false promises, etc.

FREEHOLDER. Owner of land in fee.

FRONT FOOT. The measure of land along the street frontage. Used as a unit in pricing business property.

GARNISHMENT. A court proceeding whereby a creditor attaches the wages or bank account of the debtor.

GENERAL LIEN. One which may attach to all property of a person, such as a judgment or a tax claim.

GI LOANS. The government guarantee of loans to veterans of various wars in connection with their purchase of a home with certain limitations.

GIFT DEED. A deed for which the consideration is love and affection, rather than money.

GIFT TAX (FEDERAL). A tax on gifts over a certain amount, with exemptions.

GOING PUBLIC. A corporation whose stock had previously not been made available to the general public and is now being offered to the public for the first time. Ford Motor Company stock was held by the Ford family for many years before going public.

GOOD WILL. The intangible value that a business has built up over a period of time.

GRANT DEED. Instrument used to convey title to land. Carries implied warranties.

GRANTEE. One who acquires title to property by deed.

GRANTING CLAUSE. Clause in deed stating "I grant" or "I convey." Essential to a valid deed.

GRANTOR. One who conveys title to property by deed.

GROSS INCOME. Total income from a business or property before deducting expenses.

GUARANTEE OF TITLE. An opinion on the condition of title based upon a search of the official records, and backed by a fund to compensate in case of oversight or negligence.

HARMLESS ERROR RULE. When a trial court judge commits an error and on appeal, the appellate court says that the trial court judge's error did not materially affect the defendant's rights or the law of the case.

HEAD OF A FAMILY. One who is responsible for dependents. Not necessarily a married person.

HEIRS. Those who by law obtain property upon death of another, either by will or by operation of law.

HOLDER IN DUE COURSE. One who in good faith takes a note for value and without knowledge of any defects, in the course of business.

HOLDING COMPANY. A super corporation which owns or controls such a dominant interest in one or more other corporations that it is able to dictate their policies.

HOLOGRAPHIC WILL. A will entirely handwritten and signed by the testator or testatrix.

HOMESTEAD. A home upon which a declaration of homestead has been recorded. Gives certain protection against judgments.

HYPOTHECATE. To pledge property as security for a debt, but retaining its use. As in connection with a trust deed loan.

IMPLIED WARRANTY. A warranty assumed by law to exist in an instrument although not specifically stated. As in a grant or warranty deed.

IMPOUNDS. Monthly payments by the mortgage borrower to pay for annual taxes and insurance premiums to mortgage company.

IMPROVEMENTS. Things built on land which becomes part of it.

IMPROVEMENT ACTS. State laws providing for the installation of improvements in certain districts, such as street widening and paving, installation of sewer lines and storm drains, etc. The cost is usually assessed against the properties directly benefited.

INCOMPETENT. One who is unable to manage his affairs because of feeblemindedness, senility, insanity, etc.

INDEMNITY. Guarantee against loss—as by an insurance policy. Same as a guarantor.

INDORSEMENT. A name signed on the back of a check or note.

INGRESS. A means of entering a property without trespassing.

INHERENTLY DANGEROUS. A type of danger that is an essential part of something, requiring special precautions to be taken to prevent injury.

INHERIT. To obtain property as an heir.

INHERITANCE TAX (STATE). Tax on estate of deceased resident.

INJUNCTION. An order of a court to restrain against certain acts in connection with a pending lawsuit or one adjudicated.

INSOLVENT. Inability of a person to pay his debts. Where liabilities exceed assets.

INSTALLMENT NOTE. A note which provides for payment of a certain part of the principal at stated intervals.

INSTRUMENT. A document in writing creating certain rights to its parties or transferring them.

INTEREST. The rental charge for the use of money.

INTEREST TABLE. A table giving the amount of annual interest on various sums of money at different rates of interest.

INTERSTATE COMMERCE. Involving two or more states in the United States.

INTESTATE. Death without leaving a will. The dead person is called a testator or testatrix, indicating male or female.

INTRASTATE COMMERCE. Involving one state only.

INVITEE. One who has a perfect legal right to be on the premises where the accident occurred.

INVOLUNTARY BANKRUPTCY PETITION. A document filed by three or more creditors in federal bankruptcy court, claiming that the debtor person, firm or corporation is unable to pay their bills when due and is therefore bankrupt.

INVOLUNTARY LIEN. A lien placed against property without the necessity of the owner's consent. Taxes and assessments are examples.

IOU (I OWE YOU). A written acknowledgment of a sum of money owed to another, in spite of its informality. Similar to a note except that a note contains a promise to repay at a certain time.

IRREVOCABLE. That which cannot be recalled or revoked.

IRRIGATION DISTRICT. A district created by law to furnish water. It is a quasi-political district having governing features similar to counties and cities.

JOINT NOTE. A note signed by more than one person. All have equal responsibility for payment and must be sued together.

JOINT AND SEVERAL NOTE. Same as joint note, but makers may be sued either jointly or individually in event of default.

JOINT TENANCY. Equal ownership by two or more persons under four essential unities. It features right of survivorship. If one dies, his interest goes to the survivor or survivors.

JOINT TENANCY DEED. A deed which names grantees as joint tenants. Very popular for married couples.

JUDGMENT. A court's final decree. Often involves awarding a payment of money.

JUDGMENT PROOF. Applies to persons who have no assets to satisfy a judgment for money.

JUNIOR LIEN. A lien which is subordinate to another lien which has prior claim on the security. The prior lien holders can collect before the junior lien is satisfied.

JURISDICTION. The right given by law by which courts, commissions, etc. enter into and decide cases.

KEY MAN INSURANCE. Life insurance protection paid for by management to cover the cost of replacing a key man in the organization. Sometimes used as a fringe benefit for the key man.

LACHES. Sleeping on your rights which results in failure to secure legal relief because of waiting too long.

LAND CONTRACT. An agreement whereby land is sold, usually on an installment basis, and buyer does not receive a deed until the contract is paid out.

LAND DESCRIPTIONS. A description of land recognized by law. One based on government survey or surveys based on it.

LANDLORD'S LIEN. The landlord's right to hold the tenants property as security for unpaid rent.

LAST CLEAR CHANCE. The final opportunity to avoid the accident, even though the other party was on the wrong side of the road.

LATENT DEFECTS. A defect being unknown and not discoverable by inspection.

LEGALEZE. The author's slang definition of technical language used by certain judges and members of the legal profession.

LEGAL HYBRID. A cooperative apartment because the stockholder-tenant possesses both stock in the corporation as well as a lease with the corporation.

LEGAL RATE OF INTEREST. Varies in different states. Used to be about 6 percent or 7 percent, however, some states have increased the legal rates recently.

LESSEE. A renter under a lease.

LESSOR. A landlord or owner who has leased his property.

LETTERS OF CREDIT. A written authorization from your bank to other banks all over the world permitting you to draw money to be charged against your bank. Your bank will then charge your account.

LIABLE. Responsible under the law.

LICENSEE. A person who is authorized to be on the construction project. A license gives one the right to walk over another's land.

LIEN. An encumbrance against property making it liable for a debt.

LIFE ESTATE. Rights to use property for a lifetime only.

LIMITED PARTNERSHIP. A type of partnership with limited liability.

LINE OF CREDIT. The amount of money your bank will permit you to borrow.

LIQUID ASSETS. Those readily convertible to cash.

LIQUIDATE. To sell off property at best available price in order to secure cash.

LIQUIDATED DAMAGES. Extent of damages agreed upon in a contract in event of default.

LIS PENDENS. A recorded notice to advise persons interested in certain property that a lawsuit is pending which may affect title to it.

LISTING. A contract authorizing a real estate broker to buy, sell or lease certain land under specified terms and conditions.

LITIGANTS. All parties to a lawsuit.

MAJORITY. The age at which a young man or lady becomes an adult, according to law.

MARGINAL RELEASES. Entry on the margin of an official record book showing that a claim has been paid.

MARKETABLE TITLE. Title to real estate which is free and clear from any reasonable objections.

MARKET PRICE. The going price of equivalent properties based upon recent sales.

MARKET VALUE. The best price a property would bring in dollars if freely advertised for sale for a reasonable time, to find a buyer who is fully informed on the possible uses of the property.

MASTER PLAN. A general plan for future physical development of a community.

MATERIAL FACT. A fact, which if known to the parties, might seriously affect their decisions in a transaction.

MECHANIC'S LIEN. A lien right provided by law whereby persons who have furnished labor or materials may make legal claim against the property for their money.

MENACE. Use of threats to induce one to enter into a contract.

MESNE PROFITS. Mesne means intermediate. Profits from a property during a period when a rightful owner is wrongfully deprived of the earnings.

METES AND BOUNDS. A method of describing the boundary lines of a parcel of land.

MILLER ACT. A federal law that protects all who furnish labor or materials used on a federal building project. Federal law prohibits the filing of a lien against the government, thus this substitute protection by way of the Miller Act.

MINERAL, OIL AND GAS LICENSE. Special license to deal in such lands.

MINORS. Young men and women who have not reached a legal age to vote or to enter into legal contracts. The laws of each state will vary.

MISDEMEANOR. A lesser crime than a felony. Sentences are to county jail for less than one year or a fine or both.

MONEY LEFT ON THE TABLE. Difference in money between the successful bidder on a highway construction job and the second low bidder.

MONTH TO MONTH TENANCY. When rent is paid by the month. The usual arrangement for renting houses.

MORATORIUM. A law suspending liability for paying a debt and granting more time for payment.

MORTALITY TABLES. Established procedures for determining life expectancy. Important in lawsuits to determine how long the claimant would have lived if he had not been struck by the car.

MORTGAGE. An instrument which makes property security for the payment of a loan.

MORTGAGEE. One who lends money secured by a mortgage.

MORTGAGOR. An owner who borrows money on a note secured by a mortgage.

MULTIPLE LISTING. A cooperative listing for the sale of real estate taken by a real estate board, which permits any member of their group to find a buyer. Brokerage fees are then split.

MUTUAL CONSENT. An essential to a valid contract.

MUTUAL FUNDS. An organization that invests other person's money and charges for their services. They claim to be experts in the field of investments.

MUTUAL MISTAKE OF FACT. An error or misunderstanding of a material fact that exonerates both parties to a contract.

NATIONAL BANK. Chartered and authorized to engage in the banking business by the Federal authorities. Each account is federally insured by an agency of the government.

NEGLIGENCE. Doing something wrong or failing to do something that is required to be done.

NEGOTIABLE INSTRUMENT. Those which are commonly transferred by endorsement in the course of trade—such as checks and drafts.

NEGOTIABLE NOTE. One capable of being assigned in the ordinary course of business.

NET INCOME. Remaining income from business or property after proper charges and expenses are deducted.

NET LISTING. One which provides that real estate agent gets his commission over and above a net sum to the seller.

NET WORTH. The difference between your assets and liabilities. What you own and what you owe.

NEW YORK STOCK EXCHANGE. The largest association of stockbrokers in the world where trading in securities is accomplished under an organized system.

NONPAR VALUE OF STOCK. Corporate stock that is issued without placing any value on the shares. Usually occurs in a family corporation.

NONPROFIT CORPORATION. A company that is organized for purposes other than earning money. It could be educational or charitable. Happens to some construction firms on an involuntary basis.

NOTARY PUBLIC. Person authorized by law to take acknowledgements and oaths.

NOTICE OF ABANDONMENT. Notice filed when work is discontinued on an unfinished job.

NOTICE OF COMPLETION. Document filed to give public notice that a building job is completed.

NOTICE OF DEFAULT. Notice filed by owner of a trust deed with the county recorder that borrower has defaulted and foreclosure proceedings may be started.

NOTICE OF INTENDED SALE. A notice to be recorded when a business is sold, to give notice to creditors and to the public.

NOTICE OF NON-RESPONSIBILITY. A notice provided by law, which when recorded, is designed to relieve an owner from liability for work or materials used on his property without his authorization.

NOTICE TO QUIT. A three day notice to a delinquent tenant to pay up or surrender possession of the premises.

NSF CHECKS. Abbreviation for "Not Sufficient Funds." Returned by the issuer's bank because there is not enough money on deposit to cover the check. The popular slang expression is "A HOT CHECK."

NSL (NO STOCKHOLDERS LIABILITY). A state corporation that limits the liability of its stockholders to the original amount they contributed to the new corporation.

OBLIGEE. The owner of the property and the building to be constructed is called the "obligee" under the bond issued by the bonding company.

OBLIGOR. The bonding company that issues the construction bond is called the "obligor."

OFF-SALE LICENSE. State liquor license issued to sellers of "packaged goods" to be taken from the premises.

OFFSET STATEMENT. Statement of an owner or lien holder as to present status of a lien—the remaining principal balance on the note, interest due, etc.

ON-SALE LICENSE. License to sell alcoholic beverages for consumption on the premises, such as a cocktail bar or beer hall.

OPEN LISTING. A nonexclusive listing given to one or more real estate brokers. It may be oral or written and the first agent to get owner's acceptance to an offer earns the commission.

OPTION. A written instrument which, for a consideration, gives one the right to buy or lease a property within a stated time on the terms set forth.

OPTIONEE. One who secures an option right.

OPTIONOR. An owner who gives an option.

ORAL. Verbal or spoken; not in writing.

ORIGINAL OR PRIME CONTRACTOR. The contractor who contracts with the owner to do the overall building job for an agreed price.

ORIGINATION FEE. A charge made by a financial institution in connection with starting a new loan.

OUTLAWED CLAIM. A claim is outlawed or barred by the statute of limitations when the claimant delays bringing suit beyond the time limit allowed by law.

OVER THE COUNTER. Corporate stocks that are not listed on any stock exchange, but are sold by stock brokers "over the counter." The companies involved are usually smaller and it is not always easy to sell stock over the counter because the buyers are not as numerous as when stock is sold through a large stock exchange.

OVERHEAD. The standard expenses of operating a place of business that are not chargeable to a particular part of the work.

OWNER'S EQUITY. What a property is worth over and above the liens against it.

PAID-IN CAPITAL. The amount of cash with which the new corporation is going to start their business.

PAROL EVIDENCE. Oral or verbal.

PARTIAL RELEASE CLAUSE. Clause in a mortgage or trust deed which provides for removal of certain property from the effect of the lien upon payment of an agreed sum. Subdividers must have these if their tract is subject to a "blanket lien."

PARTIAL SATISFACTION. An acknowledgment in writing that a part of a claim or judgment has been paid. Usually filed with the clerk of the court.

PARTNERSHIP. A contract between two or more persons to unite their property, labor or skill, or some of them, in prosecution of some joint or lawful business and to share profits and losses in certain proportions.

PARTY WALL. One built on the dividing line of property for use of both owners.

PAR VALUE. The value placed on new stock about to be issued. This is determined by its incorporators.

PATENT. An original conveyance of lands by the federal government. Title is granted by letters patent.

PAYOR AND PAYEE. The payor pays the sum due on a note, and the payee receives the money.

PERCENTAGE LEASE. A lease providing for rental based on the dollar volume of business done. Usually based on gross sales with an agreed minimum rental.

PERFORMANCE BOND. A guarantee that the contractor will perform the contract and also pay all bills for labor and material.

PERSONAL PROPERTY. Moveable property; that which is not real property.

PLAINTIFF. One who brings a civil lawsuit.

PLAN OF ARRANGEMENT. A written proposal filed in federal bankruptcy court by an insolvent debtor for an extension of time and a compromise payment of his obligations.

PLANS AND SPECS. Abbreviation for "plans and specifications." These are usually prepared by an architect and are vital to the construction of any structure.

PLEDGE. A deposit of personal property to secure a debt.

POINTS. In the money lending business, a point is one percent of the amount of the loan. Bonuses and commissions are often expressed in "points."

POLICE POWER. The power vested in the state to enact and enforce laws for the order, safety, health, morals and general welfare of the public.

POSTNUPTIAL PROPERTY SETTLEMENT AGREEMENT. A written agreement between husband and wife, after their marriage, specifying their property rights for the past and future.

POWER OF ATTORNEY. Authority given in writing by one person for another to act for him.

POWER OF SALE. A right given to a trustee to sell property under deed of trust if the borrower defaults.

PREFERENTIAL PAYMENT. Money paid by an insolvent debtor to a creditor in violation of bankruptcy law because it was made within four months of the filing of his bankruptcy petition. The money has to be returned to the trustee in bankruptcy for benefit of all of the creditors.

PREFERRED STOCK. Corporate stock that is entitled to priority over common stock in the distribution of profits.

PRENUPTIAL PROPERTY SETTLEMENT AGREEMENT. Written contract between husband and wife to be, before their marriage, dividing and agreeing to their ownership of property in the event of death or divorce.

PREPAYMENT PENALTY. A charge for paying off a mortgage balance ahead of schedule where the mortgage so specifies.

PRESCRIPTION. A means of obtaining title to property by long open possession under some claim, in defiance of owner's rights.

PRESUMPTION. A fact assumed by law which must be proved to the contrary.

PRIMA FACIE. On its face; presumptive.

PRIMARY MORTGAGE MARKET. Making original loans.

PRINCIPAL. One who employs an agent.

PRIOR IN TIME IS PRIOR IN RIGHT. Rights established by prompt recording, ahead of others who failed to record their lien or mortgage or other instrument promptly.

PRIORITY. Being first in rank, time or place.

PRIVITY OF CONTRACT. Lack of agreement, understanding or connection between the parties involved in the dispute.

PROBATE COURT. A special court that handles estates of persons that have died. Also all disputes that involve estates.

PROBATE SALE. Sale to liquidate the estate of a deceased person.

PROMISSORY NOTE. Written promise to pay a sum of money at a definite future time.

PROOF OF CLAIM. A special form that is filed in probate or bankruptcy court to substantiate a claim for money claimed to be owed by the decedent or bankrupt debtor.

PROPERTY. In general, anything capable of ownership.

PROPERTY MANAGEMENT. A branch of the real estate business.

PROPERTY SETTLEMENT AGREEMENT. A written agreement usually used in connection with a pending divorce action. It divides the property rights of the husband and wife as well as other matters.

PRORATION. In any transaction involving land, to divide taxes, interest, etc., proportionately between the parties, as in closing an escrow.

PROXY FIGHTS. A battle for control of a corporation by the owners of its stock granting to another the right to vote his stock.

PUBLIC LIABILITY INSURANCE. Protection against claims for the injury or death of one or more persons.

PUBLIC UTILITY. A private company giving public service, such as water, gas or electricity.

PURCHASE MONEY MORTGAGE. One given as part of the purchase price when buying property. The note it secures is given to the seller instead of cash to meet the required down payment.

QUASI-PUBLIC CORPORATION. A corporation which has been given certain powers of a private nature. The local gas company is given the right to exercise eminent domain.

QUIET TITLE. A lawsuit to determine status of title; to remove a cloud on the title.

QUITCLAIM DEED. Deed by which the grantor releases any claim or interest in a property he may possess. It says in effect, "Whatever interest, if any I have, I give to you."

QUORUM. The number of persons required to be present for corporate business to be legally transacted. For a stockholder's meeting, a quorum means a majority of the voting stock issued and not a majority of the actual bodies of the stockholders.

RANGE. A strip of land running north and south and six miles wide, established by government survey.

RECONVEYANCE. Transfer of title to a former owner, as when a trustee under a deed of trust reconveys title when the note is paid in full.

REDEMPTION. Reacquiring property lost through foreclosure within the prescribed time limit.

REFEREE IN BANKRUPTCY. An attorney who is appointed by the federal judges of his district to act as an official to preside over bankruptcy court. He is sort of a junior federal judge with limited power in his bankruptcy court.

RELEASE CLAUSE. Provision in a trust deed or mortgage to release portions of the land from the lien upon payment of an agreed amount of money. Subdividers, who sell individual lots, are required to have these.

REQUEST FOR NOTICE OF DEFAULT. Acknowledged request filed with the county recorder by holder of a junior lien so he may be notified of actions of prior lien holders.

RESCISSION OF CONTRACT. To set aside or annul a contract, either by mutual consent or by court order.

RESERVATION. A right withheld by a grantor when conveying property.

RESIDENCE. Sounds simple like where do you live. In the courtroom it can become very technical and very important. It is a matter of intent.

RES IPSA LOQUITOR. Translated from latin it means "the thing speaks for itself." A legal doctrine that eliminates the vital requirement of proof of negligence under certain circumstances.

RESOLUTION. A written approval of the board of directors of a corporation authorizing its officials to take some action of importance. Examples could be, who is authorized to sign corporate checks or the purchase of real estate by a corporation, etc.

RESTRICTION. A limitation on the use of property, usually imposed by a previous grantor.

RETAINAGE. The portion of a percentage of the monthly payments made by the owner to the contractor for construction work completed. It is withheld until the construction contract has been completed.

REVERSIONARY INTEREST. The right to an estate or its residue after present possession is terminated. As with a life estate.

RIGHT OF FIRST REFUSAL. The choice of buying an interest or land itself under specified terms and conditions. If the right is not exercised, then the owner is privileged to sell to others.

RIGHT OF SURVIVORSHIP. The right of a joint tenant to the interest of a deceased joint tenant.

RIGHT OF WAY. An easement to pass over, or maintain services, on property or a particular part thereof.

RIPARIAN RIGHT. Rights of a landowner to use the water on, under or adjacent to his land.

RUNNING DESCRIPTION. Tracing the boundaries of a tract by giving distances, angles and points around the edges. A metes and bounds description.

SANDWICH LEASE. A sublease which is subject to an original lease, the sublessee having further sublet the property. He holds an "in between" lease.

SATISFACTION. An instrument executed by a lien holder declaring that the debt has been paid. When recorded, it discharges the lien from the records.

SEAL–CORPORATION. A round metal device that contains the name of the corporation, the state of its incorporation and the date it was incorporated. Required on all real estate matters.

SEC (SECURITIES EXCHANGE COMMISSION). A federal agency that polices the sale of stocks and other securities to residents of states other than the home state of the corporation involved. Their requirements are tough, however, they effectively protect the gullible public from phony speculative investments.

SECONDARY MORTGAGE MARKET. The dealing in trust deeds and mortgages already in existence.

SECTION OF LAND. A standard land measurement containing 640 acres, or one square mile.

SECURED CREDITOR. One who holds collateral as protection that his obligation will be paid. If the debtor does not pay as agreed, then the creditor can foreclose on the collateral.

SECURITY DEVICE. An instrument or contract which results in real estate being made security for money owed, such as a trust deed, real property sales contracts, etc.

SECURITY FUNDS. Funds deposited by lessee to protect lessor if a default occurs. These are trust funds.

SEPARATE PROPERTY. That property which is owned and controlled separately by either husband or wife, as distinguished from community property.

SET-BACK ORDINANCE. Local laws requiring owners, when building, to keep improvements a certain distance from lot boundaries.

SEVERALTY OWNERSHIP. Sole ownership–as by a single person.

SHERIFF'S DEED–One given by the sheriff upon court order when property is sold to satisfy a judgment.

SIGNING BY MARK. Making a mark or an *X* by a person unable to sign his name. The mark usually has to be witnessed by two persons.

SINGLE PERSON. One who has never married or whose marriage was annulled.

SOLVENT. Able to pay all debts when they become due.

SPECIAL ASSESSMENT. A legal charge against property for improvements which particularly benefit it.

SPECIAL MASTER. A person appointed by the judge to take charge of and sell property at a public sale and report the results of the sale to the court for necessary approval. It is usually an attorney.

SPECIFIC LIEN. A lien affecting one particular property.

SPECIFIC PERFORMANCE. Court order requiring a person to do what he has agreed to do in his contract.

SPOUSE. Either one of a married couple.

STATE CHARTERED BANK. A financial institution that is incorporated and approved by its own state. It is not subject to federal regulatory bodies.

STATUS QUO. The existing state of affairs. In a dispute, leaving the parties in the same position they were in originally.

STATUTE. A law enacted by a legislative body.

STATUTE OF FRAUDS. A state law requiring certain agreements to be in writing to be enforceable at law.

STATUTE OF LIMITATIONS. A state law limiting the time in which certain court actions may be brought.

STATUTORY DEDICATION. Surrendering land for public use when required by law; as for streets in a subdivision.

STOCK CERTIFICATES. Written or printed evidence of ownership of a certain number of shares or interest in a corporation.

STOCKHOLDER-SHAREHOLDER. There is no difference between the two terms. One who owns stock in a corporation.

STOCKHOLDER'S MEETING. Meetings called by a corporation for the purpose of electing directors and transacting other business requiring the consent of the stockholders.

STOCK OPTION PLAN. Offers to key employees and officials of a corporation to buy stock at an agreed price on an optional basis. If the price of the stock goes up, you exercise your option, otherwise you forget it.

STOCK SPLITS. Dividing up of the outstanding shares of a corporation into a greater number of units. Sometimes done when the price of a stock is too high for the average investor to buy.

STOP PAYMENT ORDER. Written instructions to your bank not to pay a certain check that you issued but has not yet been presented for payment.

STRAIGHT NOTE. One payable in a lump sum and not in installments.

SUBCONTRACTOR. A builder or contractor who enters into an agreement with the prime contractor to build some part of the entire structure. The plumber, electrician, roofer, heating, and air conditioning are typical examples.

SUBDIVISION MAPS. When approved by the governing body and recorded they are the basis for good legal description.

"SUBJECT TO" A MORTGAGE. Language used when buyer does not assume personal liability for payment of a mortgage or trust deed note against a property he buys.

SUBLEASE. A lease given when the original lessee in turn sublets.

SUBMISSION AGREEMENT. A written provision that if a dispute arises, which the parties are unable to settle, the matter will be referred to a board of arbitrators.

SUBORDINATION CLAUSE. Clause in a junior mortgage or trust deed enabling the first lien to keep its priority in case of renewal or refinancing.

SUBPOENA. A court order commanding a person to appear in court at a designated time and place. Failure to appear could constitute contempt of court.

SUBROGATION. The right to stand in another's shoes by virtue of paying him money on his claim. This occurs when the insurance company pays for the repairs on your car and tries to collect their loss from the other party involved in the accident.

SUBSTANTIAL PERFORMANCE. When a certain portion of a construction contract has been completed. There is no set percentage of completion required, it will vary with the facts of each case.

SURETY. One who becomes a guarantor for another person.

TANGIBLE PROPERTY. Personal property which has substance and can be manually delivered from one person to another.

TAX DEED. One given when land is sold by the state for non-payment of taxes.

TAXES (REAL ESTATE). A levy on property by political subdivisions, such as county, city, school districts to pay for government administration and services.

TENANCY IN COMMON. Ownership of equal or unequal undivided interests in property by two or more persons, without right of survivorship.

TENANCY AT SUFFERANCE. Occurs when a lease expires and owner permits tenant to continue in possession on a temporary basis. Usually one month at a time.

TENANT IN PARTNERSHIP. Interest in property held as a partner.

TERMITES. Wood devouring insects. Enemies of home owners.

TESTATOR AND TESTATRIX. One who makes a will. Testator is a man. Testatrix a woman.

THIRD PARTY BENEFICIARY. A laborer, materialman, or subcontractor who is protected by a performance bond taken out by the general contractor, even though they are not named in the bond.

TIGHT MONEY. A situation that exists when the demand for money is greater than the supply. Banks turn down applications for loans by good customers because they are temporarily out of loanable funds.

"TIME IS OF THE ESSENCE." Necessary provision in contracts. Contemplates prompt performance by the parties within the time limits set forth.

TITLE. Evidence of ownership and lawful possession.

TITLE INSURANCE. Protection to a property owner against loss because of defective title. Policies are written by title companies and cover all usual hazards.

TITLE SEARCH. An accurate check of the courthouse records to determine if title to land has been affected by the filing of any instrument. This work is usually done by experienced personnel of the title companies.

TOPOGRAPHY. The character of the land's surface, such as level or hilly.

TORT. A civil wrong other than a crime. An automobile accident claim is a good example.

TOWNSHIP. A unit of land six miles square, or thirty-six miles. Established by government survey.

TRADE NAME. A name used by someone engaged in business, other than his own personal name. Jones Motor Company or Pacific Motor Company are both trade names for John Jones, the owner.

TREASURY STOCK. Corporate stock that has been issued and paid for, but has later been reacquired by the corporation by purchase, donation, forfeiture, or other means.

TRESPASSER. One who enters upon the lands of another unlawfully.

TRUST DEED (DEED OF TRUST). A conveyance of title to a trustee to be held until a loan secured by a note is paid, at which time title is reconveyed.

TRUST FUNDS. Money that belongs to another that is being held for a particular purpose. Should be kept separate and apart from the holder's regular funds.

TRUSTEE. A person or corporation which holds title in trust pending repayment of an obligation or the rendering of a service. In connection with trust deed, holds title until note is paid in full.

TRUSTEE IN BANKRUPTCY. An official appointed by the Referee in Bankruptcy to take charge of the bankrupt person's estate.

TRUSTEE'S DEED. One given by a trustee when foreclosed property is sold.

TRUSTOR. Borrower on a trust deed note.

TURNKEY JOB. An agreement to complete a structure for a fixed price.

ULTRA VIRES. Acts of a corporation that are beyond its legal powers as provided in its charter.

UMPIRE. Not the baseball variety. A person selected by a board of arbitrators to decide the matter in controversy when the arbitrators are unable to agree.

UNDIVIDED INTEREST. A partial interest in a whole property, merged with the interest of others.

UNDUE INFLUENCE. Taking advantage of a person because of his weakness or distress.

UNIFORM COMMERCIAL CODE. A comparatively new law requiring filings with the Secretary of State of security devices making personal property loans secured liens.

UNIFORM SIMULTANEOUS DEATH ACT. Where husband and wife died in a joint disaster leaving community property, and evidence indicates that they died simultaneously, then one-half of the property goes to the husband's heirs and the other one-half to the wife's heirs.

UNILATERAL CONTRACT. One which imposes an obligation on one party only; exchange of a promise for an act.

UNISSUED STOCK. Corporate stock that has been authorized but has not been issued.

UNIT OWNER. A person who buys an apartment in a condominium.

UNITED STATES SUPREME COURT. The highest court in the United States. Its nine justices decide litigation that involve constitutional questions. They decide what matters their court will hear.

UNITIES. Essentials such as to a joint tenancy, the unities being time, title, interest, and possession.

UNLAWFUL DETAINER. Failure of a tenant to vacate after being notified that he is default.

UNSECURED CREDITOR. One who has extended credit without obtaining any collateral as security. Typical examples could be the drug store; the grocery store; the department store; the dress shop; and all of the public utilities.

URBAN PROPERTY. City property.

URBAN RENEWAL AND REDEVELOPMENT. Plan to improve substandard areas in populated communities.

USE TAX. A sales tax on goods purchased from out of state.

USURY. Charging a high and illegal rate of interest.

VA LOANS. A mortgage loan made to a service veteran which is insured or guaranteed by the Veterans Administration.

VALUATION. Appraising. Estimating the worth of property in money.

VEHICULAR TRAFFIC. Street or highway traffic.

VENDEE. The buyer.

VENDOR. The seller.

VERBAL LISTING. A listing not reduced to writing.

VERIFICATION. Confirmation of the truth of a document by sworn statement.

VEST. To bestow upon, such as title to property.

VETERANS ADMINISTRATION. A federal governmental agency, which among other services to veterans, insures or guarantees repayment of home loans borrowed by veterans.

VETERANS TAX EXEMPTION. A property tax exemption given to certain qualified veterans or their widows.

VOID. Having no binding effect at law.

VOIDABLE. That which may be declared void, but which is not void until so adjudged by a court.

VOLUNTARY LIEN. A lien placed on property through the voluntary act of the owner, such as when he makes a mortgage loan.

WAGE EARNER'S PLAN. A petition filed in federal bankruptcy court by a debtor who is financially involved, for an extension of time in which to pay his debts in full. A sincere attempt to avoid an ordinary type of bankruptcy.

WAIVE. To relinquish; to surrender the right to require anything.

WAREHOUSE RECEIPT. A written instrument that represents that certain goods are in the hands of a warehouseman. It is a symbolical representation of the property itself.

WARRANTY DEED. A deed which recites certain warranties that are guaranteed by the seller or grantor.

WASTE. Abuse of property by a tenant or someone holding a temporary interest, such as a life estate, which results in a loss of value.

WATER TABLE. Depth of natural underground water from the surface.

WOMEN'S LIBERATION MOVEMENT. Latest demands by the female of the species for equal pay and greater equality with men.

WORKMANLIKE MANNER. An artisan performing his chores in a skillful manner. The test is what type of work would be done in his own area and not in New York or Chicago.

WORKMEN'S COMPENSATION INSURANCE. Protection furnished by the employer for benefit of his employees in the event they are injured or killed on the job.

WRIT. A written document issued by a court commanding a person to do certain acts, or sometimes to refrain from doing them.

WRITING OFF A BAD DEBT. Cancellation of a debt when the creditor is convinced that he cannot collect from the debtor. The creditor is then entitled to a tax credit due to his loss in writing off his chances to collect his money.

WRIT OF EXECUTION. A court order that property be attached and sold to pay a judgment.

ZONING. Control of the use of land by county or city authorities; power to limit property to specific use.

# *INDEX*

# INDEX


A

Abandonment, notice of, 122, 205
Abstract of judgment, 205
Abstract of title, 15-16, 205
Accommodation paper, 109-10, 205
  seven out of ten are losers, 110
Accountants, 7
  fees, 139
Act of God, 32, 205
Adverse possession, 202, 206
American Arbitration Association, 169
American Institute of Architects, 169
American Stock Exchange, 74, 206
Anticipatory breach, 34, 206
Appraisal, 140, 206
Arbitration, 48, 163, 206
  American Arbitration Association, 169
  American Institute of Architects, 169
  appraisement, 165
  arbitration agreements, 167-70
    law of contracts control, 167
  arbitration costs divided equally, 174
Arbitration *(cont.)*
    deductible as business expense, 174
  arbitration statutes, 166-67
  arbitrators, 48, 163
    lien for fee, 174
  attorneys and disputants, 165
  award, 164, 174-75
    award is written and signed, 175
    horsesense and arbitration, 164
    informal proceedings, 164
    judges encourage arbitration, 164, 166
    uniform arbitration act, 164
  controversy of any type may be arbitrated, 172
  corporations may arbitrate, 173
  court reporter eliminated, 170-71
  death terminates pending arbitration, 172
  definition, 163, 206
  Federal Arbitration Act, 167
  filing a lawsuit waives arbitration, 172
  jury trial waived, 167
  lawsuits may be switched to arbitration, 171

Arbitration *(cont.)*
majority rule, 173
partners may arbitrate their personal disputes, 173
reference by a judge, 165-66
right to lien, 170
scope of arbitration, 169
subcontractor not bound by arbitration clause in assigned contract, 172
submission to arbitration, 170
umpire, 173
United States Arbitration Act, 171
controls interstate commerce and maritime contracts, 171
what may be arbitrated, 172
witnesses are not subpoenaed, 174
Architects:
agent of owner, 197
architecture and engineering, 196
certificate of completion, 28
certification of work completed, 28
definitions, 195, 206
draftsman, 196
duties, 195
fees, 197-98
licensed and regulated, 196
lien rights, 199
negligence, 199
overseeing construction, 28, 195
plans and specifications, 21, 195
referee for owner and contractor, 195
responsibility, 27, 195
who owns plans, 197
Assignment, 37-39, 206
Associated General Contractors, 170
Assumption of risk, 60
Attorneys:
advice, 19
arbitration, 165, 169
fees, 3, 4, 87
in fact, 207
mechanics' lien, 119
Attractive nuisance doctrine, 60, 91, 207
Automobile insurance:
collision, 91
comprehensive, 91
fire and theft, 91
inadequate coverage, 91
medical expenses, 91
property damage, 91
public liability, 91
Award, 164, 174-75, 207

**B**

Bankruptcy, 70
abandonment of assets, 150-51
alimony not dischargeable, 153
arrangement proceedings:
assets greater than liabilities, 159
Chapter XI for going businesses, 159-60
extension of time, 159
50 percent creditor approval and doors stay open, 159-60
new creditors receive priority, 160
assets must be listed, 157-58
attorney for the trustee in bankruptcy, 150
attorney is not required, 151
fees, 152
child support not dischargeable, 153
conservation of assets, 149-51
creditors need not be listed, 156-57
rights of omitted creditor, 157
definition, 149, 207
discharge in bankruptcy, 153, 156
once every six years, 153
exemptions, 158
state exemption laws prevail, 158
Federal Bankruptcy Act, 154-55
federal laws control, 149
filing fees, 152
final meeting of creditors, 154
first meeting of creditors, 153
the bankrupt must appear in person, 153
fraud claims not dischargeable, 154
husband and wife filing jointly, 152
information required, 152
involuntary bankruptcy, 154-55
labor claims, 155-56
Louis, Joe, 150
materialmens' claims, 155-56
Nader, Ralph, 150
objections to discharge, 153
open to public, courtrooms, 153
payments to creditors within four months under suspicion, 155
order to disgorge, 155
Pennsylvania Railroad, 151
preferential payments, 155
proof of claim, 153
Referee in Bankruptcy, 150-53
relief from debts, 149-50
reviving a bankrupt obligation, 156
secured creditors, 152
subcontractors' claims, 155-56
taxes not dischargeable, 117, 153-54

Bankruptcy *(cont.)*
  Truman, President Harry, 149
  trustee in bankruptcy, 150, 153, 155
  unsecured creditors, 152
  voluntary petition, 150-51, 207
  Wage Earners Plan, 158, 160
    called debtor instead of bankrupt, 160
    debtor fails to pay and Wage Earners Plan is converted into bankruptcy, 161
    three years time to pay creditors, 161
    time extension for wage earners, 160
    trustee handles all money, 160
    trustee makes partial payments, 160
  wages and garnishments, 150-51
  where to file, 150
  willful judgment not dischargeable, 154
  wiping a bankruptcy off of the records, 157
Banks:
  certified financial statements, 140
  collateral, 178
  deposit slips in duplicate, 108
  escrow agents, 142
  line of credit, 138
  overdraft, 106
  personal and corporate liability, 139
  tight money, 138
Benny, Jack, 69
Bid, 27-28, 51, 207
Bill of sale, 207
Bills of lading:
  definition, 111, 207
  not a negotiable instrument, 111
Bills and notes:
  definition, 99, 207
  foreign or inland, 106
  parties, 104
Bondable:
  employees, 26, 106, 207
Bonds:
  definition, 207
  discharge of surety, 186-88
  fidelity, 212
  fraud, 185
  indemnity, 182
  labor and material, 182
  Miller Act, the, 189-93
  performance, 182
  subrogation, 188-89
  surety, 177
  surety company rights, 192-93
Bookkeeper, 14
Building code, 141
Building and construction contracts, 19-43
Building inspector, 2
Building permits, 2, 18
Business failures, 6
Bylaws, 208

C

Canadian law, 90
Cashier's check:
  definition, 104-5, 208
CD (Certificate of deposit), 110, 208
Certified check:
  definition, 105, 208
Change orders, 20, 26, 208
Chaplin, Charlie, 70
Chattel mortgage, 133, 208
Checks:
  cashier's, 104
  certified, 105
  definition, 105
  deposit slips in duplicate, 108
  endorsement stamp, 14
  kiting of checks, 107
  lost or stolen checks, 108
  negotiable instruments, 104
  postdated checks, 106
  prompt deposits, 14, 108
  prompt stamp endorsement, 108
  stop payment order, 105
  travelers checks, 106
  uncollected funds, 106
Church corporations, 77
Claim of lien forms, 120-21
Cleanup crew, 20
Cloud on the title, 116, 208
Coca Cola Bottling Company, 3, 46
Commercial land, 16
Common stock, 73, 209
  par value, 75-76
Community property, 48, 209
Completion, notice of, 122, 124
Condominiums:
  bylaws, the, 129
  death and passing of title, 131
  declaration, the, 129
  definition, 127-28, 209
  difference between cooperative apartments and condominiums, 130-31
  fee simple title, 128
  mortgage payments and taxes, 131-32
  sale of interest, 131
  statutory provisions, 129

Condominiums *(cont.)*
  tenant in common, 128
  unit owners, 128
  warranty deed, 128,
Conglomerate corporations, 75, 209
Construction business:
  name, 2-4
  trade name, 3
Consulting Engineers Council, 170
Contingent fee, 87, 119, 209
Contracts:
  accidental destruction, 33
  anticipatory breach, 34
  assignment, 37-39
  building and construction, 19
  canceling, 5
  corporation, 123-24
  cost plus, 23
  damages, 35
  death, 32-33
  definition, 19, 209
  divisible and indivisible, 33
  fraud, 41
  insurance, 33
  liquidated damages, 30
  minors, 42-43
  partnership, 123
  penalties, 30
  public liability insurance, 59
  purchase land, 5
  rescission, 39-41
  substantial performance, 29-32
  surety bond, 26
  time is of the essence, 30, 40-41
  turnkey job, 22
  verbal, 19
  workmen's compensation insurance, 59
Contractors:
  agent of property owner, 118-19
  definition, 209
  discharge of surety, 186-88
  fraud, 185
  indemnity bond, 182
  labor and material bond, 182
  licensing and regulation, 18
  Miller Act, the, 189-93
  performance bond, 182
  subrogation, 188-89
  surety bond, 177
  surety company's rights, 192-93
  who and what is covered, 183-86
Contractors licensing board:
  examination, 2
Contributory negligence, 91, 209
Cooperative apartments:
  assignment of lease, 130
  coops and coop apartments, 129
  death and title to stock, 131
  definition, 128-30, 209
  differences between condominiums and coops, 130-31
  leases, 129
    death terminates, 131
  mortgage payments and taxes, 131-32
  nonprofit corporation, 130
  promoter, 130
  sale of interest, 131
  stockholders' meetings, 130
  stockholder-tenant, 129
  stock issued, 130
  sublet, no right to, 130
Corporations:
  amending charter, 69
  annual reports, 72
  arbitration, 173
  articles of incorporation, 64, 81-84
  bankruptcy, 70
  board of directors, 80-81
  bonds and debentures, 80-81
  book value, 75
  bylaws, 68, 71, 208
  capital, 72
  capital stock, 72, 76
  certificates of stock, 76, 111
  church corporations, 71
  common stock, 73
  conglomerate corporations, 75
  construction contracts, 122-23
  convertible stock, 74
  corporate charter, 68-69, 71, 209
  corporate life, 68
  corporate seal, 69
  cost to form, 64
  death does not terminate, 49, 64
  de facto and de jure, 68
  definition, 63, 210
  dividends, 79
  domicil, 70
  dummy incorporators, 67
  family corporation, 66, 68
  family foundations, 77
  franchise, 69
  going public, 66
  holding company, 65
  legal entity, 66, 70
  merger, 209

Corporations *(cont.)*
minutes, 210
mutual fund, 75
nonpar value, 76
nonprofit, 64
NSL (no stockholders liability), 65
over the counter stock, 74
par value, 75-76
personal liability, 64
preferred stock, 73
proxy, 72, 78-79
public or private, 64
quasi-public, 64
quorum, 78
resolution of board of directors, 70
SEC (Securities Exchange Commission), 73
stockholders, 65, 72, 77-78
stock option plans, 76
stock splits, 79-80
transfer of stock, 77
treasury stock, 74
ultra vires acts, 65
unissued stock, 73
Cost plus contract, 23, 32, 189, 210
Council of Mechanical Specialty Contracting Industries, 170
Court trials:
judge, 3
jury, 3
CPA (Certified Public Accountant), 7, 210
fees, 139

D

Damages, 35, 210
proof of, 120
Death, 32-33
Debentures, 111, 210
Dedication, 16, 18, 210
Deed, 210
grant, 213
joint tenancy, 215
quitclaim, 220
tenancy in common, 222
warranty, 128, 224
Defaulting contractor, 27
Deficiency judgment, 126, 210
Dempsey, Jack, 69
Deposition, 87, 210
Deposit slips:
bank, in duplicate, 108
Discounting bills, 37, 211
Discounting commercial paper, 104
Diversity of citizenship, 167, 211
Draft:
definition, 104
Draftsman, 196, 211

E

Easement, 201, 211
Egress, 202, 211
Eminent domain, 15, 65, 211
Employees:
annual vacations, 109
bonded, 209
cost of, 109
handle company money, 109
not bondable, 109
Engineering:
engineering and architecture, 196,
Entity, 45
Equipment:
construction, 4
office, 4
Equity of redemption, 211
Escheat, 211
Escrow agent:
banks, 142
definition, 202-3, 211
fees, 203
instructions, 203
Executor (Executrix), 212
Exemption statutes, 212
Expenses:
accounting, 4,
attorney, 4,
business, 4
Expert witnesses, 21-22, 212
Extras, 212

F

Family corporation, 66, 168, 212
Federal Savings and Loan Association:
certified financial statement, 140
line of credit, 138
personal and corporate liability, 139
tight money, 138
Fidelity bond:
cost of, 109
definition, 212
employees, 109
Finance companies:
charges regulated, 112

Financial statements:
  CPA, 7
  form, 8-13
  what are they, 7, 140, 212
Fire insurance, 89-90
Fixtures:
  definition, 204, 213
  trade fixtures, 204
Foreclosures:
  death of mortgagor, 139
  deficiency judgment, 145
  definition, 133, 143, 213
  foreclosure statutes, 143-44
  judgment, 144
  mortgagor abscounds, 139
  public sale, 145
  redemption rights, 145-46
  special master, 145
  title search, 144
Fraud, 41, 185, 213

G

Gambling debts, void, 113
General lien, 213
Glossary, 205-25
Goldwyn, Samuel, 20
Government construction jobs, 189-93
Governmental agencies:
  low bidder, 27
  union contractors, 32
Guarantor, 177

H

Harmless error rule, 165, 213
Holder in due course, 103, 213
Holding company, 65, 67, 213
Home builder, 16, 18, 51
  line of credit, 138
  tight money, 138
Homestead, 213

I

Implied warranty, 27, 214
Impounds, 214
Income tax:
  capital gain, 14
  Louis, Joe, 150
  ordinary income, 14
Industrial Accident Commission, 87
Ingress, 202, 214
Innocent purchaser, 103
Installment note, 214
Insurance:
  accidental destruction, 33
  agent, 90
  automobile, 92
  cost, 90
  fire and theft, 92
  inadequate coverage, 91
  public liability, 59, 89, 91
  surety bonds, 178
  workmen's compensation, 59, 85
Insurance agent, 90, 93
Interest, 214
Interim certificate:
  definition, 111
  not negotiable, 111
Internal Revenue Service, 1, 7, 77, 108
  audit, 108
  bad debt write-off, 150
Interpleader, 38-39
Intestate, 214
Involuntary lien, 214
IOU (I OWE YOU), 100, 214

J

Joint business venture, 50
Joint tenancy deed, 215
Judgment:
  can be renewed indefinitely, 126
  definition, 215
  transcript, filing of, 117
  valid for about six years, 126

K

Key man insurance, 46-49, 215
Kiting checks, 107

L

Labor lien, 115, 118
  bankruptcy claim, 155-56
Labor and material bond, 182
Labor unions:
  membership, 32
  strikes, 32
Laches, 40, 215
Land development:
  checking title, 15
  commercial, 16
  dedication, 16, 18

Land development *(cont.)*
partial releases, 16
zoning, 15
Landlord and tenant, 122
landlord's lien, 215
Latent defects, 28, 215
Law merchant:
definition, 99
Laws:
each state different, 115
Lease:
breaking a lease, 5
definition, 5
office and yard, 5
option to renew, 5
Leasing motor vehicles, 93
Legaleze, 215
Letter of credit, 111, 215
License, 201
Licensee, 61, 215
Liens:
attorney's fees, 119
court allowance, 119
completion of building, 124
contractor and negligence, 116
defective materials, 116
defenses for owner, 116
deficiency judgment, 126
definition, 115, 215
enforcement of lien, 126
foreclosure lawsuit, 126
forms, 120-21
garageman's lien, 117
how to file, 119
interest allowed, 125
involuntary lien, 117
income tax lien, 117
property tax lien, 117
judgment lien, 117
labor lien, 115, 118
legal description of land requirement, 124
materialmen's lien, 115, 118
nonresponsibility, notice of, 123
owner's liability, 117
pawnbroker's lien, 117
priority of liens, 125
right to lien, 119-20
storage lien, 117
time to file, 119, 122
voluntary lien:
mortgage lien, 117
where to file, 119
who may file, 115
Limited partnership, 49-50, 215
Liquidated damages, 30, 37, 216
Location:
office, 4,
yard, 4
Louis, Joe, 150
Low bidder, 27

M

Master plan, 15, 216
Materialmen's liens, 115, 118
bankruptcy claims, 155-56
delivery to job site, 122
Mechanics' liens:
attorney's fees allowed, 119
definition, 18, 118-19, 216
interest allowed, 125
priority of liens, 125
time to file, 119, 122
Miller Act, the, 189-93, 216
Minors, 42-43, 216
Money:
what is it, 104
Money left on the table, 28, 216
Money orders:
definition, 111
not negotiable instruments, 111
Mortgage:
assignment or sale of note, 142
certified financial statement, 140
death of borrower, 139
definition, 134, 216
foreclosure, 143-44
impounds for taxes and fire insurance, 144-45
interest, 134
mortgage note, 134, 141
origination fee, 134
parties, 139
pay ahead of schedule, 143
points, 134
prepayment penalties, 143
recording, 141
release, 141
second mortgage, 141
statute of limitations, 141-42
title search, 144
who may borrow, 139
Mortgage liens:
mortgage recorded before construction, 125
priorities of liens, 125
Murder Incorporated, 67

Mutual Fund Corporation, 75, 217
Mutual mistake of fact, 41, 217

N

Nader, Ralph, 150
National bank, 217
National Society of Professional Engineers, 170
Negligence:
contributory, 91
definition, 21, 61, 217
proof, 91
Negotiable Instruments:
accommodation paper, 109-10
assignment, 108
bonds, 111
CD (certificate of deposit), 110
debentures, 111
definition, 99, 217
discounting, 104
holder in due course, 103
innocent purchaser for value, 103
money, 104
requirements, 104
stock certificates, 111
substitute for money, 103
Net to me listing:
land or home sale, 203, 217
New York Stock Exchange, 14, 74, 109, 217
Nonprofit corporations, 64, 217
Nonresponsibility, notice of, 123, 217
Notes:
attorney's fees, 100
debentures, 111
installment, 100
lump sum payment, 100
real estate mortgage, 100

O

Obligee, 178, 217
Obligor, 178, 217
Occupational license, 2
Option, 218
Overhead:
accounting, 4,
attorney, 4
business expenses, 4,
definition, 218
insurance, 85-97
leasing motor vehicles, 93
renting equipment, 93
Owner:
contractor agrees to satisfy, 28-29
discharge of surety, 186-88
fraud, 185
indemnity bond, 182
labor and material, 182
liability of, 182
obligee, 184
performance, 182
subrogation, 188-89
surety company rights, 192-93
who and what is covered, 183-86

P

Partnership:
arbitration, 48, 173
community property, 48
death terminates partnership, 48-49
definition, 45, 218
dissolution, 49
formation, 46
how long, 47
joint business venture, 50
key man insurance, 46-49
limited partnerships, 49-50
trade names, 46-47
verbal, 46
Party wall, 202, 218
Pawnbrokers:
charges regulated, 112
Penalties, 30, 37
Pennsylvania Railroad, 151
Performance bond:
definition, 26, 51, 182, 218
fraud, 185
Miller Act, the, 189-93
surety company rights, 192-93
who and what is covered, 183-86
Personal property liens:
garageman, 117
pawnbroker, 117
storage, 117
Petty cash fund, 107
Planning commission, 15
Plans and specifications:
definition, 218
plans and specifications, 22
prepared by architect, 21
who owns them, 197
Postdated checks, 102
Preferred stock, 73, 219

Prescription, title by, 202, 219
Prior in time is prior in right, 141, 219
Private corporation, 64-65
Privity of contract, 56, 219
Profit:
  change orders, 22
Promissory note:
  debentures, 111
  definition, 99-100, 219
  installment, 100
  penalties, 102
Proof of claim:
  bankruptcy, 153
  probate court, 219
Proof of damages, 120
Proximate cause, 60
Proxy fights, 72, 78-79, 219
Public construction work, 189-93
Public corporations, 64-65
Public liability, 59
  definition, 89, 220
  inadequate coverage, 91

Q

Qualified experts, 21-22
Quasi-public corporations, 64-65, 220
Quiet title, 220
Quitclaim deed, 220

R

Real estate mortgage note:
  definition, 100
  penalties, 102
  prepayment without penalty, 100
Rescission, 39-41, 220
Res ipsa loquitur, 57, 220
Resolution, 220
Retainage, 23, 188, 220
Retainer fee, 87

S

Savings and loan association, 67, 212
  certified financial statement, 140
  line of credit, 138
  personal and corporate liability, 139
  tight money, 138
SEC (Securities Exchange Commission), 73, 221
Shareholders-stockholders, 65, 222
Spot Zoning, 15
State chartered bank, 221
State chartered savings and loan association, 67, 212
Statute of limitations, 141-42, 221
Statutes:
  each state different, 115, 221
Stock certificates, 111, 222
Stop payment order:
  checks, 105
  procedure, 105, 222
Subcontractors:
  bankruptcy claims, 155-56
  definition, 222
  discharge of surety, 186-88
  fraud, 185
  labor and material, 182
  liens, right to file, 122
  Miller Act, the, 189-93
  performance bond, 182
  privity of contract, 56
  public liability insurance, 59
  subcontractors and contractors, 58
  subrogation, 188-89
  surety company rights, 192-93
  third party beneficiary, 178, 185
  who and what is covered, 183-86
Subpoena, 222
Subrogate, 184
Subrogation, 188-89, 222
Substantial performance, 29-32, 222
Surety bonds:
  definition, 177
  discharge of surety, 186-88
  fraud, 185
  indemnity bond, 182
  labor and material, 187
  liability, 182
  Miller Act, the, 189-93
  obligee and obligor, 178
  performance, 182
  subrogation, 188-89
  surety company rights, 192-93
  third party beneficiary, 178, 185
  who and what is covered, 183-86
Surety company, 23, 26, 177-78, 182
  discharge of, 186-88
  fraud, 185
  Miller Act, the, 189-93
  rights, 192-93
  subrogation, 188-89
  who and what is covered, 183-86
Suretyship, 178, 222
  fraud, 185

T

Taxes:
- avoidance, 64
- building permits, 2
- capital gain, 14
- evasion, 64
- occupational, 2
- ordinary income, 14
- real estate, 222
- sales, 2
- school, 2

Testator and testatrix, 223
Third party beneficiary 178, 185, 223
Time is of the essence, 30, 40-41, 55, 223
Title insurance, 15-16, 223
Title by prescription, 202
Title search, 144, 223
Tort:
- attractive nuisance, 91
- definition, 85, 223

Trade acceptance, 104
Trade fixtures, 204
Trade name, 46-47, 223
Travellers checks, 106
Treasury stock, 223
Trespasser, 91, 223
Trials:
- judge or jury, 3

Truman, President Harry, 149
Trust deed (Deed of trust):
- certified financial statement, 140
- death of borrower, 139
- deed of reconveyance, 147
- default procedures, 147
- deficiency claims not allowed, 148
- definition, 146, 223
- foreclosure statutes, 148
- impounds for taxes and fire insurance, 144-45
- interest, 147
- judgment, 144
- no redemption rights, 146
- note, 147
- notice of sale, 147
- origination fee, 134
- parties, 146
- points, 134
- posting notice of sale, 148
- power of sale, 147
- prepayment penalties, 143
- release, 147

Trust deed *(cont.)*
- second trust deed, 147-48
- title search, 144
- trustee's deed, 148
- who may borrow, 139

Truth in lending, 112
Turnkey job, 22, 223

U

Unavoidable accidents, 85
Uncollected funds, 106
Uniform commercial code, 38, 102-3, 223
Uniform negotiable instrument act, 102
Union shop, 32
United States Arbitration Act, 171
United States Supreme Court, 183, 224
Unlicensed contractors, 18
Usury:
- definition, 111, 224
- penalties, 112
- public interest, 112
- truth in lending, 112

V

Verbal contracts, 20, 27

W

Warehouse receipt:
- definition, 111, 224
- not a negotiable instrument, 111

Warranty deed, 128, 224
Woolworth Company, 5
Workmanlike manner, 21, 53-54, 116, 225
Workmen's compensation:
- amount of compensation, 88
- attorney's fees, 87
- filing fees, 86
- industrial accidents, 87
- insurance, 85, 225
- medical testimony, 22
- safe place to work, 89
- self-insured, 88
- subcontractors, 59

Z

Zoning:
- construction office and yard, 4, 140
- definition, 225